Soil Microbiology, Ecology, and Biochemistry

Soil Microbiology, Ecology, and Biochemistry

Fifth Edition

Edited by

Eldor A. Paul
Natural Resource Ecology Laboratory and
Department of Soil and Crop Sciences
Colorado State University
Fort Collins, CO, USA

Serita D. Frey
Center for Soil Biogeochemistry and Microbial Ecology
Department of Natural Resources & the Environment
University of New Hampshire
Durham, NH, USA

ELSEVIER

Elsevier
Radarweg 29, PO Box 211, 1000 AE Amsterdam, Netherlands
The Boulevard, Langford Lane, Kidlington, Oxford OX5 1GB, United Kingdom
50 Hampshire Street, 5th Floor, Cambridge, MA 02139, United States

ISBN: 978-0-12-822941-5

For information on all Elsevier publications visit our website at
https://www.elsevier.com/books-and-journals

Publisher: Peter B. Linsley
Acquisitions Editor: Rakhshan Rizwan
Senior Editorial Project Manager: Sara Valentino
Publishing Services Manager: Shereen Jameel
Project Manager: Vishnu T. Jiji
Senior Designer: Greg Harris

Printed in India

Last digit is the print number: 9 8 7 6 5 4 3 2 1

Contents

Contributors

Steven D. Allison, Department of Ecology and Evolutionary Biology, University of California, Irvine, CA, USA; Department of Earth System Science, University of California, Irvine, CA, USA

R. Balestrini, National Research Council, Institute for Sustainable Plant Protection (CNR-IPSP), Torino, Italy

Sreejata Bandopadhyay, Department of Microbiology and Molecular Genetics, Michigan State University, East Lansing, MI, USA

Pierre Barré, CNRS, École Normale Supérieure, PSL Université, Laboratoire de Géologie, Paris, France

Jennifer M. Bhatnagar, Department of Biology, Boston University, Boston, MA, USA

V. Bianciotto, National Research Council, Institute for Sustainable Plant Protection (CNR-IPSP), Torino, Italy

Christopher B. Blackwood, Department of Biological Sciences, Kent State University, Kent, OH, USA

Eoin L. Brodie, Ecology Department, Climate and Ecosystem Sciences Division, Lawrence Berkeley National Laboratory, Berkeley, CA, USA; Department of Environmental Science, Policy and Management, University of California, Berkeley, CA, USA

Claire Chenu, Université Paris-Saclay, INRAE, Agro Paris Tech, UMR Ecosys, Palaiseau, France

Brian Chung, Department of Earth System Science, University of California, Irvine, CA, USA

D.C. Coleman, Odum School of Ecology, University of Georgia, Athens, GA, USA

Elizabeth Duan, Department of Ecology and Evolutionary Biology, University of California, Irvine, CA, USA

Joanne B. Emerson, Department of Plant Pathology, College of Agricultural and Environmental Sciences, University of California, Davis, CA, USA

Serita D. Frey, Center for Soil Biogeochemistry and Microbial Ecology, Department of Natural Resources & the Environment, University of New Hampshire, Durham, NH, USA

Emmanuel Frossard, Department of Environmental Systems Science, ETH Zürich, Switzerland

S. Geisen, Department of Plant Science, Laboratory of Nematology, Wageningen University, Wageningen, the Netherlands

Kevin Geyer, Young Harris College, Young Harris, GA, USA

S. Ghignone, National Research Council, Institute for Sustainable Plant Protection (CNR-IPSP), Torino, Italy

P.M. Groffman, Cary Institute of Ecosystem Studies, Millbrook, NY and Advanced Science Research Center, City University of New York, NY, USA

Richard J. Heck, School of Environmental Sciences, University of Guelph, Guelph, ON, Canada

William R. Horwath, Department of Land, Air, and Water Resources, University of California, Davis, CA, USA

Ellen Kandeler, Institute of Soil Science and Land Evaluation, Soil Biology Department, University of Hohenheim, Stuttgart, Germany

Ulas Karaoz, Ecology Department, Climate and Ecosystem Sciences Division, Lawrence Berkeley National Laboratory, Berkeley, CA, USA

Michael A. Kertesz, School of Life and Environmental Sciences, The University of Sydney, Australia

Yakov Kuzyakov, Department of Soil Science of Temperate and Boreal Ecosystems, Büsgen Institute, Georg August University of Göttingen, Göttingen, Lower Saxony, Germany

E. Lumini, National Research Council, Institute for Sustainable Plant Protection (CNR-IPSP), Torino, Italy

William B. McGill, University of Northern British Columbia, Prince George, BC, Canada

A. Mello, National Research Council, Institute for Sustainable Plant Protection (CNR-IPSP), Torino, Italy

John C. Moore, Natural Resource Ecology Laboratory; Department of Ecosystem Science and Sustainability, Colorado State University, Fort Collins, CO, USA

Sherri J. Morris, Department of Biology, Bradley University, Peoria, IL, USA

Nathaniel Mueller, Natural Resource Ecology Laboratory; Department of Ecosystem Science and Sustainability; Department of Soil and Crop Sciences, Colorado State University, Fort Collins, CO, USA

Eldor A. Paul, Natural Resource Ecology Laboratory and Department of Soil and Crop Sciences, Colorado State University, Fort Collins, CO, USA

Alain F. Plante, University of Pennsylvania, Philadelphia, PA, USA

G.P. Robertson, Department of Plant, Soil, and Microbial Sciences and W.K. Kellogg Biological Station, Michigan State University, MI, USA

Cornelia Rumpel, CNRS, UMR IEES, Paris, France

Ashley Shade, Department of Microbiology and Molecular Genetics; Department of Plant, Soil and Microbial Sciences, Michigan State University, East Lansing, MI, USA

F. Sillo, National Research Council, Institute for Sustainable Plant Protection (CNR-IPSP), Torino, Italy

Maura Slocum, University of Pennsylvania, Philadelphia, PA, USA

D. Lee Taylor, Department of Biology, University of New Mexico, Albuquerque, NM, USA

Charlotte Védère, Université Paris-Saclay, INRAE, Agro Paris Tech, UMR Ecosys, Palaiseau, France

R. Paul Voroney, School of Environmental Sciences, University of Guelph, Guelph, ON, Canada

D.H. Wall, Department of Biology and School of Global Environmental Sustainability, Colorado State University, Fort Collins, CO, USA

Wally Xie, Center for Complex Biological Systems, University of California, Irvine, CA, USA

E. Zampieri, National Research Council, Institute for Sustainable Plant Protection (CNR-IPSP), Torino, Italy

Preface

In this fifth edition of *Soil Microbiology, Ecology, and Biochemistry* we retain the term "Soil Microbiology" for historical purposes, but the organisms in soil (its biota) include the soil fauna and interactions with the other major living soil component, the plant roots. The readers of this text come from a wide variety of backgrounds and training. All editions have therefore included a discussion of soil as a habitat for organisms and their reactions. Background information on physiological processes, biochemistry, and ecology is included in what is hoped to be a readable format. The huge amount of information in the wide variety of fields has resulted in the recent move to edited, multiauthor chapters. The electronic version has as many readers as the printed version.

The size of the soil biota ranges widely. Knowledge about the smallest, the viruses, is still very limited, with soil biota also being affected by them. The recent, and still present, COVID-19 epidemic has shown that viruses, if uncontrolled, can have devastating global effects. The individual, micron-sized, fungal hypha often are part of large, underground, macroscopic structures and aboveground fruiting bodies. The soil fauna also range widely in size and function. Many soil biochemical reactions, such as the decomposition of plant residues, require the presence of extracellular enzymes working at nanometer scales either as excretions or release during cellular death. Interactions of populations with the environment and other biota are characteristic of soil biology, with the very extensive mycorrhizal associations between fungi and plants and symbiotic nitrogen-fixing systems being prime examples of plant-biota interactions.

Our field has always been, and is still, very much methods driven. The field of microbiology was initiated by the discovery of the microscope. Understanding the study of nutrient cycling and soil organic matter dynamics depended on the availability of tracers. Computers have allowed us to process large amounts of data, to mathematically model, and to rapidly disseminate our data concepts and hypotheses. We therefore have methods-application chapters on physiology, molecular biology, modeling, and soil organic matter studies. The five editions of this book coincide with the advance of molecular biology and its application to our field. This edition reports the great advance in our knowledge of the occurrence and distribution of the vast, soil biotic populations, some understanding of the controlling factors, and a revamp of the classification of the organisms involved. We look forward to a much greater understanding in the coming years. There still is much to be learned.

This volume documents the results of the biochemical processes occurring at nanometer scales, the organisms at micrometer and meter scales, and their importance in influencing ecosystems and even landscape and global scales in the vast range of soils occurring in both disturbed and native systems. The soil biota function as nature's garbage disposal, mineralizing plant residues and cycling nutrients. However, this process is incomplete in that soil organic matter, nature's greatest natural resource, consists of the constituents of microbial bodies and altered plant residues protected by metal-mineral interactions, physical protection, and environmental factors. A significant portion of the present increase of atmospheric CO_2 is attributable to the decomposition of soil organic matter, made available to decomposition, by disruption of the soil habitat by cultivation for food production. Incomplete oxidation of plant residues in wet, cold tundra soils is attributable to the frozen conditions and inability of oxidative enzymes to work due to a lack of oxygen. Increased temperatures in the Arctic and associated dryness could result in a huge potential source of atmospheric CO_2 by microbial activity. Moist, partially anaerobic soils without the appropriate organisms and their oxidative enzymes together with time have been said to produce the vast amounts of oil and coal deposits now being oxidized by fossil fuel burning and causing global warming.

This edition again seeks to continue to coalesce new, basic, scientific knowledge in an understandable format for students, researchers, and its many other readers. This must be based on a good background of our history, documentation and interpretation of recent knowledge, and a look to the future. Application of this knowledge for the maintenance of stable ecosystems, food security, and our response to global change is vitally important. This volume therefore has a concluding chapter on the potential use of *Soil Microbiology, Ecology, and Biochemistry* knowledge to promote conservation, optimization of ecosystem processes, and food security and to improve, or at least minimize, the negative alteration in the natural processes so important to our earth and humankind.

Eldor and Serita

Acknowledgment

This volume could not have been produced without the much-appreciated and thorough editorial assistance of Laurie Richards, Research Administrator and Institutional Editor, Natural Resource Ecology Laboratory and Department of Ecosystem Science and Sustainability, Colorado State University, Fort Collins, CO, USA.

Chapter 1

Continuing our excellence in soil microbiology, ecology, and biochemistry and using it to achieve a sustainable future

Eldor A. Paul
Natural Resource Ecology Laboratory and Department of Soil and Crop Sciences, Colorado State University, Fort Collins, CO, USA

Chapter outline

1.1 The scope and challenges

The field of soil microbiology, ecology, and biochemistry continues to advance in knowledge and importance as our science and planet undergo a series of opportunities and challenges. We maintain the word soil microbiology in our title for historical continuity. The soil fauna are integral parts of the soil biota and are thus included, along with their processes, in this book. The interaction between organisms and their environment (ecology) is central to our concepts. The third component of the title (biochemistry) relates to the mechanisms of the biota with special reference to the important product of their activity: soil organic matter (SOM).

Sediment and aquatic systems, although having different physical environments, have closely interrelated controls. Their biochemical reactions are similar and one cannot understand one without a good knowledge of the other, as seen in texts on Aquatic Microbiology and Ecology (Gilbert and Kinat, 2016), Geomicrobiology (Ehrlich and Newman, 2016), and Biogeochemistry (Schlesinger and Bernhardt, 2013).

Our field continues to have an exciting present and great future based on basic knowledge and unifying concepts. This edition stresses the impact of that knowledge on the needs of our society (Fig. 1.1).

Factors involved include:

1. Scientific Knowledge – Education. We hope that the readers of this volume will use its information to increase their basic scientific knowledge and ensure its wide distribution through education and

Soil Microbiology, Ecology, and Biochemistry. https://doi.org/10.1016/B978-0-12-822941-5.00001-6

FIGURE 1.1 The interacting factors in soil microbiology, ecology, and biochemistry that affect our understanding and its application. The historical organism pictures are from the first (Lőhnis and Fred, 1923) English textbook in our field: (a) N-fixing rhizobia, (b) *Aspergillus* spore head, (c) *Spirillum*, (d) sporulating yeasts, (e) bacterial spores, and (f) protozoa.

outreach at a time when such information is very much needed to not only maintain our current environment but also improve it for future generations.

2. Ecosystem Functioning – Soil Health. It is now recognized that ecosystem functioning and soil health, whether in agricultural fields, rural and urban landscapes, grasslands, forests, or aquatic interfaces, are dependent on their soil-biota resources (Wall, 2004). All these ecosystems require a good balance of plant growth, diverse biota, and soil matrix-SOM interactions. Both natural and managed ecosystems are affected by human disturbance, global change, natural disasters and pollution.
3. Food Security – Living Space. The human population has increased to such an extent that it affects nearly all environments across the globe. Members of our society need not only to be fed but also to maintain an appropriate level of human health within a pleasant environment. The devastating impact resulting from the COVID virus in 2020–2022 is an example of what a single organism that is not controlled by normal ecosystem processes can do to human health, the health care system, and our scientific field. Plant diseases, caused by a variety of organisms, are a common threat but are typically held in check by plant breeding and crop rotations that involve soil biota, biodiversity changes. Diseases pose a constant drain on food production and can become pandemic, as in the historic example of the Irish potato famine.
4. Global Change – Climate Security. The interaction of physical, chemical, and biotic controls in the atmosphere, hydrosphere, and lithosphere has produced a variety of soil types and complex soil

habitats, which are the backbone of our existence. These vary across space and time but, fortunately in the recent past, have maintained a relatively stable, steady state that allowed for study and interpretation. These controls affect different ecosystems to various degrees, such as those observed in flood, fire, and drought-sensitive areas. Temperatures in the Arctic are increasing at a more rapid rate than that observed in other regions across the globe. Because of their unique biota and SOM-water interactions, this can have major feedbacks on atmospheric CO_2 and global temperatures and requires more intensive study.

5. Nutrient Management – Soil Organic Matter. Soil biota are counted on to decompose great quantities of plant residues and waste. Plastics and other materials, which are foreign to soil and aquatic biota, must be controlled such that the biota can preserve an environmental balance. The use of fertilizers and genetically improved crops has greatly expanded the human food supply on a limited, cultivatable-soil area. Fertilizer is now overapplied in many regions of the world, especially on high-value crops (Chapter 14). Increasing our knowledge as to the impact these practices have on the environment requires more concerted educational programs, in addition to possible regulation. More dependence on natural, ecosystem processes and soil biota-SOM interactions to supply season-long, available nutrients is a research-communications challenge as we respond to the added stress of climate change.

The readers of *Soil Microbiology, Ecology, and Biochemistry* will increasingly be called upon to help provide the basic information required for biologically sustainable ecosystem services at a reasonable cost (Cheeke, Coleman, and Wall, 2013). Older literature has been stressed in this chapter in the hope that readers will take a trip to the library and peruse some of the wealth of great knowledge contained therein. It is important to study this literature to ensure we do not repeat past mistakes and to sharpen our thinking and hypotheses.

The diverse soil biota and their interactions with mineral components of the complex soil matrix provide the habitat (Chapter 2) that allows many organisms to survive in a multitude of soil habitats (Chapter 8) for longer periods of time than one could surmise from annual plant carbon (C) inputs. This provides biodiversity and ecosystem stability but presents a challenge when one tries to apply knowledge of this field to the management of soil processes so necessary for our existence (Chapter 11).

1.2 A brief history of our field

Major advances in science that occurred in the past were often based on need, an indigenous background knowledge (Warkentin, 2006), and a receptive community. Times of relative food security and prosperity allowed the development of scholars in Asia before the Aramaic, Greek, and Roman civilizations. Scholars in the Chinese courts of the Yao Dynasty from 2357 to 2261 BCE studied soils for taxation purposes (Coleman, Crossley, and Hendrix, 2004). The ancient Vedic Indian literature classified soils for color, erosion, land use, and human health implications. They described Earthworms as "Angels of the Earth." Aristotle (384–322 BCE) in Greece called them "The Earth's intestines." The Hebrew word for soil was Adama, from which Adam, the first human documented in their theological literature, is derived.

Microbiological processes were recognized and used extensively without knowledge of their causation. Inscriptions found on an Egyptian tomb in 2400 BCE showed that beer and bread production required a starter (inoculum) and preincubation. Around 2000 years ago, Roman agriculture recognized that legumes in rotation enriched soils and that swamps were inhabited by minute animals that caused human diseases (Warkentin, 2006). Not as well recognized is the scientific knowledge of the Aztecs and Maya in Central and South America. The Aztecs differentiated soils by topography, color, water and organic content, genesis, typical plants, fertility, and farming practices using the word "thazolli" for humus-rich soil (Warkentin, 2006).

The stagnation of knowledge that occurred in Europe from the fall of the Roman Empire in the CE 5th century to the Renaissance of the 15th century came from the firmly held religious doctrines that the world was governed by an outside source rather than knowledge obtained from the laws that control our world. The development of free-thinking and the printing press aided the advancement and wide distribution of knowledge which is critical to increasing our understanding of the world around us.

1.2.1 The biota

The first evidence of life consisted of microbial accretions developed 3.7 billion years ago in an anaerobic environment. One theory for life involves reactions in hot, alkaline, hydrogen, methane, and ammonia interacting with acidic carbonics in the presence of iron—nickel sulfides (Ehrlich and Newman, 2016) that can act as energy sources (Chapter 15). Past research has also demonstrated oxidation-reduction and biogeochemical cycling by specific groups of organisms of manganese, chromium, arsenic, mercury, and selenium, often with environmentally toxic, oxidation states (Bolin and Cook, 1983). The possibility of constituents of life arriving from outside our Earth also exists.

There is a balance between the asking of important questions and the availability of methods to test hypotheses. Some fungi (especially fruiting bodies used for food) were large enough to be seen and identified without magnification. In 1665, Hooke described fossilized protozoa and two species of living microfungi (Atlas, 1984; Bardell, 1988). Early authors noted fungus-root interactions that were later identified as mycorrhiza (Waksman, 1932). The development of the microscope by Leeuwenhoek (1632—1723) allowed him to describe small "animalcules" (probably bacteria) in water containing decaying vegetation and thus initiated the recognition of the microscopic world. The Linnaeus classification of 1743 perhaps foretold the difficulties in bacterial classification, until the recent rise of genomic approaches (Chapter 6), where all animalcules seen by Leeuwenhoek were placed in the genus *Chaos*. Waksman's 1932 text on soil microbiology provides a good history of the early development of this field. Chapter 3 describes our present knowledge of the bacteria and archaea, with Chapter 4 describing the fungi and Chapter 5 the fauna.

Early chemists, such as Liebig and Berzelius, described decomposition and the evolution of CO_2 as a chemical process, stating it was a slow process of combustion by interaction with the oxygen of the atmosphere. They went so far as to say that yeasts were a noncrystalline precipitate similar to alumina and that fermentation was a mysterious catalytic force. Plant physiologists, such as Dumas and Boussingault, argued that plants and animals were the only living organisms. Darwin (1837) described earthworm processes affecting soil formation in his classic paper "On the Formation of Mould," published in Darwin (1881). For this, he should probably be recognized as the father of soil ecology. Pasteur (1830—1890) documented that life consisted of a third group of living forms, the microorganisms, that brought about the mineralization of plant residues, converting them into forms available for plant growth. He also developed the medical-pathological nature of microorganisms, leading to the present fields of medical microbiology and plant pathology. Pasteur also recognized the effect of physical controls on microbial metabolism, noting that anaerobic conditions led to different growth by-products than did growth under oxygen (Waksman, 1932).

Waksman (1932) described the turn of the 19th to 20th century as the first golden age of soil microbiology. History will determine if the application of metagenomics, automated instrumentation, tracers, and modeling from 1990 to 2020 will be called the second golden age. The development of the gelatin and agar plate made possible the isolation of some representative bacteria responsible for the major nutrient transformations. This included nitrification, sulfur oxidation, and anaerobic N fixation. Winogradsky (1949) described his much earlier isolation of two nitrifiers that utilized CO_2 as a C source, thus establishing the concept of autotrophic growth. He also established the concept of slow growth on a resistant substrate (Autochthonous-K) and faster growing organism on more available substrates (Zymogenous-R) still in use today. Beijerinck developed the nutrient culture method in which growth on specific culture media could increase a microbial population of interest and isolated the symbiotic N-fixing bacteria in legumes, as well as S-oxidizing bacteria.

The establishment of soil microbiology as an independent science was greatly enhanced when Lő;hnis (1910) published his textbook *Handbuch der Landwirtschaftliche Bacterologie*. This book went through 19 editions and was translated into four languages. The fungi and protozoa were included as "other" organisms. The English version (Lő;hnis and Fred, 1923) contains the most easily obtainable, early color images of soil biota, some of which are shown in Fig. 1.1. Conn's development of direct microscopic counts for soils in 1918 led to the realization that culturable organisms only accounted for a small proportion of the bacteria seen under the microscope. This is still of great importance today to recognize that modern molecular techniques (Chapter 6) basically describe resting populations.

Detailed studies by Waksman on plant residue decomposition in soils were followed by three textbooks on soil microbiology and two on soil humus (Waksman, 1938). His isolation of the antibiotic streptomycin from a soil actinomycete led to a 1952 Nobel Prize to complement Virtanen's 1935 Nobel Prize for identifying the plant-microbial interactions in symbiotic N fixation.

Studies of soil DNA in the 1970s (Torsvik, Sorheim, and Goksoyr, 1996) were initially oriented toward developing methods to answer the question, "How do we measure what's out there?" Chapters 3 and 4 in this volume show that molecular techniques have matured and now allow a measure of biotic occurrence and genetic controls on microbial activity in a wide range of soils. Studies of soil fauna also benefit from molecular techniques as we recognize the importance of fauna to soil food webs, soil health to ecosystem stability, and bioengineering (Chapter 5).

1.2.2 Soil ecology

The word "ecology," describing the interaction between organisms and their environment, was coined by Haeckel in 1869 (Smith and Smith, 2001) and is derived from the Greek word "*oikos*," meaning home or place to live. Studies initially concentrated on plant and animal biogeography, a continuation of the work of 18th- and 19th-century plant and animal explorers. Odum (1971) and his later editions of *Fundamentals of Ecology* attributed the term "ecosystem science" to Tansley in 1935. Ecosystems, defined as a biological community of interacting organisms and their physical environment, are an excellent way of delineating a specific community and its environment for study (Odum, 1969).

Early in the 20th century, soil biota studies concentrated on the identification of soil biota capable of growing in laboratory media. The fauna, plant-microbial interactions in the rhizosphere, mycorrhizae, and symbiotic N fixation were studied (Paul, 2007). These were later integrated with the ecological-biochemical tracer studies of the 1960s and 1970s. Alexander's (1961) textbook "*Soil Microbiology*" contained sections on soil ecology and pesticide degradation and recognized the interplay of organisms, soil characteristics, and biochemistry. *Fundamentals of Soil Ecology* (Coleman, Callaham, and Crossley, 2018) provides further background. Chapter 10, in this volume, should be referred to for background and information on ecological principles.

Plant-soil biota interactions are of significant environmental importance and thus generate considerable interest (Chapter 11). Although earlier researchers had noted the occurrence of fungal-root associations, it remained for Frank in 1885 to coin the term "mycorrhiza" and to differentiate between ecto- and endomycorrhiza. In 1877, Pfeffer recognized the symbiotic nature of the association (Paul, 2007). The mycorrhiza later became a significant subdiscipline (Harley and Smith, 1983) with a discussion of infection, growth, and C and mineral nutrition as described in Chapter 11 of this volume.

Methodology and concepts came together on a multinational basis in the International Biological Program in the mid-1960s to early 1970s (Van Dyne, 1969). This program, focused on studying the interactive components of the world's ecosystems, required the integration of soil science and soil biology, along with ecology, meteorology, and modeling (Van Dyne, 1969). This period also coincided with the introduction of computers with their ability to store and manage data and to model the interactions of life with the environment (Jenkinson and Rayner, 1977; Russell, 1964). This, together with mathematical descriptions of the effects of producers and consumers in ecosystems (Olson, 1963), initiated the modeling of ecosystems, landscapes, and the Earth system (Chapter 16). The ability to quantify and model life processes and interactions with the environment in mathematical terms is central to understanding the full spatial range of processes from subcellular kinetics to predictive, global climate models (Ojima, 1992). Computers, automated instrument control, data management, concept development, and future predictions using modeling will continue to be an important part of the work of nearly every reader of this volume.

1.2.3 Soil biochemistry

Pasteur's study of "Ferments" led the chemist Buchner to show that fermentation could occur in a pressed juice of yeasts, without living cells, establishing the field of biochemistry. Soil biochemistry includes the study of the physiology of organisms, enzymes, and SOM. The physiology of the biota must be understood to interpret biotic processes in soil. Chapter 7 discusses physiological and biochemical methods for the study of soil biota and their function. Chapter 9 provides the background required to use these methods and interpret their results on an ecosystem basis. Enzymes mediate all of life's interactions and have thus been a subject in the study of soils for as long as bacteria. Skujins (1967) quoted what he described as the first study of enzymes in soils by Woods in 1889: "I have also determined by experiment that the oxidizing enzymes, especially the peroxidases, may occur in soil and as a rule are not destroyed by the ordinary bacteria of decay. These enzymes enter the soil through the decay of roots and other parts of the plants that contain them." Notice that only bacteria were recognized as decay organisms, and plants were believed to be the source of enzymes. Skujins (1967) further reports that extensive soil enzyme results in the early 20th century were based on the use of catalases and proteases. By the 1930s, the literature recommended the extraction of enzymes to alleviate soil interference. The 1950s brought new methodology and many new papers. Today, automation continues to expand soil enzymatic studies. The editors of this treatise agree with Skujins' (1967) statement that "Although valuable insights into enzyme reactions in soils have been obtained, most of this information is, however, difficult to evaluate in terms of its importance to agriculture and its relation to nutrient cycles in soil." Burns et al. (2013) generally agreed with the above, stating that genetic tools, such as transcriptomics, the expression of enzyme-coding genes, and emerging proteomics will be most useful (see Chapter 6). The editors also feel that the promise of soil proteomics is yet to be realized.

An early written recognition of soil and aquatic biotic effects on the decay process that is still relevant today was cited by Lő;hnis and Fred (1923): "without this natural phenomenon, the world would soon be awash with a terrible amount of organic residues and waste products." Plant and animal-residue decomposition by the "soil cleanup crew" recycles nutrients in the C and N cycle and provides the C and energy for the development of the soil biota. In doing this they remove much of the residue but leave behind their bodies and some undecayed, chemically altered products. These are protected by further chemical interactions and by interactions with the soil matrix either in aggregates or in mineral particles, such as silt, clay, and sesquioxides. This results in the development of our world's greatest natural resource, SOM.

In Dokuchaev's (1883) study of soils in a Russian climate gradient, five factors of soil formation were developed: climate, parent material, organisms, topography, and time. These are so important to the definition of ecology that ecologists recognize Dokuchaev as one of their forefathers (Feller et al., 2003). Sage (1995) suggested another control, postulating that the near-simultaneous rise of plant agriculture in South America, Asia, the Mediterranean, and Africa approximately 12,000 years ago coincided with the ice recession in the Pleistocene when the atmospheric CO_2 required for photosynthesis, and thus plant growth, rose from 200 to 270 ppm. The recent rise in CO_2 to greater than 425 ppm has probably been a part of the green revolution that is presently being counted on to feed our expanding human population.

Müller in his 1889 book *Natural Forms of Humus* classified forest-soil horizons on their organic matter and earthworm content (Feller et al., 2003). His classifications of "mull" as nutrient-rich, organomineral soil, often with earthworms, and "mor" as a soil with a separate organic horizon overlaying a leached, nutrient-poor mineral horizon are still in use today. SOM, with the total global stock as SOM-C representing twice that of the atmosphere and living plants combined, is a vital part of the soil habitat (Wiesmeier et al., 2019).

It is difficult to characterize SOM in its insoluble, mineral-associated, natural forms. In 1786, Achard (Stevenson, 1982) extracted peat with sodium hydroxide and found darker materials in deeper, more highly decomposed horizons. Subsequent studies with basic solutions identified numerous fractions, but the primary ones were nonextractable, clay-associated materials called humin, base- and acid-soluble materials identified as fulvic acids, and base-soluble, acid-insoluble humic acids (Kononova, 1961). In the early 1900s Schreiner and Shorey questioned the use of strong extractions. Using then-available chemical identification techniques, they identified 40 specific organic compounds in an alcohol extract. However, the proportion of the SOM identified was too small to be useful (Waksman, 1938). The materials extracted with bases were criticized by Waksman (1938) as being artifacts of the extraction technique. However, they were used and defended for many years (Aiken et al., 1985).

Components such as amino acids, lipids, carbohydrates, and aromatic compounds that constitute the majority of the SOM, with an overall C content averaging 58%, were measured over the next 80 years (Stevenson, 1994). In Chapter 13 of this volume Chenu et al. describe other fractionation techniques such as densiometric and aggregate analyses, as well as instrumental approaches, such as computerized X-ray imagery, nuclear magnetic resonance (NMR), and nanoscale secondary ion mass spectrometry (nano-SIMS) that can be conducted on whole soils. Horwath (Chapter 12) discusses concepts related to SOM composition and dynamics.

The application of radioactive ^{14}C, found in the atmosphere and measured by radiocarbon dating, or added as plant constituents after artificial labeling, made possible the measurement of root production and the turnover of SOM components (Paul, 2016). Plant residues and other organic materials protected by aggregation are from 3 to 12 years in age. The majority of SOM, such as the nonhydrolyzable-residue fraction, humic acids, or clay-associated materials, are generally thousands of years old. The stable isotope ^{13}C, which replaced ^{14}C in residue addition experiments due to radioactivity dangers, is continuing to supply very useful information on the mineralogical, physiochemical, and microbiological controls on SOM dynamics (Kő;gel-Knaber and Kleber, 2012).

Nitrogen in proteins, enzymes, and nucleic acids is a fundamental component of all life. The determination of how legumes fix atmospheric N constitutes one of the more interesting scientific approaches in the field of soil microbiology. Lawes, Gilbert, and Pugh (1861) correctly supported Boussingault's claim that N_2 fixation involved microbial activity in root nodules rather than plant absorption of NH_3 from the atmosphere, as claimed by the chemist Liebig. Early studies also showed N to be a major stabilizing constituent of SOM, with the observation that the amino acid content, although very different in most soils of the world, had a composition that was similar in all soils (Stevenson, 1982).

The isolation of bacteria responsible for nitrification and denitrification during the Golden Age of Microbiology demonstrated the microbial nature of the N cycle (Waksman, 1932). Work in the 1920s and 1930s determined the dynamics of C and N during residue decomposition rates in nature and in the laboratory. It was determined that decomposition proceeded with no shortage or accumulation of mineral N at a residue C:N ratio of $\sim$25 (Alexander, 1961). Forest residues, comprised of slowly decomposable lignin, accumulate mineral N during decomposition at wider C:N ratios (Waksman, 1932). Although Waksman promoted a lignin-protein theory in which the interaction in soil between these two plant constituents formed decay-resistant SOM, it was later determined that lignin, although somewhat resistant, does not form a significant part of SOM (Paul, 2016).

The ability to measure the stable isotope ^{15}N with mass spectrometers (Broadbent and Norman, 1946) revolutionized the study of the N cycle. It was shown that only a portion of N added as fertilizer is utilized by plants, with an equal amount entering the soil at a slow release rate over time (Clark and Rosswall, 1981). Extensive N losses through denitrification and NO_3 leaching were quantified. Nitrogen fixation was shown to be significant primarily in symbiotic systems with the exception of a few C_4, primarily tropical, grasses. The Lőhnis N cycle (Fig. 1.2) has been expanded to include symbiotic and asymbiotic biological N fixation, microbial N utilization, plant N uptake, SOM decomposition, nitrification, and denitrification (Chapter 14). Genomic analysis has shown the occurrence of N-fixation, nitrification, and denitrification genes in microbial taxa and soils where they would not be expected. The production of the radiative gas N_2O during denitrification is of importance to temperature changes and ozone depletion in the upper atmosphere. Controlling this process will continue to be a challenge, as is the NO_3 pollution of aquatic systems. Nitrogen also moves via dust and volatile NH_3, often from cattle operations, and is contaminating what we used to believe were pristine areas, such as high mountain lakes (Olesky et al., 2020).

1.3 Putting it all together

The interaction between scientific knowledge, concepts, hypotheses, methods, and service to society (Fig. 1.3) is a significant part of our science. Both methodology and concepts have always been important. Should one study the most important question that has a possibility of being answered? An example of methodology is the development of the microscope that allowed us to see microorganisms and initiate our field. Culture techniques made possible the identification of representative biota responsible for the C, N,

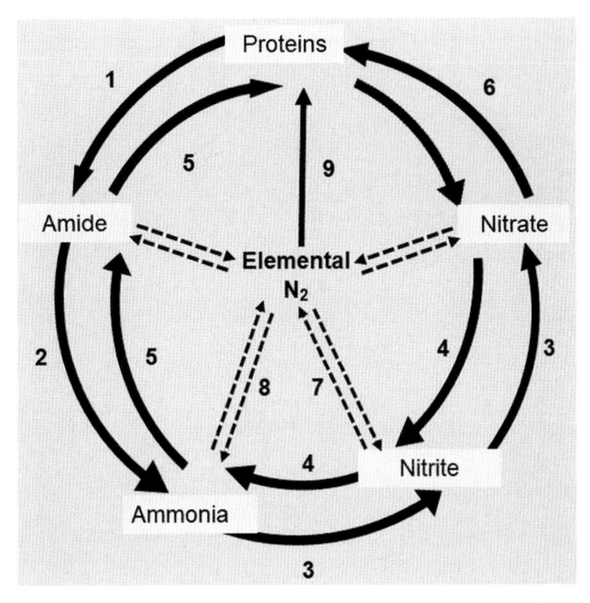

FIGURE 1.2 **The process nitrogen cycle.** Translated from the German. Redrawn from Löhnis and Fred, 1923.

S, P, and metal cycles. The organisms counted by culture techniques, however, accounted for a very small portion of those seen under the microscope. Metagenomics can increase our understanding of microbial communities and, to some extent, their interactions. However, molecular techniques can also have limitations in determining what part of the population is active and what is resting or protected by soil-matrix interactions.

Soil biotic diversity is of special importance when related to global change. There is genetic redundancy in the decomposer biota; the biota associated with plants and with elemental cycles are more restricted. Will plant-associated biota move with plants as they move with changes in temperature and moisture? Will invasive species, many with aggressive root systems, and their associated biota be more problematic with climate change?

Ecosystem services (Woodmansee et al., 2021) provided by soil biota face a number of challenges that include the following:

1. Biota that mediate decomposition and the C cycle are not infallible. Today's plastics are relatively insensitive to microbial attack.
2. The "soil cleanup crew" purifies our soils and waters from pathogens, antibiotics, etc., that enter these systems via animal and human waste. How do we ensure that the existing biota are effective; can we produce new and more effective ones?
3. The rotation of crops in many agricultural systems allows the large, diverse soil population to outcompete many root faunal and fungal pathogens. Research on the organisms involved and their possible management continues to be a challenge.

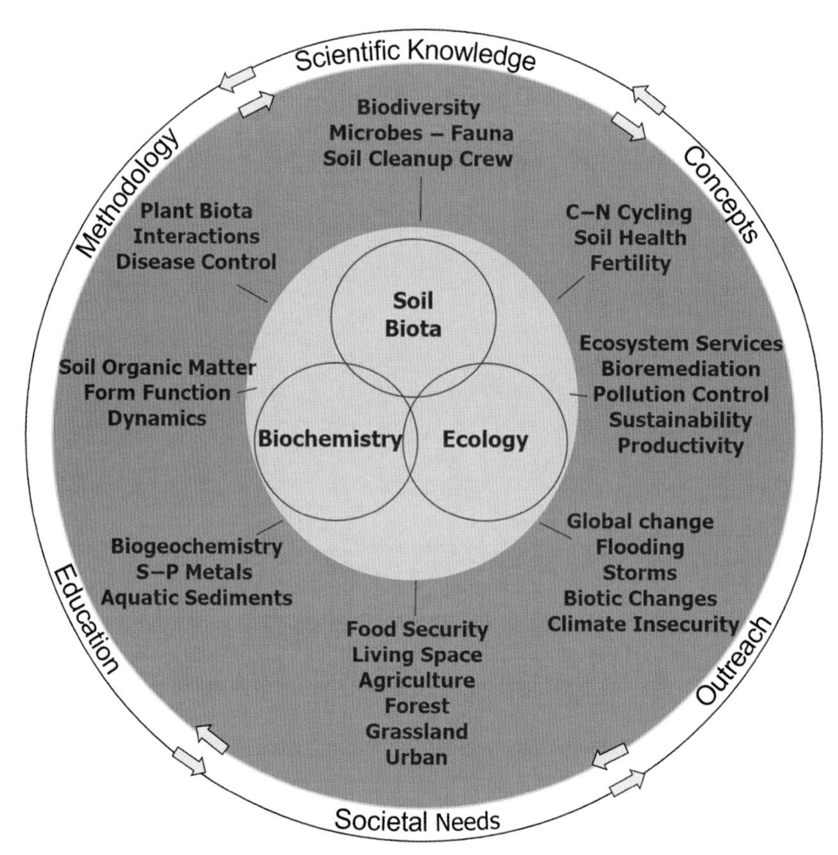

FIGURE 1.3 The interactions and processes in developing scientific knowledge and its application to education and societal needs.

4. Arbuscular mycorrhizal fungi are effective at P uptake; ectomycorrhizal fungi participate in the C cycle. Can we improve this?
5. The fact that the majority of C in most soils is thousands of years old is an essential part of its role in nutrient release and ecosystem stability. Its response to climate change is especially important for C storage, especially in the Arctic, where large amounts of decomposition-sensitive SOM are stored at low, but rapidly rising, temperatures.

Chapter 17 provides answers to some of these questions, with additional findings on the interactions and approaches for the application of knowledge to the societal needs prevalent today, with greater implications predicted for the future. We have been efficient at adopting new methods when available, as shown by the use of C dating to measure the dynamics of SOM (Paul et al., 1964), the measurement of soil DNA (Torsvik, Sorheim, and Goksoyr, 1996), tracers for understanding the N cycle (Chapter 14), and modeling for improving regional and global predictions (Chapter 16). Can we help develop new ones? Modern instrumentation for studying SOM tends to be expensive and limited to a few institutions.

The soil biota from viruses through the archaea, bacteria, fungi, fauna, and plant roots represent the greatest genetic diversity on Earth. We need more information on how their activities affect life's processes in terrestrial ecosystems, as well as their interactions with aquatic and sedimentary ecosystems. The N cycle, so important in litter decomposition, microbial growth, and plant growth inputs and losses, especially with our present needs in food security and global change, still requires extensive and increased study. An increased knowledge of the N cycle is also required to help control the damage of reactive N to many of the world's ecosystems, which is attributable to fertilizer N and high-intensity cattle operations. Because of the high energy requirement, biotic symbionts don't fix more N than needed and thus do not pollute the environment. A worthwhile challenge would be to move the complex, N-fixation system from legumes, actinorhizal-Frankia systems, or other recently discovered hosts into crops such as corn, wheat, or rice.

The authors of this textbook hope that you find the information contained herein useful for stimulating new concepts and methods in our field and in the numerous other fields of research addressing related basic scientific knowledge, teaching, and societal needs.

References

Aiken, G.R., McKnight, D.M., Wershaw, R.L., MacCarthy, P., 1985. Humic Substances in Soil, Sediment and Water. Wiley, New York.

Alexander, M., 1961. Soil Microbiology. John Wiley and Sons, New York.

Atlas, R.M., 1984. Microbiology: Fundamentals and Application. MacMillan, New York.

Bardell, D., 1988. The discovery of microorganisms by Robert Hooke. ASM News 54, 182−185.

Bolin, B., Cook, R.B. (Eds.), 1983. The Major Biogeochemical Cycles and Their Interaction. John Wiley and Sons, New York.

Broadbent, F.E., Norman, A.G., 1946. Some factors affecting the availability of organic nitrogen in soils. A preliminary report. Soil Sci. Am. Proc. 1, 261−267.

Burns, R.G., Deforest, J.L., Marxsen, J., Sinsabaugh, R.L., Stromberger, M.E., Wallenstein, M.D., 2013. Soil enzymes in a changing environment: Current knowledge and future directions. Soil Biol. Biochem. 53, 216−234.

Cheeke, T.C., Coleman, D.C., Wall, D.H. (Eds.), 2013. Microbial Ecology in Sustainable Agroecosystems. CRC Press, Boca Raton.

Clark, F.E., Rosswall, T. (Eds.), 1981. Terrestrial nitrogen cycles: Processes, ecosystem strategies, and management impacts. Ecol. Bull. 33, 671−691.

Coleman, D.C., Callaham, M.A., Crossley, D.A., 2018. Fundamentals of Soil Ecology, Third Ed. Elsevier, Academic Press, New York.

Coleman, D.C., Crossley, D.A., Hendrix, P.F., 2004. Fundamentals of Soil Ecology, Second Ed. Elsevier, Academic Press, New York.

Darwin, C., 1837. On the formation of mould. Proc. Geol. Soc. Lond. 2, 574−576.

Darwin, C., 1881. The Formation of Vegetable Mould Through the Action of Worms, With Observation of Their Habits. John Murray, London.

Dokuchaev, V.V., 1883. The Russian chernozem report to the free economic society. Imperial University of St, Petersburg, St. Petersburg.

Ehrlich, H.I., Newman, D.K., 2016. Geomicrobiology, Sixth Ed. CRC Press, Boca Raton, Florida.

Feller, C., Brown, G.G., Blanchart, E., Delaporte, P., Chernanskii, S.S., 2003. Charles Darwin and the natural sciences: Various lessons from the past and to the future. Agric. Ecosyst. Environ. 99, 29−59.

Gilbert, P.M., Kanat, T.H. (Eds.), 2016. Aquatic Microbial Ecology and Biogeochemistry: A Dual Perspective. Springer International, Zurich.

Harley, J.L., Smith, S.E., 1983. Mycorrhizal Symbiosis. Academic Press, New York.

Jenkinson, D.S., Rayner, J.H., 1977. Turnover of soil organic matter in some classical Rothamsted experiments. Soil Sci 123, 293−305.

Kőgel-Knabner, I., Kleber, M., 2012. Mineralogical, physiochemical and biological controls on soil organic matter. In: Huang, P.M., Li, Y., Sumner, M.E. (Eds.), Handbook of Soil Science: Resource Management and Environmental Impacts. CRC Press, Boca Raton, 7-1−7-21.

Kononova, M.M., 1961. Soil Organic Matter. Its Nature, Its Role in Soil Formation and Soil Fertility. Pergamon, Oxford.

Lawes, J.B., Gilbert, J.F., Pugh, E., 1861. On the sources of the nitrogen of vegetation; with special reference to the question whether plants assimilate free or uncombined nitrogen. Philos. Trans. R Soc. Lond. 431−577. https://doi.org/10.1098/rstl.1861.0024.

Lőhnis, F., 1910. Handbuch der Landwirtschaftlichen Bacteriologie. Borntrecher, Berlin.

Lőhnis, F., Fred, E.B., 1923. Textbook of Agricultural Bacteriology. McGraw Hill, New York.

Odum, E.P., 1969. The strategy of ecosystem development. Science 164, 262−270.

Odum, E.P., 1971. Principles of Ecology, Third Ed. Saunders, Philadelphia.

Ojima, D. (Ed.), 1992. Modeling the Earth System, Volume 3. UCAR, Boulder, Colorado.

Olesky, I.A., Baron, J.S., Leavitt, P., Spaulding, S.A., 2020. Nutrients and warming interact to force mountain lakes into unprecedented ecological states. Proc. Royal Soc. B 257, 20200304.

Olson, J.S., 1963. Energy storage and balance of producers and decomposers. Ecology 44, 322−331.

Paul, E.A. (Ed.), 2007. Soil Microbiology, Ecology and Biochemistry, Third Ed. Academic Press, San Diego.

Paul, E.A., 2016. The nature and dynamics of soil organic matter: Plant inputs, microbial transformations, and organic matter stabilization. Soil Biol. Biochem. 98, 109−126.

Paul, E.A., Campbell, C.A., Rennie, D.A., McCallum, K.J., 1964. Investigation of the dynamics of soil humus using carbon dating techniques. 8th Intl. Congr. Soil Sci. Bucharest, Trans. 3, 201−209.

Russell, J.S., 1964. Mathematical expression of seasonal changes in soil organic matter. Nature 5, 161−162.

Sage, R.F., 1995. Was low atmospheric CO_2 during the Pleistocene a limiting factor or the origin of agriculture? Glob. Change Biol. 1, 93−106.

Schlesinger, W.H., Bernhardt, E.S., 2013. Biogeochemistry: Analysis of Global Change. Elsevier, Amsterdam.

Skujins, J.J., 1967. Enzymes in soil. In: McLaren, A.D., Peterson, G.H. (Eds.), Soil Biochemistry, Vol. 1. Marcel Dekker, New York, pp. 371−416.

Smith, R.L., Smith, T.H., 2001. Ecology and Field Biology, Sixth Ed. Benjamin Cummings, San Francisco.

Stevenson, F.J., 1982. Nitrogen in agricultural soils, Volume 22. Am. Soc. Agron. Madison, Wisconsin.

Stevenson, F.J., 1994. Humus Chemistry, Second Ed. Wiley, New York.

Torsvik, V., Sorheim, R., Goksoyr, I., 1996. Total bacterial diversity in soil and sediment communities: A review. J. Ind. Microbiol. 17, 170−178.

Van Dyne, G.M., 1969. The Ecosystem Concept in Natural Resource Development. Academic Press, New York.

Waksman, S.A., 1932. Principles of Soil Microbiology. Williams and Wilkins, Baltimore.

Waksman, S.A., 1938. Humus: Origin, Chemical Composition and Importance in Nature, Second Ed. Williams and Wilkins, Baltimore.

Wall, D.H., 2004. Sustaining Biodiversity and Ecosystem Services in Soils and Sediments. Island Press, Washington, DC.

Warkentin, B.P., 2006. Footprints in the Soil: People and Ideas in Soil History. Elsevier, Amsterdam.

Wiesmeier, M., Urbanski, L., Hobley, E., Lang, B., von Lütz, M., Wollschläge, U.V., et al., 2019. Soil organic carbon storage as a key function of soils. A review of drivers and indicators at various scales. Geoderma 333, 149−162.

Winogradsky, S.N., 1949. Microbiologie du Sol: Problems et Methods. Mason, Paris.

Woodmansee, R., Moore, J., Ojima, D., Richards, L. (Eds.), 2021. Natural Resource Management Reimagined: Using the Systems Ecology Paradigm. Ecology, Biodiversity and Conservation. Cambridge University Press, Cambridge. https://doi.org/10.1017/9781108655354.

Chapter 2

The habitat of the soil biota

R. Paul Voroney*, Richard J. Heck*, and Yakov Kuzyakov†
*School of Environmental Sciences, University of Guelph, Guelph, ON, Canada; †Department of Soil Science of Temperate and Boreal Ecosystems, Büsgen Institute, Georg August University of Göttingen, Göttingen, Lower Saxony, Germany

Chapter outline

2.1 Introduction

Planet Earth is experiencing a prolonged period of warming 12 to 20 millennials-long, during which massive continental glaciers have melted, and exposed geological parent materials have allowed the formation of soils and peatlands. Remnants of the past are evident in receding mountain top glaciers, melting tundra permafrost, and exposed polar landscapes. The current concern for humanity is that this changing climate, which is now compounded by anthropogenic activities, will be associated with increased extremes in weather events, such as prolonged periods of drought and torrential precipitation events. The soil resource is being exploited through intense management of food, fiber, and bioenergy production for a global population that is projected to exceed 9 billion by 2050 and may not be sustainable without significant intervention and restoration. It is widely accepted that conversion of the natural prairie grasslands, forests, and peatlands for agricultural crop production and animal grazing has resulted in soil degradation: erosion, environmental contamination, and losses in soil organic matter (SOM) and biodiversity.

Soil Microbiology, Ecology, and Biochemistry. https://doi.org/10.1016/B978-0-12-822941-5.00002-8

Soils provide many essential ecosystem services including: (1) water purification and storage, (2) water and nutrient provisioning to support plant growth, (3) organic matter and nutrient storage, (4) decomposition of organic residues derived from plants and soil organisms, (5) mineral nutrient solubilization from parent materials, and (6) habitat provisioning for biodiverse populations of organisms that promote both soil resilience and plant health. As the central component of the Earth's critical zone, soils deserve special status due to their role in regulating the Earth's environment.

The principles of regional and global soil distributions, as well as the heterogeneity at macroscales, were well characterized in the 19th and 20th centuries. Soil development was related to climate, parent materials, vegetation, relief, and time and explained by the dominance of individual soil forming processes (Gerasimov, 1984) and pedogenic thresholds (Chadwick and Chorover, 2001). At the microscale, however, soils are characterized by extreme heterogeneity of physical, chemical, and biological properties that have not been well studied.

Because soils provide such a tremendous range of habitats, they support an enormous biotic biomass globally, with an estimated 2.6×10^{29} prokaryotic cells alone, and harbor much of the Earth's genetic diversity. A single gram of soil can contain kilometers of fungal hyphae and more than 10^9 bacterial and archaeal cells and organisms belonging to tens of thousands of different species. Zones of aeration may be only millimeters from areas poorly aerated. Areas near the soil surface may be enriched with decaying organic matter and other accessible nutrients, whereas the subsoil may be nutrient poor. The variance of temperature and water content of surface soils is much greater than that of subsoils. The soil solution in some pores may be acidic, yet in others basic, or vary in salinity depending on soil mineralogy, location within the landscape, and biological activity. The microenvironment of the surfaces of soil particles, where nutrients and organisms are concentrated, is different from that of the soil solution. The habitat of a particular soil organism also includes the biological component of the habitat that influences the growth, activities, interactions, and survival of other organisms associated with this space (other microorganisms, fauna, plants, and animals).

The primary soil properties influencing the gross behavior and activity of soil organisms within the bulk soil matrix are known, that is, effects of moisture and aeration, temperature, and nutrient availability. However, the principles, processes, and consequences of the distribution of soil properties at the microscale remain underinvestigated. A limitation is the lack of methods currently available that enable study of the detailed activity of soil organisms in situ at the microscale level of the soil microhabitat. It is common practice to pass soil samples through a 2-mm sieve to homogenize the sample before study, thus disrupting soil structure and the distribution of soil minerals and pore spaces.

2.2 Soil formation

Soils (pedosphere) develop at the interface where organisms (biosphere) interact with rocks and minerals (lithosphere), water (hydrosphere), and air (atmosphere), with the mineral parent material governing the inherent fertility and climate regulating the intensity of these interactions. The interactions of rock and parent material with factors such as temperature, rainfall, elevation, latitude, and exposure to sun and wind, over broad geographical regions with similar environmental conditions and characteristic plant communities, have evolved into the current terrestrial biomes with their associated soils (Fig. 2.1).

Soils derived primarily from weathered rocks and minerals, referred to as mineral soils, dominate the terrestrial landscape and extend from the northerly boreal forests to the great prairie grasslands, savannahs, dry forests, tropical forests, and deserts. When plant residues are submerged in water for prolonged

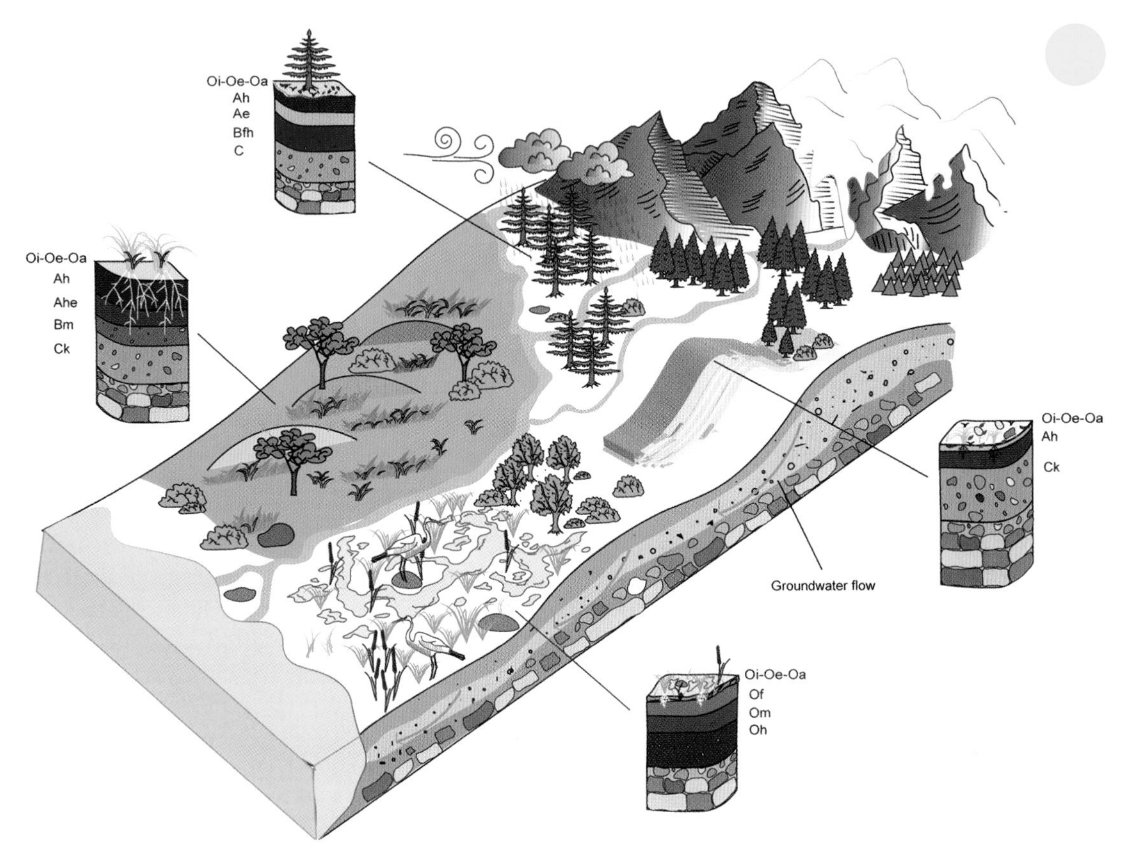

FIGURE 2.1 The concept of soil formation and development as independent natural bodies, each possessing unique properties resulting from parent material, climate, living matter, and topography interacting over time. *(With permission from M. Mesgar.)*

periods, biological decay is slowed. These accumulations of organic matter at various stages of decomposition go on to form organic soils, which include peatlands, bogs, fens, and other wetland types. When accompanied by mean annual freezing temperatures <0°C, they form the tundra permafrost. Soils can also be formed in coastal tidal marshes or inland water areas supporting plant growth where areas are periodically submerged.

The parent material of mineral soils can be the residual material weathered from solid rock masses or the loose, unconsolidated materials that often have been transported from one location and deposited in another by such processes as glaciation, sedimentation, and erosion. The disintegration of rocks into smaller mineral particles is a physical-chemical process brought about by cycles of heating and cooling; freezing and thawing; and also by abrasion from wind, water, and ice masses. Chemical and biochemical weathering processes are enhanced by the presence of water, oxygen, and the organic compounds resulting from biological activity. These reactions convert primary minerals, such as feldspars and micas, into secondary minerals, such as silicate clays and oxides of aluminum, iron, and silica. Soluble constituent elements in inorganic forms provide nutrients to support the growth of various organisms and plants.

The initial colonizers of soil parent materials are usually organisms capable of both photosynthesis and atmospheric N_2 fixation and capable of producing organic acids to promote solubilization of constituent elements in minerals, such as phosphate-containing apatite. Root-bacterial/fungal/actinomycete associations with early plants assist with supplying nutrients and water. Products of the biological decay of organic residues accumulate in the surface soil as a nutrient-rich reserve of SOM that gives soils their unique physical and chemical properties.

Soil organisms, together with plants, constitute one of the five interactive factors responsible for soil formation. By 1880, Russian and Danish soil scientists had developed the concept of soils as independent natural bodies, each possessing unique properties resulting from parent material, climate, topography, and living matter, all interacting over time (Dokuchaev, 1883). The approach to describing soil genesis in the landscape and as a unique biochemical product of organisms participating in the genesis of their own habitat was formalized by Hans Jenny in 1941 in his classic equation of soil forming factors (Jenny, 1941).

$$\text{Soil} = f[\text{Parent material, climate, living organisms, topography, time}]$$

Soils in the formerly glaciated areas of the northern hemisphere have developed over the past 10,000 to 20,000 years. In other areas, notably the tropics, soil formation has occurred over hundreds to thousands of years depending on deposition and erosion processes. Thawing of permafrost soils is currently exposing organic matter that accumulated in the Pleistocene and the Holocene.

2.2.1 Mineral soil profile

During formation, mineral soils develop horizontal layers, or horizons, with a variance in appearance (Fig. 2.2). The horizons within a soil profile vary in thickness depending on the intensity of the soil forming factors, although it can be difficult to distinguish their boundaries. Uppermost layers of mineral soils are most altered, whereas deeper layers are more similar to the original parent material. The alterations of the upper-most parent material during soil formation involve: (1) decay of organic matter from plant residues and roots and accumulating as dark-colored humus; the organic matter-enriched horizons nearest the soil surface are called A horizons; (2) eluviation by water of soluble and colloidal inorganic and organic constituents from surface soils to varying depths in the profile; and (3) accumulation of inorganic and organic precipitates in the subsurface. These underlying, enriched layers are referred to as B horizons. Together, the A and B horizons, which are most altered during formation, are referred to as the solum. The C horizons, relatively unweathered parent material, are below the solum or recognized horizons, low in organic matter and nutrients, and intermittently deficient in O_2. Nevertheless, this region of the profile annually receives dissolved organic matter and nutrients leached from the solum. In native grasslands and forests, litter-derived organic materials accumulate on the soil surface as Oi-Oe-Oa horizons containing plant residues in varying states of decay and range from recognizable plant tissues at the surface to extensively altered and amorphous, humified organic matter below.

The vadose zone is the parent material extending from the surface soil downward to where it reaches the water table and becomes saturated with groundwater. The thickness of the vadose zone can fluctuate considerably during the season, depending on landscape topography, soil texture, soil water content, and height of the water table. When the water table is near the surface, as in wetlands, it may be narrow or

FIGURE 2.2 Profile of a Spodosol formed on an acidic parent material and under mixed forest vegetation (see Global Soil Biodiversity for extensive images of soil profiles). *(With permission from M. Mesgar.)*

nonexistent. But in arid or semiarid areas, where soils are well drained, the vadose zone can extend for many meters.

2.2.2 Organic soil profile

Organic soils often contain 35 to 40% organic matter (occasionally >80%) and show evidence of horizons that are differentiated by the degree of plant residue decay (Fig. 2.3). The surface horizon is comprised of recognizable plant residues, mainly from the previous growing season, that have undergone partial decay. Deeper in the profile, residues are further fragmented and decayed, yet some resistant plant residue components can be identified. This is defined as the Om horizon. Below this horizon is the organic matter that is most decomposed and is largely comprised of amorphous, dark-colored humus. For this reason, this horizon is referred to as the Oh horizon. Organic soils of sub-Arctic and Arctic regions with tundra vegetation are typically permanently frozen within 1 to 2 m of the surface.

Humans have had a negative effect on soil formation due to agricultural practices (Amundson and Jenny, 1991; Kuzyakov and Zamanian, 2019). The clearing of native vegetation, often by slash-and-burning to expose the surface soil and tillage to prepare soils for planting, degrade soils by promoting erosion and enhancing losses of SOM. Drainage of wetlands and other organic-rich soils for agriculture has led to extensive SOM losses with significant wind erosion. However, humans have also improved soil conditions through the installation of irrigation systems and addition of plant nutrients. Where topsoil has

FIGURE 2.3 **Profile of an organic soil.** *(With permission from Y. Kuzyakov.)*

been excavated at surface mine sites, areas of exposed parent material amended with nutrients and organic matter have successfully restored plant productivity. For reclaimed areas left on their own, it could take hundreds of years to form a few centimeters of topsoil.

2.3 Nature of soil components

2.3.1 Mineral particles and inorganic precipitates

The inorganic components, including primary and secondary minerals and inorganic precipitates, represent 90 to 95% of the soil solids by weight in mineral horizons. They have an important role in affecting soil physical and chemical properties because of their sizes, surface areas, and chemical reactivity. The larger mineral particles include stones, gravels, sands, and coarse silts that are primarily derived from rock and mineral fragments.

While particles >2 mm in diameter may affect the physical attributes of a soil, the fine-earth fraction, the individual mineral particles ≤2 mm in diameter, are major controllers of microbial and plant life. The fine earth fraction of soil mineral particles ranges in size over four orders of magnitude: from 2.0 mm to smaller than 0.002 mm in diameter (Table 2.1). Sand-sized particles are individually large enough (2.0−0.05 mm) to be seen by the naked eye and feel gritty when rubbed between the fingers in a moist state. Smaller, silt-sized particles (0.05−0.002 mm) are microscopic and feel smooth and slippery even when wet. Together, the sand and silt-sized particles are predominantly comprised of primary minerals; common examples include: quartz, feldspars, muscovite, biotite, pyroxenes, micas, amphiboles, and olivines (Table 2.2). Clay-sized particles, the smallest of the mineral particles (<0.002 mm), seen only with the aid of an electron microscope, are secondary minerals that have an important role in defining the chemical properties of the soil. The abundant secondary minerals are aluminosilicates, such as montmorillonite; vermiculite; illite; and kaolinite; oxides of gibbsite, goethite, and birnessite; and amorphous materials, such as imogolite and allophane, and sulfur- and carbonate-containing minerals.

The proportions of sand-, silt-, and clay-sized particles are referred to as soil texture, and terms such as sandy loam, silty clay, and clay loam are textural classes used to identify the soil's texture. When

TABLE 2.1 Physicochemical properties of organic and mineral particulates in soils.

Particle	Diameter (mm)	Cation exchange capacity ($cmol_c\ kg^{-1}$)	Specific surface area ($m^2\ g^{-1}$)
Coarse sand	2.00–0.50	<1	0.002
Medium sand	0.50–0.25	<1	0.005
Fine sand	0.25–0.05	<1	0.016
Coarse silt	0.05–0.02	<1	0.06
Fine silt	0.02–0.002	<1	0.2
Clay	<0.002	20–150	50–800
Humus	<0.002	150–500	500–900

investigating a field site, considerable insight into the physical and chemical properties of the soil can be inferred from its texture (e.g., soil water characteristics, nutrient retention, susceptibility to compaction, ease of cultivation, engineering stability, and erosion). Thus it is often one of the first properties to be measured.

2.3.2 Organic matter

The surface of mineral soils contains an accumulation of living biomass, dead and decomposing organic material, and humus. SOM typically accounts for only 1 to 10% of the soil mass, though its effects on soil physical and chemical properties can exceed that of minerals by promoting water and nutrient retention, and through its role in preserving soil structure. The living microbial biomass and larger organisms account for 3 to 5% of SOM, of which microorganisms account for 80 to 90% and fauna 10 to 20%. While that may not appear like much, on a hectare basis, it is equivalent in living biomass to ~12,000 kg, that is, to 2 African elephants or 200 sheep. The larger recognizable remains of plant, animal, and other soil organisms can be separated from soils using hand-picking and sieving techniques (particulate organic matter), and using density floatation (light fraction organic matter). In agricultural soils the residues of annual crop growth can account for 8 to 15% of SOM. Most of the remaining SOM is adsorbed onto the surfaces of decomposing residues and fine-silt and clays as organomineral clusters. About 50% or more of SOM consists of chemically known biomolecules derived from decaying plant residues and soil organisms. The residues of dead microbial biomass (i.e., necromass), in particular, can account for most of this component (Miltner et al., 2012). The remaining 30 to 60% of the total SOM consists of brown to black-colored, colloidal, and amorphous humus, which accumulates over centuries and provides a nutrient-rich reservoir to sustain soil organisms and provide soil fertility. Because it is difficult to separate the components of living organisms from the by-products of their decomposition, or to isolate organic matter separately from the mineral particles, the precise chemical nature of SOM is far from being completely understood. Indeed, it is considered to be composed of the most chemically complex organic compounds on Earth.

TABLE 2.2 Mineral composition of soil parent materials.

Mineral	Chemical formula of mineral	Nutrient content
Primary minerals in igneous and metamorphic rocks		
Quartz	SiO_2	None
Potassium feldspars	$KAlSi_3O_8$	16.92% K_2O
Plagioclase feldspars	Continuous solid solution series between $NaAlSi_3O_8$ (albite) and $CaAl_2Si_2O_8$ (anorthite)	1% CaO in albite, 10.36% CaO in anorthite
Muscovite (mica)	$KAl_2AlSi_3O_{10}(OH)_2$	11.81% K_2O
Biotite (mica)	$K(Mg,Fe^{II})_3AlSi_3O_{10}(OH)_2$	10.86 % K_2O, 23.24% MgO, 8.29% FeO
Chlorite	$(Fe,Mg,Al)_6(Si,Al)_4O_{10}(OH)_8$	25.39% MgO, 15.09% FeO
Hornblende (amphibole)	$Ca_2(Mg,Fe,Al)_5(Si,Al)_8O_{22}(OH)_2$	19.63% MgO, 13.66% CaO, 2.43% FeO
Augite (pyroxene)	$(Ca,Na)(Mg,Fe,Al)(Si,Al)_2O_6$	21.35% CaO, 15.35% MgO, 6.08% FeO
Tourmaline	$(Na,Ca)(Li,Mg,Al)(Al,Fe,Mn)_6(BO_3)_3Si_6O_{18}(OH)_4$	20.46% FeO, 0.91% B_2O_3
Forsterite (olivine)	Mg_2SiO_4	57.29% MgO
Almandine (garnet)	$Fe_3Al_2(SiO_4)_3$	43.30% FeO
Apatite	$Ca_{10}(PO_4)_6(OH,F,Cl)_2$	55.07% CaO, 41.82% P_2O_5
Ilmenite	$FeTiO_3$	47.35% FeO
Magnetite	Fe_3O_4	31.03% FeO, 68.97% Fe_2O_3
Primary minerals in sedimentary rocks		
Name	**Chemical formula**	**Nutrient content**
Calcite	$CaCO_3$	56.03% CaO
Dolomite	$CaMg(CO_3)_2$	30.41% CaO, 21.86% MgO
Gypsum	$Ca(SO_4)_2 \cdot 2(H_2O)$	32.57% CaO, 46.50% SO_3
Halite	NaCl	(60.66% Cl)

2.3.3 Soil colloids

Soil humus and clay minerals play a major role in regulating the physical and chemical properties of soils (Fig. 2.4). Both have high surface areas due to their small size; thus they are extremely reactive with their surrounding environment. Humus colloids have an abundance of a highly complex mixture of carboxylic- and phenolic-acid functional groups. Because of the variable charge of these functional groups, both anions and cations can bind to colloids depending on the pH of the soil. Clay minerals also have both

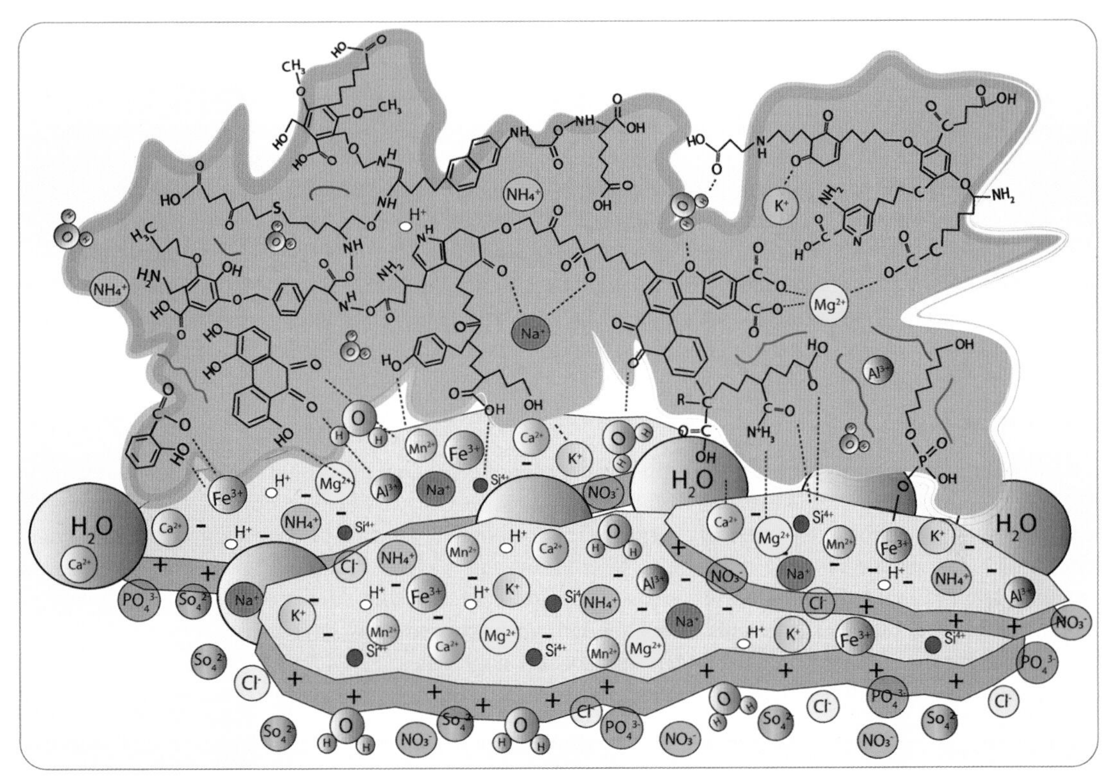

FIGURE 2.4 **Soil colloids: clays have intrinsically charged surfaces; clay mineral edges and organic matter functional groups have variable charges depending on soil pH.** *(With permission from M. Mesgar.)*

positive and negatively charged surfaces and likewise can bind to cations and anions. Intrinsic imperfections in the crystal structure of clays render their platey surfaces negatively charged, attracting cations from soil solution and concentrating them at their surfaces. The broken edges of clay minerals expose structural deficiencies that can be either positively or negatively charged, depending on soil pH. At low soil pH, anions will be adsorbed to these edges, and at neutral and higher pH, they will bind cations.

2.3.4 Aggregates in mineral soils

Most of the organic matter in surface soil becomes intimately associated with mineral soil particles as the plant fragments undergo decay and the resulting microbial products become glued to the surfaces of the finer mineral particles, the clays and fine silts. When individual mineral particles become bound together by these organic constituents and inorganic cements (calcium and magnesium carbonates, iron and aluminum oxides, silicates), they form larger, stable, and distinct units known as aggregates or peds. Together with soil texture, the size, nature, and arrangement of soil aggregates are of particular importance to the structure or microarchitecture of the soil because of their influence on soil pore size distribution, resistance to compaction, and resistance to soil erosion. Aggregates tend to be resistant to mechanical breakdown, for example, from the impact of rainfall or slaking (i.e., rapid rewetting of dry soil) or from freezing and thawing.

Due to large surface areas and electrostatically charged surfaces, both humus and clay minerals play particularly important roles in aggregate formation. Thus loamy or clayey soils are usually strongly aggregated, whereas sandy and silty soils are weakly aggregated. A network of living plant roots and fibrous organic matter physically enmesh the clusters of smaller aggregates into larger aggregates. Where they exist, casts deposited by earthworms can play a significant role in the formation of aggregates. Two size classes are commonly used to describe soil aggregation based on their resistance to break down and their associations with organic matter. Macroaggregates are >250 μm in diameter, and microaggregates are 35 to 250 μm in diameter. Macroaggregates that are 1 to 2 mm in diameter give the soil a crumb-like structure that is ideal for both plant root growth and for promoting soil aeration.

2.4 Pore space

Space-forming pores, properly termed voids, are between the solid (mineral and organic) components of soil and vary in size and connectivity. Total soil pore space can vary widely for a variety of reasons, including soil mineralogy, bulk density, organic matter content, and disturbance. On a volume basis, mineral soils are about 35 to 55% pore space, whereas organic soils are 80 to 90% pore space. Pore space can range from as low as 25% for compacted subsoils in the lower vadose zone to more than 60% in well-aggregated clay-textured surface soils. Even though sand-textured soils have a higher mean pore size, they tend to have less total pore space than do clay soil aggregates.

Soil pore space is defined as the percentage of the total soil volume occupied by soil pores:

$$\%\ \text{pore space} = [\text{pore volume}/\text{soil volume}] \times 100 \qquad \text{(Eq. 2.1)}$$

Direct measurements of soil pore volume are difficult to perform, but estimations can be obtained from data on soil bulk density and soil particle density, using the following formulas:

$$\text{Soil bulk density}\left(D_b,\ \text{Mg m}^{-3}\right) = \text{soil mass(Mg)}/\text{soil bulk volume}\left(\text{m}^3\right),\ \text{and} \qquad \text{(Eq. 2.2)}$$

$$\text{Soil particle density}\left(D_p,\ \text{Mg m}^{-3}\right) = \text{soil mass(Mg)}/\text{soil particle volume}\left(\text{m}^3\right) \qquad \text{(Eq. 2.3)}$$

(assumed to be 2.65 Mg m^{-3} for silicate minerals, but can be as high as 3.25 Mg m^{-3} for iron-rich tropical soils, and as low as 1.3 Mg m^{-3} for volcanic soils and organic soils),

$$\%\ \text{pore space} = 100 - \left[\left(D_b / D_p\right) \times 100\right] \qquad \text{(Eq. 2.4)}$$

where:

$$D_b = \text{soil bulk density},\ \text{Mg m}^{-3}\ \text{and} \qquad \text{(Eq. 2.5)}$$

$$D_p = \text{soil particle density},\ \text{Mg m}^{-3} \qquad \text{(Eq. 2.6)}$$

Although total pore space is important, the size and interconnection of the pores are key in determining the habitability of the soil and activity of soil organisms (Kravchenko and Gubera, 2017). This feature of the soil structure is largely a function of texture, aggregation, and location in the soil profile. Pore architecture affects the fluxes of air (Ball, 2013; Rappoldt and Crawford, 1999) and water (Or et al., 2007) in soil and regulates mobility of nutrients, resources, and microorganisms.

TABLE 2.3 Soil macropore and micropore distribution in soils across a textural gradient (% of total soil pore space).

Pore category	Pore function	Pore diameter (μm)	Sandy loam	Loam	Clay loam
Macropores	Aeration, water infiltration, drainage	>10	75	42	35
Micropores	Water retention	<10	25	58	65

Total pore space is usually divided into two size classes, macropores and micropores, largely based on their ability to retain water left after drainage under the influence of gravity (Table 2.3). Macropores are those larger than ~10 μm in diameter and allow for: (1) rapid diffusion of air and water vapor, (2) rapid water infiltration and drainage, and (3) access to nutrients. However, exposed surfaces of these pores dry quickly and restrict the activity of organisms living within water films. Macropores can occur as the spaces between individual sand and coarse silt grains in coarse-textured soils, and in the interaggregate pore space of well-aggregated loam- and clay-textured soils. Macropores can also be created by roots, earthworms, insects, and other soil organisms. Soil pores less than 10 μm in diameter are referred to as micropores and are important for retention of water for plants and for providing an aqueous habitat of sufficient water film thickness needed for the development of microbial biofilms. Water flow and diffusion of gases and nutrients in micropores is slow, typically <0.025 cm/hour. Thus the immediate environment for soil microorganisms remains relatively constant. While larger micropores, together with smaller macropores, can accommodate plant root hairs, microfauna and microorganisms, pores smaller than ~5 μm in diameter are not habitable by most organisms except viruses. They may even restrict diffusion of exoenzymes and nutrients, thereby inhibiting the uptake of otherwise bioavailable substrates. Although surface soils are typically about 50% pore space on a volume basis, only a quarter to a half of this pore space may be habitable by soil microorganisms. The larger pores drain relatively quickly and are too dry, and the smaller pores are too restricting in size.

The restricted size of the pores contained within smaller aggregates can slow accessibility of the associated organic matter to microbial decay and can thus restrict interactions of soil organisms, thereby protecting against predation by fauna. Larger aggregates facilitate water infiltration and soil aeration and remain intact as long as the soil is not disturbed by intensive tillage or water erosion. The pore space contained in the smaller aggregates making up the larger aggregates, referred to as intraaggregate pore space, is important for providing soil water retention. The pore space surrounding aggregates, collectively referred to as the interaggregate pore space, is where plant roots and larger fragments of plant residues are found, and its large channels promote water infiltration and aeration.

Aggregation is important for controlling microbial activity and SOM turnover in surface soils because the interaggregate pore space allows for the exchange of gases (CO_2, CH_4, and N_2O) generated by soil microorganisms and plant roots with oxygen and other gases in the atmosphere (Ball, 2013). It is also the site receiving annual deposits of plant residues. However, the larger pores in aggregates dry quickly, so they would only be a habitat for organisms able to live outside of thin water-films. In this way they may also regulate the accessibility of particulate organic matter to decay by soil microorganisms. By determining the nature and pore-size distribution within soil, aggregation can give fine-textured, clayey- and loamy-textured soils the beneficial pore space characteristics of sandy soils for aeration, water infiltration, and drainage.

2.4.1 Microscale of soil solids and pore space

The work of Kubiëna in the 1930s contributed significantly to our understanding of the nature of soil solid and pore space distributions at the microscopic scale (Kubiëna, 1938). Much of this early research was based on the examination of thin sections (25 μm thick) of intact blocks of soil. Adaptation of advancements in the acquisition and computer-assisted analysis of digital imagery during the past quarter century have led to the quantitative spatial analysis of soil components, morphology, and organization (Bullock et al., 1985; Stoops, 2021). However, thin sections represent only a single slice of soil and it is practically impossible to accurately extrapolate observations to three dimensions. Recent developments in microcomputerized X-ray tomography (CT scanning) allow for the study of the properties of the soil's intact three-dimensional structure (Fig. 2.5). These systems have high-resolution capabilities (10−30 μm), which allow differentiation of solids for quantifying the distribution of organic and mineral materials. The technology is also able to distinguish air-filled and water-filled pore space, and to identify the activities of larger soil organisms, for example, the biopores created by earthworms, ants, and

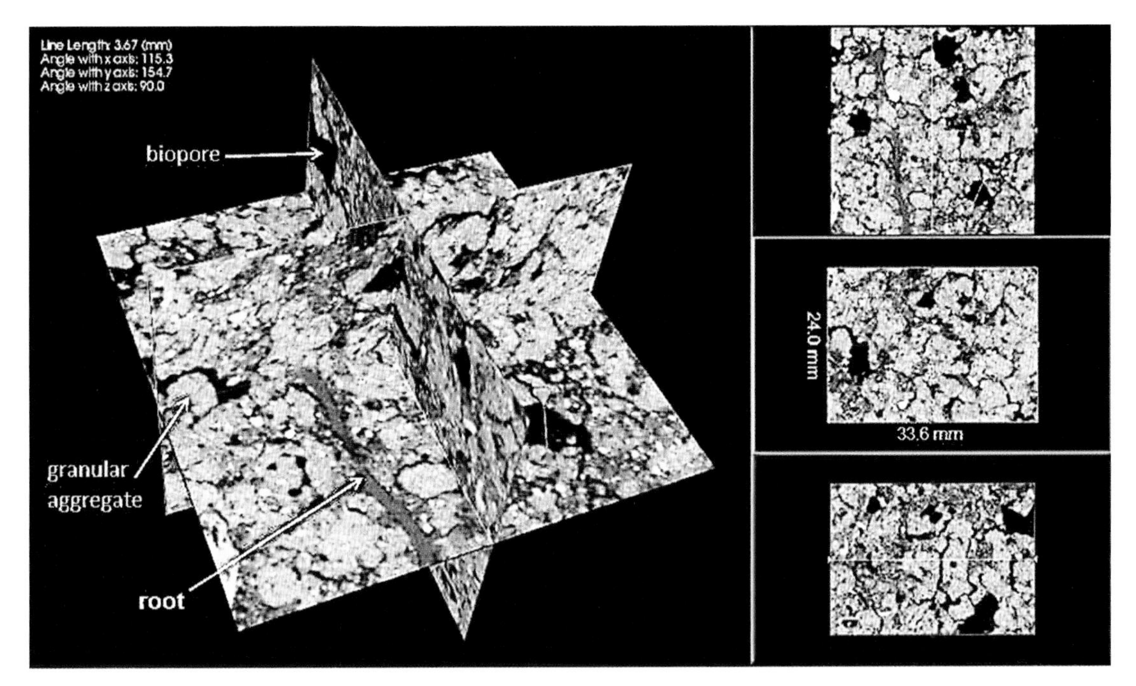

FIGURE 2.5 X-ray CT image of forest mull Ah horizon from Nepean, Ontario, Canada. The three images on the right correspond to the orthogonal planes in the main image. Voxel size of imagery is 40 μm; dimensions of the full image are 33.6 mm (width) × 33.6 mm (length) × 24.0 mm (height). By convention, air-filled pore space is dark and solid materials are lighter in tone. Diameter of the measured biopore, an earthworm channel, is ~3.7 mm.

termites. Highly attenuating features, such as iron oxide nodules, appear bright in the imagery, while features with low attenuation capability, such as pore space, appear dark. Though distinguishing microorganisms from soil particles with this technology is still limited, technologies, such as transmission electron microscopy, are capable of resolutions down to 0.150 μm. By overlaying images obtained from electron microscopy with nanoscale secondary ion mass spectrometry (NanoSIMS), two-dimensional examination of the chemistry of soil particle surfaces is possible (Fig. 2.6). Clusters of clay-sized minerals with patches of newly formed organic matter have been observed. These organomineral clusters were located on rough, not smooth, mineral surfaces, suggesting that cavities provide suitable microhabitats for microorganisms.

2.5 Soil hotspots

While the distribution of organisms and activity within the bulk soil matrix is relatively uniform, there are localized regions where population numbers and activity are several orders of magnitude higher. These areas are located wherever there is an abundance of readily available organic substrates, namely the mineral surface soil—litter boundary and in the rhizosphere, the soil most intimately associated with plant roots. A relatively new concept has been proposed for microbial hotspots and hot moments that explains spatial heterogeneity in soil biological activity at the microscale (Kuzyakov and Blagodatskaya, 2015).

Microbial hotspots can be defined as small soil volumes with faster process rates and more intensive interactions between nutrient pools compared to that in the mean bulk soil. This definition emphasizes the dynamic properties of hotspots, specifically the intensity of processes, relative to that of typical bulk soil measurements. The concept of microbial hotspots is based on localization of microorganisms and their functions in isolated, small volumes of soil where the main limitation for microbial growth, available organic carbon (C), is drastically reduced by intensive inputs of organic substances. These localized, high

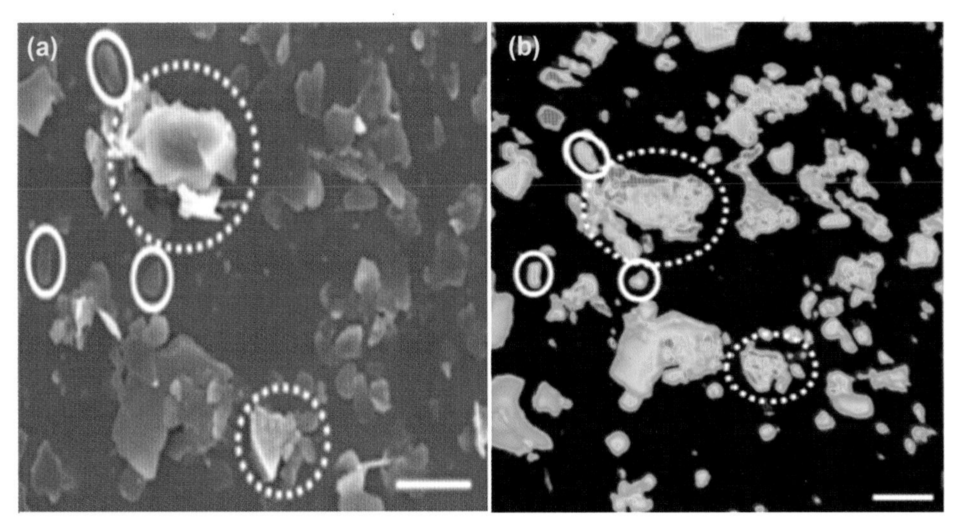

FIGURE 2.6 Organomineral clusters observed under transmission electron microscopy (a) and (b) nanoscale secondary ion mass spectrometry (NanoSIMS) to reveal clay-sized clusters, OM on mineral surfaces and patches of new litter-derived OM on rough mineral cluster surfaces. *(From Vogel et al., 2014.)*

C inputs have commonly attributed to above- and belowground plant residues; however, there are additional localized areas of high C inputs. Based on the sources of high inputs of labile organics and their localization in soil, four distinct hotspot groups have been described (Fig. 2.7). (1) The detritusphere is the surface soil volume containing dead and decomposing residues of plants and soil organisms. Detritusphere C inputs consist of recalcitrant highly polymeric, but also some labile, low-molecular-weight organics deposited as litter, primarily on the soil surface and on root death. (2) The rhizosphere is the soil volume surrounding the root that is strongly affected by root function and growth. Within the rhizosphere, labile root exudates and other less decomposable rhizodeposits are produced during seasonal root growth throughout the soil profile (Hinsinger et al., 2009; Jones et al., 2004; Kuzyakov and Razavi, 2019; Ma et al., 2017). (3) Biopores are tunnels burrowed in soils by earthworms and invertebrates and are formed by roots (Brown et al., 2000; Schrader et al., 2007; Tiunov and Scheu, 2004). Labile and recalcitrant organic C, originally from the detritusphere and rhizosphere and processed within the hindgut of earthworms (drilosphere) and other soil organisms, is deposited as linings in these biopores. Biopores can provide continuous tunnels extending from the soil surface for lengths of a meter or more down the soil profile and are contained in aboveground mounds, for example, of termites and ants. (4) Aggregate surfaces are stable soil volumes that receive fluxes of dissolved and particulate organic substrates leached and transported from the detritusphere, from the C-rich Ah horizon, and partly from the rhizosphere (Kaiser and Kalbitz, 2012). Because the detritusphere and rhizosphere derive their C directly from plant C

FIGURE 2.7 Microbial hotspots possessing faster processing rates and more intensive interactions of nutrients between pools compared to that of the mean bulk soil; (a) plant litter (detritusphere), (b) living plant roots (rhizosphere), (c) biopores and clusters of soil aggregates, and (d) drilosphere. *(With permission from M. Mesgar.)*

sources that are rich with readily available organic substrates, the organism abundance, activity, and soil volume affected are much higher than in biopores, aggregate surfaces that receive C that has already been processed by other organisms. The abundance of microorganisms in the detritusphere and rhizosphere is 3 to 10 times larger compared to average soil conditions. Typical bulk soil measurements of microbial activity reflect the mean activity of such hotspots, along with locations where the microorganisms are nearly absent. Therefore the actual contrast in activity between hotspot locations and the bulk soil matrix is probably much larger.

Other locations within the soil have been suggested as microbial hotspots. One example is the spermosphere, the immediate soil volume surrounding plant seeds where interactions between soil organisms and germinating seeds take place (Schiltz et al., 2015). It is expected that additional microbial hotspots in soil will be identified and characterized, especially when related to processes under O_2 limitations that cause greenhouse gas emissions (e.g., CH_4 and N_2O production).

Localized microbial activity in bulk soil can be contrasted with areas of higher process rates using various visualization approaches. Zymography, as an example, is a semiquantitative, two-dimensional imaging technique allowing analysis of the spatial distribution of enzyme activities on soil and root surfaces (Guber et al., 2018; Spohn et al., 2013; Spohn and Kuzyakov, 2014). This approach can be used, for example, to identify specific sites within the soil matrix with localized hydrolytic activity. Studies of hydrolytic enzymes (i.e., hydrolases), including phosphatases, β-glucosidase, chitinase, and cellobiohydrolase within the rhizosphere, have shown activities two to five times higher when compared to root-free soil (Fig. 2.8). These differences are even more pronounced in the detritusphere where there is spatially variable litter distribution and large amounts of decomposing litter (Liu et al., 2017; Ma et al., 2021). When CT scanning techniques are combined with two-dimensional zymography, the relationships of soil extracellular enzyme activity and soil pore size can be determined (Kravchenko et al., 2019).

Contrasts of hotspot properties in the rhizosphere can be visualized using planar optodes, a two-dimensional imaging technique for spatial analysis of localized soil properties such as O_2 and CO_2 concentrations, pH, temperature, and exoenzymes, among others (Li et al., 2019). For instance, planar optodes of acidification due to root release of H^+ show that the soil pH near roots is up to two pH units lower than that of bulk soil (Fig. 2.8).

Hotspots are not restricted to just the level of microbial groups (bacteria, fungi, protozoans, etc.) but cascade to the entire soil food web encompassing the micro- and macrofauna (Pausch et al., 2016). Some of these organisms, such as wood lice, spend their entire lives in the detritusphere comminuting plant litter into smaller fragments. Collembola and mites live in runways and near-surface soil pores, grazing on the microorganisms that are decomposing plant litter. Others, such as earthworms, ants, and termites, are capable of burrowing into the soil and bringing plant residues with them. Their burrowing activities produce secondary microbial hotspots located in the linings of biopores, which can reach deeper soils and thereby raise microbial activity in the whole soil profile.

The boundaries between hotspots and low-activity bulk soil have no sharp edges. For example, the detritusphere, which spans the surface litter; the Oi, Oe, and Oa horizons; and upper mineral topsoil (i.e., Ah horizon), is diffuse and difficult to distinguish due to the activities of the meso- and macrofauna mixing the mineral matrix and organic layers. Similarly, in the rhizosphere, microbial and enzyme activities extending from the root surface decrease within millimeters; therefore it does not correspond to a fixed soil volume (Kuzyakov and Razavi, 2019).

For statistical separation of hotspots from the bulk soil environment, a "Mean + 2SD" approach has been proposed (Fig. 2.9). The hotspot boundary is where the activity is higher than that of the mean bulk

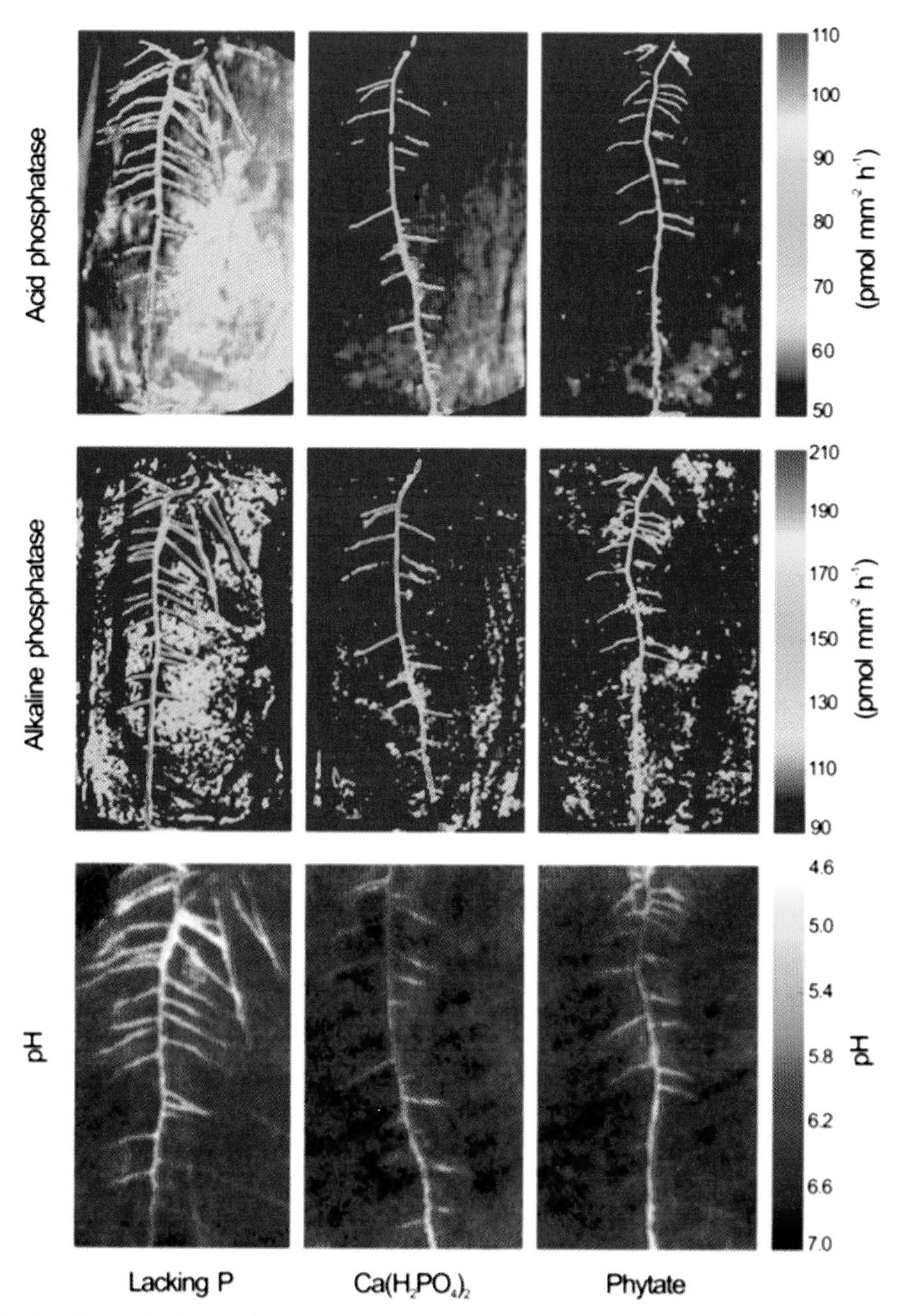

FIGURE 2.8 Lentils growing in rhizoboxes and zymography, a two-dimensional imaging technique, showing the spatial distribution of (a) acid phosphatase and (b) alkaline phosphatase. Side color maps are proportional to the enzyme activities (pmol cm^{-2} hour^{-1}). *(From Razavi et al., 2016.) Two-dimensional imaging obtained by planar optodes of soil pH in the rhizosphere of lupine grown in P-deficient soil unamended* (left), *amended with an inorganic P source* (middle), *or amended with an organic P source* (right). *Acidification due to root release of H^+ shows that the soil pH near roots is up to 2 pH units lower than that of the bulk soil (Ma et al., 2019.)*

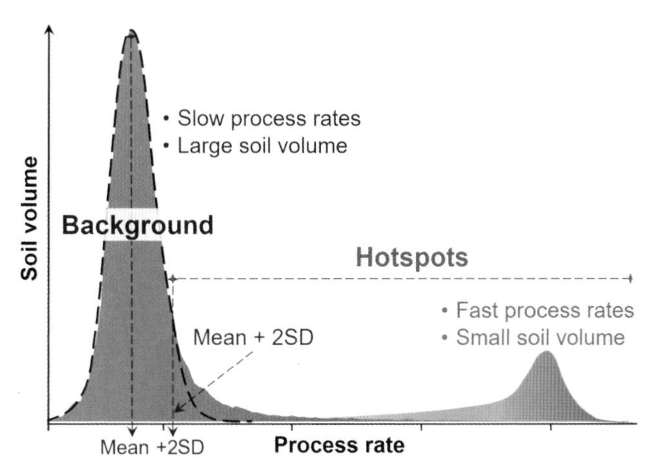

FIGURE 2.9 Concept of microbial hotspots in soil: hotspots are small soil volumes (*red*) with faster process rates and intensive interactions compared to the average soil conditions (background, *blue*). Statistical separation of hotspots from the background activity in bulk soil is presented by the Mean + 2SD approach. *(Modified from Bilyera et al., 2020; Kuzyakov and Blagodatskaya, 2015.)*

soil by two standard deviations. A computer program (Root-O-Mat) has been developed for separation of hotspots from background soil activities to identify hotspots and to estimate the area influenced by various enzymes (Tegtmeier et al., 2021).

It is well known that the fluxes of nutrients are intensive in hotspots and are based on faster decomposition rates associated with higher localized C inputs (e.g., leaf litter, rhizodeposits). Consequently, overall pool sizes, measured as contents of elements or specific substances, remain nearly the same or at least their increase is much less than the acceleration of the fluxes in the hotspots. This new concept suggests that a sharp research focus in future studies should be of localized site-specific process dynamics.

2.6 Soil water

Dependent on size, soil organisms live in or on the water-films adsorbed to particle surfaces. Water content determines availability to organisms, as well as soil aeration status, the nature and amount of soluble materials, osmotic pressure, and the pH of the soil solution. Water acts physically as an agent of transport by mass flow and as a medium through which reactants diffuse to and from sites of reaction. Chemically, water acts as a solvent and a reactant in chemical and biological reactions. Of special significance in the soil system in particular is the fact that water adsorbs strongly to itself and to soil particle surfaces by hydrogen bonding and dipole interactions.

Soil water content is expressed on a mass or volume basis. Gravimetric soil water content is the mass of water in the soil, measured as the mass loss in a soil dried at 105°C (oven-dry weight) and is expressed per unit mass of oven-dry soil. Volumetric soil water content is the volume of water per unit volume of soil. Soil water is also described in terms of its potential free energy, based on the concept of matric, osmotic, and gravitational forces affecting water potential. Soil water potential is expressed in units of pascals (Pa), or more commonly kilopascals (kPa), with pure water as a reference having a potential of

0 kPa. Matric forces are attributed to the adhesive or adsorptive forces of water attraction to surfaces of mineral and organic particles and to cohesive forces or attraction to itself. These forces reduce the free energy status of the water in soil to a negative value. Solutes dissolved in soil solution also contribute to a reduction in the free energy of water and give rise to an osmotic potential with a negative soil water potential. Combined, the matric and osmotic forces are responsible for the retention of water in soils and act in all directions. Fundamentally though, they act against gravitational force that tends to draw water downward and out of the soil. Gravitational force is usually positive.

When the gravitational force draining water downward is counterbalanced by the matric and osmotic forces holding onto water, the soil is said to be at field capacity or at its water holding capacity. This will occur after irrigation, after heavy rainfall, or at spring thaw, which leaves the soil saturated and having a soil water potential = 0 kPa. Gravitational forces drain away water in excess of that which can be retained by matric + osmotic forces, leaving the soil after 1 to 2 days at field capacity. By definition, the field capacity for loamy and clay loam soils is a soil water potential of −33 kPa and for sandy soils −10 kPa.

In the laboratory. the water holding capacity of a soil is determined by first saturating the soil (12−24 hours) and then, with care to minimize evaporation, allowing all the free water to drain (usually one to 2 days, but up to a week for clayey-textured soils). Typically, laboratory incubation studies used to simulate aerobic conditions are conducted at a gravimetric soil water content of 60 to 70% soil water holding capacity. Water retention or soil water content at a given soil water potential is a function of the sizes of pores present in the soil, or pore size distribution. Soils of different textures have very different water contents even though they have the same water potential. An important property of water influencing its behavior in soil pores and voids is surface tension. Based on matric forces and properties of surface tension, the maximum diameter of pores filled with water at a given soil water potential can be estimated using the Young-Laplace equation:

$$\text{Water film thickness}(\mu\text{m}) = 150/\text{Soil water potential (kPa)} \quad \text{(Eq. 2.5)}$$

Soil water potential(kpa)

Thus pores greater than 10 μm in diameter will be drained in soils at field capacity, and particle surfaces will be covered with a water film ~5 μm thick (Fig. 2.10).

Soil water characteristics are difficult to assess in laboratory and field studies. A computer program (SPAW) can be used to estimate soil water characteristics based on soil texture and organic matter content, which are commonly measured physical and chemical soil properties.

Soil water potential determines the energy that an organism must expend to obtain water from the soil solution. Generally, aerobic microbial activity in soil is considered optimal over soil water potentials ranging from about −50 to −150 kPa, which is 30 to 50% of the soil's total pore space depending on its texture and bulk density (Table 2.4). Aerobic activity decreases as soil becomes wetter and eventually saturated due to restricted O_2 diffusion. When greater than 60% of the pore space is water filled, the activity of microorganisms able to use alternative electron acceptors increases (e.g., anaerobic denitrifiers), resulting in the reduction of oxides of nitrogen, sulfur, and iron compounds.

As the soil dries and water potential decreases, water films on soil particles become thinner and more disconnected, restricting substrate and nutrient diffusion and increasing the concentration of salts in soil solution. Although many plants grown for agricultural purposes wilt permanently when the soil water potential reaches −1500 kPa, rates of soil microbial activity are less affected as the relative humidity within the soil remains high; respiration rates can still be ~90% of maximum. Studies in both laboratory

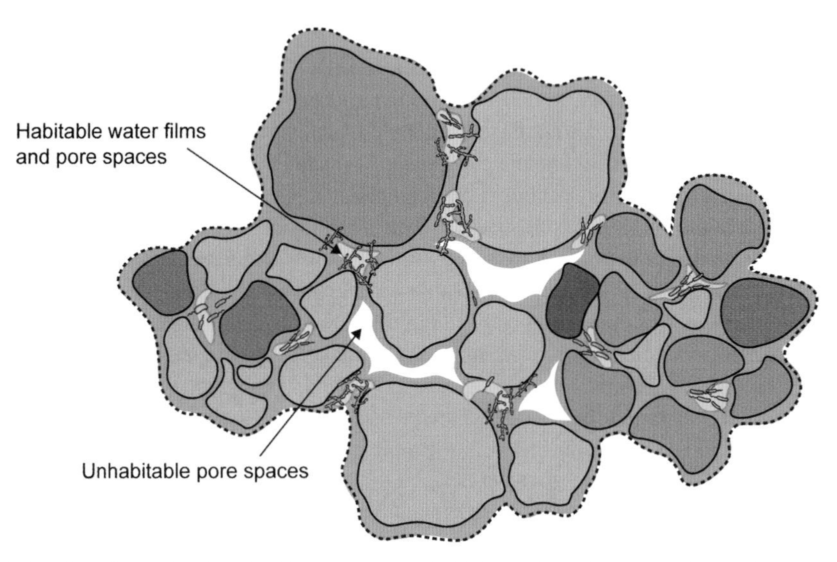

FIGURE 2.10 Water films 5 μm thick in soils at field capacity (−33 kPa) covering individual particles and contained in aggregates; air-filled pores >10 μm in diameter. *(Image with permission from M. Mesgar.)*

TABLE 2.4 Soil water potential, water film thickness, and volumetric soil water content (%) (100 × volume of water/volume of soil) across a textural gradient.

Soil water characteristic	Soil water potential (kPa)	Water film thickness (μm)	Sandy loam	Loam	Clay loam
Saturation	0	Filled	44	49	54
	−10	15	16	36	43
Field capacity	−33	5	11	28	35
	−100	1.5	9	21	32
	−500	0.3	6	14	23
Wilting point	−1500	0.1	4	8	16
Air-dry	−3100	0.04	4	6	11

incubations and in the field have reported that the logarithm of soil matric potential is a good predictor of the effect of water characteristics on soil respiration.

Some microorganisms are able to adapt to low soil water potentials by accumulating osmolytes (amino acids and polyols) or by altering the permeability of their outer membrane. The exopolysaccharide coatings of biofilms and microbial mats further benefit survival as they help retain water. However, rapid changes in soil water potential associated with drying/rewetting cause microbes to undergo osmotic shock and induce cell lysis. A flush of activity by the surviving biota, known as the Birch effect, results from mineralization of the labile cell constituents that are released.

Different microbial communities are responsible for microbial activity over the range of water potentials commonly found in soils. Protozoa are active at water potentials near field capacity in water films ≥ 5 μm thick, whereas microorganisms can be active at lower water potentials due to their size and association with the surfaces of soil particles. Fungi are generally considered to be more tolerant of lower soil water potentials than are bacteria, presumably because soil bacteria are relatively immobile and rely on diffusion processes for nutrition. Sulfur and ammonium oxidizers, typified by *Thiobacillus* and *Nitrosomonas* species, respectively, are less tolerant of water stress than are the ammonifiers typified by *Clostridium* and *Penicillium*. Ammonium may accumulate in droughty soils at water potentials where ammonifiers are still active, but nitrifiers are limited. Microalgae, protozoans, and nematodes that live in water films 5 to 10 μm thick are inactive at low water potentials. The general decline in microbial activity at soil water potentials lower than -500 kPa can be explained as due to limited diffusion of soluble substrates to microbes and to restricted microbial mobility in water films <1 μm thick.

2.7 Soil aeration

Molecular diffusion dominates the transport of gases in soil. Diffusion through continuous air-filled macropores maintains gaseous exchange between the atmosphere and the soil, and diffusion through water films of varying thickness maintains the exchange of gases with soil microorganisms and microfauna. Diffusivity through both pathways can be described by Fick's law:

$$J = D\,\mathrm{dc}/\mathrm{dx} \qquad \text{(Eq. 2.6)}$$

where J is the rate of gas diffusion (g cm^{-2} sec^{-1}), D is the diffusion coefficient for soil air and for water (cm^{2} sec^{-1}), c is the gas concentration (g cm^{-3}), x is the distance (cm), and dc/dx is the concentration gradient. The gaseous diffusion coefficient in soil air is much smaller than that in the atmosphere because of the limited fraction of total pore volume occupied by continuous air-filled pores and by pore tortuosity, where soil particles and water increase the mean path length available for diffusion. For soil air, it is referred to as the effective diffusion coefficient, D_e, and is a function of air-filled porosity. Likewise, tortuosity due to particulate material reduces rates of gaseous diffusion in soil solution. As shown in Table 2.5, diffusion of gases in water is ~1/10,000 of that in air. Thus gaseous diffusion through a 5 μm water film would take the same time as diffusion through a 5 cm air-filled pore.

2.8 Soil solution chemistry

An understanding of soil solution chemistry, which provides the environment for soil organisms, needs to take into account the nature and quantity of its major components: water, dissolved organic matter and inorganic constituents, and O_2 and CO_2. The biogeochemistry of the soil solution is largely determined by acid-base and redox reactions. Consequently, the thermodynamic activities of protons and electrons in soil solution define the chemical environment that controls biotic activity. Conceptually, both can be considered as flowing from regions of high concentration to regions of low concentration. Soil microbial activity has a profound effect on regulating this flow.

TABLE 2.5 Temperature effects on gaseous diffusion in (A) air and in (B) water, and (C) gas solubility in water.

A. Gaseous Diffusion Coefficients in Air (cm^2/sec)				
Temperature (°C)	N_2	O_2	CO_2	N_2O
0	0.148	0.178	0.139	0.179
10	0.157	0.189	0.150	0.190
20	0.170	0.205	0.161	0.206
30	0.180	0.217	0.172	0.218
B. Gaseous Diffusion Coefficients in Water ($\times 10^{-4}$ cm^2/sec)				
Temperature (°C)	N_2	O_2	CO_2	N_2O
0	0.091	0.110	0.088	0.111
10	0.130	0.157	0.125	0.158
20	0.175	0.210	0.167	0.211
30	0.228	0.275	0.219	0.276
C. Solubility Coefficients (volume of dissolved gas relative to volume of water, cm^3/cm^3)				
Temperature (°C)	N_2	O_2	CO_2	N_2O
0	0.0235	0.0489	1.713	1.30
10	0.0186	0.0380	1.194	1.01
20	0.0154	0.0310	0.878	0.71
30	0.0134	0.0261	0.665	0.42

2.8.1 Soil pH

Protons supplied to the soil from atmospheric and organic sources react with bases contained in aluminosilicates, carbonates, and other mineralogical and SOM constituents. In a humid climate, with excess precipitation and sufficient time, base cations (Na^+, K^+, Ca^{2+}, Mg^{2+}) will be exchanged from mineral and organic constituents by H^+ and leached from the surface soil. The presence of calcite and clay minerals, such as smectites which are saturated with base cations, retards the rate of acidification. Continued hydrolysis results in the formation of the secondary minerals, kaolinite, gibbsite, and goethite, and a soil solution buffered between pH 3.5 and 5. Semiarid and arid conditions lead to an opposite trend, a soil solution buffered at an alkaline pH. Soil pH influences a number of factors affecting microbial activity, such as solubility and ionization of inorganic and organic soil solution constituents, which in turn affect soil enzyme activity. There are a large number of both organic and inorganic acids found in soils, yet the majority of these acids are relatively weak.

Measurements of soil-solution pH provide important data for predicting potential microbial reactions and enzyme activity in soil. Though easily measured in a soil suspension with a pH electrode, interpretation of its effects on microbial processes is complicated, largely because concentrations of cations adsorbed to the surfaces of negatively charged soil colloids are 10 to 100 times higher than those of the soil solution. For enzymes sorbed to colloid surfaces, their apparent pH optimum is 1 to 2 pH units higher than if they were not sorbed. An example of this is soil urease activity, which has an apparent pH optimum of 8.5 to 9.0 in soil, which is about two pH units higher than optimal urease activity measured in solution.

2.8.2 Soil redox

The most reduced material in the biosphere is the organic matter contained in living biomass. Organic matter in soils ranges from total dominance, as in peatlands, to the minor amounts found in young soils or at depth in the vadose zone. The metabolic activity of soil organisms produces electrons during the oxidation of organic matter. These electrons must be transferred to an electron acceptor, with O_2 being the most abundant in freely drained, aerobic soils. The O_2 contained in soil air or present in soil solution can be consumed within hours depending on the activity of soil organisms and is replenished by O_2 diffusion. If O_2 consumption rates by soil organisms are high due to an abundant supply of readily decomposable organic C, or if O_2 diffusion into the soil is impeded because of waterlogging or restricted macroporosity, soil solution O_2 concentrations decrease. When all available dissolved O_2 is consumed, the solution changes from aerobic (oxic) to anaerobic (anoxic). Microbial activity will then be controlled by the movement of electrons to alternative electron acceptors, as well as by air diffusion rates into the hotspots of activity.

Development of anaerobic conditions results in a shift in the activity of the soil microbial populations, with the activity of aerobic and facultative organisms which dominate well-drained oxic soils decreasing and the activity of obligate anaerobic and fermentative organisms increasing. This switch in electron acceptors promotes the reduction of several important elements in soil, including nitrogen, manganese, iron, and sulfur, in a process known as anaerobic respiration, and of CO_2 by methanogenesis. Evidence of O_2 limiting conditions are mottles, often seen in the subsoil, whereby rusty-colored iron oxide (Fe^{3+}) is reduced to grayish-colored (gleyed) Fe^{2+} (Fig. 2.11).

2.9 Soil temperature

Many physical, chemical, and biological processes that occur in soil are influenced by temperature. Increasing temperature enhances mineralization of SOM or decomposition of plant residues by increasing rates of physiological reactions and by accelerating diffusion of soluble substrates in soil. An increase in temperature can also induce a shift in the composition of the biotic community. Whereas rates of molecular diffusion always increase with increasing temperature, solubility of gases in soil solution does not, and it can even decrease, thereby slowing biotic activity (Table 2.5).

The relation between a chemical reaction rate and temperature was first proposed by Arrhenius:

$$k = A\, e^{-\mathrm{Ea}/\mathrm{RT}} \qquad \text{(Eq. 2.7)}$$

The constant A is called the frequency factor and is related to the frequency of molecular collisions, Ea is the activation energy or energy required to initiate the reaction, R is the gas constant R and has a value

FIGURE 2.11 Mottles providing evidence of oxidized and reduced iron. *(Photo with permission from M. Mesgar.)*

of 8.314×10^{-3} kJ mol^{-1} T^{-1}, e is the base of the natural logarithm, T is the absolute temperature (in °K), and k is the specific reaction rate constant ($time^{-1}$).

Conversion of Eq. 2.7 to natural logarithmic form gives:

$$\ln k = (-Ea/RT) + \ln A \qquad \text{(Eq. 2.8)}$$

By determining the value of k over a moderate range of soil temperatures, the plot of ln k versus 1/T results in a linear relationship providing the activation energy is constant over the temperature interval; Ea is obtained from the slope of the line, and A, the frequency factor, from the intercept.

A similar equation can be used to describe temperature effects on enzyme activity:

$$k_{(cat)} = k(K_B T/h)e^{-\Delta G\#/RT} \qquad \text{(Eq. 2.9)}$$

where $k_{(cat)}$ is the reaction rate ($time^{-1}$), κ is the transmission coefficient, k_B is the Boltzmann constant, h is the Planck constant, ΔG# is the activation energy, R is the universal gas constant, and T is the absolute temperature (in K). The transmission coefficient varies significantly with the viscosity of the soil solution, which increases by a factor of almost 2 over a temperature drop from 20°C to near 0°C. In both instances a change in temperature results in an exponential change in the reaction rate, the magnitude of which is a function of the activation energy. Note also that there is an inverse relationship between reaction rate and activation energy.

The relationship between temperature and biologically mediated processes is complicated as individual species differ in the optimal temperature response, different microbial communities are active as temperatures change, and microorganisms are able to acclimate by altering their physiology and cellular mechanisms, membrane fluidity and permeability, and structural flexibility of enzymes and proteins.

The relative sensitivity of soil microbial activity to temperature can be expressed as a Q_{10} function, which is the proportional change in activity associated with a 10°C temperature change

$$Q_{10} = (k_2/k_1)^{[(10/T2-T1)]} \quad \text{(Eq. 2.10)}$$

where k_2/k_1 are the rate constants for a microbial process under study at temperatures differing by 10°C. It is generally accepted that a Q_{10} of ~2 can be used to describe the temperature sensitivity of soil biochemical processes, such as respiration, over the mesophilic temperature range (20–45°C); that is, microbial activity at 30°C is twofold higher than it is at 20°C. At temperatures beyond 45°C, microbial community composition shifts from mesophilic to thermophilic, and microorganisms adapt by increasing concentrations of saturated fatty acids in their cytoplasmic membranes and by production of heat-stable proteins. Soil chemical reaction rates increase, often very sharply at low temperatures, with increases in temperature due to increased molecular interactions. Since individual species differ in their optimal response to temperature, quite different microbial communities dominate activity over the range of soil temperatures.

Microorganisms that have an upper growth temperature limit of 20°C, commonly referred to as psychrophiles, are capable of growth at low temperatures by adjusting upward both the osmotic concentration of their cytoplasmic constituents to permit cell interiors to remain unfrozen and the proportion of unsaturated fatty acids in their cytoplasmic membrane. A common adaptive feature of psychrophiles to low temperatures is that their enzymes have much lower activation energies and much higher (up to 10-fold) specific activities than do those of mesophiles, resulting in a reaction rate, $k_{(cat)}$, that is largely independent of temperature. Although microbial activity slows at lower temperatures, rates are significantly higher and more sensitive to temperature changes than those predicted from studies over the mesophilic temperature range. Researchers studying decomposition of SOM, soil respiration, and N mineralization have reported values for Q_{10} increasing to near 8 to 10 with a soil temperature increase from −5 to +5°C (Kirschbaum, 2013). Díaz-Raviña et al. (1994) reported Q_{10} values between 3.7 and 6.7 for the 0 to 10°C interval for thymidine incorporation and between 5.0 and 13.9 for acetate incorporation for a soil bacterial community.

Low temperatures are common over vast areas of the Earth and include soils in temperate, polar, and alpine regions where mean annual temperatures are <5°C. Soils in these environments contain a great diversity of cold-adapted microorganisms able to thrive even at subzero temperatures and to survive repeated freeze/thaw events. Though global warming may result in only a few degrees of temperature change, it is predicted to dramatically increase microbial decay rates of the huge organic matter reserves stored in the tundra, which will boost emissions of CO_2 and CH_4.

Very few soils maintain a uniform temperature in their upper layers. Variations may be either seasonal or diurnal. Because of the high specific heat of water, wet soils are subject to lower diurnal temperature fluctuations than are dry soils. Among factors affecting the rate of soil warming, the intensity and reflectance of solar irradiation are critical. The aspect (south versus north facing slopes), steepness of the slope, degree of shading, and surface cover (vegetation, litter, mulches) determine effective solar irradiation. Given the importance of soil temperature in controlling soil processes, models of energy movement into the surface soil profile have been developed. They are based on physical laws of soil heat transport and thermal diffusivity and include empirical parameters related to the temporal (seasonal) and sinusoidal variations in the diurnal pattern of near-surface air temperatures. The amplitude of the diurnal soil temperature variation is greatly dampened with profile depth.

2.10 Environmental factors, temperature, and moisture interactions

Soil moisture and temperature are the critical factors affected by climate regulating soil biological activity. This control is affected by changes in the underlying rates of enzyme-catalyzed reactions and sizes of the substrate organic and inorganic pools. Where water is nonlimiting, biological activity depends primarily on temperature, and standard Arrhenius theory can be used to predict temperature effects. But as soils dry, moisture is more controlling of biological processes than is temperature. These two environmental influences do not regulate microbial activity in linear relationships but display complex, nonlinear interactions that reflect the individual responses of the various microbial communities that develop, and their associated enzyme systems (Davidson and Janssens, 2006).

The interaction of temperature, moisture, and organisms is exemplified by the current concerns about the effects of climate change on soil biology. A hundred years ago, Swedish scientist Svante Arrhenius asked the important question, "Is the mean temperature of the ground in any way influenced by the presence of the heat-absorbing gases in the atmosphere?" He went on to become the first person to investigate the effect that doubling atmospheric CO_2 would have on global climate. The question was debated throughout the early part of the 20th century and is a main concern of Earth system scientists today. The Earth's surface is warming for several reasons, among which are increased emissions of greenhouse gases from combustion of fossil fuels, industrial pollution, deforestation, and soil disturbance due to past and current agricultural management practices. Global temperatures have increased by ~0.5°C over the past 100 years and are expected to increase another 1°C to 6°C by 2100. Although this represents only a few degrees of temperature change, global warming will dramatically increase microbial decay rates of the organic matter stored in boreal forests and tundra regions, estimated to contain ~30% of global soil C (Kirschbaum, 1995). The permafrost thaw rate has more than tripled over the past half-century.

The critical concern is that SOM decomposition in a warmer Earth will be stimulated to a greater extent than is net plant productivity, the C input to SOM, due to increased nutrient cycling. Theory suggests that the resistant constituents of SOM, with high activation energies, would become more decomposable at higher temperatures (Davidson and Janssens, 2006). Degradation of cellulose, hemicellulose, and other components of SOM by extracellular enzymes is the rate-limiting step in O_2 emissions. However, feedback mechanisms characteristic of all biogeochemical cycles may dampen the effects of temperature changes.

Soils are complex, and many environmental constraints affect decomposition reactions by altering organic matter (substrate) concentrations at the site at which all decomposition occurs, that of the enzyme reaction site. We must also consider decomposition rates at the enzyme affinity level: Michaelis-Menten models of enzyme kinetics and energy yield, as discussed in Chapter 9. The kinetic and thermodynamic properties of extracellular enzymes and their responses to environmental factors are now being considered in models of the effects of global warming on carbon cycling. Changes in microbial community structure (Chapter 3) will also have profound influences. The goal of this chapter is to provide an environmental boundary of the soil habitat and a description of its fundamental physical, chemical, and biological properties. With this as a foundation, later chapters in this volume explore, in detail, information about organisms, their ecology, biochemistry, and interactions.

References

Amundson, R., Jenny, H., 1991. The place of humans in the state factor theory of ecosystems and their soils. Soil Sci. 151, 99–109.

Ball, B.C., 2013. Soil structure and greenhouse gas emissions: a synthesis of 20 years of experimentation. Eur. J. Soil Sci. 64, 357–373.

Bilyera, N., Kuzyakova, I., Guber, A., Razavi, B.S., Kuzyakov, Y., 2020. How "hot" are hotspots: statistically localizing the high-activity areas on soil images. Rhizosphere 16, 100259. https://doi.org/10.1016/j.rhisph.2020.100259.

Brown, G.G., Barois, I., Lavelle, P., 2000. Regulation of soil organic matter dynamics and microbial activity in the drilosphere and the role of interactions with other edaphic functional domains. Eur. J. Soil Biol. 36, 177–198.

Bullock, P., Fedoroff, N., Jongerius, A., Stoops, G., Tursina, T., Babel, U., 1985. Handbook for Soil Thin Section Description. Waine Research Publications, Wolverhampton.

Chadwick, O.A., Chorover, J., 2001. The chemistry of pedogenic thresholds. Geoderma 100, 321–353. https://doi.org/10.1016/S0016-7061(01)00027-1.

Davidson, E.A., Janssens, I.A., 2006. Temperature sensitivity of soil carbon decomposition and feedbacks to climate change. Nature 440 (7081), 165–173.

Díaz-Raviña, M., Frostegård, Å., Bååth, E., 1994. Thymidine, leucine and acetate incorporation into soil bacterial assemblages at different temperatures. FEMS Microbiol. Ecol. 14, 221–232.

Dokuchaev, V.V., 1883. The Russian Chernozem Report to the Free Economic Society. Imperial University of St. Petersburg, St. Petersburg, Russia.

Gerasimov, I., 1984. The system of basic genetic concepts that should be included in modern Dokuchayevian soil science. Sov. Geogr. 25, 1–14.

Guber, A., Kraychenko, A., Razavi, B.S., Uteau, D., Peth, S., Blagodatskaya, E., et al., 2018. Quantitative soil zymography: mechanisms, processes of substrate and enzyme diffusion in porous media. Soil Biol. Biochem. 127, 156–167. https://doi.org/10.1016/j.soilbio.2018.09.030.

Hinsinger, P., Bengough, A.G., Vetterlein, D., Young, I.M., 2009. Rhizosphere: biophysics, biogeochemistry and ecological relevance. Plant Soil 321, 117–152.

Jenny, H., 1941. Factors of Soil Formation: A System of Quantitative Pedology. McGraw-Hill, New York.

Jones, D.L., Hodge, A., Kuzyakov, Y., 2004. Plant and mycorrhizal regulation of rhizodeposition. New Phytol. 163, 459–480.

Kaiser, K., Kalbitz, K., 2012. Cycling downwards – dissolved organic matter in soils. Soil Biol. Biochem. 52, 29–32. https://doi.org/10.1016/j.soilbio.2012.04.002.

Kirschbaum, M.U.F., 1995. The temperature dependence of soil organic-matter decomposition, and the effect of global warming on soil C storage. Soil Biol. Biochem. 27, 753–760.

Kirschbaum, M.U.F., 2013. Seasonal variations in the availability of labile substrate confound the temperature dependence of organic matter decomposition. Soil Biol. Biochem. 57, 568–576. https://doi.org/10.1016/j.soilbio.2012.10.012.

Kravchenko, A.N., Gubera, A.K., 2017. Soil pores and their contributions to soil carbon processes. Geoderma 287, 31–39. https://doi.org/10.1016/j.geoderma.2016.06.027.

Kravchenko, A.N., Gubera, A.K., Razavi, B.S., Koestel, J., Blagodatskaya, E.V., Kuzyakov, Y., 2019. Spatial patterns of extracellular enzymes: combining X-ray computed micro-tomography and 2D zymography. Soil Biol. Biochem. 135, 411–419. https://doi.org/10.1016/j.soilbio.2019.06.002.

Kubiëna, W.L., 1938. Micropedology. Collegiate Press, Ames.

Kuzyakov, Y., Blagodatskaya, E.V., 2015. Microbial hotspots and hot moments in soil. Soil Biol. Biochem. 83, 184–199. https://doi.org/10.1016/j.soilbio.2015.01.025.

Kuzyakov, Y., Razavi, B.S., 2019. Rhizosphere size and shape: temporal dynamics and spatial stationarity. Soil Biol. Biochem. 135, 343–360. https://doi.org/10.1016/j.soilbio.2019.05.011.

Kuzyakov, Y., Zamanian, K., 2019. Reviews and syntheses: agropedogenesis – humankind as the sixth soil-forming factor and attractors of agricultural soil degradation. Biogeosciences 16, 4783–4803. https://doi.org/10.5194/bg-16-4783-2019.

Li, C., Ding, S., Yang, L., Zhu, Q., Chen, M., Tsang, D., et al., 2019. Planar optode: a two-dimensional imaging technique for studying spatial-temporal dynamics of solutes in sediment and soil. Earth Sci. Rev. 197, 102916. https://doi.org/10.1016/j.earscirev.2019.102916.

Liu, S.B., Razavi, B.S., Su, X., Maharjan, M., Zarebanadkouki, M., Blagodatskaya, E., et al., 2017. Spatio-temporal patterns of enzyme activities after manure application reflect mechanisms of niche differentiation between plants and microorganisms. Soil Biol. Biochem. 112, 100–109. https://doi.org/10.1016/j.soilbio.2017.05.006.

Ma, X., Razavi, B.S., Holz, M., Blagodatskaya, E., Kuzyakov, Y., 2017. Warming increases hotspot areas of enzyme activity and shortens the duration of hot moments in the root detritusphere. Soil Biol. Biochem. 107, 226–233. https://doi.org/10.1016/j.soilbio.2017.01.009.

Ma, X., Mason-Jones, K., Liu, Y., Blagodatskaya, E., Kuzyakov, Y., Guber, A., et al., 2019. Coupling zymography with pH mapping reveals a shift in lupine phosphorus acquisition strategy driven by cluster roots. Soil Biol. Biochem. 135, 420–428. https://doi.org/10.1016/j.soilbio.2019.06.001.

Ma, X., Liu, Y., Shen, W., Kuzyakov, Y., 2021. Phosphatase activity and acidification in lupine and maize rhizosphere depend on P availability and root properties: coupling zymography with planar optodes. Appl. Soil Ecol. 167, 104029. https://doi.org/10.1016/j.apsoil.2021.104029.

Miltner, A., Bombach, P., Schmidt-Brücken, B., Kästner, M., 2012. SOM genesis: microbial biomass as a significant source. Biogeochemistry 111, 41–55.

Or, D., Smets, B.F., Wraith, J.M., Dechesne, A., Friedman, S.P., 2007. Physical constraints affecting bacterial habitats and activity in unsaturated porous media – a review. Adv. Water. Res. 30 (6–7), 1505–1527.

Pausch, J., Kramer, S., Scharroba, A., Scheunemann, N., Butenschoen, O., Kandeler, E., et al., 2016. Small but active – pool size does not matter for carbon flow in belowground food webs. Funct. Ecol. 30, 479–489. https://doi.org/10.1111/1365-2435.12512.

Rappoldt, C., Crawford, J.W., 1999. The distribution of anoxic volume in a fractal model of soil. Geoderma 88 (3–4), 329–347.

Razavi, B.S., Blagodatskaya, E., Kuzyakov, Y., 2016. Temperature selects for static soil enzyme systems to maintain high catalytic efficiency. Soil Biol. Biochem. 97, 15–22. https://doi.org/10.1016/j.soilbio.2016.02.018.

Schiltz, S., Gaillard, I., Pawlicki-Jullian, N., Thiombiano, B., Mesnard, F., Gontier, E., 2015. A review: what is the spermosphere and how can it be studied? J. Appl. Microbiol. 119, 1467–1481. https://doi.org/10.1111/jam.12946.

Schrader, S., Rogasik, H., Onasch, I., Jegou, D., 2007. Assessment of soil structural differentiation around earthworm burrows by means of X-ray computed tomography and scanning electron microscopy. Geoderma 137, 378–387.

Spohn, M., Carminati, A., Kuzyakov, Y., 2013. Soil zymography – a novel in situ method for mapping distribution of enzyme activity in soil. Soil Biol. Biochem. 58, 275–280. https://doi.org/10.1016/j.soilbio.2012.12.004.

Spohn, M., Kuzyakov, Y., 2014. Spatial and temporal dynamics of hotspots of enzyme activity in soil as affected by living and dead roots-a soil zymography analysis. Plant Soil 379, 67–77.

Stoops, G., 2021. Guidelines for Analysis and Description of Soil and Regolith Thin Sections. Soil Science Society of America, Inc. and John Wiley & Sons, Inc.

Tegtmeier, J., Dippold, M.A., Kuzyakov, Y., Spielvogel, S., Loeppmann, S., 2021. Root-o-Mat: a novel tool for 2D image processing of root-soil interactions and its application in soil zymography. Soil Biol. Biochem. 157, 108236. https://doi.org/10.1016/j.soilbio.2021.108236.

Tiunov, A.V., Scheu, S., 2004. Carbon availability controls the growth of detritivores (*Lumbricidae*) and their effect on nitrogen mineralization. Oecologia 138, 83–90.

Vogel, C., Mueller, C.W., Höschen, C., Buegger, F., Heister, K., Schulz, S., et al., 2014. Submicron structures provide preferential spots for carbon and nitrogen sequestration in soils. Nature. Commun 5, 2947. https://doi.org/10.1038/ncomms3947.

Supporting online material

The Global Soil Biodiversity Atlas: Atlas Introduction — Global Soil Biodiversity Initiative

Soil Stories, The Whole Story: http://youtube.com/watch?v? Ego6LI-IjbY; see online supplemental material at: http://booksite.elsevier.com/9780124159556.

The Five Factors of Soil Formation: http://www.youtube.com/watch?v=bTzslvAD1Es; see online supplemental material at: http://booksite.elsevier.com/9780124159556.

Hans Jenny Memorial Lecture in Soil Science – The Genius of Soil: http://www.youtube.com/watch?v=y3q0mg54Li4; see online supplemental material at: http://booksite.elsevier.com/9780124159556.

Soil horizons – see Web Soil Survey: https://www.nrcs.usda.gov/wps/portal/nrcs/site/soils/home.

Maps of soils located throughout the United States and descriptions of their compositions are available at this website: https://casoilresource.lawr.ucdavis.edu/gmap/.

How to test your soil texture: https://www.youtube.com/watch?v=fufeaLBLGlk; see online supplemental material at: http://booksite.elsevier.com/9780124159556.

Estimation of soil water characteristics based on soil texture and organic matter content. see online supplemental material at. http://hydrolab.arsusda.gov/soilwater/Index.htm. http://booksite.elsevier.com/9780124159556.

Soil water characteristics explained:

https://www.nrcs.usda.gov/wps/ portal/nrcs/detailfull/national/water/manage/drainage/?cid=stelprdb1045310.

Chapter 3

Soil bacteria and archaea

Sreejata Bandopadhyay*,[1] and Ashley Shade†,[2]
*Department of Microbiology and Molecular Genetics, Michigan State University, East Lansing, MI, USA; †Department of Microbiology and Molecular Genetics; Department of Plant, Soil and Microbial Sciences, Michigan State University, East Lansing, MI, USA

Chapter outline

1. Present address: Soil Ecosystems Science Team, Biological Systems Science, Earth and Biological Science Directorate, Pacific Northwest National Laboratory, Richland, WA, United States
2. Present address: Université de Lyon, France; Research Group on "Bacterial efflux and Environmental Resistance", CNRS, INRAe, Ecole Nationale Véterinaire de Lyon and Université Lyon 1, UMR 5557 Écologie Microbienne, Villeurbanne Cedex, France

Soil Microbiology, Ecology, and Biochemistry. https://doi.org/10.1016/B978-0-12-822941-5.00003-X

3.1 General introduction to bacteria and archaea

3.1.1 Overview

The lineages of microbes that compose the bacteria and archaea are the most numerically abundant organisms on Earth. They collectively offer greater breadth of phylogenetic and functional diversity than all of the eukaryotic lineages combined, and by magnitudes. Soil bacteria and archaea are key sources and transformers of organic carbon (C) and nitrogen (N), and thus are important contributors to global biogeochemical cycles. Consequently, bacterial and archaeal populations in soil help maintain proper functioning of terrestrial ecosystems.

Bacteria and archaea are two phylogenetic domains that belong to a classification previously known as prokaryotes. Prokaryotes were grouped based on their differences from eukaryotes, which include organisms such as algae, protozoa, fungi, plants, and animals. Most prokaryotes are unicellular, while eukaryotes include both unicellular and multicellular organisms. Also, prokaryotes lack cellular compartments that separate from the cytoplasm via membranes, while eukaryotes have compartmentalized organelles that carry out specific functions. Prokaryotes are further distinguished by their nuclear division by fission and the presence of cell walls that contain peptidoglycan. Ribosomes, the translational machinery of the cell, are smaller in prokaryotes (70S) than in eukaryotes (80S). Some functions, such as atmospheric N fixation and methane production, are uniquely prokaryotic. However, there are many exceptions to many of the earlier attributed "defining" features of prokaryotes. Phylogenetic analysis has confirmed that the bacteria and archaea have divergent evolutionary histories that place them in entirely separate domains of life. The term "prokaryotes" is now considered imprecise by many, and even old-fashioned, because it does not accurately reflect evolutionary relationships.

In this chapter, the cellular, metabolic, and functional characteristics of soil bacteria and archaea will be described to provide insights into how these organisms are adapted and responsive to the terrestrial environment. This chapter will also discuss the major phylogenetic lineages of soil bacteria and archaea in soils, in addition to their traits, abundances, distributions, and their broader context within the universal tree of life.

3.1.2 Historical and recent advances in understanding the tree of life

To understand soil bacteria and archaea, it is first necessary to understand their phylogenetic context within the universal tree of life. The understanding of the universal tree of life continues to advance rapidly. Most of the recent changes concern the expansion of the known diversity of lineages within the bacteria and archaea, and their phylogenetic relationships to each other and to the eukaryotes. By sequencing the 16S ribosomal RNA (rRNA) gene as a biomarker and then using those sequences to build the first universal phylogeny, Woese et al. (1990) discovered that the archaea were a separate domain from bacteria, despite their shared unicellularity and lack of organelles (Fig. 3.1a). This discovery eventually led to the acceptance that "prokaryotes" were not monophyletic. More recently, advances in sequencing technology and high-throughput methods permitted a deep and expansive perspective of bacterial and archaeal diversity, revealing divergent lineages that had eluded discovery because the common 16S rRNA gene primers did not capture them. These lineages include the candidate phyla radiation (CPR), which is a large, evolutionary radiation of candidate bacterial phyla whose members are mostly uncultivable and have been understood through advances in metagenomics (Hug et al., 2016). Genome-centric metagenomics (Parks et al., 2017) and single-cell (Stepanauskas et al., 2017; Woyke et al., 2010) approaches

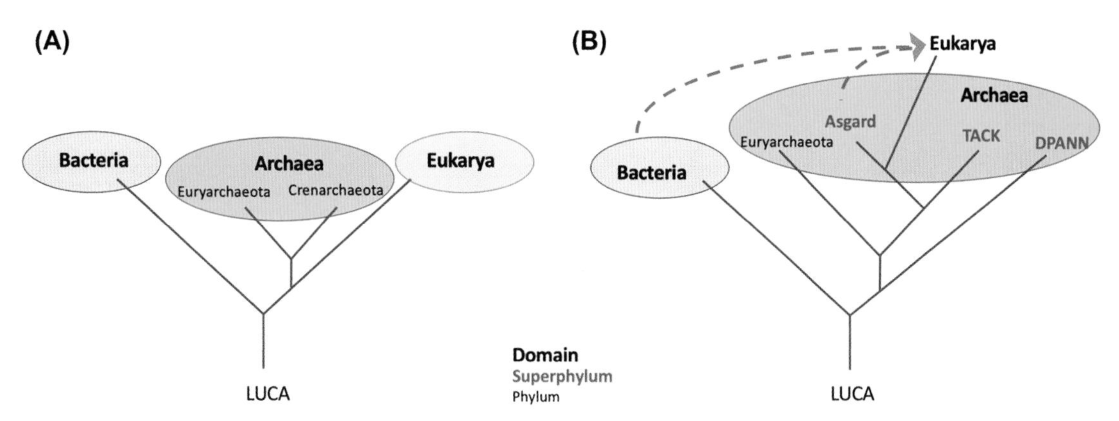

FIGURE 3.1 (A) A simplified depiction of the original universal tree of life as proposed by Woese et al. (1990) based on available 16S rRNA gene sequences available at the time. The original tree contained three domains of life (bacteria, archaea, and eukarya) and the domain archaea was further delineated into "kingdoms:" Euryarchaeota and Crenarchaeota. (b) A simplified depiction of the current view of the universal tree of life with two domains (bacteria and archaea), with eukarya deriving from the *Asgard* superphylum within the archaea. *LUCA*, Last Universal Common Ancestor.

continue to uncover phylogenetic novelty within bacteria and archaea. Regardless, all new knowledge affirms the centrality of bacteria and archaea in understanding the evolution of life and their dominance in both numbers and diversity.

Knowledge about the universal tree of life and microbial lineages therein continues to change rapidly. There are two major points of note. The first concerns the domains of life. Recent evidence from the discovery of the Lokiarchaeota and analyses of related lineages within the Thaumarchaeota, Aigarchaeota, Crenarchaeota, Korarchaeota (TACK) group suggest that Woese's three-domain tree (Woese et al., 1990), which included the bacteria, archaea, and eukarya, is no longer supported. Instead, there is likely a two-domain tree consisting of bacteria and archaea, with eukarya as a lineage deriving from the Asgard superphylum within the archaea, and merged with bacterial endosymbiont(s) (Fig. 3.1b). It is hypothesized that the symbiogenesis of Alphaproteobacteria with an Asgard archaeon derived a eukaryotic cell, with the former becoming the mitochondria and the latter serving as its host. Though the current view of the universal tree of life is yet unsolved and debated, our understanding of the tree of life originating from the Last Universal Common Ancestor (LUCA) continues to change with improved molecular tools and observations.

The second point is that the model of the progressive, branching universal tree of life falls short because it does not completely acknowledge the widespread importance of horizontal transfer of genes or the importance of the co-mingling of previously separate organisms into one biological unit. For example, the eukaryotic energy-capturing organelles, chloroplasts, and mitochondria were previously free-living Cyanobacteria and Alphaproteobacteria, respectively, that were engulfed as endosymbionts within a pre-eukaryotic cell, likely an archaeon. Bacteria and archaea can transfer or gain new genes via plasmids (conjugation), viruses (transduction), or uptake of free DNA from the environment (transformation). Thus the universal tree of life is more analogous to a "tangled web" than to a tree.

A final note is that phylogenetic relationships can be inferred from any gene sequence, and in soil microbiology and ecology, it can be useful to resolve relationships using phylogenetic markers or conserved functional genes. The intergenic spacer (ITS) is often used for yeasts and other fungi, and the 18S rRNA genes are common phylogenetic markers for microeukaryotes or invertebrates, such as nematodes. RNA polymerase (e.g., rpoB) and rRNA proteins (e.g., rpl) have also been used as alternatives to the 16S rRNA gene for bacterial and archaeal classification. Functional genes, such as those providing resistance to antibiotics, involved in N fixation or in transformations in elemental cycles (ammonia oxidation, carbon assimilation) are also used to build phylogenies and can offer more focus and resolution for particular questions of interest to environmental or agricultural research.

3.2 Metabolism and physiology

3.2.1 Bacterial and archaeal cell characteristics

The external structures surrounding bacterial and archaeal cells are called cell walls (Fig. 3.2). These rigid structures protect the cells from osmotic lysis. The structure of the cell wall is used to phenotypically classify bacteria and archaea and enables distinction between gram-positive (henceforth Gr+) and gram-negative bacteria (henceforth Gr−). Gr+ cell walls (Fig. 3.2a) consist of a single layer of peptidoglycan surrounding the cytoplastic membrane. Peptidoglycan (murein) is an essential component that preserves cell integrity by withstanding turgor. It helps maintain a distinct cell shape and serves as a framework for anchoring other external cellular components such as proteins and teichoic acids. The Gr− cell wall (Fig. 3.2b) is more complex than the Gr+ cell wall, with the peptidoglycan being much thinner (5−10% of total cell wall) and surrounded by an outer membrane enclosing a periplasmic space. The periplasmic space is a separate cellular compartment, which is important for protein transport, protein folding, protein oxidation, electron transport, nutrient acquisition, and toxin exclusion. The outer membrane is different from the cell membrane due to the presence of large molecules called lipopolysaccharides, which

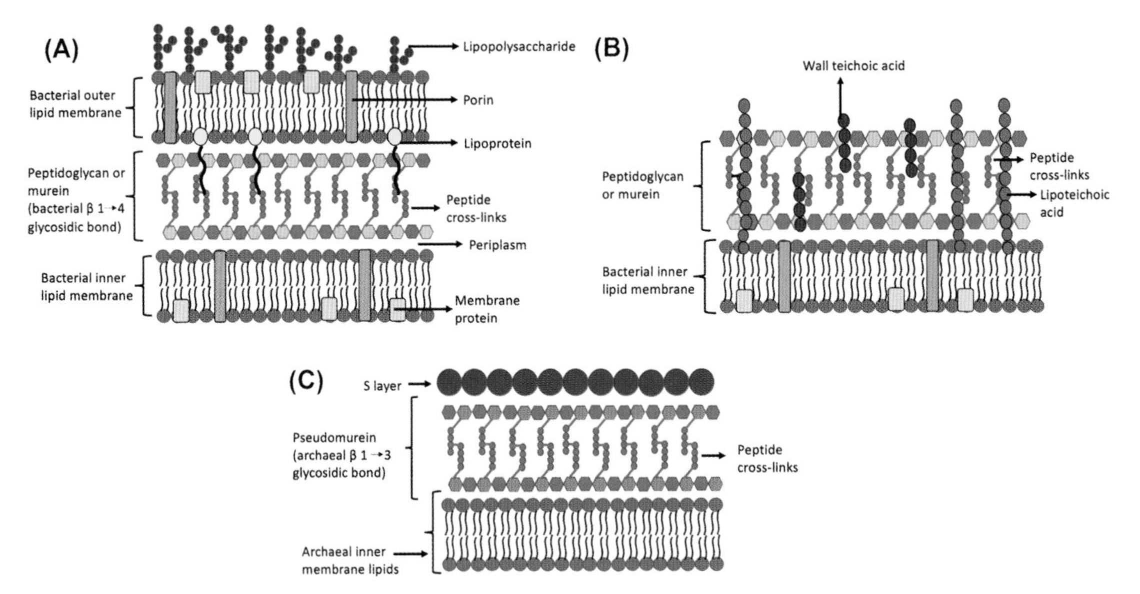

FIGURE 3.2 Cell wall characteristics of (a) gram-negative (Gr−) bacteria, (b) gram-positive (Gr+) bacteria, and (c) archaea.

project from the cell into the outside environment. Peptidoglycan consists of linear glycan strands cross-linked by short peptides. The glycan strands are made of alternating N-acetyl glucosamine (NAG) and N-acetyl muramic acid (NAM) residues linked by β-1-4 bonds and resulting in long chains of glycan strands. A tetrapeptide, comprised of L-alanine, D-glutamine, L-lysine, or meso-diaminopimelic acid and D-alanine, extends off the NAM sugar unit, forming a lattice-like structure. The tetrapeptides are sometimes directly cross-linked: the D-alanine on one tetrapeptide binding to the L-lysine on another tetrapeptide. Alternatively, there is a cross-bridge of five amino acids in many Gr+ bacteria, such as glycine (called pentapeptide bridge), that connects the tetrapeptides. The cross-linking increases the strength of the peptidoglycan. Apart from a thick peptidoglycan layer (90% of total cell wall), Gr+ cell walls usually contain teichoic acids, which encompasses a wide-ranging family of cell surface glycopolymers containing phosphodiester-linked polyol repeat units (Ward, 1981). There are two types of teichoic acids: lipoteichoic acids (LTA) and wall teichoic acids (WTA). The LTA are anchored to the plasma membrane and extend from the cell surface to the peptidoglycan layer, whereas the WTA are covalently attached to the peptidoglycan and extend beyond the cell wall (Vollmer et al., 2008).

The archaeal cell wall (Fig. 3.2c) contains pseudomurein instead of peptidoglycan. Pseudomurein is a polymer that helps to maintain cell shape and protects the cell. It is similar to bacterial peptidoglycan but typically consists of L-N-acetyltalosaminuronic acid and D-N-acetylglucosamine with a β-1,3 linkage instead of NAM linked to NAG, such as in bacterial peptidoglycan (Klingl, 2014). Furthermore, the amino acid cross-links in pseudomurein consist of L-amino acids (glutamic acid, alanine, lysine) versus D-amino acids in bacterial peptidoglycan. Commonly, the archaeal cell wall also contains a glycoprotein S-layer forming a distinct symmetrical, two-dimensional pseudocrystalline array on the cell surface (Klingl, 2014). This protein array is typically anchored in the cytoplasmic membrane with the help of stalk-like structures, forming a quasiperiplasmic space. The S-layer symmetry is often used as a taxonomic trait, with distinct symmetries being unique for certain archaeal groups, such as the S-layer of *Nitrososphaera viennensis* belonging to the phylum Thaumarchaeota. S-layers in halophilic archaea are often glycosylated, which increases protein stability and could also contribute to thermal stabilization in extreme temperatures (Klingl, 2014). As mentioned in Klingl (2014), further information on S-layer proteins, their genetic background, and distinguishing features can be obtained from such reviews as Albers and Meyer (2011); Claus et al. (2001) and König et al. (2007), whereas other focused studies on S-layer proteins of mesophilic and extremely thermophilic archaea, and mesophilic, thermophilic, and extremely thermophilic methanococci, can be found in Akca et al. (2002) and Claus et al. (2002).

In terms of nuclear material, the bacterial cell generally contains one chromosome. However, multiple chromosomes are also known to exist in some bacteria (Teyssier et al., 2004; Volff and Altenbuchner, 2000). In contrast to the eukaryotic nucleus, bacterial cells do not possess a separate compartment hosting the nuclear material. Instead, the nuclear chromatin is present as a highly condensed structure, occupying a region in the cellular cytoplasm. This structure is considered equivalent to the eukaryotic nucleus and hence is called the nucleoid. The structure of the bacterial chromatin is complex, with the first insights into its architecture dating back to the 1970s. Worcel and Burgi (1972) found that the DNA in the nucleoids was compacted by supercoiling, with several nicks needed to relax the chromatin versus only one single-stranded break needed to relax a plasmid. This observation paved the way for the discovery of small, dynamically positioned topological domains in bacterial cells, the positions of which could be comparable in cells sharing the same developmental stage but would appear randomly when averaged over a mixed population (Thanbichler et al., 2005). These domains provide several advantages for bacterial cells; e.g., breaks in the bacterial chromatin, caused by DNA damage or repair, relax only one single domain without affecting the other domains (Postow et al., 2004). This is critical because even slight changes in the superhelicity of the chromosome could be lethal to the cell (Gellert et al., 1976; 1977). Eukaryotic cells wrap their DNA around proteins known as histones, whereas most prokaryotes do not have histones, with the exception of certain species in domain archaea. However, there are some histone-like proteins which are nucleoid-associated polypeptides, which help modulate the structure of bacterial chromatin.

Bacterial cells also produce a wide range of pigments, including carotenoids, prodigiosin (red), melanin (indolic polymers), violacein (violet), and pyocyanin (blue). Pigment production is most commonly seen in *Actinobacteria*, such as the genera *Streptomyces, Rhodococcus, Nocardia,* and *Micromonospora*, among others. Pigment-producing bacterial strains can be found in varying ecological niches, such as soil (Zhu et al., 2007), rhizosphere soil (Peix et al., 2005), desert sand (Liu et al., 2009), and other environments. Use of biotechnology to harness these natural pigments as colorants is of wide industrial application and provides an advantage over synthetic and plant-derived pigments (Narsing Rao et al., 2017).

3.2.2 Cell morphology and its plasticity

There are many possible shapes of bacterial and archaeal cells, ranging from circular to rod-shaped, to spiral, to stalked cells. The shape and size of bacterial and archaeal cells are important determinants of nutrient uptake in a cell, which can be assessed by the surface area to volume ratio. The surface area to volume ratio can also determine the fitness of bacterial cells under environmental stressors, such as substrate limitation. For example, under substrate limiting conditions, the higher surface area to volume ratio of spiral-shaped bacteria, such as *Spirillum*, can place them at a competitive advantage over rod-shaped bacteria, such as *Bacillus*. Cell shape and size also impact prey selection by protozoa and flagellates (as reviewed in Young, 2006).

Bacterial and archaeal cell shape and size are most often not fixed and can fluctuate based on environmental conditions, physiological needs, and growth phase. This cell shape-changing phenomenon is called morphological plasticity. An example of morphological plasticity is in the symbiotic nitrogen-fixing (N_2-fixing) bacterium *Rhizobium*, which is known for its ability to form root nodules on leguminous crops. Initially, the bacterium attaches to the roots of the plant in its flagellated form. The *Rhizobium* penetrates the root hairs forming a tubular structure known as an infection thread. Cortical cell divisions of the root lead to the formation of a nodule. The nodules are eventually colonized by the *Rhizobium* with an altered cell morphology. The developing root nodules contain masses of rapidly dividing, unflagellated, large, irregularly shaped branching cells called bacteroids. As the *Rhizobium* depends on the host plant for survival, the bacteria in turn help the host plant by fixing N.

Though bacteria and archaea are generally unicellular, several examples exist within these domains of life that exhibit chain-like or filamentous multicellular forms. In streptococci and cyanobacteria filamentous growth is observed as chains of cells. Actinobacteria can form continuous filaments similar to hyphae but also exhibit a wide variety of other growth forms, ranging from single-celled rods and cocci to mycelial structures (Prosser and Tough, 1991). The Actinobacteria *Streptomyces* spp. can develop mycelial structures growing as branched hyphae comparable to those of filamentous fungi.

Some bacterial multicells can exhibit social behavior, as in the case of the *Myxococcus* spp., which forms cooperative multicell swarms that move by contact-dependent social motility (Munoz-Dorado et al., 2016). Many filamentous cyanobacteria have "division of labor" across two differentiated cell types, such that each separately carry out oxygenic photosynthesis and N fixation, an oxygen-sensitive process. In the cyanobacterium, *Anabaena* sp., 5−10% of the filamentous cells are specialized heterocysts for N fixation (Zhang et al., 2016). Another example of cellular division of labor can be seen in *Streptomyces* spp. that have vegetative and reproductive growth stages. Vegetative hyphae consume complex organic compounds, and reproductive hyphae produce spores that can tolerate nutrient starvation (Zhang et al., 2016).

Soil is a highly heterogenous environment, containing a range of environmental gradients and microhabitats. A number of studies have suggested that bacterial filamentous morphology can provide survival advantages in response to environmental stress, such as antibiotic exposure and protist predators, among others (Justice et al., 2008). Filamentous growth forms also provide bacteria a mechanism to access microhabitats that have soluble organic compounds, which are otherwise inaccessible to nonmotile and nonfilamentous soil bacteria. In the soil-dwelling bacterium, *Burkholderia pseudomallei*, filamentation is associated with an improved survival strategy (Chen et al., 2005). When exposed to antibiotics, the filamented *B. pseudomallei* form remains more viable than the bacillary form. This protection lasts until the antibiotic becomes inactive. Interestingly, it has been seen that after the removal of antibiotic, when cell division is restored, the daughter cells derived from the filamented form remain viable with continued cell division capacity despite reexposure to other classes of antibiotics (Chen et al., 2005). This suggests that the developmental changes associated with filamentation are also conferred onto the daughter cells, thus providing multigenerational survival benefits (Justice et al., 2008). The mechanism by which this phenotype is transferred to the next generation in this bacterium is not yet explored.

Starvation can have major morphological effects on bacterial and archaeal cell size and shape. Starving cells often have relatively smaller cell size and decreased numbers of ribosomes, in addition to other changes in cell characteristics. Some starved cells can transition to a dormant state known as a viable but nonculturable (VBNC) form. The VBNC form can be associated with morphological changes in some species (Baker et al., 1983; Liu et al., 2017; Rollins and Colwell, 1986; van Teeseling et al., 2017); such as many Gr− pathogens change from rod to coccoid forms. In some cases, these morphological adaptations can be correlated with the regulation of expression of cell wall genes. Even though the induction of VBNC form is associated with remodeling of the cell wall, the relevance of this remodeling to morphogenesis is not clear.

Starvation conditions can induce the expression of a suite of starvation genes in bacteria. The stringent response, which is widely conserved across bacteria, controls adaptation to nutrient limitation and is activated by different starvation and stress signals (Boutte and Crosson, 2013). During starvation, substrates at low concentrations become available to bacterial cells via upregulation of genes that encode high-affinity nutrient uptake systems with broad substrate specificity. Starvation responses, like the stringent response, increase the survival of vegetative cells exposed to environmental stress. The starvation survival response of non−spore-forming Gr− bacteria, such as *Escherichia coli*, *Salmonella typhimurium*, and *Vibrio* spp., as well as Gr+ bacteria, such as *Micrococcus luteus*, have been well characterized (Watson et al., 1998). Under prolonged starvation, *M. luteus* persists in a dormant state and does not form colonies on agar plates. Interestingly, the recovery of dormant cells can increase 100-fold when incubated in spent growth medium from like cells growing in exponential phase, suggesting that a recovery signal made by the growing cells can promote resuscitation from dormancy (Kaprelyants and Kell, 1993). Some of these observations are particularly relevant for a soil environment because the majority of the soil microbial populations are dormant at a given time and have to cope with nutrient-limiting conditions.

Diverse bacterial species can form metabolically inert spores in response to starvation, which is another morphological adaptation of microbial cells. Internal spores, or endospores, are produced by lineages within the phylum Firmicutes, such as bacilli and clostridia, whereas exospores are produced by Actinobacteria. After the initial response to nutrient deprivation, the peptidoglycan in the spores of certain bacterial species, such as *Bacillus subtilis*, is remodeled to form a specific peptidoglycan known as the cortex, which has a much lesser degree of crosslinking and fewer peptide stems (Tan and Ramamurthi, 2014; van Teeseling et al., 2017).

Aside from their advantage to survive temporary nutrient limitation, microbial spores can also resist a suite of other environmental stresses, including extreme temperatures, desiccation, radiation, and pressure. Exospores are not as resistant to environmental stresses as endospores but are very resistant to desiccation, enabling their survival in soil. Their hardiness promotes their passive dispersal to potentially more favorable environments; both exospores and endospores can effectively disperse through the atmosphere. Additionally, aerial hyphae produced by Actinobacteria can support dispersal of their exospores.

3.2.3 Fundamentals of archaeal and bacterial metabolism

Given the wide phylogenetic diversity observed in bacteria and archaea, it is not surprising that both have a wide spectrum of metabolic and physiological capabilities. The physiological diversity observed informs how these domains can occupy the widest range of soil habitats and provide specialized soil functions. Though many bacteria and archaea share some metabolism with eukaryotes (e.g., aerobic respiration and oxygenic photosynthesis), they also uniquely contribute select functional capabilities, such as anaerobic respiration, prokaryotic fermentation, lithotrophy, photoheterotrophy, and methanogenesis (archaea), among others. A brief description of the various metabolic processes in bacteria and archaea is provided in Table 3.1.

One challenge to uncovering the metabolic diversity of soil bacteria and archaea is that the majority of soil microbial lineages are not yet culturable; ideally, physiology is best studied using pure cultures. This knowledge gap spurred by "missing" cultured isolates has been circumvented to some extent using technologies that provide insights into the functional potentials of uncultured bacteria and archaea. An example of technology enabling insights into uncultured isolates is the discovery that some archaea are capable of ammonia oxidation, which was deduced using metagenome sequencing (Treusch et al., 2005; Venter et al., 2004). These ammonia-oxidizing archaeal candidates are now classified within the Thaumarchaeota, with ongoing efforts to bring more representatives into culture for direct interrogation (Bartossek et al., 2012). For example, the physiology of ammonia oxidation has been confirmed for several archaea in pure culture, such as *Nitrososphaera viennensis* isolated from Viennese garden soil (Tourna et al., 2011), and *Nitrosopumilus koreensis* from Korean agricultural soil (Jung et al., 2011). Further use of new techniques, such as genome-centric analyses (e.g., metagenome-assembled genomes, single-cell genomics), will be instrumental in both uncovering the functional diversity of complex soil microbiomes and providing insights into the conditions that will promote the laboratory isolation and cultivation of missing lineages.

The distinction of soil microbes based on their physiology and metabolism is critical to understand their ecological roles and, consequently, ecosystem functions. There are several strategies for classification of microbial physiology. One such strategy is the energy source, e.g., light (photoautotrophs) or

TABLE 3.1 Typical Metabolic Processes Prevalent in Bacteria and Archaea

Type of Metabolism	Processes Involved in Metabolism
Aerobic respiration	Respiration is carried out using oxygen as the terminal electron acceptor
Anaerobic respiration	Respiration is carried out using other alternatives of terminal electron acceptors to oxygen such as nitrate and sulfate
Fermentative processes	Bacteria are characterized by distinct fermentative pathways where the end product of the Embden-Meyerhof pathway, pyruvate, can be converted to products such as lactic acid
Archaean methanogenesis	In this process archaeal methanogens use carbon dioxide as terminal electron acceptor and produce methane as byproduct
Chemolithotrophy	This type of metabolism uses inorganic compounds as electron donors such as hydrogen gas, sulfur compounds (sulfide, sulfur), nitrogen compounds (ammonium, nitrite), and iron
Photolithotrophy	Photolithotrophs use light energy and synthesize microbial biomass entirely from inorganic molecules
Photoheterotrophy	Photoheterotrophs use light as energy source and organic compounds as carbon source
Anoxygenic photosynthesis	This process occurs in anaerobic conditions and uses reduced inorganic electron donors such as hydrogen sulfide, hydrogen, or ferrous ion

chemical energy (chemotrophs). Chemical energy also relates to carbon source. For example, the source of chemical energy could be from CO_2 (autotrophs or lithotrophs) or organic compounds (organotrophs or heterotrophs). Examples of photoautotrophs include cyanobacteria, green sulfur bacteria, and purple nonsulfur bacteria. Organisms that oxidize reduced forms of N are examples of chemoautotrophs, such as the nitrifiers *Nitrosomonas* and *Nitrobacter*. Though these are broad classifications, it is important to bear in mind that not all carbon source–dependent physiological classifications can be tied to ecosystem function, in part because many microbes have complicated physiologies that change according to the availability of energy sources in the environment.

Oxygen requirement is yet another important factor determining microbial survival and provides another classification strategy applied to microbes. This is especially relevant in terrestrial ecosystems because soil is a highly heterogenous system. The transport of oxygen in soil is mainly by diffusion. Oxygen diffusion depends on the physical properties of the soil, with soil porosity being the most important factor. Soil porosity can be attributed to air-filled or water-filled pores, with larger pores showing more rapid movement of gas than smaller pores. Thus macropores are more aerated than micropores (Neira et al., 2015). However, there are barriers to oxygen transport in soils, such as soil compaction and water saturation, of which flooding poses a critical barrier. In regard to their oxygen requirement, microbes can be classified as either obligate or facultative aerobes or anaerobes. Obligate aerobes include chemoheterotrophs (e.g., *Rhizobium*) and chemoautotrophs (e.g., *Thiobacillus*), which use oxygen as the terminal electron acceptor. Aerobic soil microbes require both oxygen and water. Thus, in soils at field capacity where both oxygen and water are readily available to microbes, soil respiration via aerobic metabolism is at a maximum. However, low-oxygen concentrations can exist in isolated pore regions causing the development of anaerobic microsites in aerobic soils. Thus the oxygen concentration and availability can vary widely over small spatial scales. Conversely, obligate anaerobes cannot grow in the presence of molecular oxygen due to the toxic effects and thus they depend on other electron acceptors. Obligate anaerobes are known to possess fermentative metabolic pathways, in which organic compounds are reduced to organic acids and alcohols. One important class of soil obligate aerobes, called microaerophiles, typically grow best at low oxygen tension. Notably, high-energy phosphate is more readily available in microbial aerobic metabolism as compared to anaerobic metabolism, in part due to the buildup of partially oxidized products during microbial growth. Facultative organisms are more versatile because they can thrive both in presence and absence of oxygen. When oxygen becomes limiting, facultative organisms can switch from aerobic to anaerobic metabolism quite rapidly and can use other substitutes in place of oxygen as their terminal electron acceptor. For example, denitrifiers, such as *Pseudomonas aeruginosa*, can substitute oxygen with nitrate as their terminal electron acceptor and reduce it to nitrous oxide or free N.

Bacteria and archaea use enzymes to convert substrates to products that are ultimately assimilated by the microorganisms. Although enzymes dictate the wide suite of substrates that can be degraded by bacteria and archaea, other factors also play a role, such as microbial competition and substrate availability. Enzymes that act on substrates can be either extracellular or intracellular, although the breakdown of environmental substrates typically requires use of extracellular enzymes to catabolize complex polymers into simpler units (Burns et al., 2013). These smaller units (oligomers or monomers) are taken up by cells and then acted upon by intracellular enzymes to catabolize the unit into even simpler constituents that then provide energy through metabolic pathways within the cell. Some examples of intracellular and extracellular enzymes that are involved in substrate utilization in soils are listed in Table 3.2. Extracellular soil enzymes produced by bacteria play a major role in organic matter decomposition and elemental cycling and respond to changes in soil management much earlier than other detectable indicators of soil health (Das and Varma, 2011). Carbon is critical for soil functioning and plant productivity, and management practices that impact the pools of soil organic carbon affect overall soil health, which is a combination of soil physical, chemical, and biological parameters. Several management practices are known to increase soil organic carbon and reduce carbon loss into the atmosphere, including conservation tillage practices, crop residue management, use of cover crops, and the addition of organic amendments, such as manure and compost, among others (Paustian et al., 2019). Some of the common extracellular enzymes that are important in soil organic carbon and N cycling include cellubiosidase, glucosidase, urease, leucine aminopeptidase, and N-acetyl-β-glucosaminidase, whose primary functions correspond to cellulose, sugar, urea, protein, and chitin degradation, respectively. Certain autotrophic bacteria can carry out oxidation of their substrates, such as the nitrifying bacteria *Nitrosomonas* (ammonia to hydroxylamine using ammonia monooxygenase) and *Nitrobacter* (nitrite to nitrate using nitrite oxidoreductase). Other examples include methanotrophs that use methane monooxygenase, e.g., *Methylomonas* and *Methylococcus*. While some of these enzymes are always expressed (constitutive), others are only expressed in certain conditions (inducible).

3.2.4 Growth kinetics and substrate availability

To understand the competitive ability of a bacterium under certain conditions of substrate availability, we need to know their comparative kinetics of growth and the relationship between substrate concentration

TABLE 3.2 Examples of Enzymes and Bacteria Involved in Organic Substrate Utilization

Classification	Degradative Enzymes Used for Organic Substrate Utilization	Distribution of Enzymes/Examples of Soil Bacteria With Ecologically Significant Activity of These Enzymes
Hydrolase	Cellulase (cellulose → glucose subunits)	Species of *Bacillus, Cellulomonas,* and *Pseudomonas*
	Protease (protein → amino acids)	Widespread among soil bacteria and archaea, but species of *Pseudomonas* and *Flavobacterium* are strongly proteolytic
	Urease (urea → ammonia + carbon dioxide)	About 50% of heterotrophic soil bacteria are ureolytic
	Amylase and glucosidase (starch → glucose)	Species of *Bacillus, Pseudomonas,* and *Chromobacterium*
	Ligninase (lignin → aromatic subunits)	Although lignin degradation is primarily the domain of the white rot fungi, species of *Arthrobacter, Flavobacterium,* and *Pseudomonas* are sometimes involved
	Pectinase (pectin → galacturonic acid subunits)	Species of *Arthrobacter, Pseudomonas,* and *Bacillus* (some species possess all of the pectinase enzymes— polygalacturonase, pectate lyase, pectin lyase, and pectin esterase). Many plant pathogens possess pectinase to assist in plant host penetration
	Phosphatase (phosphate esters → phosphate)	About 30% of heterotrophic soil bacteria possess phosphatase enzymes
	Sulfatase (sulfate esters → sulfate)	Much fewer possess sulfatase
	Invertase (sucrose → fructose + glucose)	Particularly active in saprotrophic soil bacteria such as species of *Acinetobacter* and *Bacillus*
	Chitinase (chitin → amino sugar subunits)	The actinomycetes *Streptomyces* and *Nocardia*
Lyase	Amino acid decarboxylase	Both aromatic and nonaromatic amino acid decarboxylases are found in a wide range of soil bacteria synthesizing amino acids
Oxidoreductase	Glucose oxidase (glucose → CO_2)	Ubiquitous enzyme among soil bacteria

Adapted from Killham and Prosser (2015) with permission, published in Soil Microbiology, Ecology and Biochemistry, Fourth Edition, Chapter 3 The Bacteria and Archaea, pp. 62–63, Copyright Elsevier (2015).

and the specific growth rate of organisms. This approach to understanding the competitive behavior of soil microbes was first addressed by Sergei Winogradsky 1924, a pioneer in soil microbiology. At high substrate concentrations, a population C can outcompete populations A and B, whereas, at low substrate concentrations, population A will outcompete B and C (Fig. 3.3). Population B is not the best competitor at any substrate concentration. In such a situation the soil bacteria that exhibit the characteristics of population C, with high growth rates and greater competitiveness at high substrate concentrations, are called zymogenous. Population A, with lower specific growth rates and higher substrate affinity, are

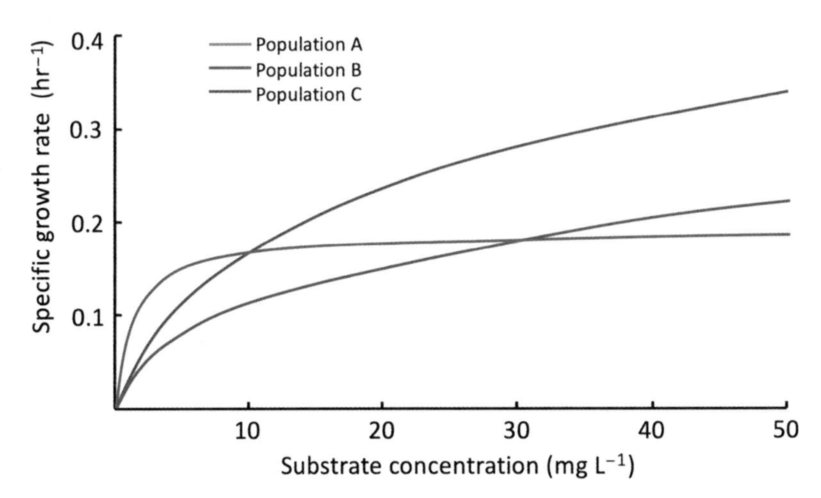

FIGURE 3.3 A representation of enzyme kinetics elucidating the relationship between substrate concentration and specific growth rate of microorganisms. *(Redrawn with permission from Killham and Prosser (2015), published in Soil Microbiology, Ecology and Biochemistry, Fourth Edition, Chapter 3 The Bacteria and Archaea, page number 65, Copyright Elsevier (2015).)*

termed autochthonous. These populations are more prevalent in soil microniches where pores are smaller and nutrient availability is low, whereas zymogenous organisms thrive better in nutrient-rich environments, such as rhizosphere compartments, which have a good supply of carbon-rich substrates in root exudates, such as glucose.

Related concepts to that of zymogeny and autochthony are those of copiotrophy and oligotrophy, respectively, as applied typically to heterotrophic microorganisms. The term *oligotroph* has been in use since the 1900s and could be considered equivalent to autochthonous microorganisms, as depicted by Winogradsky (1924). Oligotrophs are primarily found in environments with low levels of nutrients and can cope within a nutrient-limited environment. Copiotrophs thrive in nutritionally rich environments and use resources rapidly when available. Both oligotrophs and copiotrophs can survive in nutrient-poor environments, but only oligotrophs will persist in chronic starvation conditions and may not be the best competitors in nutrient-rich environments. In essence, copiotrophs could be equivalent to zymogenous organisms (Koch, 2001).

Notably, relationships have been observed between the number of copies of the 16S rRNA gene that a microbial heterotroph has in its genome and its trophic status as either a copiotroph or oligotroph (Klappenbach et al., 2000; Roller et al., 2016). The concept is that copiotrophs are selected for fast growth when resources are plentiful and therefore maintain relatively more copies of their ribosomal operons to enable rapid response and protein translation given newly available resources. Oligotrophs maintain relatively fewer copies. While there are notable exceptions to this trend among particular lineages of bacteria (e.g., spore-forming Firmicutes) (Bishop and Doi, 1966), and likely variation in what constitutes "more" versus "fewer" copies across different ecosystems, calculating a mean 16S rRNA gene copy number for communities has been applied to soils (Kearns and Shade, 2018) and other systems (Nemergut et al., 2016) to gain an understanding of how the heterotrophic strategy in the system trends over time or treatment. The ability to link heterotrophic strategy to a commonly sequenced phylogenetic biomarker like the 16S rRNA gene can add value to such microbiome studies (Stoddard et al., 2015); however, more work is needed to provide precise copy numbers for underdescribed microbial lineages and improve precision (Louca et al., 2018).

3.3 Microbial adaptations to the terrestrial environment

3.3.1 Water availability

Abiotic stress conditions, such as drought, have a strong effect on soil biota (Leng and Hall, 2019). Drought causes osmotic stress, increases soil heterogeneity, limits nutrient mobility and access, and increases soil oxygen (de Vries et al., 2020). All these factors can ultimately cause decreases in soil microbial biomass (Jansson and Hofmockel, 2020; Naylor and Coleman-Derr, 2017). Microbes can resist these alterations in water availability depending on the presence of certain "response traits" that protect against desiccation. Among such traits are morphological adaptations, such as the thick peptidoglycan cell wall in Gr+ bacteria, as well as spore formation and dormant phenotypes (Martiny et al., 2012; Schimel et al., 2007; Xu and Coleman-Derr, 2019). Such traits have also coevolved in diverse organisms, most notably in Gr+ actinomycetes (Martiny et al., 2017). Consequently, these organisms are often referred to as stress-tolerant strategists (Malik et al., 2020). Filamentous growth forms also have advantages under conditions of low hydraulic conductivity typical in drought conditions. The ability to grow in a filamentous form is an essential strategy for soil microbes to connect air-filled pores in water-limiting soil conditions (Wolf et al., 2013). Most soil bacterial morphologies, such as rods, spherical coccus, or spirals, are unable to bridge air-filled pores common in dried soils, with the filamentous Actinobacteria being an exception to this rule. Thus filamentous bacteria can enhance growth and expansion under conditions of low hydraulic conductivity. This helps access microhabitats that may not be accessible to other nonmotile bacteria (Wolf et al., 2013).

3.3.2 Dynamic soil chemical gradients

Movement toward a chemical attractant in soil or chemotaxis is crucial for cell-to-plant communication in the soil environment; for instance, in certain high-cell-density hotspots, such as the rhizosphere, where root exudates can act as signaling molecules for microbial recruitment. The movement of cells toward or away from an attractant is linked to the bacterial flagellum. The bacterial flagellum (plural flagella) consists of a long, helically shaped protein called flagellin anchored to the cytoplasmic membrane and extending through the cell wall. Bacteria and archaea are mostly motile, with the most common mechanism of movement referred to as "swimming." This is achieved via rotation of one (polar), two, or many (peritrichous) flagella in bacteria, whereas in archaea, this is accomplished with archaella. Archaella is similar to flagella in that it also helps serve as a rotating organelle driving cellular motility through a fluid medium. However, both flagella and archaella are very different and do not seem to have common ancestry, the archaella being more related to the bacterial pili (long extracellular polymers in Gr− bacteria that are critical virulence factors for pathogens) than to flagella (Khan and Scholey, 2018). Rotation of the flagellum is achieved via a proton motive force generated across the cell membrane and consequently leads to cell movement. Archaella consist of archaellins and have different structures than flagella. Flagellar motility is useful in soil environments for unicellular organisms searching for nutrients. Other mechanisms that bacteria employ for motility in soil environments include axial filaments in spirochetes, which enables cellular movement by flexing and spinning. Others, such as cyanobacteria and mycobacteria (belonging to phylum Actinobacteria), can move by gliding over surfaces (Harshey, 2003;

Killham and Prosser, 2015; McBride, 2001; Wilde and Mullineaux, 2015). Water availability is important in this context as microbes utilize water as a medium or channel for gliding and movement across heterogeneous soil environments.

Motility is important in soils and sediments where steep, complex, and dynamic chemical gradients can be observed. Recent evidence suggests that some bacteria can sense chemical gradients over the lengths of their bodies in contrast to the standard notion that emphasizes that this is not the case (Mitchell, 2002). For environmental bacteria, motility is generally considered energetically expensive and is used sparingly, especially in nutrient-limiting conditions.

3.3.3 Microniches: soil aggregates and microhabitats

Soil is a highly heterogeneous system with several small distinct habitat types. Soil texture and structure together form soil microhabitats and play a key role in influencing the distribution and activity of microbial populations. Microbes on the other hand act as aggregation agents and influence the soil structure. Distribution of bacteria in soil occurs throughout pores of varying sizes and within soil aggregates. These microhabitats exist in different soil layers, with the upper layer typically drier with more organic material, which provides microniches adequate for growth of microbes. The role of microbes in the spatial and temporal dynamics of soil aggregation has been described by Chenu and Cosentino (2011), where they discuss the formation, stabilization, and destruction of aggregates by microbes and the consequences of the evolution of soil structure on soil microbial activity (Watteau and Villemin, 2018). Thus there is a close overlap of soil functions and processes with the soil physical structure and soil microbial habitats.

Microniches are occupied by microcolonies of organisms. This spatial segregation among populations reduces competitive interactions and results in overall high soil bacterial diversity (Chenu and Cosentino, 2011). That said, development of microniches in soil also limits resource availability, which might be compartmentalized. Several morphological adaptations in bacteria and fungi help access these resources that are otherwise inaccessible, particularly in unsaturated soils. In saturated soils water films connect the microhabitats with the connectivity dependent on the geometry of the pore network. The level of connectivity between the microhabitats via water films determines the accessibility of resources by a given organism (Young et al., 2008). Under conditions where water is not limiting, bacteria rely on the water films or water-filled pores for passive or active motility. Filamentous growth forms are particularly useful to bridge air-filled gaps in water-limiting soil conditions. This places fungi at a competitive advantage compared to most bacteria due to their ability to form hyphae and bridge these pore spaces to access different microniches in soil (Wolf et al., 2013). Most growth forms of bacteria, such as coccoid, rod, or spirals, are not adapted to spread over large distances separating microhabitats. However, some motile bacteria can comigrate with other bacteria along the surface of fungal hyphae promoting their spread in unsaturated soils. An exception to this rule is posed by Actinobacteria, which forms filamentous growth forms and thus, this adaptation can help this group of bacteria to potentially outcompete other groups in water-limiting conditions in soil. Numerous sources from recent literature demonstrate the ubiquity of Actinobacteria under drought conditions in soil (Xu et al., 2018).

Apart from water, soil microniches also vary in pH, temperature, redox potential, nutrient sources, and pollutant concentrations (Young and Crawford, 2004). All of these factors can contribute to bacterial diversity in soil. Even though horizontal gene transfer processes are considered rare events in soil, the formation of soil aggregates and biofilms could lead to increased transfer of conjugative plasmids within soil bacterial communities (Molin and Tolker-Nielsen, 2003). The high cell densities observed in the microhabitats could also increase the chances of cell-to-cell contact, which is essential for conjugation (Heuer and Smalla, 2012). Spatially isolated microhabitats will inhibit plasmid-mediated gene transfer between their populations unless the transfer of plasmids is assisted by water channels, as previously described. Given these soil habitat complexities, such as spatial isolation, added with the reduced metabolic activity of a significant portion of the soil bacterial community, plasmids are frequently detected in bacterial isolates from soil "hotspots." Hotspots are areas in soil of high microbial activity, such as the rhizosphere with sufficient nutrients. In these hotspots the metabolic activity of mating partners and proliferation of transconjugants is increased. As pointed out by Heuer and Smalla (2012), studies show that hotspots of HGT exist in soil (van Elsas et al., 2003), and nutrient availability can increase the likelihood of gene transfer events. The authors also point out that soil microniches can be considered a potential hotspot where bacteria have enhanced access to substrates that are adsorbed to clay particles or soil organic matter. The added advantage of HGT in the microniches could be attributed to increased fitness to those bacteria with plasmids carrying resistance genes or pollutant degradation genes.

3.4 Major lineages of bacteria and archaea in soil

3.4.1 Overview

Here, we present some of the dominant soil bacterial and archaeal groups. These major lineages were selected based on their reported functional roles in the soil environment and contributions to terrestrial ecosystems. For each phylum, we mention their general characteristics, environmental preferences, and special features, in addition to references from the literature for further reading.

For the general characteristics, we include outcomes of a Gram stain test, which reveals insights into the cell wall structure, where Gr^{+} cells typically have a single and thick peptidoglycan layer that retains a purple color after being stained and washed with alcohol, while Gr− cells typically have both outer and inner peptidoglycan membranes that do not retain color after the alcohol wash. We also include, if known, the percent of guanine-cytosine nucleotide (GC) content in the genome, which reveals the general stability of the base pairing due to the number of hydrogen bonds required. Low GC content genomes are relatively less thermostable, while those with high GC content are relatively more thermostable. The GC content is relevant for understanding utility of common molecular techniques, e.g., polymerase chain reaction for those lineages, and also for determining the phylogenetic signal. Three categories can be established based on available data on mean GC content of different bacterial and archaeal phyla: GC-rich phyla, such as Actinobacteria with greater than 60% GC content; GC intermediate phyla, such as Crenarchaeota and Proteobacteria with GC content ranging from 49% to 57%; and GC poor phyla, such as Bacteroidetes and Firmicutes whose GC content ranges from 30% to 46%. It is important to bear in mind that both phylogeny and environment considerably affect nucleotide composition, with GC content varying significantly as a function of environment (Reichenberger et al., 2015). The GC content reported below for each phylum is using complete genomes of candidates from each phylum. We also include any general morphological characteristics, such as typical cell shape(s). Finally, we include the relative abundances of the lineages in soils and their biogeographic distributions across different soils.

We have focused on the major and most common bacterial and archaeal lineages by key soil functions, abundance, or biogeographic distributions in soils and acknowledge that there are many additional lineages present in soil that we do not mention. We use the Earth Microbiome Project (EMP) terrestrial dataset (Thompson et al., 2017) to show the distribution of soil samples collected across the

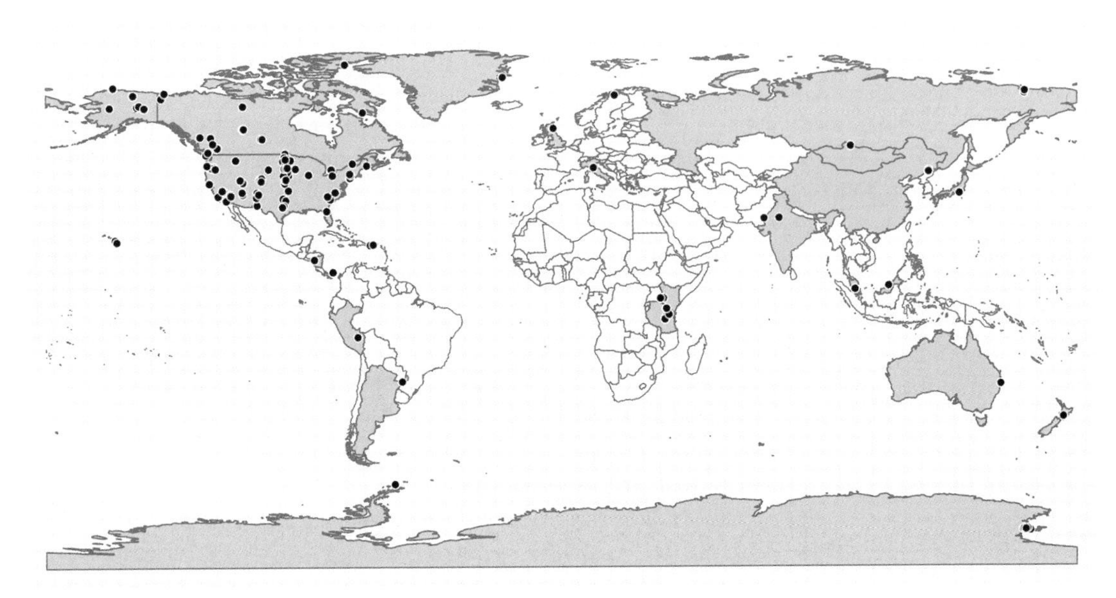

FIGURE 3.4 Global distribution of soil samples collected from the Earth Microbiome Project (EMP). EMP soil samples originated in countries that are shaded in *gray*. *Solid black circles* with *white outline* show the sampling locations. These soil samples were analyzed to determine the dominant soil microbial phyla across terrestrial environments. Input data and metadata files from the EMP can be found at the remote server here: ftp://ftp.microbio.me/emp/release1.

globe, along with details on the areas and samples (Fig. 3.4; Table 3.3). We use this data to summarize the occupancy and average relative abundances of major soil bacterial and archaeal phyla (Fig. 3.5). Occupancy is defined as the proportion of samples that show the presence of a phylum. We also show their contributions to globally distributed soil samples within different biomes (Fig. 3.6). EMP soil samples originated from the following countries and United States territories: Antarctica, Argentina, Australia, Canada, China, Greenland, India, Italy, Japan, Kenya, Malaysia, Mongolia, New Zealand, Nicaragua, Panama, Peru, Puerto Rico, Russia, Scotland, Sweden, Tanzania, and the United States. While there is a notable bias towards U.S. soil samples in the EMP dataset, the Proteobacteria was the most dominant phylum across all biomes. Acidobacteria, Actinobacteria, Verrucomicrobia, and Bacteroidetes were common and in high relative abundance. We see substantial phylum-level variation in relative abundance across biomes, but with the same major phyla represented (Fig. 3.6). Even though soils harbor the most diverse and numerically largest microbial communities known, there are lineages characteristic of soil, as well as biome-distinctive expectations, even at the broad phylum level of taxonomic resolution.

3.4.2 Archaea

3.4.2.1 Crenarcheaota

The Crenarchaeota is one of the most abundant archaeal phyla in terrestrial environments, with a mean relative abundance of 0.5% of total prokaryotic phyla detected in terrestrial soils. They are widespread

TABLE 3.3 Major Geographic Areas From Where the Earth Microbiome Project Samples Are Derived. Area Name, Number of Samples Originating From That Area, and Major Characteristic Features of Soil Samples From the Respective Areas Are Provided

Area	Number of Samples	Characteristic Feature
Antarctica	44	Antarctica is Earth's southernmost continent. Samples from Antarctica represented here are characterized as part of polar desert biomes (some samples containing oil contamination) and tundra biomes. Subzero temperatures are characteristic of this area. The soils classified as tundra have pH ranging from 5 to 10, with the majority having pH between 9 and 10.
Argentina	4	Argentina is a country in South America. The samples referred to here are characterized as part of grassland biomes with slightly acidic pH ranging from 5 to 6.
Australia	253	The samples collected from Australia are all classified as part of cropland biomes with cultivated habitats. Since human interference is present in these areas, they can also be considered part of anthropogenic terrestrial biome. Most samples represented here have pH between 4 and 5.
Canada	55	Canada is a country in the northern part of North America. The samples referred to here are a mix from several classified biomes such as tundra, montane shrubland, temperate grassland, temperate mixed and coniferous forest, cropland biomes, and dense settlement areas. Some samples have acidic pH ranging from 3 to 5.
China	22	China is an east Asian country. The samples from this area represent the montane shrubland biome. All samples are from mountains and range from pH 3–6.
Greenland	20	Greenland is the world's largest island and is located between the Arctic and Atlantic oceans. All samples from Greenland are characterized as part of tundra biome.
India	2	India is a south Asian tropical country with diverse climate ranging from hot and humid northeastern areas to cooler northern areas. It also hosts arid desert areas in the northwestern part of the country. The samples shown here are characterized as part of the desert biome in India. The pH values of these two soils are 6.4 and 8.5.
Italy	48	The samples collected from Italy encompass agricultural areas and are classified as part of cropland biome. Since agriculture requires human intervention the samples can also be characterized as part of anthropogenic terrestrial biome.
Japan	309	Japan is an east Asian country and all samples represented here come from alluvial paddy field soils, hence classified as cropland biome.
Kenya	120	Kenya is an east African country. Samples represented here encompass a variety of biomes such as rangeland, cropland, and tropical moist broadleaf forest biomes. The samples from tropical moist broadleaf forests are from national parks, whereas other samples reflect either farm or agricultural soils.
Malaysia	53	Malaysia is a southeast Asian country. Samples referred to here are from tropical moist broadleaf forest (few samples within this group originating from oil palm plantations) and other forest biomes.
Mongolia	228	Mongolia is an east Asian country. All samples from this area are categorized as part of montane grassland biome containing mostly steppe soils and some fluvisols. Steppes are characterized by a semiarid or temperate climate with huge seasonal variations in temperatures.

Continued

TABLE 3.3 Major Geographic Areas From Where the Earth Microbiome Project Samples Are Derived. Area Name, Number of Samples Originating From That Area, and Major Characteristic Features of Soil Samples From the Respective Areas Are Provided—cont'd

Area	Number of Samples	Characteristic Feature
New Zealand	23	Samples from New Zealand can be characterized as part of the cropland biome and are all from pasture areas with slightly acidic pH ranging from 5 to 6.
Nicaragua	60	Nicaragua is a Central American country. The samples referred to here are from coffee plantations and thus are part of the cropland biome.
Panama	41	Panama is also a Central American country and all samples referred to here are characterized as part of the tropical moist broadleaf forest biome. Samples depicted are all slightly acidic or neutral ranging from pH 4 to 7.
Peru	4	Peru is a country in South America and is home to a part of the Amazon rainforest. The four samples represented here are part of the forest biome with pH ranging from 3 to 6.
Puerto Rico	26	Puerto Rico is a Caribbean island and is characterized by mountains and tropical rainforests. All samples referred to here are part of the tropical moist broadleaf forest biome.
Russia	69	Russia is a transcontinental country spanning eastern Europe to northern Asia. All samples referred to here are part of the tundra biome with pH ranging from acidic to neutral (4–7).
Scotland	14	Scotland is a country of the United Kingdom. All samples referred to here are part of the cropland biome consisting of agricultural soils. The soil pH ranges from 4 to 7.
Sweden	2	Sweden is a Scandinavian country characterized by coastal islands, inland lakes, boreal forests, and mountains. Both samples referred to here are from a tundra biome.
Tanzania	123	Tanzania is a country in East Africa and is known for its expansive wilderness areas. All samples represented here are part of the shrubland biome.
United States	1598	The samples from the United States belong to diverse biomes ranging from high-temperature tropical coniferous forest biomes to tundra biomes. Samples also belong to tropical shrubland biome, temperate grassland biome, cropland biome, and anthropogenic terrestrial biome. The pH values of these soil samples are broad ranging from 3 to 9.

across soil samples with an occupancy of around 88%. These organisms are Gr− and morphologically diverse with rod, cocci, or filamentous forms. The GC content of members within this phylum is around 46 to 50% and thus falls within the GC intermediate range (Reichenberger et al., 2015). Though most of the members that have been cultured are characterized as thermophilic, many are also acidophilic occurring in acidic thermal springs and submarine hydrothermal vents. Nonthermophilic Crenarchaeota are found in marine picoplankton, freshwater sediments, and soils; they have been reported to account for as much as 2% of microbial rRNA in soils. This lineage of soil Crenarchaeota is phylogenetically distinct

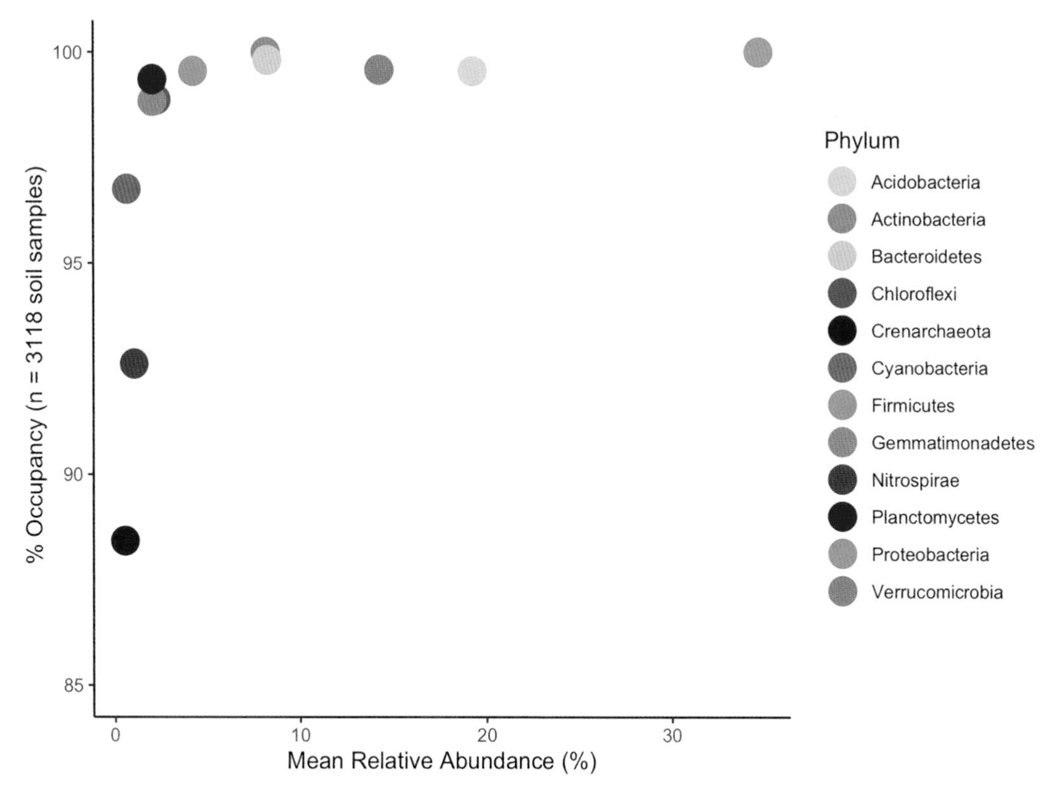

FIGURE 3.5 Mean relative abundance and occupancy of the dominant soil bacterial and archaeal phyla across terrestrial environments. Data derived from 3118 samples from the Earth Microbiome Project (EMP) are presented. Occupancy is defined as the proportion of samples that show the presence of a phylum. Data were subsampled to 10,000 sequences per sample to calculate the mean relative abundance. Note that the *y*-axis starts at 85% occupancy. The standard error of the mean relative abundance across all samples for each phylum was less than 0.5%.

from the hyperthermophiles (McLain, 2005). Some of the previously classified mesophilic Crenarchaeota are now part of another branch of the domain archaea called Thaumarchaeota (Ren et al., 2019).

3.4.2.2 Euryarachaeota

Euryarchaeota is another archaeal phylum that is highly diverse. It accounts for a smaller fraction of the total soil bacterial and archaeal community, with a relative abundance of 0.3% with an occupancy of about 32%. They can be either Gr+ or Gr− and can be found in both terrestrial and marine environments. Candidates belonging to this phylum come in a range of shapes, both cocci and rod shaped. Data obtained from complete genomes from members of this phylum indicate that the GC content of this phylum is around 63% (Li and Du, 2014). This phylum includes methanogens, salt-loving, or halophilic bacteria, as well as thermophilic aerobes and anaerobes. Though candidates of this phylum were initially thought to be restricted to these environments, recent studies demonstrate that they are also found in lower-temperature, acidic environments.

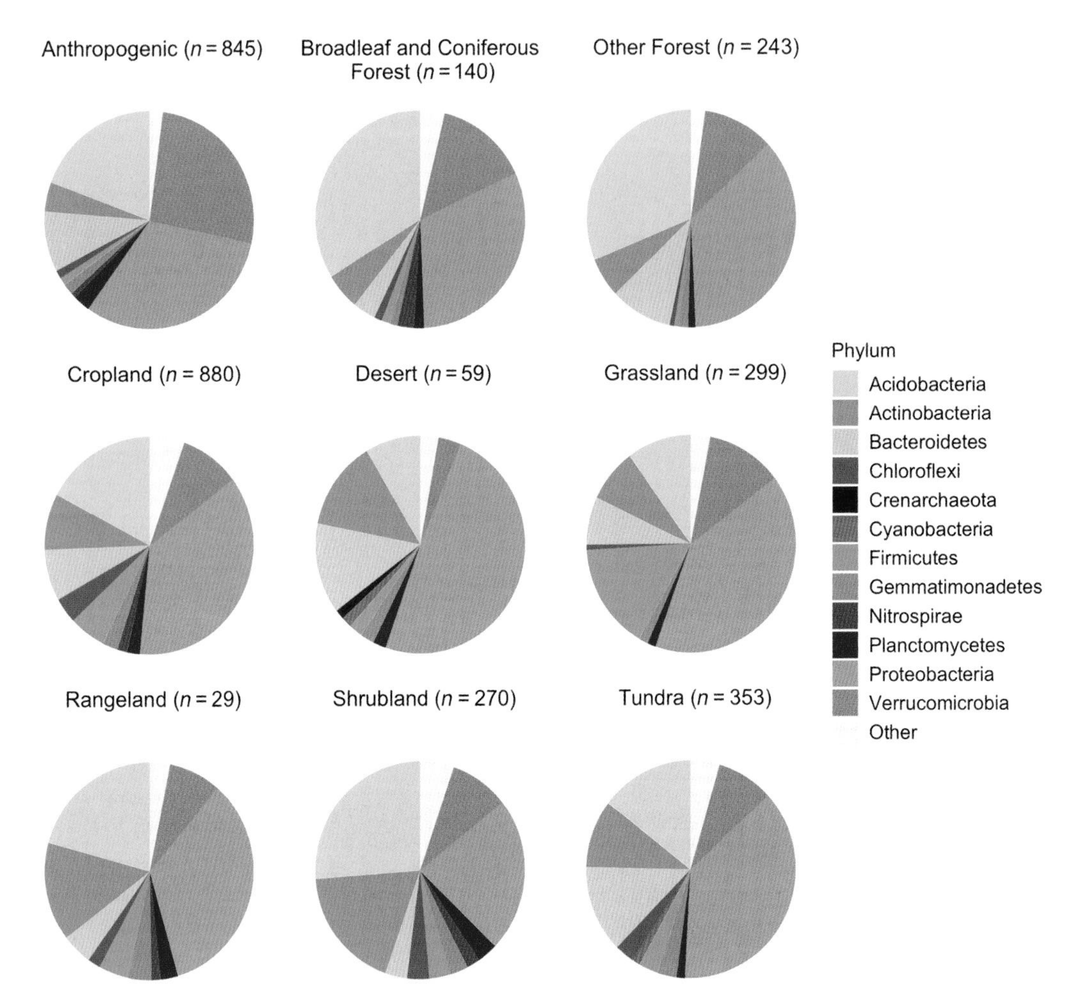

FIGURE 3.6 Pie charts showing the distribution of the dominant soil phyla across terrestrial biomes expressed as the mean relative abundance (%); namely, anthropogenic (n = 845 soil samples), broadleaf and coniferous forest (n = 140 soil samples), other forest (n = 243 soil samples), cropland (n = 880 soil samples), desert (n = 59 soil samples), grassland (n = 299 soil samples), rangeland (n = 29 soil samples), shrubland (n = 270 soil samples), and tundra (n = 353 soil samples).

3.4.3 Bacteria

3.4.3.1 Acidobacteria

The Acidobacteria are found in a wide range of environments including but not limited to the soil environment. They have acidophilic or anaerobic metabolism. This phylum is ubiquitous and highly abundant in soils, with a mean relative abundance of 19% and an occupancy of greater than 99% across all terrestrial soil samples. The genome GC content is around 57–60% (Eichorst et al., 2018) and candidates of this phylum stain Gr−. The phylum Acidobacteria has 26 subdivisions (Kielak et al., 2016), with the vast majority of cultivated isolates affiliated to subdivision 1 (class Acidobacteria). Members of this division are widely distributed in soils as well as other habitats (Sait et al., 2006). They are heterotrophic bacteria with differing oxygen requirements for different species; for example, most species are aerobic, microaerophilic, while some are facultative anaerobes. Although there is minor evidence, genomic studies have revealed that decomposition and utilization of natural polymers, such as extracellular polymeric substance (EPS), chitin, and cellulose, are important features of this phylum and require further study. The ability of Acidobacteria to successfully adapt to harsh soil conditions can be attributed to the increased number of high affinity transporters, utilization of a range of different carbohydrate substrates, and resistance to antibiotics, among other factors. Due to the strong negative correlation between the abundance of Acidobacteria and the concentration of soil organic carbon, it is postulated that some members of this phylum may be oligotrophic (Kielak et al., 2016), if not all (Fierer et al., 2007).

3.4.3.2 Actinobacteria

The phylum Actinobacteria are Gr+ bacteria characterized by high GC nucleotide content in their genomes, roughly around 62% (Reichenberger et al., 2015). They are also one of the most widely distributed phyla in soils, with a relative abundance of 8% and an occupancy of 100% across all terrestrial soil samples. They are aerobic, heterotrophic organisms and are considered major decomposers. Actinobacteria are one of the rare bacterial groups that form mycelial structures and hence, often have a competitive advantage over other growth forms of bacteria such as cocci and rods for nutrient acquisition in soil microniches and accessing less explored soil compartments. Actinobacteria possess many properties that make them suitable candidates in bioremediation; they have tremendous metabolic versatility and can be present under extreme conditions, such as high temperature, low moisture, and nutrient starvation. Their special role in bioremediation also stems from the presence of large genomes, production of secondary metabolites, high resistance to toxins, and ability to degrade a variety of complex polymers ranging from small to long chain hydrocarbons, among other pollutants (Trögl et al., 2018). Thus they are known to play important roles in the removal of xenobiotics, such as heavy metals, petroleum contaminants, and plastic fragments, in terrestrial environments. Key soil members within this phylum include *Streptomyces*, *Micromonospora*, *Cellulomonas*, and *Nocardia*. *Streptomyces* are also known to degrade cellulose, which is a component of plant cell walls, using the cellulase enzyme complex. While some representatives from this group are important human and animal pathogens, such as *Mycobacterium* sp., only few members are plant pathogens, such as *Rhodococcus* sp.

3.4.3.3 Bacteroidetes

The Bacteroidetes phylum primarily consists of Gr− bacteria with a GC content of approximately 46% (Reichenberger et al., 2015). They have a relative abundance of nearly 8% in terrestrial soils, with an occupancy of around 99% across all soil samples. Their metabolisms can range from aerobic to anaerobic with a facultative anaerobic metabolism. Some members, such as those belonging to class Sphingobacteria, are common in soils, whereas members of class Flavobacteria are less common, with class Bacteroidetes being absent from most soils (Janssen, 2006). Cultured representatives of this phylum include *Flavobacterium*, *Hymenobacter*, and *Pedobacter*, indicating that this phylum contains more readily culturable candidates than other dominant soil phyla.

3.4.3.4 Cyanobacteria

Cyanobacteria are photosynthetic prokaryotes found in varying habitats extending from soil to fresh water and oceans. The relative abundance of this phylum is approximately 0.6%, with an occupancy of approximately 97% across all terrestrial soil samples. These are Gr− bacteria that generate a significant portion of oxygen in the Earth's atmosphere. The GC content of Cyanobacteria ranges from 34% to 55%. Newly discovered isolates of the marine cyanobacterium, *Synechococcus*, have been shown to possess some of the smallest genomes while having some of the highest GC contents of approximately 60% (Lee et al., 2019). The filamentous terrestrial cyanobacterium *Nostoc* shows a GC content of around 40% (Hirose et al., 2016).

3.4.3.5 Chloroflexi

The phylum Chloroflexi consists of a diverse group of organisms, including anoxygenic photoautotrophs, aerobic chemoheterotrophs, thermophilic organisms, and anaerobic organisms. They are known to carry out reductive dehalogenation of organic chlorinated compounds as a means to obtain energy. This phylum has a mean relative abundance of around 2%, with occupancy of around 99%. Organisms from this phylum are mostly Gr− but lack the lipopolysaccharide outer membrane characteristic of Gr− bacteria. The average GC content of members of this phylum is around 56% (Zhang and Gao, 2017). Over the last two decades, this phylum has expanded in membership due to discovery of new species and reassignment of existing species from other bacterial phyla such as Actinobacteria. The phylum currently consists of six classes, i.e., Chloroflexia, Thermomicrobia, Dehalococcoidetes, Anaerolineae, Caldilineae, and Ktedonobacteria (Gupta, 2013). Species belonging to the class *Chloroflexia* have photosynthetic ability, with most of the photosynthetic genera belonging to the order Chloroflexales. Several species within Chloroflexi have available genome sequences as well.

3.4.3.6 Firmicutes

The phylum Firmicutes consists of low-GC-content bacteria, with the average content estimated at 43% (Reichenberger et al., 2015). These are Gr+ organisms found in soil and water environments. Their mean relative abundance is around 4%, with occupancy of greater than 99% across all terrestrial soil samples. Their metabolism can be either aerobic or anaerobic. Their morphology ranges from cocci to rod shaped, with some members able to form endospores. Endospores are highly resistant spores with a tough outer coat that is formed as a state of dormancy to protect a vegetative cell from harsh environmental conditions, such as desiccation, heat, and radiation, among others. Increased abundances of transcripts belonging to phylum Firmicutes have been reported in soils under drought conditions (Xu et al., 2018). Some examples of soil Firmicute lineages include *Clostridium*, *Peptococcus*, *Bacillus*, and *Mycoplasma*. A few *Bacillus* species, such as *B. thuringiensis* and *B. anthracis*, can cause diseases in insects and animals/humans, respectively, whereas several species of *Clostridium* cause toxin-mediated diseases in humans. *Bacillus subtilis*, on the other hand, have plant growth-promoting characteristics and are an important rhizobacteria that suppress plant disease.

3.4.3.7 Gemmatimonadetes

Gemmatimonadetes are one the most abundant soil bacterial phyla and are found in several other natural habitats, such as freshwater and sediments. This phylum is widespread in grassland, prairie, and pasture soil, and also eutrophic lake sediments and alpine soils. The relative abundance of this phylum in soil is around 2%, with an occupancy of around 99%. Candidates from this phylum are Gr− and are known to be adapted to low soil moisture conditions, with evidence that they prefer dryer soils. Even though this phylum is ubiquitous in many terrestrial environments, only a few cultured representatives exist. Of those, *Gemmatimonas aurantiaca* strain T-27 was isolated from activated sludge from a sewage treatment system in 2003 and is regarded as the type of species for the phylum. It is Gr− and able to grow via anaerobic and aerobic respiration. Another cultured species within this phylum is *Gemmatimonas phototrophica* strain AP64 and was isolated from a shallow freshwater desert lake in north China. This organism has a special feature in that it possesses bacterial photosynthetic reaction centers. Given the prevalence of this phylum in soil, it is critical to expand the cultured soil representatives to better understand the physiology and function of this important phylum.

3.4.3.8 Nitrospirae

The phylum Nitrospirae is less abundant in terrestrial ecosystems compared with other phyla, such as Bacteroidetes, Actinobacteria, and Verrucomicrobia. It is present at an occupancy of 93% across all terrestrial soil samples and a mean relative abundance of 1%. This phylum contains one class, order, and family within it, namely Nitrospira, Nitrospirales, and Nitrospiraceae, respectively. They are Gr− and are found in curved, spiral, or vibrioid shapes. Metabolically diverse, genera within this phylum include nitrifiers, dissimilatory sulfate reducers, and magnetotactic forms. Members of this phylum are mostly aerobic chemolithotrophs; however, thermophilic, obligately acidophilic and anaerobic members are also present, such as the genus *Thermodesulfovibrio* (Garrity et al., 2001). The genus *Nitrospira* plays an important role in nitrification as a nitrite-oxidizing bacterium and thus is often found in close association with ammonia-oxidizing bacteria or archaea, which convert ammonia to nitrite. Some recently discovered species within this genus are also known to carry out both steps in nitrification by themselves and are thus called complete ammonia oxidizers or commamox organisms (Daims and Wagner, 2018). Draft genome sequences of two *Nitrospira*-like strains assembled from metagenomic data revealed that the average GC content of these strains was between 54.9% and 59.2%. One of these strains was identified as a commamox organism, whereas the other was a nitrite-oxidizing bacterium (Camejo et al., 2017).

3.4.3.9 Planctomycetes

This phylum constitutes a part of the PVC (Planctomycetes, Verrucomicrobia, Chlamydiae) superphylum. They are present at a mean relative abundance of 2% in soils with an occupancy of around 99% across soil samples. Comparative genomics of four genera from the family Isosphaeraceae show GC content ranging from 62% to 66% (Ivanova et al., 2017). Members of this phylum possess a notable compartmentalized cell plan, including a pirellulosome compartment that contains a condensed nucleoid and ribosomes surrounded by an intracytoplasmic membrane, in addition to a ribosome-free paryphoplasm compartment between the intracytoplasmic membrane and cytoplasmic membrane (Lee et al., 2009; Pinos et al., 2016). This organization differs from classical bacterial organization, and the presence of this internal membrane structure that divides and compartmentalizes the cytoplasm makes *Planctomycetes* unique among prokaryotes. Aside from soil, they are also found in water environments. They are obligate aerobes. Their morphological features include the presence of flagellated swarmer cells with cell walls lacking peptidoglycan. In contrast, these bacteria have proteinaceous cell walls, such as in archaea, and contain pit-like surfaces or crateriform structures. They are known to invest energy to transport proteins, RNA, and metabolites through internal membranes. The nucleic acid, which is present in one compartment called nucleoid, is dense and fibrillary. It is also similar to the nucleoid of primitive eukaryotes and very unlike typical bacterial structures. Examples include *Planctomyces*, *Pasteuria*, and *Isocystis pallida*, among others (Op den Camp et al., 2007).

3.4.3.10 Proteobacteria

Proteobacteria are perhaps of the prototypical Gr− bacterial lineage. They are found in almost all soil samples with a mean relative abundance of 35% and occupancy of 100% across all soil samples. The average GC content of this phylum is around 56.4% (Reichenberger et al., 2015). Their metabolisms range from heterotrophic metabolism to chemolithotrophs and chemophototrophs. Proteobacteria can be obligate anaerobes (Marín, 2014) or aerobic, photosynthetic organisms with some N fixers. They are often motile with flagella or can possess gliding capabilities. Proteobacteria constitute a diverse phylum in soil environments, including several classes that also harbor substantial phylogenetic diversity and numerous lineages within them. Four different well-characterized classes fall under this phylum: Alphaproteobacteria, Betaproteobacteria, Gammaproteobacteria, and Deltaproteobacteria. The class *Alphaproteobacteria* is known to thrive at very low nutrient levels and can possess unusual morphological structures, such as stalks and buds. Some agriculturally relevant bacteria capable of symbiotic N fixation with plants also fall under this category. Some examples of genera within Alphaproteobacteria include *Agrobacterium*, which is a plant pathogen, and *Nitrobacter*, which converts nitrite to nitrate in terrestrial nitrification processes. The type order for Alphaproteobacteria is Caulobacterales. Betaproteobacteria is a highly diverse group of organisms composed of chemolithoautotrophs, photoautotrophs, and heterotrophs. Some examples of genera within this class include ammonia-oxidizing *Nitrosomonas*, sulfur-oxidizing *Thiobacillus*, and *Spirillum*. The type order for Betaproteobacteria is Burkholderiales. The largest class in terms of well-documented species is that of Gammaproteobacteria, the type order of which is Pseudomonadales, which includes the well-known genera *Pseudomonas* and N-fixing Azotobacter. *Pseudomonas* is also capable of biodegradative activities, such as degradation of cellulose. Deltaproteobacteria are typically known as predators and contribute to the anaerobic part of the sulfur cycle. Examples from this class include myxobacteria, which are predominantly found in soil and have characteristically large genomes. *Desulfovibrio* is an example of a sulfate-reducing bacteria belonging to this class. The type order is Myxococcales, which includes organisms with self-organizing abilities.

3.4.3.11 Verrucomicrobia

The phylum Verrucomicrobia is widespread, is high in relative abundance across soils, and is consistently reported in the literature as within the top three phyla along with Proteobacteria and *Acidobacteria* in a given soil. The mean relative abundance of this phylum in soil is 14%, with an occupancy of around 100%. The cultivated members of this phylum are Gr and are either coccoid or rod-shaped (Schlesner et al., 2006). They are also found in other environments, such as freshwater and marine waters, and also in extremely acidophilic environments, such as from hot springs. Verrucomicrobia forms part of the PVC superphylum. Similar to members of the phylum Planctomycetes, members of Verrucomicrobia possess a compartmentalized cell plan with a pirellulosome compartment. Very few cultured representatives are known for this phylum, but a high-quality, representative genome from a Verrucomicrobia population, designated as genus *Udaeobacter*, was assembled from metagenome reads in high completeness from a soil that was highly enriched with it. This metagenome-assembled genome revealed *Udaeobacter* to be an aerobic oligotroph capable of using few carbon substrates, including chitobiose, pyruvate, and glucose.

However, it was predicted to synthesize starch and glycogen, suggesting ample storage capacity during times of resource limitation. It was predicted to be auxotrophic for essential vitamins, like B12, and branched/aromatic amino acids, and it had multiple putative predicted amino acid transporter genes. Finally, the *Udaeobacter* had relatively small genome size, which suggested a trade-off of "versatility for efficiency" (Brewer et al., 2016). Overall, the genome assembly of the *Udaeobacter* reveals the potential ways that Verrucomicrobia lineages may be making a living in diverse soils and the adaptations that have led to their prevalence. It also provides an example of how, when enrichment and isolation are lacking, genomic information can be especially insightful to better understand the ecology and biology of yet-uncultivated bacteria and archaea.

3.5 Phylogeny and function in the bacteria and archaea

3.5.1 Microbial functions and phylogenetic signal

Microbial phylogenetic diversity is often used as a proxy for microbial functional diversity because of the reasonable assumption that evolutionary diversity creates trait differentiation. Historically, the term "guild" was used to refer to a group of microorganisms that could perform the same metabolic functions, such as N fixation, ammonia oxidation, or sulfate reduction. In some cases these guilds were aligned with phylogenetic clades in that the functions had evolved and were conserved within particular lineages. Thus clades refer to a group of organisms that are monophyletic and reflect vertical evolutionary history in a bifurcating phylogenetic tree. In other cases similar functions evolved independently or were transferred across lineages by horizontal gene transfer (Top et al., 2000). While it is shown that some microbial functions or responses to a specific stressor are well predicted by phylogeny (Isobe et al., 2019; Martiny et al., 2015; Wallenstein and Hall, 2012), others are more ambiguous. In addition, there are many microbial functions that are yet un-annotated or underannotated, and therefore robust analyses of the relationship between phylogeny and function are not yet possible. For well-annotated functions, phylogeny-informed approaches can be used to predict functional traits from phylogenetic marker genes (Asshauer et al., 2015; Langille et al., 2013).

Especially for bacteria and archaea, functional genes cannot necessarily be assumed to have been transferred vertically from mother to daughter cell. In addition to vertical transmission, horizontal gene transfers (HGT) between nonkin microbial cells are important events that can shape the functional profile of a microbial community. HGT, though commonly known for its spread of antibiotic resistance, has recently been recognized to play a role in the diversification of bacterial genomes. HGT can lead to acquisition of new genetic information not only from one bacterial species to another but also across domains (e.g., to bacteria from archaea or eukaryotes). In some cases of HGT loss of function is an outcome, in addition to gaining new genetic material. The potential for the transfer of genetic material between the domains of life can permit bacteria to expand into new niches and gain a competitive edge. Nelson-Sathi et al. (2015) reported that inter-domain gene transfer is asymmetric, as gene transfer from bacteria to archaea is reported to be far more frequent than from archaea to bacteria. The transfer of genetic information at distinct evolutionary transitions indicates that genes for metabolic functions are acquired in the higher archaeal taxa from bacteria. Thus, in addition to the vertical transmission of genes, lateral gene transfers are also likely to affect bacterial and archaeal community structure and dynamics of ecosystems. However, complex natural environments make it difficult to study these interactions in situ, and thus it remains challenging to authentically reproduce the in situ biology when using laboratory approaches (van de Guchte, 2017).

When we consider evolutionary time on the phylogenetic tree, soil microbial communities include both lineages that have diverged very early on ("ancient divergence") and those with relatively recent divergence, which can influence functional trait diversity within the community (Groussin et al., 2017; Herrera, 1992; Valiente-Banuet et al., 2006). Traits such as oxygenic photosynthesis and methane oxidation, which require more complex molecular machinery, are typically found conserved deeply in the phylogeny, whereas other traits, such as utilization of simple organic compounds (e.g., root exudates and fresh plant residue), are conserved at shallow depths. As a result, even though certain traits could be randomly distributed across phylogenies due to gene gains and losses, evolutionary trait conservatism is seen across the tree of life, with close relatives having more similar traits than distantly related ones.

3.5.2 Functional redundancy and resilience

Functional redundancy occurs when more than one species can perform the same or similar functions in their community, and thus one species can be substituted for the other with little or no effect on ecosystem functioning (Rosenfeld, 2002). The root of this concept extends back to that of ecological guilds, as mentioned above.

Due to inherently high phylogenetic and functional diversity, soil microbial communities are expected to have high functional redundancy, such that changes to the composition of the microbial community do not impact the functional performance (Morris and Blackwood, 2015), especially for broad functions like aerobic heterotrophy. However, for specific physiologies ("narrow" microbial processes), there may be a lower likelihood of redundancy within a community due to the expectation that relatively fewer members have the function, and among those that do, there can be variability in process rates. Subsequently, this variability can have consequences for carbon, N, and other major ecosystem transformations (Schimel and Schaeffer, 2012). The challenge is that many process rates often are difficult to measure in situ and at the population level, and so experiments to test hypotheses of functional redundancy may be difficult or impossible to execute with available methods. However, it is possible that variation in microbial species' process rates, even when at a limiting step, may be negligible at the ecosystem level.

Redundant species are expected to support ecosystem functional resilience to disturbance or perturbation (Rosenfeld, 2002). Microbiomes that have high diversity and multiple redundant members are more likely to retain a particular function after stress, because the likelihood of the presence of a stress-resistant member after disturbance is greater (Konopka, 2009). Microbial communities are generally sensitive to disturbances, and if the community does not recover quickly, rates of ecosystem processes can also remain altered (Allison and Martiny, 2008). The observation of different functional outcomes for different microbiome compositions suggests that the composition of microbial taxa in a community can affect function, at least in some contexts.

Apart from total species richness, the extent to which some communities are more likely to be resilient after a disturbance may also depend on the abundance of generalist and specialist species within each functional group. Functions that are more general, broad, or measured in aggregate, such as organic matter decomposition and community-level respiration, often remain unchanged after shifts in microbial communities, whereas more narrow, specialized functions, such as N fixation and xenobiotic degradation, may be less resilient (Singh et al., 2014).

The potential for microbiome functional overlap and redundancy can be assessed at several different levels of confidence and degrees of insight (Table 3.4). The high-throughput generation of microbial sequencing data can provide insights into a microbiome's functional potential but cannot directly address

TABLE 3.4 Examples of Assessment of Microbial Functional Overlap, Complementarity, Differences, and Redundancy in an Ecosystem. Microbiome Members Can Be Compared by Various Methods to Inform Their Degree of Functional Overlap or Redundancy, From a Comparison of Their Relevant Genes and Genetic Pathways to More Precise Analysis of the Rates of Their Functional Output Under Varying Conditions

Knowledge of Functions Shared Among Taxa	Description	Type of Information Needed	Supporting Microbial/ Molecular Data	Example
Overlap in functional potential	Taxa have the same potential to perform the function	Genetic/genomic	PCR/qPCR of functional gene targets (Meta)genome sequencing	Two taxa have the same genetic pathways for nitrogen fixation; their potential to use the pathways is unknown.
Complementarity in function	Taxa perform the function, but under different conditions (environmental or temporal partitioning)	Genetic/genomic investigations over a longitudinal series or different conditions, optionally and usefully paired with functional assays	PCR/qPCR of functional genes or transcripts (Meta)genome sequencing (Meta)transcript sequencing	One taxon upregulates their nitrogen fixation pathway in early season, the other in late season
Differences in efficiency	Taxa perform the function under the same condition but at different rates	Taxon-discriminating activity/ output analyses paired with rate assays over time in a controlled setting or shared environment	Transcript sequencing Protein/enzyme analysis Metabolite quantification	Two taxa fix nitrogen under the same field condition but one fixes 2× faster than the other
Functional *redundancy*	Taxa perform the function under the same condition and with the same rates	Taxon-discriminating activity/ output analyses paired with rate assays over time in a controlled setting or shared environment	Transcript sequencing Protein/enzyme analysis Metabolite quantification	Two taxa fix nitrogen under the same field condition and at rates that are statistically indistinguishable

function without precise measurements. Multiple, complementary methods that include both functional assays and microbiome assessment are needed to understand the extent and importance of functional redundancy in an ecosystem, for a particular key process or elemental cycle.

3.6 Summary

In this chapter, we introduce the soil bacteria and archaea, their basic metabolisms and physiologies relevant to the terrestrial environment, the major lineages inhabiting soils, and considerations for understanding how their community composition relates to ecosystem functions. In the first section we focus on understanding how the two domains of bacteria and archaea fit within the universal tree of life and how this view has changed over time. We discuss how Woese's three-domain tree (Woese et al., 1990) of bacteria, archaea, and eukarya is no longer supported. Instead, there is strong evidence of a two-domain tree of life consisting of bacteria and archaea, with eukarya branching out from archaea that merged with bacterial endosymbionts. We introduce shortcomings of targeting the most common 16S rRNA genes in that certain lineages can elude discovery with these gene primers, for instance, the CPR. We talk about the importance of using functional genes aside from the 16SrRNA gene to infer phylogenetic relationships to better understand specific ecological questions.

In the second section we focus on the metabolism and physiology of soil bacteria and archaea. We discuss their cell characteristics, such as cell wall structure of gram-positive (Gr+), gram-negative (Gr−) bacterial cells and archaeal cells, nuclear material, and pigments, and highlight differences in cellular constituents across bacteria and archaea where appropriate. In terms of cellular morphology, we emphasize the plasticity exhibited by bacterial and archaeal cells, meaning the inherent ability of bacterial and archaeal cells to change shape and size depending on environmental and physiological needs. Often morphological adaptation can help adapt to environmental stress such as antibiotics and predation. We discuss these in detail with examples in Section II. We touch base on morphological adaptations to starvation, such as formation of spores and dormant states, i.e., the VBNC form. Metabolic pathways that are unique to bacteria and archaea are mentioned, such as anaerobic metabolism, fermentation, and methanogenesis (archaea). We emphasize the importance of pure cultures to understand phylogeny but acknowledge that there are missing cultured representatives that hinder these investigations. However, with the advent of metagenomics and use of metagenome-assembled genomes and single-cell genomics, there is hope to overcome these challenges and uncover novel functions of soil microbiomes. Strategies for classification of microbial physiology based on energy source (light energy, chemical energy) and oxygen requirement (aerobes, anaerobes, microaerophiles) are elaborated in detail. Enzymes that aid in the transformation of soil carbon pools and metabolic processes are discussed, as well as the significance of extracellular enzymes as an important indicator of soil health because of their sensitivity to management. We introduce concepts such as copiotrophy (better competitors in nutrient-rich conditions) and oligotrophy (better competitors in chronic starvation conditions), and the importance and relevance of using phylogenetic markers as links to the heterotrophic strategy of microbes.

Next, we discuss microbial adaptations that help them thrive in terrestrial ecosystems. Microbes adapt to varying environmental conditions in several ways, such as low water availability and chemical gradients. To combat water limitation, some common strategies include the formation of thick peptidoglycans, dormant phenotypes, and filamentous growth forms, whereas to adapt to the chemical gradients present in a soil and to aid motility, bacteria and archaea use flagella and archaella, respectively. Soil is a very heterogeneous system and consists of small aggregates and microhabitats. These microniches play a critical role in overall soil functions and activities. Here again, filamentation is a widely adopted microbial survival strategy, providing access to resources that may be compartmentalized in microaggregates and, as such, may not be readily accessible to most bacterial growth forms, such as coccoid or rod-shaped bacteria.

In the next section we present the major lineages of soil bacteria and archaea that are most abundant and have the highest occupancy across a diverse suite of soil samples collected globally as part of the EMP initiative. Using standardized metadata from the EMP, we provide the distribution of major bacterial and archaeal phyla across diverse terrestrial biomes and show the dominance of Proteobacteria, Acidobacteria, Actinobacteria, and Verrucomicrobia across all biomes. We discuss key characteristics of each lineage and its known contributions or functions in the terrestrial environment.

Finally, we conclude the chapter with a focus on how phylogeny and function can inform our understanding of the soil bacteria and archaea. We introduce concepts such as guilds and clades, mechanisms other than vertical transmission for inheritance of functional genes such as horizontal gene transfer and its role in transfer of genetic information across the domains of life, and functional redundancy that stems from high phylogenetic and functional diversity in soils.

Taken together, we cover several aspects of the ecology of soil bacteria and archaea. We highlight key features in their cellular characteristics, physiology, and metabolism and shed light on how our understanding of these two domains of life is rapidly evolving. New high-throughput techniques and metagenomic approaches will shape our understanding of the diverse suite of functions that these two domains are capable of and uncover novel functions in the future. The two-domain view of the universal tree of life is a critical point that will lead to the reassessment of the evolution of life in the terrestrial biome. This chapter serves as a cornerstone for subsequent chapters that delve deeper into the processes, patterns, and consequences of bacteria, archaea, and other microbes in soil ecosystems.

3.7 Further reading and resources

This textbook chapter provides an introductory primer on bacteria and archaea in the context of soil science. There are additional resources for students who are beginners in the field of microbiology and wish to learn more detail. Some popular textbooks include *Brock Biology of Microorganisms* (15th edition; authors: Michael M. Madigan, Kelly S. Bender, Daniel H. Buckley, W. Matthew Sattley, David A. Stahl); *Prescott's Microbiology* (10th edition; authors: Joanne Willey, Linda Sherwood, and Christopher J. Woolverton); *Microbiology: Principles and Explorations* (9th edition; authors: Jacquelyn G. Black and Laura J. Black); and *Microbiology: Laboratory Theory and Applications* (4th edition; authors: Michael J. Leboffe and Burton E. Pierce). The Brock edition balances information on cutting-edge research with important concepts necessary to understand the field of microbiology and includes a strong emphasis on ecology, evolution, and metabolism. *Prescott's Microbiology* textbook is a beginner's guide to microbiology and provides comprehensive introduction to all major areas of microbiology. The laboratory textbook is relevant for students wishing to learn basic techniques required in a microbiology lab. It is designed for students taking an introductory-level microbiology lab course.

References

Akca, E., Claus, H., Schultz, N., Karbach, G., Schlott, B., Debaerdemaeker, T., et al., 2002. Genes and derived amino acid sequences of S-layer proteins from mesophilic, thermophilic, and extremely thermophilic methanococci. Extremophiles 6, 351–358.

Albers, S.V., Meyer, B.H., 2011. The archaeal cell envelope. Nat. Rev. Microbiol. 9, 414–426.

Allison, S.D., Martiny, J.B., 2008. Colloquium paper: resistance, resilience, and redundancy in microbial communities. Proc. Natl. Acad. Sci. 105, 11512.

Asshauer, K.P., Wemheuer, B., Daniel, R., Meinicke, P., 2015. Tax4Fun: predicting functional profiles from metagenomic 16S rRNA data. Bioinformatics 31, 2882–2884.

Baker, R.M., Singleton, F.L., Hood, M.A., 1983. Effects of nutrient deprivation on *Vibrio cholerae*. Appl. Environ. Microbiol. 46, 930–940.

Bartossek, R., Spang, A., Weidler, G., Lanzen, A., Schleper, C., 2012. Metagenomic analysis of ammonia-oxidizing archaea affiliated with the soil group. Front. Microbiol. 3 (208), 1–6.

Bishop, H.L., Doi, R.H., 1966. Isolation and characterization of ribosomes from *Bacillus subtilis* spores. J. Bacteriol. 91, 695–701.

Boutte, C.C., Crosson, S., 2013. Bacterial lifestyle shapes stringent response activation. Trends Microbiol. 21, 174–180.

Brewer, T.E., Handley, K.M., Carini, P., Gilbert, J.A., Fierer, N., 2016. Genome reduction in an abundant and ubiquitous soil bacterium "*Candidatus* Udaeobacter copiosus. Nat. Microbiol. 2, 16198.

Burns, R.G., DeForest, J.L., Marxsen, J., Sinsabaugh, R.L., Stromberger, M.E., Wallenstein, M.D., et al., 2013. Soil enzymes in a changing environment: current knowledge and future directions. Soil Biol. Biochem. 58, 216–234.

Camejo, P.Y., Santo Domingo, J., McMahon, K.D., Noguera, D.R., 2017. Genome-enabled insights into the ecophysiology of the comammox bacterium "*Candidatus* Nitrospira nitrosa. Msystems 2, e00059–17.

Chen, K., Sun, G.W., Chua, K.L., Gan, Y.H., 2005. Modified virulence of antibiotic-induced Burkholderia pseudomallei filaments. Antimicrob. Agents Chemother. 49, 1002–1009.

Chenu, C., Cosentino, D., 2011. Microbial regulation of soil structural dynamics. In: Ritz, K., Young, I. (Eds.), The Architecture and Biology of Soils: Life in Inner Space. CAB International, Wallingford, Oxfordshire, pp. 37–70.

Claus, H., Akca, E., Schultz, N., Karbach, G., Schlott, B., Debaerdemaeker, T., et al., 2001. Surface (Glyco-) Proteins: Primary Structure and Crystallization under Microgravity Conditions. Proceedings of the First European Workshop on Exo/Astrobiology. ESA Publication Division, Noordwijk, The Netherlands, pp. 313–320.

Claus, H., Akca, E., Debaerdemaeker, T., Evrard, C., Declercq, J.P., Konig, H., 2002. Primary structure of selected archaeal mesophilic and extremely thermophilic outer surface layer proteins. Syst. Appl. Microbiol. 25, 3–12.

Daims, H., Wagner, M., 2018. Nitrospira. Trends Microbiol. 26, 462–463.

Das, S.K., Varma, A., 2011. Role of enzymes in maintaining soil health. In: Shukla, G., Varma, A. (Eds.), Soil Enzymology. Springer, Berlin, Heidelberg.

de Vries, F.T., Griffiths, R.I., Knight, C.G., Nicolitch, O., Williams, A., 2020. Harnessing rhizosphere microbiomes for drought-resilient crop production. Science 368, 270–274.

Eichorst, S.A., Trojan, D., Roux, S., Herbold, C., Rattei, T., Woebken, D., 2018. Genomic insights into the Acidobacteria reveal strategies for their success in terrestrial environments. Environ. Microbiol. 20, 1041–1063.

Fierer, N., Bradford, M.A., Jackson, R.B., 2007. Toward an ecological classification of soil bacteria. Ecology 88, 1354–1364.

Garrity, G.M., Holt, J.G., Spieck, E., Bock, E., Johnson, D.B., Spring, S., et al., 2001. Phylum BVIII. nitrospirae phy. nov. In: Boone, D.R., Castenholz, R.W., Garrity, G.M., et al. (Eds.), Volume One, the Archaea and the Deeply Branching and Phototrophic Bacteria. Springer, New York, pp. 451–464.

Gellert, M., O'Dea, M.H., Itoh, T., Tomizawa, J., 1976. Novobiocin and coumermycin inhibit DNA supercoiling catalyzed by DNA gyrase. Proc. Natl. Acad. Sci. 73, 4474–4478.

Gellert, M., Mizuuchi, K., O'Dea, M.H., Itoh, T., Tomizawa, J.I., 1977. Nalidixic acid resistance: a second genetic character involved in DNA gyrase activity. Proc. Natl. Acad. Sci. 74, 4772–4776.

Groussin, M., Mazel, F., Sanders, J.G., Smillie, C.S., Lavergne, S., Thuiller, W., et al., 2017. Unraveling the processes shaping mammalian gut microbiomes over evolutionary time. Nat. Commun. 8, 1–12.

Gupta, R.S., 2013. Molecular markers for photosynthetic bacteria and insights into the origin and spread of photosynthesis. Adv. Bot. Res. 66, 37–66.

Harshey, R.M., 2003. Bacterial motility on a surface: many ways to a common goal. Annu. Rev. Microbiol. 57, 249–273.

Herrera, C.M., 1992. Historical effects and sorting processes as explanations for contemporary ecological patterns: character syndromes in Mediterranean woody plants. Am. Nat. 140, 421–446.

Heuer, H., Smalla, K., 2012. Plasmids foster diversification and adaptation of bacterial populations in soil. FEMS Microbiol. Rev. 36, 1083–1104.

Hirose, Y., Fujisawa, T., Ohtsubo, Y., Katayama, M., Misawa, N., Wakazuki, S., et al., 2016. Complete genome sequence of cyanobacterium *Nostoc* sp. NIES-3756, a potentially useful strain for phytochrome-based bioengineering. J. Biotechnol. 218, 51–52.

Hug, L.A., Baker, B.J., Anantharaman, K., Brown, C.T., Probst, A.J., Castelle, C.J., et al., 2016. A new view of the tree of life. Nat. Microbiol. 1, 16048.

Isobe, K., Allison, S.D., Khalili, B., Martiny, A.C., Martiny, J.B.H., 2019. Phylogenetic conservation of bacterial responses to soil nitrogen addition across continents. Nat. Commun. 10, 2499.

Ivanova, A.A., Naumoff, D.G., Miroshnikov, K.K., Liesack, W., Dedysh, S.N., 2017. Comparative genomics of four isosphaeraceae planctomycetes: a common pool of plasmids and glycoside hydrolase genes shared by *Paludisphaera borealis* PX4^T, *Isosphaera pallida* IS1B^T, *Singulisphaera acidiphila* DSM 18658T, and Strain SH-PL62. Front. Microbiol. 8, 412. https://doi.org/10.3389/fmicb.2017.00412.

Janssen, P.H., 2006. Identifying the dominant soil bacterial taxa in libraries of 16S rRNA and 16S rRNA genes. Appl. Environ. Microbiol. 72, 1719.

Jansson, J.K., Hofmockel, K.S., 2020. Soil microbiomes and climate change. Nat. Rev. Microbiol. 18, 35–46.

Jung, M.Y., Park, S.J., Min, D., Kim, J.S., Rijpstra, W.I., Sinninghe Damste, J.S., et al., 2011. Enrichment and characterization of an autotrophic ammonia-oxidizing archaeon of mesophilic crenarchaeal group I.1a from an agricultural soil. Appl. Environ. Microbiol. 77, 8635–8647.

Justice, S.S., Hunstad, D.A., Cegelski, L., Hultgren, S.J., 2008. Morphological plasticity as a bacterial survival strategy. Nat. Rev. Microbiol. 6, 162–168.

Kaprelyants, A.S., Kell, D.B., 1993. Dormancy in stationary-phase cultures of micrococcus luteus: Flow cytometric analysis of starvation and resuscitation. Appl. Environ. Microbiol. 59, 3187–3196.

Kearns, P.J., Shade, A., 2018. Trait-based patterns of microbial dynamics in dormancy potential and heterotrophic strategy: case studies of resource-based and post-press succession. ISME J. 12, 2575–2581.

Khan, S., Scholey, J.M., 2018. Assembly, functions and evolution of archaella, flagella and cilia. Curr. Biol. 28, R278–R292.

Kielak, A.M., Barreto, C.C., Kowalchuk, G.A., van Veen, J.A., Kuramae, E.E., 2016. The ecology of acidobacteria: moving beyond genes and genomes. Front. Microbiol. 7, 1–16.

Killham, K., Prosser, J.I., 2015. The bacteria and archaea. In: Paul, E.A. (Ed.), Soil Microbiology, Ecology and Biochemistry, fourth ed. Elsevier, London.

Klappenbach, J.A., Dunbar, J.M., Schmidt, T.M., 2000. rRNA operon copy number reflects ecological strategies of bacteria. Appl. Environ. Microbiol. 66, 1328–1333.

Klingl, A., 2014. S-layer and cytoplasmic membrane – exceptions from the typical archaeal cell wall with a focus on double membranes. Front. Microbiol. 5, 624.

Koch, A.L., 2001. Oligotrophs versus copiotrophs. Bioessays 23, 657–661.

König, H., Rachel, R., Claus, H., 2007. Proteinaceous surface layers of archaea: ultrastructure and biochemistry. In: Cavicchioli, R. (Ed.), Archaea: Molecular and Cellular Biology. ASM Press, Washington, D.C., pp. 315–340

Konopka, A., 2009. What is microbial community ecology? ISME J. 3, 1223–1230.

Langille, M.G., Zaneveld, J., Caporaso, J.G., McDonald, D., Knights, D., Reyes, J.A., et al., 2013. Predictive functional profiling of microbial communities using 16S rRNA marker gene sequences. Nat. Biotechnol. 31, 814–821.

Lee, K.C., Webb, R.I., Janssen, P.H., Sangwan, P., Romeo, T., Staley, J.T., et al., 2009. Phylum Verrucomicrobia representatives share a compartmentalized cell plan with members of bacterial phylum Planctomycetes. BMC Microbiol. 9 (1), 5.

Lee, M.D., Ahlgren, N.A., Kling, J.D., Walworth, N.G., Rocap, G., Saito, M.A., et al., 2019. Marine *Synechococcus* isolates representing globally abundant genomic lineages demonstrate a unique evolutionary path of genome reduction without a decrease in GC content. Environ. Microbiol. 21, 1677–1686.

Leng, G., Hall, J., 2019. Crop yield sensitivity of global major agricultural countries to droughts and the projected changes in the future. Sci. Total Environ. 654, 811–821.

Li, X.Q., Du, D., 2014. Variation, evolution, and correlation analysis of C+G content and genome or chromosome size in different kingdoms and phyla. PLoS One 9, e88339.

Liu, J., Zhou, R., Li, L., Peters, B.M., Li, B., Lin, C.W., et al., 2017. Viable but non-culturable state and toxin gene expression of enterohemorrhagic *Escherichia coli* O157 under cryopreservation. Res. Microbiol. 168, 188–193.

Liu, M., Peng, F., Wang, Y., Zhang, K., Chen, G., Fang, C., 2009. *Kineococcus xinjiangensis* sp. nov., isolated from desert sand. Int. J. Syst. Evol. Microbiol. 59, 1090–1093.

Louca, S., Doebeli, M., Parfrey, L.W., 2018. Correcting for 16S rRNA gene copy numbers in microbiome surveys remains an unsolved problem. Microbiome 6, 41.

Malik, A.A., Martiny, J.B.H., Brodie, E.L., Martiny, A.C., Treseder, K.K., Allison, S.D., 2020. Defining trait-based microbial strategies with consequences for soil carbon cycling under climate change. ISME J. 14, 1–9.

Marín, I., 2014. Proteobacteria. In: Amils, R., Gargaud, M., Cernicharo Quintanilla, J., Cleaves, H.J., Irvine, W.M., et al. (Eds.), Encyclopedia of Astrobiology. Springer, Berlin and Heidelberg.

Martiny, A., Treseder, K., Pusch, G., 2012. Phylogenetic conservatism of functional traits in microorganisms. ISME J. 7 (4), 830–838.

Martiny, J.B., Jones, S.E., Lennon, J.T., Martiny, A.C., 2015. Microbiomes in light of traits: a phylogenetic perspective. Science 350, aac9323.

Martiny, J.B., Martiny, A.C., Weihe, C., Lu, Y., Berlemont, R., Brodie, E.L., et al., 2017. Microbial legacies alter decomposition in response to simulated global change. ISME J. 11, 490–499.

McBride, M.J., 2001. Bacterial gliding motility: multiple mechanisms for cell movement over surfaces. Annu. Rev. Microbiol. 55, 49–75.

McLain, J.E.T., 2005. Archaea. In: Hillel, D. (Ed.), Encyclopedia of Soils in the Environment. Elsevier, Oxford, pp. 88–94.

Mitchell, J.G., 2002. The energetics and scaling of search strategies in bacteria. Am. Nat. 160, 727–740.

Molin, S., Tolker-Nielsen, T., 2003. Gene transfer occurs with enhanced efficiency in biofilms and induces enhanced stabilisation of the biofilm structure. Curr. Opin. Biotechnol. 14, 255–261.

Morris, S.J., Blackwood, C.B., 2015. The ecology of the soil biota and their function. In: Paul, E.A. (Ed.), Soil Microbiology, Ecology and Biochemistry, fourth ed. Academic Press, Boston, pp. 273–309.

Munoz-Dorado, J., Marcos-Torres, F.J., Garcia-Bravo, E., Moraleda-Munoz, A., Perez, J., 2016. Myxobacteria: moving, killing, feeding, and surviving together. Front. Microbiol. 7, 781.

Narsing Rao, M.P., Xiao, M., Li, W.J., 2017. Fungal and bacterial pigments: secondary metabolites with wide applications. Front. Microbiol. 8, 1113.

Naylor, D., Coleman-Derr, D., 2017. Drought stress and root-associated bacterial communities. Front. Plant Sci. 8, 2223.

Neira, J., Ortiz, M., Morales, L., Acevedo, E., 2015. Oxygen diffusion in soils: understanding the factors and processes needed for modeling. Chil. J. Agric. Res. 75, 35–44.

Nelson-Sathi, S., Sousa, F.L., Roettger, M., Lozada-Chavez, N., Thiergart, T., Janssen, A., et al., 2015. Origins of major archaeal clades correspond to gene acquisitions from bacteria. Nature 517, 77–80.

Nemergut, D.R., Knelman, J.E., Ferrenberg, S., Bilinski, T., Melbourne, B., Jiang, L., et al., 2016. Decreases in average bacterial community rRNA operon copy number during succession. ISME J. 10, 1147–1156.

Op den Camp, H.J.M., Jetten, M.S.M., Strous, M., 2007. Anammox. In: Bothe, H., Ferguson, S.J., Newton, W.E. (Eds.), Biology of the Nitrogen Cycle. Elsevier, Amsterdam, pp. 245–262.

Parks, D.H., Rinke, C., Chuvochina, M., Chaumeil, P.A., Woodcroft, B.J., Evans, P.N., et al., 2017. Recovery of nearly 8,000 metagenome-assembled genomes substantially expands the tree of life. Nat. Microbiol. 2, 1533–1542.

Paustian, K., Larson, E., Kent, J., Marx, E., Swan, A., 2019. Soil C sequestration as a biological negative emission strategy. Front. Clim. 1, 8.

Peix, A., Berge, O., Rivas, R., Abril, A., Velazquez, E., 2005. *Pseudomonas argentinensis* sp. nov., a novel yellow pigment-producing bacterial species, isolated from rhizospheric soil in Cordoba, Argentina. Intl. J. Syst. Evol. Microbiol. 55, 1107–1112.

Pinos, S., Pontarotti, P., Raoult, D., Baudoin, J.P., Pagnier, I., 2016. Compartmentalization in PVC super-phylum: evolution and impact. Biol. Direct 11, 38.

Postow, L., Hardy, C.D., Arsuaga, J., Cozzarelli, N.R., 2004. Topological domain structure of the Escherichia coli chromosome. Genes Dev. 18, 1766–1779.

Prosser, J.I., Tough, A.J., 1991. Growth mechanisms and growth kinetics of filamentous microorganisms. Crit. Rev. Biotechnol. 10, 253–274.

Reichenberger, E.R., Rosen, G., Hershberg, U., Hershberg, R., 2015. Prokaryotic nucleotide composition is shaped by both phylogeny and the environment. Genome Biol. Evol. 7, 1380–1389.

Ren, M., Feng, X., Huang, Y., Wang, H., Hu, Z., Clingenpeel, S., et al., 2019. Phylogenomics suggests oxygen availability as a driving force in Thaumarchaeota evolution. ISME J. 13, 2150–2161.

Roller, B.R., Stoddard, S.F., Schmidt, T.M., 2016. Exploiting rRNA operon copy number to investigate bacterial reproductive strategies. Nat. Microbiol. 1, 16160.

Rollins, D.M., Colwell, R.R., 1986. Viable but nonculturable stage of *Campylobacter jejuni* and its role in survival in the natural aquatic environment. Appl. Environ. Microbiol. 52, 531–538.

Rosenfeld, J.S., 2002. Functional redundancy in ecology and conservation. Oikos 98, 156–162.

Sait, M., Davis, K.E., Janssen, P.H., 2006. Effect of pH on isolation and distribution of members of subdivision 1 of the phylum *Acidobacteria* occurring in soil. Appl. Environ. Microbiol. 72, 1852–1857.

Schimel, J., Balser, T.C., Wallenstein, M., 2007. Microbial stress-response physiology and its implications for ecosystem function. Ecology 88, 1386–1394.

Schimel, J.P., Schaeffer, S.M., 2012. Microbial control over carbon cycling in soil. Front. Microbiol. 3, 348.

Schlesner, H., Jenkins, C., Staley, J.T., 2006. The phylum Verrucomicrobia: a phylogenetically heterogeneous bacterial group. In: The Prokaryotes. Springer, New York.

Singh, B.K., Quince, C., Macdonald, C.A., Khachane, A., Thomas, N., Al-Soud, W.A., et al., 2014. Loss of microbial diversity in soils is coincident with reductions in some specialized functions. Environ. Microbiol. 16, 2408–2420.

Stepanauskas, R., Fergusson, E.A., Brown, J., Poulton, N.J., Tupper, B., Labonte, J.M., et al., 2017. Improved genome recovery and integrated cell-size analyses of individual uncultured microbial cells and viral particles. Nat. Commun. 8 (1), 84.

Stoddard, S.F., Smith, B.J., Hein, R., Roller, B.R., Schmidt, T.M., 2015. rrnDB: improved tools for interpreting rRNA gene abundance in bacteria and archaea and a new foundation for future development. Nucleic Acids Res. 43, D593–D598.

Tan, I.S., Ramamurthi, K.S., 2014. Spore formation in *Bacillus subtilis*. Environ. Microbiol. Rep. 6, 212–225.

Teyssier, C., Marchandin, H., Jumas-Bilak, E., 2004. The genome of alpha-proteobacteria: complexity, reduction, diversity and fluidity. Can. J. Microbiol. 50, 383–396.

Thanbichler, M., Wang, S.C., Shapiro, L., 2005. The bacterial nucleoid: a highly organized and dynamic structure. J. Cell. Biochem. 96, 506–521.

Thompson, L.R., Sanders, J.G., McDonald, D., Amir, A., Ladau, J., Locey, K.J., et al., 2017. A communal catalogue reveals Earth's multiscale microbial diversity. Nature 551, 457–463.

Top, E., Moënne-Loccoz, Y., Pembroke, T., Thomas, C., 2000. Phenotypic traits conferred by plasmids. In: Thomas, C.M. (Ed.), The Horizontal Gene Pool: Bacterial Plasmids and Gene Spread. Harwood Academic Publishers, London, pp. 249–285.

Tourna, M., Stieglmeier, M., Spang, A., Konneke, M., Schintlmeister, A., Urich, T., et al., 2011. *Nitrososphaera viennensis*, an ammonia oxidizing archaeon from soil. Proc. Natl. Acad. Sci. 108, 8420–8425.

Treusch, A.H., Leininger, S., Kletzin, A., Schuster, S.C., Klenk, H.P., Schleper, C., 2005. Novel genes for nitrite reductase and Amo-related proteins indicate a role of uncultivated mesophilic crenarchaeota in nitrogen cycling. Environ. Microbiol. 7, 1985–1995.

Trögl, J., Esuola, C.O., Křiženecká, S., Kuráň, P., Seidlová, L., Veronesi-Dáňová, P., et al., 2018. Biodegradation of high concentrations of aliphatic hydrocarbons in soil from a petroleum refinery: implications for applicability of new actinobacterial strains. Appl. Sci. 8, 1855.

Valiente-Banuet, A., Rumebe, A.V., Verdú, M., Callaway, R.M., 2006. Modern quaternary plant lineages promote diversity through facilitation of ancient tertiary lineages. Proc. Natl. Acad. Sci. 103, 16812–16817.

van de Guchte, M., 2017. Horizontal gene transfer and ecosystem function dynamics. Trends Microbiol. 25, 699–700.

van Elsas, J.D., Turner, S., Bailey, M.J., 2003. Horizontal gene transfer in the phytosphere. New Phytol. 157, 525–537.

van Teeseling, M.C.F., de Pedro, M.A., Cava, F., 2017. Determinants of bacterial morphology: from fundamentals to possibilities for antimicrobial targeting. Front. Microbiol. 8, 1264.

Venter, J.C., Remington, K., Heidelberg, J.F., Halpern, A.L., Rusch, D., Eisen, J.A., et al., 2004. Environmental genome shotgun sequencing of the Sargasso Sea. Science 304, 66–74.

Volff, J.N., Altenbuchner, J., 2000. A new beginning with new ends: linearisation of circular chromosomes during bacterial evolution. FEMS Microbiol. Lett. 186, 143–150.

Vollmer, W., Blanot, D., de Pedro, M.A., 2008. Peptidoglycan structure and architecture. FEMS Microbiol. Rev. 32, 149–167.

Wallenstein, M.D., Hall, E.K., 2012. A trait-based framework for predicting when and where microbial adaptation to climate change will affect ecosystem functioning. Biogeochemistry 109, 35–47.

Ward, J.B., 1981. Teichoic and teichuronic acids: biosynthesis, assembly, and location. Microbiol. Rev. 45, 211–243.

Watson, S.P., Clements, M.O., Foster, S.J., 1998. Characterization of the starvation-survival response of *Staphylococcus aureus*. J. Bacteriol. 180, 1750–1758.

Watteau, F., Villemin, G., 2018. Soil microstructures examined through transmission electron microscopy reveal soil-microorganisms interactions. Front. Environ. Sci. 6, 1–10.

Wilde, A., Mullineaux, C.W., 2015. Motility in cyanobacteria: polysaccharide tracks and Type IV pilus motors. Mol. Microbiol. 98, 998–1001.

Winogradsky, S., 1924. Sur la microflora autochtone de la terre arable. Comptes rendus hebdomadaires des seances de l'Academie des Sciences (Paris) D. 178, 1236–1239.

Woese, C.R., Kandler, O., Wheelis, M.L., 1990. Towards a natural system of organisms: proposal for the domains archaea, bacteria, and eucarya. Proc. Natl. Acad. Sci. 87, 4576–4579.

Wolf, A.B., Vos, M., de Boer, W., Kowalchuk, G.A., 2013. Impact of matric potential and pore size distribution on growth dynamics of filamentous and non-filamentous soil bacteria. PLoS One 8, e83661.

Worcel, A., Burgi, E., 1972. On the structure of the folded chromosome of *Escherichia coli*. J. Mol. Biol. 71, 127–147.

Woyke, T., Tighe, D., Mavromatis, K., Clum, A., Copeland, A., Schackwitz, W., et al., 2010. One bacterial cell, one complete genome. PLoS One 5, e10314.

Xu, L., Coleman-Derr, D., 2019. Causes and consequences of a conserved bacterial root microbiome response to drought stress. Curr. Opin. Microbiol. 49, 1–6.

Xu, L., Naylor, D., Dong, Z., Simmons, T., Pierroz, G., Hixson, K.K., et al., 2018. Drought delays development of the sorghum root microbiome and enriches for monoderm bacteria. Proc. Natl. Acad. Sci. 115, E4284–E4293.

Young, I.M., Crawford, J.W., 2004. Interactions and self-organization in the soil-microbe complex. Science 304, 1634–1637.

Young, I.M., Crawford, J.W., Nunan, N., Otten, W., Spiers, A., 2008. Microbial distribution in soils: physics and scaling. Adv. Agron. 100, 81–121.

Young, K.D., 2006. The selective value of bacterial shape. Microbiol. Mol. Biol. Rev. 70, 660–703.

Zhang, G., Gao, F., 2017. Quantitative analysis of correlation between AT and GC biases among bacterial genomes. PLoS One 12, e0171408.

Zhang, Z., Claessen, D., Rozen, D.E., 2016. Understanding microbial divisions of labor. Front. Microbiol. 7, 2070.

Zhu, H.H., Guo, J., Yao, Q., Yang, S.Z., Deng, M.R., Phuong, L.T.B., et al., 2007. *Streptomyces vietnamensis* sp. nov., a streptomycete with violet blue diffusible pigment isolated from soil in Vietnam. Int. J. Syst. Evol. Microbiol. 57, 1770–1774.

Chapter 4

Fungi in soil: a rich community with diverse functions

D. Lee Taylor* and Jennifer M. Bhatnagar[†]
*Department of Biology, University of New Mexico, Albuquerque, NM, USA; †Department of Biology, Boston University, Boston, MA, USA

Chapter outline

Soil Microbiology, Ecology, and Biochemistry. https://doi.org/10.1016/B978-0-12-822941-5.00004-1

4.1 Introduction

While plants dominate global and terrestrial biomass, fungi likely rank second only to plants in terrestrial biomes and constitute the major fraction of biomass in soil (Bar-On et al., 2018). From an evolutionary perspective, there is now considerable circumstantial evidence that fungi were instrumental in both the colonization of land by the ancestors of terrestrial plants (Pirozynski and Malloch, 1975; Simon et al., 1993) and the termination of carbon (C) deposition into geological reserves (i.e., fossil fuels, Floudas et al., 2012). The traits that underlie these features illustrate why fungi play such an important role in soils. Most fungi interact intimately with both living and dead organisms, especially plants. The mycorrhizal symbiosis with plant roots is thought to have permitted aquatic plants to transition into the challenging terrestrial habitat. Mycorrhizal and other fungal interactions with living plants (pathogens, endophytes) may be highly specific or generalized, with outcomes that vary among taxa but influence the structure and function of plant communities. Fungi have a profound influence on biogeochemical cycles through their growth habits, which include external digestion of food resources using a powerful arsenal of degradative enzymes and secondary metabolism. It was the innovation and diversification of polyphenolic-degrading enzyme machinery among the white-rot basidiomycete fungi that may have halted the accumulation of undecayed plant materials during the carboniferous (Floudas et al., 2012). The filamentous habit common to the majority of soil-dwelling fungi allows them to bridge gaps between pockets of soil water and nutrients, force their way into substrates such as decaying wood, and redistribute C, minerals, and water through the soil. Filamentous growth may underlie the abilities of some fungi to withstand soil water deficits and cold temperatures that are beyond the tolerance of bacteria and archaea. Fungi constitute large fractions of living and dead soil biomass, particularly in forested habitats. Their growth and production of cell wall materials lead to the creation and stabilization of soil aggregates, which are key elements of soil structure. Rates of production and turnover of fungal biomass have important consequences for C cycling and long-term sequestration in soil. In this chapter we summarize recent advances in our understandings of the phylogeny, biodiversity, and ecology of fungi of relevance to their diverse roles in soil environments. Of particular note is the accumulation of massive phylogenomic datasets that provide new insights into deep-level relationships and early evolution of the fungal kingdom, as well as key changes in gene repertoires across lineages and time. In addition, new insights have been acquired into how fungal communities assemble and how these processes impact ecosystem function, such as soil C storage.

4.2 Diversity and evolutionary relationships of the true fungi

4.2.1 The origin of fungi

The fungi are recognized as a kingdom (see online Supplemental material Section S4.1 for a refresher on the taxonomic hierarchy of life). Although sometimes loosely referred to as microbes, fungi are eukaryotes and most are multicellular. The superkingdom of eukaryotes that includes fungi and animals is the Opisthokonta. The common ancestor of all Opisthokonta was almost certainly a heterotrophic, single-celled, phagotrophic protist (Berbee et al., 2020). While the relationships of the earliest branches of the eukaryotes remain to be resolved with certainty, it is clear that the Opisthokonta dates back to the early origins of eukaryotes some ~1.6 billion years ago (Cerón-Romero et al., 2022). This superkingdom has been further subdivided into two groups: the Holozoa (animal lineage) and Holomycota (fungal lineage).

There is now strong evidence that a small group of protists within the Opisthokonta, including Nuclearia and the cellular slime mold Fonticula, are the closest sister group to the true fungi, or Eumycota (Liu et al., 2009; Steenkamp et al., 2006). There is also support for monophylly of the Eumycota (James et al., 2006; Li et al., 2021), although relationships among basal lineages near the divergence between nucle-arids and fungi are not yet certain.

About 30 years ago, a system of five fungal phyla achieved universal recognition based, in part, on initial rDNA phylogenies (Bruns et al., 1991). The five phyla were the Chytridiomycota (water molds), Zygomycota (bread molds), Glomeromycota (arbuscular mycorrhizal fungi [AMF]), *Ascomycota* (cup fungi), and Basidiomycota (club fungi). Later multilocus, molecular analyses suggested that neither the Zygomycota nor the Chytridiomycota is a monophyletic group (James et al., 2006; Liu et al., 2009; O'Donnell et al., 2001), although there remains uncertainty concerning the Zygomycota (Li et al., 2021; Spatafora et al., 2016). Furthermore, this five-phylum system ignored diverse taxa now known to be basal fungi, including Rozella. Current taxonomic ranks and best estimates of deep branch orders are in flux, but consensus for several key points is growing. Recent phylogenomic studies and related taxonomic revisions recognize 12 to 20 phylum-level lineages organized into a few subkingdom supergroups (James et al., 2020; Li et al., 2021; Voigt et al., 2021). Most recognize a constellation of obscure, ancient, mostly parasitic phyla within the supergroup Opisthosporidia, including Rozella, Microsporidia, and Aphelids. A second supergroup of basal fungi is sometimes named the Chytridiomyceta and includes the Chy-tridiomycota, Monoblepharidomycota, and Neocallimastigomycota, which are all flagellated and single to few-celled. Additionally, nonflagellated ancient lineages are currently recognized as discrete phyla that are not lumped within the preceding two supergroups and include the Mucoromycota, Zoopagomycota, and Dikarya. As described further below, only the Chytridiomycota and Mucoromycota are known to have prominent roles in soil among these early branching lineages. A third, more recently evolved su-pergroup is the Dikarya, which includes the majority of important soil fungi belonging to the phyla Entorrhizomycota, Ascomycota, and Basidiomycota. Molecular phylogenies confirm the long-held view that the Oomycota, which includes the important filamentous plant pathogens *Pythium* and *Phytophthora*, belong to the superkingdom Stramenopiles (heterokonts), while most slime molds (i.e., except *Fonticula*) belong to the superkingdom Amoebozoa, both outside the Opisthokonta. Even extremely data-rich phylogenomic analyses have yet to resolve with certainty the relationships at the base of the fungal tree (James et al., 2020; Li et al., 2021), so we can expect further rearrangements and optimization of high-order taxonomy in the years to come. Current understandings of the major evolutionary lines of fungi, along with a few exemplar taxa and their trophic niches, are presented in Figs. 4.1–4.3. (See Fig. S4.1 and Table S4.1 in online Supplemental material for help in making sense of some confusing terminology in fungal systematics.) Good sources for the most up-to-date taxonomic hierarchy for the fungi include Wijayawardene et al. (2020) and Index Fungorum (www.indexfungorum.org).

A number of shared, derived traits are characteristic of fungi, although no single unique trait is shared by all fungi (James et al., 2006, 2020; Stajich et al., 2009). Fungi depend on organic compounds for C, energy, and electrons, with most species taking up these resources via osmotrophy. Although many fungi can fix CO_2 using enzymes of central anabolic cycles (e.g., pyruvate carboxylase), they are considered heterotrophs in a broad sense. Chitin, a polymer of N-acetylglucosamine, is a feature of the cell wall matrix of most fungi, although quite a few parasitic lineages, especially within the Apistophelida, have life stages that lack cell walls (e.g., *Rozella*), and some groups have little or no chitin in their walls (e.g., ascomycete yeasts). The ancestral state for the true fungi includes a mobile, flagellated meiospore (zoospore) stage; flagella appear to have been lost several times through the evolution of the fungi and

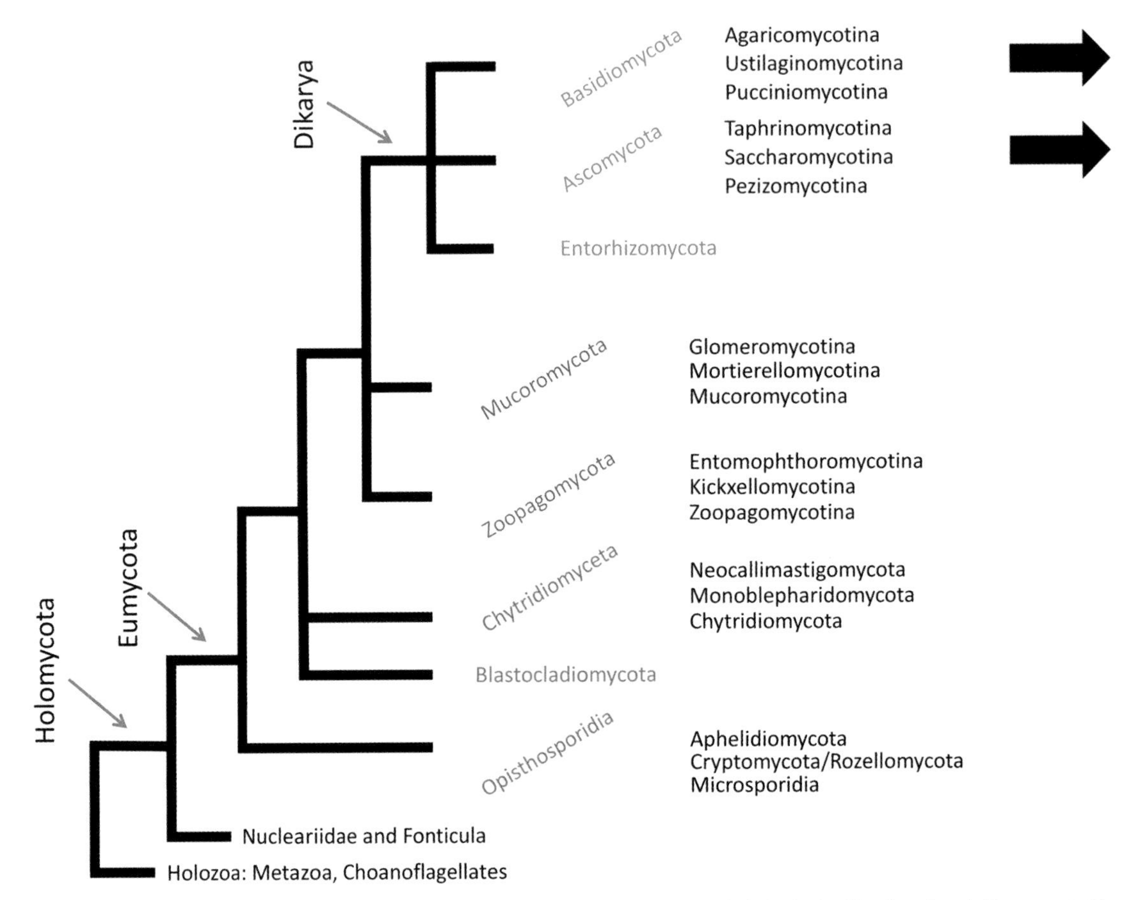

FIGURE 4.1 Overview of fungal phylogeny. Currently recognized phyla and subphyla in the kingdom fungi (Eumycota). The Holozoa are an outgroup within the Opisthokonta, while the Nucleariidae are thought to be the closest relatives of the fungi. Exemplar taxa and their ecological roles are provided on the right.

today exist only in several early diverging lineages, most prominently within the Chytridiomyceta (James et al., 2020). It is now clear that the oldest lineages of fungi were aquatic. Most fungi also use the sugar trehalose as an energy store, display apical growth, and have spindle-pole bodies rather than centrioles (with the exception of ancient, flagellated lineages). While fungi have a very long history of intimate association with land plants and their green-algal ancestors, the most ancient lineages of the Apistophelida lack the unique enzymes, such as pectinases, that are suited to attacking plant cell walls (Berbee et al., 2020). Hence the common ancestor of the Eumycota was likely not plant associated. These patterns are logical given that the origin of Eumycota has been estimated at about 1.3 billion years based on several sophisticated molecular clock studies (James et al., 2020). This origin predates the evolution of pectin-cellulosic walls in the charophycean algal ancestors of land plants (Viridiplantae). The common ancestor of all fungi, and indeed all Opistokonts, was likely a swimming, flagellated, single-celled, phagotrophic protist (Berbee et al., 2020). Indeed, nuclearids are phagotrophs (organisms lacking cell

FIGURE 4.2 Currently recognized subphyla and classes within the phylum Basidiomcota. Several subphyla and classes that contain only a few, rarely encountered species are not shown. Exemplar taxa and their ecological roles are provided on the right.

walls that engulf relatively large food items via phagocytosis). Some extant members of Opisthophelida/ Opisthosporidia still exhibit phagotrophy (i.e., uptake of food particles via encasement and engulfment of membrane-bound spheres). In contrast, in the next more recently evolved cluster of ancient fungal lineages, the "chytrids" (Blastocladiomycota and Chytriomyceta, Fig. 4.1), we see a sharp transition to osmotrophy (i.e., the direct uptake of very small molecules such as sugars and amino acids through the cell wall and into the cytoplasm via transporters). Osmotrophy requires that an organism grow into its food and digest external resources into sufficiently small molecules outside its cells and is the only option available for organisms with stiff cell walls that therefore cannot use phagotrophy.

4.2.2 Major lineages in soil

The most abundant and diverse fungal phyla in the majority of soils belong to the more recently evolved Ascomycota and Basidiomycota. However, a number of taxa from older lineages, particularly the

FIGURE 4.3 Currently recognized subphyla and classes within the phylum Ascomycota. Several subphyla and classes that contain only a few, rarely encountered species are not shown. Exemplar taxa and their ecological roles are provided on the right.

Mucoromycotina and Mortierellomycotina, are also abundant in many soils. Even members of the ancient Chytridiomycota are often recovered in soil metabarcoding studies (Taylor et al., 2014; Tedersoo et al., 2014), though at low abundances. As discussed in the next section, whether these low abundances may be due in part to negative biases of current primer sets is yet to be determined. The trophic roles of chytrids are poorly understood, although it has been suggested that they may decompose allochthonous plant materials in some unvegetated high-elevation soils where they comprise a higher fraction of the fungal community than in less extreme environments (Freeman et al., 2009; Schmidt et al., 2012).

Many taxa of zygospore-forming fungi belonging to the Mortierellomycotina and Mucoromycotina are found primarily in soil. Some are clearly decomposers, with apparent preferences for labile C sources (e.g., members of the cosmopolitan soil genus *Mortierella*). However, in the last decade a diverse array of *Mucoromycotina* related to *Endogone* have been documented forming arbuscular mycorrhizal (AM)–like coils within roots based on combined molecular diagnostics and detailed microscopy (Bidartondo et al., 2011; Field et al., 2015; Humphreys et al., 2010). The first records came from some of the oldest living lineages of nonvascular plants and led to the hypothesis that relatives of these fungi, which were thought to be older than true AMF of the Glomeromycotina, may have been the symbionts that aided plants in colonizing land rather than AMF. In support of this hypothesis these fungi exchange P and N in return for C, demonstrating a true mutualistic mycorrhizal role (Field et al., 2015). On the other hand, these fungi often cooccur with AMF in nonvascular plants (Field et al., 2016) and have more recently been found in vascular and even flowering plants (Albornoz et al., 2020). Furthermore, very recent phylogenomic studies suggest that the Glomeromycotina is not sister to the Dikarya and, instead, is likely affiliated with the Mucoromycota (Li et al., 2021; Spatafora et al., 2016). Hence mycorrhizal Mucoromycotina may or may not predate AM Glomeromycotina. The class Glomeromycotina encompasses all fungi that form true arbuscular mycorrhizae, as well as the enigmatic, algal symbiont Geosiphon (Gehrig et al., 1996).

The subkingdom Dikarya is comprised of the most recent and derived "crown" phyla Ascomycota and Basidiomycota. A recent surprise has been the proposed addition of a new phylum, the Entorrhizomycota, to the Dikarya (Bauer et al., 2015). This new phylum is comprised of just a few species of filamentous root pathogens. These fungi share more cellular and ultrastructural features with Basidiomycota, including dikaryotic vegetative cells, tripartite septa, spindle-pole-body structure, and meiospores formed in tetrads. Yet current multigene phylogenies place them basal to Ascomycota + Basidiomycota (Riess et al., 2019). However, no full Entorrhizomycota genomes are yet available, and the placement of this phylum relative to the Ascomycota and *Basidiomycota* remains uncertain.

The Ascomycota constitute the most species-rich fungal phylum, accounting for roughly 75% of described fungal species. These species fall into three subphyla, as follows: (1) Taphrinomycotina, which encompasses the fission yeasts (Schizosaccharomyces), the animal pathogen Pneumocystis, the unusual dimorphic plant pathogen *Taphrina*, the root-associated, sporocarp-forming, filamentous genus *Neolecta*, and the newly erected class of filamentous soil fungi, the Archaeorhizomyces; (2) Saccharomycotina include the budding yeasts, such as *Saccharomyces*, *Debaromyces*, *Pichia*, *Candida*, and others, many of which are found in soils, presumably decomposing labile organic materials both aerobically and anaerobically; and (3) Pezizomycotina which accounts for the greatest phylogenetic, species, and functional diversity within the Ascomycota. The majority of lichen-forming fungi fall within this lineage (there are a few basidiomycete lichens), as do a few ectomycorrhizal fungal (EMF) taxa, all dark-septate endophytes (DSE), essentially all ericoid-mycorrhizal (ERM) species, and a wide spectrum of endophytes, pathogens, and saprotrophs. Recent phylogenomic studies support each of these subphyla as monophyletic and suggest that the Sacharomycotina and Pezizomycotina are sister groups (Shen et al., 2020).

The diverse phylum Basidiomycota is also divided into three well-supported subphyla. The branching order of these subphyla, which likely radiated in a short period of time, remains unresolved even using advanced phylogenomic data and methods (Li et al., 2021). The subphylum Pucciniomycotina includes all rust fungi, an economically important group of plant pathogens, as well as some yeasts that are common in soil, yet rarely studied, such as *Sporobolomyces* and Leucosporidium. The second subphylum, Ustilaginomycotina, is also predominantly comprised of plant pathogens, the smuts. Similar to the rusts, the Ustilagomycotina includes several yeasts, such as *Malassezia* (a skin pathogen that is frequently recovered from soils) and *Arcticomyces* (a cold-soil yeast). The Agaricomycotina includes the vast majority of filamentous Basidiomycota, including all mushroom-forming taxa. These encompass nearly every conceivable soil niche (except thermophiles and psychrophiles), accounting for all brown rot and white rot fungi, as well as most EMF taxa.

4.2.3 Extremophiles and novel lineages

While fungi do not equal the extremes of heat, salinity, acidity, or UV tolerance of the most extremophilic of the prokaryotes, certain fungal taxa surpass most other eukaryotes in tolerance to these and other extreme conditions, including cold and aridity. Fungi are found in essentially all biomes and habitats, both terrestrial and aquatic. Some of these habitats are extreme, such as soils in the Dry Valleys of Antarctica, on Arctic glaciers, and in high molarity salterns. A key question is the extent to which these taxa are active in these hostile environments as opposed to surviving in highly resistant, dormant spore stages following introduction by wind or other vectors (Bridge and Spooner, 2012; Pearce et al., 2009). Nonetheless, there is clear evidence that some of these extreme environment fungi are indeed active in these habitats. Phylogenetically diverse yeasts dominate the isolates obtained from many extreme habitats. For example, species of the dimorphic basidiomycete yeast *Cryptococcus* can be dominant in glacial habitats, permafrost, marine sediments, and unvegetated Antarctic Dry Valley soils (Bridge and Spooner, 2012; Buzzini et al., 2012; Connell et al., 2006). In the marine study these yeasts were detected by culture-independent DNA and RNA methods, the latter strongly suggesting in situ metabolic activity (Edgcomb et al., 2011). Ascomycota yeasts, such as *Pichia*, *Debaromyces*, *Candida*, *Metschnikowia*, and *Aureobasidium pullulans*, are also found in extremely cold and/or saline habitats (Butinar et al., 2011; Cantrell and Baez-Félix, 2010; Cantrell et al., 2011; Gunde-Cimerman et al., 2003; Zalar et al., 2008). The so-called "meristematic," "microcolonial," or "black yeasts" are distributed among several lineages of Sordariomycetes, Eurotiomycetes, and Dothideomycetes (Onofri et al., 2000; Selbmann et al., 2005; Sterflinger et al., 2012) and commonly occur in extreme habitats; e.g., in canyon walls of the Antarctic Dry Valleys. Some species grow on or within rocks (endolithic) in both hot and cold deserts and at high elevations. Some members of this group display moderate halotolerance and extreme drought tolerance or can grow at pHs down to zero (Starkey and Waksman, 1943). Convergent evolution of strong melanization, slow growth, isodiametric meristematic cells (i.e., cells that can reinitiate growth when dislodged from the colony), and other putative stress-related features is seen among the black yeasts (Selbmann et al., 2005; Sterflinger et al., 2012). Several filamentous Ascomycota are also noteworthy extremophiles. Species of *Geomyces* (*Leotiomycetes*; now placed in teleomorph *Pseudogymnoascus*) have been recorded from marine habitats as well as cold soils (Arenz and Blanchette, 2011; Bridge and Spooner, 2012; Richards et al., 2012; Waldrop et al., 2008). One isolate was reported to be metabolically active down to $-35°C$ (Panikov and Sizova, 2007).

Lichens are also formed predominantly by filamentous taxa of Ascomycota and are found across the array of extreme hot, cold, dry, and saline environments (discussed above), often playing the role of chief primary producer by virtue of photosynthetic activities of their cyanobacterial or algal photobionts (Vitt, 2007). Lichens are also important and widespread in less extreme terrestrial habitats (Feuerer and Hawksworth, 2007). These same classes of Ascomycota include species with the highest heat tolerance seen in the Eukaryota. Thermophilic and thermotolerant fungi share several key convergent traits. Their spores usually do not germinate below temperatures of 45°C, even though the mycelium can grow at lower temperatures (Maheshwari et al., 2000). These fungi, which belong to several orders within the Ascomycota (Sordariales, Eurotiales, Onygenales), one of which is within the early diverging lineages (*Mucorales*), have been recovered from diverse soils in both hot and cold regions. Within the Ascomycota, closely related species may be mesophiles and thermophiles, although many members of the Chaetomiaceae (Sordariales) are thermophilic. The primary niches of thermophiles are concentrated aggregations of moist, well-oxygenated organic material that self-heat due to intensive respiration during decomposition (Maheshwari et al., 2000).

In the case of black yeasts, thermophiles, and lichens, there is no question of their activity and adaptation to extremes since they can be observed actively growing under these extreme conditions. This is also true of some of the cold-tolerant taxa, such as Geomyces, which can be observed growing in a dense mat across permanently frozen ice lenses in the Fox Permafrost Tunnel near Fairbanks, Alaska, United States (Waldrop et al., 2008). It seems likely that some of the other extremophiles described above, such as yeasts isolated from glacial habitats, are also indigenous taxa that are adapted to these extreme environments since they display tolerance to extreme conditions in the laboratory (Arenz and Blanchette, 2011; Butinar et al., 2011; Buzzini et al., 2012; Gunde-Cimerman et al., 2003; Onofri et al., 2000; Selbmann et al., 2005). Some cosmopolitan taxa detected in extreme environments, such as species of *Penicillium* and *Aspergillus*, may be present only as inactive spores. A recent study used RNA-based metabarcoding to isolate the putatively active fraction of the fungal community and compared it to the total community, derived from standard DNA metabarcoding in high-latitude cold soils (Cox et al., 2019). Differences between the RNA and DNA profiles revealed overrepresentation of endemic taxa, particularly chytrids, in the active community and overrepresentation of cosmopolitan, wind-dispersed Ascomycota in the DNA community, supporting the hypothesis that some of these cosmopolitan taxa do not actually grow under extreme conditions (Cox et al., 2019).

Although members of the Dikarya dominate records for extremophilic fungi, members of the early diverging fungal lineages have been reported from marine habitats using culture-independent methods (Comeau et al., 2016; Richards et al., 2012). Culture-independent approaches have also revealed a preponderance of chytrid lineages in soils at globally distributed, high-elevation sites that are above the vegetated zone (Freeman et al., 2009; Gleason et al., 2010). It has been proposed that these fungi may derive nutrients from algae and/or pollen transported by wind from distant locations; their presence in marine habitats might also be due to trophic linkages with algae (Richards et al., 2012).

4.2.4 Fungal structure and growth

The body of a fungal individual may contain one type of nucleus with only a single set of chromosomes — a haploid growth form. Alternatively, the cells of two different haploid individuals may fuse (plasmogamy). In most fungi this fusion occurs only immediately before nuclear fusion (karyogamy) and meiosis (see Fig. S4.1 in online Supplemental material for an exemplar fungal life cycle). However, in some fungi there is a brief (phylum Ascomycota) or prolonged (phylum Basidiomycota) stage in which the two different nuclei multiply in a synchronized fashion; this phase of the life cycle is termed dikaryotic. While not representing any particular taxon, the features shown in Fig. 4.4 are characteristic of the Dikarya. As a mycelium grows, hyphae branch at regulated intervals in response to external and internal signals (Fig 4.5a). In many fungi cytoplasm is retracted from older parts of the mycelium, leaving walled-off empty cells. The newly formed thin and soft hyphal tip extends due to turgor pressure. The growing tip is the area of most active enzyme secretion and nutrient uptake. In the more ancient fungal groups the filaments in which the nuclei are housed do not contain cross-walls (septa), while in other groups septa divide hyphal filaments into distinct cells (Fig 4.5b). Cross walls or septa separate individual cells (numbers of nuclei are usually variable in Ascomycota but are more often fixed in Basidiomycota). In the two most recently evolved phyla the Ascomycota and Basidiomycota, nuclear division, and

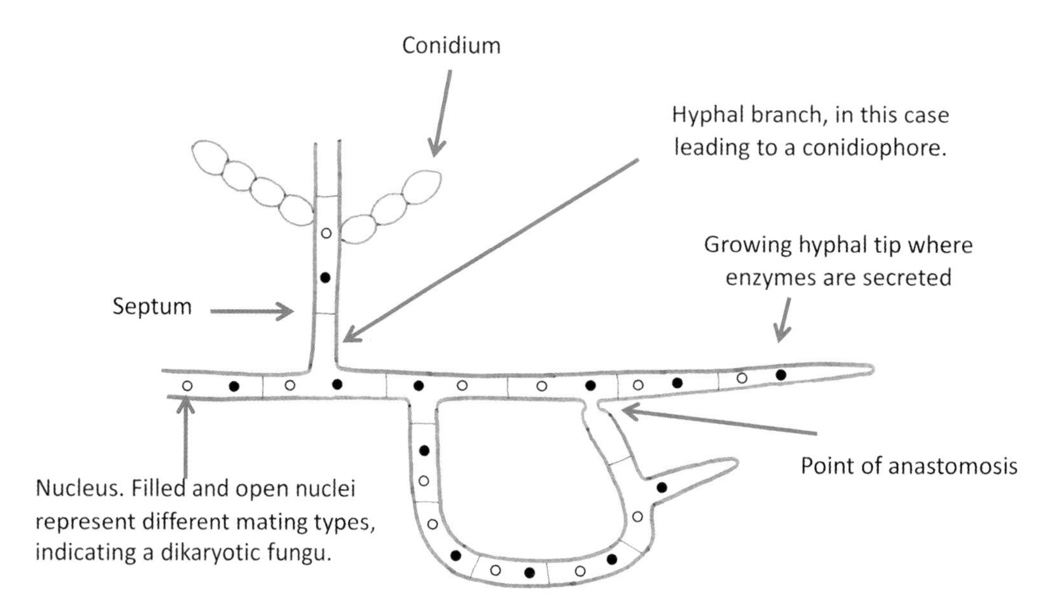

FIGURE 4.4 Filamentous fungal growth form typical of fungi in soil. Cartoon depicts a mycelium growing from left to right.

apportionment to the cells comprising the hyphae are tightly regulated (see Fig. 4.5b, c, d), while in groups without septa nuclei flow freely through the entire mycelium (e.g., in phylum Glomeromycota).

Fungal mycelia generally grow radially as fractal networks in soil, wood, and litter (Bolton and Boddy, 1993). Fungi alter hyphal development in response to environmental conditions to minimize cost:benefit ratios in terms of C or nutrient capture versus expenditure on growth. Very fine feeder hyphae are elaborated in resource-rich patches, while nutrient-poor areas are less densely colonized by hyphae specialized for efficient searching and nutrient transport. The transport hyphae may aggregate into tightly woven bundles called cords, strands, or rhizomorphs depending on their developmental structure. All these aggregations provide larger diameter transport tubes. However, species vary considerably in hyphal growth patterns, the size and structure of transport networks, and the resulting foraging strategies (Agerer, 2001; Boddy, 1999; Donnelly et al., 2004).

4.3 Diversity and biogeography

Fungi generally dominate microbial biomass and activity (i.e., respiration) in soil organic horizons, particularly in forests (Joergensen and Wichern, 2008). Bacterial:fungal ratios tend to be lower in acidic, low-nutrient soils with recalcitrant litter and high C:N ratios (Fierer et al., 2009), while bacteria are increasingly prominent in high−N and P, saline, alkaline, and anaerobic (waterlogged) soils (Joergensen and Wichern, 2008).

Fungal biomass varies widely within and across biomes in relation to plant litter composition, root density, and nutrient availability. Fungi may comprise up to 20% of the mass of decomposing plant litter.

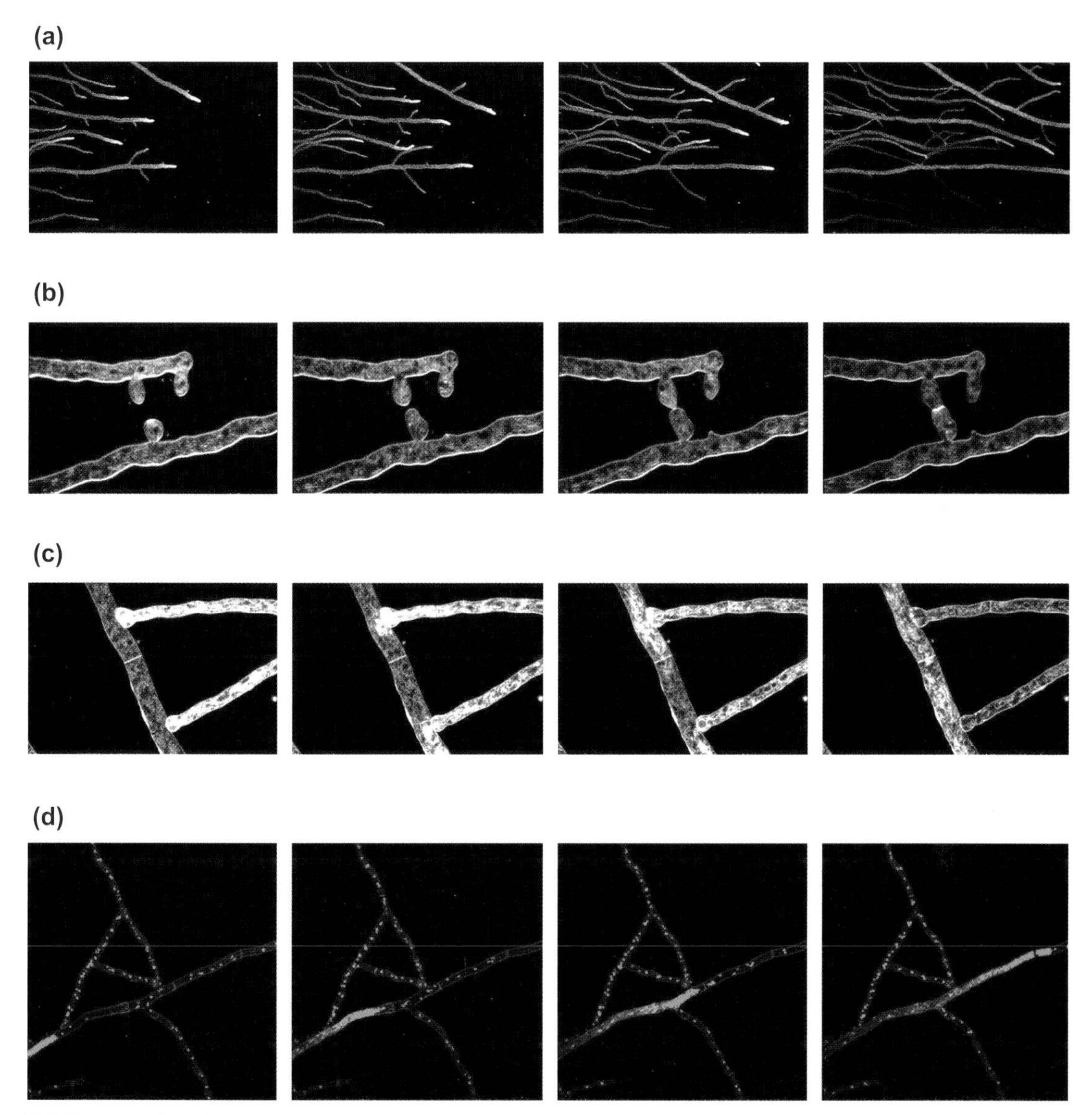

FIGURE 4.5 Mycelial structure and growth. (a) Time series showing tip growth and branching as mycelium spreads across substrate. Left image is the earliest in the series. *(Images courtesy of Patrick Hickey.)* (b) Time series showing anastomosis (fusion) of hyphal branches followed by septum formation. *(Images courtesy of Patrick Hickey.)* (c) Time series showing mixing of cytoplasmic contents following anastomosis. *(Images courtesy of Patrick Hickey.)* (d) Time series showing migration of cytoplasm and nuclei through septa. *(Images courtesy of Patrick Hickey.)*

In biomes dominated by ectomycorrhizal plants extraradical mycelia may comprise 30% of the microbial biomass and 80% of the fungal biomass (Högberg and Högberg, 2002). Although fungal abundance and ratios of fungal to bacterial biomass tend to increase as soil pH decreases (Högberg et al., 2007; Joergensen and Wichern, 2008; Rousk et al., 2009), studies suggest that fungal distributions are more influenced by N and P availability than pH per se (Fierer et al., 2009; Lauber et al., 2009). Estimates of fungal biomass turnover in soils are on the order of months (i.e., 130–150 days; Rousk and Baath, 2011), which is comparable to studies of fungal growth rates on submerged litter (e.g., Gulis et al., 2008).

Fungi are present in all soils and are a prominent biological component of most. At broad phylogenetic scales, the aphorism "everything is everywhere" does seem to apply to fungi: beyond soils, they are also found in nearly every other habitat on Earth. High-throughput sequencing (HTS) is providing remarkable new insights into the local to global distribution of fungal species (Bahram et al., 2018; Talbot et al., 2014; Taylor et al., 2014; Tedersoo et al., 2014, 2021). However, a consensus as to whether endemism or global distribution accounts for the larger fraction of species in typical communities remains elusive. What we do know with certainty is that some species occur over only a limited range, from regional to continental, while other species are clearly able to disperse across the globe. Patterns with respect to distribution ranges and species diversity across latitudes and functional guilds are also beginning to emerge.

4.3.1 Estimates of species richness

The true diversity of the Eumycota is uncertain and controversial. To date, there are roughly 148,000 described taxa (Index Fungorum). New species are being described at a rapid rate (Cheek et al., 2020), limited primarily by a dearth of fungal taxonomists rather than a lack of fungi in need of description (Blackwell, 2011; Hawksworth, 2012). Mycologists widely acknowledge that the true number of species on Earth vastly exceeds the number that has been described (Bass and Richards, 2011). A variety of approaches have been used to estimate how many species of fungi may exist. The approach that has received the most attention has been to census both plants and fungi at focal sites within a region to derive species ratios of fungal to plant taxa. These regional ratios are then multiplied by the estimated numbers of plants on Earth to arrive at an estimate for fungal richness. Hawksworth compiled data from well-studied sites in the United Kingdom and obtained a ratio of six fungi per vascular plant species, giving rise to a widely cited global estimate of 1.5 million fungi (Hawksworth, 1991). Updates by Hawksworth have raised the estimate to 2.2 to 3.8 million (Hawksworth and Lücking, 2017). Several molecular surveys of fungi in soil have applied the fungus:plant ratio method and yielded estimates of 5 to 6 million species (O'Brien et al., 2005; Taylor et al., 2014). Estimates based on thorough censuses from single plant species have yielded estimates as high as 10 million (Cannon, 1997). However, there is also evidence that fungus:plant ratios vary geographically and with latitude and are lower in the tropics (Tedersoo et al., 2014). Hence fungus:plant ratios may not be the best path to accurately estimate fungal diversity. It is also important to note that soil-based estimates largely overlook aquatic fungi, insect pathogens, gut symbionts, and many others (Blackwell, 2011). PCR-based surveys also likely overlook divergent taxa due to primer biases. Thus we can confidently state that estimates of total fungal species diversity on Earth of 2 to 3 million are likely to fall well below the true number but also that an immense amount of additional work will be required to approach a reliable tally.

4.3.2 Fungal dispersal and biogeography

Fungal biogeography and assembly of local communities are determined, in part, by dispersal. Fungi in soils disperse, albeit slowly, through growth of their mycelial networks. They disperse more rapidly and over larger distances through movement of various propagules. A hallmark of fungi is the production of propagules in the forms of meiotic or mitotic single-celled resting structures (zygospores, ascospores, basidiospores, conidia, etc.) (see Fig. S4.1 and Section 4.2.4) capable of surviving harsh conditions and dispersing great distances by air. Thick-walled vegetative cells (e.g., monilioid cells) or aggregations of such cells, such as the canon ball—like sclerotia formed by Cenococcum and *Thanatephorus*, can also be important propagules. Meiotic basidiospores are very small and are released into the air by fungi that form mushrooms and by other plant pathogenic fungi such as rusts and smuts. However, studies of mushrooms have shown that, much like plant seeds, the vast majority of basidiospores fall to the ground within centimeters of the fruiting body (Galante et al., 2011). Studies of EMF colonizing individual pine islands amid a sea of non-EMF plants have shown that distance to forest edge and airborne dispersal capabilities of spores of different fungi have strong effects on potential and actual colonization (Peay et al., 2007, 2010, 2012). On the other hand, recent molecular surveys and meta-analyses have shown that a core set of Ascomycota are both dominant in many soils and have global species distributions (Egidi et al., 2019; Větrovský et al., 2019). These were DNA-based whole-soil surveys, so the degree to which the incidence of these global taxa can be attributed to inactive spores or relic DNA (Carini et al., 2017) versus active mycelium is an open question (Cox et al., 2019).

A single taxon defined by traditional morphological and anatomical characters may encompass two or more subgroups that are distinct when phylogenetic or biological species concepts are applied; these are cryptic species. Cryptic species nested within traditional taxonomic species often have much narrower geographic distributions. Molecular studies are revealing cryptic species within widespread species complexes in a number of EMF and decomposer fungi (Aldrovandi et al., 2015; Carlsen et al., 2011; Feng et al., 2016; Geml et al., 2008; Grubisha et al., 2012; Haight et al., 2016; James et al., 1999; Taylor et al., 2006). Some of these studies also reveal finer host-specificity than previously recognized. It seems reasonable to expect that the capacity for aerial dispersal interacts with host and habitat specificity to influence the patterns of successful dispersal that are observed in nature. In contrast to EMF and decomposers, plant pathogens are notorious for very wide dispersal capabilities. However, this perspective may be driven in part by our alarm at the devastation that can ensue when a rare dispersal event (often human-mediated) carries a virulent pathogen to a novel, susceptible host. Native elms and chestnut trees were effectively lost to North America due to the introduction of virulent Dutch elm disease (*Ophiostoma novo-ulmi*), possibly from Asia, and chestnut blight (*Cryphonectria parasitica*) from Japan. However, the survival of these trees for millennia before arrival of these pathogens again underlines that cross-continental jumps are rare.

4.4 Fungal communities

In mainstream ecology the term "community" refers to the set of sympatric, metabolically active organisms that either interact or can potentially interact (Krebs, 1978). We know relatively little about when, where, and how fungi interact with other organisms in soil, aside from conspicuous manifestations, such as mycorrhizal colonization of plant roots or nematode-trapping fungi, and a few recent discoveries of fungal-bacteria interactions (Deveau et al., 2018; Romaní et al., 2006). Fungi in soil vary at least four orders of magnitude in size. Single-celled yeasts may be 3 to 10 μm in diameter. In contrast, a single mycelial individual of the white-rot, root pathogen *Armillaria gallica* spans >37 hectares with a predicted mass greater than 400,000 kg (Anderson et al., 2018; Smith et al., 1992). In some areas of mycology careful attention has been paid to the spatial definition of community. In particular, researchers studying wood, litter, and dung decay have recognized that fungal species must colonize, grow, and reproduce within the confines of a particular substrate, leading to the designation of unit communities (Cooke and Rayner, 1984). For example, the fungi that occupy a single, isolated leaf might constitute a unit community. The application of the unit community perspective to soil is difficult due to the complex distribution of resources and lack of distinct spatial boundaries. From a practical perspective, soil fungal communities are usually sampled at the plot scale (e.g., 5−50 m diameter), which is defensible in terms of the potential for filamentous individuals to interact, given the size range of sizes of fungal genets.

4.4.1 Abiotic drivers

Like all terrestrial organisms, the distributions and abundances of fungal species are influenced by history (e.g., dispersal, plate tectonics), climate, and numerous environmental factors, especially edaphic factors. There is increasing evidence that soil fungal communities are influenced by climatic conditions on both geographic and temporal scales. For example, the correlation of soil fungal communities is stronger with temperature than with latitude across the five bioclimatic subzones of the Arctic (Timling et al., 2014). Moisture, measured as mean annual precipitation (MAP), was one of the strongest predictors of community composition in the global soil studies by Tedersoo and colleagues (Bahram et al., 2018; Tedersoo et al., 2014, 2021). In communities of EMF, climate explained 58% of the variance in richness and 41% in community composition in 68 sites across North America (Steidinger et al., 2020). In concert with patterns of fungal species composition, climate and pH appear to be the major drivers of the relative dominance of EMF versus AMF or nonmycorrhizal symbioses in forests worldwide (GFBI consortium et al., 2019). Ectomycorrhizal symbioses increase in dominance with distance from the equator in a pattern that aligns with decreasing decomposition coefficient (an index based on moisture and temperature). A recent meta-analysis of fungal communities in soil globally supports the preeminence of climate as a driver (Větrovský et al., 2019). However, the relative importance of temperature versus moisture and how these factors influence the physiology and performance of different fungi remain outstanding questions. Moreover, the degree to which climate influences fungal communities directly or indirectly via differences in vegetation remains to be elucidated.

Most studies have undertaken sampling in the summer at higher latitudes or the equivalent at peak growing season (e.g., wet season) in tropical latitudes, yet most belowground activity likely occurs at other times. For example, midlatitude, high-elevation alpine sites at Niwot Ridge in Colorado showed that microbial (including fungal) biomass peaked in late winter under the snowpack (Lipson and Schmidt, 2004; Schadt, 2003). Enzyme activities also peaked in winter following leaf-drop in a deciduous forest, with saprotrophs increasing in biomass through the winter (Voříšková et al., 2014). Given that up to 10 months of the year are snow covered in boreal, Arctic, and Antarctic regions, activity under snow could strongly influence annual biogeochemical fluxes (Sturm et al., 2005). In temperate biomes (Averill et al., 2019; Voříšková et al., 2014), and even in cold-dominated boreal and alpine biomes, fungal communities in soil have been shown to shift predictably across seasons (Schadt, 2003; Taylor et al., 2010).

Other more extreme and less predictable disturbances also perturb the biomass and composition of fungal communities in soil. The disturbance that has received the greatest attention is fire. Ectomycorrhizal communities are strongly impacted by fire as a consequence of direct heat injury and consumption of fungal biomass, as well as death or injury to host plants, loss of organic matter, and changes in soil chemistry (e.g., transient increases in N availability, higher pH). Studies showing mild effects of fire have usually been conducted in habitats with frequent low-severity burns that do not kill the host trees (Dahlberg et al., 2001; Smith et al., 2004; Stendell et al., 1999). Where hotter, stand-replacing fires occur, impacts on ECM fungi are stronger (Baar et al., 1999; Cairney and Bastias, 2007). In general, fire reduces fungal diversity and preferentially removes those taxa that have strong preferences for the litter layer. When mycorrhizal hosts are killed, vegetation succession is reset and suites of so-called early-stage fungi are the dominant colonizers (see next section; Taylor et al., 2010; Treseder et al., 2004; Visser, 1995). There are also a number of soil fungi that respond positively to fire, the so-called pyrophilous fungi (Glassman et al., 2016; Hughes et al., 2020). Most of these appear to be saprotrophic taxa that benefit from new substrates and possibly reduced competition from established mycorrhizal and decomposer taxa.

Other well-studied abiotic disturbances include land management, both forest and agricultural, as well as pollution. For example, agricultural tillage strongly reduces the diversity of AMF (Helgason et al., 1998). Heavy metal contamination (Colpaert and van Assche, 1987; Op De Beeck et al., 2015) and nitrogen (N) deposition (Lilleskov et al., 2002; van der Linde et al., 2018) also reduce diversity and reshape fungal communities. Disturbances that occur at smaller spatial scales also impact fungal communities. For example, fungal communities in paired vegetated versus cryoturbated microsites (frost boils) in the Arctic are highly distinct (Timling et al., 2014). At the micrometer scale of hyphae and mycelia, soil fungi face constant impacts ranging from fresh litter inputs to dying roots to fluctuation in moisture to grazing by soil fauna such as collembolans.

At local to global scales, soil fungi respond to environmental gradients in factors such as pH and moisture (Bahram et al., 2018; Tedersoo et al., 2014). A series of papers report sharp differences in fungal communities as a function of soil horizon (Bahram et al., 2015; Dickie et al., 2002; Taylor and Bruns, 1999; Taylor et al., 2014). Soil pH, moisture content, and nutrient levels (particularly N) are also correlated with community composition in soil (Cox et al., 2010; Taylor et al., 2000; Toljander et al., 2006; van der Linde et al., 2018) but are relatively weak drivers compared to soil horizon, except in cases of extreme gradients such as gaseous ammonia pollution from a fertilizer plant (Lilleskov et al., 2002). On the other hand, at regional to continental scales, geographic distance was shown to play a larger role than edaphic factors or horizon in structuring soil fungi in North American coniferous forests (Talbot et al., 2014).

Not surprisingly, edaphic factors are strongly correlated with the composition of both mycorrhizal and saprotrophic fungal communities. Among EMF, there is evidence that mycelial exploration type (i.e., hyphal growth and foraging patterns; Agerer, 2001) may be related to strategies for N acquisition from different sources (Hobbie and Agerer, 2010; Taylor et al., 2000). Taylor et al. (2000) found higher capacities for growth on media containing only organic N sources among fungal isolates from the pristine end of an N-deposition gradient across Europe compared with those from the polluted end of the gradient where mineral N is more available. Corresponding to the aforementioned pattern, EMF taxa that produce potent peroxidases increase in abundance with latitude and declining soil mineral N availability (Argiroff et al., 2022). Less attention has been paid to how pH or moisture may underlie habitat preferences other than the well-known high levels of enzyme activity, and resulting competitive dominance, exhibited by ericoid mycorrhizal fungi under acidic conditions (Read, 1991; Read and Perez-Moreno, 2003). Several papers have suggested that soil K and Ca levels are predictive of the composition of EMF communities associated with *Alnus* species (Roy et al., 2013; Tedersoo et al., 2009). The physiological traits that underlie preferences of soil fungi for particular abiotic conditions are mostly unknown, although soil C:N ratios, decomposition rates, and host mycorrhizal type appear to exert selection that alters the relative roles of mycorrhizal versus saprotrophic fungi in the liberation of macronutrients from soil organic matter (SOM) (Argiroff et al., 2022; Bahram et al., 2018; GFBI consortium et al., 2019; Kyaschenko et al., 2017; Mayer et al., 2021).

4.4.2 Biotic drivers

Vegetation is the most important biotic factor that influences the composition of fungal communities in soil. Most fungi consume living or dead plant material for their primary energy source, with a large fraction of fungi displaying some degree of specialization toward living or dead tissue of particular plant lineages or functional groups. Thus plant community composition plays a critical role in determining which soil fungi are present at a site. It has been well-documented that EMF range from genus-specific, or even species-specific, to quite generalist, associating with both angiosperms and gymnosperms (Molina et al., 1992). Broader associations appear to be the norm in EMF, but host-specialist fungi can be important players, such as in the reciprocally specific associations of alder (Kennedy et al., 2011; Roy et al., 2013). While the several hundred species of AMF found in >200,000 vascular plant species have been assumed to have low specificity, a more complicated picture is emerging. Much like mycorrhizal fungi, decomposer fungi display a range of specialization toward their substrates. The genetic, physiological, or ecological basis for such specialization are not known in most cases. Certain white and brown rot wood decay fungi are found on wide arrays of both angiosperm and gymnosperm hosts. Yet many fungi in these guilds favor either angiosperms or gymnosperms and may even prefer families or genera within these lineages.

Historically, only the plant-to-fungus direction of influence received much recognition. However, there is increasing evidence that these influences are reciprocal: the spectrum of species present, i.e., the plant microbiome, and their relative abundances can also have major impacts on plant community composition (Rúa et al., 2016; van der Heijden et al., 1998). Specific plant pathogens can dramatically reduce or eliminate particular host species, altering plant community composition (Packer and Clay, 2000; Reynolds et al., 2003). Plant-soil-feedbacks (Bever, 2003, 1994; Kulmatiski et al., 2008), which often involve the buildup of host-specific pathogens in soil, can alter coexistence and competitive dynamics within plant communities.

Competition among fungal species also plays an important role in structuring communities. Studies have demonstrated that the arrival order of ECM species can shift competitive dominance in colonization of seedling root systems (Kennedy et al., 2009). These priority effects have been shown to occur for fungal decomposer communities in wood (Fukami et al., 2010). In this latter example the order of species arrivals also affected the progress of decay, suggesting that competition and community assembly may have ecosystem consequences (see also Argiroff et al., 2022; Kyaschenko et al., 2017). These interactions likely involve indirect exploitative competition for resources as well as various forms of direct interference competition. Further evidence for competition in soil fungi comes from statistical analyses of patterns of cooccurrence and avoidance. In general, cooperating/synergistic species should cooccur more often than expected by chance, while antagonistic species should cooccur less often than expected by chance; the latter situation is called a "checkerboard" pattern (Stone and Roberts, 1990). In a study of pine ECM communities avoidance patterns suggestive of competition were more common than were cooccurrence patterns (Koide et al., 2004). Wood decay fungi are well-known for overt signs of competitive interactions, as many use combative strategies, including production of bioactive compounds as well as direct invasion and lysis of opposing hyphae (Boddy, 2006; Cooke and Rayner, 1984).

Soil fungal communities undergo succession, likely driven by a combination of species interactions and changing resource/environmental conditions, analogous to patterns known from prokaryotes and plants. Patterns of succession are complicated in soil fungal communities due to the wide range of relevant spatial and temporal scales. In vegetated ecosystems succession of EMF occurs in tandem with plant succession. These observations led to the classification of early-stage and late-stage EMF taxa (Deacon et al., 1983). Early-stage species are the first to colonize young tree seedlings in habitats with no mature trees (e.g., old-fields), while late-stage species are typical of mature forests. These changes in fungal communities may be driven by changes in the C-provisioning capacity of trees as they grow (late-stage fungi tend to produce larger mycelial mats and may have larger C demands) or due to changes in the soil environment, particularly the build-up of a well-decomposed organic horizon (late-stage fungi appear to have a greater capacity to degrade complex organic polymers). Studies of fungal communities in soil beyond only ECM taxa also demonstrate strong shifts in composition in concert with plant successional stage (Taylor et al., 2010).

A microhabitat in which fungal succession has been well studied is coarse woody debris. As with EMF communities, wood-decomposer fungi are territorial, usually occupying contiguous patches of substrate to the exclusion of other species. A series of studies have demonstrated that the first colonists of wood are present at low levels in the living tree and grow actively once the dead wood has dried beyond a certain threshold (Chapela and Boddy, 1988; Parfitt et al., 2010). These fungi are primarily soft rot members of the Ascomycota, such as *Xylaria*. These taxa are quickly followed by white and brown rot fungi in the Basidiomycota. Observations of fruit body formation over time on downed logs, as well as direct molecular analyses of the wood itself, agree that Basidiomycota-dominated communities follow a predictable series of species appearance and disappearance. Carbohydrate and phenolic composition, as well as C:N:P ratios, change significantly as decomposition progresses, which likely provides a basis for niche-differentiation and successional patterns among decomposers. However, chemical composition does not entirely explain the successional patterns. For example, certain wood-rot Basidiomycota follower species seem to occur only when another species of Basidiomycota has previously colonized the log. Whether these patterns arise from direct species interactions or from changes in the chemical environment imparted by the primary species (i.e., facilitation in the sense of classical succession theory) is unknown.

Successional patterns in leaf litter are similar but faster than in wood. The earliest colonizers are often found within the attached, senescent leaves as endophytes, primarily Sordariomycetes and Dothideomycetes (Ascomycota) (Snajdr et al., 2011). R-selected sugar fungi that are not present as endophytes, such as *Mortierella*, may also play important roles in the earliest stages of decay when labile carbohydrates are readily available. Once litter is exposed to the moist forest floor, rhizomorph and cord-forming Basidiomycota can aggressively colonize leaves. However, in arid lands it appears that various drought-tolerant, melanized Ascomycota predominate over Basidiomycota as both plant symbionts and decomposers (Porras-Alfaro et al., 2011).

4.5 Fungal traits

Trait-based perspectives have been extremely important in plant and animal ecology (Chapin et al., 1996; Grime, 1974, 1977). Considerable effort has been put into building trait databases for plants and animals (Kattge et al., 2011). Efforts to apply similar perspectives to microbial ecology have gained momentum (Aguilar-Trigueros et al., 2015; Crowther et al., 2014; Lajoie and Kembel, 2019; Lustenhouwer et al., 2020; Romero-Olivares et al., 2021; Zanne et al., 2020a, 2020b). The trait-based perspective seeks to link specific phenotypic characteristics of an organism to its performance (fitness) in a particular environment. Through assembling data on numerous traits across species and environments, ecologists seek to discern tradeoffs between trait values and combinations of traits (strategies) that perform well under the specific conditions. This perspective can improve mechanistic understandings of fundamental ecological processes, such as community assembly, and can also improve capacity to predict future community composition and function (Lajoie and Kembel, 2019; Zanne et al., 2020a), such as under climate change.

Efforts to build trait databases are underway for fungi (Nguyen et al., 2016; Põlme et al., 2020; Zanne et al., 2020b), particularly soil fungi, although there are considerable challenges. In introductory biology textbooks fungi are often divided between parasites, decomposers, and mutualistic symbionts. These coarse trophic categories can then be further divided into numerous guilds, such as AMF versus EMF symbiotic fungi. Distinguishing systems that are dominated by one mycorrhizal type versus another provide valuable predictive power with respect to key soil processes such as C storage, decomposition rates, and productivity (Bennett et al., 2017; Clemmensen et al., 2013; Read, 1991). At the same time, variation in ecological strategies among fungal species belonging to a single guild can be immense. This is well illustrated by recent advances in our understandings of fungal wood decay brought by genomics and biochemical techniques (Eastwood et al., 2011; Riley et al., 2014).

4.5.1 Structural traits

Broad categories of traits of interest in ecology span morphology to biochemistry to genes. For soil fungi, mycelial growth traits (described above) vary widely among taxa and in response to internal and external stimuli. Because fungi must grow to both find and digest their food, mycelial traits are clearly essential to their fitness and ecosystem functions. Among the key traits are hyphal diameters, growth rates, and branching and anastomosis frequencies (Fricker et al., 2017). How individuals alter these traits in response to environmental stimuli such as resource distribution and abundance and the presence of competitors will likely contribute insights into fundamental ecological strategies of fungi. Unfortunately, fitness of filamentous fungi is extremely difficult to measure (Pringle and Taylor, 2002), particularly in natural settings like soil, where distinguishing the boundaries of a single, physiologically integrated individual is nigh to impossible. However, the imprint of natural selection as well as developmental and genetic constraints can be inferred through study of trait combinations that do and do not occur in extant species. We thus expect tradeoffs in trait expression due to resource limitation. For example, Lehman et al. (2019) studied a range of mycelial traits in phylogenetically diverse isolates from the same habitat and uncovered correlations between long internodes (distance between branches) and wide hyphae. Mycelial network architecture was most complex in members of the *Mucoromycotina*, but trait values varied widely, even among closely related taxa.

The morphology of fungal reproductive structures is obviously of great importance to dispersal and fitness. With respect to sexual structures in the *Dikarya*, there are clear tradeoffs between above- and belowground fruiting strategies. Mushrooms have little capacity to restrict water loss (Lilleskov et al., 2009), which explains why aboveground (epigeous) mushrooms are seen nearly exclusively in moist habitats or seasons. In semiarid to arid environments taxa that rely on epigeous fruiting are largely replaced by taxa that fruit belowground. This includes taxa with resupinate (stalkless, mat-like crusts) or gastroid (sealed, semispherical) sexual structures. These taxa rely less on wind for dispersal and many have evolved to attract animal vectors such as mites (Lilleskov and Bruns, 2005) and small mammals (Fogel and Trappe, 1978). Most famous among these are the highly sought edible truffles, which are gastroid fruitbodies produced by ectomycorrhizal Ascomycota in the genus *Tuber*. Not surprisingly, belowground fruiters have much shorter average dispersal distances than their epigeous counterparts (Kivlin et al., 2014; Kjøller and Bruns, 2003), illustrating a key tradeoff.

There is recent evidence showing that Ascomycota with forcibly ejected ascospores are precisely apportioned to minimize drag (Roper et al., 2008), which aids spores in escaping the zone of still air immediately adjacent to the sporocarp. That simultaneous ejection of ascospores across the surface further increases dispersal by creating a small wind current (Roper et al., 2010). Asexual spores are also key components to the dispersal and survival of some fungi. Conidia are small (one- to few-celled) haploid, asexual spores produced by many Dikarya. These spores are often dispersed by wind but also by water (aquatic hyphomycetes) or insects. Spores play critical roles in surviving disturbances or harsh conditions, not just in dispersal. Emphasizing their role in survival, durable resting (metabolically inactive) structures of soil fungi have sometimes been called resistant propagules. In addition to sexual spores and conidia, resistant propagules include thick-walled, inflated hyphal segments called chlamydospores as well as aggregations of cells in highly protected, tough structures called sclerotia. Sclerotia are seen in many plant-pathogenic fungi and serve various roles, such as overwintering in soil while awaiting the next generation of host plants. The diverse structures that soil fungi use to disperse and withstand harsh conditions are clearly key traits with respect to understanding the movement and population dynamics of fungi. Improved knowledge of these traits should aid in applying fundamental ecological theory to soil fungi. For example, do competition-colonization tradeoffs help explain distinct fungal strategies and community assembly (Smith et al., 2018)? Are there tradeoffs in spore size versus number, as well known for seeds in vascular plants (Zanne et al., 2020a)?

4.5.2 Elemental stoichiometry

Considerable work in animal macroecology has focused on scaling laws in relation to body size and metabolism (Brown, 1995). From a theoretical point of view, such principles should apply equally to fungi and animals. Unfortunately, taking the necessary measurements is impractical for most fungi, in part due to difficulties in measuring body size. However, a related body of theory holds more promise in fungi, namely ecological stoichiometry, which relates tissue elemental ratios to nutrient acquisition (Sterner and Elser, 2017). In general, fungi have higher tissue N contents and lower C:N ratios than the plant tissues from which most obtain their nutrition due to the higher N content of chitin as compared to cellulose. However, studies of C:N:P contents and ratios in terrestrial fungi have revealed wide variation (Lodge, 1987; Zhang and Elser, 2017), making it difficult to infer ecological strategies from stoichiometric traits of soil fungi. A recent analysis of saprotrophic versus EMF mushrooms growing in the same habitats across steep soil N and P gradients did reveal some strong guild-level patterns despite considerable variation among species within a guild (Kranabetter et al., 2019). More specifically, EMF had consistently lower N and P contents than saprotrophs, where nutrient contents were strongly correlated with their availability in soil. One inference was that mycorrhizal fungi have lower N and P because they provide any excess to their hosts, while saprotrophs may benefit from accumulating these resources. Variation in nutrient contents within guilds may point the way toward elucidating further axes of niche partitioning. The same can be said of stable isotope ratios, which differ predictably between EMF and saprotrophic taxa (Hobbie et al., 2001). *Laccaria bicolor* is an ectomycorrhizal fungus that often displays an unusual ^{15}N natural abundance, likely because it supplements its N supply by killing and consuming soil collembola (Klironomos and Hart, 2001).

4.5.3 Genes and enzymes

One of the most important categories of trait variation in soil fungi from the perspective of ecosystem function and nutrient cycling is the production of extracellular enzymes. All osmotrophic fungi produce some extracellular enzymes to break down extracellular molecules into subunits that can be absorbed. However, the range of enzymes produced, and likely the controls over production, vary widely across fungal guilds and taxa. Some fungi, such as AMF, have limited arrays of hydrolytic and oxidative enzymes. At the other extreme, white-rot fungi in the Basidiomycota express complex batteries of enzymes, including polyphenol peroxidases, that allow them to attack recalcitrant, lignocellulosic polymers from woody plants. Analyses of enzyme potentials from soil have been widely used to interrogate the functional potentials of soils under various conditions (Sinsabaugh et al., 2002; Sinsabaugh, 2010; Snajdr et al., 2011; Talbot et al., 2013, 2015; Talbot and Treseder, 2012). In such analyses the activities of the entire microbial community are summed, providing little insight into trait variation among taxa. However, studies of pure cultures and single-fungus, ectomycorrhizal root tips have provided valuable information about the enzymatic activities of specific saprotrophic and mycorrhizal fungi (Courty et al., 2005; Kirk and Cullen, 1998; Rineau and Courty, 2011; Ruess et al., 2019).

An exciting extension of physiological and biochemical assays has been the acceleration of genome and transcriptome sequencing across diverse fungal lineages (Grigoriev et al., 2014). Together with biochemical studies, accumulating genomes suggest that hard and fast distinctions between mycorrhizal fungi and decomposers, as well as decomposer categories within wood decay fungi (white vs. brown vs. soft rotters), do not hold up. Some mycorrhizal fungi retain oxidative genes inherited from the common ancestor of the Agaricomycotina (Floudas et al., 2012), presumably as a mechanism for obtaining N to trade to their host trees (Bödeker et al., 2014), while white and brown rot fungi display a wide spectrum of hydrolytic and oxidative gene contents (Eastwood et al., 2011; Riley et al., 2014).

4.6 Ecosystem functions

Soil fungi regulate the cycling of C, N, and other elements in ecosystems through their activities as decomposers, pathogens, endophytes, and mutualists of other organisms. Filamentous fungi support plant production through mycorrhizal associations that enhance the acquisition of water and nutrients, while fungal endophytes of plant roots can confer plant resistance to thermal and drought stress and reduce herbivory (Porras-Alfaro and Bayman, 2011). Fungi also support C and N fixation by algae and cyanobacteria through lichen associations (Liu et al., 2021) and in arid and polar regions, the formation of biotic crusts. Biotic crusts mediate soil-atmosphere exchange of greenhouse gases (such as CO_2), as well as water infiltration and stabilization of surface soils against erosion (Pointing and Belnap, 2012).

Fungi mediate nearly every aspect of organic matter production, decomposition, and sequestration in soil. As major drivers of surface organic matter (i.e., plant litter) decomposition, fungi mediate the creation of SOM, as well as the balance between its sequestration in aggregates or mineral-associated particles and its mineralization into CO_2. From a microbial perspective, SOM can be defined as the point at which the cost of further decomposition is not energetically favorable without additional inputs of more labile material (Moorhead et al., 2013). SOM can develop from the decay of surface leaf and stem litter or from the growth and turnover of roots, their associated fungi, and other microorganisms (Clemmensen et al., 2013; Wilson et al., 2009). Fungi can also protect SOM from decomposition through both chemical and physical means. Filamentous fungi promote formation of large macroaggregates in soil by binding soil particles with hypha and produce cell wall materials that act as adhesives (Willis et al., 2013). Aggregate formation promotes soil C sequestration by providing physical protection from decomposers and their degradative enzymes (Wilson et al., 2009). Paradoxically, fungi are also major decomposers of SOM on land, generating a large fraction of extracellular enzyme activity in soil (Fernandes et al., 2022; López-Mondéjar, 2020). In the process they release large amounts of major growth-supporting elements from dead organic matter (Joergensen and Wichern, 2008).

4.6.1 Carbon and nutrient cycling

Soil fungi control the cycling of C and other nutrients (primarily N and P) through both exchange with plants (Averill et al., 2019) and their biochemical interactions with SOM. Root symbionts (e.g., mycorrhizal fungi), endophytes, and epiphytes can promote plant growth by providing nutrients like N and P in exchange for C, and by ameliorating abiotic stressors, such as high salinity, temperature extremes, drought, and metal toxicity (Hashem et al., 2018). Root-associated fungi can act as biopesticides that deter microbial and insect pathogens of roots (Müller and Ruppel, 2014; Rodriguez et al., 2009). Mycorrhizal fungi often serve as conduits of plant C to soil, providing photosynthate to other free-living members of the soil microbiome (e.g., saprotrophs) (Gorka et al., 2019). By contrast, saprotrophic fungi feed on plant litter and microbial necromass as their main source of C. Saprotrophic, mycorrhizal, and all other aerobic soil fungi release a portion of their acquired C into the atmosphere as CO_2 and recycle the rest into fungal molecules in biomass and exudates. This process feeds nearby fungi and other organisms in soil. In some northern forest ecosystems the majority of soil organic C may be sequestered within fungal biomass (Clemmensen et al., 2013) due to chemical and physical protection of fungal molecules from decomposition (Fernandez and Koide, 2014).

One of the greatest impacts of soil fungi on the Earth system is through regulating the balance between C sequestration in the biosphere and C release into the atmosphere as CO_2. Root-associated fungi can consume up to 20% of plant C, with additional C consumed by fungal pathogens and parasites. Through the process of decomposition, saprotrophic fungi are responsible for releasing up to 36 Pg C from soils into the atmosphere as CO_2 annually, while root-associated fungi are estimated to release an additional 1.3 to 5.4 Pg C as CO_2 (Dighton and White, 2017). Through these processes, fungi are second only to plants in directly cycling and sequestering C in most terrestrial ecosystems. However, fungi also indirectly influence plant C cycling through their activities in nutrient cycling and disease.

4.6.1.1 Nutrient exchange with plants

Soil fungi facilitate uptake of N, P, other elements, and water by plant roots either by direct association and exchange (e.g., mycorrhizal fungi) or by decomposing and mineralizing nutrients from SOM for root uptake (Schimel and Bennett, 2004). Both strategies are widespread; over 90% of plant families, comprising approximately 250,000 plant species, engage in mycorrhizal symbioses. An estimated 50,000 species of fungi are mycorrhizal (van der Heijden et al., 2015), with the vast majority of the remaining 3.5 million fungal species having saprotrophic capabilities. Both mycorrhizal and saprotrophic functional groups have positive effects on plant nutrition, photosynthesis rates, and biomass accumulation. Mycorrhizal fungi can trigger stress resistance metabolism in plants such that they are capable of living in extreme — and otherwise deadly — environmental conditions (Bunn et al., 2009). Mycorrhizal fungi are even known to connect plants within a community (GFBI consortium et al., 2019) allowing the movement of nutrients between coexisting plants (Gorzelak et al., 2015). However, the rate of C and nutrient exchange between plants and soil fungi depends on environmental conditions, the identity of the fungus (Kiers et al., 2011), with the molecular controls over the metabolic exchange between fungi and plants just beginning to be revealed (Liao et al., 2014).

Mycorrhizal fungi can provide up to 80% of plant N and 90% of plant P (van der Heijden et al., 2015). Nevertheless, careful experiments with isotopic tracers in the AM symbiosis have established that the identity of the fungus, including both its taxonomy and nutrient delivery traits, can determine the amount of nutrients and C exchanged between soil and plants. Multiple studies have shown that plants can preferentially allocate more C to roots associated with a fungus delivering more P to the plant (Ji and Bever, 2016) and that conversely, C allocation can trigger N and P transfer from fungus to host plant (Fellbaum et al., 2012). Because nutrient uptake and transfer from AMF to plant hosts can increase foliar P and deplete soil P pools (van der Heijden et al., 1998), higher rates of C and nutrient exchange among strong plant-fungal mutualists could drive faster rates of soil nutrient cycling and plant productivity at the ecosystem level. Preferential plant C allocation to nutrient-generous AMF is predicted as part of biological market theory, in which plants and fungi can discriminate resource transfer between partners depending on the rate of transfer. Similar activities among EMF and their plant hosts have recently been revealed (Bogar et al., 2019). Nevertheless, resource exchange in mycorrhizal symbiosis does not always follow predictions from this theory, especially in the case of mycoheterotrophic plants or when plant health is compromised and mycorrhizal fungi turn parasitic (Walder and van der Heijden, 2015).

Despite the commonality of mycorrhizal symbioses in nature, C and nutrient exchange between the plant and fungal partners appears to be extremely chemical specific. In EMF-pine associations C is transferred to the fungus primarily as sugar alcohols (e.g., mannitol). Compatibility between plant and fungus — and therefore, the ability to develop mycorrhizal cellular structures and exchange molecules in the root — seems to be controlled by secreted proteases (Tang et al., 2021), which are small secreted proteins that have molecular properties similar to proteins associated with fungal virulence (Liao et al., 2014, 2016). It is still unclear how these molecular processes vary across fungal and plant taxa.

4.6.1.2 Enzymes and the decomposition of biopolymers

To fuel their growth, fungi must generate small molecular weight molecules (i.e., substrates) by enzymatically degrading complex organic matter (i.e., biopolymers) outside their cells. Enzymes released into the environment to acquire nutrients, or as a result of cell lysis, can be beyond the control of that organism (Fig. 4.1) such that a substantial fraction of biogeochemical cycling in soils is a legacy of fungal enzymes that are spatially and temporally displaced from their origins (Burns et al., 2013). However, some fungi secrete extracellular enzymes within mucopolysaccharides that are reabsorbed after the decomposition of organic matter. This strategy could allow fungi to reuse or reduce the loss of C and N used to construct extracellular enzymes and might also allow fungi to control the decomposition process and soil biogeochemical cycling more carefully than previously thought.

Fungi are considered principal degraders of plant cell wall material, especially during the early stages of litter decomposition in soil when their filamentous growth form and their capacity to secrete a variety of enzymes (primarily glycosidases and oxidases) allows them to bore through the cellular structure of plant litter. The capacity to partially or wholly degrade cellulose, especially after it has been decrystallized, is widespread in fungi, including basal lineages. Ascomycota and Basidiomycota have the widest genetic and ecological capacity for cell wall decomposition, which is facilitated by the synergistic expression of a variety of other polysaccharide-degrading enzymes. The ecological capacity of the Glomeromycota to degrade cellulose and other cell wall polysaccharides appears more limited, but several studies indicate greater capacity than once thought (Kowalchuk, 2012; Talbot et al., 2008).

Production of laccases and other phenol oxidases is also widely distributed among Ascomycota, Basidiomycota, and Glomeromycota (Baldrian, 2006). In some organisms, mainly saprotrophs, these enzymes primarily degrade lignin and other secondary metabolites in plant cell walls, often indirectly by producing small reactive oxidants known as redox mediators (Rabinovich et al., 2004). Other saprotrophs oxidatively degrade SOM to obtain chemically protected C, N, and P (Burns et al., 2013). In many taxa a large but indeterminate portion of oxidative activity is related to morphogenesis (e.g., melanin production), detoxification, and oxidative stress (Baldrian, 2006; Sinsabaugh, 2010) rather than nutrient acquisition. However, once released into the environment through biomass turnover, these activities also contribute to the oxidative potential of soils, which catalyzes nonspecific degradation reactions that contribute to the loss of SOM (Burns et al., 2013) as well as condensation reactions that can reform SOM (Sinsabaugh, 2010).

Peroxidases, with greater oxidative potential than laccases, are produced by some members of the Ascomycota and Basidiomycota. The most widely distributed enzyme, Mn peroxidase, acts indirectly on organic compounds by generating diffusible Mn^{+3}. The contribution of Mn peroxidase to soil C dynamics is highlighted by manipulation studies showing that Mn availability can limit the decomposition of plant litter (Trum et al., 2011). Some Basidiomycota, principally wood-rotting fungi, produce lignin peroxidase, which directly oxidizes aromatic rings (Rabinovich et al., 2004).

While most hydrolytic and oxidative enzymes are broadly distributed across taxonomic groups, functional groups of fungi (i.e., fungi that have different primary C sources) have unique enzyme profiles (Zanne et al., 2020a). For example, the distribution of extracellular enzymatic capacity is the basis for the traditional soft rot/brown rot/white rot ecological classification of wood rot decomposer fungi. In enzymatic terms these classifications refer to organisms that primarily attack cell wall polysaccharides (soft rot), those that deploy lower redox potential laccases in addition to glycosidases (brown rot), and those that deploy high redox potential laccase-mediated or peroxidase systems capable of effectively depolymerizing lignin (white rot). Interestingly, recent phylogenomic studies suggest that the common ancestor of the Agaricomycotina (basidiomycete class containing all EMF, brown rot, and white rot fungi) was capable of white rot such that fungi in all these functional groups retain some capacity for recalcitrant SOM breakdown (Bödeker et al., 2009, 2016; Lindahl and Tunlid, 2015). The evolution of brown rot appears to have occurred independently several times, in part through the loss of Mn-peroxidase genes.

EMF were once thought to have little saprotrophic capability compared to wood and litter decay fungi, but there is growing evidence that at least some EMF can attack a wide range of organic compounds (Pritsch and Garbaye, 2011; Read and Perez-Moreno, 2003; Talbot et al., 2013) such that these EMF are now considered a class of decomposers (Lindahl and Tunlid, 2015). EMF extracellular enzymes seem largely targeted toward N-acquisition based both on actual activity (Talbot and Treseder, 2010) and on genomic sequence data (Martin et al., 2008; Zak et al., 2019). However, some EMF have retained the capacity to generate oxidative enzymes (such as peroxidases) and Fenton chemistry, which can depolymerize lignin, cellulose, and hemicellulose (Bödeker et al., 2016; Lindahl and Tunlid, 2015; Nicolás et al., 2019; Op De Beeck et al., 2015; Rineau et al., 2013; Shah et al., 2016), suggesting that the ability to decompose litter C to some extent may be common among EMF. For example, the relative abundance of *Cortinarus acutus*, an EMF species capable of producing Mn peroxidases, was linked to a 33% decrease in C stocks in the organic soil horizon of a Swedish boreal forest (Lindahl et al., 2021). Nevertheless, the selective pressures that drive the adaptive evolution of enzyme activities likely differ for EMF versus saprotrophic fungi such that enzyme activities generated by EMF may be under different environmental controls (e.g., plant physiology) than those of free-living saprotrophs.

4.6.1.3 Soil nitrogen cycling

Soil fungi are involved in many key steps of the soil N cycle. While no fungi have been discovered that conduct N fixation, fungi are adept at decomposing and taking up N-rich molecules from soil, potentially because decomposition and symbiosis with plants require large nutrient (e.g., N, P) investment by the fungus. The potential to use chitin as an N source is widespread among fungi (Geisseler et al., 2010), potentially because fungi also decompose chitin internally during the process of hyphal extension (Gooday et al., 1986). However, proteins and their degradation products are the largest source of organic N in soils and it is likely that most saprotrophic and biotrophic fungi obtain the majority of their N from degradation of peptides in soil (Hofmockel et al., 2010; Sinsabaugh and Shah, 2011). In contrast to their role in P acquisition, the role of AMF in N acquisition is poorly resolved (Veresoglou et al., 2012). The enzymatic capacity of EMF to mine N from SOM has received more attention (Pritsch and Garbaye, 2011; Talbot et al., 2008). The role of fungi in the N cycle was once considered primarily assimilatory, where fungi assimilated inorganic N and N-containing organic molecules to support the production of new fungal biomass and supply plant hosts. However, recent studies have shown that denitrification pathways are widespread among *Ascomycota* and are responsible for a large fraction of nitrous oxide efflux, especially in arid soils (Shoun et al., 2012; Spott et al., 2011).

While fungi often mine soil for N, anthropogenic N deposition and addition has been shown to decrease fungal diversity, biomass, and respiration in soil (Knorr et al., 2005; Saiya-Cork et al., 2002; Treseder, 2008) for multiple potential reasons. Increased N deposition is associated with decreases in laccase and peroxidase activities in litter and soil, leading to slower decomposition and increased soil C sequestration (Gallo et al., 2004; Sinsabaugh et al., 2002). Nitrogen deposition also reduces the need for plants to engage in mycorrhizal symbiosis such that mycorrhizal fungi often decline in abundance with N deposition (Treseder, 2004), which may contribute to observed decreases in overall fungal biomass and activity in soil. Another contributing factor may be the difference in biomass C:N ratio of fungi and bacteria, where high N levels might promote bacterial growth relative to fungi (Strickland and Rousk, 2010; Van Der Heijden et al., 2008). Fungi can have extremely wide-ranging C:N:P ratios that can vary with both environmental conditions and across lineages (Zhang and Elser, 2017). Nevertheless, many fungal species have high C:N ratios (in the range of 13–60) (Strickland and Rousk, 2010) that exceed C:N ratios in biomass of bacteria (Danger et al., 2016). One review noted a positive relationship between fungal dominance (as indicated by qPCR) and soil C:N ratios across biomes (Fierer et al., 2009). Nitrogen deposition is typically composed of nitrate, ammonium, and organic N (Holland et al., 1999). There is evidence that these molecules may also suppress fungal activity in soil through various mechanisms (Fog, 1988).

Despite the suppressive effect of N deposition on fungal growth and activity, soil fungi also have the capability to process nitrate into more reduced N forms that can be exported from soil systems as greenhouse gas. Nitrate reduction has historically been considered specific to prokaryotes in soil, yet nitrate reduction was reported in eukaryotes as early as the 1980s (Finlay et al., 1983) and nitrous oxide production was reported in soil fungi during the 1990s (Shoun and Tanimoto, 1991). The ability of fungi to reduce nitrite often occurs under anoxic (or less oxic) conditions and may be performed through the mitochondria. However, there is some suggestion that fungi have acquired the ability to reduce nitric oxide (NO) to nitrous oxide (N_2O) through horizontal gene transfer (Kamp et al., 2015). Forty-three percent of the over 380 soil fungal isolates tested have been observed to produce measurable amounts of N_2O in culture (Jirout, 2015; Mothapo et al., 2013; Takaya, 2002). These strains represent a diversity of lineages in the Ascomycota, Basidiomycota, and Zygomycota. Species that release N_2O also encompass a diversity of functional types, including EMF, plant pathogens, and saprotrophic fungi (Kamp et al., 2015). Studies using selective inhibition of fungi in soil have shown that fungi can contribute 10% to 89% of N_2O emissions from soil, with the highest contributions under moderately reducing to weakly oxidizing conditions, such as under intense cattle farming (Jirout, 2015).

4.6.1.4 Soil phosphorus cycling

Extracellular phosphatase production is common in soil fungi, indicating that many taxa are capable of releasing phosphate from organic P sources. For fungi, it appears that most phosphatases released for extracellular P acquisition have acidic pH optima, while those intended for intracellular reactions have optima at neutral to alkaline pH (Plassard et al., 2011). Inositol phosphates, produced mostly as P storage products by plants, account for half of soil organic P (Menezes-Blackburn et al., 2013; Plassard et al., 2011). The abundance of these compounds has more to do with their resistance to degradation than with their rate of production. Phytate (inositol hexaphosphate) is crystalline such that specific enzymes (phytases) are needed to hydrolyze the phosphate. Ascomycota are considered the best producers of phytase, but some Glomeromycota and Basidiomycota also produce them (Menezes-Blackburn et al., 2013; Plassard et al., 2011). Like all extracellular phosphatases, enzyme expression is induced by P deficiency.

Mineral phosphate is also important to fungi and plants. In alkaline soil calcium phosphates may be abundant, while weathered acid soils may have high concentrations of iron and aluminum phosphates. Some fungi, particularly ectomycorrhizal Basidiomycota, solubilize phosphate from mineral sources using low-molecular-weight organic acids, such as oxalate (Courty et al., 2010; Plassard et al., 2011).

4.6.1.5 Soil carbon cycling

Soil fungi release C from the biosphere through plant litter C and SOM decomposition and respiration of CO_2, but they also sequester CO_2 by producing recalcitrant C molecules that can be stored in soil (King, 2011) or by recycling nutrients to promote growth of other organisms, including plants, invertebrates, and animals. Fungi are responsible for 27% to 95% of CO_2 respiration from soils across ecosystems, averaging 60% (+/− 9−12%) of total respiration from aerobic soils based on selective inhibition (Joergensen and Wichern, 2008). Somewhat paradoxically, however, the more active fungi are in SOM decomposition, the more capacity they have for sequestering C in soil. Across soil types, fungal biomass contributes 68%

to 76% microbial C on average in soils, with the highest contribution in leaf litter layers (Joergensen and Wichern, 2008). Fungal necromass can decompose more slowly than bacterial necromass (Joergensen and Wichern, 2008; Strickland and Rousk, 2010), in part because fungal molecules like melanin (and other hydrophobic compounds) may decompose slowly (Fernandez and Kennedy, 2016). Hydrophobins play important roles in the hydrophobicity of spores and other cell surfaces in fungi (King, 2011), so they may also contribute to slow turnover of fungal biomass in soil.

Different species and functional groups of soil fungi will contribute to stable soil C stocks in different biomes. In boreal forests much of the C stabilized to decomposition (i.e., over 100 years old) can be attributed to mycelium of EMF (Clemmensen et al., 2013), which are the major fungal symbionts of the dominant vegetation of the region (Read et al., 2004). In systems dominated by AMF, such as temperate grasslands, savannah, and some tropical forests (Treseder and Cross, 2006), stable soil C in aggregates may develop from AMF-derived C compounds like glomalin, a heat-shock protein (Rillig and Steinberg, 2002; Rillig et al., 2010; Wright and Upadhyaya, 1996). Some functional traits related to C sequestration potential may vary more across taxa or lineage than across functional groups of fungi. For example, members of the Glomeraceae can produce less extraradical mycelium than taxa within the Gigasporaceae (Maherali and Klironomos, 2007). Glomalin production varies among AMF taxa (Treseder and Turner, 2007).

Fungi can also contribute to the release of other greenhouse gases, such as methane. Microbial production of methane in soils is mainly conducted by prokaryotes (methanogenic archaea), yet the presence of fungi can increase methane yields from methanogenic archaea (Beckmann et al., 2011). Fungi have also been reported to release methane during aerobic respiration (Lenhart et al., 2012) through the activity of mitochondria. These activities indicate that fungal metabolism in soil could be an important component of nutrient loss from ecosystems via greenhouse gas emissions beyond CO_2.

4.6.2 Bioremediation

The filamentous growth habit and enzymatic versatility of fungi can also be adapted to treat waste streams and remediate soils contaminated with organic pollutants or toxic metals (Harms et al., 2011; Strong and Claus, 2011). The most effective pollutant remediators belong to the phyla Ascomycota and Basidiomycota and include many EMF species. Most of this capacity is related to the production of a broad spectrum of extracellular laccases and peroxidases with varying redox potentials, pH optima, and substrate specificity that oxidatively modify or degrade aliphatic and aromatic pollutants, including halogenated compounds. In addition, some fungi produce nitroreductases and reductive dehalogenases that further contribute to the degradation of explosive residues and halogenated contaminants. Intracellularly, many fungi have cytochrome P450 oxidoreductases that can mitigate the toxicity of a broad range of compounds. The toxicity of metal contaminants can be mitigated by translocation and sequestration in chemically inaccessible complexes. Improving the bioremediation capabilities of EMF is of particular interest because the C supply from host plants may support fungal growth into contaminated hotspots and stimulate cometabolic reactions (Policelli et al., 2020).

4.7 Soil fungi and global change

Soil fungi are affected by global change — including changes in climate and other environmental conditions over the longer-term — but also feedback to some of these changes through their biogeochemical cycling activities as pathogens, mutualists of other organisms, and free-living decomposers (saprotrophs). Fungi can be affected directly by a shift in their abiotic environment, or indirectly through climate change impacts on their biotic resource pools or on other members of the soil microbial community with which they interact. Any climate changes that alter resource availability will not only alter total fungal activity in the environment but also affect species composition due to varying resource preferences and competitive abilities among fungal species (Hawkes and Keitt, 2015). Fungal communities also change because of the dependence of fungal species on the presence and activity of others, e.g., through processes such as cross-feeding, cheating, and commensalism. One of the strongest indirect impacts of climate change on fungal communities could be through the effect on bacterial communities, which can act as population controllers of soil fungi (Romaní et al., 2006). These direct and indirect impacts of global change on soil fungal communities can often be related to the tolerance of specific species to new environmental conditions (e.g., response traits). The feedback of these community changes to ecosystem biogeochemistry is a result of the emergent biogeochemical functions of the new fungal community, which can be shaped by its most active members.

Experiments have applied global change manipulations to soils in the field and in the laboratory that are regularly applied at a level comparable to the effects that have been observed over the past decades or that are expected to occur over the next century. Most of these manipulative experiments apply changes almost immediately, simulating an event even more extreme than the most extreme weather documented in the last century. These manipulations often reveal where the boundaries of species niches and survivability lie in the face of future worst-case global change scenarios, rather than the response to gradual, chronic, longer-term global changes.

4.7.1 Climate change effects on soil fungi

One of the most dramatic and widespread human-induced changes in Earth's condition has been rising air temperatures due to increased greenhouse gas emissions. Climate warming affects many other aspects of climate, including precipitation regimes, surface humidity, and extreme temperature events, as well as the amount of snow and ice on land. These cascading effects of warming often have larger impacts on soil fungi than climate warming alone, yet fungi do not have a uniformly negative response to each of these individual and interacting factors.

4.7.1.1 Elevated CO_2

Elevated levels of atmospheric CO_2 have overall positive effects on growth and C acquisition of soil fungi, yet seem to be strongest for plant root associates. Recent meta-analyses of field-based elevated CO_2 experiments show that increasing atmospheric CO_2 (100−500 ppm above ambient) tends to increase both fungal biomass and richness of fungal communities (Zhou et al., 2020) in soil, although the trends are not significant. By contrast, elevated CO_2 consistently increases the biomass of mycorrhizal fungi in soil. Ectomycorrhizal fungi increase biomass by 19% and AMF increase biomass by 84% under elevated CO_2 (Garcia et al., 2008; Treseder, 2004). These results suggest that most CO_2 effects on fungi may be indirect, through changes in plant growth, and by inducing plant C allocation belowground into the rhizosphere where root symbionts presumably have first access to the supplied C. A recent meta-analysis documented that under elevated CO_2, ectomycorrhizal forests increase C in plant biomass but reduce C in soils. Alternatively, AM grasslands show an increase in C inputs, particularly to the more stable, difficult-to-decompose, mineral-associated organic matter (MAOM) (Terrer et al., 2021). The authors hypothesize that these trends result from EMF more efficiently, leveraging plant C to mine soil for N, which accelerates soil C losses but increases photosynthetic capacity of the host plant. Variations in AMF responses to CO_2 enrichment have been observed to depend on both plant and fungal species, perhaps due to fundamentally different C exchange relationships between mutualistic mycorrhizal fungi and their hosts (Johnson et al., 2005).

4.7.1.2 Temperature changes

Fungal physiology is sensitive to temperature in a way that can impact growth and nutrient-cycling activity of individual fungi. Fungal metabolism increases with temperature as enzyme-catalyzed reaction rates increase (Allison et al., 2018; Gasch et al., 2000; Mohsenzadeh et al., 1998), up to an optimum temperature which, for most fungi, is just under 30°C (Kerry, 1990). Beyond this optimum, metabolic activity decreases due to biochemical phenomena such as protein denaturation (Boddy et al., 2014). Both yeasts and filamentous fungi can acclimate to warming by reducing C use efficiency (CUE), which increases the amount of CO_2 respired per unit C uptake (Allison et al., 2018; Bradford and Crowther, 2013). Recent studies have also revealed that fungi can evolve in response to repeat warming events, increasing respiration rates and reproduction (Romero-Olivares et al., 2015), and labile C consumption through beta-glucosidase activity (Finestone et al., 2022), while reducing the ability to decompose recalcitrant substrates in soil, such as aromatic organic matter (Anthony et al., 2021). These observations suggest that fungi could feed back to climate change by increasing soil respiration, at least initially; however, this impact could be offset by increased soil C storage if fungi fail to recover the ability to degrade older, more recalcitrant SOM.

At the community level, warming can decrease total fungal abundance in soil (DeAngelis et al., 2015; Morrison et al., 2019), yet the effects are often small, potentially because warming effects can vary based on initial soil temperature and moisture regimes. Warming tends to reduce the abundance of fungi and lichen in soils that are dry due to habitat type (Allison and Treseder, 2008; Ferrenberg et al., 2015) or long-term warming (DeAngelis et al., 2015). This is true especially when temperatures are reaching the limit for growth (Bárcenas-Moreno et al., 2009) but can also occur in dry habitats with small increases in temperature (e.g., +0.5°C). High summer temperatures are associated with low yields of *T. melanosporum* fruit bodies (i.e., truffles) and reduced soil moisture availability (Boddy et al., 2014). Based on these observations, one study projected decreased truffle yields across most of the Mediterranean Basin based on the expected increasing temperatures and decreasing precipitation in the coming decades (Büntgen et al., 2012). By contrast, warming can increase fungal abundance in soils that have sufficient moisture, such as tundra soils. A meta-analysis by Chen et al. (2015) found that warming significantly increased fungal abundances in tundra soils and histosols, rising 9.5% and 31% above control soils, respectively. Both soils experience low mean annual temperatures (−2.4°C for tundra and 3.1°C for histosols) and store large amounts of C for fungi that can be metabolized quickly to generate biomass. This indicates that fungi in historically cold soils show higher sensitivity to warming, which is possibly contingent on resource availability.

Warming also appears to select for certain lineages or groups of fungi, reshaping community composition and, potentially, its functional capabilities. Warming generally increases fungal diversity across ecosystems, with the greatest effects in forests (Zhou et al., 2020). In their global meta-analysis Gang et al. (Gang et al., 2019) found that biomass of saprotrophic fungi increased with experimental warming, while that of AMF (based on PLFA) showed no significant change. A recent publication in a northern hardwood forest found an increase in brown rot fungi and plant pathogenic fungi in soil, a response that correlated with an increase in peroxidase activity in soil and a decrease in plant photosynthetic rate, respectively (Garcia et al., 2020). This shift in fungal functional groups was driven by changes in the relative abundance of dominant fungal genera; changes in AMF were associated with changes in the dominant genus *Glomus*. Changes in the relative abundance of brown rot fungi in soil with warming were driven by increases in the dominant brown rot genera *Cerinosterus* and *Amylocystis*. Another recent study in a different northern hardwood forest found that community composition of saprotrophic and AMF shifted with ~10 years of continuous soil warming (Anthony et al., 2021). Heated plots were hyperdominated by certain fungi (e.g., *Russula* species and *Mortierella gemmifera*), which showed a 10-fold increase in heated soil, while several other warming-sensitive taxa declined in heated plots.

Fungi that survive and thrive in warmed soils seem to reallocate resources away from biomass accumulation and towards maintenance of metabolic rate (i.e., rate-yield trade-off), potentially contributing to the often observed increase in soil C losses with warming. Recent -omics studies show that soil fungi under 10 years of continuous soil warming have greater rRNA gene copy numbers without a parallel increase in biomass production (Anthony et al., 2021) and express more genes for cell metabolic maintenance than for carbohydrate decomposition control soils (Romero-Olivares et al., 2019). Similarly, it was recently reported that under longer-term warming, eukaryotic genes coding for CAZymes are lower under long-term warming than control soils (Pold et al., 2016). The accumulation of examples where fungi trade off growth for metabolic rate under warming suggests that soil fungi invest in metabolic maintenance to ensure survival under the stress of increased temperatures (Bennett and Lenski, 2007). Under warming-only experiments, soil fungi appear to evolve to trade off growth for activity of oxidative enzymes that decompose more recalcitrant C substrates, such as those in soil humus (Pold et al., 2015, 2016). Interestingly, a recent study found that in heated soils, there was a lower negative cooccurrence and higher positive cooccurrence among fungal community members compared to unheated plots (Anthony et al., 2021), suggesting that warming leads to selection for soil fungi that either collaborate more than fungi under cooler temperatures or share similar environmental preferences.

Paradoxically, climate warming can generate colder soils in some regions, forcing soil fungi to contend with the stress of a wider range in annual soil temperatures. Rising air temperatures increase soil temperatures during the growing season but also decrease the fraction of precipitation occurring as snow in winter in northern latitudes, which decreases snowpack formation, increases snowmelt, and increases the frequency of soil freeze/thaw events (Campbell et al., 2010). This counterintuitive effect of climate change — warmer soils in the growing season but colder soils in winter — has unusual effects on soil fungi that we are just beginning to understand. Soil freeze/thaw cycles may exert particularly strong selection for taxa that can survive extreme fluctuations in temperature, moisture, and physical structure (Ostroumov and Siegert, 1996), as well as changes in plant C availability belowground due to root damage (Sanders-DeMott et al., 2018). Fungi originating from polar climates, for example, have developed mechanisms to survive extreme conditions, including higher concentrations of sugars, alcohols, lipids and fatty acids, or antifreeze proteins in their cells compared to their mesophilic relatives (Robinson, 2001). There is new evidence that soil fungi can rapidly evolve in response to freeze/thaw cycles, potentially shaping forest biogeochemical cycling over the longer term. Common-garden experiments with fungal cultures have shown that soil fungi evolved under the combined impact of soil warming during the growing season and increased frequency of freeze/thaw cycles in winter to have inherently higher cellulase activity but lower acid phosphatase activity than fungi that are evolving under warming alone (Finestone et al., 2022).

Repeated soil freeze/thaw cycles may exacerbate soil C losses under these conditions as severely stressed microbial cells are more easily decomposed in soils than unstressed cells (Crowther et al., 2015). By contrast, consistently lower acid phosphatase activities of soil communities, both initially (Sorensen et al., 2018) and over the longer term (Finestone et al., 2022), suggest a decoupling of C and P decomposition from SOM by fungi that would not occur with warming alone.

4.7.1.3 Precipitation changes

Changes in precipitation can affect fungal biomass and community composition through their effects on soil moisture. Fungi are generally considered more resistant to desiccation than bacteria (de Vries et al., 2012) but less resilient in the face of desiccation as they grow slower in soils compared to bacteria. Nevertheless, global meta-analyses have shown that soil fungal biomass can respond positively to precipitation (Blankinship et al., 2011) and reduced soil moisture (i.e., drought) can severely reduce the abundance of fungi in laboratory-based experiments (Waring and Hawkes, 2015). Both temperature and moisture impact fungal physiology. Under moisture stress, some fungi produce trehalose, which protects fungal cell membranes from desiccation, freezing, and heat shock (Treseder and Lennon, 2015). This trait is commonly found among fungi in the Arctic (Gunde-Cimerman et al., 2003), where water availability is low, as in xerotolerant lichenicolous fungi (Mittermeier et al., 2015). Interestingly, recent work shows that soil fungal diversity responds negatively to increased soil moisture, declining with increased precipitation (Zhou et al., 2020). In addition, fluctuating water potential of soil can cause turnover in fungal community composition (Kaisermann et al., 2015), indicating selection against many species by rainfall or drought. In soils that were experimentally dried and warmed in an Alaskan boreal forest, soil fungi orders known to house lichenized, corticioid, melanized, high-sporulating, endophytic, pathogenic, and xerotolerant fungi were significantly more abundant in the warmed treatment compared to controls (Romero-Olivares et al., 2019).

Similar to warming effects, moisture effects on soil fungi depend on initial environmental conditions. One study found that fungi in tropical soils that experience historical drought are more tolerant of future extreme drought conditions (Waring and Hawkes, 2015). By contrast, fungal communities in soils that are seasonally dry often have negative responses to precipitation, i.e., reductions in biodiversity and total biomass of fungi (Hawkes et al., 2011). These effects are thought to occur when the new environment falls outside the range of conditions previously experienced (Hawkes and Keitt, 2015), because taxa may have evolved a particular range of niche optima best suited to the initial environment. Modeling experiments show that a greater range of niche optima in diverse communities leads to a greater chance of these species showing resiliency to environmental change (Hawkes and Keitt, 2015). Fungal communities under fluctuating precipitation regimes may therefore have a diversity of taxa with specific physiologies (specialists) to tolerate a wider range of moisture conditions.

Certain functional guilds tolerate drought more than others. In contrast to saprotrophs, mycorrhizal fungi can receive water from the host tree through hydraulic lift (nocturnal water transfer from the tree to the associated mycorrhizal symbiont) (Querejeta et al., 2003) and transfer this water to their sporocarps (Lilleskov et al., 2009). In addition, root-associated fungal symbionts often respond to soil drying through increased root colonization (Rudgers et al., 2014; Talbot et al., 2008). This may be driven by increased nutrient limitation of plants due to low nutrient diffusion rates in dry soil or by the need for plants to reduce water losses through root systems (Finlay et al., 2008). AMF can down-regulate transcription of genes coding for aquaporins in plant roots, which may serve as a mechanism for both water-conservation and tolerance of salt-stress (Finlay et al., 2008). Fungal endophytes also confer drought tolerance to many plant species (Kivlin et al., 2013), with endophyte colonization of roots observed to increase under experimental warming that inadvertently dries soils (Rudgers et al., 2014). By contrast, EMF showed greater variability in their response to drought (Talbot et al., 2008).

In addition to a functional guild effect, there appears to be a phylogenetic signal to fungal drought responses. Melanin may be an adaptive response to prevent cells from desiccation (Horikoshi et al., 2010). Generally, more derived fungi are more resilient to changes in water availability as they have drought-resistant biochemistry (e.g., melanized spores) that basal lineages lack (Treseder et al., 2014). Older phyla of soil fungi dominate soils of low-latitude, high-moisture ecosystems (Treseder et al., 2014), potentially due to their zoospore stage of development that requires high-moisture conditions. Indeed, basal lineages like the *Chytridiomycota* drop out in drought experiments in the field (Waring and Hawkes, 2015). These traits may constrain the phylogenetic distribution of taxa around the globe, such that ranges of entire fungal clades could be determined by precipitation regimes.

4.7.2 Other global change effects on soil fungi

Global changes beyond climate change — including pollution, fire frequency and severity, vegetation shifts, and species introductions — often have more pronounced immediate impacts on soil fungi and their function than climate changes per se.

4.7.2.1 Fire

Soil fungal abundances are reduced under fire (Holden et al., 2013) regardless of fire severity or habitat type; however, certain lineages recover quickly suggesting an adaptive response to the disturbance created by fire. *Pyronema* sp. and several EMF species are known to be fire-adapted, but their postfire recovery depends on the intensity and frequency of fire (Dove and Hart, 2017; Glassman et al., 2016). These EMF may be able to capitalize on the N (ammonium) that increases in soil after fire (Wan et al., 2001) and transfer it to their hosts (Claridge et al., 2009). Postfire fungal communities can establish from fire-adapted propagules present in the soil, dispersal from adjacent areas, or propagules present in less affected areas, such as deeper soil horizons (Policelli et al., 2020).

4.7.2.2 Plant community change

While soil fungal abundances are reduced under plant harvest disturbance, such as clearcutting forest (Jones et al., 2003) in tropical forests (Holden and Treseder, 2013), shifts in plant community composition can impact fungal abundance, community composition, and activity belowground. Nonnative plant invasions are increasing worldwide (Seebens et al., 2018) and can impact soil fungi through shifts in plant chemistry (often conserved at the family level), the plant microbiome, or other species- and community-specific plant traits. For example, invasions by EMF-associating trees are associated with coinvasion by their EMF symbionts such that nonnative EMF have been introduced to many countries in the Southern Hemisphere (Policelli et al., 2019; Pringle et al., 2009; Vellinga et al., 2009). In eastern North America invasion by garlic mustard shifts belowground fungal communities over the longer term (Lankau, 2011) due to the nonmycorrhizal status of garlic mustard and the production of secondary metabolites (glucosinolates) that can be toxic to fungi (Rodgers et al., 2008; Stinson et al., 2006). Garlic mustard appears to target mycorrhizal fungi, shifting the soil fungal community toward dominance by saprotrophic and plant pathogenic fungi (Anthony et al., 2017). In some cases plant invasion effects are exacerbated by other concomitant abiotic stressors. A recent study found that fungi dominant in warmed soils were sensitive to invasion, whereas fungi dominant in control soils were less responsive (Anthony et al., 2020). The authors also found a positive correlation between fungal community response to invasion and mean annual temperature, over a temperature gradient of 4°C, providing additional support for the idea that warming increases the invasibility of soil fungal communities.

4.7.2.3 Pollution

Organic and inorganic toxins can impact community composition of soil fungi. One major pollutant of interest has been N, with an increase in availability to soil fungi (and other soil organisms) since the mid-20th century, primarily through the application of N-based fertilizer to agricultural lands (Galloway et al., 2008). Of all the global change factors applied to soil, N appears to be the most consistent in its impact on fungi. Across ecosystems, N reduces fungal biomass and diversity in soil (Zhou et al., 2020). The strength of the effect can vary, with strongest impacts on historically low-N soil and weakest impacts on historically polluted, high-N ground (Cox et al., 2010). One hypothesis is that N effects can largely be explained by pH shifts, because increased ammonium deposition can promote the process of nitrification by bacteria, which decreases soil pH (Averill and Waring, 2018). While many soil fungi thrive at low pH (<5), it has been observed that N addition tends to acidify soil across experiments. This reduction in pH is correlated with a reduction in soil fungal diversity (Zhou et al., 2020). Similar to other global change factors, N addition to soil selects for specific taxa, particularly more nitrophilic taxa that show lower levels of hyphal production and N-acquiring enzyme activities (i.e., extracellular peptidases and proteases; van der Linde et al., 2018). These types of shifts in fungal communities, combined with overall declines in fungal abundance, under N fertilization may be associated with decreases in decomposition and soil C accumulation (Frey et al., 2014).

Other soil contaminants can have strong effects on fungal communities belowground. While organic contaminants can be degraded, metals need to be physically removed or immobilized (Policelli et al., 2020). Some EMF species can be negatively affected by high levels of heavy metals, reducing the number of sporocarps produced as heavy metal concentrations in soil increase (Rühling and Söderström, 1990). However, several common EMF species — including *Amanita*, *Paxillus*, *Pisolithus*, *Scleroderma*, *Suillus*, and *Rhizopogon* — accumulate high metal concentrations in the extramatrical mycelium (Khan et al., 2000). In these cases EMF can reduce metal toxicity to their host plants (Wilkins, 1991).

Tropospheric ozone (O_3), which absorbs UV radiation, has negative chemical effects on organisms, including fungi. Ozone tends to reduce conidia germination, increase rates of hyphae death, and promote production of reactive oxygen species (ROS) that damage fungal tissues (Savi and Scussel, 2014). Total fungal biomass (Bao et al., 2015) and fungal/bacterial ratios in soil can decline under elevated O_3 (Li et al., 2015). However, effects of O_3 on fungal communities in the field can be delayed (Cotton et al., 2015), potentially when O_3 affects fungi indirectly through damaging plants. For example, colonization of tomato by the AMF *Glomus fasciculatum* was reduced after only nine weeks of exposure to elevated O_3 (McCool and Menge, 1984) and microbial communities of meadow mesocosms were unaffected by elevated O_3 after 2 months, showing response after 2 years (Kanerva et al., 2008).

4.7.3 Effects of multiple interacting global changes

Global change is caused by various factors (Aber et al., 2001). Nevertheless, the impact of multiple global change stressors on microbial communities is seldom tested — only 20% of studies have examined more than one factor, and only 1% have examined more than two factors (Rillig et al., 2019). One interesting study on soil fungi in laboratory microcosms examined the effects of an increasing number of global change factors in combination, including temperature, drought, resource availability, chemical toxicants, and microplastics (Rillig et al., 2019). When exposed to multiple factors at once, soil fungal communities showed a decrease in diversity and an increase in similarity (regardless of the global change factor) and were primarily composed of generalist stress-tolerant fungi while losing Basidiomycota. These soils experienced water repellency at an increased level relative to single-factor global changes and far greater than the effect predicted from a purely additive response of each change. Soils with multiple global change manipulations also exhibited severely reduced CO_2 flux. However, the type of global change factor was a better predictor of soil fungal function and community composition than the number of manipulations alone. Long-term soil warming in combination with nonnative plant invasion significantly increased relative abundances of saprotrophic fungi and fungal genes encoding for hydrolytic enzymes (Anthony et al., 2020), suggesting that the stress of multiple interacting global changes on certain fungal functions, such as plant C decomposition, can be remediated by sufficient C substrate supply.

Acknowledgments

This material is based in part upon work supported by the National Science Foundation through awards DEB-1354674 and OPP-1603710 to DLT and DEB-1457695 to JMB as well as DOE awards DE-SC0020403 and DE-SC0022194 to JMB. Sincere thanks to Patrick Hickey for providing the photomicrographs shown in Fig. 4.5.

References

Aber, J., Neilson, R.P., McNulty, S., Lenihan, J.M., Bachelet, D., Drapek, R.J., 2001. Forest processes and global environmental change: predicting the effects of individual and multiple stressors. Bioscience 51, 735–751.

Agerer, R., 2001. Exploration types of ectomycorrhizae. Mycorrhiza 11, 107–114.

Aguilar-Trigueros, C.A., Hempel, S., Powell, J.R., Anderson, I.C., Antonovics, J., Bergmann, J., et al., 2015. Branching out: towards a trait-based understanding of fungal ecology. Fungal Biol. Rev. 29, 34–41.

Albornoz, F.E., Hayes, P.E., Orchard, S., Clode, P.L., Nazeri, N.K., Standish, R.J., et al., 2020. First cryo-scanning electron microscopy images and X-ray microanalyses of mucoromycotinian fine root endophytes in vascular plants. Front. Microbiol. 11, 2018.

Aldrovandi, M.S.P., Johnson, J.E., OMeara, B., Petersen, R.H., Hughes, K.W., 2015. The *Xeromphalina campanella/kauffmanii* complex: species delineation and biogeographical patterns of speciation. Mycologia 107, 1270–1284.

Allison, S.D., Treseder, K.K., 2008. Warming and drying suppress microbial activity and carbon cycling in boreal forest soils. Glob. Change Biol. 14, 2898–2909.

Allison, S.D., Romero-Olivares, A.L., Lu, L., Taylor, J.W., Treseder, K.K., 2018. Temperature acclimation and adaptation of enzyme physiology in *Neurospora discreta*. Fungal Ecol. 35, 78–86.

Anderson, J.B., Bruhn, J.N., Kasimer, D., Wang, H., Rodrigue, N., Smith, M.L., 2018. Clonal evolution and genome stability in a 2500-year-old fungal individual. Proc. R. Soc. B 285, 20182233.

Anthony, M.A., Frey, S.D., Stinson, K.A., 2017. Fungal community homogenization, shift in dominant trophic guild, and appearance of novel taxa with biotic invasion. Ecosphere 8, e01951.

Anthony, M.A., Stinson, K.A., Moore, J.A.M., Frey, S.D., 2020. Plant invasion impacts on fungal community structure and function depend on soil warming and nitrogen enrichment. Oecologia 194, 659–672.

Anthony, M.A., Knorr, M., Moore, J.A.M., Simpson, M., Frey, S.D., 2021. Fungal community and functional responses to soil warming are greater than for soil nitrogen enrichment. Elem. Sci. Anthr. 9, 000059.

Arenz, B., Blanchette, R., 2011. Distribution and abundance of soil fungi in Antarctica at sites on the Peninsula, Ross Sea Region and McMurdo Dry Valleys. Soil Biol. Biochem. 43, 308–315.

Argiroff, W.A., Zak, D.R., Pellitier, P.T., Upchurch, R.A., Belke, J.P., 2022. Decay by ectomycorrhizal fungi couples soil organic matter to nitrogen availability. Ecol. Lett. 25, 391–404.

Averill, C., Waring, B., 2018. Nitrogen limitation of decomposition and decay: how can it occur? Glob. Change Biol. 24, 1417–1427.

Averill, C., Cates, L.L., Dietze, M.C., Bhatnagar, J.M., 2019. Spatial vs. temporal controls over soil fungal community similarity at continental and global scales. ISME J. 13, 2082–2093.

Baar, J., Horton, T., Kretzer, A., Bruns, T., 1999. Mycorrhizal colonization of *Pinus muricata* from resistant propagules after a stand-replacing wildfire. New Phytol. 143, 409–418.

Bahram, M., Peay, K.G., Tedersoo, L., 2015. Local-scale biogeography and spatiotemporal variability in communities of mycorrhizal fungi. New Phytol. 205, 1454–1463.

Bahram, M., Hildebrand, F., Forslund, S.K., Anderson, J.L., Soudzilovskaia, N.A., Bodegom, P.M., et al., 2018. Structure and function of the global topsoil microbiome. Nature 560, 233–237.

Baldrian, P., 2006. Fungal laccases: occurrence and properties. FEMS Microbiol. Rev. 30, 215–242.

Bao, X., Yu, J., Liang, W., Lu, C., Zhu, J., Li, Q., 2015. The interactive effects of elevated ozone and wheat cultivars on soil microbial community composition and metabolic diversity. Appl. Soil Ecol. 87, 11–18.

Bar-On, Y.M., Phillips, R., Milo, R., 2018. The biomass distribution on Earth. Proc. Natl. Acad. Sci. 115, 6506–6511.

Bárcenas-Moreno, G., Gómez-Brandón, M., Rousk, J., Bååth, E., 2009. Adaptation of soil microbial communities to temperature: comparison of fungi and bacteria in a laboratory experiment. Glob. Change Biol. 15, 2950–2957.

Bass, D., Richards, T.A., 2011. Three reasons to re-evaluate fungal diversity "on Earth and in the ocean". Fungal Biol. Rev. 25, 159–164.

Bauer, R., Garnica, S., Oberwinkler, F., Riess, K., Weiß, M., Begerow, D., 2015. *Entorrhizomycota*: a new fungal phylum reveals new perspectives on the evolution of fungi. PLoS One 10, e0128183.

Beckmann, S., Krüger, M., Engelen, B., Gorbushina, A.A., Cypionka, H., 2011. Role of Bacteria, Archaea and Fungi involved in methane release in abandoned coal mines. Geomicrobiol. J. 28, 347–358.

Bennett, A.F., Lenski, R.E., 2007. An experimental test of evolutionary trade-offs during temperature adaptation. Proc. Natl. Acad. Sci. 104, 8649–8654.

Bennett, J.A., Maherali, H., Reinhart, K.O., Lekberg, Y., Hart, M.M., Klironomos, J., 2017. Plant-soil feedbacks and mycorrhizal type influence temperate forest population dynamics. Science 355, 181–184.

Berbee, M.L., Strullu-Derrien, C., Delaux, P.-M., Strother, P.K., Kenrick, P., Selosse, M.-A., et al., 2020. Genomic and fossil windows into the secret lives of the most ancient fungi. Nat. Rev. Microbiol. 18 (12), 717–730.

Bever, J.D., 1994. Feedback between plants and their soil communities in an old field community. Ecology 75, 1965–1977.

Bever, J.D., 2003. Soil community feedback and the coexistence of competitors: conceptual frameworks and empirical tests. New Phytol. 157, 465–473.

Bidartondo, M.I., Read, D.J., Trappe, J.M., Merckx, V., Ligrone, R., Duckett, J.G., 2011. The dawn of symbiosis between plants and fungi. Biol. Lett. 7, 574–577.

Blackwell, M., 2011. The Fungi: 1, 2, 3... 5.1 million species? Am. J. Bot. 98, 426–438.

Blankinship, J.C., Niklaus, P.A., Hungate, B.A., 2011. A meta-analysis of responses of soil biota to global change. Oecologia 165, 553–565.

Boddy, L., 1999. Saprotrophic cord-forming fungi: meeting the challenge of heterogeneous environments. Mycologia 91 (1), 13–32.

Boddy, L., 2006. Interspecific combative interactions between wood-decaying basidiomycetes. FEMS Microbiol. Ecol. 31, 185–194.

Boddy, L., Büntgen, U., Egli, S., Gange, A.C., Heegaard, E., Kirk, P.M., et al., 2014. Climate variation effects on fungal fruiting. Fungal Ecol. 10, 20–33.

Bödeker, I.T.M., Nygren, C.M.R., Taylor, A.F.S., Olson, Å., Lindahl, B.D., 2009. ClassII peroxidase-encoding genes are present in a phylogenetically wide range of ectomycorrhizal fungi. ISME J. 3, 1387–1395.

Bödeker, I.T.M., Clemmensen, K.E., de Boer, W., Martin, F., Olson, Å., Lindahl, B.D., 2014. Ectomycorrhizal *Cortinarius* species participate in enzymatic oxidation of humus in northern forest ecosystems. New Phytol. 203, 245–256.

Bödeker, I.T.M., Lindahl, B.D., Olson, Å., Clemmensen, K.E., 2016. Mycorrhizal and saprotrophic fungal guilds compete for the same organic substrates but affect decomposition differently. Funct. Ecol. 30, 1967–1978.

Bogar, L., Peay, K., Kornfeld, A., Huggins, J., Hortal, S., Anderson, I., et al., 2019. Plant-mediated partner discrimination in ectomycorrhizal mutualisms. Mycorrhiza 29, 97–111.

Bolton, R.G., Boddy, L., 1993. Characterization of the spatial aspects of foraging mycelial cord systems using fractal geometry. Mycol. Res. 97, 762–768.

Bradford, M.A., Crowther, T.W., 2013. Carbon use efficiency and storage in terrestrial ecosystems. New Phytol. 199, 7–9.

Bridge, P., Spooner, B., 2012. Non-lichenized Antarctic fungi: transient visitors or members of a cryptic ecosystem? Fungal Ecol. 5, 381–394.

Brown, J., 1995. Macroecology. University of Chicago Press.

Bruns, T.D., White, T.J., Taylor, J.W., 1991. Fungal molecular systematics. Annu. Rev. Ecol. Syst. 22, 525–564.

Bunn, R., Lekberg, Y., Zabinski, C., 2009. Arbuscular mycorrhizal fungi ameliorate temperature stress in thermophilic plants. Ecology 90, 1378–1388.

Büntgen, U., Kauserud, H., Egli, S., 2012. Linking climate variability to mushroom productivity and phenology. Front. Ecol. Environ. 10, 14–19.

Burns, R., DeForest, J., Marxsen, J., Sinsabaugh, R., Stromberger, M., Wallenstein, M., et al., 2013. Soil enzyme research: current knowledge and future directions. Soil Biol. Biochem. 58, 216–234.

Butinar, L., Strmole, T., Gunde-Cimerman, N., 2011. Relative incidence of ascomycetous yeasts in Arctic coastal environments. Microb. Ecol. 61, 832–843.

Buzzini, P., Branda, E., Goretti, M., Turchetti, B., 2012. Psychrophilic yeasts from worldwide glacial habitats: diversity, adaptation strategies and biotechnological potential. FEMS Microbiol. Ecol. 82, 217–241.

Cairney, J., Bastias, B., 2007. Influences of fire on forest soil fungal communities. Can. J. For. Res. 37, 207–215.

Campbell, J.L., Ollinger, S.V., Flerchinger, G.N., Wicklein, H., Hayhoe, K., Bailey, A.S., 2010. Past and projected future changes in snowpack and soil frost at the Hubbard Brook Experimental Forest, New Hampshire, USA. Hydrol. Process. 24, 2465–2480.

Cannon, P.F., 1997. Diversity of the Phyllachoraceae with special reference to the tropics. In: Hyde, K.D. (Ed.), Biodiversity of Tropical Microfungi. Hong Kong University Press, pp. 255–278.

Cantrell, S.A., Baez-Félix, C., 2010. Fungal molecular diversity of a Puerto Rican subtropical hypersaline microbial mat. Fungal Ecol. 3, 402–405.

Cantrell, S.A., Dianese, J.C., Fell, J., Gunde-Cimerman, N., Zalar, P., 2011. Unusual fungal niches. Mycologia 103, 1161–1174.

Carini, P., Marsden, P.J., Leff, J.W., Morgan, E.E., Strickland, M.S., Fierer, N., 2017. Relic DNA is abundant in soil and obscures estimates of soil microbial diversity. Nat. Microbiol. 2, 16242.

Carlsen, T., Engh, I.B., Decock, C., Rajchenberg, M., Kauserud, H., 2011. Multiple cryptic species with divergent substrate affinities in the *Serpula himantioides* species complex. Fungal Biol. 115, 54–61.

Cerón-Romero, M.A., Fonseca, M.M., de Oliveira Martins, L., Posada, D., Katz, L.A., 2022. Phylogenomic analyses of 2,786 genes in 158 lineages support a root of the eukaryotic tree of life between Opisthokonts (Animals, Fungi and their microbial relatives) and all other lineages. Genome. Biol. Evol. 14, evac119.

Chapela, I., Boddy, L., 1988. Fungal colonization of attached beech branches. I. Early stages of development of fungal communities. New Phytol. 39–45.

Chapin, F.S., Bret-Harte, M.S., Hobbie, S.E., Zhong, H., 1996. Plant functional types as predictors of transient responses of arctic vegetation to global change. J. Veg. Sci. 7, 347–358.

Cheek, M., Nic Lughadha, E., Kirk, P., Lindon, H., Carretero, J., Looney, B., et al., 2020. New scientific discoveries: plants and fungi. Plants. People. Planet 2, 371–388.

Chen, J., Luo, Y., Xia, J., Jiang, L., Zhou, X., Lu, M., et al., 2015. Stronger warming effects on microbial abundances in colder regions. Sci. Rep. 5, 18032.

Claridge, A.W., Trappe, J.M., Hansen, K., 2009. Do fungi have a role as soil stabilizers and remediators after forest fire? For. Ecol. Manag. 257, 1063–1069.

Clemmensen, K.E., Bahr, A., Ovaskainen, O., Dahlberg, A., Ekblad, A., Wallander, H., et al., 2013. Roots and associated fungi drive long-term carbon sequestration in boreal forest. Science 339, 1615–1618.

Colpaert, J.V., van Assche, J.A., 1987. Heavy metal tolerance in some ectomycorrhizal fungi. Funct. Ecol. 1, 415.

Comeau, A.M., Vincent, W.F., Bernier, L., Lovejoy, C., 2016. Novel chytrid lineages dominate fungal sequences in diverse marine and freshwater habitats. Sci. Rep. 6, 30120.

Connell, L., Redman, R., Craig, S., Rodriguez, R., 2006. Distribution and abundance of fungi in the soils of Taylor Valley, Antarctica. Soil Biol. Biochem. 38, 3083–3094.

Cooke, R.C., Rayner, A.D.M., 1984. Ecology of Saprophytic Fungi. Longman.

Cotton, T.E.A., Fitter, A.H., Miller, R.M., Dumbrell, A.J., Helgason, T., 2015. Fungi in the future: interannual variation and effects of atmospheric change on arbuscular mycorrhizal fungal communities. New Phytol. 205, 1598–1607.

Courty, P.-E., Pritsch, K., Schloter, M., Hartmann, A., Garbaye, J., 2005. Activity profiling of ectomycorrhiza communities in two forest soils using multiple enzymatic tests: methods. New Phytol. 167, 309–319.

Courty, P.-E., Buée, M., Diedhiou, A.G., Frey-Klett, P., Le Tacon, F., Rineau, F., et al., 2010. The role of ectomycorrhizal communities in forest ecosystem processes: new perspectives and emerging concepts. Soil Biol. Biochem. 42, 679–698.

Cox, F., Barsoum, N., Lilleskov, E.A., Bidartondo, M.I., 2010. Nitrogen availability is a primary determinant of conifer mycorrhizas across complex environmental gradients. Ecol. Lett. 13, 1103–1113.

Cox, F., Newsham, K.K., Robinson, C.H., 2019. Endemic and cosmopolitan fungal taxa exhibit differential abundances in total and active communities of Antarctic soils. Environ. Microbiol. 21, 1586–1596.

Crowther, T.W., Maynard, D.S., Crowther, T.R., Peccia, J., Smith, J.R., Bradford, M.A., 2014. Untangling the fungal niche: the trait-based approach. Front. Microbiol. 5, 579.

Crowther, T.W., Sokol, N.W., Oldfield, E.E., Maynard, D.S., Thomas, S.M., Bradford, M.A., 2015. Environmental stress response limits microbial necromass contributions to soil organic carbon. Soil Biol. Biochem. 85, 153–161.

Dahlberg, A., Schimmel, J., Taylor, A., Johannesson, H., 2001. Post-fire legacy of ectomycorrhizal fungal communities in the Swedish boreal forest in relation to fire severity and logging intensity. Biol. Conserv. 100, 151–161.

Danger, M., Gessner, M.O., Bärlocher, F., 2016. Ecological stoichiometry of aquatic fungi: current knowledge and perspectives. Fungal Ecol. Aquatic. Fungi 19, 100–111.

de Vries, F.T., Liiri, M.E., Bjørnlund, L., Bowker, M.A., Christensen, S., Setälä, H.M., et al., 2012. Land use alters the resistance and resilience of soil food webs to drought. Nat. Clim. Change 2, 276–280.

Deacon, J., Donaldson, S., Last, F., 1983. Sequences and interactions of mycorrhizal fungi on birch. Plant Soil 71, 257–262.

DeAngelis, K.M., Pold, G., Topçuoğlu, B.D., van Diepen, L.T.A., Varney, R.M., Blanchard, J.L., et al., 2015. Long-term forest soil warming alters microbial communities in temperate forest soils. Front. Microbiol. 6, 104.

Deveau, A., Bonito, G., Uehling, J., Paoletti, M., Becker, M., Bindschedler, S., et al., 2018. Bacterial–fungal interactions: ecology, mechanisms and challenges. FEMS Microbiol. Rev. 42, 335–352.

Dickie, I.A., Xu, B., Koide, R.T., 2002. Vertical niche differentiation of ectomycorrhizal hyphae in soil as shown by T-RFLP analysis. New Phytol. 156, 527–535.

Fungal communities and climate change. In: Dighton, J., White, J.F. (Eds.), 2017. The Fungal Community. CRC Press.

Donnelly, D.P., Boddy, L., Leake, J.R., 2004. Development, persistence and regeneration of foraging ectomycorrhizal mycelial systems in soil microcosms. Mycorrhiza 14, 37–45.

Dove, N.C., Hart, S.C., 2017. Fire reduces fungal species richness and in situ mycorrhizal colonization: a meta-analysis. Fire Ecol. 13, 37–65.

Eastwood, D.C., Floudas, D., Binder, M., Majcherczyk, A., Schneider, P., Aerts, A., et al., 2011. The plant cell wall–decomposing machinery underlies the functional diversity of forest fungi. Science 333, 762–765.

Edgcomb, V.P., Beaudoin, D., Gast, R., Biddle, J.F., Teske, A., 2011. Marine subsurface eukaryotes: the fungal majority. Environ. Microbiol. 13, 172–183.

Egidi, E., Delgado-Baquerizo, M., Plett, J.M., Wang, J., Eldridge, D.J., Bardgett, R.D., et al., 2019. A few *Ascomycota* taxa dominate soil fungal communities worldwide. Nat. Commun. 10, 2369.

Fellbaum, C.R., Gachomo, E.W., Beesetty, Y., Choudhari, S., Strahan, G.D., Pfeffer, P.E., et al., 2012. Carbon availability triggers fungal nitrogen uptake and transport in arbuscular mycorrhizal symbiosis. Proc. Natl. Acad. Sci. 109, 2666–2671.

Feng, B., Wang, X.-H., Ratkowsky, D., Gates, G., Lee, S.S., Grebenc, T., et al., 2016. Multilocus phylogenetic analyses reveal unexpected abundant diversity and significant disjunct distribution pattern of the Hedgehog Mushrooms (Hydnum L.). Sci. Rep. 6, 25586.

Fernandes, M.L.P., Bastida, F., Jehmlich, N., Martinović, T., Větrovský, T., Baldrian, P., et al., 2022. Functional soil mycobiome across ecosystems. J. Proteomics 252, 104428.

Fernandez, C.W., Kennedy, P.G., 2016. Revisiting the "Gadgil effect": do interguild fungal interactions control carbon cycling in forest soils? New Phytol. 209, 1382–1394.

Fernandez, C.W., Koide, R.T., 2014. Initial melanin and nitrogen concentrations control the decomposition of ectomycorrhizal fungal litter. Soil Biol. Biochem. 77, 150–157.

Ferrenberg, S., Reed, S.C., Belnap, J., 2015. Climate change and physical disturbance cause similar community shifts in biological soil crusts. Proc. Natl. Acad. Sci. 112, 12116–12121.

Feuerer, T., Hawksworth, D.L., 2007. Biodiversity of lichens, including a world-wide analysis of checklist data based on Takhtajan's floristic regions. Biodivers. Conserv. 16, 85–98.

Field, K.J., Rimington, W.R., Bidartondo, M.I., Allinson, K.E., Beerling, D.J., Cameron, D.D., et al., 2015. First evidence of mutualism between ancient plant lineages (Haplomitriopsida liverworts) and Mucoromycotina fungi and its response to simulated Palaeozoic changes in atmospheric CO_2. New Phytol. 205, 743–756.

Field, K.J., Rimington, W.R., Bidartondo, M.I., Allinson, K.E., Beerling, D.J., Cameron, D.D., et al., 2016. Functional analysis of liverworts in dual symbiosis with *Glomeromycota* and *Mucoromycotina* fungi under a simulated Palaeozoic CO_2 decline. ISME J. 10, 1514–1526.

Fierer, N., Strickland, M.S., Liptzin, D., Bradford, M.A., Cleveland, C.C., 2009. Global patterns in belowground communities. Ecol. Lett. 12, 1238–1249.

Finestone, J., Templer, P.H., Bhatnagar, J.M., 2022. Soil fungi exposed to warming temperatures and shrinking snowpack in a northern hardwood forest have lower capacity for growth and nutrient cycling. Front. For. Glob. Change 5. https://doi.org/10.3389/ffgc.2022.800335.

Finlay, B.J., Span, A.S.W., Harman, J.M.P., 1983. Nitrate respiration in primitive eukaryotes. Nature 303, 333–336.

Finlay, R.D., Lindahl, B.D., Taylor, A.F., 2008. Responses of mycorrhizal fungi to stress. In: Avery, S.V., Stratford, M., Van West, P. (Eds.), British Mycological Society Symposia Series. Elsevier, pp. 201–219.

Floudas, D., Binder, M., Riley, R., Barry, K., Blanchette, R.A., Henrissat, B., et al., 2012. The Paleozoic origin of enzymatic lignin decomposition reconstructed from 31 fungal genomes. Science 336, 1715–1719.

Fog, K., 1988. The effect of added nitrogen on the rate of decomposition of organic matter. Biol. Rev. 63, 433–462.

Fogel, R., Trappe, J.M., 1978. Fungus consumption (mycophagy) by small animals. Northwest Sci. 52, 1–31.

Freeman, K., Martin, A., Karki, D., Lynch, R., Mitter, M., Meyer, A., et al., 2009. Evidence that chytrids dominate fungal communities in high-elevation soils. Proc. Natl. Acad. Sci. 106, 18315–18320.

Frey, S.D., Ollinger, S., Nadelhoffer, K., Bowden, R., Brzostek, E., Burton, A., et al., 2014. Chronic nitrogen additions suppress decomposition and sequester soil carbon in temperate forests. Biogeochemistry 121, 305–316.

Fricker, M.D., Heaton, L.L.M., Jones, N.S., Boddy, L., 2017. The mycelium as a network. In: Heitman, J., Howlett, B.J., Crous, P.W., Stukenbrock, E.H., James, T.Y., Gow, N.A.R. (Eds.), The fungal kingdom, pp. 335–367.

Fukami, T., Dickie, I.A., Paula Wilkie, J., Paulus, B.C., Park, D., et al., 2010. Assembly history dictates ecosystem functioning: evidence from wood decomposer communities. Ecol. Lett. 13, 675–684.

Galante, T.E., Horton, T.R., Swaney, D.P., 2011. 95% of basidiospores fall within 1 m of the cap: a field-and modeling-based study. Mycologia 103, 1175–1183.

Gallo, M., Amonette, R., Lauber, C., Sinsabaugh, R., Zak, D., 2004. Microbial community structure and oxidative enzyme activity in nitrogen-amended north temperate forest soils. Microb. Ecol. 48, 218–229.

Galloway, J.N., Townsend, A.R., Erisman, J.W., Bekunda, M., Cai, Z., Freney, J.R., et al., 2008. Transformation of the nitrogen cycle: recent trends, questions, and potential solutions. Science 320, 889–892.

Gang, F., Haorui, Z., Shaowei, L., Wei, S., 2019. A meta-analysis of the effects of warming and elevated CO_2 on soil microbes. J. Resour. Ecol. 10, 69.

Garcia, M.O., Ovasapyan, T., Greas, M., Treseder, K.K., 2008. Mycorrhizal dynamics under elevated CO_2 and nitrogen fertilization in a warm temperate forest. Plant Soil 303, 301–310.

Garcia, M.O., Templer, P.H., Sorensen, P.O., Sanders-DeMott, R., Groffman, P.M., Bhatnagar, J.M., 2020. Soil microbes trade-off biogeochemical cycling for stress tolerance traits in response to year-round climate change. Front. Microbiol. 11, 616.

Gasch, A.P., Spellman, P.T., Kao, C.M., Carmel-Harel, O., Eisen, M.B., Storz, G., et al., 2000. Genomic expression programs in the response of yeast cells to environmental changes. Mol. Biol. Cell 11, 4241−4257.

Gehrig, H., Schüßler, A., Kluge, M., 1996. *Geosiphon pyriforme*, a fungus forming endocytobiosis with Nostoc (Cyanobacteria), is an ancestral member of the glomales: evidence by SSU rRNA Analysis. J. Mol. Evol. 43, 71−81.

Geisseler, D., Horwath, W.R., Joergensen, R.G., Ludwig, B., 2010. Pathways of nitrogen utilization by soil microorganisms—a review. Soil Biol. Biochem. 42, 2058−2067.

Geml, J., Tulloss, R.E., Laursen, G.A., Sazanova, N.A., Taylor, D.L., 2008. Evidence for strong inter-and intracontinental phylogeographic structure in *Amanita muscaria*, a wind-dispersed ectomycorrhizal basidiomycete. Mol. Phylogenet. Evol. 48, 694−701.

Glassman, S.I., Levine, C.R., DiRocco, A.M., Battles, J.J., Bruns, T.D., 2016. Ectomycorrhizal fungal spore bank recovery after a severe forest fire: some like it hot. ISME J. 10, 1228−1239.

Gleason, F.H., Schmidt, S.K., Marano, A.V., 2010. Can zoosporic true fungi grow or survive in extreme or stressful environments? Extremophiles 14, 417−425.

Gooday, G.W., Humphreys, A.M., McIntosh, W.H., 1986. Roles of chitinases in fungal growth. In: Muzzarelli, R., Jeuniaux, C., Gooday, G.W. (Eds.), Chitin in Nature and Technology. Springer, Boston, MA, pp. 83−91.

Gorka, S., Dietrich, M., Mayerhofer, W., Gabriel, R., Wiesenbauer, J., Martin, V., et al., 2019. Rapid transfer of plant photosynthates to soil bacteria via ectomycorrhizal hyphae and its interaction with nitrogen availability. Front. Microbiol. 10, 168.

Gorzelak, M.A., Asay, A.K., Pickles, B.J., Simard, S.W., 2015. Inter-plant communication through mycorrhizal networks mediates complex adaptive behaviour in plant communities. AoB Plants 7, plv050.

Grigoriev, I.V., Nikitin, R., Haridas, S., Kuo, A., Ohm, R., Otillar, R., et al., 2014. MycoCosm portal: gearing up for 1000 fungal genomes. Nucleic Acids Res. 42, D699−D704.

Grime, J.P., 1974. Vegetation classification by reference to strategies. Nature 250, 26−31.

Grime, J.P., 1977. Evidence for the existence of three primary strategies in plants and its relevance to ecological and evolutionary theory. Am. Nat. 111, 1169−1194.

Grubisha, L.C., Levsen, N., Olson, M.S., Taylor, D.L., 2012. Intercontinental divergence in the *Populus*-associated ectomycorrhizal fungus, Tricholoma populinum. New Phytol. 194, 548−560.

Gulis, V., Suberkropp, K., Rosemond, A.D., 2008. Comparison of fungal activities on wood and leaf litter in unaltered and nutrient-enriched headwater streams. Appl. Environ. Microbiol. 74, 1094−1101.

Gunde-Cimerman, N., Sonjak, S., Zalar, P., Frisvad, J.C., Diderichsen, B., Plemenitaš, A., 2003. Extremophilic fungi in arctic ice: a relationship between adaptation to low temperature and water activity. Phys. Chem. Earth. Parts. ABC. 28, 1273−1278.

Haight, J.-E., Laursen, G.A., Glaeser, J.A., Taylor, D.L., 2016. Phylogeny of *Fomitopsis pinicola*: a species complex. Mycologia 108, 925−938.

Harms, H., Schlosser, D., Wick, L.Y., 2011. Untapped potential: exploiting fungi in bioremediation of hazardous chemicals. Nat. Rev. Microbiol. 9, 177−192.

Hashem, A., Abd_Allah, E.F., Alqarawi, A.A., Egamberdieva, D., 2018. Arbuscular Mycorrhizal fungi and plant stress tolerance. In: Egamberdieva, D., Ahmad, P. (Eds.), Plant Microbiome: Stress Response, Microorganisms for Sustainability. Springer, pp. 81−103.

Hawkes, C.V., Keitt, T.H., 2015. Resilience vs. historical contingency in microbial responses to environmental change. Ecol. Lett. 18, 612−625.

Hawkes, C.V., Kivlin, S.N., Rocca, J.D., Huguet, V., Thomsen, M.A., Suttle, K.B., 2011. Fungal community responses to precipitation. Glob. Change Biol. 17, 1637−1645.

Hawksworth, D.L., 1991. The fungal dimension of biodiversity: magnitude, significance, and conservation. Mycol. Res. 95, 641−655.

Hawksworth, D., 2012. Global species numbers of fungi: are tropical studies and molecular approaches contributing to a more robust estimate? Biodivers. Conserv. 21 (9), 2425−2433.

Hawksworth, D.L., Lücking, R., 2017. Fungal diversity revisited: 2.2 to 3.8 million species. Microbiol. Spectr. 5 (4), 10.

Helgason, T., Daniell, T., Husband, R., Fitter, A., Young, J., 1998. Ploughing up the wood-wide web? Nature 394, 431−431.

Hobbie, E.A., Agerer, R., 2010. Nitrogen isotopes in ectomycorrhizal sporocarps correspond to belowground exploration types. Plant Soil 327, 71–83.

Hobbie, E.A., Weber, N.S., Trappe, J.M., 2001. Mycorrhizal vs saprotrophic status of fungi: the isotopic evidence. New Phytol. 150, 601–610.

Hofmockel, K.S., Fierer, N., Colman, B.P., Jackson, R.B., 2010. Amino acid abundance and proteolytic potential in North American soils. Oecologia 163, 1069–1078.

Högberg, M.N., Högberg, P., 2002. Extramatrical ectomycorrhizal mycelium contributes one-third of microbial biomass and produces, together with associated roots, half the dissolved organic carbon in a forest soil. New Phytol. 154, 791–795.

Högberg, M.N., Högberg, P., Myrold, D.D., 2007. Is microbial community composition in boreal forest soils determined by pH, C-to-N ratio, the trees, or all three? Oecologia 150, 590–601.

Holden, S., Treseder, K., 2013. A meta-analysis of soil microbial biomass responses to forest disturbances. Front. Microbiol. 4, 163.

Holden, S.R., Gutierrez, A., Treseder, K.K., 2013. Changes in soil fungal communities, extracellular enzyme activities, and litter decomposition across a fire chronosequence in Alaskan boreal forests. Ecosystems 16, 34–46.

Holland, E.A., Dentener, F.J., Braswell, B.H., Sulzman, J.M., 1999. Contemporary and pre-industrial global reactive nitrogen budgets. Biogeochemistry 46, 7–43.

Horikoshi, K., Antranikian, G., Bull, A.T., Robb, F.T., Stetter, K.O., 2010. Extremophiles Handbook. Springer Science & Business Media.

Hughes, K.W., Matheny, P.B., Miller, A.N., Petersen, R.H., Iturriaga, T.M., Johnson, K.D., et al., 2020. Pyrophilous fungi detected after wildfires in the Great Smoky Mountains National Park expand known species ranges and biodiversity estimates. Mycologia 112, 677–698.

Humphreys, C.P., Franks, P.J., Rees, M., Bidartondo, M.I., Leake, J.R., Beerling, D.J., 2010. Mutualistic mycorrhiza-like symbiosis in the most ancient group of land plants. Nat. Commun. 1, 103.

James, T.Y., Porter, D., Hamrick, J.L., Vilgalys, R., 1999. Evidence for limited intercontinental gene flow in the cosmopolitan mushroom, Schizophyllum commune. Evolution 53, 1665–1677.

James, T.Y., Kauff, F., Schoch, C.L., Matheny, P.B., Hofstetter, V., Cox, C.J., et al., 2006. Reconstructing the early evolution of Fungi using a six-gene phylogeny. Nature 443, 818–822.

James, T.Y., Stajich, J.E., Hittinger, C.T., Rokas, A., 2020. Toward a fully resolved fungal tree of life. Annu. Rev. Microbiol. 74, 291–313.

Ji, B., Bever, J.D., 2016. Plant preferential allocation and fungal reward decline with soil phosphorus: implications for mycorrhizal mutualism. Ecosphere 7, e01256.

Jirout, J., 2015. Nitrous oxide productivity of soil fungi along a gradient of cattle impact. Fungal Ecol. 17, 155–163.

Joergensen, R., Wichern, F., 2008. Quantitative assessment of the fungal contribution to microbial tissue in soil. Soil Biol. Biochem. 40, 2977–2991.

Johnson, N.C., Wolf, J., Reyes, M.A., Panter, A., Koch, G.W., Redman, A., 2005. Species of plants and associated arbuscular mycorrhizal fungi mediate mycorrhizal responses to CO_2 enrichment. Glob. Change Biol. 11, 1156–1166.

Jones, M., Durall, D., Cairney, J., 2003. Ectomycorrhizal fungal communities in young forest stands regenerating after clearcut logging. New Phytol. 157, 399–422.

Kaisermann, A., Maron, P.A., Beaumelle, L., Lata, J.C., 2015. Fungal communities are more sensitive indicators to non-extreme soil moisture variations than bacterial communities. Appl. Soil Ecol. 86, 158–164.

Kamp, A., Høgslund, S., Risgaard-Petersen, N., Stief, P., 2015. Nitrate storage and dissimilatory nitrate reduction by eukaryotic microbes. Front. Microbiol. 6, 1492.

Kanerva, T., Palojärvi, A., Rämö, K., Manninen, S., 2008. Changes in soil microbial community structure under elevated tropospheric O_3 and CO_2. Soil Biol. Biochem. 40, 2502–2510.

Kattge, J., Díaz, S., Lavorel, S., Prentice, I.C., Leadley, P., Bönisch, G., et al., 2011. Try – a global database of plant traits. Glob. Change Biol. 17, 2905–2935.

Kennedy, P.G., Peay, K.G., Bruns, T.D., 2009. Root tip competition among ectomycorrhizal fungi: are priority effects a rule or an exception? Ecology 90, 2098–2107.

Kennedy, P.G., Garibay-Orijel, R., Higgins, L.M., Angeles-Arguiz, R., 2011. Ectomycorrhizal fungi in Mexican *Alnus* forests support the host co-migration hypothesis and continental-scale patterns in phylogeography. Mycorrhiza 21, 559–568.

Kerry, E., 1990. Effects of temperature on growth rates of fungi from subantarctic Macquarie Island and Casey, Antarctica. Polar Biol. 10, 293–299.

Khan, A.G., Kuek, C., Chaudhry, T.M., Khoo, C.S., Hayes, W.J., 2000. Role of plants, mycorrhizae and phytochelators in heavy metal contaminated land remediation. Chemosphere 41, 197–207.

Kiers, E.T., Duhamel, M., Beesetty, Y., Mensah, J.A., Franken, O., Verbruggen, E., et al., 2011. Reciprocal rewards stabilize cooperation in the mycorrhizal symbiosis. Science 333, 880–882.

King, G.M., 2011. Enhancing soil carbon storage for carbon remediation: potential contributions and constraints by microbes. Trends Microbiol. 19, 75–84.

Kirk, T.K., Cullen, D., 1998. Enzymology and molecular genetics of wood degradation by white-rot fungi. In: Young, R.A., Akhtar, M. (Eds.), Environmentally Friendly Technologies for the Pulp and Paper Industry. Wiley, pp. 273–307.

Kivlin, S.N., Emery, S.M., Rudgers, J.A., 2013. Fungal symbionts alter plant responses to global change. Am. J. Bot. 100, 1445–1457.

Kivlin, S.N., Winston, G.C., Goulden, M.L., Treseder, K.K., 2014. Environmental filtering affects soil fungal community composition more than dispersal limitation at regional scales. Fungal Ecol. 12, 14–25.

Kjøller, R., Bruns, T.D., 2003. *Rhizopogon* spore bank communities within and among California pine forests. Mycologia 95, 603–613.

Klironomos, J.N., Hart, M.M., 2001. Food-web dynamics: animal nitrogen swap for plant carbon. Nature 410, 651–652.

Knorr, M., Frey, S., Curtis, P., 2005. Nitrogen additions and litter decomposition: a meta-analysis. Ecology 86, 3252–3257.

Koide, R.T., Xu, B., Sharda, J., Lekberg, Y., Ostiguy, N., 2004. Evidence of species interactions within an ectomycorrhizal fungal community. New Phytol. 165, 305–316.

Kowalchuk, G.A., 2012. Bad news for soil carbon sequestration? Science 337, 1049–1050.

Kranabetter, J.M., Harman-Denhoed, R., Hawkins, B.J., 2019. Saprotrophic and ectomycorrhizal fungal sporocarp stoichiometry (C : N : P) across temperate rainforests as evidence of shared nutrient constraints among symbionts. New Phytol. 221, 482–492.

Krebs, C., 1978. Ecology: The Experimental Analysis of Distribution and Abundance. Harper and Row.

Kulmatiski, A., Beard, K.H., Stevens, J.R., Cobbold, S.M., 2008. Plant–soil feedbacks: a meta-analytical review. Ecol. Lett. 11, 980–992.

Kyaschenko, J., Clemmensen, K.E., Karltun, E., Lindahl, B.D., 2017. Below-ground organic matter accumulation along a boreal forest fertility gradient relates to guild interaction within fungal communities. Ecol. Lett. 20, 1546–1555.

Lajoie, G., Kembel, S.W., 2019. Making the most of trait-based approaches for microbial ecology. Trends Microbiol. 27, 814–823.

Lankau, R.A., 2011. Resistance and recovery of soil microbial communities in the face of Alliaria petiolata invasions. New Phytol. 189, 536–548.

Lauber, C.L., Hamady, M., Knight, R., Fierer, N., 2009. Pyrosequencing-based assessment of soil pH as a predictor of soil bacterial community structure at the continental scale. Appl. Environ. Microbiol. 75, 5111.

Lehmann, A., Zheng, W., Soutschek, K., Roy, J., Yurkov, A.M., Rillig, M.C., 2019. Tradeoffs in hyphal traits determine mycelium architecture in saprobic fungi. Sci. Rep. 9, 14152.

Lenhart, K., Bunge, M., Ratering, S., Neu, T.R., Schüttmann, I., Greule, M., et al., 2012. Evidence for methane production by saprotrophic fungi. Nat. Commun. 3, 1046.

Li, Q., Yang, Y., Bao, X., Liu, F., Liang, W., Zhu, J., et al., 2015. Legacy effects of elevated ozone on soil biota and plant growth. Soil Biol. Biochem. 91, 50–57.

Li, Y., Steenwyk, J.L., Chang, Y., Wang, Y., James, T.Y., Stajich, J.E., et al., 2021. A genome-scale phylogeny of the kingdom Fungi. Curr. Biol. 31 (18), 1653–1665.

Liao, H.-L., Chen, Y., Bruns, T.D., Peay, K.G., Taylor, J.W., Branco, S., et al., 2014. Metatranscriptomic analysis of ectomycorrhizal roots reveals genes associated with Piloderma-Pinus symbiosis: improved methodologies for assessing gene expression in situ. Environ. Microbiol. 16, 3730–3742.

Liao, H.-L., Chen, Y., Vilgalys, R., 2016. Metatranscriptomic study of common and host-specific patterns of gene expression between pines and their symbiotic ectomycorrhizal fungi in the genus *Suillus*. PLoS Genet. 12, e1006348.

Lilleskov, E.A., Bruns, T.D., 2005. Spore dispersal of a resupinate ectomycorrhizal fungus, *Tomentella sublilacina*, via soil food webs. Mycologia 97, 762–769.

Lilleskov, E.A., Fahey, T.J., Horton, T.R., Lovett, G.M., 2002. Belowground ectomycorrhizal fungal community change over a nitrogen deposition gradient in Alaska. Ecology 83, 104–115.

Lilleskov, E.A., Bruns, T.D., Dawson, T.E., Camacho, F.J., 2009. Water sources and controls on water-loss rates of epigeous ectomycorrhizal fungal sporocarps during summer drought. New Phytol. 182, 483–494.

Lindahl, B.D., Tunlid, A., 2015. Ectomycorrhizal fungi – potential organic matter decomposers, yet not saprotrophs. New Phytol. 205, 1443–1447.

Lindahl, B.D., Kyaschenko, J., Varenius, K., Clemmensen, K.E., Dahlberg, A., Karltun, E., et al., 2021. A group of ectomycorrhizal fungi restricts organic matter accumulation in boreal forest. Ecol. Lett. 24, 1341–1351.

Lipson, D., Schmidt, S., 2004. Seasonal changes in an alpine soil bacterial community in the Colorado Rocky Mountains. Appl. Environ. Microbiol. 70, 2867–2879.

Liu, Y., Steenkamp, E.T., Brinkmann, H., Forget, L., Philippe, H., Lang, B.F., 2009. Phylogenomic analyses predict sistergroup relationship of nucleariids and fungi and paraphyly of zygomycetes with significant support. BMC Evol. Biol. 9, 272.

Liu, Y.-R., Eldridge, D.J., Zeng, X.-M., Wang, J., Singh, B.K., Delgado-Baquerizo, M., 2021. Global diversity and ecological drivers of lichenised soil fungi. New Phytol. 231, 1210–1219.

Lodge, D.J., 1987. Nutrient concentrations, percentage moisture and density of field-collected fungal mycelia. Soil Biol. Biochem. 19, 727–733.

López-Mondéjar, R., 2020. Metagenomics and stable isotope probing reveal the complementary contribution of fungal and bacterial communities in the recycling of dead biomass in forest soil. Soil Biol. Biochem. 148, 107875.

Lustenhouwer, N., Maynard, D.S., Bradford, M.A., Lindner, D.L., Oberle, B., Zanne, et al., 2020. A trait-based understanding of wood decomposition by fungi. Proc. Natl. Acad. Sci. 117, 11551–11558.

Maherali, H., Klironomos, J.N., 2007. Influence of phylogeny on fungal community assembly and ecosystem functioning. Science 316, 1746–1748.

Maheshwari, R., Bharadwaj, G., Bhat, M.K., 2000. Thermophilic fungi: their physiology and enzymes. Microbiol. Mol. Biol. Rev. 64, 461–488.

Martin, F., Aerts, A., Ahrén, D., Brun, A., Danchin, E.G.J., Duchaussoy, F., et al., 2008. The genome of *Laccaria bicolor* provides insights into mycorrhizal symbiosis. Nature 452, 88–92.

Mayer, M., Rewald, B., Matthews, B., Sandén, H., Rosinger, C., Katzensteiner, K., et al., 2021. Soil fertility relates to fungal-mediated decomposition and organic matter turnover in a temperate mountain forest. New Phytol. 231, 777–790.

McCool, P., Menge, J., 1984. Interaction of ozone and mycorrhizal fungi on tomato as influenced by fungal species and host variety. Soil Biol. Biochem. 16, 425–427.

Menezes-Blackburn, D., Jorquera, M.A., Greiner, R., Gianfreda, L., de la Luz Mora, M., 2013. Phytases and phytase-labile organic phosphorus in manures and soils. Crit. Rev. Environ. Sci. Technol. 43, 916–954.

Mittermeier, V.K., Schmitt, N., Volk, L.P.M., Suárez, J.P., Beck, A., Eisenreich, W., 2015. Metabolic profiling of alpine and Ecuadorian lichens. Molecules 20, 18047–18065.

Mohsenzadeh, S., Saupe-Thies, W., Steier, G., Schroeder, T., Fracella, F., Ruoff, P., et al., 1998. Temperature adaptation of house keeping and heat shock gene expression in *Neurospora crassa*. Fungal Genet. Biol. 25, 31–43.

Molina, R., Massicotte, H., Trappe, J.M., 1992. Specificity phenomena in mycorrhizal symbioses: community-ecological consequences and practical implications. In: Allen, M.F. (Ed.), Mycorrhizal Functioning: An Integrative Plant-Fungal Process. Chapman and Hall, pp. 357–423.

Moorhead, D.L., Lashermes, G., Sinsabaugh, R.L., Weintraub, M.N., 2013. Calculating co-metabolic costs of lignin decay and their impacts on carbon use efficiency. Soil Biol. Biochem. 66, 17–19.

Morrison, E.W., Pringle, A., van Diepen, L.T.A., Grandy, A.S., Melillo, J.M., Frey, S.D., 2019. Warming alters fungal communities and litter chemistry with implications for soil carbon stocks. Soil Biol. Biochem. 132, 120–130.

Mothapo, N.V., Chen, H., Cubeta, M.A., Shi, W., 2013. Nitrous oxide producing activity of diverse fungi from distinct agro-ecosystems. Soil Biol. Biochem. 66, 94–101.

Müller, T., Ruppel, S., 2014. Progress in cultivation-independent phyllosphere microbiology. FEMS Microbiol. Ecol. 87, 2–17.

Nguyen, N.H., Song, Z., Bates, S.T., Branco, S., Tedersoo, L., Menke, J., et al., 2016. FUNGuild: an open annotation tool for parsing fungal community datasets by ecological guild. Fungal Ecol. 20, 241–248.

Nicolás, C., Martin-Bertelsen, T., Floudas, D., Bentzer, J., Smits, M., Johansson, T., et al., 2019. The soil organic matter decomposition mechanisms in ectomycorrhizal fungi are tuned for liberating soil organic nitrogen. ISME J. 13, 977–988.

Onofri, S., Fenice, M., Cicalini, A.R., Tosi, S., Magrino, A., Pagano, S., et al., 2000. Ecology and biology of microfungi from Antarctic rocks and soils. Ital. J. Zool. 67, 163–167.

Op De Beeck, M., Ruytinx, J., Smits, M.M., Vangronsveld, J., Colpaert, J.V., Rineau, F., 2015. Belowground fungal communities in pioneer Scots pine stands growing on heavy metal polluted and non-polluted soils. Soil Biol. Biochem. 86, 58–66.

Ostroumov, V.E., Siegert, C., 1996. Exobiological aspects of mass transfer in microzones of permafrost deposits. Adv. Space Res. 18, 79–86.

O'Brien, H., Parrent, J., Jackson, J., Moncalvo, J., Vilgalys, R., 2005. Fungal community analysis by large-scale sequencing of environmental samples. Appl. Environ. Microbiol. 71, 5544–5550.

O'Donnell, K., Lutzoni, F.M., Ward, T.J., Benny, G.L., 2001. Evolutionary relationships among mucoralean fungi (*Zygomycota*): evidence for family polyphyly on a large scale. Mycologia 93 (2), 286–297.

Packer, A., Clay, K., 2000. Soil pathogens and spatial patterns of seedling mortality in a temperate tree. Nature 404, 278–281.

Panikov, N.S., Sizova, M.V., 2007. Growth kinetics of microorganisms isolated from Alaskan soil and permafrost in solid media frozen down to −35 C. FEMS Microbiol. Ecol. 59, 500–512.

Parfitt, D., Hunt, J., Dockrell, D., Rogers, H.J., Boddy, L., 2010. Do all trees carry the seeds of their own destruction? PCR reveals numerous wood decay fungi latently present in sapwood of a wide range of angiosperm trees. Fungal Ecol. 3, 338–346.

Pearce, D.A., Bridge, P.D., Hughes, K.A., Sattler, B., Psenner, R., Russell, N.J., 2009. Microorganisms in the atmosphere over Antarctica. FEMS Microbiol. Ecol. 69, 143–157.

Peay, K.G., Bruns, T.D., Kennedy, P.G., Bergemann, S.E., Garbelotto, M., 2007. A strong species–area relationship for eukaryotic soil microbes: island size matters for ectomycorrhizal fungi. Ecol. Lett. 10, 470–480.

Peay, K.G., Garbelotto, M., Bruns, T.D., 2010. Evidence of dispersal limitation in soil microorganisms: isolation reduces species richness on mycorrhizal tree islands. Ecology 91, 3631–3640.

Peay, K.G., Schubert, M.G., Nguyen, N.H., Bruns, T.D., 2012. Measuring ectomycorrhizal fungal dispersal: macroecological patterns driven by microscopic propagules. Mol. Ecol. 21, 4122–4136.

Pirozynski, K., Malloch, D., 1975. The origin of land plants: a matter of mycotrophism. Biosystems 6, 153–164.

Plassard, C., Louche, J., Ali, M.A., Duchemin, M., Legname, E., Cloutier-Hurteau, B., 2011. Diversity in phosphorus mobilisation and uptake in ectomycorrhizal fungi. Ann. For. Sci. 68, 33–43.

Pointing, S.B., Belnap, J., 2012. Microbial colonization and controls in dryland systems. Nat. Rev. Microbiol. 10, 551–562.

Pold, G., Melillo, J.M., DeAngelis, K.M., 2015. Two decades of warming increases diversity of a potentially lignolytic bacterial community. Front. Microbiol. 6, 480.

Pold, G., Billings, A.F., Blanchard, J.L., Burkhardt, D.B., Frey, S.D., Melillo, J.M., et al., 2016. Long-term warming alters carbohydrate degradation potential in temperate forest soils. Appl. Environ. Microbiol. 82, 6518–6530.

Policelli, N., Bruns, T.D., Vilgalys, R., Nuñez, M.A., 2019. Suilloid fungi as global drivers of pine invasions. New Phytol. 222, 714–725.

Policelli, N., Horton, T.R., Hudon, A.T., Patterson, T.R., Bhatnagar, J.M., 2020. Back to roots: the role of ectomycorrhizal fungi in boreal and temperate forest restoration. Front. For. Glob. Change 3, 97.

Põlme, S., Abarenkov, K., Henrik Nilsson, R., Lindahl, B.D., Clemmensen, K.E., Kauserud, H., et al., 2020. FungalTraits: a user-friendly traits database of fungi and fungus-like stramenopiles. Fungal Divers. 105, 1–16.

Porras-Alfaro, A., Bayman, P., 2011. Hidden fungi, emergent properties: endophytes and microbiomes. Annu. Rev. Phytopathol. 49, 291–315.

Porras-Alfaro, A., Herrera, J., Natvig, D.O., Lipinski, K., Sinsabaugh, R.L., 2011. Diversity and distribution of soil fungal communities in a semiarid grassland. Mycologia 103, 10–21.

Pringle, A., Taylor, J.W., 2002. The fitness of filamentous fungi. Trends Microbiol. 10, 474–481.

Pringle, A., Bever, J.D., Gardes, M., Parrent, J.L., Rillig, M.C., Klironomos, J.N., 2009. Mycorrhizal symbioses and plant invasions. Annu. Rev. Ecol. Evol. Syst. 40, 699–715.

Pritsch, K., Garbaye, J., 2011. Enzyme secretion by ECM fungi and exploitation of mineral nutrients from soil organic matter. Ann. For. Sci. 68, 25–32.

Querejeta, J., Egerton-Warburton, L., Allen, M., 2003. Direct nocturnal water transfer from oaks to their mycorrhizal symbionts during severe soil drying. Oecologia 134, 55–64.

Rabinovich, M., Bolobova, A., Vasil'chenko, L., 2004. Fungal decomposition of natural aromatic structures and xenobiotics: a review. Appl. Biochem. Microbiol. 40, 1–17.

Read, D.J., 1991. Mycorrhizas in ecosystems. Experientia 47, 376–391.

Read, D.J., Perez-Moreno, J., 2003. Mycorrhizas and nutrient cycling in ecosystems – a journey towards relevance? New Phytol. 157, 475–492.

Read, D.J., Leake, J.R., Perez-Moreno, J., 2004. Mycorrhizal fungi as drivers of ecosystem processes in heathland and boreal forest biomes. Can. J. Bot. 82, 1243–1263.

Reynolds, H.L., Packer, A., Bever, J.D., Clay, K., 2003. Grassroots ecology: plant-microbe-soil interactions as drivers of plant community structure and dynamics. Ecology 84, 2281–2291.

Richards, T.A., Jones, M.D., Leonard, G., Bass, D., 2012. Marine fungi: their ecology and molecular diversity. Annu. Rev. Mar. Sci. 4, 495–522.

Riess, K., Schön, M.E., Ziegler, R., Lutz, M., Shivas, R.G., Piątek, M., et al., 2019. The origin and diversification of the Entorrhizales: deep evolutionary roots but recent speciation with a phylogenetic and phenotypic split between associates of the Cyperaceae and Juncaceae. Org. Divers. Evol. 19, 13–30.

Riley, R., Salamov, A.A., Brown, D.W., Nagy, L.G., Floudas, D., Held, B.W., et al., 2014. Extensive sampling of basidiomycete genomes demonstrates inadequacy of the white-rot/brown-rot paradigm for wood decay fungi. Proc. Natl. Acad. Sci. 111, 9923–9928.

Rillig, M.C., Steinberg, P.D., 2002. Glomalin production by an arbuscular mycorrhizal fungus: a mechanism of habitat modification? Soil Biol. Biochem. 34, 1371–1374.

Rillig, M.C., Mardatin, N.F., Leifheit, E.F., Antunes, P.M., 2010. Mycelium of arbuscular mycorrhizal fungi increases soil water repellency and is sufficient to maintain water-stable soil aggregates. Soil Biol. Biochem. 42, 1189–1191.

Rillig, M.C., Ryo, M., Lehmann, A., Aguilar-Trigueros, C.A., Buchert, S., Wulf, A., et al., 2019. The role of multiple global change factors in driving soil functions and microbial biodiversity. Science 366, 886–890.

Rineau, F., Courty, P.-E., 2011. Secreted enzymatic activities of ectomycorrhizal fungi as a case study of functional diversity and functional redundancy. Ann. For. Sci. 68, 69–80.

Rineau, F., Shah, F., Smits, M.M., Persson, P., Johansson, T., Carleer, R., et al., 2013. Carbon availability triggers the decomposition of plant litter and assimilation of nitrogen by an ectomycorrhizal fungus. ISME J. 7, 2010–2022.

Robinson, C.H., 2001. Cold adaptation in arctic and antarctic fungi. New Phytol. 151, 341–353.

Rodgers, V.L., Stinson, K.A., Finzi, A.C., 2008. Ready or not, garlic mustard is moving in: *Alliaria petiolata* as a member of Eastern North American forests. Bioscience 58, 426–436.

Rodriguez, R.J., White Jr., J.F., Arnold, A.E., Redman, R.S., 2009. Fungal endophytes: diversity and functional roles. New Phytol. 182, 314–330.

Romaní, A.M., Fischer, H., Mille-Lindblom, C., Tranvik, L.J., 2006. Interactions of bacteria and fungi on decomposing litter: differential extracellular enzyme activities. Ecology 87, 2559–2569.

Romero-Olivares, A.L., Taylor, J.W., Treseder, K.K., 2015. Neurospora discreta as a model to assess adaptation of soil fungi to warming. BMC Evol. Biol. 15, 198.

Romero-Olivares, A.L., Meléndrez-Carballo, G., Lago-Lestón, A., Treseder, K.K., 2019. Soil metatranscriptomes under long-term experimental warming and drying: fungi allocate resources to cell metabolic maintenance rather than decay. Front. Microbiol. 10, 1914.

Romero-Olivares, A.L., Morrison, E.W., Pringle, A., Frey, S.D., 2021. Linking genes to traits in fungi. Microb. Ecol. 82, 145–155.

Roper, M., Pepper, R.E., Brenner, M.P., Pringle, A., 2008. Explosively launched spores of ascomycete fungi have drag-minimizing shapes. Proc. Natl. Acad. Sci. 105, 20583–20588.

Roper, M., Seminara, A., Bandi, M.M., Cobb, A., Dillard, H.R., Pringle, A., 2010. Dispersal of fungal spores on a cooperatively generated wind. Proc. Natl. Acad. Sci. 107, 17474–17479.

Rousk, J., Baath, E., 2011. Growth of saprotrophic fungi and bacteria in soil. FEMS Microbiol. Ecol. 78, 17–30.

Rousk, J., Brookes, P.C., Baath, E., 2009. Contrasting soil pH effects on fungal and bacterial growth suggest functional redundancy in carbon mineralization. Appl. Environ. Microbiol. 75, 1589–1596.

Roy, M., Rochet, J., Manzi, S., Jargeat, P., Gryta, H., Moreau, P.-A., et al., 2013. What determines *Alnus* -associated ectomycorrhizal community diversity and specificity? A comparison of host and habitat effects at a regional scale. New Phytol. 198, 1228–1238.

Rúa, M.A., Antoninka, A., Antunes, P.M., Chaudhary, V.B., Gehring, C., Lamit, L.J., et al., 2016. Home-field advantage? evidence of local adaptation among plants, soil, and arbuscular mycorrhizal fungi through meta-analysis. BMC Evol. Biol. 16, 122.

Rudgers, J.A., Kivlin, S.N., Whitney, K.D., Price, M.V., Waser, N.M., Harte, J., 2014. Responses of high-altitude graminoids and soil fungi to 20 years of experimental warming. Ecology 95, 1918–1928.

Ruess, R.W., Swanson, M.M., Kielland, K., McFarland, J.W., Olson, K.D., Taylor, D.L., 2019. Phosphorus mobilizing enzymes of *Alnus*-associated ectomycorrhizal fungi in an Alaskan boreal floodplain. Forests 10, 554.

Rühling, Å., Söderström, B., 1990. Changes in fruitbody production of mycorrhizal and litter decomposing macromycetes in heavy metal polluted coniferous forests in north Sweden. Water. Air. Soil. Pollut. 49, 375–387.

Saiya-Cork, K.R., Sinsabaugh, R.L., Zak, D.R., 2002. The effects of long term nitrogen deposition on extracellular enzyme activity in an Acer saccharum forest soil. Soil Biol. Biochem. 34, 1309–1315.

Sanders-DeMott, R., Sorensen, P.O., Reinmann, A.B., Templer, P.H., 2018. Growing season warming and winter freeze–thaw cycles reduce root nitrogen uptake capacity and increase soil solution nitrogen in a northern forest ecosystem. Biogeochemistry 137, 337–349.

Savi, G., Scussel, V., 2014. Inorganic compounds at regular and nanoparticle size and their anti-toxigenic fungi activity. J. Nanotechnol. Res 97, 589–598.

Schadt, C.W., 2003. Seasonal dynamics of previously unknown fungal lineages in tundra soils. Science 301, 1359–1361.

Schimel, J.P., Bennett, J., 2004. Nitrogen mineralization: challenges of a changing paradigm. Ecology 85, 591–602.

Schmidt, S., Naff, C., Lynch, R., 2012. Fungal communities at the edge: ecological lessons from high alpine fungi. Fungal Ecol. 5, 443–452.

Seebens, H., Blackburn, T.M., Dyer, E.E., Genovesi, P., Hulme, P.E., Jeschke, J.M., et al., 2018. Global rise in emerging alien species results from increased accessibility of new source pools. Proc. Natl. Acad. Sci. 115, E2264–E2273.

Selbmann, L., De Hoog, G.S., Mazzaglia, A., Friedmann, E.I., Onofri, S., 2005. Fungi at the edge of life: cryptoendolithic black fungi from Antarctic desert. Stud. Mycol. 51, 1–32.

Shah, F., Nicolás, C., Bentzer, J., Ellström, M., Smits, M., Rineau, F., et al., 2016. Ectomycorrhizal fungi decompose soil organic matter using oxidative mechanisms adapted from saprotrophic ancestors. New Phytol. 209, 1705–1719.

Shen, X.-X., Steenwyk, J.L., LaBella, A.L., Opulente, D.A., Zhou, X., Kominek, J., et al., 2020. Genome-scale phylogeny and contrasting modes of genome evolution in the fungal phylum *Ascomycota*. Sci. Adv. 6, eabd0079.

Shoun, H., Tanimoto, T., 1991. Denitrification by the fungus *Fusarium oxysporum* and involvement of cytochrome P-450 in the respiratory nitrite reduction. J. Biol. Chem. 266, 11078–11082.

Shoun, H., Fushinobu, S., Jiang, L., Kim, S.-W., Wakagi, T., 2012. Fungal denitrification and nitric oxide reductase cytochrome P450nor. Philos. Trans. R. Soc. B Biol. Sci. 367, 1186–1194.

Simon, L., Bousquet, J., Levesque, R.C., Lalonde, M., 1993. Origin and diversification of endomycorrhizal fungi and coincidence with vascular land plants. Nature 363, 67–69.

Sinsabaugh, R.L., 2010. Phenol oxidase, peroxidase and organic matter dynamics of soil. Soil Biol. Biochem. 42, 391–404.

Sinsabaugh, R.L., Follstad Shah, J.J., 2011. Ecoenzymatic stoichiometry of recalcitrant organic matter decomposition: the growth rate hypothesis in reverse. Biogeochemistry 102, 31–43.

Sinsabaugh, R., Carreiro, M., Repert, D., 2002. Allocation of extracellular enzymatic activity in relation to litter composition, N deposition, and mass loss. Biogeochemistry 60, 1–24.

Smith, M.L., Bruhn, J.N., Anderson, J.B., 1992. The fungus Armillaria bulbosa is among the largest and oldest living organisms. Nature 356, 428–431.

Smith, J.E., McKay, D., Niwa, C.G., Thies, W.G., Brenner, G., Spatafora, J.W., 2004. Short-term effects of seasonal prescribed burning on the ectomycorrhizal fungal community and fine root biomass in ponderosa pine stands in the Blue Mountains of Oregon. Can. J. For. Res. 34, 2477–2491.

Smith, G.R., Steidinger, B.S., Bruns, T.D., Peay, K.G., 2018. Competition–colonization tradeoffs structure fungal diversity. ISME J. 12, 1758–1767.

Snajdr, J., Cajthaml, T., Valášková, V., Merhautová, V., Petránková, M., Spetz, P., et al., 2011. Transformation of Quercus petraea litter: successive changes in litter chemistry are reflected in differential enzyme activity and changes in the microbial community composition. FEMS Microbiol. Ecol. 75, 291–303.

Sorensen, P.O., Finzi, A.C., Giasson, M.-A., Reinmann, A.B., Sanders-DeMott, R., Templer, P.H., 2018. Winter soil freeze-thaw cycles lead to reductions in soil microbial biomass and activity not compensated for by soil warming. Soil Biol. Biochem. 116, 39–47.

Spatafora, J.W., Chang, Y., Benny, G.L., Lazarus, K., Smith, M.E., Berbee, M.L., et al., 2016. A phylum-level phylogenetic classification of zygomycete fungi based on genome-scale data. Mycologia 108, 1028–1046.

Spott, O., Russow, R., Stange, C.F., 2011. Formation of hybrid N_2O and hybrid N_2 due to codenitrification: first review of a barely considered process of microbially mediated N-nitrosation. Soil Biol. Biochem. 43, 1995–2011.

Stajich, J.E., Berbee, M.L., Blackwell, M., Hibbett, D.S., James, T.Y., Spatafora, J.W., et al., 2009. Primer—the fungi. Curr. Biol. 19, R840.

Starkey, R.L., Waksman, S.A., 1943. Fungi tolerant to extreme acidity and high concentrations of copper sulfate. J. Bacteriol. 45, 509–519.

Steenkamp, E.T., Wright, J., Baldauf, S.L., 2006. The protistan origins of animals and fungi. Mol. Biol. Evol. 23, 93–106.

GFBI consortium, Steidinger, B.S., Crowther, T.W., Liang, J., Van Nuland, M.E., Werner, G.D.A., et al., 2019. Climatic controls of decomposition drive the global biogeography of forest-tree symbioses. Nature 569, 404–408.

Steidinger, B.S., Bhatnagar, J.M., Vilgalys, R., Taylor, J.W., Qin, C., Zhu, K., et al., 2020. Ectomycorrhizal fungal diversity predicted to substantially decline due to climate changes in North American Pinaceae forests. J. Biogeogr. 47, 772–782.

Stendell, E., Horton, T., Bruns, T., 1999. Early effects of prescribed fire on the structure of the ectomycorrhizal fungus community in a Sierra Nevada ponderosa pine forest. Mycol. Res. 103, 1353–1359.

Sterflinger, K., Tesei, D., Zakharova, K., 2012. Fungi in hot and cold deserts with particular reference to microcolonial fungi. Fungal Ecol. 5, 453–462.

Sterner, R.W., Elser, J.J., 2017. Ecological Stoichiometry. Princeton University Press.

Stinson, K.A., Campbell, S.A., Powell, J.R., Wolfe, B.E., Callaway, R.M., Thelen, G.C., et al., 2006. Invasive plant suppresses the growth of native tree seedlings by disrupting belowground mutualisms. PLoS Biol. 4, e140.

Stone, L., Roberts, A., 1990. The checkerboard score and species distributions. Oecologia 85, 74–79.

Strickland, M.S., Rousk, J., 2010. Considering fungal: bacterial dominance in soils—Methods, controls, and ecosystem implications. Soil Biol. Biochem. 42, 1385–1395.

Strong, P.J., Claus, H., 2011. Laccase: a review of its past and its future in bioremediation. Crit. Rev. Environ. Sci. Technol. 41, 373–434.

Sturm, M., Schimel, J., Michaelson, G., Welker, J.M., Oberbauer, S.F., Liston, G.E., et al., 2005. Winter biological processes could help convert Arctic tundra to shrubland. Bioscience 55, 17–26.

Takaya, N., 2002. Dissimilatory nitrate reduction metabolisms and their control in fungi. J. Biosci. Bioeng. 94, 506–510.

Talbot, J.M., Treseder, K.K., 2010. Controls over mycorrhizal uptake of organic nitrogen. Pedobiologia 53, 169–179.

Talbot, J.M., Treseder, K.K., 2012. Interactions among lignin, cellulose, and nitrogen drive litter chemistry-decay relationships. Ecology 93, 345–354.

Talbot, J., Allison, S., Treseder, K., 2008. Decomposers in disguise: mycorrhizal fungi as regulators of soil C dynamics in ecosystems under global change. Funct. Ecol. 22, 955–963.

Talbot, J.M., Bruns, T.D., Smith, D.P., Branco, S., Glassman, S.I., Erlandson, S., et al., 2013. Independent roles of ectomycorrhizal and saprotrophic communities in soil organic matter decomposition. Soil Biol. Biochem. 57, 282–291.

Talbot, J.M., Bruns, T.D., Taylor, J.W., Smith, D.P., Branco, S., Glassman, S.I., et al., 2014. Endemism and functional convergence across the North American soil mycobiome. Proc. Natl. Acad. Sci. 111, 6341–6346.

Talbot, J.M., Martin, F., Kohler, A., Henrissat, B., Peay, K.G., 2015. Functional guild classification predicts the enzymatic role of fungi in litter and soil biogeochemistry. Soil Biol. Biochem. 88, 441–456.

Tang, N., Lebreton, A., Xu, W., Dai, Y., Yu, F., Martin, F.M., 2021. Transcriptome profiling reveals differential gene expression of secreted proteases and highly specific gene repertoires involved in Lactarius–Pinus symbioses. Front. Plant Sci. 12, 714393.
Taylor, D.L., Bruns, T.D., 1999. Community structure of ectomycorrhizal fungi in a Pinus muricata forest: minimal overlap between the mature forest and resistant propagule communities. Mol. Ecol. 8, 1837–1850.
Taylor, A.F.S., Martin, F., Read, D.J., 2000. Fungal diversity in ectomycorrhizal communities of Norway spruce (Picea abies [L.] Karst.) and Beech (Fagus sylvatica L.) along north-south transects in Europe. In: Schulze, E.-D. (Ed.), Carbon and Nitrogen Cycling in European Forest Ecosystems: With 106 Tables, Ecological Studies. Springer, pp. 343–365.
Taylor, J.W., Turner, E., Townsend, J.P., Dettman, J.R., Jacobson, D., 2006. Eukaryotic microbes, species recognition and the geographic limits of species: examples from the kingdom Fungi. Philos. Trans. R. Soc. B Biol. Sci. 361, 1947–1963.
Taylor, D.L., Herriott, I.C., Stone, K.E., McFarland, J.W., Booth, M.G., Leigh, M.B., 2010. Structure and resilience of fungal communities in Alaskan boreal forest soils. Can. J. For. Res. 40, 1288–1301.
Taylor, D.L., Hollingsworth, T.N., McFarland, J.W., Lennon, N.J., Nusbaum, C., Ruess, R.W., 2014. A first comprehensive census of fungi in soil reveals both hyperdiversity and fine-scale niche partitioning. Ecol. Monogr. 84, 3–20.
Tedersoo, L., Suvi, T., Jairus, T., Ostonen, I., Polme, S., 2009. Revisiting ectomycorrhizal fungi of the genus *Alnus*: differential host specificity, diversity and determinants of the fungal community. New Phytol. 182, 727–735.
Tedersoo, L., Bahram, M., Põlme, S., Kõljalg, U., Yorou, N.S., Wijesundera, R., et al., 2014. Global diversity and geography of soil fungi. Science 346, 1256688.
Tedersoo, L., Mikryukov, V., Anslan, S., Bahram, M., Khalid, A.N., Corrales, A., et al., 2021. The Global Soil Mycobiome consortium dataset for boosting fungal diversity research. Fungal Divers. 111, 573–588.
Terrer, C., Phillips, R.P., Hungate, B.A., Rosende, J., Pett-Ridge, J., Craig, M.E., et al., 2021. A trade-off between plant and soil carbon storage under elevated CO2. Nature 591, 599–603.
Timling, I., Walker, D.A., Nusbaum, C., Lennon, N.J., Taylor, D.L., 2014. Rich and cold: diversity, distribution and drivers of fungal communities in patterned-ground ecosystems of the North American Arctic. Mol. Ecol. 23, 3258–3272.
Toljander, J., Eberhardt, U., Toljander, Y., Paul, L., Taylor, A., 2006. Species composition of an ectomycorrhizal fungal community along a local nutrient gradient in a boreal forest. New Phytol. 170, 873–883.
Treseder, K.K., 2004. A meta-analysis of mycorrhizal responses to nitrogen, phosphorus, and atmospheric CO_2 in field studies. New Phytol. 164, 347–355.
Treseder, K.K., 2008. Nitrogen additions and microbial biomass: a meta-analysis of ecosystem studies. Ecol. Lett. 11, 1111–1120.
Treseder, K.K., Cross, A., 2006. Global distributions of arbuscular mycorrhizal fungi. Ecosystems 9, 305–316.
Treseder, K.K., Lennon, J.T., 2015. Fungal traits that drive ecosystem dynamics on land. Microbiol. Mol. Biol. Rev. 79, 243–262.
Treseder, K.K., Turner, K.M., 2007. Glomalin in ecosystems. Soil Sci. Soc. Am. J. 71, 1257–1266.
Treseder, K., Mack, M., Cross, A., 2004. Relationships among fires, fungi, and soil dynamics in Alaskan Boreal Forests. Ecol. Appl. 14, 1826–1838.
Treseder, K.K., Maltz, M.R., Hawkins, B.A., Fierer, N., Stajich, J.E., McGuire, K.L., 2014. Evolutionary histories of soil fungi are reflected in their large-scale biogeography. Ecol. Lett. 17, 1086–1093.
Trum, F., Titeux, H., Cornelis, J.-T., Delvaux, B., 2011. Effects of manganese addition on carbon release from forest floor horizons. Can. J. For. Res. 41, 643–648.
van der Heijden, M.G.A., Klironomos, J.N., Ursic, M., Moutoglis, P., Streitwolf-Engel, R., Boller, T., et al., 1998. Mycorrhizal fungal diversity determines plant biodiversity, ecosystem variability and productivity. Nature 396, 69–72.
Van Der Heijden, M.G., Bardgett, R.D., Van Straalen, N.M., 2008. The unseen majority: soil microbes as drivers of plant diversity and productivity in terrestrial ecosystems. Ecol. Lett. 11, 296–310.
van der Heijden, M.G.A., Martin, F.M., Selosse, M.-A., Sanders, I.R., 2015. Mycorrhizal ecology and evolution: the past, the present, and the future. New Phytol. 205, 1406–1423.
van der Linde, S., Suz, L.M., Orme, C.D.L., Cox, F., Andreae, H., Asi, E., et al., 2018. Environment and host as large-scale controls of ectomycorrhizal fungi. Nature 558, 243–248.
Vellinga, E.C., Wolfe, B.E., Pringle, A., 2009. Global patterns of ectomycorrhizal introductions. New Phytol. 181, 960–973.
Veresoglou, S.D., Chen, B., Rillig, M.C., 2012. Arbuscular mycorrhiza and soil nitrogen cycling. Soil Biol. Biochem. 46, 53–62.

Větrovský, T., Kohout, P., Kopecký, M., Machac, A., Man, M., Bahnmann, B.D., et al., 2019. A meta-analysis of global fungal distribution reveals climate-driven patterns. Nat. Commun. 10, 5142.

Visser, S., 1995. Ectomycorrhizal fungal succession in jack pine stands following wildfire. New Phytol. 129, 389–401.

Vitt, D.H., 2007. Estimating moss and lichen ground layer net primary production in tundra, peatlands, and forests. In: Fahey, T.J., Knapp, A.K. (Eds.), Principles and Standards for Measuring Primary Production, Long-Term Ecological Research Network Series. Oxford University Press, New York, pp. 82–105.

Voigt, K., James, T.Y., Kirk, P.M., Santiago, A.L.C.M. de A., Waldman, B., Griffith, G.W., et al., 2021. Early-diverging fungal phyla: taxonomy, species concept, ecology, distribution, anthropogenic impact, and novel phylogenetic proposals. Fungal Divers. 109, 59–98.

Voříšková, J., Brabcová, V., Cajthaml, T., Baldrian, P., 2014. Seasonal dynamics of fungal communities in a temperate oak forest soil. New Phytol. 201, 269–278.

Walder, F., van der Heijden, M.G.A., 2015. Regulation of resource exchange in the arbuscular mycorrhizal symbiosis. Nat. Plants 1, 1–7.

Waldrop, M., White, R., Douglas, T., 2008. Isolation and identification of cold-adapted fungi in the fox permafrost tunnel, Alaska. Proc. NICOP II, 1887–1891.

Wan, S., Hui, D., Luo, Y., 2001. Fire effects on nitrogen pools and dynamics in terrestrial ecosystems: a meta-analysis. Ecol. Appl. 11, 1349–1365.

Waring, B.G., Hawkes, C.V., 2015. Short-term precipitation exclusion alters microbial responses to soil moisture in a wet tropical forest. Microb. Ecol. 69, 843–854.

Wilkins, D.A., 1991. The influence of sheating (ecto-)mycorrhizas of trees on the uptake and toxicity of metals. Agric. Ecosyst. Environ. 35, 245–260.

Willis, A., Rodrigues, B., Harris, P., 2013. The ecology of arbuscular mycorrhizal fungi. Crit. Rev. Plant Sci. 32, 1–20.

Wilson, G.W., Rice, C.W., Rillig, M.C., Springer, A., Hartnett, D.C., 2009. Soil aggregation and carbon sequestration are tightly correlated with the abundance of arbuscular mycorrhizal fungi: results from long-term field experiments. Ecol. Lett. 12, 452–461.

Wright, S.F., Upadhyaya, A., 1996. Extraction of an abundant and unusual protein from soil and comparison with hyphal protein of arbuscular mycorrhizal fungi. Soil Sci. 161, 575–586.

Zak, D.R., Pellitier, P.T., Argiroff, W.A., Castillo, B., James, T.Y., et al., 2019. Exploring the role of ectomycorrhizal fungi in soil carbon dynamics. New Phytol. 223, 33–39.

Zalar, P., Gostincar, C., De Hoog, G., Ursic, V., Sudhadham, M., Gunde-Cimerman, N., 2008. Redefinition of *Aureobasidium pullulans* and its varieties. Stud. Mycol. 61, 21–38.

Zanne, A.E., Abarenkov, K., Afkhami, M.E., Aguilar-Trigueros, C.A., Bates, S., Bhatnagar, J.M., et al., 2020a. Fungal functional ecology: bringing a trait-based approach to plant-associated fungi. Biol. Rev. 95, 409–433.

Zanne, A.E., Powell, J.R., Flores-Moreno, H., Kiers, E.T., van 't Padje, A., Cornwell, W.K., 2020b. Finding fungal ecological strategies: is recycling an option? Fungal Ecol. 46, 100902.

Zhang, J., Elser, J.J., 2017. Carbon:nitrogen:phosphorus stoichiometry in fungi: a meta-analysis. Front. Microbiol. 8, 1281.

Zhou, Z., Wang, C., Luo, Y., 2020. Meta-analysis of the impacts of global change factors on soil microbial diversity and functionality. Nat. Commun. 11, 1–10.

Supplemental material

S4.1 A primer on fungal systematics

While kingdoms were long recognized as the most encompassing categories in the taxonomic hierarchy of life, the landmark work of Karl Woese, showing that two groups of prokaryotes (microbial organisms without nuclei) were equally divergent from one another as from eukaryotes (single-celled and multicellular organisms with nuclei and additional organelles), led to the establishment of domain as the deepest division (Woese and Fox 1977). There are three domains: the prokaryotic domains of archaea and bacteria, and the *Eukaryota*. Thus the formal Linnaean hierarchy proceeds as follows: domain, kingdom, phylum (= division for animalia), class, order, family, genus, and species. However, this limited set of rankings does not allow recognition of all important groupings; hence informal rankings above (e.g., "superkingdom") and below (e.g., "subphylum") these formal levels are often used in systematics. Exemplar taxonomic classifications downloaded from the NCBI taxonomy pages (www.ncbi.nlm.nih.gov/taxonomy) for two soil-dwelling fungi are shown in Table S4.1. The ranks at the top are the broadest and proceed through successively narrower groupings until reaching the species. Notice that several unranked taxonomic groupings have been used for these fungi due to insufficient resolution provided by the traditional Linnaean levels alone.

Bacteria and archaea are subject to the International Code of Nomenclature of Bacteria (ICNB; www.the-icsp.org/), while formal taxonomy in the kingdom fungi falls under the International Code of Nomenclature (ICN) for algae, fungi, and plants (www.iapt-taxon.org/nomen/main.php). Molecular, physiological, and ultrastructural data are increasingly useful in efforts to ascertain the boundaries of species and higher taxonomic ranks, but technical descriptions of the morphology of reproductive structures are still required to name (or rename) a fungal taxon. In most cases these descriptions relate to structures produced in the sexual phase of the fungal lifecycle in which meiosis occurs (see Section II),

TABLE S4.1 Two Examples of Formal Fungal Classification

Taxonomic Rank	Classification for *Tuber melanosporum*	Classification for *Agaricus bisporus*
Superkingdom	*Eukaryota*	*Eukaryota*
Unranked	*Opisthokonta*	*Opisthokonta*
Kingdom	Fungi (=*Eumycota*)	Fungi (=*Eumycota*)
Subkingdom	*Dikarya*	*Dikarya*
Phylum	*Ascomycota*	*Basidiomycota*
Unranked	*Saccharomyceta*	
Subphylum	*Pezizomycotina*	*Agaricomycotina*
Class	*Pezizomycetes*	*Agaricomycetes*
Subclass		*Agaricomycetidae*
Order	*Pezizales*	*Agaricales*
Family	*Tuberaceae*	*Agaricaceae*
Genus	*Tuber*	*Agaricus*
Species	*Tuber melanosporum*	*Agaricus bisporus*
Common name	Perigord truffle	Button mushroom

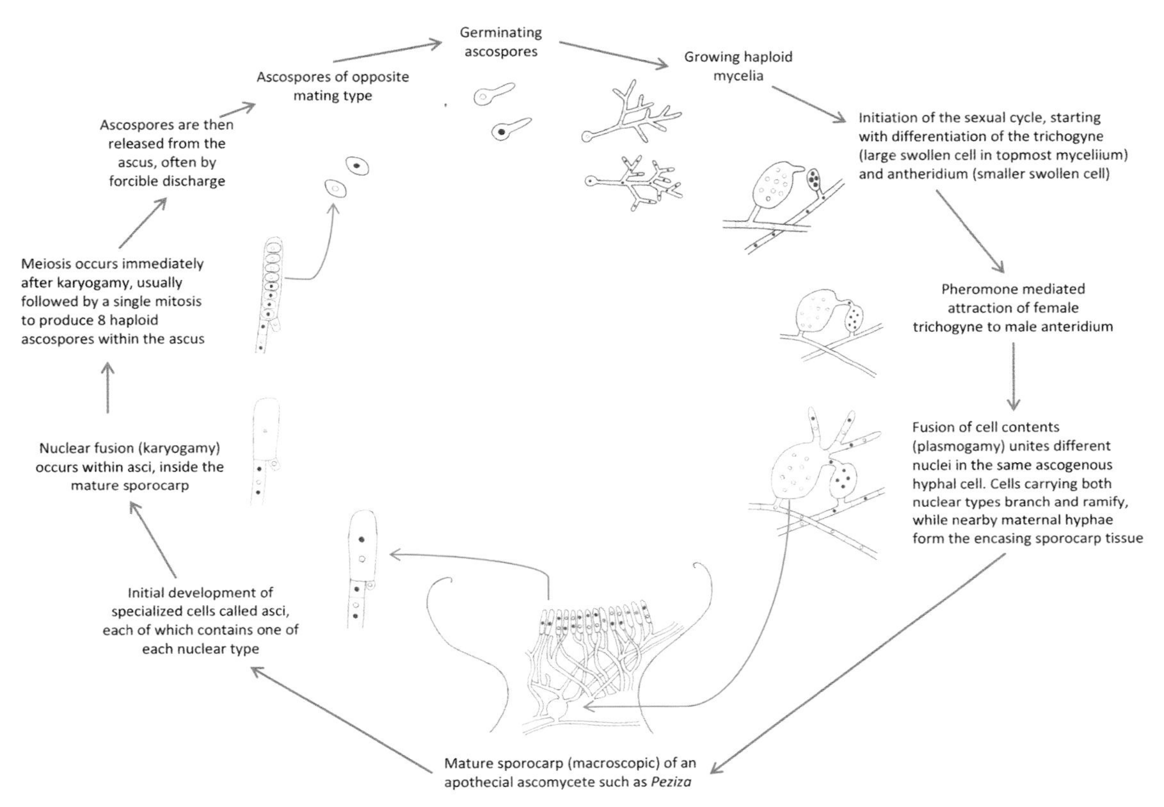

FIGURE S4.1 A typical life cycle for a fungus in the *Ascomycota*.

such as mushrooms of the *Agaricomycetes* or morels of the *Ascomycota*. Species names based on a sexual sporocarp are called teleomorphs. If a fungus produces asexual reproductive structures (e.g., conidia and conidiophores, Fig. S4.1) and no sexual structures are known for this species, the Latin species description may be based upon these asexual forms; in this case the species name is called the anamorph. Many fungi have been named based on sexual sporocarps collected in the wild without any accompanying pure culture isolate. Many other fungi have been named based on isolates brought into pure culture. Often these isolates produce asexual reproductive forms readily in culture but rarely or never produce sexual structures (e.g., molds of the *Ascomycota*). As a consequence of these two strategies for naming new fungi, a number of species had both a teleomorphic and an anamorphic name attached to them. The connections between teleomorphs and anamorphs are slowly being discovered over time as more sterile fungi are induced to produce sexual structures in culture and molecular diagnostics are brought to bear. Furthermore, the sexual reproductive structures are historically the basis upon which fungi are placed in higher taxonomic ranks, e.g., class *Agaricomycetes* or phylum *Basidiomycota*. Hence, the asexual fungi known only by their anamorphs were traditionally placed in an artificial class, the *Deuteromycetes*. It turns out that most members of the *Deuteromycetes* belong to the *Ascomycota* in terms of their shared evolutionary history; however, there are some anamorphic names spread among the other fungal phyla (e.g., *Rhizoctonia*, a widespread soil fungus in the *Basidiomycota*). These aspects of fungal systematics have caused great confusion for biologists from other disciplines seeking to use or interpret fungal names. Recently, a movement under the banner "one fungus, one name" was successful in their lobbying to abolish the two-name, anamorph-teleomorph system (Taylor, 2011). Every fungus will now have only one formally recognized name. In addition, nonphylogenetic classifications, such as Deuteromycetes, are no longer recognized. These changes will help make fungal systematics more digestible for ecologists, soil scientists, evolutionary biologists, and others. However, some familiarity with these outdated terms may be helpful, since they will be widely encountered in the literature. Some additional terms that are fading from use are hyphomycete, which is a synonym for Deuteromycetes, and microfungi, which refers to fungi that only produce very small sexual or asexual reproductive structures, i.e., filamentous soil fungi other than mushroom-forming *Basidiomycota*. Microfungi are mostly members of the *Ascomycota*; the term encompasses common soil molds, such as *Trichoderma*, *Aspergillus*, and *Penicillium*.

S4.2 Fungal life cycles

In general, fungi spend the majority of their life cycle in a haploid state. Haploid spores produced by meiosis or mitosis disperse. If conditions are conducive, the spore will then germinate and a hypha will begin to grow (unless the fungus is a yeast). A fungal body or thallus develops through progressive growth and branching of this original hypha (Figs. 4.5 and S4.1). In many fungi asexual reproduction by mitotic production of conidia or other mitospores can occur at any time during growth of the mycelium (usually triggered by specific growth conditions). For sexual reproduction to occur, karyogamy to generate a transient diploid stage must take place. Most fungi are heterothallic, meaning they cannot mate with themselves but must instead mate with a different individual of an opposing mating type. When a haploid hypha encounters a hypha or conidium of another individual of the same species, but different mating type, it may engage in a series of steps leading to sexual reproduction. In brief, the cytoplasms fuse by anastomosis – a process called plasmogamy. This places the two different nuclei within the same cell (see Fig. S4.1). Next, a complex developmental process unfolds, often involving synchronized mitotic replication of the two nuclei within the developing sporocarp. Finally, the two different nuclei fuse (karyogamy) within a specialized reproductive cell (e.g., the ascus in *Ascomycota* and the basidium in *Basidiomycota*), followed by meiosis and production of haploid spores. There are exceptions to immediate meiosis; for example, the zygospore in members of the former *Zygomycota* remains diploid until this resting spore is triggered to germinate. The exception to dominance of the haploid stage occurs in the *Basidiomycota*, in which mycelia are more likely to be dikaryotic. Here, the difference is that compatible hyphae of opposing mating types anastomose early in the life cycle, rather than late in the life cycle immediately before sexual reproduction. However, as with other fungi, these nuclei do not fuse until immediately before meiosis. Hence cells of the vegetative hyphae of most *Basidiomycota* contain pairs of unfused haploid nuclei from the two parents, a situation that is termed "dikaryotic" (Fig. S4.1). Many basidiomycete fungi have bumps (actually a small hyphal loop) over the septa that separate cells called clamp connections. Clamp connections maintain exactly one nucleus of each type per cell. A third exception to the life cycle patterns described above concerns so-called homothallic species. These are species that do not require nuclei to have opposing mating types and so are able to mate with themselves and undergo sexual reproduction.

S4.3 Glossary of terms

Term	Definition
-ales	Ending for taxa at the ordinal level.
-cetes	Ending for taxa at the class level.
-eae	Ending for taxa at the familial level.
-ota	Ending for taxa at the phylum level.
-otina	Ending for taxa at the subphylum level.
anamorph	The asexual phase of the life cycle of a fungus. Can also refer to the Latin name for a fungal species based on a description of the asexual stage; e.g., anamorph = *Fusarium moniliforme*, teleomorph = *Gibberella fujikuroi*.
anastomosis	The fusion of two fungal hyphae to form a unified cytoplasm. Occurs often within the mycelium of a single fungal individual; can sometimes occur among different genotypes within a species.
ascocarp	A conglomeration of asci and surrounding tissue forming the sexual reproductive structure of fungi in the *Ascomycota*.
ascospore	A single-celled haploid spore produced by meiosis within an ascus of a fungus in the *Ascomycota*.
ascus (singular), asci (plural)	A specialized sack-shaped reproductive cell of fungi in the *Ascomycota* in which meiosis occurs and ascospores are produced.
basal fungal lineages = early-diverging fungal lineages	Refers to major evolutionary lineages at the class to phylum level toward the base of the fungal tree of life. Most of these lineages were once placed within either the *Chytridiomycota* or the *Zygomycota* but now stand on their own due to the polyphyly of the aforementioned phyla. Relationships among these lineages remain to be resolved.
basidiocarp	A conglomeration of basidia and surrounding tissue forming the sexual reproductive structure of fungi in the *Ascomycota*.
basidiospore	A single-celled haploid spore produced by meiosis within a basidium of a fungus in the *Basidiomycota*.
basidium (singular), basidia (plural)	A specialized club-shaped reproductive cell of fungi in the *Basidiomycota* in which meiosis occurs and basidiospores are produced.
black yeasts = microcolonial yeasts = meristematic yeasts	A polyphyletic assemblage
brown rot	A type of wood decay in which lignin is only partially depolymerized and its C is not consumed. The cellulose and hemicellulose are attacked using hydrogen peroxide and the cellulosic C is mostly consumed. This type of decay leaves behind blocky brown material. Carried out by a spectrum of fungi in the *Basidiomycota*.

Continued

—cont'd

Term	Definition
clamp connection	A hump-shaped outgrowth forming a channel from one cell to the next, bypassing a septum; found only in some species of *Basidiomycota*.
codenitrification	A process in which additional amines react with intermediates in normal denitrification, resulting in N-N compounds with each N atom coming from a different source. This process can alter the dynamics of N turnover and loss from soil systems but is poorly understood or quantified.
conidium (singular), conidia (plural)	A single-celled dispersal unit produced mitotically (i.e., not a sexual product).
cryptic species	A situation in which multiple phylogenetic or biological species occur within a single morphologically demarcated species.
Deuteromycete (= Fungi Imperfecti)	An outdated taxonomic term to encompass asexual, anamorphic fungal species of uncertain evolutionary affiliation.
dikaryon	A fungal individual that contains two types of haploid nuclei, one from each parent.
dimorphic	A fungus that is able to switch between a single-celled yeast growth form and a multicellular filamentous growth form.
dissimilatory denitrification	The reduction of nitrate or nitrite to gaseous forms (N_2O, N_2), resulting in loss of N back to the atmosphere. Nitrate is used in place of oxygen as the electron acceptor to carry out respiration.
DSE	Dark septate endophyte; melanized fungi growing inside of roots, may also extend beyond the root and may be mildly beneficial or mildly harmful to the plant.
extracellular enzyme	An enzyme that is targeted for export across the cell membrane and cell wall so that it can function outside the cytoplasm; often involved in degradation of organic polymers, such as starch, cellulose, proteins, or lignin.
glomalin	A complex glycoprotein compound found in soil thought to be secreted by arbuscular mycorrhizal fungi. Glomalin contributes to aggregate formation and soil stabilization.
glycosidase = glycosyl hydrolase	A hydrolytic enzyme that contributes to depolymerization of sugar polymers, such as cellulose and chitin. This is a very large group containing many families of enzymes; in fungi many are secreted (i.e., extracellular). This class includes xylanases, chitinases, beta-glucosidases, cellobiohydrolases, and endo-glucanases, among others.
heterothallic	A species of fungus that requires the joining of nuclei from different mating types for sexual reproduction.

—cont'd

Term	Definition
homothallic	A species of fungus identical nuclei (i.e., not from different mating types) can fuse and undergo sexual reproduction.
hypha (singular), hyphae (plural)	An individual filament or thread formed of tube-like cells attached end to end; hyphae are the basic building blocks of all multicellular fungal structures, such as mushrooms.
hyphomycete (= soil microfungi)	An umbrella term for asexual filamentous fungi in soils, particularly ascomycetous molds.
karyogamy	The fusion of two different haploid nuclei from compatible mating types to form the diploid stage; followed immediately by meiosis in the vast majority of fungi.
laccase	Copper-containing oxidative enzymes that may function inside the cytoplasm as part of various developmental processes or may be secreted to attack polyphenolics, such as lignin.
mold (= mould)	An evolutionarily heterogeneous group of fast-growing fungi that rapidly produce masses of asexual spores soon after colonizing a high-energy substrate, such as bread, fruit, or dung.
monophyletic	A group that encompasses all taxa descended from a single common ancestor; synonymous with natural evolutionary group and clade.
mycelium	A constellation of interconnected hyphae belonging to a single fungal individual.
mycoparasite	A fungus that preys upon (i.e., infects) other living fungi.
phosphatase	A hydrolytic enzyme that cleaves phosphoric acid monoesters to release phosphate. Secreted by many fungi to aid in P scavenging.
phytase	A type of phosphatase that cleaves phosphate groups from phytic acid.
plasmogamy	The fusion of cytoplasmic material by the joining of the cell walls of two compatible fungal individuals; usually precedes karyogamy and meiosis.
polyphyletic	An assemblage of taxa that does not include all descendants from a single common ancestor; rather, taxa derived from different common ancestors are placed in the same grouping, such as yeasts.
psychrophile	Cold-loving organisms that can grow below 5°C with a maximum growth rate below 20°C.
psychrotolerant	Cold-tolerant organisms that can grow below 5°C with a maximum growth rate above 20°C.

Continued

—cont'd

Term	Definition
rhizomorph	A developmentally complex tube structure formed through the coordinated growth or a large number of hyphae. Similar to cords but includes a cap that resembles a root cap rather than a diffuse mycelial front. Most common in *Basidiomycota*.
rust	Common name for plant pathogens of the order *Pucciniales, Basidiomycota*. Rusts typically have complex life cycles that may include five different spore stages and alternation between two unrelated host species. The common name comes from the typical orange-brown color of the telial spore stage.
sclerotium (singular), sclerotia (plural)	A conglomeration of thick-walled, resistant cells produced by mitosis.
septum (singular), septa (plural)	Cell wall material laid down within a hypha to divide the filament into discrete cells; has a septal pore in the middle that controls movement of structures, such as nuclei from cell to cell.
smut	Common name for a class of fungi in the *Basidiomycota*, most of which are plant pathogens. The life cycles of smuts are not as complex as those of rusts, with completion of the life cycle usually requiring only one host. The common name comes from the gray-black, moldy teliospore stage produced within the flowering tissue of hosts, such as maize (corn).
soft rot	A type of wood decay in which lignin is only partially depolymerized and its C is not consumed. Hyphae secrete cellulase leading to microcavities in the wood. Carried out primarily by members of the *Ascomycota*.
sporocarp	A conglomeration of fungal cells in which meiosis occurs and from which spores are released, e.g., a mushroom. Analogous to a plant flower.
strand = cord	An aggregation of hyphae into a larger-diameter, tube-like transport structure. Similar to rhizomorphs but with a diffuse mycelial front. Most common in *Basidiomycota*.
teleomorph	The sexual phase of the life cycle of a fungus. Can also refer to the Latin name for a fungal species based on a description of the sexual stage, e.g., teleomorph = *Talaromyces spiculisporus*, anamorph = *Penicillium lehmanii*.
thermophile	Heat-loving organisms that can grow at 45–60°C.
truffle	Species of fungi that produce macroscopic sexual sporocarps belowground. Divided into the true truffles (e.g., *Tuber*) that belong to the *Ascomycota* and the false truffles that belong to other phyla, especially the *Basidiomycota*.

—cont'd

Term	Definition
white rot	A type of wood decay in which lignin is completely broken down and its C is consumed. This type of decay leaves behind a stringy/powdery white residue primarily composed of crystalline cellulose. Carried out by a limited number of species in the *Basidiomycota*.
yeast	Any fungus with cells that separate after budding (e.g., *Saccharomyces*) or splitting (e.g., *Schizosaccharomyces*); note that the term does not describe a phylogenetically united group of fungi.
zygospore	A single-celled diploid resting spore resulting from fusion of compatible nuclei of opposite mating types that is produced within a Zygosporangium; characteristic of fungi that were once placed in the *Zygomycota*. Meiosis occurs when the zygospore is triggered to germinate.

See also:
www.mushroomexpert.com/glossary.html
botit.botany.wisc.edu/toms_fungi/

Supplemental references

Taylor, J.W., 2011. One fungus = one name: DNA and fungal nomenclature 20 years after PCR. IMA Fungus 2, 113. https://doi.org/10.5598/imafungus.2011.02.02.01.

Woese, C.R., Fox, G.E., 1977. Phylogenetic structure of the prokaryotic domain: the primary kingdoms. Proc. Natl. Acad. Sci. 74, 5088–5090.

Chapter 5

Soil fauna: occurrence, biodiversity, and roles in ecosystem function

D.C. Coleman*, S. Geisen[†], and D.H. Wall[‡]

*Odum School of Ecology, University of Georgia, Athens, GA, USA; [†]Department of Plant Science, Laboratory of Nematology, Wageningen University, Wageningen, the Netherlands; [‡]Department of Biology and School of Global Environmental Sustainability, Colorado State University, Fort Collins, CO, USA

Chapter outline

5.1 Introduction

Animals, a key group of heterotrophs in soils, shape bacterial and fungal biomass, activity, diversity, and community composition (Hättenschwiler et al., 2005). Many soil animals, which are mostly invertebrates and often referred to as "fauna," regulate nutrient cycling at ecosystem and global scales by feeding directly on plant materials and other organic substrates. The fragmentation or comminution of these materials enhances their decomposition. Comminution increases the surface area of plant structural materials and exposes cytoplasm, thereby enabling greater access by microbes. Decomposition is further

Soil Microbiology, Ecology, and Biochemistry. https://doi.org/10.1016/B978-0-12-822941-5.00005-3

accelerated, as the feeding activity of soil animals often results in the translocation of nutrients such as nitrogen (N) and phosphorus (P) from the soil to the substrate in the form of fecal material. Invertebrates consume bacteria and fungi, thereby directly cycling C and N and disseminating microbes from one organic source to another, as many microbes adhere to invertebrate exoskeletons and cuticles and survive passage through their digestive tracts (Coleman et al., 2012; Meier and Honegger, 2002).

Soil animals exist in food webs containing several trophic levels (Moore and de Ruiter, 2012). Some are herbivores, since they feed directly on roots of living plants, but most subsist on dead plant matter (saprophytes), the living microorganisms associated with it, or a combination of the two. Still others are carnivores or parasites. The structure of these food webs is complex, with many "missing links" only slowly being described (Geisen et al., 2019a; Scharroba et al., 2012; Scheu and Setälä, 2002; Walter et al., 1991).

5.2 Overview of faunal biodiversity in soils

The nature and extent of biodiversity in soils is impressively large (Coleman and Whitman, 2005; Fierer and Lennon, 2011; Jeffery et al., 2010; Whitman et al., 1998). Current estimates suggest that for most soil organism groups, a maximum of 10% of the existent species diversity is described (André et al., 2002; Geisen et al., 2019a). A potentially high degree of functional specialization may underlie this enormous diversity of soil life (Heemsbergen et al., 2004, Wurst and van der Putten, 2007). Species with similar biologies and morphologies (Coleman et al., 1993, 2004, 2018; Hendrix et al., 1986; Hunt et al., 1987) or ecological functions (Geisen et al. 2019a) are often grouped together for purposes of integration at the system level.

Soil fauna also may be characterized by the portion of their life cycle that is spent in the soil. Transient species, exemplified by the ladybird beetle, hibernate in the soil but otherwise live in the plant stratum. Gnats (Diptera) are temporary residents of the soil given that the adult stages live aboveground. Their eggs are laid in the soil and their larvae feed on decomposing organic debris. In some situations dipteran larvae are important scavengers. Cutworms are temporary soil residents, whose larvae feed on seedlings by night. Some nematodes that parasitize insects and beetles spend all or part of their life cycle in the soil. Periodic residents spend their life histories belowground, with adults, such as velvet mites, emerging to reproduce. Soil food webs are linked to aboveground systems, making trophic analyses much more complicated than in one subsystem alone (van der Putten et al., 2013; Wardle et al., 2004). Even permanent residents of the soil may be adapted to life at various depths within the soil profile.

A generalized classification by length and width illustrates a commonly used device for separating the soil fauna into size classes: microfauna, mesofauna, macrofauna, and megafauna. This classification encompasses the range from smallest to largest, i.e., from about 1 to 2 μm for some protists to 2 m for giant Australian earthworms (Moreira et al., 2009). Both faunal body lengths and widths are commonly related to their microhabitats (Fig. 5.1). Protists and the microfauna (rotifers, tardigrades, nematodes) inhabit water films. The mesofauna inhabit existing air-filled pore spaces and are largely restricted to existing spaces. Macrofauna have the ability to create their own spaces through their burrowing activities and, like the megafauna, can significantly influence gross soil structure (Lavelle and Spain, 2001; van Vliet and Hendrix, 2003).

The body size range among the soil fauna (grazers and predators) affects soil processes at a range of spatial scales. Three levels of participation have been suggested (Lavelle et al., 1995; Wardle, 2002): (1) "Ecosystem engineers," such as earthworms, termites, or ants, alter the physical structure of the soil itself, influencing rates of nutrient and energy flow (Jones et al., 1994); (2) "Litter transformers," the microarthropods, fragment decomposing litter and improve its availability to microbes; and (3) "Grazers and predators" directly consume bacteria, fungi (i.e., protist and nematode grazers), and microfauna (e.g., nematode and mite predators) and thus contribute significantly to the soil food web. These three levels operate on different size, spatial, and time scales (Fig. 5.2; Wardle, 2002).

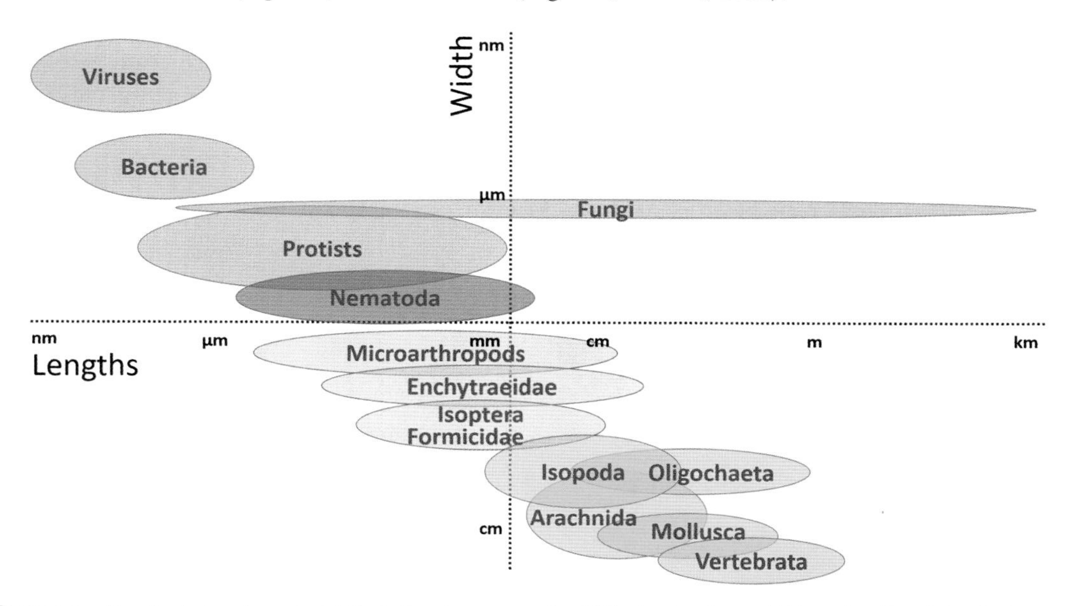

FIGURE 5.1 Size (body lengths and width) of all major groups of soil life showing the multidimensionality of size for large-scale categorization of soil organisms. Colours denote current large-scale classifications; Gray: microorganisms; Green: microfauna; Blue: mesofauna; Orange: macrofauna. *(Redrawn from Decaëns, 2010 and Swift et al., 1979.)*

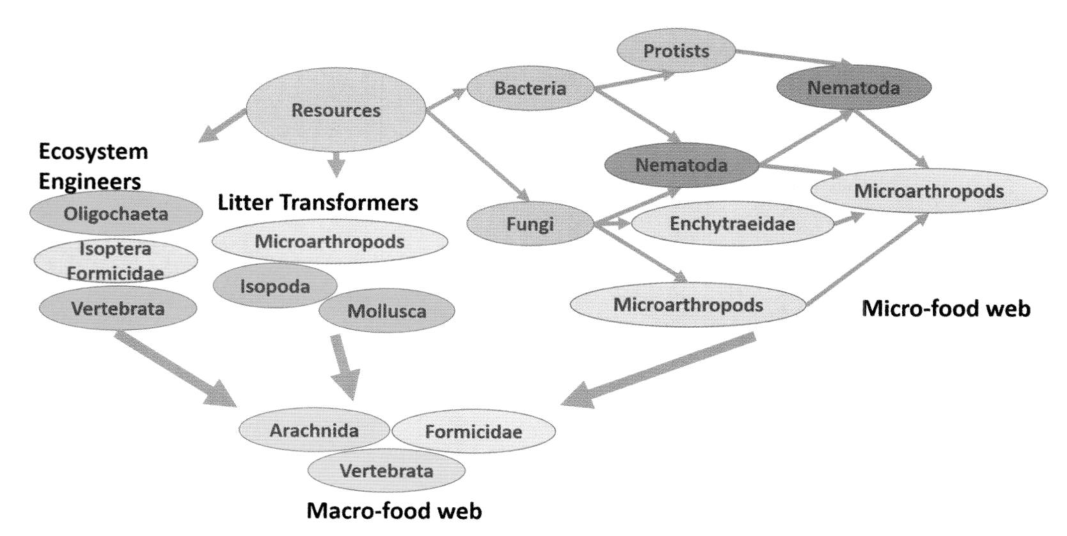

FIGURE 5.2 Organization of the soil food web into three categories – ecosystem engineers, litter transformers, and micro-food webs. *(Redrawn from Lavelle et al., 1995 and Wardle, 2002.)*

5.3 Protists: microbes among "fauna"

Protists can easily be called the problematic ones. They are not animals (Fig. 5.3) — hence no longer called protozoa (Adl et al., 2012) — but microbes (Caron et al., 2008). Similar to other microbes, a common feature of soil protists is that they can form highly resistant stages, called cysts, to survive adverse conditions (Geisen et al., 2018a). These cysts can remain for extended periods in the environment, as recently shown for those that were revived from millennial-old permafrost samples (Shmakova et al., 2016). Major taxonomic and phylogenetic revisions mainly based on molecular sequence information have resulted in major reshufflings of the eukaryotic tree of life (Adl et al., 2019) (Fig. 5.3). These insights have shown that many photosynthetic

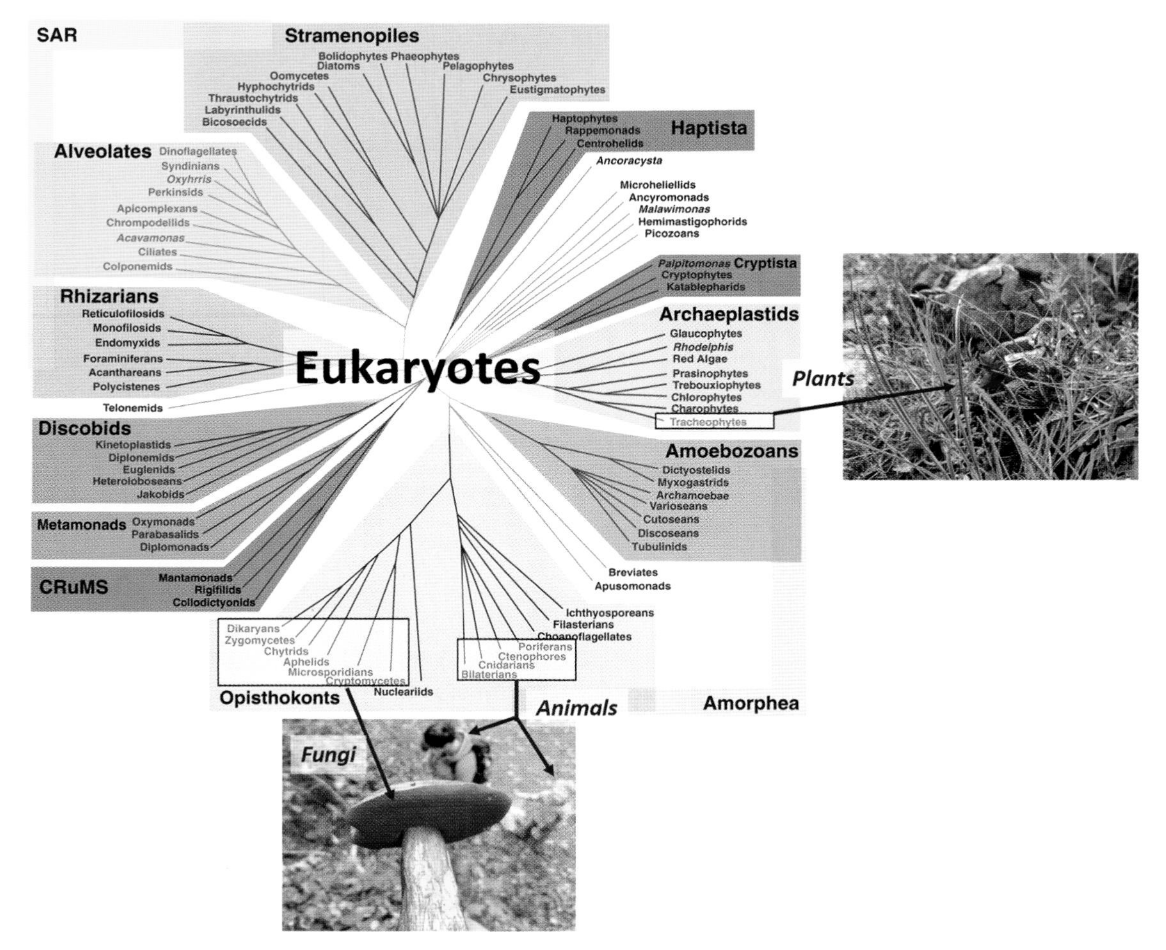

FIGURE 5.3 Overview of the phylogenetic diversity of eukaryotes. Multicellular organisms commonly considered to host most eukaryotic diversity, including animals, fungi, and plants, are shown in boxes. However, the remaining sections of the tree are made up of single-celled protists. *(Modified from Keeling and Burki, 2019.)*

organisms ("algae" *sensu lato*) are phylogenetically intermingled with heterotrophic organisms ("protozoa" *sensu lato*) and that mixotrophs can live photo- and heterotrophically (Geisen et al., 2018a). Commonly used morphologically distinguishable groups, amoebae and flagellates, were shown to be paraphyletic (Geisen et al., 2018a) (Fig. 5.3). Today, morphotype-based investigations are rarely performed, and although they may still provide ecologically relevant information on protistan community changes (Geisen et al., 2014), they have mostly been replaced by phylogenetically informed analyses based on sequencing data (Geisen and Bonkowski, 2018). This phylogenetic view shows that most cultivated soil protists are placed in the eukaryotic supergroups Amoebozoa (most amoebae), Rhizaria (most flagellates and amoeba-flagellates), Discoba (amoebae and flagellates), and Alveolata (ciliates) (Adl et al., 2019; Geisen et al., 2018a). Sequence-based analyses have also revealed several other groups as being common soil protists, such as parasites in the Apicomplexa (Bates et al., 2013; Geisen et al., 2015; Oliverio et al., 2020). Apicomplexans represent up to 50% of all protists in some tropical forests, leading Mahé et al. (2017) to speculate that protists might be key controllers of animal biodiversity. Phototrophic protists that are major carbon fixers are also common in soils (Oliverio et al., 2020).

5.3.1 Methods to study soil protists

Different techniques can be applied to study soil protists (Geisen and Bonkowski, 2018), including traditional morphology-based investigations and molecular techniques. Morphological investigations suffer from profound biases as the most abundant soil protist groups cannot directly be observed due to their small size and transparent body shapes, and their primary use is to determine and quantify ciliates and testate amoebae (Foissner, 1999). Cultivation-based methods, such as the most probable number technique used in many studies, focus only on well-cultivable taxa (Berthold and Palzenberger, 1995; Darbyshire et al., 1974). The last decades have brought in a wealth of novel molecular tools to study protists, with high-throughput sequencing as the current method of choice (Geisen and Bonkowski, 2018). Yet, further study is needed to standardize and optimize the information gained from molecular protistan surveys, as issues including primer biases do not allow a cumulative investigation of soil protistan diversity (Geisen et al., 2019a). There are currently no molecular approaches available that provide biomass or abundance data, essential information necessary to answer many ecological questions beyond diversity estimates (Fig. 5.4)

5.3.2 Ecology of protists

Soil protists are diverse in many ways; their sizes can range from a few micrometers to many centimeters as in the case of network-forming myxomycetes as the largest cells on earth (Geisen et al., 2017). Protist sizes can reflect adaptations to specific environments and indicate specific habitat niches. Smaller taxa inhabit all soil pore sizes and can penetrate into tiny pores in the soil matrix, while larger taxa, including testate amoebae and ciliates, are found in larger soil pores in the upper soil or litter layers. Smaller protists are more abundant than larger ones, with small flagellates (but also amoebae) dominating soil protistan communities (Finlay et al., 2000). Many factors, other than soil pore sizes, determine protist distribution and community composition. Soil moisture appears to be the main determinant of protistan communities at small and large spatial scales (Bates et al., 2013; Geisen et al., 2014; Oliverio et al., 2020). Other factors such as pH and organic carbon content also contribute to the structure of soil protistan communities (Dupont et al., 2016; Oliverio et al., 2020). The use of fertilization for soil management has directly

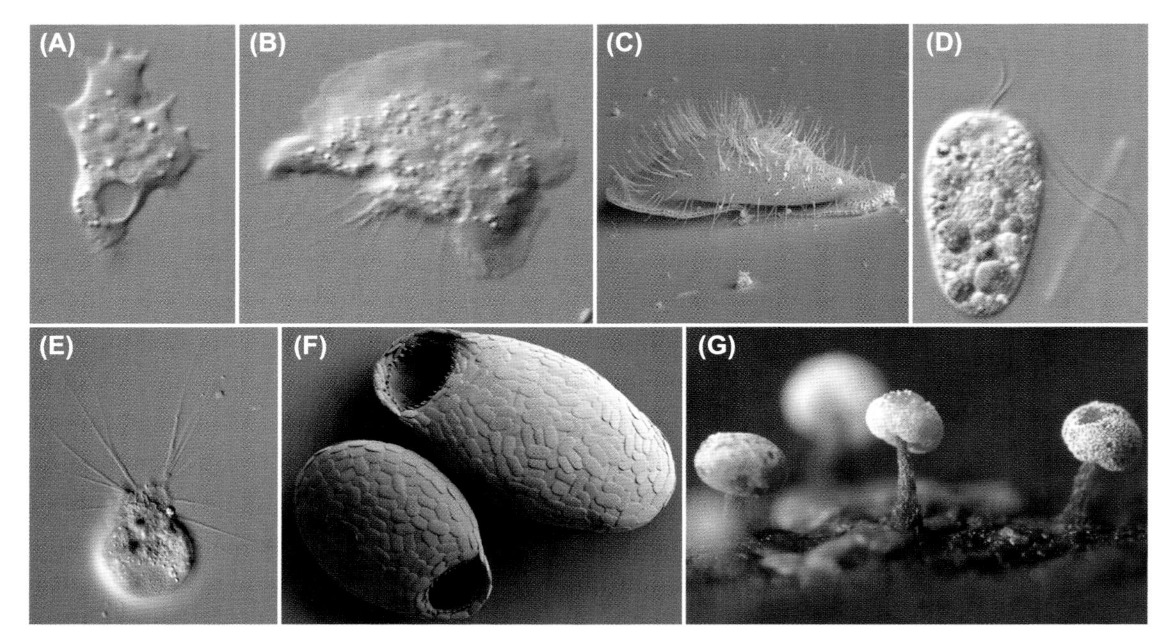

FIGURE 5.4 Morphology of common soil protists. (a–c) amoebae; (a): *Acanthamoeba* sp., (b): *Flamella* sp.; (c): *Cochliopodium vestitum*); (d): the flagellate *Viridiraptor invadens*; (e–f): testate amoebae (e): *Lecythium hyalinum*; (f): *Corythion* sp; (g): a slime mold that builds fungal-like multicellular fruiting structures from aggregated single-celled amoebae. *(Pictures with permission from Stefan Geisen (a–b), Eckhard Völcker, Steffen Clauß (c–f), and Maries Elemans (g).)*

affected soil protists (Lentendu et al., 2014), resulting in significant changes over those found in other groups of soil microorganisms (Zhao et al., 2019). Plants shape protistan communities (Hünninghaus et al., 2019; Sapp et al., 2018), with protists being the soil biodiversity group that increases the most with an increase in plant diversity (Scherber et al., 2010).

5.3.3 Ecological relevance of protists

Protists are not only influenced by plants, but in turn, affect plant performance. In fact, the ecology of protists is immense, such as shown by their diversity of feeding modes. While the majority of protists feed on bacteria, others feed on fungi, and are omnivores, phototrophs, animal parasites, or plant pathogens (Dupont et al., 2016; Geisen, 2016; Oliverio et al., 2020). Through these trophic interactions, protists serve as an immensely important intermediate level for nutrient transfer to higher trophic levels (de Ruiter et al., 1995). Through trophic feeding interactions, protists liberate nutrients, particularly N, which acts as a fertilizer to boost plant growth (Bonkowski, 2004). However, protists select for their preferred prey items (Geisen et al., 2016). However, protists are not *per se* plant-beneficial as several plant pathogenic protistan groups exist, among them plasmodiophorids and oomycetes (Schwelm et al., 2018). These protists have traditionally been considered fungi due to the presence of convergent features, such as the formation of hyphae and presence of easily dispersed fruiting bodies (Beakes et al., 2012). Due to different physiological and biochemical features, including nonchitinous cell walls as well as those based on phylogenetic analyses, plasmodiophorids and oomycetes are now unambiguously placed as nonfungal organisms among protists (Adl et al., 2019). Protists affect plant growth depending on the presence of keystone species (pathogens) and the overall community composition. For example, the community composition of protists may be a reliable determinant of plant performance at the seedling establishment stage, indicating that certain protistan communities can guarantee plant health while others can predict plant disease during later stages of plant growth (Xiong et al., 2020).

Protists generally increase microbial activity and therefore may stimulate other microbial-driven ecological processes in soils (Geisen et al., 2018a). For instance, decomposition, and therefore C release to the atmosphere, is potentially increased through protist predation of microbes (Geisen et al., 2020). Heterotrophic protists may even contribute to the increase of the stable soil C fraction through predation on bacteria and fungi, as microbial-derived C represents the major fraction of recalcitrant C in soils (Liang et al., 2019). Specific protist species, as well as the composition of protistan communities, might also serve as responsive indicators for environmental change or soil quality, as they often respond more dramatically to changes compared to other groups of soil microorganisms including bacteria and fungi (Foissner, 1999; Zhao et al., 2019).

5.4 Microfauna

Soils, litter layers, and moss layers or biocrusts, especially those with high periodic water saturation, are environments in which soil microfauna inhabit thin water films. Nematodes, which are the most abundant microfauna group in soils, are covered in more detail below. Rotifers and tardigrades can reach abundances of above 100×10^4 m^{-2} to a depth of 10 cm in forest topsoils (Ito and Abe, 2001; Sohlenius, 1979). Both groups contain members of diverse functional feeding groups that have different mouth and ingestion apparatuses (Wallace, 2002). Tardigrades feed on bacteria, yeasts, and protists but also on rotifers and plants (Schill et al., 2011). Tardigrades can be major predators of nematodes (Hyvönen and Persson, 1996), including root-feeding nematodes, that can suppress nematode-induced root damage (Sánchez-Moreno et al., 2008). Tardigrades are preyed on by microarthropods (Hyvönen and Persson, 1996). Tardigrades, rotifers, and nematodes have a matchless ability to survive the most severe environmental conditions — for example, desiccation in soils, as they have the capacity to enter into an anhydrobiotic stage (Iglesias Briones et al., 1997; Kutikova, 2003). This adaptation allows microfauna to be present even in the driest soils that only sporadically receive water input. The ecology of rotifers (see below) and tardigrades in soils is still poorly understood (Wallace, 2002).

5.4.1 Rotifera

These small (0.05—3 mm long) fauna are typically only found in soils and other environments containing an abundance of water films (Segers, 2007). Rotifers exist in leaf litter, mosses, lichens, and even more extreme environments, such as wet soils of the Antarctic Dry Valleys (Segers, 2007; Treonis et al., 1999).

Rotifers are characterized by being transparent and having a three-part body: (1) an anterior ciliary structure or a ‘crown’ with cilia, (2) a body wall that appears to be pseudo-segmented (lorica), and (3) a feeding apparatus with strong muscles and jaws. More than 90% of soil rotifers are in the order Bdelloidea, or worm-like rotifers. Life history features include the construction of a shell, which may have particles of debris and/or fecal material adhering to it. Some rotifers will use the empty shells of Testacea, the thecate amoebae, to survive. The parthenogenic Bdelloidea are vortex feeders, creating currents of water that bring food particles, such as unicellular algae or bacteria, to the mouth for ingestion. The importance of these organisms is largely unknown, except as having roles in biogeochemical cycling. They are extremely diverse globally (Robeson et al., 2009, 2011), reaching abundances exceeding 10^5 m^{-2} in moist, organic soils (Wallwork, 1970). Rotifers are extracted from soil samples and enumerated using methods similar to those used for nematodes (see the following section), but morphological identification has been stymied because they can only be identified while alive and active (Segers, 2007).

5.4.2 Nematoda

The phylum Nematoda contains aquatic nematodes or roundworms, which are among the most numerous and diverse of the multicellular organisms found in any ecosystem. It has been estimated that four of every five animals on Earth are nematodes (Bongers and Ferris, 1999), with the total number of nematodes in soils estimated to be $4.4 \pm 0.64 \times 10^{20}$ (van den Hoogen et al., 2019). Nematodes are important belowground parasites and pathogens of plants with effects on net primary productivity (NPP) and organic matter decomposition (Wall et al., 2012). Nematodes have a very early evolutionary origin among the Eukarya, likely in the Cambrian explosion about 550 million years ago (Blaxter et al., 1998; dos Reis et al., 2015).

The overall body shape is cylindrical, tapering at the ends. Nematode body plans are characterized by a "tube within a tube" (alimentary tract/the body wall). They have a complete digestive system or an alimentary tract, consisting of a stoma or stylet, pharynx (or esophagus), and intestine and rectum, which opens externally at the anus. The reproductive structures are complex, and sexes are generally dimorphic. Many species are parthenogenetic, producing only females. Nematodes are highly diverse, with estimated species numbers at a million (Geisen et al., 2019b). Due to their enormous abundances, high species richness, and functional diversity (Fig. 5.5), nematodes are used as indicators of soil quality (Bongers, 1990; Domene et al., 2011; Ferris and Bongers, 2006; Ferris et al., 2001; Vervoort et al., 2012; Wilson and Kakouli-Duarte, 2009).

5.4.2.1 Methods to study soil nematodes

Nematodes are extracted from soils using a variety of techniques, either active or passive, each having distinct advantages and disadvantages (Geisen et al., 2019b). The most used approaches take advantage of their aquatic nature, along with gravity, and require ideally hundreds of grams of soil submerged in water to provide reliable information on nematode communities. The Baermann funnel method has many modifications and consists of soil being placed on a tissue or filter paper on a screen (sieve or window screen) at the large mouth of a funnel that is hanging vertically. The funnel is filled with water and closed off with a rubber tube clamped at the funnel base. The nematodes actively move through the soil and down through the water to the funnel base, where they are collected for examination (Hooper, 1970). These methods provide unbiased information on the activity of organisms, as they are based on the animal's movement.

Following extraction, nematodes are identified morphologically under an inverted or dissection microscope. While enumeration and even functional placement can be done after a short training, taxonomic identification to family or genus level remains an expert task and requires a substantial commitment of many hours per sample. Alternatively, qPCR approaches can be conducted to estimate abundances of targeted taxa with specific primers (Vervoort et al., 2012). Use of high-throughput

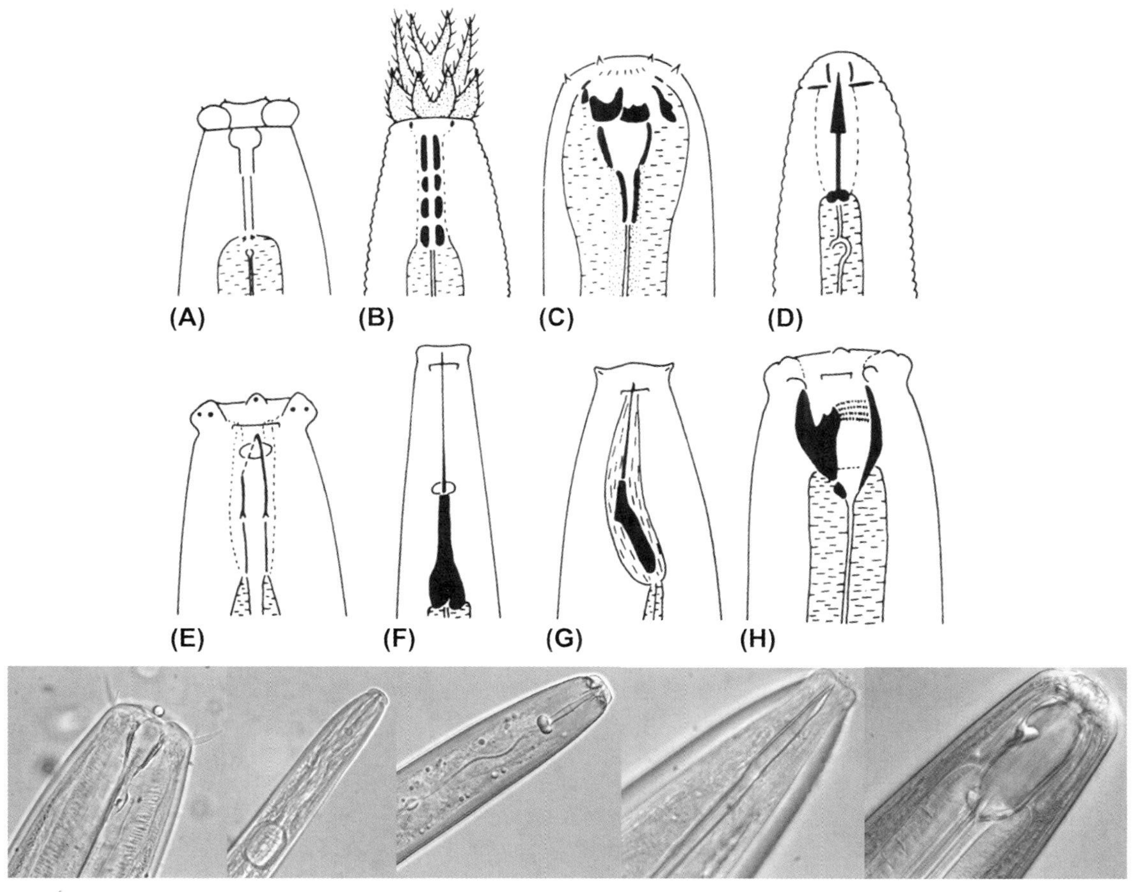

FIGURE 5.5 Head structures of a range of soil nematodes. (a) *Rhabditis* (bacterial feeding); (b) *Acrobeles* (bacterial feeding); (c) *Diplogaster* (bacterial feeding, predator); (d) tylenchid (plant feeding, fungal feeding, predator); (e) *Dorylaimus* (feeding undetermined, omnivore); (f) *Xiphinema* (plant feeding); (g) *Trichodorus* (plant feeding); (h) *Mononchus* (predator). Lower row shows real images of a bacterial-feeding *Epitobrilus steineri*, a fungal-feeding *Aphelenchoides* sp., a plant-feeding *Pratylenchus crenatus*, an omnivore *Dorylaimus stagnalis*, and a predator *Mononchus* sp. *(With permission from Yeates and Coleman, 1982.)*

sequencing for high-resolution taxonomic profiling of nematode communities is increasingly done (Geisen et al., 2018b). There are still some biases with this approach, as sequence abundances cannot be used directly to infer species abundances (Griffiths et al., 2018); however, relative sequence abundance closely reflects relative species biomasses (Schenk et al., 2019). Nematode functioning can be assessed based on a priori knowledge of feeding preferences, growth rates, etc. More exact information on trophic positions can be obtained using fatty acid markers or profiles (Kühn et al., 2018). Stable isotope-based techniques are used to predict feeding patterns, especially when combined with high-resolution taxonomic identification tools such as high-throughput sequencing techniques (Crotty et al., 2012; Geisen et al., 2019b). The method of choice to determine nematode communities depends on the availability of experts, time, and techniques. Ideally, a combination of methods is employed to obtain information on abundances, community structure, and feeding preferences of soil nematode communities.

5.4.2.2 Nematodes in soil food webs

Soil nematode abundance generally decreases with increasing soil depth and distance from plants, as many soil nematodes are largely concentrated in the litter layer or rhizosphere, or where organic matter accumulates. Nematodes also move vertically in soils toward plant roots, but the distance of movement is dependent on nematode species, in addition to soil type, temperature, and moisture. In deserts nematodes are associated with plant roots to depths of 15 m, as are mites and other biota (Freckman and Virginia, 1989); the nematode *Halicephalobus mephisto* was recovered from soils 3 km deep (Borgonie et al., 2011).

Nematodes have several dispersal methods and are transported with plants, soils, and other invertebrates, and by wind and water; some of these methods involve survival mechanisms. Wallace (1959) noted that local movement of nematodes was optimum when soil pores were half drained of free water. Elliott et al. (1980) noted that the limiting factor for nematode survival often hinges on the availability and size of soil pore necks, which enable passage between soil pores. Yeates et al. (2002) measured the movements, growth, and survival of three genera of bacterial-feeding soil nematodes in undisturbed soil cores maintained on soil pressure plates. Interestingly, nematodes showed significant reproduction in water-filled pores above 1 μm in diameter.

Nematodes feed on a wide range of foods, with general trophic groupings including bacterial feeders, fungal feeders, plant feeders, predators, and omnivores. In most cases these trophic groups can easily be determined using morphological features, particularly anterior (stomal or mouth) structures (Fig. 5.5) (Moore and de Ruiter, 2012; Yeates and Coleman, 1982; Yeates et al., 1993). Plant-feeding nematodes have a hollow stylet that pierces cell walls of higher plants. Some species are facultative plant feeders, feeding occasionally on plant roots or root hairs. Others, recognized for their damage to agricultural crops and forest plantations that include root-knot and cyst nematodes, are obligate parasites of plants and feed internally or externally on plant roots. The effect nematodes have on plants is generally species specific and host plant specific and can include alterations in root architecture, water transport, plant metabolism, or all of these.

5.4.2.3 Ecology of soil nematodes

We are rapidly increasing our understanding of the main determinants of the biogeography of soil nematodes. Griffiths and Caul (1993) found that nematodes migrated to "hot spots" of decomposing grass residues where there were considerable amounts of labile substrates and microbial food sources. At the global level and at large spatial scales, soil texture, sand content, and soil organic matter (SOM) levels are key determinants of nematode abundances and community composition (Song et al., 2017; van den Hoogen et al., 2019). The presence of plants obviously is the key determinant for the abundance of plant parasites (van den Hoogen et al., 2019), while even the identity of plant species affects abundances and nematode community composition (Wilschut et al., 2019). Like most soil organisms, nematodes are enriched in topsoils and the rhizosphere compared to root-free bulk soil (Ingham et al., 1985). Nematodes can occur at a depth of 15 m in deserts (Freckman and Virginia, 1989) and can even be found in soils of several kilometers' depth (Borgonie et al., 2011).

Evidence suggests nematodes have similar patterns to larger aboveground animals, with their diversity peaking in the tropics (Li et al., 2020; Porazinska et al., 2012). Other studies reported highest nematode diversity in temperate zones (Song et al., 2017), or with nematode diversity being similar in tropical and polar regions (Kerfahi et al., 2016).

Nematode communities are useful bioindicators of soil quality (Bongers, 1990; Ferris and Bongers, 2006; Ferris et al., 2001; Vervoort et al., 2012). Intense management increases the ratio of bacteria to fungi leading to higher abundances of bacterivorous nematodes. Disturbance, such as plowing, especially decreases larger nematode species, such as predators, that reproduce more slowly than their smaller bacterivore counterparts, leading to an even stronger reduction in the ratio between bacterivorous taxa to omnivores and predators. As such, it was shown that biomass of trophic groups declines and that the diversity of nematodes is reduced by intensive agricultural practices (Tsiafouli et al., 2015). However, this seems to depend on the abiotic surroundings, as agricultural practices can even stabilize nematode communities and increase diversity in harsh environments (Li et al., 2020). Some plant parasitic nematodes are endemic to a region or country and, when dispersed with soil elsewhere, may result in crop damage and quarantine regulations to prohibit the species. For example, the potato cyst nematode, *Globodera rostochiensis*, originated in Central America and was transported with potatoes globally. Quarantine regulations, *Globodera*-resistant potato varieties, and noncrop rotations have successfully limited the nematode in the United States to areas in New York. The causative agent of the pine wilt disease, *Bursaphelenchus xylophilus*, a native of the United States that spread in the 1990s with wood shipped to Europe, is a quarantined pest in many countries.

5.5 Mesofauna

5.5.1 Microarthropods

Many microarthropods feed on fungi and nematodes, thereby linking microfauna and microbes with the mesofauna (Chamberlain et al., 2006). Microarthropods in turn are prey for macroarthropods, such as spiders, beetles, ants, and centipedes, thus bridging a connection of nutrients cycled to the macrofauna and to further identification of food webs in soils. Large numbers of microarthropods (mainly mites and collembolans) are found in most types of soils, and like nematodes, a square meter of forest floor may contain hundreds of thousands of individuals representing thousands of species. They have a significant impact on decomposition processes and are important reservoirs of biodiversity.

Mites and collembola are characterized as mesofauna and live in soil air-filled pore spaces. They are typically enumerated by extracting them from soil samples using drying rather than water to drive the organisms out of the soil. Variations of the Tullgren funnel, which uses heat to desiccate the sample and force the arthropods into a collection fluid, or flotation in solvents or saturated sugar solutions, are followed by filtration (Edwards, 1991).

Microarthropod densities vary seasonally within and between different ecosystems. Temperate forest floors with large organic matter content support high numbers of microarthropods (33−88,000 m^{-2}), with coniferous forests having in excess of 130,000 m^{-2}. Tropical forests, where the organic layer is thin, contain fewer microarthropods (Coleman et al., 2004, 2018; Seastedt, 1984). Tillage, fire, and pesticide applications typically reduce populations, but recovery may be rapid, with microarthropod groups responding differently. These effects of disturbance on abundance and diversity can be variable for many groups of soil animals, such as nematodes, collembola, and earthworms. In the spring forest leaf litter may develop large populations of collembolan "snow fleas" (*Hypogastrura nivicola* and related species).

Considerable progress has been made in determining the details of trophic (feeding) interactions in microarthropods, particularly for the oribatid mites. Stable isotope techniques have provided new information on the diet of mites over time — the ^{13}C and ^{15}N stable isotope signatures of mite gut contents and tissues enable the assignment of oribatid mites into feeding guilds (Pollierer et al., 2009; Schneider et al., 2004). An alternative approach uses a morphological analysis of mite chelicerae (chewing mouth parts). In cases of ambiguous trophic relationships the dietary preferences are then further resolved through isotope analyses (Perdomo et al., 2012). Chamberlain et al. (2006) used a combination of fatty acid gut analysis with compound-specific C isotope analysis to show that two collembolan species consumed nematodes. The study of the fate of DNA from two prevalent species of soil nematodes, eight species of oribatid mites, and one Mesostigmatid determined that these organisms acted as either predators or scavengers on soil nematodes (Heidemann et al., 2011). More recently, Crotty and Adl (2019), using ^{13}C and ^{15}N enrichment, established trophic cascades in microcosms containing an Ascomycetous fungus and single or multiple species of fungivorous microarthropods (the collembolan *Lepidocyrtus curvicollis*, an astigmatid *Tyrophagus putrescentiae*, and the oribatid *Oribatula tibialis*). A mesostigmatid predator *Hypoaspis aculeifer* provided an upper trophic level in this trophic cascade.

The cold hardiness of certain microarthropods, such as collembola in the genus *Tullbergia mediantarctica*, is impressive. Fox (2020) notes that this springtail has survived more than 30 onslaughts of ice sheets over the rocky slopes of the Transantarctic Mountains in far southern Antarctica. They have survived in temperature regimes occasionally colder than −40°C, with their genomes unchanged over 5 million years' time. Perhaps *Tullbergia* reproduce parthenogenetically, thus enabling them to persist successfully without the complications of mating in an extreme environment.

5.5.2 Enchytraeids

Species from 19 of 28 genera of enchytraeids, small unpigmented oligochaetes, are found in soil. The remainder occur primarily in marine and freshwater habitats (Jeffery et al., 2010, Schmelz and Collado, 2010a,b; van Vliet, 2000). The Enchytraeidae are thought to have arisen in cool temperate climates, where they are commonly found in moist soils rich in organic matter. Identification of enchytraeid species is difficult, but genera may be identified by observing internal structures through the transparent body wall of specimens mounted on slides. Molecular tools are slowly finding their way into enchytraeid species identification, revealing profound numbers of cryptic species. However, molecular barcode regions are not fully assigned, and diversity analyses currently remain based on morphological tools (Schmelz et al., 2017).

The Enchytraeidae are typically 10 to 20 mm in length and are anatomically similar to the earthworms, except for the miniaturization and rearrangement of features. They possess setae (with the exception of one genus) and a clitellum in segments XII and XIII, which contain both male and female pores. Sexual reproduction in enchytraeids is hermaphroditic and functions similar to that in earthworms. Cocoons may contain one or more eggs, and maturation of newly hatched individuals ranges from 65 to 120 days depending on species and environmental temperature (van Vliet, 2000). Enchytraeids also display asexual strategies of parthenogenesis and fragmentation, which enhance their probability of new habit colonization (Dósza-Farkas, 1996).

5.5.2.1 Methods to study enchytraeids

Enchytraeids are typically sampled in the field using cylindrical soil cores of 5 to 7.5 cm in diameter. Large numbers of replicates may be needed for a sufficient sampling due to the clustered distribution and spatial heterogeneity of enchytraeid populations (van Vliet, 2000). Extractions are often done with a wet-funnel technique, similar to the Baermann funnel extraction used for nematodes. In this case soil cores are submerged in water in the funnel and exposed for several hours to a heat and light source from above; enchytraeids move downward and are collected in the water below.

5.5.2.2 Distribution and abundance of enchytraeids

Enchytraeids are distributed globally from subarctic to tropical regions. Keys to the common genera are presented by Schmelz and Collado (2010a,b) and Schmelz et al. (2017). Enchytraeid densities range from 1000 to a high of 140,000 individuals m^{-2}, depending on ecosystem type and management practice (van Vliet et al., 1995). Vertical distributions of enchytraeids in soil are related to organic horizons. Up to 90% of populations may occur in the upper layers in forest and no-tillage agricultural soils (Davidson et al., 2002). Seasonal trends in enchytraeid population densities appear to be primarily associated with moisture and temperature regimes (van Vliet, 2000).

In more acidic soils with reduced earthworm abundances, enchytraeids have significant effects on SOM dynamics and on soil physical structure. Enchytraeids affect soil structure by producing fecal pellets, which, depending on the animal size distribution, may enhance aggregate stability in the 600 to 1000 μm aggregate size fraction. In organic horizons these pellets are composed mainly of fine, highly decomposed organic matter particles, but in mineral soils, organic matter and mineral particles may be mixed into fecal pellets with a loamy texture. Davidson et al. (2002) estimated that enchytraeid fecal pellets constituted nearly 30% of the volume of the surface horizon in a Scottish grassland soil. Encapsulation or occlusion of organic matter into these structures may reduce decomposition rates by protecting organic matter from microbial attack.

5.5.2.3 Food sources of enchytraeids

Enchytraeids ingest both mineral and organic particles, although typically of smaller size ranges than those of earthworms. Finely divided plant materials, often enriched with fungal hyphae and bacteria, are a principal portion of the diet of enchytraeids. Microbial tissues are probably the fraction most readily assimilated because enchytraeids lack the gut enzymes to digest more recalcitrant SOM (Jeffery et al., 2010; van Vliet, 2000). Didden (1990, 1993) suggested that enchytraeids feed predominantly on fungi, at least in arable soils, and classified a community as 80% microbivorous and 20% saprovorous. The mixed microbiota that occur on decaying organic matter, either litter or roots, are probably an important part of the diet of these creatures. The remaining portions of organic matter, after the processes of ingestion, digestion, and assimilation, are excreted and become part of the slow-turnover pool of SOM. Mycorrhizal hyphae have been found in the fecal pellets of enchytraeids from pine litter (Ponge, 1991). Enchytraeids probably consume and further process larger fecal pellets and castings of soil macrofauna, such as collembolans and earthworms (Rusek, 1985; Zachariae, 1964). Thus fecal contributions by soil-dwelling invertebrates provide feedback mechanisms affecting the abundance and diversity of other soil-dwelling animals.

5.6 Macrofauna

5.6.1 Macroarthropods

Larger insects, spiders, myriapods, and others are considered together under the appellation "macroarthropods." Typical body lengths range from about 10 mm to as much as 15 cm for centipedes (Shelley, 2002). The group includes a mixture of various arthropod classes, orders, and families. Like the microarthropods, the macroarthropods are defined more by the methods used to sample them than by measurements of body size. Large soil cores (10 cm diameter or greater) may be appropriate for euedaphic (dwelling within the soil) species. Arthropods can be recovered using flotation techniques (Edwards, 1991); however, hand sorting of soils and litter, though time consuming, yields better estimates of population size than flotation. Capture-mark-recapture methods have been used to estimate population sizes of selected macroarthropod species, but the method is problematic due to inadequate recoveries of organisms during resampling (Southwood, 1978). Pitfall traps have been widely used to sample litter- and surface-dwelling macroarthropods. This method collects arthropods that fall into cups, filled with preservative (e.g., antifreeze), and whose rims are flush with the soil surface. Absolute population estimates are difficult to obtain with pitfall traps, but the method yields comparative estimates when used with caution. Many of the macroarthropods are members of the group termed "cryptozoa," a group consisting of animals that dwell beneath stones or logs, under bark, or in cracks and crevices. Cryptozoans typically emerge at night to forage, and some are attracted to artificial lights. The cryptozoa fauna is poorly defined but remain useful for identifying a group of invertebrate species with similar patterns of habitat utilization. Due to the well-established expertise on morphological species identification, molecular tools are only slowly emerging to study cryptozoans (Deiner et al., 2017).

Termites and ants in particular are important movers of soil, depositing parts of lower strata on top of the litter layer (Fig. 5.6). Emerging nymphal stages of cicadas may be numerous enough to disturb soil structure. Larval stages of soil-dwelling scarabaeid beetles sometimes churn the soil in grasslands. These and other macroarthropods are part of the group that has been termed ecological engineers (Jones et al., 1994). Some macroarthropods participate in both above- and belowground parts of terrestrial ecosystems. Many macroarthropods are transient or temporary soil residents and thus form a connection between food chains in the "green world" of foliage and the "brown world" of the soil. Caterpillars descending to the soil to pupate or migrating armyworm caterpillars are prey to ground-dwelling spiders and beetles. Macroarthropods may have a major influence on the microarthropod portion of belowground food webs. Collembola, among other microarthropods, are important food items for spiders, thus providing a macro- to microfauna connection. Other macroarthropods, such as cicadas, emerging from soil may serve as prey for some vertebrate animals (Lloyd and Dybas, 1966), including endangered vertebrate species (small mammals) (Decaëns et al., 2006), thus providing a link to the larger megafauna. Among the macroarthropods, there are many litter-feeding species, such as millipedes, that are significant consumers of leaf, grass, and wood litter. The decomposition of vertebrate carrion is largely accomplished through the actions of soil-dwelling insects (Payne, 1965).

Litter-feeding millipedes harbor a gut microbiota that is distinctly different from that of the ingested organic substrate of leaf litter. Using denaturing gradient gel electrophoresis (DGGE), Knapp et al. (2009) found a stable, indigenous microbial community in the gut of the millipede *Cylindroiulus fulviceps* in abandoned alpine pastureland. Its gut microbiota was dominated by Gamma- and Delta-proteobacteria and resisted dietary changes even during a varied dietary intake.

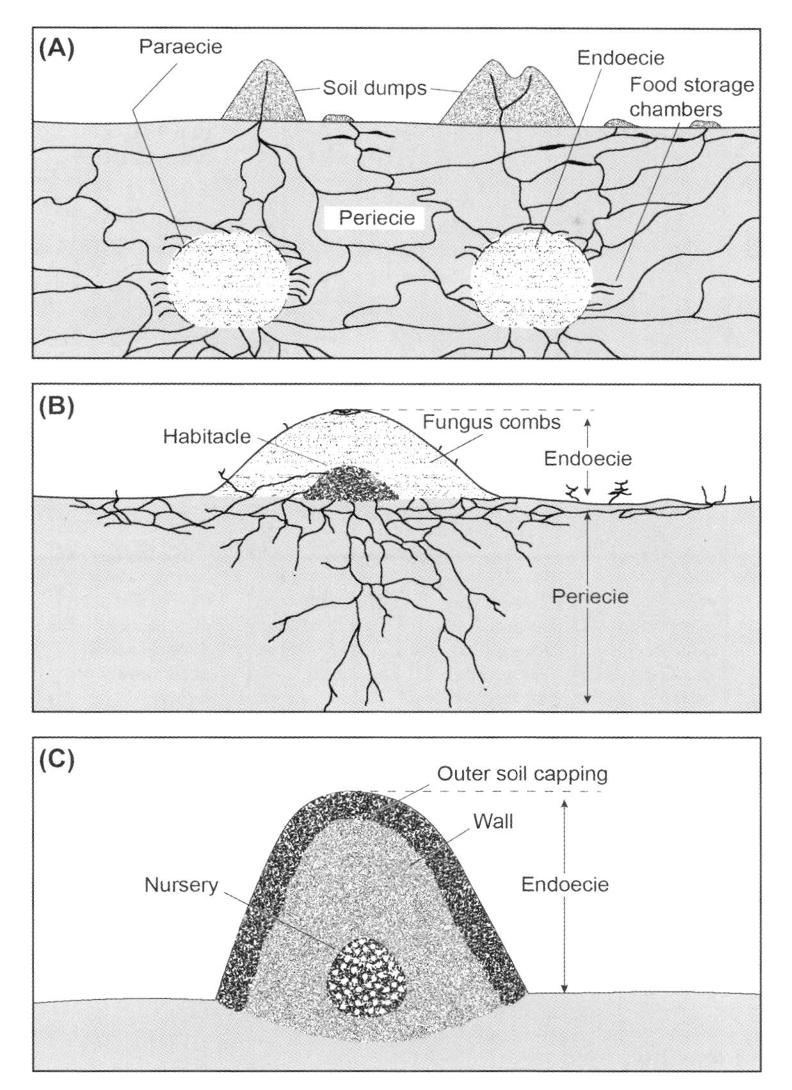

FIGURE 5.6 Termite mounds. Diagrammatic representation of different types of concentrated nest systems: (a) *Hodotermes mossambicus*, (b) *Macrotermes subhyalinus*, and (c) *Nasutitermes exitiosus*. *(With permission from Lee and Wood, 1971.)*

A more general aspect of arthropod feeding biology is addressed by the study of enzymatic roles in the digestion of plant cell walls, in particular, cellulose and lignin. Considered over evolutionary time, insects were more active in the "brown world" of decomposition of plant tissues on or in the soil, until the development of herbivory, possibly coinciding with the origin of the Angiosperms c. 160 million years ago. Some of the earliest orders of insects to arise, such as the Isoptera (Termites), have approached the problem of development of cellulases by developing symbiotic associations with protists in their hindguts

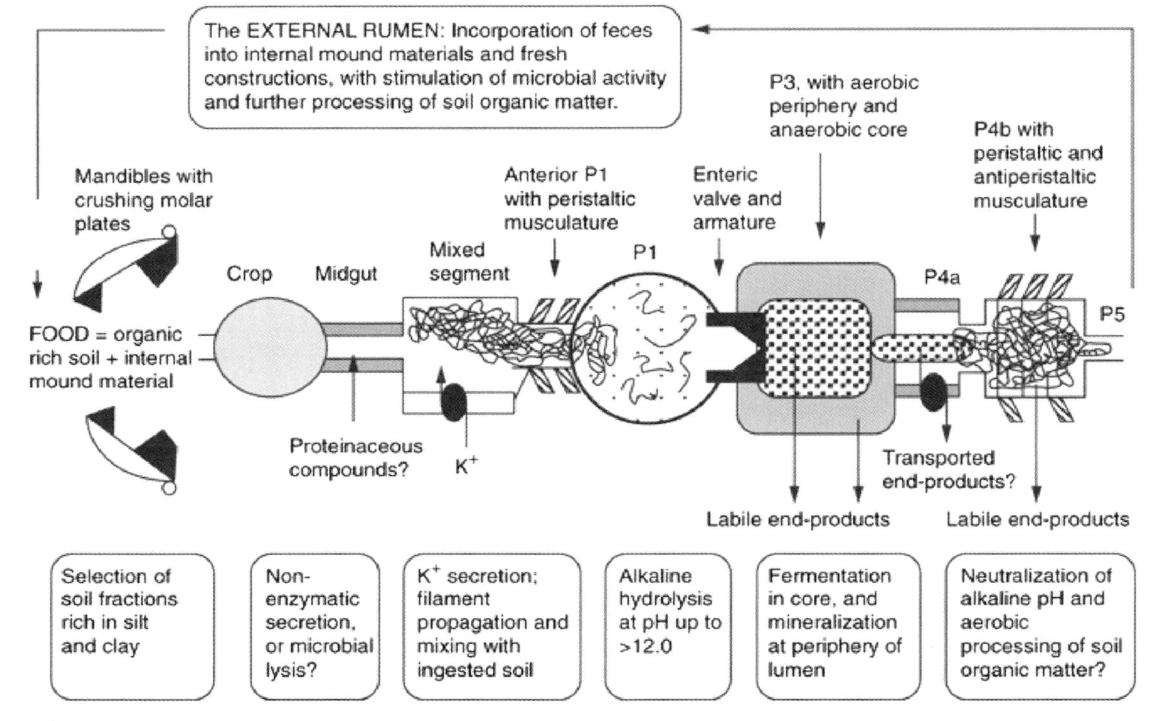

FIGURE 5.7 Hypothesis of gut organization and sequential processing in soil-feeding Cubitermes-clade termites. The model emphasizes the role of filamentous prokaryotes, the extremely high pH reached in the P1, and the existence of both aerobic and anaerobic zones within the hindgut. *(From Brauman et al. (2000).)*

(the more primitive termite families) or their own endogenous cellulases (Termitidae) (Calderón-Cortés et al., 2012) (Fig. 5.7). Significantly, the combined synthesis and activity of cellobiohydrolases and xylanases effectively degrade lignocellulose in the termite hindgut (Gilbert, 2010). The development of endogenous cellulases and in some cases laccases occurred in numerous insect families, including several in the order Coleoptera, and a few Lepidoptera and Diptera (Calderón-Cortés et al., 2012).

5.6.2 Oligochaeta (earthworms)

Earthworms are the most familiar and, with respect to soil processes, often the most important of the soil fauna. The importance of earthworms arises from their influence on soil structure (e.g., aggregate or crumb formation, soil pore formation) and the breakdown of organic matter (e.g., fragmentation, burial, and mixing of plant residues) (Carpenter et al., 2008).

Invasive earthworms are now colonizing less disturbed and even pristine forested habitats of the glaciated portions of North America (Cameron et al., 2007; Frelich et al., 2006; Hale et al., 2005). The history of road construction in northern boreal forests is a strong predictor of the extent of earthworm invasion, and the distance of a particular forested habitat from active agriculture is directly related to earthworm invasion (Cameron and Bayne, 2009). The modern era of earthworm research began with

Darwin's (1881) book, "The Formation of Vegetable Mold through the Actions of Worms, with Observations of Their Habits," which called attention to the beneficial effects of earthworms. Since then, a vast literature has established the importance of earthworms as biological agents in soil formation, organic litter decomposition, and redistribution of organic matter in the soil (Edwards, 1998; Hendrix, 1995; Lavelle and Spain, 2001).

Earthworms are classified within the phylum Annelida, class Oligochaeta. Species within the families Lumbricidae and Megascolecidae are ecologically the most important in North America, Europe, Australia, and Asia. Some of these species have been introduced worldwide by human activities and now dominate the earthworm fauna in many temperate areas. Any given locality may be inhabited by only native species, all exotic species, a combination of native and exotic species, or no earthworms at all. Relative abundance and species composition of local fauna depend greatly on soil, climate, vegetation, topography, land use history, and especially past invasions by exotic species.

5.6.2.1 Earthworm distribution and abundance

Earthworms are native to habitats where soil water and temperature are favorable for at least part of the year, while in North American forests they have now been introduced and become invasive. They are most abundant in forests and grasslands of temperate and tropical regions and least so in arid and frigid environments, such as deserts, tundra, or polar regions. Earthworm densities in a variety of habitats worldwide range from 10 to 2000 individuals m^{-2}, the highest values occurring in fertilized pastures and the lowest in acid or arid soils (coniferous or sclerophyll forests). Typical densities from temperate deciduous or tropical forests and certain arable systems range from 100 to over 400 individuals m^{-2}, representing a range of from 4 to 16 g dry mass m^{-2}. Earthworms are also highly diverse, with at least 7000 described species and many more awaiting description (Phillips et al., 2019). Intensive land management (especially soil tillage and application of toxic chemicals such as soil and plant pesticides) often reduces the density of earthworms or may completely eliminate them. Conversely, degraded soils converted to conservation management often show increased earthworm densities after a few years (Curry et al., 1995; Edwards and Bohlen, 1996).

5.6.2.2 Biology and ecology

Earthworms are often grouped into functional categories based on their morphology, behavior, and feeding ecology and their microhabitats within the soil (Lavelle, 1983; Lee, 1985). Epigeic and epiendogeic species are often polyhumic, meaning they prefer organically enriched substrates and utilize plant litter on the soil surface and C-rich upper layers of mineral soil. Polyhumic endogeic species inhabit mineral soil with high organic matter content (3%), such as the rhizosphere, while meso- and oligohumic endogeic species inhabit soil with moderate (1−3%) and low (1%) organic matter contents, respectively. Anecic species exploit both the surface litter as a source of food and the mineral soil as a refuge. The familiar *Lumbricus terrestris* is an example of an anecic species, constructing burrows and pulling leaf litter down into them. The American log worm (*Bimastos parvus*) exploits leaf litter and decaying logs with little involvement in the soil, making it an epigeic species. Epigeic species promote the breakdown and mineralization of surface litter, whereas anecic species incorporate organic matter deeper into the soil profile and facilitate aeration and water infiltration through their formation of burrows.

5.6.2.3 Influence on ecosystems and soil processes

Earthworms, as ecosystem engineers (Lavelle et al., 1998), have pronounced effects on soil structure as a consequence of their burrowing activities as well as their ingestion of soil and production of castings (Lavelle and Spain, 2001; van Vliet and Hendrix, 2003). Casts are produced after earthworms ingest mineral soil or particulate organic matter, mix and enrich them with organic secretions in the gut, and then deposit the material as a slurry lining their burrows or as discrete fecal pellets. Excretion of fecal pellets can occur within or upon the soil, depending on the species. Turnover rates of soil through earthworm casting range from 40–70 t ha^{-1} y^{-1} in temperate grasslands (Bouché, 1983) to 500–1000 t ha^{-1} y^{-1} in tropical savannas (Lavelle et al., 1992). While in the earthworm gut, casts are colonized by microbes that begin to break down SOM. As casts are deposited in the soil, microbial colonization and activity continue until readily decomposable compounds are depleted. Mechanisms of cast stabilization include organic bonding of particles by polymers secreted by earthworms and microbes, mechanical stabilization by plant fibers and fungal hyphae, and stabilization due to wetting and drying cycles and age-hardening effects (Tomlin et al., 1995). Mineralization of organic matter in earthworm casts and burrow linings produces zones of nutrient enrichment compared to bulk soil. These zones, referred to as the "drilosphere," are often sites of enhanced activity of plant roots and other soil biota (Lavelle et al., 1998). Plant growth-promoting substances have also been suggested as constituents of earthworm casts. Many earthworm castings from commercial vermicomposting operations are sold commercially as soil amendments to improve soil physical properties and enhance plant growth (Edwards and Shipitalo, 1998).

Earthworm burrowing in soil creates macropores of various sizes, depths, and orientations, depending on species and soil type. Burrows range from about 1 to 10 mm in diameter and constitute among the largest of soil pores. Continuous macropores resulting from earthworm burrowing may enhance water infiltration by functioning as bypass flow pathways through soils. These pores may or may not be important in solute transport, depending on soil water content, the nature of the solute, and chemical exchange properties of the burrow linings (Edwards and Shipitalo, 1998). Earthworm burrows in agricultural systems can lead to groundwater contamination with pesticides and fertilizers (Edwards and Shipitalo, 1998).

Despite the many beneficial effects of earthworms on soil processes, some aspects of earthworm activities may be undesirable (Lavelle et al., 1998; Parmelee et al., 1998). Detrimental effects include: (1) removing and burying of surface residues that would otherwise protect soil surfaces from erosion; (2) producing fresh casts that increase erosion and surface sealing; (3) increasing compaction of surface soils by decreasing SOM, particularly for some tropical species; (4) riddling irrigation ditches, making them leaky; (5) increasing losses of soil N through leaching and denitrification; and (6) increasing soil C loss through enhanced microbial respiration. Earthworms may transmit pathogens, either as passive carriers or as intermediate hosts, raising concerns that some earthworm species could be a vector for the spread of certain plant and animal diseases. The net result of positive and negative effects of earthworms, or any other soil biota, determines whether they have detrimental impacts on ecosystems (Lavelle et al., 1998). An effect, such as mixing of O (litter [L], fermentation [F], and humification [H] zones, also called Oi, Oe, and Oa zones, respectively) and A horizons (top of the mineral soil surface, containing the preponderant amount of organic matter), may be considered beneficial in one setting (e.g., urban gardens) and detrimental in another (e.g., native forests) (Edwards, 1998).

5.6.3 Formicidae (ants)

Formicidae, the ants, are probably the most significant family of soil insects due to the very large influence they have on soil structure. Ants are numerous, diverse, and widely distributed from arctic to tropical ecosystems. Ant communities contain many species, even in desert areas (Whitford, 2000), and local species diversity is especially large in tropical areas. Populations of ants are equally numerous. About one-third of the animal biomass of the Amazonian rainforest is composed of ants and termites, with each hectare containing in excess of 8 million ants and 1 million termites (Hölldobler and Wilson, 1990). Furthermore, ants are social insects, living in colonies with several castes.

Ants are major predators of small invertebrates. Their activities reduce the abundance of other predators such as spiders and carabid beetles (Wilson, 1987). Ants are ecosystem engineers, moving large volumes of soil, much as earthworms do (Hölldobler and Wilson, 1990). Ant influences on soil structure are particularly important in deserts where earthworm densities are low. Given the large diversity of ants, identification to species is problematic for many, but Wheeler and Wheeler (1990) offer keys to subfamilies and genera of the Nearctic ant fauna. See also Ellison et al. (2012).

5.6.4 Isoptera (termites)

Along with earthworms and ants, termites are the third major earth-moving group of invertebrates. Termites are social insects with a well-developed caste system. Through their ability to digest wood, they have become economic pests of major importance in some regions of the world (Bignell and Eggleton, 2000; Lee and Wood, 1971). Termites are highly successful, constituting up to 75% of the insect biomass and 10% of all terrestrial animal biomass in the tropics (Bignell, 2000; Wilson, 1992). While termites are mainly tropical in distribution, they occur in temperate zones as well. Termites have been called the tropical analogs of earthworms since they reach large abundances in the tropics and process large amounts of litter. Termites in the primitive families, such as Kalotermitidae, possess a gut flora of protists, which enables them to digest cellulose. Their normal food is wood that has come into contact with soil. Most species of termites construct runways of soil and some are builders of spectacular mounds. Members of the phylogenetically advanced family Termitidae do not have protistan symbionts but possess a formidable array of microbial symbionts (bacteria and fungi) that enable them to process and digest the humified organic matter in tropical soils (Bignell, 1984; Breznak, 1984; Pearce, 1997). A generalized sequence of events in a typical Termitinae soil-feeder gut emphasizes the role of filamentous prokaryotes, the extremely high pH reached in the front of the hindgut, and the existence of both aerobic and anaerobic zones within the hindgut (Brauman et al., 2000).

Three nutritional categories include wood-feeding species, plant- and humus-feeding species, and fungus growers. The last group lacks intestinal symbionts and depends upon cultured fungi for nutrition. Termites have an abundance of unique microbes living in their guts. One study of bacterial microbiota in the gut of the wood-feeding termite *Reticulitermes speratus* found 268 phylotypes of bacteria (16S rRNA genes, amplified by PCR), including 100 clostridial, 61 spirochetal, and 31 Bacteroides-related phylotypes (Hongoh et al., 2003). More than 90% of the phylotypes were found for the first time, but it is unknown if they are active and participating in wood decay. Other phylotypes were monophyletic clusters with sequences recovered from the gut of other termite species. Cellulose digestion in termites, which was once considered to be solely due to the activities of fungi and protists and occasionally bacteria, has now been demonstrated to be endogenous to termites. Endogenous cellulose-degrading enzymes occur in the midguts of two species of higher termites in the genus *Nasutitermes* and in the Macrotermitinae (which cultivate basidiomycete fungi in elaborately constructed gardens) as well (Bignell, 2000).

In contrast to C-degradation by termites, only prokaryotes are capable of producing nitrogenase to fix N_2, which occurs in the organic matter-rich, microaerophilic milieu of termite guts. Some termite genera have bacteria that fix relatively small amounts of N, but others, including *Mastotermes* and *Nasutitermes*, fix from 0.7 to 21 g N g^{-1} fresh wt. day^{-1}. This equals 20 to 61 μg N per colony per day, which would double the N content if N_2 fixation was the sole source of N and the rate per termite remained constant (N content of termites assumed to be 11% on a dry weight basis) (Breznak, 2000). Some insight into the roles of soil-feeding termites in the terrestrial N cycle has been provided by Ngugi and Brune (2012), who measured significant denitrification in the hindguts of two genera of soil-feeding termites ranging from 0.4 to 3.9 nmol h^{-1} (g fresh wt.)$^{-1}$ N_2O, providing direct evidence that soil-feeding termites are a hitherto unrecognized source of this greenhouse gas in tropical soils. For an extensive exposition of the role of termites in the dynamics of SOM and nutrient cycling in ecosystems worldwide, refer to Bignell and Eggleton (2000). An overview of omics research in termites is covered by Scharf (2015).

A profound demonstration of the potential for termites to regulate ecosystem stability/resilience is described by Ashton et al. (2019), where termites were responsible for mitigating the effects of drought on an entire tropical forest ecosystem. In this study the presence of termites was associated with higher soil moisture during drought conditions, which in turn was associated with faster decomposition rates (and attendant nutrient supply rates), ultimately resulting in greater seedling survival among rainforest plants. These results emphasize the importance of soil fauna in the stability (or even persistence) of an ecosystem under stress.

5.7 Roles of soil fauna in ecosystems and societal impacts

Evidence on how soil fauna will respond to environmental change and how shifts in soil faunal abundance, diversity, and community composition will influence aboveground processes has advanced considerably (see Cheeke et al., 2012; de Vries et al., 2012, 2020; Rillig et al., 2020; Wall et al., 2012). Loss of species due to varying management practices, soil erosion, pollution, and urbanization, and the resulting effects on ecosystem function and services are becoming widely recognized and are related to larger issues of biodiversity loss, desertification, and elevated greenhouse gas concentrations (Koch et al., 2013). This has resulted in international attention to the importance of soils beyond agricultural soils and, in particular, to soil biodiversity, as now summarized in the FAO report on soil biodiversity (FAO et al., 2020). We are now able to answer basic questions about soil fauna including such topics as whether there are cosmopolitan vs. endemic fauna species; what the local and global biogeographic distribution and range of faunal species is; what the factors influencing the distribution of key species are; and whether the loss of species affects ecosystem function (Veresoglou et al., 2015; Wall et al., 2012). Food web ecology, with its emphasis on community assembly and disassembly, has the potential to act as an integrating concept across conservation biology and community and ecosystem ecology as well as for provision of ecosystem services (de Vries et al., 2012; Moore and de Ruiter, 2012; Thompson et al., 2012; Wall et al., 2012).

Global experiments and syntheses have continued to address the quantification of the role of soil fauna in ecosystem processes and, in particular, have led to increased evidence for the faunal contribution to C cycling. Global multisite experiments show that soil fauna are key regulators of decomposition rates at biome and global scales (Makkonen et al., 2012; Powers et al., 2009; Wall et al., 2008). García-Palacios et al. (2013) conducted a meta-analysis on 440 litterbag studies across 129 sites to assess how climate, litter quality, and soil invertebrates affect decomposition. This analysis showed that fauna were responsible for a ~27% enhancement, on average, of litter decomposition across global and biome scales.

Agricultural practices affect many of the key functional and structural attributes of ecosystems in several ways: the transformation of mature ecosystems into ones that are in a managed developmental state is induced by tillage operations and other activities, such as applying fertilizers and pesticides. These manipulations have the potential to shift the elemental balance of a system, decrease species diversity, and alter the soil food web (Cheeke et al., 2012; Moore and de Ruiter, 2012). Conventional tillage practices alter the distribution of organic material and affect the rate of formation of micro- and macroaggregates in the soil profile. This has a profound effect on the turnover rates of organic matter that is associated with the aggregates (Elliott and Coleman, 1988; O'Brien and Jastrow, 2013; Six et al., 2004), as well as affecting ecosystem services (Cheeke et al., 2012). de Vries et al. (2012) showed that grassland, fungal-based food webs were more resilient than agricultural fields with bacterial-dominated food webs and provided evidence for management options that enhanced ecosystem services. Cock et al. (2012) provide evidence for manipulating soil invertebrates to benefit agriculture and to enhance ecosystem services, such as biological control and C sequestration. Their focus includes successful case studies in sustainable agriculture. Koch et al. (2013) bring attention to the global policy impact of land degradation and loss of soils, biodiversity and ecosystem services, and its implications for food security. Soil fauna continue to be an exciting research field, linking aboveground systems to belowground diversity and function, as well as to environmental issues at local to global scales (Cock et al., 2012; Wall, 2004; Wall et al., 2012).

5.8 Summary

Soil fauna may be considered very efficient means to assist microbes in colonizing and extending their reach into the horizons of soils worldwide. Their roles as colonizers, comminutors, and engineers within soils have been emphasized, but new technologies and global environmental issues are yielding new questions about how to manipulate soil fauna for the long-term sustainability of soils. The demand for taxonomic specialists for all groups of soil biota is increasing as we currently recognize that molecular information alone is insufficient for many studies. Stable isotope technologies are being used with more precision to reveal the transfer of C and N through soil food webs and elucidate the role of each trophic group. This technique is clarifying that soil faunal species hitherto thought to be within one trophic group are not, thus leading to research about the structure and resilience of soil food webs. Information on the biogeography of soil fauna, their latitudinal gradient patterns, and their relationship to aboveground hot spots and to land management strategies, as well as their taxonomic status and natural history, will be critical for understanding how bacteria, archaea, fungi, protists, and soil invertebrates interact and respond to multiple global changes (Fierer et al., 2012; Wall et al., 2001, 2008; Wu et al., 2011). For example, soil invertebrates can be invasive species, which, depending on the species, can affect soil C sequestration, soil fertility, and plant and animal health, resulting in economic and ecosystem change. We feel protection of known biodiversity in ecosystems clearly must include the rich pool of soil species. This is because data for some of these species individually and collectively indicate tight connections to biodiversity aboveground, major roles in ecosystem processes, and provision of ecosystem benefits for human well-being (Wall, 2004; Wall et al., 2005; Wardle et al., 2004). Additionally, climate envelopes (e.g., niche dimensions of climate tolerance ranges) can be developed as more is known about soil species identity, activity, and geographic ranges, aiding in the development of land management practices and projections for the functioning of tomorrow's ecosystems under climate change. For further reading on the roles of fauna in soil processes, see Coleman (2008), Coleman et al. (2012; 2018), Geisen et al. (2019a; b), Nielsen (2019), and Wall et al. (2012).

References

Adl, S.M., Simpson, A.G.B., Lane, C.E., Lukeš, J., Bass, D., Bowser, S.S., et al., 2012. The revised classification of eukaryotes. J. Eukaryot. Microbiol. 59, 429–514.

Adl, S.M., Bass, D., Lane, C.E., Lukeš, J., Schoch, C.L., Smirnov, A., et al., 2019. Revisions to the classification, nomenclature, and diversity of eukaryotes. J. Eukaryot. Microbiol. 66, 4–119.

André, H.M., Ducarme, X., Lebrun, P., 2002. Soil biodiversity: myth, reality or conning? Oikos 96, 3–24.

Ashton, L.A., Griffiths, H.M., Parr, C.L., Evans, T.A., Didham, R.K., Hasan, F., et al., 2019. Termites mitigate the effects of drought in tropical rainforest. Science 363 (6423), 174–177. https://doi.org/10.1126/science.aau9565.

Bates, S.T., Clemente, J.C., Flores, G.E., Walters, W.A., Parfrey, L.W., Knight, R., et al., 2013. Global biogeography of highly diverse protistan communities in soil. ISME J. 7, 652–659.

Beakes, G.W., Glockling, S.L., Sekimoto, S., 2012. The evolutionary phylogeny of the oomycete "fungi". Protoplasma 249, 3–19.

Berthold, A., Palzenberger, M., 1995. Comparison between direct counts of active soil ciliates (Protozoa) and most probable number estimates obtained by Singh's dilution culture method. Biol. Fertil. Soils 19, 348–356.

Bignell, D.E., 1984. The arthropod gut as an environment for microorganisms. In: Anderson, J.M., Rayner, A.D.M., Walton, D.W.H. (Eds.), Invertebrate-Microbial Interactions. Cambridge University Press, Cambridge, pp. 205–227.

Bignell, D.E., 2000. Introduction to symbiosis. In: Abe, T., Bignell, D.E., Higashi, M. (Eds.), Termites: Evolution, Sociality, Symbioses, Ecology. Kluwer Academic, Dordrecht, pp. 189–208.

Bignell, D.E., Eggleton, P., 2000. Termites in ecosystems. In: Abe, T., Bignell, D.E., Higashi, M. (Eds.), Termites: Evolution, Sociality, Symbioses, Ecology. Kluwer Academic, Dordrecht, pp. 363–387.

Blaxter, M.L., De Ley, P., Garey, J.R., Liu, L.X., Scheldeman, P., Vierstraete, A., et al., 1998. A molecular evolutionary framework for the phylum Nematoda. Nature 392, 71–75.

Bongers, T., 1990. The maturity index: an ecological measure of environmental disturbance based on nematode species composition. Oecologia 83, 14–19.

Bongers, T., Ferris, H., 1999. Nematode community structure as a bioindicator in environmental monitoring. Trends Ecol. Evol. 14, 224–228.

Bonkowski, M., 2004. Protozoa and plant growth: the microbial loop in soil revisited. New Phytol. 162, 617–631.

Bonkowski, M., Clarholm, M., 2012. Stimulation of plant growth through interactions of bacteria and protozoa: testing the auxiliary microbial loop hypothesis. Acta Protozool. 51, 237–247.

Borgonie, G., Garcia-Moyano, A., Litthauer, D., Bert, W., Bester, A., van Heerden, E., et al., 2011. Nematoda from the terrestrial deep subsurface of South Africa. Nature 474, 79–82.

Bouché, M.B., 1983. The establishment of earthworm communities. In: Satchell, J.E. (Ed.), Earthworm Ecology: From Darwin to Vermiculture. Chapman & Hall, London, pp. 431–448.

Brauman, A., Bignell, D.E., Tayasu, I., 2000. Soil-feeding termites: biology, microbial associations and digestive mechanisms. In: Abe, T., Bignell, D.E., Higashi, M. (Eds.), Termites: Evolution, Sociality, Symbioses. Kluwer Academic Publishers, Dordrecht, pp. 233–259.

Breznak, J.A., 1984. Biochemical aspects of symbiosis between termites and their intestinal micro-biota. In: Anderson, J.M., Rayner, A.D.M., Walton, D.W.H. (Eds.), Invertebrate-Mirobial Interactions. Cambridge University Press, Cambridge, UK, pp. 209–231.

Breznak, J., 2000. Ecology of prokaryotic microbes in the guts of wood- and litter-feeding termites. In: Abe, T., Bignell, D.E., Higashi, M. (Eds.), Termites: Evolution, Sociality, Symbioses, Ecology. Kluwer Academic Publishers, Dordrecht, pp. 209–231.

Calderón-Cortés, N., Quesada, M., Watanabe, H., Cano-Comacho, H., Oyama, K., 2012. Endogenous plant cell wall digestion: a key mechanism in insect evolution. Annu. Rev. Ecol. Evol. Syst. 43, 45–71.

Cameron, E.K., Bayne, E.M., 2009. Road age and its importance in earthworm invasion of northern boreal forests. J. Appl. Ecol. 46, 28–36.

Cameron, E.K., Bayne, E.M., Clapperton, M.J., 2007. Human-facilitated invasion of exotic earthworms into northern boreal forests. Ecoscience 14, 482–490.

Caron, D.A., Worden, A.Z., Countway, P.D., Demir, E., Heidelberg, K.B., 2008. Protists are microbes too: a perspective. ISME J. 3, 4–12.

Carpenter, D., Hodson, M.E., Eggleton, P., Kirk, C., 2008. The role of earthworm communities in soil mineral weathering: a field experiment. Mineral. Mag. 72, 33–36.

Chamberlain, P.M., Bull, I.D., Black, H.I.J., Ineson, P., Evershed, R.P., 2006. Collembolan trophic preferences determined using fatty acid distributions and compound-specific stable carbon isotope values. Soil Biol. Biochem. 38, 1275–1281.

Cheeke, T., Coleman, D.C., Wall, D.H. (Eds.), 2012. Microbial Ecology in Sustainable Agroecosystems. CRC Press, Boca Raton.

Cock, M.J., Biesmeijer, J.C., Cannon, R.J., Gerard, P., Gillespie, D., Jiménez, J., et al., 2012. The positive contribution of invertebrates to sustainable agriculture and food security. CAB. Rev. Perspect. Agric. Vet. Sci. Nutr. Nat. Resour. 7, 1–27.

Coleman, D.C., 2008. From peds to paradoxes: linkages between soil biota and their influences on ecological processes. Soil Biol. Biochem. 40, 271–289.

Coleman, D.C., Whitman, W.B., 2005. Linking species richness, biodiversity and ecosystem function in soil systems. Pedobiologia 49, 479–497.

Coleman, D.C., Hendrix, P.F., Beare, M.H., Cheng, W., Crossley Jr., D.A., 1993. Microbial and faunal dynamics as they affect soil organic matter dynamics in subtropical agroecosystems. In: Paoletti, M.G., Foissner, W., Coleman, D.C. (Eds.), Soil Biota and Nutrient Cycling Farming Systems. CRC Press, Boca Raton, pp. 1–14.

Coleman, D.C., Crossley Jr., D.A., Hendrix, P.F., 2004. Fundamentals of Soil Ecology. Elsevier, San Diego.

Coleman, D.C., Vadakattu, G., Moore, J.C., 2012. Soil ecology and agroecosystem studies: a dynamic world. In: Cheeke, T., Coleman, D.C., Wall, D.H. (Eds.), Microbial Ecology in Sustainable Agroecosystems. CRC Press, Boca Raton, pp. 1–21.

Coleman, D.C., Callaham, M.A., Crossley Jr., D.A., 2018. Fundamentals of Soil Ecology, third ed. Elsevier, Academic Press, London.

Crotty, F.V., Adl, S.M., 2019. Competition and predation in soil fungivorous microarthropods using stable isotope ratio mass spectrometry. Front. Microbiol. 10, 1274. https://doi.org/10.3389/fmicb.2019.01274.

Crotty, F.V., Adl, S.M., Blackshaw, R.P., Murray, P.J., 2012. Protozoan pulses unveil their pivotal position within the soil food web. Microb. Ecol. 63, 905–918.

Curry, J.P., Byrne, D., Boyle, K.E., 1995. The earthworm population of a winter cereal field and its effects on soil and nitrogen turnover. Biol. Fertil. Soils 19, 166–172.

Darbyshire, J.F., Whitley, R.E., Graebes, M.P., Inkson, R.H.E., 1974. A rapid micromethod for estimating bacterial and protozoan populations in soil. Rev. Ecol. Biol. Soil 11, 465–475.

Darwin, C., 1881. The Formation of Vegetable Mould, through the Action of Worms. John Murray, London.

Davidson, D.A., Bruneau, P.M.C., Grieve, I.C., Young, I.M., 2002. Impacts of fauna on an upland grassland soil as determined by micromorphological analysis. Appl. Soil Ecol. 20, 133–143.

Decaëns, T., 2010. Macroecological patterns in soil communities. Glob. Ecol. Biogeogr. 19, 287–302.

Decaëns, T., Jimenez, J.J., Gioia, C., Measey, G.J., Lavelle, P., 2006. The values of soil animals for conservation biology. Eur. J. Soil Biol. 42, S23–S38.

de Ruiter, P.C., Neutel, A.M., Moore, J.C., 1995. Energetics, patterns of interaction strengths, and stability in real ecosystems. Science 269, 1257–1260.

de Vries, F.T., Liiri, M.E., Bjornlund, L., Bowker, M.A., Christensen, S., Setälä, H.M., et al., 2012. Land use alters the resistance and resilience of soil food webs to drought. Nat. Clim. Change 2, 276–280.

de Vries, F.T., Griffiths, R.I., Knight, C.G., Nicolitch, O., Williams, A., 2020. Harnessing rhizosphere microbiomes for drought-resilient crop production. Science 368, 270–274.

Deiner, K., Bik, H.M., Mächler, E., Seymour, M., Lacoursière-Roussel, A., Altermatt, F., et al., 2017. Environmental DNA metabarcoding: transforming how we survey animal and plant communities. Mol. Ecol. 26, 5872–5895.

Didden, W.A.M., 1990. Involvement of Enchytraeidae (Oligochaeata) in soil structure evolution in agricultural fields. Biol. Fertil. Soils 9, 152–158.

Didden, W.A.M., 1993. Ecology of Enchytraeidae. Pedobiologia 37, 2–29.

Domene, X., Chelinho, S., Campana, P., Natal-da-Luz, T., Alcañiz, J.M., Andrés, P., et al., 2011. Influence of soil properties on the performance of *Folsomia candida*: implications for its use in soil ecotoxicology testing. Environ. Toxicol. Chem. 30, 1497–1505.

dos Reis, M., Thawornwattana, Y., Angelis, K., Telford, M.J., Donoghue, P.C.J., Yang, Z., 2015. Uncertainty in the timing of origin of animals and the limits of precision in molecular timescales. Curr. Biol. 25, 2939–2950.

Dósza-Farkas, K., 1996. Reproduction strategies in some enchytraeid species. In: Dosza-Farkas, K. (Ed.), Newsletter on Enchytraeidae. Eotvos Lorand University, Budapest.

Dupont, A.O., Griffiths, R.I., Bell, T., Bass, D., 2016. Differences in soil micro-eukaryotic communities over soil pH gradients are strongly driven by parasites and saprotrophs. Environ. Microbiol. 18, 2010–2024.

Edwards, C.A., 1991. The assessment of populations of soil-inhabiting invertebrates. Agric. Ecosyst. Environ. 34, 145–176.

Edwards, C.A. (Ed.), 1998. Earthworm Ecology. St. Lucie Press, Boca Raton.

Edwards, C.A., Bohlen, P.J., 1996. Earthworm Biology and Ecology. Chapman & Hall, London.

Edwards, W.M., Shipitalo, M.J., 1998. Consequences of earthworms in agricultural soils: aggregation and porosity. In: Edwards, C.A. (Ed.), Earthworm Ecology. CRC Press, Boca Raton, pp. 147–161.

Elliott, E.T., Coleman, D.C., 1988. Let the soil work for us. Ecol. Bull. Copenhagen 39, 23–32.

Elliott, E.T., Anderson, R.V., Coleman, D.C., Cole, C.V., 1980. Habitable pore-space and microbial trophic interactions. Oikos 35, 327–335.

Ellison, A.M., Gotelli, N.J., Farnsworth, E.J., Alpert, G.D., 2012. A Field Guide to the Ants of New England. Yale University Press, New Haven.

FAO, ITPS, CBD, GSBI, EC, 2020. State of Knowledge of Soil Biodiversity – Status, Challenges and Potentialities, Full Report. FAO, Rome.

Ferris, H., Bongers, T., 2006. Nematode indicators of organic enrichment. J. Nematol. 38, 3–12.

Ferris, H., Bongers, T., de Goede, R.G.M., 2001. A framework for soil foodweb diagnostics: extension of the nematode faunal analysis concept. Appl. Soil Ecol. 18, 13–29.

Fierer, N., Lennon, J.T., 2011. The generation and maintenance of diversity in microbial communities. Am. J. Bot. 98, 439–448.

Fierer, N., Leff, J.W., Adams, B.J., Nielsen, U.N., Bates, S.T., Lauber, C.L., et al., 2012. Cross-biome metagenomic analyses of soil microbial communities and their functional attributes. PNAS USA 109, 21390–21395.

Finlay, B.J., Black, H.I.J., Brown, S., Clarke, K.J., Esteban, G.F., Hindle, R.M., et al., 2000. Estimating the growth potential of the soil protozoan community. Protist 151, 69–80.

Foissner, W., 1999. Soil protozoa as bioindicators: pros and cons, methods, diversity, representative examples. Agric. Ecosyst. Environ. 74, 95–112.

Fox, D., 2020. Extreme survivor. Sci. Am. 322 (4), 50–57.

Freckman, D.W., Virginia, R.A., 1989. Plant-feeding nematodes in deep-rooting desert ecosystems. Ecology 70, 1665–1678.

Frelich, L.E., Hale, C.M., Scheu, S., Holdsworth, A.R., Heneghan, L., Bohlen, P.J., et al., 2006. Earthworm invasion into previously earthworm-free temperate and boreal forests. Biol. Invasions 8, 1235–1245.

Gao, Z., Karlsson, I., Geisen, S., Kowalchuk, G., Jousset, A., 2019. Protists: puppet masters of the rhizosphere microbiome. Trends Plant Sci. 24, 165–176.

García-Palacios, P., Maestre, F.T., Kattge, J., Wall, D.H., 2013. Climate and litter quality differently modulate the effects of soil fauna on litter decomposition across biomes. Ecol. Lett. 16, 1045–1053.

Geisen, S., 2016. The bacterial-fungal energy channel concept challenged by enormous functional versatility of soil protists. Soil Biol. Biochem. 102, 22–25.

Geisen, S., Bonkowski, M., 2018. Methodological advances to study the diversity of soil protists and their functioning in soil food webs. Appl. Soil Ecol. 123, 328–333.

Geisen, S., Bandow, C., Römbke, J., Bonkowski, M., 2014. Soil water availability strongly alters the community composition of soil protists. Pedobiologia 57, 205–213.

Geisen, S., Tveit, A.T., Clark, I.M., Richter, A., Svenning, M.M., Bonkowski, M., et al., 2015. Metatranscriptomic census of active protists in soils. ISME J. 9, 2178–2190.

Geisen, S., Koller, R., Hünninghaus, M., Dumack, K., Urich, T., Bonkowski, M., 2016. The soil food web revisited: diverse and widespread mycophagous soil protists. Soil Biol. Biochem. 94, 10–18.

Geisen, S., Mitchell, E.A.D., Wilkinson, D.M., Adl, S., Bonkowski, M., Brown, M.W., et al., 2017. Soil protistology rebooted: 30 fundamental questions to start with. Soil Biol. Biochem. 111, 94–103.

Geisen, S., Snoek, L.B., ten Hooven, F.C., Duyts, H., Kostenko, O., Bloem, J., et al., 2018a. Integrating quantitative morphological and qualitative molecular methods to analyze soil nematode community responses to plant range expansion. Methods Ecol. Evol. 9, 1366–1378.

Geisen, S., Mitchell, E.A.D., Adl, S., Bonkowski, M., Dunthorn, M., Ekelund, F., et al., 2018b. Soil protists: a fertile frontier in soil biology research. FEMS. Microbiol. Revs. 42, 293–323.

Geisen, S., Vaulot, D., Mahé, F., Lara, E., de Vargas, C., Bass, D., 2019a. A User Guide to Environmental Protistology: Primers, Metabarcoding, Sequencing, and Analyses. bioRxiv, p. 850610.

Geisen, S., Briones, M.J.I., Gan, H., Behan-Pelletier, V.M., Friman, V.P., de Groot, G.A., et al., 2019b. A methodological framework to embrace soil biodiversity. Soil Biol. Biochem. 136, 107536.

Geisen, S., Hu, S., dela Cruz, T.E.E., Veen, G.F., 2020. Protists as catalyzers of microbial litter breakdown and carbon cycling at different temperature regimes. ISME J. 15, 618–621. https://doi.org/10.1038/s41396-020-00792-y.

Gilbert, H.J., 2010. The biochemistry and structural biology of plant cell wall decomposition. Plant. Physiol 153, 444–455.

Griffiths, B.S., Caul, S., 1993. Migration of bacterial-feeding nematodes, but not protozoa, to decomposing grass residues. Biol. Fertil. Soils 15, 201–207.

Griffiths, B.S., de Groot, G.A., Laros, I., Stone, D., Geisen, S., 2018. The need for standardization: exemplified by a description of the diversity, community structure and ecological indices of soil nematodes. Ecol. Indic. 87, 43–46.

Hale, C.M., Frelich, L.E., Reich, P.B., Pastor, J., 2005. Effects of European earthworm invasion on soil characteristics in northern hardwood forests of Minnesota, USA. Ecosystems 8, 911–927.

Hättenschwiler, S., Tiunov, A.V., Scheu, S., 2005. Biodiversity and litter decomposition in terrestrial ecosystems. Annu. Rev. Ecol. Evol. Syst. 36, 191–218.

Heemsbergen, D., Berg, M.P., Van Hall, J., van Hal, J.R., Faber, J.H., Verhoef, H.A., 2004. Biodiversity effects on soil processes explained by inter-specific functional trait dissimilarity. Science 306, 1019–1020.

Heidemann, K., Scheu, S., Ruess, L., Maraun, M., 2011. Molecular detection of nematode predation and scavenging in oribatid mites: laboratory and field experiments. Soil Biol. Biochem. 43, 2229–2236.

Hendrix, P.F. (Ed.), 1995. Earthworm Ecology and Biogeography in North America. CRC Press, Boca Raton.

Hendrix, P.F., Parmalee, R.W., Crossley Jr., D.A., Coleman, D.C., Odum, E.P., Groffman, P.M., 1986. Detritus food webs in conventional and no-tillage agroecosystems. Bioscience 36, 374–380.

Hölldobler, B., Wilson, E.O., 1990. The Ants. Belknap Press, Cambridge, MA.

Hongoh, Y., Ohkuma, M., Kudo, T., 2003. Molecular analysis of bacterial microbiota in the gut of the termite *Reticulitermes speratus* (Isoptera; Rhinotermitidae). Fems. Microbiol. Ecol. 44, 231–242.

Hooper, D.J., 1970. Extraction of free-living stages from soil. In: Southey, J.F. (Ed.), Laboratory Methods for Work with Plant and Soil Nematodes. Ministry of Agriculture, Fisheries and Food, London, UK, pp. 5–30.

Hünninghaus, M., Dibbern, D., Kramer, S., Koller, R., Pausch, J., Schloter-Hai, B., 2019. Disentangling carbon flow across microbial kingdoms in the rhizosphere of maize. Soil Biol. Biochem. 134, 122–130.

Hunt, H.W., Coleman, D.C., Ingham, E.R., Ingham, R.E., Elliott, E.T., Moore, J.C., et al., 1987. The detrital food web in a shortgrass prairie. Biol. Fertil. Soils 3, 57–68.

Hyvönen, R., Persson, T., 1996. Effects of fungivorous and predatory arthropods on nematodes and tardigrades in microcosms with coniferous forest soil. Biol. Fertil. Soils 21, 121–127.

Iglesias Briones, M.J., Ineson, P., Piearce, T.G., 1997. Effects of climate change on soil fauna; responses of enchytraeids, Diptera larvae and tardigrades in a transplant experiment. Appl. Soil Ecol. 6, 117–134.

Ingham, R.E., Trofymow, J.A., Ingham, E.R., Coleman, D.C., 1985. Interactions of bacteria, fungi, and their nematode grazers: effects on nutrient cycling and plant growth. Ecol. Monogr. 55, 119–140.

Ito, M.T., Abe, W., 2001. Micro-distribution of soil inhabiting tardigrades (Tardigrada) in a sub-alpine coniferous forest of Japan. Zool. Anz. 240, 403–407.

Jeffery, S., Gardi, C., Jones, A., Montanarella, L., Marmo, L., Miko, L., et al., 2010. European Atlas of Soil Biodiversity. European Commission, Publications Office of the European Union, Luxembourg.

Jones, C.G., Lawton, J.H., Shachak, M., 1994. Organisms as ecosystem engineers. Oikos 69, 373–386.

Keeling, P.J., Burki, F., 2019. Progress towards the tree of eukaryotes. Curr. Biol. 29, R808–R817.

Kerfahi, D., Tripathi, B.M., Porazinska, D.L., Park, J., Go, R., Adams, J.M., 2016. Do tropical rain forest soils have greater nematode diversity than High Arctic tundra? A metagenetic comparison of Malaysia and Svalbard. Glob. Ecol. Biogeogr. 25, 716–728.

Knapp, B.A., Seeber, J., Podmirseg, S.M., Rief, A., Meyer, E., Insam, H., 2009. Molecular fingerprinting analysis of the gut microbiota of *Cylindroiulus fulviceps* (Diplopoda). Pedobiologia 52, 325–336.

Koch, A., McBratney, A., Adams, M., Field, D., Hill, R., Crawford, J., et al., 2013. Soil security: solving the global soil crisis. Glob. Policy 4, 434–441.

Kühn, J., Richter, A., Kahl, T., Bauhus, J., Schöning, I., Ruess, L., 2018. Community level lipid profiling of consumers as a tool for soil food web diagnostics. Methods Ecol. Evol. 9, 1265–1275.

Kutikova, L.A., 2003. Bdelloid Rotifers (Rotifera, Bdelloidea) as a component of soil and land biocenoses. Biol. Bull. Russ. Acad. Sci. 30, 271–274.

Lavelle, P., 1983. The structure of earthworm communities. In: Satchell, J.E. (Ed.), Earthworm Ecology: From Darwin to Vermiculture. Chapman & Hall, London, pp. 449–466.

Lavelle, P., Spain, A.V., 2001. Soil Ecology. Kluwer Academic, Dordrecht.

Lavelle, P., Blanchart, E., Martin, A., Spain, A.V., Martin, S., 1992. Impact of soil fauna on the properties of soils in the humid tropics. In: Lal, R., Sanchez, P.A. (Eds.), Myths and Science of Soils of the Tropics. SSSA Special Publications, Madison, WI, pp. 157–185.

Lavelle, P., Lattaud, C., Trigo, D., Barois, I., 1995. Mutualism and biodiversity in soils. Plant Soil 170, 23–33.

Lavelle, P., Pashanasi, B., Charpentier, F., Gilot, C., Rossi, J.P., Derouard, L., et al., 1998. Large-scale effects of earthworms on soil organic matter and nutrient dynamics. In: Edwards, C.A. (Ed.), Earthworm Ecology. St. Lucie Press, Columbus, Ohio, pp. 103–122.

Lee, K.E., 1985. Earthworms: Their Ecology and Relationships with Soils and Land Use. Academic Press, Sydney.

Lee, K.E., Wood, T.G., 1971. Termites and Soils. Academic Press, London.

Lentendu, G., Wubet, T., Chatzinotas, A., Wilhelm, C., Buscot, F., Schlegel, M., 2014. Effects of long-term differential fertilization on eukaryotic microbial communities in an arable soil: a multiple barcoding approach. Mol. Ecol. 23, 3341–3355.

Li, X., Zhu, H., Geisen, S., Bellard, C., Hu, F., Li, H., et al., 2020. Agriculture erases climate constraints on soil nematode communities across large spatial scales. Glob. Change Biol. 26, 919–930.

Liang, C., Amelung, W., Lehmann, J., Kästner, M., 2019. Quantitative assessment of microbial necromass contribution to soil organic matter. Glob. Change Biol. 25, 3578–3590.

Lloyd, M., Dybas, H.S., 1966. The periodical cicada problem. I. Population ecology. Evolution 20, 133–149.

Mahé, F., de Vargas, C., Bass, D., Czech, L., Stamatakis, A., Lara, E., et al., 2017. Parasites dominate hyperdiverse soil protist communities in Neotropical rainforests. Nat. Ecol. Evol. 1, 0091.

Makkonen, M., Berg, M.P., Handa, I.T., Hättenschwiler, S., van Ruijven, J., van Bodegom, P.M., et al., 2012. Highly consistent effects of plant litter identity and functional traits on decomposition across a latitudinal gradient. Ecol. Lett. 15, 1033–1041.

Meier, F.A., Honegger, R., 2002. Faecal pellets of lichenivorous mites contain viable cells of the lichen-forming ascomycete *Xanthoria parietina* and its green algal photobiont, *Tebouxia arboricola*. Biol. J. Linn. Soc. 76, 259–268.

Moore, J.C., De Ruiter, P.C., 2012. Soil food webs in agricultural ecosystems. In: Cheeke, T., Coleman, D.C., Wall, D.H. (Eds.), Microbial Ecology of Sustainable Agroecosystems. CRC Press, Boca Raton, pp. 63–88.

Moreira, F.M.S., Huising, E.J., Bignell, D.E., 2009. A Handbook of Tropical Soil Biology: Sampling and Characterization of Belowground Biodiversity. Earthscan, London, UK.

Ngugi, D.K., Brune, A., 2012. Nitrate reduction, nitrous oxide formation, and anaerobic ammonia oxidation to nitrite in the gut of soil-feeding termites (*Cubitermes* and *Ophiotermes* spp.). Environ. Microbiol. 14, 860–871.

Nielsen, U.N., 2019. Soil Fauna Assemblages. Cambridge University Press, Cambridge, U.K.

O'Brien, S.L., Jastrow, J.D., 2013. Physical and chemical protection in hierarchical soil aggregates regulates soil carbon and nitrogen recovery in restored perennial grasslands. Soil Biol. Biochem. 61, 1−13.

Oliverio, A.M., Geisen, S., Delgado-Baquerizo, M., Maestre, F.T., Turner, B.L., Fierer, N., 2020. The global-scale distributions of soil protists and their contributions to belowground systems. Sci. Adv. 6, eaax8787.

Parmelee, R.W., Bohlen, P.J., Blair, J.M., 1998. Earthworms and nutrient cycling processes: integrating across the ecological hierarchy. In: Edwards, C.A. (Ed.), Earthworm Ecology. St. Lucie Press, Boca Raton, pp. 123−143.

Payne, J.A., 1965. A summer carrion study of the baby pig, *Sus scrofa* Linnaeus. Ecology 46, 592−602.

Pearce, M.J., 1997. Termites: Biology and Pest Management. CAB International, Wallingford, UK.

Perdomo, G., Evans, A., Maraun, M., Sunnucks, P., Thompson, R., 2012. Mouthpart morphology and trophic position of microarthropods from soils and mosses are strongly correlated. Soil Biol. Biochem. 53, 56−63.

Phillips, H.R.P., Guerra, C.A., Bartz, M.L.C., Briones, M.J.I., Brown, G., Crowther, T.W., et al., 2019. Global distribution of earthworm diversity. Science 366, 480.

Pollierer, M.M., Langel, R., Scheu, S., Maraun, M., 2009. Compartmentalization of the soil animal food web as indicated by dual analysis of stable isotope ratios (N-15/N-14 and C-13/C-12). Soil Biol. Biochem. 41, 1221−1226.

Ponge, J.F., 1991. Food resources and diets of soil animals in a small area of Scots pine litter. Geoderma 49, 33−62.

Porazinska, D.L., Giblin-Davis, R.M., Powers, T.O., Thomas, W.K., 2012. Nematode spatial and ecological patterns from tropical and temperate rainforests. PLoS One 7, e44641.

Powers, J.S., Montgomery, R.A., Adair, E.C., Brearley, F.Q., DeWalt, S.J., Castanho, C.T., et al., 2009. Decomposition in tropical forests: a pan-tropical study of the effects of litter type, litter placement and mesofaunal exclusion across a precipitation gradient. J. Ecol. 97, 801−811.

Rillig, M.C., Ryo, M., Lehmann, A., Aguilar-Trigueros, C.A., Buchert, S., Wulf, A., et al., 2020. The role of multiple global change factors in driving soil functions and microbial biodiversity. Science 366, 886−890.

Robeson, M.S., Costello, E.K., Freeman, K.R., Jeremy Whiting, J., Adams, B., Martin, A.P., et al., 2009. Environmental DNA sequencing primers for eutardigrades and bdelloid rotifers. BMC Ecol. 9, 25.

Robeson, M.S., King, A.J., Freeman, K.R., Birky Jr., C.W., Martin, A.P., Schmidt, S.K., 2011. Soil rotifer communities are extremely diverse globally but spatially autocorrelated locally. PNAS USA 108, 4406−4410.

Rusek, J., 1985. Soil microstructures − contributions on specific soil organisms. Quaest. Entomol. 21, 497−514.

Sánchez-Moreno, S., Ferris, H., Guil, N., 2008. Role of tardigrades in the suppressive service of a soil food web. Agric. Ecosyst. Environ. 124, 187−192.

Sapp, M., Ploch, S., Fiore-Donno, A.M., Bonkowski, M., Rose, L.E., 2018. Protists are an integral part of the *Arabidopsis thaliana* microbiome. Environ. Microbiol. 20, 30−43.

Scharf, M.E., 2015. Omic research in termites: an overview and a roadmap. Front. Genet. https://doi.org/10.3389/gene 2015.00076.

Scharroba, A., Dibbern, D., Hunninghaus, M., Kramer, S., Moll, J., Butenschoen, O., et al., 2012. Effects of resource availability and quality on the structure of the micro-food web of an arable soil across depth. Soil Biol. Biochem. 50, 1−11.

Schenk, J., Geisen, S., Kleinboelting, N., Traunspurger, W., 2019. Metabarcoding data allow for reliable biomass estimates in the most abundant animals on earth. MBMG 3, e46704.

Scherber, C., Eisenhauer, N., Weisser, W.W., Schmid, B., Voigt, W., Fischer, M., et al., 2010. Bottom-up effects of plant diversity on multitrophic interactions in a biodiversity experiment. Nature 468, 553−556.

Scheu, S., Setälä, H., 2002. Multitrophic interactions in decomposer food-webs. In: Tscharntke, B., Hawkins, B.A. (Eds.), Multitrophic Level Interactions. Cambridge University Press, Cambridge, UK, pp. 223−264.

Schill, R.O., Jönsson, K.I., Pfannkuchen, M., Brümmer, F., 2011. Food of tardigrades: a case study to understand food choice, intake and digestion. J. Zoolog. Syst. Evol. Res. 49, 66−70.

Schmelz, R.M., Collado, R., 2010a. A guide to European terrestrial and freshwater species of Enchytraeidae (Oligochaeta). Soil Org 82, 1−176.

Schmelz, R.M., Collado, R., 2010b. Checklist of taxa of Enchytraeidae (Oligochaeta): an update. Soil Org. 87, 149−153.

Schmelz, R.M., Beylich, A., Boros, G., Dózsa-Farkas, K., Graefe, U., Hong, Y., et al., 2017. How to deal with cryptic species in Enchytraeidae, with recommendations on taxonomical descriptions. Opusc. Zoolog. Budapest 48, 45−51.

Schneider, K., Migge, S., Norton, R.A., Scheu, S., Langel, R., Reineking, A., et al., 2004. Trophic niche differentiation in soil microarthropods (Oribatida, Acari): evidence from stable isotope ratios (15N/14N). Soil. Biol. Biochem. 36, 1769−1774.

Schulz-Bohm, K., Geisen, S., Wubs, E.R.J., Song, C., de Boer, W., Garbeva, P., et al., 2017. The prey's scent – volatile organic compound mediated interactions between soil bacteria and their protist predators. ISME J. 11, 817–820.

Schwelm, A., Badstober, J., Bulman, S., Desoignies, N., Etemadi, M., Falloon, R.E., et al., 2018. Not in your usual top 10: protists that infect plants and algae. Mol. Plant Pathol. 19, 1029–1044.

Seastedt, T.R., 1984. The role of microarthropods in decomposition and mineralization processes. Annu. Rev. Entomol. 29, 25–46.

Segers, H., 2007. Annotated checklist of the rotifers (Phylum Rotifera), with notes on nomenclature, taxonomy and distribution. Zootaxa 1–104.

Shelley, R.M., 2002. A synopsis of the North American centipedes of the order scolopendromorpha (Chilopoda). V. Museum. Nat. Hist. Mem. 5, 1–108.

Shmakova, L., Bondarenko, N., Smirnov, A., 2016. Viable species of Flamella (Amoebozoa: variosea) isolated from ancient arctic permafrost sediments. Protist 167, 13–30.

Six, J., Bossuyt, H., Degryze, S., Denef, K., 2004. A history of the research on the link between (micro) aggregates, soil biota, and soil organic matter dynamics. Soil Tillage Res. 79, 7–31.

Sohlenius, B., 1979. A carbon budget for nematodes, rotifers and tardigrades in a Swedish coniferous forest soil. Holarctic Ecol. 2, 30–40.

Song, D., Pan, K., Tariq, A., Sun, F., Li, Z., Sun, X., et al., 2017. Large-scale patterns of distribution and diversity of terrestrial nematodes. Appl. Soil Ecol. 114, 161–169.

Southwood, T.R.E., 1978. Ecological Methods with Particular Reference to the Study of Insect Populations. Chapman, London.

Swift, M.J., Heal, O.W., Anderson, J.M., 1979. Decomposition in Terrestrial Ecosystems. University of California Press, Berkeley.

Thompson, R.M., Brose, U., Dunne, J.A., Hall Jr., R.O., Hladyz, S., Kitching, R.L., et al., 2012. Food webs: reconciling the structure and function of biodiversity. Trends Ecol. Evol. 27, 689–697.

Tomlin, A.D., Shipitalo, M.J., Edwards, W.M., Protz, R., 1995. Earthworms and their influence on soil structure and infiltration. In: Hendrix, P.F. (Ed.), Earthworm Ecology and Biogeography in North America. CRC Press, Boca Raton, pp. 159–183.

Treonis, A.M., Wall, D.H., Virginia, R.A., 1999. Invertebrate biodiversity in Antarctic dry valley soils and sediments. Ecosystems 2, 482–492.

Tsiafouli, M.A., Thebault, E., Sgardelis, S.P., de Ruiter, P.C., van der Putten, W.H., Birkhofer, K., et al., 2015. Intensive agriculture reduces soil biodiversity across Europe. Glob. Change Biol. 21, 973–985.

van den Hoogen, J., Geisen, S., Routh, D., Ferris, H., Traunspurger, W., Wardle, D.A., et al., 2019. Soil nematode abundance and functional group composition at a global scale. Nature 572, 194–198.

van der Putten, W.H., Bardgett, R.D., Bever, J.D., Bezemer, T.M., Casper, B.B., Fukami, T., et al., 2013. Plant-soil feedbacks: the past, the present and future challenges. J. Ecol. 101, 265–276.

van Vliet, P.C.J., 2000. Enchytraeids. In: Sumner, M. (Ed.), Handbook of Soil Science. CRC Press, Boca Raton, pp. C70–C77.

van Vliet, P.C.J., Hendrix, P.F., 2003. Role of fauna in soil physical processes. In: Abbott, L.K., Murphy, D.V. (Eds.), Soil Biological Fertility – A Key to Sustainable Land Use in Agriculture. Kluwer Academic, Dordrecht, pp. 61–80.

van Vliet, P.C.J., Beare, M.H., Coleman, D.C., 1995. Population dynamics and functional roles of Enchytraeidae (Oligochaeta) in hardwood forest and agricultural systems. Plant Soil 170, 199–207.

Veresoglou, S.D., Halley, J.M., Rillig, M.C., 2015. Extinction risk of soil biota. Nat. Commun. 6, 8862–8872.

Vervoort, M.T.W., Vonk, J.A., Mooijman, P.J.W., Sven, J.J., Van den Elsen, S.J.J., Van Megen, H.H.B., et al., 2012. SSU ribosomal DNA-based monitoring of nematode assemblages reveals distinct seasonal fluctuations within evolutionary heterogeneous feeding guilds. PLoS One 7 (10), e47555.

Wall, D.H., Snelgrove, V.R., Covich, A.P., 2001. Conservation priorities for soil and sediment invertebrates. In: Soule, M.E., Orians, G.H. (Eds.), Conservation Biology: Research Priorities for the Next Decade. Island Press, Washington, DC, pp. 99–123.

Wall, D.H. (Ed.), 2004. Sustaining Biodiversity and Ecosystem Services in Soil and Sediments. SCOPE No 64. Island Press, Washington, DC.

Wall, D.H., Fitter, A., Paul, E., 2005. Developing new perspectives from advances in soil biodiversity research. In: Bardgett, R.D., Usher, M.B., Hopkins, D.W. (Eds.), Biological Diversity and Function in Soils. Cambridge University Press, Cambridge, UK, pp. 3–30.

Wall, D.H., Bradford, M.A., St John, M.G., Trofymow, J.A., Behan-Pelletier, V., Bignell, D.E., et al., 2008. Global decomposition experiment shows soil animal impacts on decomposition are climate-dependent. Glob. Change Biol. 14, 2661–2677.

Wall, D.H., Bardgett, R.D., Behan-Pelletier, V., Herrick, J.E., Jones, H., Ritz, K., et al. (Eds.), 2012. Soil Ecology and Ecosystem Services. Oxford University Press, Oxford, UK.

Wallace, H.R., 1959. The movement of eelworms in water films. Ann. Appl. Biol. 47, 366–370.

Wallace, R.L., 2002. Rotifers: exquisite metazoans. Integr. Comp. Biol. 42, 660–667.

Wallwork, J.A., 1970. Ecology of Soil Animals. McGraw-Hill, London.

Walter, D.E., Kaplan, D.T., Permar, T.A., 1991. Missing links: a review of methods used to estimate trophic links in soil food webs. Agric. Ecosyst. Environ. 34, 399–405.

Wardle, D.A., 2002. Communities and Ecosystems: Linking the Aboveground and Belowground Components. Princeton University Press, Princeton.

Wardle, D.A., Bardgett, R.D., Klironomos, J.N., Setälä, H., van der Putten, W.H., Wall, D.H., 2004. Ecological linkages between aboveground and belowground biota. Science 304, 1629–1633.

Wheeler, G.C., Wheeler, J., 1990. Insecta: hymenoptera formicidae. In: Dindal, D.L. (Ed.), Soil Biology Guide. Wiley, New York, pp. 1277–1294.

Whitford, W.G., 2000. Keystone arthropods as webmasters in desert ecosystems. In: Coleman, D.C., Hendrix, P.F. (Eds.), Invertebrates as Webmasters in Ecosystems. CAB International, Wallingford, UK, pp. 25–41.

Whitman, W.B., Coleman, D.C., Wiebe, W.J., 1998. Perspective. Prokaryotes: the unseen majority. Proc. Natl. Acad. Sci. USA 95, 6578–6583.

Wilschut, R.A., Geisen, S., Martens, H., Kostenko, O., de Hollander, M., ten Hooven, F.C., et al., 2019. Latitudinal variation in soil nematode communities under climate warming-related range-expanding and native plants. Glob. Change Biol. 25, 2714–2726.

Wilson, E.O., 1987. Causes of ecological success - the case of the ants - the 6th Tansley lecture. J. Anim. Ecol. 56, 1–9.

Wilson, E.O., 1992. The Diversity of Life. Norton, New York.

Wilson, M.J., Kakouli-Duarte, T., 2009. Nematodes as Environmental Indicators. CABI Press, Wallingford, UK.

Wu, T., Ayres, E., Bardgett, R., Wall, D., Garey, J., 2011. Molecular study of worldwide distribution and diversity of soil animals. Proc. Natl. Acad. Sci. USA 108, 17720–17725.

Wurst, S., van der Putten, W.H., 2007. Root herbivore identity matters in plant-mediated interactions between root and shoot herbivores. Basic Appl. Ecol. 8, 491–499.

Xiong, W., Song, Y., Yang, K., Gu, Y., Wei, Z., Kowalchuk, G.A., et al., 2020. Rhizosphere protists are key determinants of plant health. Microbiome 8, 27.

Yeates, G.W., Coleman, D.C., 1982. Role of nematodes in decomposition. In: Freckman, D.W. (Ed.), Nematodes in Soil Ecosystems. University of Texas Press, Austin, pp. 55–80.

Yeates, G.W., Bongers, T., Degoede, R.G.M., Freckman, D.W., Georgieva, S.S., 1993. Feeding habits in soil nematode families and genera – an outline for soil ecologists. J. Nematol. 25, 315–331.

Yeates, G.W., Dando, J.L., Shepherd, T.G., 2002. Pressure plate studies to determine how moisture affects access of bacterial-feeding nematodes to food in soil. Eur. J. Soil Sci. 53, 355–365.

Zachariae, G., 1964. Welche bedeutung haben enchytraus in waldboden? In: Jongerius, A. (Ed.), Soil Micromorphology. Elsevier, Amsterdam, pp. 57–68.

Zhao, Z.-B., He, J.-Z., Geisen, S., Han, L.-L., Wang, J.-T., Shen, J.-P., et al., 2019. Protist communities are more sensitive to nitrogen fertilization than other microorganisms in diverse agricultural soils. Microbiome 7, 33.

Chapter 6

Molecular and associated approaches for studying soil biota and their functioning

Ulas Karaoz*, Joanne B. Emerson[†], and Eoin L. Brodie*[‡]

*Ecology Department, Climate and Ecosystem Sciences Division, Lawrence Berkeley National Laboratory, Berkeley, CA, USA;
[†]Department of Plant Pathology, College of Agricultural and Environmental Sciences, University of California, Davis, CA, USA;
[‡]Department of Environmental Science, Policy and Management, University of California, Berkeley, CA, USA

Chapter outline

Soil Microbiology, Ecology, and Biochemistry. https://doi.org/10.1016/B978-0-12-822941-5.00006-5

6.1 Introduction

Soils are arguably the most diverse microbial ecosystem on Earth (Fierer and Lennon, 2011). As a dynamic, porous medium, complex gradients of water, carbon (C), and other elements combine to create a diversity of niches that support a broad range of microbial metabolisms and growth strategies. Within this context, microorganisms (bacteria, fungi, viruses, microeukaryotes) compete and cooperate with each other, as well as with plants and soil macrofauna. This breadth of metabolic diversity, complex metabolic interactions, and gradients of energy and resources make cultivation of soil microorganisms challenging. For this reason, cultivation-independent approaches to study soil microorganisms are essential complements to our understanding of the traits required for microbial fitness in the soil environment. These approaches involve the detection of molecules of biological importance as proxies for the presence, metabolic potential, or activity of soil microorganisms. Collectively, these approaches have been termed "-omics."

This chapter describes the application of diverse molecular methods to the study of soil biota and focuses on a subset of commonly applied omics techniques and their applications. These include genomics, transcriptomics, proteomics, and metabolomics, in addition to their "meta-" applications to whole communities, driven by technological developments, new computational hardware, and algorithms, including artificial intelligence. This omics information can be used to infer the traits of microorganisms using information encoded in their genomes. With the increasing ability to resolve complete genomes of soil microorganisms, new possibilities exist to explore how ecological theories apply to microbial systems across scales.

Although this chapter touches on some molecular methods used to measure microbial metabolism and physiology in situ, Chapter 7 provides a more comprehensive review. Similarly, details on methods to quantify the diversity of soil microeukaryotes and other soil fauna are provided in Chapter 4. Recognition and interest in the importance of viruses in soil microbial ecology and biogeochemical cycles have increased substantially in recent times, with new molecular tools beginning to uncover the significance of viral diversity and function. An up-to-date view on the status of this emerging field is presented here.

Issues of scale are discussed in this chapter, particularly the spatial scale at which omics approaches are commonly used. Due to the nested nature of soil compartments (Thakur et al., 2020), the scale of observation has important implications for how one interprets the properties of microbial niches, microbe-microbe interactions, and the influence of higher trophic levels and environmental drivers on microbial community composition and function. The chapter concludes with a discussion on new opportunities that promise to make molecular tools more accessible and better integrated into soil biology.

6.2 Omics in the context of soil biota

Systems biology is the study of complex biological systems using observation, manipulation, and computation. Approaches developed for microorganisms in pure culture, or for less complex microbiomes, are continuously adapted and applied to soil systems. This growing field of soil systems, or "ecosystems biology" (Bissett et al., 2013; Raes and Bork, 2008), targets the physiology, metabolism, ecology, and evolution of soil microorganisms and communities in the context of their habitat, with molecular understanding being a primary goal. Genomes provide the information, with core functions inherited over evolutionary time and shaping the capabilities of modern-day microbiomes (Jia and Whalen, 2020). Niche adaptation can occur through the evolution of new genetic traits that can be shared across microbiomes via horizontal gene transfer (HGT) mediated by mobile genetic elements, sometimes termed the "mobilome" (for a recent review of HGT, see Brito, 2021). The global analysis of all genetic information in a system is termed metagenomics. The activation and expression of genetic traits can be observed by analyzing the collective pool of expressed genes in a system (metatranscriptomics) that, due to the short half-life of messenger RNA (mRNA), provides a snapshot into the current investment strategies of soil microorganisms. Metaproteomics tracks the translation of these messages into proteins and reveals further information about the activity of specific microbial strains, as well as the metabolic costs incurred by microbes as they acquire resources, protect themselves from stress, engage in combat with local competitors, and repair damage. Microbes and microbiomes are complex chemical factories, producing materials for cellular biosynthesis and the processes described above, including a broad range of what are termed "secondary metabolites" — compounds that are not thought to be typically required for growth, but provide a competitive advantage under certain conditions, antibiotics being a good example. The analysis of these diverse chemicals is called metabolomics with growing applications to complex systems such as soil (Bahureksa et al., 2021; Swenson et al., 2015).

DNA approaches include the targeted analysis of phylogenetic or functional gene biomarkers, typically via PCR and are then sequenced to obtain a profile (or barcode) of community composition. For nontargeted DNA or shotgun sequencing, DNA extracts are sequenced and compared to existing databases to obtain profiles of community composition and function (read-based) or assembled into longer sequences (contigs) which may be further assembled into scaffolds. Contigs and scaffolds can be associated with genomes by binning; that is, grouping sequences together based on their properties and/or patterns of abundance (coverage) over space or time. This collection of sequences may be further refined into complete or near-complete genome sequences. This genome-resolved approach allows for the prediction of the role and ecological niche of an organism. The expression of genes is determined by sequencing RNA or reverse-transcribed DNA copy (cDNA). Various fractions of the total pool of RNA in soil can be enriched to target eukaryotes (polyA enrichment) or mRNA (rRNA depletion). These reads from expressed genes or RNA viruses can be mapped to contigs, scaffolds, and genomes, or assembled directly. Counting reads mapped to genes (open-reading frames) is then used to evaluate differences in gene expression between samples. Similarly, protein expression can be evaluated by counting peptides produced by fragmentation of protein extracts with the enzyme trypsin. Peptide fragments can be detected using high mass accuracy spectrometry and then compared to predicted fragment patterns from protein databases or peptide fragments predicted from genomes or metagenomes. Counts of peptides per gene are used to compare protein expression across samples. Metabolomics represent a suite of approaches to determine the metabolite composition of microbiome samples. Compositional properties of metabolites such as their elemental stoichiometry or energy density can be compared across samples. The patterns of presence-absence or abundance can be used to associate metabolites to networks where connected metabolites may have functional relatedness, e.g., as part of the same metabolic pathway or originating from the same organism. These approaches are summarized in Fig. 6.1.

With this growing suite of analytical tools to explore the composition and function of all components of soil microbiomes, there is great potential for new discoveries and change to existing paradigms. However, this will also require both data-driven discovery and hypothesis testing (Prosser, 2020). Given the challenge of interpreting complicated data from complex systems, data standardization, integration, and synthesis are key, and the use of quantitative mathematical approaches is essential. Some of these challenges and opportunities are discussed in Sections 6.7 and 6.10.

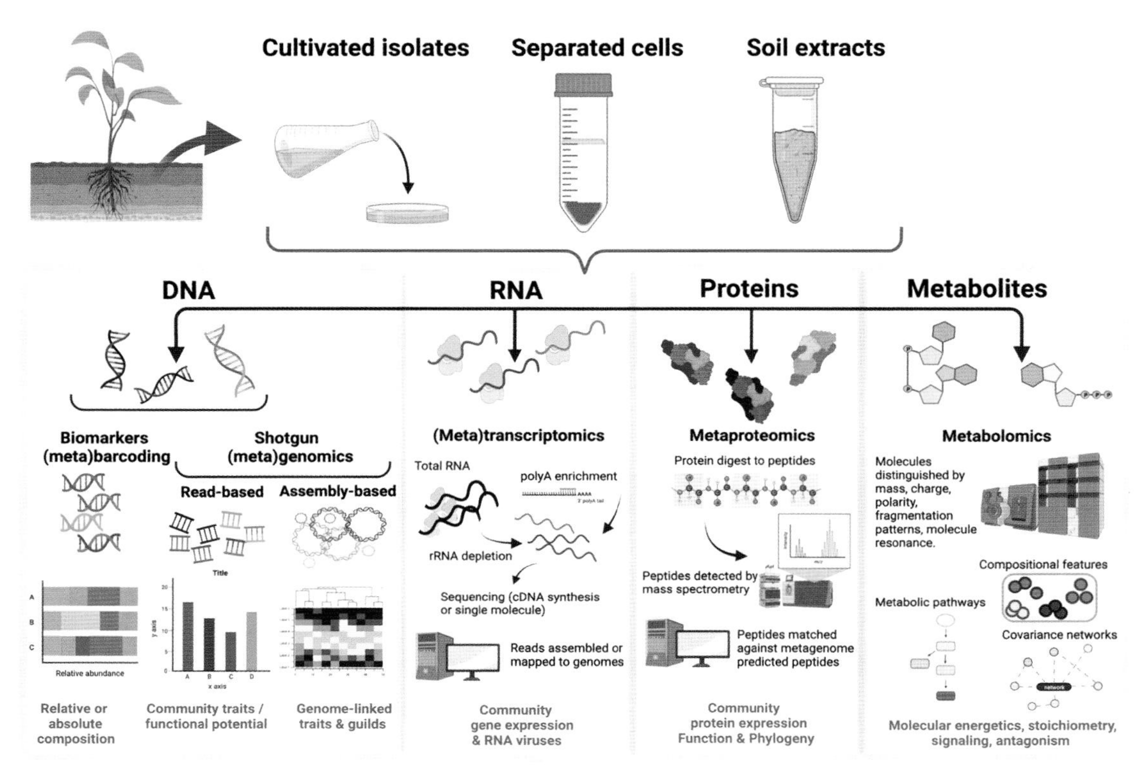

FIGURE 6.1 Application of omic approaches to soil systems. These approaches can be applied to isolated organisms or enrichment cultures, to microbial cells or viruses separated from soil and enriched through centrifugation, or to biomolecules extracted directly from soil. Analysis of DNA quantifies the composition or functional potential of a soil microbiome. RNA and protein analyses evaluate the activation of that potential under specific conditions, while the products of microbial activity (e.g., trace gases, metabolites) can be quantified to explore the consequences of changes in microbiome composition and activity for the functioning of soil. *(Created with BioRender.com.)*

6.3 Direct extraction, relic DNA, and cell separation for omic analyses

A common consideration in the molecular analysis of soils is whether to directly extract nucleic acids or attempt to extract and enrich cells to reduce the influence of extracellular DNA. Extracellular DNA can persist in soil (Pietramellara et al., 2009) and can represent a significant proportion of the extracted soil DNA. This DNA pool likely includes remnants of inactive and dead cells (termed *relic DNA*), distorts estimates of diversity (Carini et al., 2016), and may also constrain the detection of temporal changes in soil microbiome composition (Carini et al., 2020). While laboratory approaches have been developed to reduce the influence of extracellular DNA on microbial censuses (Carini et al., 2016), the alternative approach, cell extraction prior to cell lysis and DNA extraction, has shown to have limited benefit. To extract cells from the soil matrix, cells must be dislodged from their natural habitats (typically biofilms on particle surfaces). This is achieved with physical shaking (e.g., using a vortex or blender) or gentle sonication in the presence of a buffer, typically containing surfactants (i.e., Tween 20; polysorbate 20) and salts to reduce adhesion of cells and biofilms to particles. Detached cells may be separated and concentrated from soil particles by density centrifugation using Nycodenz (see Ouyang et al., 2021). The extracted cells can then be assessed for viability using differential staining methods, cultivation on growth media, or via viability PCR (Ouyang et al., 2021). Moreover, the recovery of intact cells allows the estimation of critical biophysical traits, such as cell size, shape, and surface-area-to-volume relationships (Portillo et al., 2013). Recovering intact cells also enables innovative single-cell omic studies combined with isotopes to focus on metabolically active cells (Jing et al., 2021). Despite these numerous advantages, recovering cells from soil without creating bias in terms of community composition remains a challenge due to differences in cell adhesion to minerals, cell wall rigidity (e.g., Gram positive or negative), and occurrence as biofilms or as planktonic cells. For these reasons, in addition to sample processing throughput considerations, most studies continue to use direct soil nucleic acid extraction procedures.

Direct nucleic acid extraction from soils typically involves the weakening of cell walls and membranes using detergents (e.g., SDS, CTAB), enzymes (e.g., proteinase K), or solvents (e.g., phenol, chloroform). Physical methods (grinding, blending, ultrasonication, high pressure, bead beating) are then used to release nucleic acids into chemical buffers that reduce or eliminate their enzymatic degradation (e.g., EDTA or dithiothreitol), buffer pH changes, and reduce sorption to soil minerals (e.g., phosphate buffers). This chemical milieu is then separated from soil particles and cell debris via centrifugation and the various liquid phases are recovered for purification of nucleic acids and other biomolecules. High molecular weight DNA is typically required with the advent of single molecule, long-read sequencing approaches. Gentler chemical and enzymatic cell lysis are favored, with physical lysis methods that cause DNA fragmentation typically avoided. A single extraction protocol has been developed for the analysis of proteins, metabolites, and lipids from the same soil sample (Nicora et al., 2018).

6.4 The use of biomarkers and metabarcoding for diversity studies

Biomarkers are biomolecules used to indicate the presence (current or past) of an organism. In soil systems, the most commonly used biomarkers for the study of microorganisms are the conserved ribosomal RNA (rRNA) coding genes or the internal transcribed spacer (ITS) regions that flank them. The 16S rRNA gene remains the most commonly used biomarker for bacteria and archaea, while 18S, 28S, or ITS regions are commonly analyzed for fungi and other eukaryotes. PCR primers, or sets of primers, have been designed to balance the specificity to these regions of the genome, while maintaining their flexibility (also termed degeneracy) to capture as many variants as possible and ensure a representative sampling of all members of a microbiome. The process of sampling and analyzing these biomarkers across a complex system is now frequently termed "metabarcoding," where "meta"- signifies profiling of the microbial content of numerous microbial community samples. This is made possible through the use of high-throughput sequencing, where many samples can be pooled together and analyzed in a single sequencing run taking advantage of the millions of reads these platforms produce to improve cost effectiveness. As hundreds to thousands of samples may be pooled together, a system of short DNA barcodes is incorporated into the process. Thus, microbial biomarker sequence information is computationally assigned to each sample after the sequences are obtained. By making microbiome sequencing affordable, these high-throughput PCR amplicon sequencing approaches have revolutionized the study of microbiomes in complex systems. However, not all sequences in a microbiome may be captured due to differences between primer sequences and the related regions of the organism (i.e., PCR primer bias). The most commonly used sequencing platforms (e.g., Illumina) produce millions of short reads (100–300 bp in length). For this reason, both specific and variable regions of the biomarker genes are targeted, limiting their ability to distinguish closely related microorganisms that may have different ecological roles and niches, potentially obscuring important patterns. A recent review by Francioli and colleagues (Francioli et al., 2021) provides an excellent description of potential pitfalls of metabarcoding approaches for terrestrial ecosystem studies. In the following sections, we discuss common challenges faced when interpreting high-throughput biomarker data.

6.4.1 Compositional nature of high-throughput biomarker data

A key feature of high-throughput amplicon sequencing data is that it is compositional. Compositional data add up to a fixed sum and contain information only on portions of various components (Aitchison, 1982; Pawlowsky-Glahn et al., 2015), meaning no information is contained on the actual (absolute) frequencies. As a result, standard methods for multivariate analysis result in misleading inferences for compositional data because it is assumed that each observation is independent and originates from a normal distribution. The fixed-sum-constraint (due to relativizing sequence count data) invalidates the assumption of independence between component variables as an increase in the proportion of a sequence from a specific organism naturally resulting in the decrease of others creating a false dependency. These pitfalls in relation to multivariate analysis have been noted in Pearson's seminal work (Pearson, 1897). Furthermore, it also leads to constraints on the variance-covariance matrix and invalidates standard methods for regression and multivariate analysis which depend on the assumption of multivariate normality.

Conceptually, starting with soil DNA and ending with reads, the data generation process from modern sequencing instruments is as follows. A random sample of the DNA in a soil sample is taken as a library and a random sample of the DNA in the library is sequenced. The instrument generates data up to a machine-specific capacity such that the total read count is uninformative of the absolute number of molecules in the input sample. Therefore, the read counts can only be interpreted as relative abundance of the molecules in the loaded sample after normalizing to the total count from a machine run (Lovell et al., 2020).

Many of the microbiome data analysis methods were originally adopted from analysis of macro-ecological data, where organismal counts are normalized to a constant area or volume. The development and evaluation of compositional data analysis (CoDa) methods for omics data, including amplicon metagenomics, play an active role in the field of research (Erb et al., 2020; Lovell et al., 2011; Quinn et al., 2019). Several compositional replacements of standard microbiome analysis techniques have been suggested for normalization (Aitchison, 1982), distance metrics (Aitchison et al., 2000; Silverman et al., 2017; Wong et al., 2016), ordination (Aitchison, 1983; Aitchison and Greenacre, 2002; Gloor et al., 2016), correlation (Erb and Notredame, 2016; Friedman and Alm, 2012; Kurtz et al., 2015; Schwager et al., 2017), and differential abundance testing (Fernandes et al., 2014; Hawinkel et al., 2019; Mandal et al., 2015; Thorsen et al., 2016) reviewed in Gloor et al. (2017).

6.4.2 Effect of gene copy number variation

Several biomarker genes exist in multiple copies per genome, a common biological variation; therefore, their use for estimation of taxonomic proportions based on counts of PCR amplicon variants is biased. For instance, copy numbers of the 16S rRNA gene used for bacterial and archaeal diversity studies (Acinas et al., 2005) and ITS and 18S rRNA gene regions used for fungal and eukaryotic diversity (Schoch et al., 2012; Thornhill et al., 2007) can span up to an order of magnitude (1−15 in bacteria, 1−5 in archaea). Copy number variation shows strong phylogenetic conservation (Angly et al., 2014), with a wide variance across phylogenetic scales (Větrovský and Baldrian, 2013). Available genome sequences of "close" relatives for taxa allows correction of relative abundances based on genome-based copy number estimates (Angly et al., 2014; Stoddard et al., 2015), but in practice, leads to inaccurate predictions for many clades that are distant from sequenced genomes (Louca et al., 2018).

6.4.3 Approaches for absolute quantification

The effects of compositionality can be remedied with statistical methods, yet these methods cannot be used to estimate absolute abundances from compositional data. Absolute abundances are clearly important to quantify as the number of cells determines the microbial biomass, contributions to biogeochemical process rates, and the probability of interactions with other organisms.

DNA- or cell-based techniques can be coupled with high-throughput sequencing to estimate absolute microbial abundances. DNA-based quantification methods have a long history yet their use for microbiome studies is less frequent and includes spike-ins and methods based on quantitative PCR (qPCR). A known number of cells of a species corresponding to a fixed number of biomarker gene copies, assumed not to be found in the environment of interest, can be spiked into environmental samples. The relative abundance of this species in the output reads can be calculated and used as a sample-specific calibration factor to estimate total microbial numbers (Jones et al., 2015; Stämmler et al., 2016). A similar DNA-based approach is to use synthetic DNA as the spike-in; that is, the addition of an internal standard (Harrison et al., 2021; Smets et al., 2016). Spike-ins specific to various domains of life allow both estimation of differences in absolute abundance within and across amplicon classes (i.e., domains). The primary challenge with these techniques is choosing the optimum amount of internal standards for a given environmental condition (Tkacz et al., 2018).

qPCR of genomic DNA with domain-specific universal primers (for bacteria/archaea or fungi) is a DNA-based technique to estimate total microbial numbers (Smith and Osborn, 2009). Starting with the same input material, qPCR is performed in parallel to sequencing library preparation. Per gram estimates of genomes present in the sample are calculated from qPCR threshold cycle (Ct) values and then used as a proxy for total counts. As with PCR, qPCR is subject to primer bias that results in uneven amplification of the target genes across species, as well as to the effects of the DNA extraction method on microbial community composition. Despite its limitations, qPCR is an accessible and cost-effective method for quantitative microbiome profiling. Due to the lack of a universal marker gene for viruses, qPCR cannot be used for absolute viral community quantification, but has been used to assess the proportions of specific known viral groups in environmental samples (Uyaguari-Diaz et al., 2016).

An alternative to DNA-based techniques toward absolute quantification is flow cytometric single-cell enumeration of microbial cells in the samples (Frossard et al., 2016; Props et al., 2017; Vandeputte et al., 2017). Flow cytometry requires extensive expertise to ensure reproducibility and remains especially difficult for microbial cells within complex matrices, such as sediments and soils that require mechanical or acoustic disruption of cell-to-particle associations (Emerson et al., 2017; Frossard et al., 2016; Lee et al., 2021). Unlike qPCR, which targets microbial DNA, flow cytometry counts microbial cells and exclusively intact cells. As a result, significant differences in quantification may be observed for samples containing free extracellular DNA that can be quantified both with qPCR and sequencing. Systematic comparison of qPCR and flow cytometry for quantification is scarce, with available results confirming discrepancies in estimates (Galazzo et al., 2020; Vandeputte et al., 2017). To date, there is no standardized, reliable approach for absolute quantification of viral particles in soil; however, this is typically attempted via epifluorescence microscopy, where a nucleic acid stain is added to purified soil viral particles in solution. The viral particles appear as fluorescent pinpricks that can be counted under the microscope based on protocols originally developed for marine water and sediments (Wilhelm et al., 2010; Williamson et al., 2005). Challenges associated with counting soil viral particles via epifluorescence microscopy are similar to those encountered for flow cytometry of microbial cells, including nonspecific binding of dyes to soil particles resulting in background fluorescence that can make it difficult to identify true viral particles (Emerson et al., 2017).

6.4.4 New approaches to link microbial diversity to sequencing data products

Historically, the most common approach for determining taxonomic and phylogenetic diversity based on marker gene sequencing has involved clustering sequence readouts at a sequence similarity threshold that matches a taxonomic rank (e.g., the often applied 97% 16S rRNA gene identity as a proxy for species) to obtain operational taxonomic units (OTUs), followed by selection of a single representative sequence for each OTU to represent the taxonomic rank in question. The motivations for this choice were diverse. The taxonomic resolution for single marker genes is limited, and the per-base error rates of high-throughput sequencing technologies were too high to confidently allow finer sequence discrimination between closely related sequences.

Although used in practice, these similarity thresholds are arbitrarily defined cutoffs with an aim to pick a single threshold that applies "well" to all clades per loci. There are significant differences in the optimum threshold between clades and indeed between the different regions of loci being targeted (e.g., the different 16S rRNA gene hypervariable regions; Edgar, 2018). An additional drawback for using cluster (OTU) representatives as primary data products for biodiversity studies is the difficulty in comparing results across multiple studies. The unsupervised clustering algorithms used for sequence clustering are diverse. All use significant heuristics both by their design and to deal with data volumes, resulting in a challenging comparison for diversity metrics and relative counts of taxa.

With increased accuracy of short-read sequencing, it is now feasible to catalog exact sequencing variants (ESVs), also termed amplicon sequence variants (ASVs) (Callahan et al., 2017; Tikhonov et al., 2015). These new approaches depend on effective and strict denoising algorithms (Amir et al., 2017; Callahan et al., 2016; Edgar, 2016) to come up with clusters where within cluster sequence differences match the error rate of the sequencing platform (significantly less than the traditional 3% difference). Compared to OTUs, ASVs have a higher resolution, as by design, they represent a smaller variance in base pair level differences. This can be important in biogeography studies and in resolving ecologically distinct subpopulations, especially in natural environments with high diversity (Ettinger et al., 2021; Ghannam et al., 2020; Tikhonov et al., 2015). With ASVs also comes greater reproducibility through avoidance of clustering heuristics, as well as data reusability since ASVs can be used as universal signatures to compare across studies.

The latest approaches for identifying viral species (vOTUs) rely on shotgun metagenomic (or viromic) sequencing data (discussed in subsequent sections) Amplicon-based targeting of specific viral marker genes, such as those encoding major capsid proteins, is only used for identifying specific viral groups and has thus far not been extensively applied to soil.

6.5 Meta-omics of soil biota: read-centric data analysis approaches

Meta-omics refers to creating a "parts list" of a microbial community through metagenomics and determining the role and significance of each part within specific environmental niches through metatranscriptomics, metaproteomics, and metabolomics. This information is synthesized in the context of physical or biogeochemical measurements to better understand and predict soil functioning and microbial impacts on soil ecosystems. The resolution that we consider a "part" and the assumed nature of relationships within parts drives the data analysis strategy. The two major meta-omics data analysis strategies relate to the resolution of the "part" and are termed gene-centric and genome-centric analysis workflows (Fig. 6.2). DNA extracted and isolated from microbiome samples is randomly fragmented and packaged as sequencing libraries specific to a shotgun sequencing platform, generating hundreds of thousands to millions of reads. Raw reads go through quality control to identify and remove sequencing artifacts, contaminants, and quality trim reads. QC-ed reads are assembled into longer contiguous sequences (scaffolds) and reads from each sample are mapped back to the resulting scaffolds to determine the depth of coverage of scaffolds across samples. The resulting scaffolds can be subjected to marker gene-based analysis, or further clustered into genome bins based on composition and depth of coverage profiles across samples. The resulting bins are evaluated for completeness and contamination based on patterns of conserved features of complete isolate genomes and for mis-assemblies as evaluated by read coherence. Quality filtered bins (metagenome-assembled genomes [MAGs]) are annotated for genetic elements (structural annotation, i.e., protein-coding genes, noncoding RNA) and functional roles are assigned to the inferred coding sequences.

Metagenomics, that is sequencing of all the DNA from a microbial community, lies at the heart of a typical meta-omics study. Metagenomics is often referred to as "shotgun" metagenomics, which is untargeted sequencing of all the microbial genomes in a sample after randomly breaking "shotgun" DNA into smaller pieces. Since metagenomic sequence data is generated from the genetic material of entire communities, it can be analyzed from different perspectives (e.g., taxonomic composition, metabolic potential, and phylogenetic diversity), and at different resolutions (e.g., taxonomically from broader (phylum) to more specific (strain), or at a community level).

6.5.1 Quality control of sequencing reads

The primary observable in a sequencing experiment is a sequencing read, that is, an inferred sequence of base pairs from part or all of a DNA fragment. The length (number of base pairs), error profile (probability of error across the read), and throughput (number of reads) are sequencing platform specific. Second-generation sequencing platforms (includes Illumina, Ion torrent, and the now discontinued 454 pyrosequencing) generate large numbers of short reads, while third-generation sequencing platforms (currently includes Pacific Biosciences single-molecule sequencing and Oxford Nanopore sequencing) generate a smaller number of long reads with higher error rates (Slatko et al., 2018). Independent of the platform, the first step for any metagenomic analysis is quality control, which includes identification and removal of low-quality, artifactual, and contaminant sequences.

Low-quality sequences are identified using quality scores (probability of error) for base pairs reported by the sequencer. Several tools are available for profiling (FastQC) and removal (Trimmomatic, BBtools, Sickle) of reads based on the magnitude and distribution of quality scores along the read. Artifactual sequences, such as adapter sequences, primers, and polyA tails, can be removed with cutadapt (Martin,

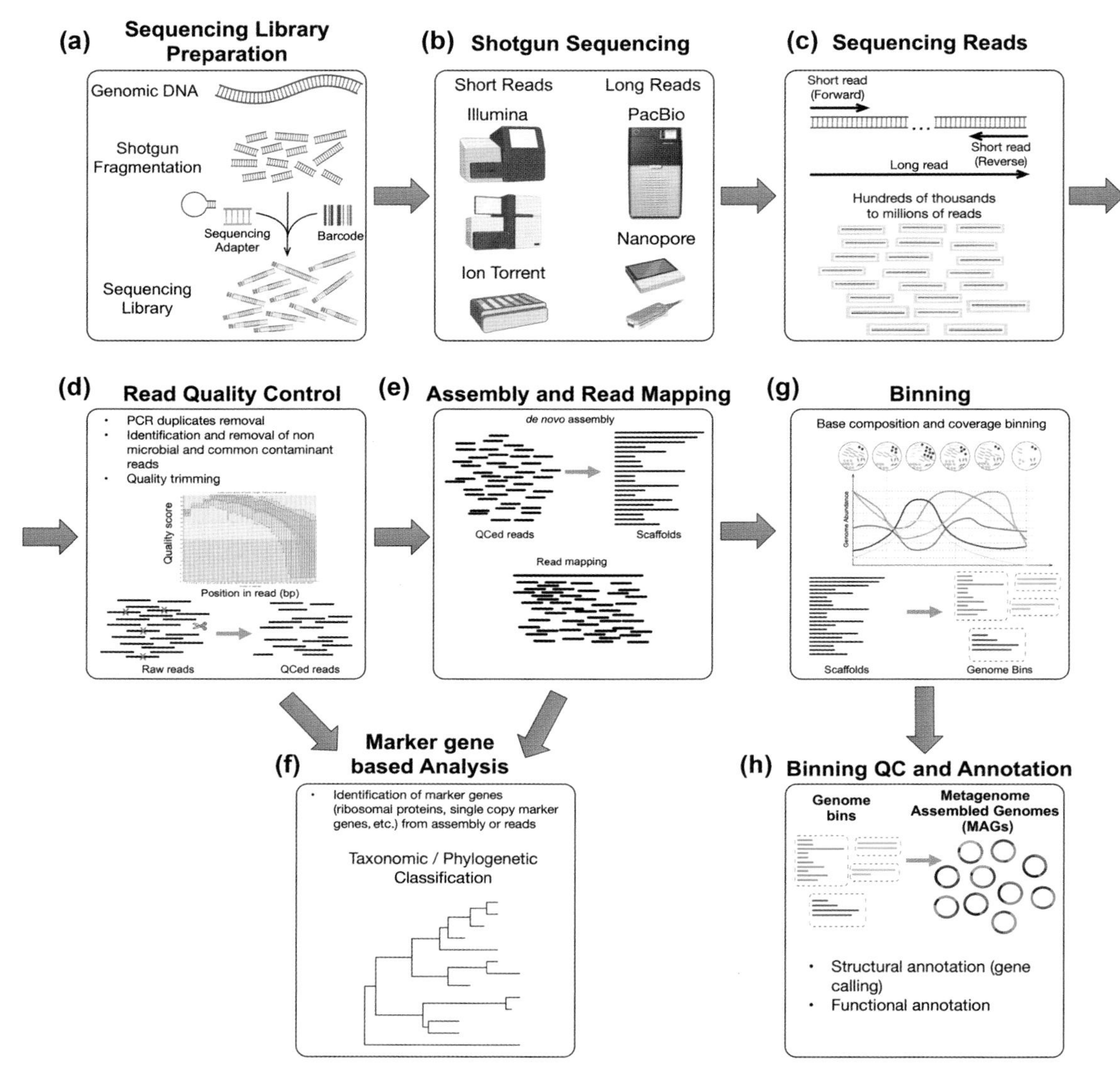

FIGURE 6.2 Genome-centric analysis of metagenome data. (a) DNA is randomly fragmented and (b) sequenced across a variety of platforms that vary in sequence length and throughput with (c) hundreds of thousands to millions of reads generated. (d) Raw reads are quality controlled (QCed). (e) QCed reads are assembled into scaffolds and sample reads are mapped back to the scaffolds to quantify coverage of scaffolds. (f) Marker gene-based analysis or (g) genome binning are carried out. (h) Bin quality control, i.e., completeness, contamination, and artifacts, is evaluated and quality-filtered bins (metagenome-assembled genomes [MAGs]) are annotated for genetic elements.

2011). If the soil sample comes from an environment with significant amounts of nonmicrobial DNA (e.g., the rhizoplane or rhizosphere) or if contaminant DNA is introduced during sample processing, the contaminant (undesired) reads should be removed before further processing. The identification of contaminant reads depends on comparison to reference databases and can be performed with tools such as FastQ Screen and BBDuk within BBTools. For post-quality control, depending on the research goals, reads can either be grouped into taxonomic bins (taxonomic classification) or can be assembled into longer contiguous sequences (assembly-based analysis).

6.5.2 Taxonomic classification with reads

Taxonomic classification is the process of inferring the taxonomic clades that make up a microbial community and their relative abundances. In comparison to marker gene amplicon sequencing, metagenome sequencing circumvents primer biases and allows sampling across all domains of life. However, as for marker genes, taxonomic classifiers depend on reference databases of sequenced isolate genomes to associate taxa with metagenome sequences. Therefore, the degree to which the reference database represents the biodiversity of the soil sample has a large impact on its accuracy. Methods for taxonomic classification can be grouped based on how the reference genome sequences are used.

Marker-gene-based methods focus on a small subset of most taxonomically informative "marker" genes (i.e., clade-specific with high discriminatory power between taxa). Since a limited number of genes rather than whole genome sequences are used, the reference database is pruned to only include the marker genes, resulting in reduced search space and computational requirements. Universal markers and clade-specific markers are both used for taxonomic classification.

Universal markers are genetic loci present in all microbes yet contain variable regions that can be used for taxonomic and phylogenetic differentiation. These include the 16S rRNA gene as well as other highly conserved genes such as RNA polymerase gene beta-subunit (rpoB) or ribosomal protein genes. Tools that use universal makers include PhylOTU (Sharpton et al., 2011), using the 16S rRNA gene only, mOTUS (Milanese et al., 2019), AMPHORA 2 (Wu and Scott, 2012), and MetaPhyler (Liu et al., 2011), which use several dozen universal marker genes. PhyloPhlAn (Beghini et al., 2021; Segata et al., 2013) relaxes the universality requirement to include hundreds of additional genes that are almost universal yet in conjunction provide high phylogenetic accuracy.

Clade-specific marker genes are defined with respect to a specific clade (i.e., species); that is, they exist only in that clade and have no sequence homology to any gene outside of it. MetaPhlAn (Beghini et al., 2021; Segata et al., 2013; Truong et al., 2015, 2017) compiles and uses more than 1 million such species-specific marker genes to estimate abundance of well-characterized taxonomic groups in a metagenome sample.

Composition-based methods leverage base pair compositional features of sequences (i.e., GC-content, codon usage bias, k-mer distribution) benchmarked as organismal fingerprints. Clade-specific (genus or species) statistical models with a defined set of features are built (training) using reference genomes which are then applied to label the metagenome sequences with clade information. Tools built around composition-based methods include Kraken 2 (Wood et al., 2019), CLARK (Ounit et al., 2015), PhyloPythiaS (Gregor et al., 2016), and LMAT (Ames et al., 2013), all of which use distribution of genomic k-mers (a string of nucleotides of length k) as features for taxonomic classification of metagenome reads.

Mapping-based methods classify reads into taxonomic groups based on read mapping to sequences of reference genomes. Reference genomes are indexed with compact indices (i.e., using FM-index based on

Burrows-Wheeler transform) that allow efficient identification of matching subsequences (mapping). Centrifuge (Kim et al., 2016), GOTTCHA (Freitas et al., 2015), and Kaiju (Menzel et al., 2016) are tools that map reads to an indexed reference database and use the taxonomic labels of the hits to assign taxonomy to the reads. Among these, Kaiju is unique in that it is a protein sequence-based classifier that leverages higher conservation of protein sequences compared to underlying DNA to achieve higher classification accuracy.

Hybrid approaches that incorporate features of composition- and homology-based strategies have also been implemented in software tools including PhymmBL (Brady and Salzberg, 2009) that combines interpolated Markov models with sequence mapping and RITA (MacDonald et al., 2012) that couples a suite of BLAST family homology search tools with a Naive Bayes classifier.

The primary challenge in taxonomic classification of reads derived from soil samples is the limited availability of diverse soil reference genomes. Compared to human-associated microbes, soil microbes are not as well studied, resulting in their underrepresentation in genome databases. Multiple clades detected in soils lack cultured representatives leading to undersampling of genomes across the tree of life. The sparsity of soil genomes in some portions of the tree of life limits the identification of phylogenetically conserved sequences beyond major lineages.

De novo metagenome assembly and binning of reads provide an alternative analysis path for soil metagenomes (Fig. 6.2). Reads are assembled into contigs or scaffolds (ordered contigs with gaps). Contigs/scaffolds can be used as inputs to taxonomic classification or functional analysis, whereas those from closely related organisms are further grouped together, a process called binning. Depending on complexity of the samples, bins will represent a single genome or a mixture of several closely related organisms. In recent years, a plethora of strategies and tools that allow binning of draft genomes from metagenome samples have been developed (covered in the next section). Downstream taxonomic or functional analysis is then performed starting with bins instead of reads or contigs/scaffolds.

6.5.3 Metagenome assembly

Sequence reads, ignoring sequencing errors, are the primary window into the contiguous segments of genomic DNA from a soil sample. Given a set of sequence reads, metagenome assembly refers to the computational process of deriving longer contiguous segments of genomic DNA supported by the read set. As such, an assembly represents a hypothesis about the identity of the genomic segments in the sample from a multitude of organisms. Here we primarily refer to de novo metagenomic assembly. That is, one in which only information from the sequenced reads is used, as opposed to a reference-guided assembly that makes use of reference sequences of closely related organisms. From a theoretical point of view, sequence assembly is a computationally hard problem with no known efficient computational solution (Medvedev et al., 2007). The task of genome assembly becomes more complex in the case of metagenome data obtained from soil where sequence reads originate from a large number of organisms, a significant portion of which can have a high sequence of similarity, yet at a different relative abundance.

There are two computational approaches to sequence assembly, each leveraging a different data structure: (1) overlap-layout-consensus (OLC) and (2) the de Bruijn graph (DBG) approach. A sequence assembler implements one or a combination of these two approaches during different stages of the assembly pipeline and also incorporates a variety of heuristic techniques. OLC is the more intuitive strategy in which overlaps (O) between all read pairs are found (stored in an O graph), on which a layout (L) — contiguous reads based on the O — is formed and a consensus contiguous sequence (contig) from overlapped laid-out reads is determined. DBG works by determining the short sequence content of reads at a fixed length k (known as k-mers: a string of nucleotides of length k), represented in a graph structure (DBG) that infers the contig sequence of the DBG (Pevzner et al., 2001). Compared to OLC, reads in DBG are replaced by k-mers, which means chopping off reads into smaller pieces and as such, is counter intuitive to the goal of sequence assembly. All assemblers that deal with large volumes of short read sequencing data use a DBG strategy. The need for k-mer (Compeau et al., 2011) rather than an overlap-based representation of input reads stems from the intractability of building overlap-graphs with large volumes of short reads. k-mer approaches consider reads one-by-one, while the overlap approach needs to sift through all read pairs, translating into a dramatic advantage in compute time and memory complexity. The "price" of the k-mer representation of reads is the loss of sequence context as a result of chopping off the reads which increases the likelihood of joining discontiguous regions from genomes with similar k-mer content.

Deploying best practices for metagenome assembly of reads generated from soil samples requires an understanding of the algorithmic trade-offs at play. First and foremost, one should observe that the cardinality, identity, diversity, and relative abundance of genomes in a sample are unknown. Consequently, the expected coverages (sampling redundancy: mean number of times a readout from a position in the genome is generated) of the target genomes for a given sequencing effort (total sequences generated for a sample) cannot be accurately predicted. Larger genomes or genomes with a lower abundance would have lower coverage compared to smaller genomes or genomes with a higher abundance. As a result, unlike DBGs representing isolate genome reads where k-mers with low coverage are assumed to be due to sequencing errors and removed to simplify the graph, the same assumption cannot be made for metagenome reads. The resulting graph is significantly more complex and traversing the graph to generate accurate contigs is more difficult.

In addition, it is difficult to detect repeats during metagenomic assembly. For isolates, assuming a uniform coverage of the genome with sequencing, repetitive DNA can be detected based on irregularities in the depth of coverage in the graph. For metagenomes, an uneven and unknown distribution of organisms rules out the use of simple coverage statistics for repeat detection. The presence of nearly identical regions from unrelated genomes (i.e., mobile genetic elements) and multiple genomes of either related or the same species presents additional complexity to hosting differences (i.e., strain variants). When inferring contigs from the graph, it is up to the user to decide whether such differences should be ignored.

Metagenome assemblers include IDBA-UD (Peng et al., 2012), MEGAHIT (Li et al., 2015), Meta-Velvet (Namiki et al., 2012), Ray Meta (Boisvert et al., 2012), metaSPAdes (Nurk et al. 2017), and MetaHipMer (Hofmeyr et al., 2020). A metagenome assembler takes sequence reads as its input and outputs a list of primarily longer sequences supported by those reads. It is a computational pipeline that typically consists of the following steps: (1) read error correction, (2) construction of the DBG, (3) inference of an assembly graph using graph simplification, and (4) traversal of the assembly graph to infer genomic fragments. While the error-correction step is optional, the subsequent steps are iterated for different k-mer sizes merging the outputs from specific iterations. The a priori decision on the size of the k-mer has a large effect on the resulting assembly. Larger k-mer sizes will disconnect the resulting graph structured while smaller k-mer sizes will result in an overly connected graph structure that is difficult to traverse. For a fixed k-mer size, the differences in building the DBG are largely on the implementation side. Much of the impactful differences stem from the strategy to pick and stagger the range of k-mers, downstream heuristics to simplify and traverse a complex DBG to generate output contigs and further, link contigs into scaffolds (ordered contigs with gaps of known length in between). Some of these assemblers are specifically engineered to deal with larger data volumes. MEGAHIT uses a succinct DBG (DBG with Burrows-Wheeler transform) to reduce storage and computational requirements. Ray Meta and MetaHipMer are designed for distributed memory systems available in a networked cluster of low-memory machines. Metagenomic assembly for the recovery of viral contigs (e.g., from viral size-fraction metagenomes, known as viromes, described below) proceeds in much the same way as described here, with metaSPADES and MEGAHIT shown to perform well for viral assembly from large datasets (Roux et al., 2017).

6.6 Meta-omics of soil biota: from potential to function

A genome or a metagenome assembly provides the blueprint for the capabilities of a microbe or microbiome. Only a fraction of this capacity is expressed at any one time, with local soil conditions at the scale of the microbe influencing which genes and pathways are expressed or silenced, which proteins are produced or degraded, and which metabolites are synthesized or consumed. Transcriptomics, proteomics, and metabolomics, and their "meta-" versions are increasingly applied to explore how the functional potential of a microbiome is activated. For gene expression, RNA is extracted from isolates, extracted cells, or directly from soil, often in combination with DNA, proteins, and metabolites. The sum of RNA in a cell (total RNA) is ~80% rRNA, molecules that provide scaffolding for ribosomal proteins to interact and form a functional ribosome. RNA that carries codes for protein synthesis is messenger RNA (mRNA) that makes up ~5% of the total RNA. To analyze mRNA, it needs to be actively enriched from the total RNA pool or rRNA needs to be depleted. For eukaryotes, mRNA is typically polyadenylated at the 3′ end, resulting in a poly(A) tail of multiple adenosine monophosphates that can be used to capture, or selectively synthesize via reverse transcription, this fraction of mRNA from the total RNA pool. This approach has been effectively used to study fungal gene expression in soil (Damon et al., 2012), while new transcriptomic markers have been identified as markers of fungal growth and respiration that together predict the efficiency of fungal biomass synthesis (Hasby et al., 2021). As most bacterial and archaeal mRNA transcripts are not polyadenylated, an approach to deplete the rRNA fraction of total RNA is preferred. Using specific nucleic acid probes, it is possible to bind and remove the majority of rRNA molecules in a mixture, leaving RNA enriched in mRNA that can then be converted to complementary DNA (cDNA) and sequenced, or sequenced directly using third generation single molecule nucleic acid sequencing instruments. This novel approach of direct RNA sequencing has not yet been applied to soil microbiomes, but its potential is reviewed in Tedersoo et al. (2021).

Metagenomics and metatranscriptomics have been successfully combined to explore niche partitioning in the rhizosphere. For example, Nuccio et al. (2020) showed that living roots, decaying roots, and a combination of the two influenced the assembly and carbohydrate-active enzyme expression of bacterial genomes. This allowed guilds (organisms occupying the same niche with similar functional roles) to be defined, representing organisms with specialist and generalist growth strategies in these contrasting soil habitats. Reducing the complexity of soil microbiomes into a tractable number of guilds can simplify the representation of microbial diversity in mathematical models of soil biogeochemistry.

While analysis of soil enzyme activity has been common place in the soil literature for over a century, the identity of the proteins and their producers has only recently been explored in situ in natural systems (Banfield et al., 2005). The analysis of all proteins extracted from soil (metaproteomics) allows the activated functions of soil organisms to be explored simultaneously regardless of their phylogeny.

However, the process for identifying proteins requires that the enzyme trypsin be used to digest proteins into smaller peptide fragments that can then be identified using high-resolution mass spectrometry, followed by a comparison to predicted peptide masses based on computationally simulated trypsin digests of proteomes from reference genomes. Because of amino acid sequence variability that can significantly change trypsin digest peptide masses, the most suitable reference genome database for peptide prediction in soil systems typically comes from the same sample. This requirement of paired metagenomes for metaproteome analysis, along with the technical difficulties of obtaining intact proteins, currently limits the number of reported applications of this powerful (proteogenomics) method. Nevertheless, over the last decade, applications of metaproteomics in soil have begun to explore plant-microbial interactions and nutrient cycling, as well as the response to climate change and soil contamination (reviewed by Tartaglia et al., 2020). With new developments in long-read sequencing that will improve microbial genome reconstruction from soil (particularly for fungi), metaproteomics is likely to become more commonly applied across the soil sciences.

In contrast to metaproteomics, metabolomic analyses have been rapidly adopted in recent years, with several distinct analytical platforms including liquid chromatography-mass spectrometry (LC-MS), gas chromatography-MS (GC-MS), nuclear magnetic resonance (NMR), and Fourier-transform ion cyclotron resonance MS (FTICR-MS) (Swenson et al., 2015; Withers et al., 2020). Each of these methods differs in their accuracy and sensitivity for detecting and identifying the complex suite of metabolites in soil microbiomes and combining them allows better insights into soil microbiome responses to global change. An excellent example is a study using multiple metabolomics platforms and genome-resolved metagenomics (discussed below) to explore peatland response to global warming (Wilson et al., 2021). The authors discovered that experimental warming resulted in an increase in plant-derived metabolites and simple sugars. Combined with measurement of changes in the oxidation state of C in dissolved organic matter (from FTICR-MS data), they provided a mechanistic explanation for why warming favored methane emissions over carbon dioxide.

6.7 The new era of genome-resolved metagenomics

The last decade of metagenomics has witnessed the development of a suite of computational techniques and tools for recovery of genomes of soil microorganisms. Genome-resolved metagenomics refer to the family of computational strategies that deliver genome-level information from metagenome data. This is a critical advance in soil biology in that we can now ask questions about how combinations of traits are linked within genomes and how genomic traits vary across gradients that span micrometer to Earth system scales (e.g., mineralogy, pH, temperature, redox, vegetation, elevation, latitude).

6.7.1 Genome binning

The output from metagenome assembly is a set of contigs. The next step is to partition this set into subsets that ideally represent genomes, strains, or species if possible; a process called genome or contig binning. As for assembly, this task can be performed leveraging reference genomes (supervised) or in an unsupervised fashion. Some of the very same tools for read-based metagenomic data analysis are in theory suitable for supervised binning of contigs (Huson et al., 2007; Jia et al., 2011; Ounit et al., 2015; Wilke et al., 2016; Wood et al., 2019). However, since these depend on the availability of diverse and high-quality representative genomes, in practice, they perform poorly for contigs constructed from complex environments. As a result, the primary focus has been on the application of unsupervised clustering for genome binning.

All unsupervised binning tools use information on base composition and relative abundance (across samples) of contigs to be binned. Base composition is an informative feature set for contig binning because the within-taxa variance in base composition signatures is smaller than that between taxa. Therefore, contigs that belong to the same taxa can be grouped based on the similarity of their base composition. Base composition is most commonly represented as frequencies of all possible tetranucleotides, as this metric encapsulates amino acid composition, synonymous codon usage, minimizes effects of codon pair bias (Coleman et al., 2008), and avoids errors in gene-calling or frame shifts.

The relative abundance of contigs across a set of samples (relative abundance profile) is another feature used as input to unsupervised binning. The rationale here is that contigs that belong to the same genome will show, on average, more similarity in relative abundance trends across samples compared to contigs from different genomes. The relative abundance profile of each contig is estimated from depth-of-coverage of reads from each of the samples used in the profile. To estimate read depth-of-coverage of assembled contigs from a sample, reads are mapped to the contigs, and depth-of-coverage for each position of the sequence is calculated from which contig coverage is estimated. Unlike the calculation of tetranucleotide frequencies, there are multiple ways to infer contig coverage, thus the choice of mapper and downstream processing of the mapping information will affect the estimates and therefore the binning. A binning tool will typically package its own choice of a read mapper and will use a default estimation of contig coverage; however, a user can also feed in a matrix of precomputed contig coverages.

The primary input to a binning tool is a contig x feature matrix where the feature vector typically consists of tetranucleotide frequencies and contig coverages. A range of binning tools to perform binning based on sequence composition and coverage are available: CONCOCT (Alneberg et al., 2013) applies modified Bayesian model selection to automatically determine the number of clusters; MaxBin2 (Wu et al., 2016) implements the Expectation-Maximization (EM) algorithm; MetaBAT (Kang et al., 2015) uses a k-medoid clustering algorithm without a need to preset the number of clusters; MetaBAT2 (Kang et al., 2019) employs graph partitioning using a label propagation algorithm; and COCACOLA (Lu et al., 2016) takes advantage of both hard and soft clustering, using L1-norm distance rather than Euclidean distance to measure the similarity of pairwise contigs, and applies nonnegative matrix factorization (NMF) to cluster contigs (Lee and Seung, 1999).

6.7.2 Metagenome-assembled genome quality assessment

The aim of genome binning is to group contigs so that each group (bin) represents a MAG or a taxonomic group. This process is imperfect with use of independent metrics necessary to evaluate the quality of the final bins. Completeness and contamination are the primary metrics used for assessing MAG quality. Completeness refers to genome completeness, an estimate of the recovered proportion of the total length of the MAG. Contamination refers to the proportion of the MAG that is estimated to belong to another closely related MAG or a MAG with a similar coverage profile. These two indices represent a fundamental tradeoff due to the nature of clustering. As one tries to reduce the contamination of a genome bin by removing contigs with "high" probability of belonging to other genome bins, completeness of the bin is compromised, and vice versa. It is important to realize that no access is available for true completeness and contamination of a MAG; thus, we depend on observable variables that are assumed to correlate with them.

The presence and absence of universal single copy "marker" genes (~50 genes) in a MAG has been used for estimating MAG completeness and contamination (Eren et al., 2015; Sharon et al., 2013; Wrighton et al., 2012). Completeness is estimated as the proportion of the marker set as recovered, while contamination is estimated as the proportion of redundant genes (i.e., multiple copies of genes from the marker set) observed in a MAG with respect to the full set. The small number and uneven distribution of the universal single-copy marker set has an impact on the accuracy of these estimates. To alleviate this, marker gene sets specific to a genome's inferred lineage within a reference genome tree can be used. CheckM is an automated tool that uses lineage-specific bacterial and archaeal marker genes to provide accurate quality estimates for MAGs (Parks et al., 2015).

6.7.3 MAG refinement, dereplication, and curation

MAGs from automatic binning can be further refined to increase completeness and decrease contamination using additional automated tools or manual curation. Automated approaches take MAGs called by multiple binning tools from the same metagenomic assembly as input and generate nonredundant refined sets of MAGs. DAS Tool (Sieber et al., 2018), Binning_refiner (Song and Thomas, 2017), and metaWRAP's Bin_refinement module (Uritskiy et al., 2018) are tools that systematically look for refinement opportunities by aggregating bins from multiple binning tools to improve the overall quality of the MAGs. Anvi'o (Eren et al., 2015), an analysis and visualization platform, uses interactive tools to allow the user to manually refine the MAGs.

6.7.4 Viral recovery

The recovery of viral genomes from soil follows much the same approach as for other microorganisms. However, as described above for separation of bacteria or fungal cells from soil to improve signal-to-noise in genomic studies, the enrichment of virus-like particles by size fractionation (e.g., recovery in the post-0.2 μm size fraction) prior to nucleic acid extraction can significantly improve the diversity of detected viruses (Santos-Medellin et al., 2021). Shotgun metagenomic sequencing data derived from the viral size fraction is known as the "virome," and although some studies use this term to mean the viral portion of the microbiome, the term "virosphere" is preferred for the viral microbiome to avoid confusion.

To recover viral population sequences from a virome or total metagenome, current best practices do not use genome binning per se. Rather, viral contigs ≥10 kbp are clustered at 95% average nucleotide identity (ANI) into viral population ("species") sequences, known as viral operational taxonomic units (vOTUs). These vOTUs can then be assigned taxonomy via network analysis based on shared predicted protein content with isolated viral genomes in the NCBI RefSeq prokaryotic viral database, using vConTACT v2.0 (Bin Jang et al., 2019). However, ~2−5% of a typical soil viral community can be assigned taxonomy, highlighting the poor representation of viral genomes in public databases. Genomic and comparative genomic analyses of vOTUs (e.g., investigations of functional annotation) can proceed as genomes from other biota. Viral community ecological analyses are performed similar to those for ASVs and other compositional data described above. The primary difference is the input data for generating the "vOTU coverage table" of viral relative abundances in each sample, which is derived through read mapping from each virome to a dereplicated set of vOTUs for a given dataset (Roux et al., 2017; Santos-Medellin et al., 2021).

6.8 Big data challenges in molecular analyses of soil biota

Omics analyses can produce data across the entirety of biomolecules from soil biota. Each omics method measures a particular type of molecule in the sample collectively. As a consequence, each type of omics generates a large amount of information simultaneously, and different types of omics experiments on the same sample generate a heterogeneous suite of large datasets. For example, through sequential or co-extractions, a soil sample can be subjected to shotgun metagenomics and transcriptomics to generate billions of DNA and RNA reads from the entire community to untargeted metabolomics (NMR, FTICR-MS or LC/MS) to generate a spectrum containing information on various metabolites. When applied to tens to hundreds of samples, each of these techniques generates a large volume of heterogeneous data that need to be analyzed and integrated in an efficient and reproducible fashion.

6.8.1 Experimental planning

To conduct a successful multi-omics study, a carefully thought-out experimental design, together with computational and statistical considerations, must be constructed. The study should begin with the question of interest, determining the final statistical model and comparisons, and then work backward to apply the basic principles of statistical design of experiments: randomization, replication, and control (Fisher, 1936). Proper randomization should first be considered for all stages of the project, such as for samples with respect to treatments, treatments with respect to plots, and sources of nonbiological errors (i.e., machines, sequencing lanes, technicians, etc.). Next, the identification of molecular features that have strong statistical support for downstream actionability necessitates high-powered experimental designs. Replication is indispensable to estimate the variability associated with the phenomenon of interest and to achieve the statistical power to detect true differences in an underlying signal. Assuming the variability with different omics methods is well-constrained, biological replicates are more important than technical replicates to increase the power for detecting differences. Finally, soil microbiomes can both modify and be modified by environmental factors; thus, environmental covariates (i.e., soil moisture, temperature, pH, conductivity, dissolved oxygen, organic matter, nutrients, and other elements) provide important context to interpret high-throughput molecular measurements and their relationships to soil biogeochemical processes.

While the increasing availability of multiple omics methods makes their application to soil samples alluring, with limited resources, the application of all types of omics to every sample should be carefully considered. The number of molecular predictors generated by multi-omics projects is orders of magnitude larger than the number of samples (i.e., big-p, little-n; $p>>n$). In addition, omics data are often noisy and have considerable batch effects. Consequently, small and well-constrained studies testing hypotheses of limited scope may have a greater potential to yield well-grounded conclusions by decreasing noise and confounding factors. Reducing the number of omics layers, that is the number of molecular features measured, while increasing the number of replicates, will often result in statistically well-supported conclusions that will advance our knowledge.

6.8.2 Computational resources

During the last decade, high-throughput multi-omics experiments have become a "big data" science requiring significant computational resources to extract knowledge from data. An average soil research laboratory is typically ill-equipped to handle data storage and computational resources required for processing omics data. A sufficiently capable compute and data storage server for omics data analysis is not only expensive to acquire and setup, but also must keep up with the dynamic requirements of different computational workflows. This often quickly leads to a suboptimal computing environment.

A popular way to meet the computational needs for large-scale omics analysis is to use high-performance computing (HPC) clusters, hosted and managed by academic and government research institutions. An HPC cluster is a collection of server-grade computers (called nodes) connected with a fast interconnect. Nodes serving different types of tasks include compute, high-memory, GPU, and data transfer nodes. The diversity of these nodes provides for the wide variation in computational needs of omics analysis. Running DBG-based metagenome assemblers requires access to large memory nodes ($\geq$500 Gb of RAM) as they need to store each k-mer present in the graph with their additional attributes. A DBG-based assembler (MetaHipMer, that can scale to thousands of compute nodes and therefore utilizes petabytes of memory), has been developed (Hofmeyr et al., 2020). MetaHipMer allows the assembly of large sets of reads by taking advantage of distributed memory HPC supercomputers, such as those resulting from deep sequencing or co-assembly of multiple samples. Contig binning and other tasks that depend on unsupervised clustering techniques will typically require large memory nodes for large datasets. On the other hand, read mapping from individual samples to contigs is an embarrassingly parallel task that can be split by sample and sets of reads, with each processed in parallel across several nodes and the results put back together. This way of parallelization works well for any task that can be split into individual sections which can thus run independently. These include database searches by sequence homology techniques.

Distributed cloud computing platforms (HPC in the cloud) have been used as an alternative HPC solution to the computational demands of big omics data, such as Amazon Elastic Compute Cloud (AWS EC2) which offers a wide-variety of on-demand compute options, priced per CPU hours. Additional analysis costs would include network transfers to and from the cloud (bandwidth priced per GB) and persistent data storage costs (priced per GB per month). Cloud computing can be a suitable approach when access to an institutional HPC is not available. While HPC in the cloud offers on-demand and flexible computing power with minimal upfront costs, for regular analysis of large omics datasets on a per unit basis, it will be costly compared to an in-house HPC.

6.9 Isotope-enabled molecular analyses of soil biota

As described in Chapter 7, stable isotope analysis of biomolecules can be a powerful approach to determine microbial substrate use in situ in soils, exploring either natural isotopic fractionation signatures or tracing the isotopic signatures of added substrates. It is estimated that only a small fraction of the soil microbiome is active at any one point in time or space (Blagodatskaya and Kuzyakov, 2013). Further, DNA from nonviable organisms may persist in soil (Carini et al., 2016) making connections between genetic content of soils and soil biogeochemical processes difficult to make without coincident measurements of gene or protein expression. A key challenge in soil microbiology has been the estimation of the active or growing fraction of the community, with stable-isotope incorporation particularly powerful for addressing this. Stable isotope-labeled water (^{18}O-H_2O) is commonly used. When added to soil, the ^{18}O is incorporated into the biomolecules of active microorganisms during anabolic reactions, increasing their buoyant density and allowing for separation of natural abundance and isotopically enriched or "heavy" molecules via density gradient centrifugation (described in Chapter 7). This approach has been used to estimate the fraction of growing microorganisms in soil (Schwartz, 2007) and also to track dynamic growth and mortality of specific microbial taxa (Blazewicz and Schwartz, 2011; Blazewicz et al., 2020). The assimilation of ^{18}O from isotope-labeled water into DNA is proportional to a taxon's replication rate, meaning growth rates can be estimated. This is possible because the change in buoyant density of DNA due to isotope incorporation can be quantified by determining the quantity of DNA from specific taxa (or genomes) in the different fractions from a density centrifugation experiment. By comparing a shift in the distribution of taxa across density fractions in unlabeled controls and samples where label has been added, the isotopic enrichment of a taxon's DNA can be predicted using models. In this way, a taxon's growth and mortality rate can be directly estimated in situ. This approach has been termed quantitative stable isotope probing or qSIP (Hungate et al., 2015). Its extension to genome-centric analysis of growth and mortality and substrate use in soil microbiomes is an exciting new direction.

6.10 From nanometer-scale cells to global-scale questions

6.10.1 The importance of scale

Although soil microorganisms are typically nanometer to micrometer scale organisms, they are the biogeochemical engines of life on Earth (Falkowski et al., 2008). However, the composition of microbial communities and their functions in soil are regulated by processes that occur from the subcellular to the global scale (Fig. 6.3). Microbial metabolism and other traits can be inferred with increasing accuracy

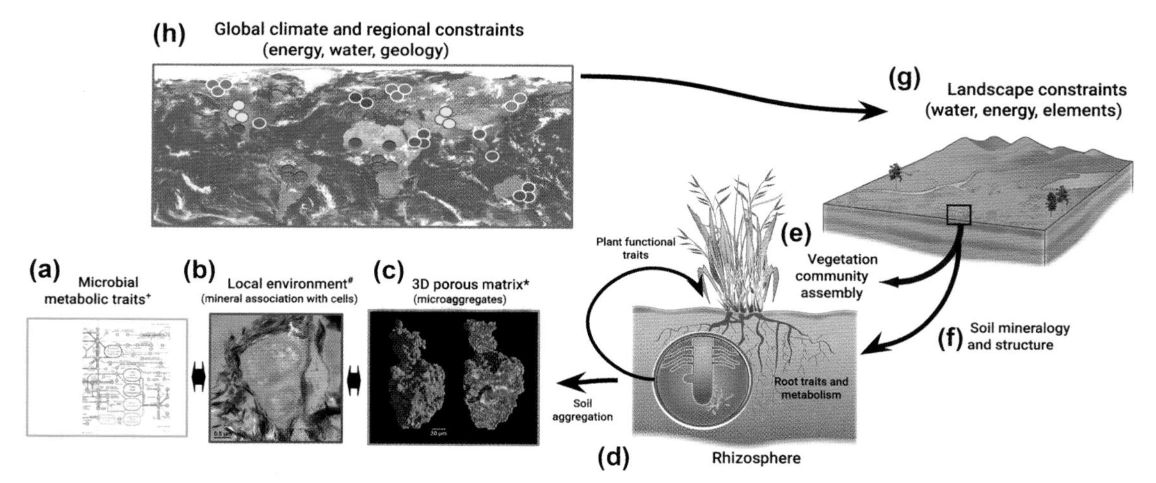

FIGURE 6.3 The composition of microbial communities and their functions in soil are regulated by processes that occur from the subcellular to the global scale. (a) Microbial metabolism and other traits, (b) the local mineral environment (scale bar is 0.5 μm), and the aggregation of soil particles (c) interact to regulate microbial metabolism and interactions in soil. In the rhizosphere (d), aggregates are dynamic, and growing roots activate subsets of the soil microbiome that interact with each other and in turn influence plant physiology. Across a landscape (e) communities of plants assemble across gradients of energy, water, and nutrients. Both plant and microbial activity interact with geology (f) and local climate (g). Meta-analyses that synthesize soil microbiome findings across the globe have the potential to identify generalizable relationships between soil, vegetation, and landscape properties and microbial phylogenetic and functional distributions (h). *(+Original image by Chakazul, used under CC 4.0 license. https://en.wikipedia.org/wiki/File:Metabolic_Metro_Map.svg. #Image by Manfred Auer [unpublished]. *Image unpublished, acquired as described in Voltolini et al., 2017. Thanks to Diana Swantek of Berkeley Lab for graphical support.)*

from omics information (Fig 6.3a). However, the local environment a microbe experiences, such as the minerals attached to a dividing microbial cell (scale bar is 0.5 μm) (Fig 6.3b), will constrain that metabolism through availability of space and resources, as well as protection from predators. The adhesion of soil particles (Fig 6.3c) creates a further scale of complexity (i.e., aggregation) and a porous matrix that regulates water, C and nutrient flux, as well as microbe-microbe interactions. In the rhizosphere (Fig 6.3d), aggregates are formed and destroyed, embedding or exposing microbes with new environmental conditions leading to new opportunities and constraints. Growing roots activate previously inactive members of the soil microbiome that then flourish, compete, cooperate, and exchange genetic material. In densely populated zones near a growing root tip with active root exudation, phage, protozoa, and some predatory bacteria control microbial populations and release nutrients that promote plant growth. Communities of plants and the expression of their traits are similarly controlled by their local environment, where communities assemble across gradients of energy, water, and nutrients (Fig 6.3e). Plant and microbial activity interact with climate properties such as precipitation and temperature to alter the nature of the geologic parent material, impacting mineral chemistry and structure, soil pH, permeability, porosity, conductivity, and water holding capacity (Fig 6.3f). This co-evolution of soils, plants, and microbes provides some predictive power. Variation in plant species can drive variation in microbial communities and microbial traits. These interactions are further impacted by variance in geology, weather, and climate (Fig 6.3g). Meta-analyses that synthesize soil microbiome findings across the globe have the potential to identify generalizable relationships between soil, vegetation, and landscape properties, in addition to microbial phylogenetic and functional distributions and their response to perturbation (Fig 6.3h). Leveraging these global research activities to produce newly synthesized knowledge requires methodological and data standardization and data FAIRification (see below for definition).

It is critical to remember that many of the approaches described in this chapter are applied to the study of soil microorganisms at a sample scale (millimeters to centimeters), which is vastly different from their body size (nanometers to micrometers) and thus the scale at which microbes interact with their habitat. As Fierer and Lennon (2011) put it, if we want to study soil microorganisms at a body-size-adjusted scale similar to how grassland ecologists would study their plant communities, microbial ecologists should be surveying microbial community composition and interactions over 100-μm^2 areas. Major technical challenges must be resolved to ensure scale-appropriate soil microbial research does not lag behind soil physical and biochemical understanding (Baveye et al., 2018). A consistent challenge has been the characterization of the spatial heterogeneity of soil microbes and their association with soil microenvironments. New approaches in parallel microbiome fields may offer some practical routes forward. For example, Sheth et al. (2019) embedded fecal material in a polymer, fracturing the polymer into roughly 30 μm^3 pieces and sequencing the microbiome composition of the fragments. This allowed microbial co-occurrence to be determined at appropriate scales, and if coupled with other analytical approaches like X-ray microtomography (Baveye et al., 2018), it may potentially reveal some aspects of the relationship of microbiome composition to the chemical and physical properties of their environment at the microbial scale. We see this both as a challenge as well as an opportunity at the interface between disciplines to establish solid foundations for our concepts and theories of microbial life in soil.

As multi-omic platforms become more accessible, there is greater potential to generalize findings across systems in a way that transitions from new observations to new knowledge across Earth's soils. Standardizing data generation, manipulation, and reporting is key to enable such syntheses. Removing barriers to data access and interpretation has clear benefits. Efforts like the Earth Microbiome Project (Gilbert et al., 2018) have driven the adoption of common nucleic acid extraction, sequencing, and analysis workflows. Community efforts like the Genomic Standards Consortium (Field et al., 2011) have driven the adoption of community data standards (e.g., Minimum Information about a Genome Sequence [MIGS]/Minimum Information about a Metagenomic Sequence/Sample [MIMS]), with the support of community ontology efforts such as the Environment Ontology (ENVO) (Buttigieg et al., 2013) to deliver "machine-actionable knowledge representation of environmental entities." These efforts are driven by the need to deliver data that is Findable, Accessible, Interoperable, and Reusable (FAIR).

6.10.2 FAIRification of multi-omics data

Soil heterogeneity across scales and its physical, chemical, and biological complexity necessitates the sharing of data across disciplines and geographies. The increasing complexity and volume of multi-omics datasets in particular make it difficult to share data in ways that enable integrative analyses. The objective of FAIR principles (Wilkinson et al., 2016) is to enable data discovery and reuse by rendering data and metadata Findable, Accessible, Interoperable, and Reusable (i.e., FAIR). These principles extend to algorithms, tools, and analysis workflows that generate all data products.

Public data repositories for storage and dissemination of the components of multi-omics data include EBI-ENA (Leinonen et al., 2011a) and NCBI-SRA (Leinonen et al., 2011b) for metagenomics and metatranscriptomics; PRIDE (Perez-Riverol et al., 2019) and ProteomeXchange (Vizcaíno et al., 2014) for metaproteomics; and Metabolomics Workbench (Sud et al., 2016), MetaboLights (Haug et al., 2019), and GNPS (Wang et al., 2016) for metabolomics. These repositories store and disseminate raw data and data products to a certain degree (i.e., MAGs), yet linking different types of multi-omics data from them in a searchable manner remains a challenge. Omics Discovery Index (Perez-Riverol et al., 2017) links datasets from different databases.

FAIR code sharing for multi-omics analysis can be enabled by platforms that use Git for version control such as GitHub and Bitbucket, and Markdown and Jupyter−based notebooks that provide dissemination of data analysis narratives. Multi-omics data analysis consists of complex multi-step workflows where for each step, several implementations (tools) to choose from exist, each with its own set of parameterizations. This necessitates machine-readable formalization of computational workflows that address FAIR principles (Wilkinson et al., 2016). Different tools and standards that describe computational workflows include Workflow Description Language (WDL; https://openwdl.org/), Common Workflow Language (CWL) (Amstutz et al., 2016), Nextflow (Di Tommaso et al., 2017), and Snakemake (Koster and Rahmann, 2012).

To further increase interoperability and reproducibility, containers are used to disseminate and deploy software components and computational environments. Conda packages is a popular repository and environment management system that provides a configuration file to deploy the computational environment anywhere. An alternative popular tool is Docker that automates the deployment of applications into lightweight, portable virtual containers, where an environment can be recreated with software dependencies to reproduce an analysis. In particular, Bioconda (The Bioconda Team et al., 2018) and Biocontainer (da Veiga Leprevost et al., 2017) are bioinformatics-specific tools.

A few efforts are particularly notable in that they encompass different aspects of FAIRification of multi-omics that spans data, tools, and workflows. NMDC (Wood-Charlson et al., 2020) is a recent effort to develop metadata standards (Vangay et al., 2021), provide curated, high-quality reference data, and open-source workflows for processing raw multi-omics data. The U.S. Department of Energy's Systems Biology Knowledgebase, or Kbase (Arkin et al., 2018), is an example of an integrated suite of analysis tools and workflows in a web-based user interface with a scalable compute infrastructure. Kbase is dedicated to improving reproducibility in plant and microbial systems biology, providing users a way to document, record, and share their data analysis workflows.

6.10.3 Answering global scale questions through meta-analyses of soil microbiomes

The recent expansion of publications that explore the global diversity of microbiomes is a product of these efforts to make data FAIR. These syntheses are illuminating the phylogenetic and functional traits of global microbiomes. These syntheses have revealed that relatively few taxa ($\sim 2\%$) comprise almost half the global soil microbiomes and in the process, generate a "most-wanted list" of organisms on which communities of researchers could focus collaborative efforts (Delgado-Baquerizo et al., 2018). For example, similar recent efforts have determined that differential abundance of traits involved in energy generation is a key signature of microbial adaptation across global biomes (Ramírez-Flandes et al., 2019), and that precipitation patterns and soil pH have contrasting impacts on bacterial and fungal phylogenetic and functional diversity (Bahram et al., 2018). Nelson et al. (2016) discovered that globally, soil microbial N cycling traits could be explained from relatively few variables (i.e., habitat type, soil C and N content), but they also discovered that dissimilatory nitrate reduction to ammonium (DNRA) was far more common in soil bacteria than expected, and therefore less studied. Such data sharing and cross-system analyses allow us to move beyond local observations in search of generalizable rules that can change dogma and spur further research directions (Fig. 6.3h). A prime example of this is the discovery that organisms that use trace gases as C and/or energy sources are commonly found and active across global soil microbiomes (Bay et al., 2021), raising the question of how important gaseous substrates are in soils where aqueous diffusion is thought to limit microbial activity. Looking across Earth's ecosystems again, recent work has demonstrated the common occurrence of large (>200 kb) phage genomes that potentially have a CRISPR-Cas capacity to manipulate their microbial hosts by silencing gene expression or even stimulating their host to target competing phage (Al-Shayeb et al., 2020). As a final example of global collaborative efforts leading to new discoveries in soil microbiology, it seems that human transformation and management of the natural environment in urban systems is homogenizing soil microbiome composition and function and depleting these soil systems of key functional guilds such as ectomycorrhizal fungi (Delgado-Baquerizo et al., 2021).

6.10.4 The need for a theoretical framework to guide measurements and integrate new knowledge

The expanded availability of standardized data and metadata, in addition to linked complementary and supporting data, is both recommended and in some cases, mandated by funding agencies. When combined with new data science tools, such as deep learning and artificial intelligence, this will surely continue to produce surprises and paradigm-shifting discoveries. Machine learning approaches are evolving and now offer the detection of high-order interactions in complex systems such as genetic networks (Basu et al., 2018) which can now be applied to soils in a manner beyond traditional statistics and the intuition of scientists.

Despite these advances, the complexity of soil microbiomes requires a common theoretical framework to integrate and distill new knowledge and to explore generalizable patterns emerging across studies. In this regard, traits (properties of organisms that explain differences in their fitness) have been used successfully to interpret and predict the distribution and activity of macro-organisms. Trait-based approaches can be used to provide a mechanistic explanation of patterns in microbial distribution and activity and to identify consistent combinations of traits that might represent the different strategies that microbes have evolved to compete in soil niches (Lajoie and Kembel, 2019; Martiny et al., 2015). As the omics approaches described in this chapter now allow us to infer traits from whole genome sequences, it may be possible to reduce the complexity of soil microbiomes by considering organisms (genomes) with similar combinations of traits to be members of microbial guilds. These guilds could then be used in numerical models to represent microbial diversity in the form of trait combinations (Allison, 2012; Malik et al., 2019; Wieder et al., 2013). Finally, because most soil organisms have yet to be cultivated, most traits inferred from omics information remain a hypothesis − that is, they are not directly observed features with direct connections to the fitness of the organism. To verify the importance of key traits, new approaches to genetically manipulate microorganisms in situ could be particularly powerful tools (Rubin et al., 2022). This is an exciting time, where theory, hypothesis development, new genetic tools, analytical technologies, and data science can converge to accelerate our understanding and rehabilitation of Earth's soils.

Acknowledgements

Eoin Brodie and Ulas Karaoz are supported by the Watershed Function Science Focus Area, funded by the Department of Energy, Office of Science, Office of Biological and Environmental Research, Environmental System Science program under contract DE-AC02-05CH11231 to Lawrence Berkeley National Laboratory. Joanne Emerson is supported by the U.S. Department of Energy (DOE), Office of Science, Office of Biological and Environmental Research (BER) Early Career Research Program under Award Number DE-SC0021198.

References

Acinas, S.G., Sarma-Rupavtarm, R., Klepac, V., 2005. PCR-induced sequence artifacts and bias: insights from comparison of two 16S rRNA clone libraries constructed from the same sample. Appl. Environ. Microbiol. 71, 8966−8969. https://doi.org/10.1128/AEM.71.12.8966-8969.2005.

Aitchison, J., 1982. The statistical analysis of compositional data. J. R. Stat. Soc. Series B Stat. Methodol. 44, 139−160. https://doi.org/10.1111/j.2517-6161.1982.tb01195.x.

Aitchison, J., 1983. Principal component analysis of compositional data. Biometrika 70, 57−65. https://doi.org/10.1093/biomet/70.1.57.

Aitchison, J., Greenacre, M., 2002. Biplots of compositional data. J. R. Stat. Soc. Series C Appl. Stat. 51, 375−392. https://doi.org/10.1111/1467-9876.00275.

Aitchison, J., Barceló-Vidal, C., Martín-Fernández, J.A., Pawlowsky-Glahn, V., 2000. Logratio analysis and compositional distance. Math. Geol. 32, 271−275. https://doi.org/10.1023/A:1007529726302.

Al-Shayeb, B., Sachdeva, R., Chen, L.-X., Ward, F., Munk, P., Devoto, A., et al., 2020. Clades of huge phages from across Earth's ecosystems. Nature 578 (7795), 425−431. https://doi.org/10.1038/s41586-020-2007-4.

Allison, S.D., 2012. A trait-based approach for modelling microbial litter decomposition. Ecol. Lett. 15, 1058−1070.

Alneberg, J., Bjarnason, B.S., de Bruijn, I., Schirmer, M., Quick, J., Ijaz, U.Z., et al., 2013. CONCOCT: clustering cONtigs on COverage and ComposiTion. Software https://github.com/BinPro/CONCOCT.

Ames, S.K., Hysom, D.A., Gardner, S.N., Lloyd, G.S., Gokhale, M.B., Allen, J.E., 2013. Scalable metagenomic taxonomy classification using a reference genome database. Bioinformatics 29, 2253−2260. https://doi.org/10.1093/bioinformatics/btt389.

Amir, A., McDonald, D., Navas-Molina, J.A., Kopylova, E., Morton, J.T., Zech Xu, Z., et al., 2017. Deblur rapidly resolves single-nucleotide community sequence patterns. mSystems 2. https://doi.org/10.1128/mSystems.00191-16.

Amstutz, P., Crusoe, M.R., Tijanić, N., Chapman, B., Chilton, J., Heuer, M., et al., 2016. Common Workflow Language, v1.0. https://doi.org/10.6084/M9.FIGSHARE.3115156.V2.

Angly, F.E., Dennis, P.G., Skarshewski, A., Vanwonterghem, I., Hugenholtz, P., Tyson, G.W., 2014. CopyRighter: a rapid tool for improving the accuracy of microbial community profiles through lineage-specific gene copy number correction. Microbiome 2, 11. https://doi.org/10.1186/2049-2618-2-11.

Bahram, M., Hildebrand, F., Forslund, S.K., Anderson, J.L., Soudzilovskaia, N.A., Bodegom, P.M., et al., 2018. Structure and function of the global topsoil microbiome. Nature 560, 233−237. https://doi.org/10.1038/s41586-018-0386-6.

Bahureksa, W., Tfaily, M.M., Boiteau, R.M., Young, R.B., Logan, M.N., McKenna, A.M., et al., 2021. Soil organic matter characterization by Fourier Transform Ion Cyclotron Resonance Mass Spectrometry (FTICR MS): a critical review of sample preparation, analysis, and data interpretation. Environ. Sci. Technol. 55, 9637−9656. https://doi.org/10.1021/acs.est.1c01135.

Banfield, J.F., Verberkmoes, N.C., Hettich, R.L., Thelen, M.P., 2005. Proteogenomic approaches for the molecular characterization of natural microbial communities. OMICS A J. Integr. Biol. 9, 301−333. https://doi.org/10.1089/omi.2005.9.301.

Basu, S., Kumbier, K., Brown, J.B., Yu, B., 2018. Iterative random forests to discover predictive and stable high-order interactions. Proc. Natl. Acad. Sci. 115, 1943−1948. https://doi.org/10.1073/pnas.1711236115.

Baveye, P.C., Otten, W., Kravchenko, A., Balseiro-Romero, M., Beckers, É., Chalhoub, M., et al., 2018. Emergent properties of microbial activity in heterogeneous soil microenvironments: different research approaches are slowly converging, yet major challenges remain. Front. Microbiol. 9, 1929. https://doi.org/10.3389/fmicb.2018.01929.

Bay, S.K., Dong, X., Bradley, J.A., Leung, P.M., Grinter, R., Jirapanjawat, T., et al., 2021. Trace gas oxidizers are widespread and active members of soil microbial communities. Nat. Microbiol. 6, 246−256. https://doi.org/10.1038/s41564-020-00811-w.

Beghini, F., McIver, L.J., Blanco-Míguez, A., Dubois, L., Asnicar, F., Maharjan, S., et al., 2021. Integrating taxonomic, functional, and strain-level profiling of diverse microbial communities with bioBakery 3. elife 10, e65088. https://doi.org/10.7554/eLife.65088.

Bin Jang, H., Bolduc, B., Zablocki, O., Kuhn, J.H., Roux, S., Adriaenssens, E.M., et al., 2019. Taxonomic assignment of uncultivated prokaryotic virus genomes is enabled by gene-sharing networks. Nat. Biotechnol. 37, 632−639. https://doi.org/10.1038/s41587-019-0100-8.

Bissett, A., Brown, M.V., Siciliano, S.D., Thrall, P.H., 2013. Microbial community responses to anthropogenically induced environmental change: towards a systems approach. Ecol. Lett. 16, 128−139. https://doi.org/10.1111/ele.12109.

Blagodatskaya, E., Kuzyakov, Y., 2013. Active microorganisms in soil: critical review of estimation criteria and approaches. Soil Biol. Biochem. 67, 192−211. https://doi.org/10.1016/j.soilbio.2013.08.024.

Blazewicz, S.J., Schwartz, E., 2011. Dynamics of ^{18}O incorporation from H_2 ^{18}O into soil microbial DNA. Microb. Ecol. https://doi.org/10.1007/s00248-011-9826-7.

Blazewicz, S.J., Hungate, B.A., Koch, B.J., Nuccio, E.E., Morrissey, E., Brodie, E.L., et al., 2020. Taxon-specific microbial growth and mortality patterns reveal distinct temporal population responses to rewetting in a California grassland soil. ISME J. 61, 911−916.

Boisvert, S., Raymond, F., Godzaridis, É., Laviolette, F., Corbeil, J., 2012. Ray Meta: scalable de novo metagenome assembly and profiling. Genome Biol. 13, R122. https://doi.org/10.1186/gb-2012-13-12-r122.

Brady, A., Salzberg, S.L., 2009. Phymm and PhymmBL: metagenomic phylogenetic classification with interpolated Markov models. Nat. Methods 6, 673−676. https://doi.org/10.1038/nmeth.1358.

Brito, I.L., 2021. Examining horizontal gene transfer in microbial communities. Nat. Rev. Microbiol. 19, 442−453. https://doi.org/10.1038/s41579-021-00534-7.

Buttigieg, P., Morrison, N., Smith, B., Mungall, C.J., Lewis, S.E., The ENVO Consortium, 2013. The environment ontology: contextualising biological and biomedical entities. J. Biomed. Semant. 4, 43. https://doi.org/10.1186/2041-1480-4-43.

Callahan, B.J., McMurdie, P.J., Rosen, M.J., Han, A.W., Johnson, A.J.A., Holmes, S.P., 2016. DADA2: high-resolution sample inference from Illumina amplicon data. Nat. Methods. https://doi.org/10.1038/nmeth.3869.

Callahan, B.J., McMurdie, P.J., Holmes, S.P., 2017. Exact sequence variants should replace operational taxonomic units in marker-gene data analysis. ISME J. 11, 2639−2643. https://doi.org/10.1038/ismej.2017.119.

Carini, P., Marsden, P.J., Leff, J.W., Morgan, E.E., Strickland, M.S., Fierer, N., 2016. Relic DNA is abundant in soil and obscures estimates of soil microbial diversity. Nat. Microbiol. 2, 16242. https://doi.org/10.1038/nmicrobiol.2016.242.

Carini, P., Delgado-Baquerizo, M., Hinckley, E.-L.S., Holland-Moritz, H., Brewer, T.E., Rue, G., et al., 2020. Effects of spatial variability and relic DNA removal on the detection of temporal dynamics in soil microbial communities. mBio 11. https://doi.org/10.1128/mBio.02776-19.

Coleman, J.R., Papamichail, D., Skiena, S., Futcher, B., Wimmer, E., Mueller, S., 2008. Virus attenuation by genome-scale changes in codon pair bias. Science 320, 1784–1787. https://doi.org/10.1126/science.1155761.

Compeau, P.E.C., Pevzner, P.A., Tesler, G., 2011. How to apply de Bruijn graphs to genome assembly. Nat. Biotechnol. 29, 987–991. https://doi.org/10.1038/nbt.2023.

da Veiga Leprevost, F., Grüning, B.A., Alves Aflitos, S., Röst, H.L., Uszkoreit, J., Barsnes, H., et al., 2017. BioContainers: an open-source and community-driven framework for software standardization. Bioinformatics 33, 2580–2582. https://doi.org/10.1093/bioinformatics/btx192.

Damon, C., Lehembre, F., Oger-Desfeux, C., Luis, P., Ranger, J., Fraissinet-Tachet, L., et al., 2012. Metatranscriptomics reveals the diversity of genes expressed by eukaryotes in forest soils. PLoS One 7, e28967. https://doi.org/10.1371/journal.pone.0028967.

Delgado-Baquerizo, M., Oliverio, A.M., Brewer, T.E., Benavent-González, A., Eldridge, D.J., Bardgett, R.D., et al., 2018. A global atlas of the dominant bacteria found in soil. Science 359, 320–325. https://doi.org/10.1126/science.aap9516.

Delgado-Baquerizo, M., Eldridge, D.J., Liu, Y.-R., Sokoya, B., Wang, J.-T., Hu, H.-W., et al., 2021. Global homogenization of the structure and function in the soil microbiome of urban greenspaces. Sci. Adv. 7, eabg5809. https://doi.org/10.1126/sciadv.abg5809.

Di Tommaso, P., Chatzou, M., Floden, E.W., Barja, P.P., Palumbo, E., Notredame, C., 2017. Nextflow enables reproducible computational workflows. Nat. Biotechnol. 35, 316–319. https://doi.org/10.1038/nbt.3820.

Edgar, R.C., 2016. UNOISE2: improved error-correction for Illumina 16S and ITS amplicon sequencing (preprint). Bioinformatics. https://doi.org/10.1101/081257.

Edgar, R.C., 2018. Updating the 97% identity threshold for 16S ribosomal RNA OTUs. Bioinformatics 34, 2371–2375. https://doi.org/10.1093/bioinformatics/bty113.

Emerson, J.B., Adams, R.I., Román, C.M.B., Brooks, B., Coil, D.A., Dahlhausen, K., et al., 2017. Schrödinger's microbes: tools for distinguishing the living from the dead in microbial ecosystems. Microbiome 5, 86. https://doi.org/10.1186/s40168-017-0285-3.

Erb, I., Notredame, C., 2016. How should we measure proportionality on relative gene expression data? Theory. Biosci. 135, 21–36. https://doi.org/10.1007/s12064-015-0220-8.

Erb, I., Gloor, G.B., Quinn, T.P., 2020. Editorial: compositional data analysis and related methods applied to genomics – a first special issue. NAR. Genom. Bioinform. 2, lqaa103. https://doi.org/10.1093/nargab/lqaa103.

Eren, A.M., Esen, Ö.C., Quince, C., Vineis, J.H., Morrison, H.G., Sogin, M.L., et al., 2015. Anvi'o: an advanced analysis and visualization platform for 'omics data. PeerJ 3, e1319. https://doi.org/10.7717/peerj.1319.

Ettinger, C.L., Vann, L.E., Eisen, J.A., 2021. Global diversity and biogeography of the *Zostera marina* mycobiome. Appl. Environ. Microbiol. 87. https://doi.org/10.1128/AEM.02795-20.

Falkowski, P.G., Fenchel, T., Delong, E.F., 2008. The microbial engines that drive Earth's biogeochemical cycles. Science 320, 1034–1039. https://doi.org/10.1126/science.1153213.

Fernandes, A.D., Reid, J.N., Macklaim, J.M., McMurrough, T.A., Edgell, D.R., Gloor, G.B., 2014. Unifying the analysis of high-throughput sequencing datasets: characterizing RNA-seq, 16S rRNA gene sequencing and selective growth experiments by compositional data analysis. Microbiome 2, 15. https://doi.org/10.1186/2049-2618-2-15.

Field, D., Amaral-Zettler, L., Cochrane, G., Cole, J.R., Dawyndt, P., Garrity, G.M., et al., 2011. The genomic standards Consortium. PLoS Biol. 9, e1001088. https://doi.org/10.1371/journal.pbio.1001088.

Fierer, N., Lennon, J.T., 2011. The generation and maintenance of diversity in microbial communities. Am. J. Bot. 98, 439–448. https://doi.org/10.3732/ajb.1000498.

Fisher, R.A., 1936. Design of experiments. BMJ 1. https://doi.org/10.1136/bmj.1.3923.554-a, 554–554.

Francioli, D., Lentendu, G., Lewin, S., Kolb, S., 2021. DNA metabarcoding for the characterization of terrestrial microbiota – pitfalls and solutions. Microorganisms 9, 361. https://doi.org/10.3390/microorganisms9020361.

Freitas, T.A.K., Li, P.-E., Scholz, M.B., Chain, P.S.G., 2015. Accurate read-based metagenome characterization using a hierarchical suite of unique signatures. Nucleic Acids Res. 43, e69. https://doi.org/10.1093/nar/gkv180.

Friedman, J., Alm, E.J., 2012. Inferring correlation networks from genomic survey data. PLoS Comput. Biol. 8, e1002687. https://doi.org/10.1371/journal.pcbi.1002687.

Frossard, A., Hammes, F., Gessner, M.O., 2016. Flow cytometric assessment of bacterial abundance in soils, sediments and sludge. Front. Microbiol. 7. https://doi.org/10.3389/fmicb.2016.00903.

Galazzo, G., van Best, N., Benedikter, B.J., Janssen, K., Bervoets, L., Driessen, C., et al., 2020. How to count our microbes? The effect of different quantitative microbiome profiling approaches. Front. Cell. Infect. Microbiol. 10, 403. https://doi.org/10.3389/fcimb.2020.00403.

Ghannam, R.B., Schaerer, L.G., Butler, T.M., Techtmann, S.M., 2020. Biogeographic patterns in members of globally distributed and dominant taxa found in port microbial communities. mSphere 5. https://doi.org/10.1128/mSphere.00481-19.

Gilbert, J.A., Jansson, J.K., Knight, R., 2018. Earth Microbiome Project and Global Systems Biology. mSystems 3. https://doi.org/10.1128/mSystems.00217-17.

Gloor, G.B., Wu, J.R., Pawlowsky-Glahn, V., Egozcue, J.J., 2016. It's all relative: analyzing microbiome data as compositions. Ann. Epidemiol. 26, 322–329. https://doi.org/10.1016/j.annepidem.2016.03.003.

Gloor, G.B., Macklaim, J.M., Pawlowsky-Glahn, V., Egozcue, J.J., 2017. Microbiome datasets are compositional: and this is not optional. Front. Microbiol. 8, 2224. https://doi.org/10.3389/fmicb.2017.02224.

Gregor, I., Dröge, J., Schirmer, M., Quince, C., McHardy, A.C., 2016. PhyloPythiaS+: a self-training method for the rapid reconstruction of low-ranking taxonomic bins from metagenomes. PeerJ 4, e1603. https://doi.org/10.7717/peerj.1603.

The Bioconda Team, Grüning, B., Dale, R., Sjödin, A., Chapman, B.A., Rowe, J., et al., 2018. Bioconda: sustainable and comprehensive software distribution for the life sciences. Nat. Methods 15, 475–476. https://doi.org/10.1038/s41592-018-0046-7.

Harrison, J.G., John Calder, W., Shuman, B., Alex Buerkle, C., 2021. The quest for absolute abundance: the use of internal standards for DNA-based community ecology. Mol. Ecol. Resour. 21, 30–43. https://doi.org/10.1111/1755-0998.13247.

Hasby, F.A., Barbi, F., Manzoni, S., Lindahl, B.D., 2021. Transcriptomic markers of fungal growth, respiration and carbon-use efficiency. FEMS Microbiol. Lett. 368, fnab100. https://doi.org/10.1093/femsle/fnab100.

Haug, K., Cochrane, K., Nainala, V.C., Williams, M., Chang, J., Jayaseelan, K.V., et al., 2019. MetaboLights: a resource evolving in response to the needs of its scientific community. Nucleic Acids Res. https://doi.org/10.1093/nar/gkz1019.

Hawinkel, S., Mattiello, F., Bijnens, L., Thas, O., 2019. A broken promise: microbiome differential abundance methods do not control the false discovery rate. Brief. Bioinform. 20, 210–221. https://doi.org/10.1093/bib/bbx104.

Hofmeyr, S., Egan, R., Georganas, E., Copeland, A.C., Riley, R., Clum, A., et al., 2020. Terabase-scale metagenome coassembly with MetaHipMer. Sci. Rep. 10, 10689. https://doi.org/10.1038/s41598-020-67416-5.

Hungate, B.A., Mau, R.L., Schwartz, E., Caporaso, J.G., Dijkstra, P., van Gestel, N., et al., 2015. Quantitative microbial ecology through stable isotope probing. Appl. Environ. Microbiol. 81, 7570–7581. https://doi.org/10.1128/AEM.02280-15.

Huson, D.H., Auch, A.F., Qi, J., Schuster, S.C., 2007. MEGAN analysis of metagenomic data. Genome Res. 17, 377–386. https://doi.org/10.1101/gr.5969107.

Jia, Y., Whalen, J.K., 2020. A new perspective on functional redundancy and phylogenetic niche conservatism in soil microbial communities. Pedosphere 30, 18–24. https://doi.org/10.1016/S1002-0160(19)60826-X.

Jia, P., Xuan, L., Liu, L., Wei, C., 2011. MetaBinG: using GPUs to accelerate metagenomic sequence classification. PLoS One 6, e25353. https://doi.org/10.1371/journal.pone.0025353.

Jing, X., Gong, Y., Xu, T., Meng, Y., Han, X., Su, X., et al., 2021. One-cell metabolic phenotyping and sequencing of soil microbiome by Raman-activated gravity-driven encapsulation (RAGE). mSystems 6. https://doi.org/10.1128/mSystems.00181-21.

Jones, M.B., Highlander, S.K., Anderson, E.L., Li, W., Dayrit, M., Klitgord, N., et al., 2015. Library preparation methodology can influence genomic and functional predictions in human microbiome research. Proc. Natl. Acad. Sci. 112, 14024–14029. https://doi.org/10.1073/pnas.1519288112.

Kang, D.D., Froula, J., Egan, R., Wang, Z., 2015. MetaBAT, an efficient tool for accurately reconstructing single genomes from complex microbial communities. PeerJ 3, e1165. https://doi.org/10.7717/peerj.1165.

Kang, D.D., Li, F., Kirton, E., Thomas, A., Egan, R., An, H., et al., 2019. MetaBAT 2: an adaptive binning algorithm for robust and efficient genome reconstruction from metagenome assemblies. PeerJ 7, e7359. https://doi.org/10.7717/peerj.7359.

Kim, D., Song, L., Breitwieser, F.P., Salzberg, S.L., 2016. Centrifuge: rapid and sensitive classification of metagenomic sequences. Genome Res. 26, 1721–1729. https://doi.org/10.1101/gr.210641.116.

Koster, J., Rahmann, S., 2012. Snakemake – a scalable bioinformatics workflow engine. Bioinformatics 28, 2520–2522. https://doi.org/10.1093/bioinformatics/bts480.

Kurtz, Z.D., Müller, C.L., Miraldi, E.R., Littman, D.R., Blaser, M.J., Bonneau, R.A., 2015. Sparse and compositionally robust inference of microbial ecological networks. PLoS Comput. Biol. 11, e1004226. https://doi.org/10.1371/journal.pcbi.1004226.

Lajoie, G., Kembel, S.W., 2019. Making the most of trait-based approaches for microbial ecology. Trends Microbiol. 27, 814–823. https://doi.org/10.1016/j.tim.2019.06.003.

Lee, D.D., Seung, H.S., 1999. Learning the parts of objects by non-negative matrix factorization. Nature 401, 788–791. https://doi.org/10.1038/44565.

Lee, J., Kim, H.-S., Jo, H.Y., Kwon, M.J., 2021. Revisiting soil bacterial counting methods: optimal soil storage and pretreatment methods and comparison of culture-dependent and -independent methods. PLoS One 16, e0246142. https://doi.org/10.1371/journal.pone.0246142.

Leinonen, R., Akhtar, R., Birney, E., Bower, L., Cerdeno-Tarraga, A., Cheng, Y., et al., 2011a. The European nucleotide archive. Nucleic Acids Res. 39, D28–D31. https://doi.org/10.1093/nar/gkq967.

Leinonen, R., Sugawara, H., Shumway, M., on behalf of the International Nucleotide Sequence Database Collaboration, 2011b. The sequence read archive. Nucleic Acids Res. 39, D19–D21. https://doi.org/10.1093/nar/gkq1019.

Li, D., Liu, C.-M., Luo, R., Sadakane, K., Lam, T.-W., 2015. MEGAHIT: an ultra-fast single-node solution for large and complex metagenomics assembly via succinct de Bruijn graph. Bioinformatics 31, 1674–1676. https://doi.org/10.1093/bioinformatics/btv033.

Liu, B., Gibbons, T., Ghodsi, M., Treangen, T., Pop, M., 2011. Accurate and fast estimation of taxonomic profiles from metagenomic shotgun sequences. BMC Genom. 12, S4. https://doi.org/10.1186/1471-2164-12-S2-S4.

Louca, S., Doebeli, M., Parfrey, L.W., 2018. Correcting for 16S rRNA gene copy numbers in microbiome surveys remains an unsolved problem. Microbiome 6, 41. https://doi.org/10.1186/s40168-018-0420-9.

Lovell, D., Müller, W., Taylor, J., Zwart, A., Helliwell, C., 2011. Proportions, percentages, PPM: do the molecular biosciences treat compositional data right? In: Pawlowsky-Glahn, V., Buccianti, A. (Eds.), Compositional Data Analysis. John Wiley & Sons, Chichester, UK, pp. 191–207. https://doi.org/10.1002/9781119976462.ch14.

Lovell, D.R., Chua, X.-Y., McGrath, A., 2020. Counts: an outstanding challenge for log-ratio analysis of compositional data in the molecular biosciences. NAR. Genom. Bioinform 2. https://doi.org/10.1093/nargab/lqaa040. lqaa040.

Lu, Y.Y., Chen, T., Fuhrman, J.A., Sun, F., 2016. COCACOLA: binning metagenomic contigs using sequence COmposition, read CoverAge, CO-alignment and paired-end read LinkAge. Bioinformatics. https://doi.org/10.1093/bioinformatics/btw290.

MacDonald, N.J., Parks, D.H., Beiko, R.G., 2012. Rapid identification of high-confidence taxonomic assignments for metagenomic data. Nucleic Acids Res. 40, e111. https://doi.org/10.1093/nar/gks335.

Malik, A.A., Martiny, J.B., Brodie, E.L., Martiny, A.C., Treseder, K.K., Allison, S.D., 2019. Defining trait-based microbial strategies with consequences for soil carbon cycling under climate change. ISME J. 14, 1–9. https://doi.org/10.1038/s41396-019-0510-0.

Mandal, S., Van Treuren, W., White, R.A., Eggesbø, M., Knight, R., Peddada, S.D., 2015. Analysis of composition of microbiomes: a novel method for studying microbial composition. Microb. Ecol. Health Dis. 26. https://doi.org/10.3402/mehd.v26.27663.

Martin, M., 2011. Cutadapt removes adapter sequences from high-throughput sequencing reads. EMBnet. J. 17, 10. https://doi.org/10.14806/ej.17.1.200.

Martiny, J.B.H., Jones, S.E., Lennon, J.T., Martiny, A.C., 2015. Microbiomes in light of traits: a phylogenetic perspective. Science 350 (6261), aac9323. https://doi.org/10.1126/science.aac9323.

Medvedev, P., Georgiou, K., Myers, G., Brudno, M., 2007. Computability of models for sequence assembly. In: Giancarlo, R., Hannenhalli, S. (Eds.), Algorithms in Bioinformatics, Lecture Notes in Computer Science. Springer Berlin Heidelberg, Berlin, Heidelberg, pp. 289–301. https://doi.org/10.1007/978-3-540-74126-8_27.

Menzel, P., Ng, K.L., Krogh, A., 2016. Fast and sensitive taxonomic classification for metagenomics with Kaiju. Nat. Commun. 7, 11257. https://doi.org/10.1038/ncomms11257.

Milanese, A., Mende, D.R., Paoli, L., Salazar, G., Ruscheweyh, H.-J., Cuenca, M., et al., 2019. Microbial abundance, activity and population genomic profiling with mOTUs2. Nat. Commun. 10, 1014. https://doi.org/10.1038/s41467-019-08844-4.

Namiki, T., Hachiya, T., Tanaka, H., Sakakibara, Y., 2012. MetaVelvet: an extension of Velvet assembler to de novo metagenome assembly from short sequence reads. Nucleic Acids Res. 40, e155. https://doi.org/10.1093/nar/gks678.

Nelson, M.B., Martiny, A.C., Martiny, J.B.H., 2016. Global biogeography of microbial nitrogen-cycling traits in soil. Proc. Natl. Acad. Sci. 113, 8033–8040. https://doi.org/10.1073/pnas.1601070113.

Nicora, C.D., Burnum-Johnson, K.E., Nakayasu, E.S., Casey, C.P., White III, R.A., Chowdhury, T.R., et al., 2018. The MPLEx protocol for multi-omic analyses of soil samples. J. Vis. Exp. (135), 57343. https://doi.org/10.3791/57343.

Nuccio, E.E., Starr, E., Karaoz, U., Brodie, E.L., Zhou, J., Tringe, S.G., et al., 2020. Niche differentiation is spatially and temporally regulated in the rhizosphere. ISME J. 14 (4), 999–1014. https://doi.org/10.1038/s41396-019-0582-x.

Nurk, S., Meleshko, D., Korobeynikov, A., Pevzner, P.A., 2017. MetaSPAdes: a new versatile metagenomic assembler. Genome Res. 27, 824–834. https://doi.org/10.1101/gr.213959.116.

Ounit, R., Wanamaker, S., Close, T.J., Lonardi, S., 2015. CLARK: fast and accurate classification of metagenomic and genomic sequences using discriminative k-mers. BMC Genom. 16, 236. https://doi.org/10.1186/s12864-015-1419-2.

Ouyang, Y., Chen, D., Fu, Y., Shi, W., Provin, T., Han, A., et al., 2021. Direct cell extraction from fresh and stored soil samples: impact on microbial viability and community compositions. Soil Biol. Biochem. 155, 108178. https://doi.org/10.1016/j.soilbio.2021.108178.

Parks, D.H., Imelfort, M., Skennerton, C.T., Hugenholtz, P., Tyson, G.W., 2015. CheckM: assessing the quality of microbial genomes recovered from isolates, single cells, and metagenomes. Genome Res. 25 (7), 1043–1055. https://doi.org/10.1101/gr.186072.114.

Pawlowsky-Glahn, V., Egozcue, J.J., Tolosana-Delgado, R., 2015. Modelling and Analysis of Compositional Data: Pawlowsky-Glahn/Modelling and Analysis of Compositional Data. John Wiley & Sons, Chichester, UK. https://doi.org/10.1002/9781119003144.

Pearson, K., 1897. Mathematical contributions to the theory of evolution. On a form of spurious correlation which may arise when indices are used in the measurement of organs. Proc. R. Soc. Lond. 60, 489–498. https://doi.org/10.1098/rspl.1896.0076.

Peng, Y., Leung, H.C., Yiu, S.-M., Chin, F.Y., 2012. IDBA-UD: a de novo assembler for single-cell and metagenomic sequencing data with highly uneven depth. Bioinformatics 28 (11), 1420–1428. https://doi.org/10.1093/bioinformatics/bts174.

Perez-Riverol, Y., Bai, M., da Veiga Leprevost, F., Squizzato, S., Park, Y.M., Haug, K., et al., 2017. Discovering and linking public omics data sets using the Omics Discovery Index. Nat. Biotechnol. 35, 406–409. https://doi.org/10.1038/nbt.3790.

Perez-Riverol, Y., Csordas, A., Bai, J., Bernal-Llinares, M., Hewapathirana, S., Kundu, D.J., et al., 2019. The PRIDE database and related tools and resources in 2019: improving support for quantification data. Nucleic Acids Res. 47, D442–D450. https://doi.org/10.1093/nar/gky1106.

Pevzner, P.A., Tang, H., Waterman, M.S., 2001. An Eulerian path approach to DNA fragment assembly. Proc. Natl. Acad. Sci. 98, 9748–9753. https://doi.org/10.1073/pnas.171285098.

Pietramellara, G., Ascher, J., Borgogni, F., Ceccherini, M.T., Guerri, G., Nannipieri, P., 2009. Extracellular DNA in soil and sediment: fate and ecological relevance. Biol. Fertil. Soils 45, 219–235. https://doi.org/10.1007/s00374-008-0345-8.

Portillo, M.C., Leff, J.W., Lauber, C.L., Fierer, N., 2013. Cell size distributions of soil bacterial and archaeal taxa. Appl. Environ. Microbiol. 79, 7610–7617. https://doi.org/10.1128/AEM.02710-13.

Props, R., Kerckhof, F.-M., Rubbens, P., De Vrieze, J., Hernandez Sanabria, E., Waegeman, W., et al., 2017. Absolute quantification of microbial taxon abundances. ISME J. 11, 584–587. https://doi.org/10.1038/ismej.2016.117.

Prosser, J.I., 2020. Putting science back into microbial ecology: a question of approach. Philos. Trans. R. Soc. B Biol. Sci. 375, 20190240. https://doi.org/10.1098/rstb.2019.0240.

Quinn, T.P., Erb, I., Gloor, G., Notredame, C., Richardson, M.F., Crowley, T.M., 2019. A field guide for the compositional analysis of any-omics data. GigaScience 8, giz107. https://doi.org/10.1093/gigascience/giz107.

Raes, J., Bork, P., 2008. Molecular eco-systems biology: towards an understanding of community function. Nat. Rev. Microbiol. 6, 693–699. https://doi.org/10.1038/nrmicro1935.

Ramírez-Flandes, S., González, B., Ulloa, O., 2019. Redox traits characterize the organization of global microbial communities. Proc. Natl. Acad. Sci. 116, 3630–3635. https://doi.org/10.1073/pnas.1817554116.

Roux, S., Emerson, J.B., Eloe-Fadrosh, E.A., Sullivan, M.B., 2017. Benchmarking viromics: an in silico evaluation of metagenome-enabled estimates of viral community composition and diversity. PeerJ 5, e3817. https://doi.org/10.7717/peerj.3817.

Rubin, B.E., Diamond, S., Cress, B.F., Crits-Christoph, A., Lou, Y.C., Borges, A.L., et al., 2022. Species- and site-specific genome editing in complex bacterial communities. Nat. Microbiol. 7, 34–47. https://doi.org/10.1038/s41564-021-01014-7.

Santos-Medellin, C., Zinke, L.A., ter Horst, A.M., Gelardi, D.L., Parikh, S.J., Emerson, J.B., 2021. Viromes outperform total metagenomes in revealing the spatiotemporal patterns of agricultural soil viral communities. ISME J. https://doi.org/10.1038/s41396-021-00897-y.

Schoch, C.L., Seifert, K.A., Huhndorf, S., Robert, V., Spouge, J.L., Levesque, C.A., et al., 2012. Nuclear ribosomal internal transcribed spacer (ITS) region as a universal DNA barcode marker for Fungi. Proc. Natl. Acad. Sci. 109, 6241–6246. https://doi.org/10.1073/pnas.1117018109.

Schwager, E., Mallick, H., Ventz, S., Huttenhower, C., 2017. A Bayesian method for detecting pairwise associations in compositional data. PLoS Comput. Biol. 13, e1005852. https://doi.org/10.1371/journal.pcbi.1005852.

Schwartz, E., 2007. Characterization of growing microorganisms in soil by stable isotope probing with $H_2{}^{18}O$. Appl. Environ. Microbiol. 73, 2541–2546. https://doi.org/10.1128/AEM.02021-06.

Segata, N., Börnigen, D., Morgan, X.C., Huttenhower, C., 2013. PhyloPhlAn is a new method for improved phylogenetic and taxonomic placement of microbes. Nat. Commun. 4, 2304. https://doi.org/10.1038/ncomms3304.

Sharon, I., Morowitz, M.J., Thomas, B.C., Costello, E.K., Relman, D.A., Banfield, J.F., 2013. Time series community genomics analysis reveals rapid shifts in bacterial species, strains, and phage during infant gut colonization. Genome Res. 23 (1), 111–120. https://doi.org/10.1101/gr.142315.112.

Sharpton, T.J., Riesenfeld, S.J., Kembel, S.W., Ladau, J., O'Dwyer, J.P., Green, J.L., et al., 2011. PhylOTU: a high-throughput procedure quantifies microbial community diversity and resolves novel taxa from metagenomic data. PLoS Comput. Biol. 7, e1001061. https://doi.org/10.1371/journal.pcbi.1001061.

Sheth, R.U., Li, M., Jiang, W., Sims, P.A., Leong, K.W., Wang, H.H., 2019. Spatial metagenomic characterization of microbial biogeography in the gut. Nat. Biotechnol. 37, 877–883. https://doi.org/10.1038/s41587-019-0183-2.

Sieber, C.M., Probst, A.J., Sharrar, A., Thomas, B.C., Hess, M., Tringe, S.G., et al., 2018. Recovery of genomes from metagenomes via a dereplication, aggregation and scoring strategy. Nat. Microbiol. 3 (7), 836–843. https://doi.org/10.1038/s41564-018-0171-1.

Silverman, J.D., Washburne, A.D., Mukherjee, S., David, L.A., 2017. A phylogenetic transform enhances analysis of compositional microbiota data. Elife 6, e21887. https://doi.org/10.7554/eLife.21887.

Slatko, B.E., Gardner, A.F., Ausubel, F.M., 2018. Overview of next-generation sequencing technologies. Curr. Protoc. Mol. Biol. 122. https://doi.org/10.1002/cpmb.59.

Smets, W., Leff, J.W., Bradford, M.A., McCulley, R.L., Lebeer, S., Fierer, N., 2016. A method for simultaneous measurement of soil bacterial abundances and community composition via 16S rRNA gene sequencing. Soil Biol. Biochem. 96, 145–151. https://doi.org/10.1016/j.soilbio.2016.02.003.

Smith, C.J., Osborn, A.M., 2009. Advantages and limitations of quantitative PCR (Q-PCR)-based approaches in microbial ecology: application of Q-PCR in microbial ecology. FEMS Microbiol. Ecol. 67, 6–20. https://doi.org/10.1111/j.1574-6941.2008.00629.x.

Song, W.-Z., Thomas, T., 2017. Binning_refiner: improving genome bins through the combination of different binning programs. Bioinformatics 33, 1873–1875. https://doi.org/10.1093/bioinformatics/btx086.

Stämmler, F., Gläsner, J., Hiergeist, A., Holler, E., Weber, D., Oefner, P.J., et al., 2016. Adjusting microbiome profiles for differences in microbial load by spike-in bacteria. Microbiome 4, 28. https://doi.org/10.1186/s40168-016-0175-0.

Stoddard, S.F., Smith, B.J., Hein, R., Roller, B.R.K., Schmidt, T.M., 2015. rrnDB: improved tools for interpreting rRNA gene abundance in bacteria and archaea and a new foundation for future development. Nucleic Acids Res. 43, D593–D598. https://doi.org/10.1093/nar/gku1201.

Sud, M., Fahy, E., Cotter, D., Azam, K., Vadivelu, I., Burant, C., et al., 2016. Metabolomics Workbench: an international repository for metabolomics data and metadata, metabolite standards, protocols, tutorials and training, and analysis tools. Nucleic Acids Res. 44, D463–D470. https://doi.org/10.1093/nar/gkv1042.

Swenson, T.L., Jenkins, S., Bowen, B.P., Northen, T.R., 2015. Untargeted soil metabolomics methods for analysis of extractable organic matter. Soil Biol. Biochem. 80, 189–198. https://doi.org/10.1016/j.soilbio.2014.10.007.

Tartaglia, M., Bastida, F., Sciarrillo, R., Guarino, C., 2020. Soil metaproteomics for the study of the relationships between microorganisms and plants: a review of extraction protocols and ecological insights. Int. J. Mol. Sci. 21, 8455. https://doi.org/10.3390/ijms21228455.

Tedersoo, L., Albertsen, M., Anslan, S., Callahan, B., 2021. Perspectives and benefits of high-throughput long-read sequencing in microbial ecology. Appl. Environ. Microbiol. 87. https://doi.org/10.1128/AEM.00626-21.

Thakur, M.P., Phillips, H.R.P., Brose, U., De Vries, F.T., Lavelle, P., Loreau, M., et al., 2020. Towards an integrative understanding of soil biodiversity. Biol. Rev. 95, 350–364. https://doi.org/10.1111/brv.12567.

Thornhill, D.J., Lajeunesse, T.C., Santos, S.R., 2007. Measuring rDNA diversity in eukaryotic microbial systems: how intragenomic variation, pseudogenes, and PCR artifacts confound biodiversity estimates. Mol. Ecol. 16, 5326–5340. https://doi.org/10.1111/j.1365-294X.2007.03576.x.

Thorsen, J., Brejnrod, A., Mortensen, M., Rasmussen, M.A., Stokholm, J., Al-Soud, W.A., et al., 2016. Large-scale benchmarking reveals false discoveries and count transformation sensitivity in 16S rRNA gene amplicon data analysis methods used in microbiome studies. Microbiome 4, 62. https://doi.org/10.1186/s40168-016-0208-8.

Tikhonov, M., Leach, R.W., Wingreen, N.S., 2015. Interpreting 16S metagenomic data without clustering to achieve sub-OTU resolution. ISME J. 9, 68–80. https://doi.org/10.1038/ismej.2014.117.

Tkacz, A., Hortala, M., Poole, P.S., 2018. Absolute quantitation of microbiota abundance in environmental samples. Microbiome 6, 110. https://doi.org/10.1186/s40168-018-0491-7.

Truong, D.T., Franzosa, E.A., Tickle, T.L., Scholz, M., Weingart, G., Pasolli, E., et al., 2015. MetaPhlAn2 for enhanced metagenomic taxonomic profiling. Nat. Methods 12, 902–903. https://doi.org/10.1038/nmeth.3589.

Truong, D.T., Tett, A., Pasolli, E., Huttenhower, C., Segata, N., 2017. Microbial strain-level population structure and genetic diversity from metagenomes. Genome Res. 27, 626–638. https://doi.org/10.1101/gr.216242.116.

Uritskiy, G.V., DiRuggiero, J., Taylor, J., 2018. MetaWRAP – a flexible pipeline for genome-resolved metagenomic data analysis. Microbiome 6, 158. https://doi.org/10.1186/s40168-018-0541-1.

Uyaguari-Diaz, M.I., Chan, M., Chaban, B.L., Croxen, M.A., Finke, J.F., Hill, J.E., et al., 2016. A comprehensive method for amplicon-based and metagenomic characterization of viruses, bacteria, and eukaryotes in freshwater samples. Microbiome 4, 20. https://doi.org/10.1186/s40168-016-0166-1.

Vandeputte, D., Kathagen, G., D'hoe, K., Vieira-Silva, S., Valles-Colomer, M., Sabino, J., et al., 2017. Quantitative microbiome profiling links gut community variation to microbial load. Nature 551, 507–511. https://doi.org/10.1038/nature24460.

Vangay, P., Burgin, J., Johnston, A., Beck, K.L., Berrios, D.C., Blumberg, K., et al., 2021. Microbiome metadata standards: report of the national microbiome data collaborative's workshop and follow-on activities. mSystems 6, e01194–20. https://doi.org/10.1128/mSystems.01194-20.

Větrovský, T., Baldrian, P., 2013. The variability of the 16S rRNA gene in bacterial genomes and its consequences for bacterial community analyses. PLoS One 8, e57923. https://doi.org/10.1371/journal.pone.0057923.

Vizcaíno, J.A., Deutsch, E.W., Wang, R., Csordas, A., Reisinger, F., Ríos, D., et al., 2014. ProteomeXchange provides globally coordinated proteomics data submission and dissemination. Nat. Biotechnol. 32, 223–226. https://doi.org/10.1038/nbt.2839.

Voltolini, M., Taş, N., Wang, S., Brodie, E.L., Ajo-Franklin, J.B., 2017. Quantitative characterization of soil micro-aggregates: new opportunities from sub-micron resolution synchrotron X-ray microtomography. Geoderma 305, 382–393. https://doi.org/10.1016/j.geoderma.2017.06.005.

Wang, M., Carver, J.J., Phelan, V.V., Sanchez, L.M., Garg, N., Peng, Y., et al., 2016. Sharing and community curation of mass spectrometry data with global natural products social molecular networking. Nat. Biotechnol. 34, 828–837. https://doi.org/10.1038/nbt.3597.

Wieder, W.R., Bonan, G.B., Allison, S.D., 2013. Global soil carbon projections are improved by modelling microbial processes. Nat. Clim. Chang. 3, 909–912. https://doi.org/10.1038/nclimate1951.

Wilhelm, S., Weinbauer, M., Suttle, C. (Eds.), 2010. Manual of Aquatic Viral Ecology. American Society of Limnology and Oceanography. https://doi.org/10.4319/mave.2010.978-0-9845591-0-7.

Wilke, A., Bischof, J., Gerlach, W., Glass, E., Harrison, T., Keegan, K.P., et al., 2016. The MG-RAST metagenomics database and portal in 2015. Nucleic Acids Res. 44, D590–D594. https://doi.org/10.1093/nar/gkv1322.

Wilkinson, M.D., Dumontier, M., Aalbersberg, I.J., Appleton, G., Axton, M., Baak, A., et al., 2016. The FAIR guiding principles for scientific data management and stewardship. Sci. Data 3, 160018. https://doi.org/10.1038/sdata.2016.18.

Williamson, K.E., Radosevich, M., Wommack, K.E., 2005. Abundance and diversity of viruses in six Delaware soils. Appl. Environ. Microbiol. 71, 3119–3125. https://doi.org/10.1128/AEM.71.6.3119-3125.2005.

Wilson, R.M., Tfaily, M.M., Kolton, M., Johnston, E.R., Petro, C., Zalman, C.A., et al., 2021. Soil metabolome response to whole-ecosystem warming at the spruce and peatland responses under changing environments experiment. Proc. Natl. Acad. Sci. 118, e2004192118. https://doi.org/10.1073/pnas.2004192118.

Withers, E., Hill, P.W., Chadwick, D.R., Jones, D.L., 2020. Use of untargeted metabolomics for assessing soil quality and microbial function. Soil Biol. Biochem. 143, 107758. https://doi.org/10.1016/j.soilbio.2020.107758.

Wong, R.G., Wu, J.R., Gloor, G.B., 2016. Expanding the UniFrac toolbox. PLoS One 11, e0161196. https://doi.org/10.1371/journal.pone.0161196.

Wood, D.E., Lu, J., Langmead, B., 2019. Improved metagenomic analysis with Kraken 2. Genome Biol. 20, 257. https://doi.org/10.1186/s13059-019-1891-0.

Wood-Charlson, E.M., Auberry, D.A., Blanco, H., Borkum, M.I., Corilo, Y.E., Davenport, K., et al., 2020. The national microbiome data collaborative: enabling microbiome science. Nat. Rev. Microbiol. 18, 313–314. https://doi.org/10.1038/s41579-020-0377-0.

Wrighton, K.C., Thomas, B.C., Sharon, I., Miller, C.S., Castelle, C.J., VerBerkmoes, N.C., et al., 2012. Fermentation, hydrogen, and sulfur metabolism in multiple uncultivated bacterial phyla. Science 337, 1661–1665. https://doi.org/10.1126/science.1224041.

Wu, M., Scott, A.J., 2012. Phylogenomic analysis of bacterial and archaeal sequences with AMPHORA2. Bioinformatics 28, 1033–1034. https://doi.org/10.1093/bioinformatics/bts079.

Wu, Y.-W., Simmons, B.A., Singer, S.W., 2016. MaxBin 2.0: an automated binning algorithm to recover genomes from multiple metagenomic datasets. Bioinformatics 32, 605–607. https://doi.org/10.1093/bioinformatics/btv638.

Chapter 7

Physiological and biochemical methods for studying soil biota and their functions

Ellen Kandeler

Institute of Soil Science and Land Evaluation, Soil Biology Department, University of Hohenheim, Stuttgart, Germany

Chapter outline

7.1 Introduction

Biological and biochemically mediated processes in soils are fundamental to terrestrial ecosystem functions because members of all trophic levels depend on the soil as a source of nutrients and energy, along with the transformation of complex into simpler organic compounds and the production of soil organic matter

Soil Microbiology, Ecology, and Biochemistry. https://doi.org/10.1016/B978-0-12-822941-5.00007-7

(SOM). Biotic processes are studied at multiple levels of resolution: (1) at the molecular level, SOM chemistry and enzymatic characteristics of degradation are investigated; (2) at the organismal level, the focus is on functional gene analyses, regulation of enzyme expression, and growth kinetics; and (3) at the community level, research concentrates on metabolism, microbial succession, and interactions between microbial and faunal communities. Results from multiple levels of resolution must be integrated to fully understand soil biotic functions and how they influence ecosystem-scale processes (Sinsabaugh et al., 2002; Thiele-Bruhn et al., 2020). The functions of soil biota are investigated by a range of methods focusing either on broad physiological properties (e.g., soil respiration, nitrogen [N] mineralization) or specific reactions (e.g., ammonia monooxygenase production by bacterial and archaeal nitrifiers). A challenge for the future is to locate these specific processes in the three-dimensional soil network and relate their activity to broader soil processes.

This chapter will focus on the important biochemical and physiological methods applied in soil microbiology and soil biochemistry. Whereas biochemical techniques (e.g., phospholipid fatty acid [PLFA] analyses) are used to assess the biomass and distribution of soil microorganisms, physiological methods are applied to understand the metabolism of individual taxa, whole microbial communities, and biogeochemical cycling in terrestrial ecosystems. Information on methods focusing specifically on faunal abundance and activity can be found in Chapter 4, and information on different methods in molecular soil ecology is discussed in Chapter 6. Material about the development and application of biochemical and physiological methods to study soil microorganisms is summarized in Table 7.1. Before starting any analysis, it is important to select the optimal experimental design and sampling strategy.

TABLE 7.1 Books and book chapters of methods in soil microbiology.

Methods in soil microbiology	References
Soil microbiology and soil biochemistry	Pepper et al. (2014); Schinner et al. (1996); Van Elsas et al. (2019)
Fungi	Frankland et al. (1991); Newell and Fallon (1991)
Digital image analysis of soil microorganisms	Daims and Wagner (2007)
Soil enzymes	Burns and Dick (2002); Dick (2011)
Tracer techniques (^{13}C, ^{14}C, ^{11}C, ^{15}N)	Boutton and Yamasaki (1996); Coleman and Fry (1991); Knowles and Blackburn (1993); Schimel (1993)
Gross nitrogen fluxes (^{15}N pool dilution)	Murphy et al. (2003)
Soil biological processes and soil organisms	Pepper et al. (2014); Robertson et al. (1999); Thiele-Bruhn et al. (2020)
Stable isotope probing techniques	Coyotzi et al. (2016); Neufeld et al. (2007)
PLFAs	Ruess and Chamberlain (2010)

7.2 Scale of investigations and collection of samples

Investigations at the plot scale are still the dominant sampling strategy for soil chemical and biological studies. A representative number of soil samples are taken from the study site (e.g., arable land, grassland, forest) and then either combined into bulk samples or treated separately. Random samples are typically combined from representative areas that are described by uniform soil type, texture, and habitat characteristics. Samples of agricultural soils are often taken from specific soil depths, often related to the depth of the plow layer (e.g., 0–20 cm), whereas samples of forest soils are taken from specific soil horizons (e.g., litter layer, organic horizon, A horizon). The same sampling procedure can be used to cover larger areas where single sampling points are selected by typical two-dimensional sampling patterns (e.g., random, transect, two-stage, and grid sampling techniques). When mapping an area with unknown spatial variability, a grid sampling pattern is recommended where samples are taken systematically at regular intervals with fixed spacing. Transect sampling is chosen when specific gradients of soil properties are expected (e.g., glacial retreat area). As an example, Waldrop et al. (2017) chose two continental-scale transects across multiple climatic, physiographic, land use, geologic, pedologic, and ecological boundaries to clarify which edaphic and/or environmental factors determine the composition and functioning of microbial communities. Random sampling is frequently applied for soil sampling of field experiments where random sampling points are chosen within a plot of interest. Grid, transect, or random sampling techniques are used for the characterization of the spatial distribution of microbiological properties at the regional scale. If soil biological data are distributed normally, the number of samples that are necessary for a given level of accuracy can be found by using the relationship:

$$N = t^2C^2/E^2, \quad \text{(Eq. 7.1)}$$

where N = the number of samples to be collected; t = student's statistic that is appropriate for the level of confidence and number of samples collected; C = the coefficient of variation (standard deviation divided by mean); and E = the acceptable error as a proportion of the mean.

Descriptions of sampling time, frequency, intensity, as well as preparation, archiving, and quality control, are given by Robertson et al. (1999). Soil biological data become more informative if supplemented by information about corresponding soil physical, chemical, and biotic factors (Table 7.2).

Spatial patterns strongly depend on the organisms studied, the characteristics of the study area, and the spacing of the samples. When topography and soil chemical and physical properties are relatively

TABLE 7.2 Physical, chemical, and biological properties that help to interpret data on the function and abundance of soil biota.

Physical and chemical soil properties		Biological soil properties
Topography	Particle size and type	Plant cover and productivity
Parent material	CO_2 and O_2 status	Vegetation history
Soil type, soil pH	Bulk density	Abundance of soil animals
Moisture status	Temperature: range and variation	Microbial biomass
Water infiltration	Rainfall: amount and distribution	Organic matter inputs and roots present

uniform, spatial patterns of soil biota are primarily structured by plants (plant size, growth form, and spacing). Therefore, simple a priori sampling designs are often inappropriate. A nested spatial sampling design is useful to explore spatial aggregation across a range of scales. For patch size estimation and mapping at a particular scale, the spatial sampling design can be optimized using simulation. To increase the statistical power in belowground field experiments and monitoring programs, exploratory spatial sampling and geostatistical analysis can be used to design a stratified sampling scheme for hotspots (Slaets et al., 2020). Pico- and nano-scale investigations are used to reveal the structure and chemical composition of organic substances and microorganisms, as well as the interactions between biota and humic substances. These analyses can identify organisms, characterize their relationships, determine their numbers, and measure the rates of physiological processes. Tremendous progress has been made at the nano-scale in visualizing nutrient uptake by individual soil microbes, element storage in microorganisms, and in the clarification of root-microbe interactions using secondary ion mass spectrometry (SIMS, see the imaging section below) (Clode et al., 2009; Wagner, 2009). Such results boost our understanding of chemical and biological processes and structures at larger scales.

Microscale investigations concentrate either on soil aggregates or on different microhabitats characterized by high turnover of organic material (e.g., rhizosphere, detritusphere, drilosphere, and the soil—litter interface). For example, microbial abundance and carbon (C) dynamics at the soil—litter interface can be studied in microcosms where moisture content and solute transport through soil are regulated by irrigation through fine needles and suction plates on the bottom of the soil cores. These microcosms have been used to study the metabolism and co-metabolism of bacterial and fungal degradation of different organic substances (Pinheiro et al., 2018; Poll et al., 2010). Since high-activity areas are heterogeneously distributed within the soil matrix, hotspots of biotic activity may make up $<10\%$ of the total soil volume but may represent $> 90\%$ of the total biological activity (Beare et al., 1995; Bernhardt et al., 2017). Consequently, upscaling of data from the small- to the plot- or regional-scale remains difficult because spatial distribution patterns are still relatively unknown.

Knowledge of the spatial dependency of soil biotic attributes helps to interpret their ecological significance at the ecosystem level. Fig. 7.1 gives an example of the spatial distribution of microbial biomass in the top 10 cm of a grassland soil at the scale of $6 \times 6\ m^2$. Kriged maps can be used to relate the spatial distribution of soil microorganisms to physical and chemical soil properties of the same site. Nevertheless, biochemical processes in the soil are dynamic, leading to variations in both space and time. Landscape-scale analyses by geostatistical methods are a useful tool for identifying and explaining spatial relationships between soil biochemical processes and site properties. The use of co-occurrence networks of sequence data is an alternative approach to move beyond the basic inventory descriptions of the composition and diversity of natural microbial communities at different scales (Barberán et al., 2012). At larger scales, Oliverio et al. (2020) used amplicon sequencing of soils from 180 locations across six continents to investigate the ecological preferences of protists and their functional contributions to belowground systems. Further model improvements, however, should focus on identifying and mapping time-space patterns using approaches such as fuzzy classification and geostatistical interpolation (Tscherko et al., 2007).

7.3 Storage and pretreatment of samples

After soil sampling, microbial communities within a sample change, regardless of the storage method. Therefore, it is important to minimize these potential changes during storage to the extent possible. Since multiple analyses are often applied to the same soil sample, it is necessary to find a compromise that

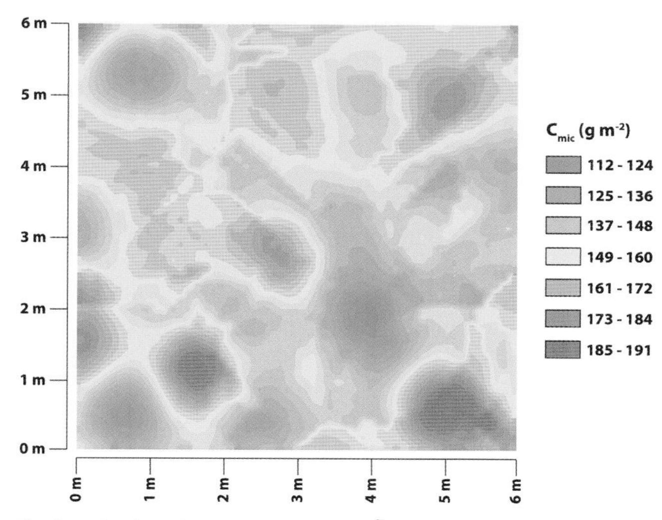

FIGURE 7.1 Spatial distribution of microbial carbon (g C_{mic} m^{-2}) in a temperate grassland. The kriged map illustrates the heterogeneity of C_{mic} at the scale of 6 × 6 m^{-2}.

reduces the potential risk of microbiological changes during storage. The best option is to perform biological analyses as soon as possible after soil sampling. Ideally, biochemical analyses should be made within 48 hours of sampling; however, moist soil can be stored for up to 3 weeks at 4°C when samples cannot be processed immediately. If longer storage periods are necessary, the samples taken to measure most soil physiological properties (soil microbial biomass, PLFAs, etc.) can be stored at −20°C. The soil is then allowed to thaw at 4°C for about 2 days before analysis. The soil disturbance incidental to sampling may in itself trigger changes in soil populations during the storage interval. Observations on a stored sample may not be representative of the undisturbed field soil. If samples are stored, care should be taken to ensure that samples do not dry out and that anaerobic conditions do not develop. Analyses are usually preceded by sieving the samples through a mesh sieve (<2 mm) to remove stones, roots, and large organic debris. Wet soil samples can be sieved either through a 5 mm mesh or gently predried before using the 2 mm mesh.

7.4 Microbial biomass

The estimation of microbial biomass is the basis of most studies in soil microbiology. Consequently, several approaches have been developed over the last 40 years to characterize different aspects of the microbial biomass. Whereas standard methods focus on the quantification of organic C bound in soil microorganisms (e.g., chloroform-fumigation extraction [CFE], substrate-induced respiration [SIR]), other methods give insight into the biomass of specific cell compounds (PLFAs, ergosterol, quinones) representing, for example, fungi, and gram-positive or gram-negative bacteria. A general aspect is to also link the microbial biomass with their energy status; for example, by measuring the ATP content or the heat output. Although the latter methods fell out of favor over the last 20 years, these measurements deserve renewed attention. Thermodynamic descriptions of soils provide a systemic view of energy and matter fluxes and their interactions with living and nonliving soil components (Bölscher et al., 2020). This new viewpoint in soil science considers energy flows and storage of SOM, living biomass, and necromass that are subject to the laws of thermodynamics. Further reading on methods focusing on soil fauna, molecular techniques, and SOM can be found in Chapters 5, 6, and 13, respectively.

7.4.1 Fumigation incubation and fumigation extraction methods

Microbial biomass is measured to give an indication of the response of soil microbiota to management, environmental change, site disturbance, and soil pollution. Two different approaches are based on CO_2 evolution. The chloroform fumigation incubation (CFI) method (Jenkinson and Powlson, 1976) exposes moist soil to ethanol-free chloroform for 24 hours to kill the indigenous microorganisms. After removal of the fumigant, a flush of mineralized CO_2 and NH_4^+ is released during a 10-day incubation. This flush is caused by soil microorganisms that have survived the fumigation and are using cell lysates produced by the fumigation as available C and energy sources. The released CO_2 is measured with an infrared gas analyzer (IRGA) or trapped in an alkaline solution and quantified by titration. Alternatively, the CO_2 of the headspace of the samples is measured by gas chromatography (GC) or automated respiration equipment (e.g., chambers linked to an IRGA). An assay of nonfumigated soil serves as a control. The amount of microbial biomass C is calculated as:

$$\text{Biomass C} = (Fc - UFc)/Kc, \qquad \text{(Eq. 7.2)}$$

where biomass C = the amount of C trapped in the microbial biomass; Fc = CO_2 produced by the fumigated soil; UFc = CO_2 produced by the nonfumigated soil sample; and Kc = fraction of the biomass C mineralized to CO_2. The Kc value is a constant, representative of the cell utilization efficiency of the fumigation procedure. This efficiency is assumed to be about 40% to 45% for many soils (e.g., a constant of 0.40−0.45). Deviations from this range are found for subsurface and tropical soils. The Kc factor of soil samples can be estimated by measuring the $^{14}CO_2$ release of labeled soil biota.

The CFE method involves the extraction and quantification of microbial constituents (C, N, S, and P) immediately following $CHCl_3$ fumigation of the soil (Brookes et al., 1985). The efficiency of extraction has to be considered. To convert extracted organic C to biomass C, the calibration factor of k_{EC} 0.45 is recommended for surface soils (Joergensen, 1996; Joergensen et al., 2011); the k_{EC} factor of soils from other environments (subsurface soils, peat soils, etc.) should be experimentally derived. Soils containing large amounts of living roots require a pre-extraction procedure of roots because these cells are also affected by the fumigation procedure. Direct extraction of C from nonfumigated and chloroform-fumigated soils with 0.5 M K_2SO_4 solution for 30 minutes is a rapid alternative method for quantifying microbial biomass C in soils (Setia et al., 2012). Results of the direct extraction method are comparable with results derived from the CFE method.

7.4.2 Substrate-induced respiration (SIR)

The SIR method estimates the amount of C held in living, heterotrophic organisms by measuring their initial respiration following the addition of an available substrate (Anderson and Domsch, 1978). In general, soil samples are amended with glucose and respiration is followed for several hours (no increase in the population of microorganisms should occur prior to measurement). The initial respiratory response is proportional to the amount of microbial C present in the soil sample. Results can be converted to mg biomass C by applying the following conversion factor, derived from the calibration of SIR to the CFI technique:

$$y = 40.04x + 0.37, \quad \text{(Eq. 7.3)}$$

where y is biomass C (mg 100 g^{-1} dry weight soil) and x is the respiration rate (mL CO_2 100 g^{-1} soil h^{-1}). Respired CO_2 can be measured by titration of the CO_2 trapped in an alkaline solution, by gas chromatographic analysis, or by using an IRGA. The addition of glucose at levels of optimum concentration to generate the maximal initial release of CO_2 must be independently determined for each soil type and should be applied accordingly to standardize the substrate induction method between different soil types. The SIR method using titrimetric measurement of CO_2 is frequently applied because it is simple, fast, and inexpensive. A disadvantage of static systems, such as SIR that use alkaline absorption of evolved CO_2, is that the O_2 partial pressure may change, causing overestimations in neutral or alkaline soils. Nevertheless, most versions of the three methods for estimating microbial biomass (FI, FE, and SIR) give similar results (Beck et al., 1997).

By using selective antibiotics, the SIR approach can also be applied to measure the relative contributions of fungi and bacteria to the soil microbial community. Glucose-induced respiration is determined in the presence of streptomycin to inhibit bacteria and in the presence of cyclohexamide to inhibit fungi. An automated IRGA system is used to continuously measure CO_2 production, and a computer program is used to calculate the bacterial:fungal respiration ratio based on the following criteria: (1) proof of no nonselective inhibition, and (2) proof of no shifts in the biosynthesis rates of bacteria and fungi (Bååth and Anderson, 2003).

7.5 Compound-specific analyses of microbial biomass and microbial community composition

Signal molecules differ in their specificity. For example, the amount of ATP extracted from soil gives a signature of all living organisms, while the ergosterol content is specific to fungi. Signal molecules, such as PLFAs or respiratory quinones, are used as indicators of the microbial community's structural diversity. PLFA quantification gives an estimate of microbial community composition at low taxonomic resolution in comparison to molecular tools based on sequencing (Chapter 6). A prerequisite for using a compound as a signal molecule is that it be unstable outside the cell such that the compound extracted from the soil is representative of living organisms only.

7.5.1 ATP and heat output as a measure of active microbial biomass

All biosynthetic and catabolic reactions within cells require the participation of adenosine triphosphate (ATP), which is sensitive to phosphatases and does not persist in soil in a free state. The adenine nucleotides (adenosine 5′-monophosphate (AMP), adenosine 5′-diphosphate (ADP), and ATP serve as energy carriers that are produced during exergonic reactions and used to drive endergonic reactions. The sum of AMP, ADP, and ATP calculates the adenylate energy charge (AEC). The detection of ATP is usually carried out with a bioluminescence assay (Schinner et al., 1996). Thermodynamic methods (e.g., heat output measurement using an isothermal microcalorimeter) provide different and complementary information to that from ATP and other techniques characterizing active microbial biomass (Thiele-Bruhn et al., 2020; Wu et al., 2014). Using isothermal microcalorimetry, Herrmann et al. (2014) showed that direct measures of energetics provide a functional link between energy flows and the composition of belowground microbial communities at a high taxonomic level.

7.5.2 Microbial membrane components and fatty acids

Lipids play an important role as sources of energy (i.e., neutral lipids), as structural components of membranes (i.e., phospholipids), and/or as storage products of cells. To date, more than 1000 different lipids have been identified. Fatty acids, the major fraction of lipids, contain either a fully saturated or unsaturated C chain containing, in most cases, one or two double bonds. Their classification is based on the number of C atoms counted from the carboxyl group (delta end) to the nearest double bond (Fig. 7.2). Alternatively, a single fatty acid is expressed as the number of C atoms counted from the terminal methyl group (ω end). The latter classification is frequently used in soil microbiology, because the position of the double bond in the fatty acid depends on the pathway of biosynthesis, revealing characteristic "ω families" (ω3, ω6, and ω9) (Ruess and Chamberlain, 2010).

Lipids can be extracted from the soil with a one-phase chloroform-methane extraction. Separation of the extracted lipids on silicic acid columns yields neutral lipids, glycolipids, and polar lipids. The neutral lipids can be further separated by high-pressure liquid chromatography (HPLC), derivatization, and GC to yield sterols and triglycerides. Glycolipids yield poly-β-hydroxybutyrate through the processes of hydrolysis and derivatization, followed by GC. Polar lipids are separated by hydrolysis, derivatization, and GC to yield phospholipid phosphate, phospholipid glycerol, PLFAs, and ether lipids.

A simple approach to characterize microbial communities is the extraction and subsequent methylation of lipids from soils to release their respective fatty acid methyl esters (FAMEs). Since the extracted lipids derive not only from cellular storage compounds and membranes of living organisms but also from

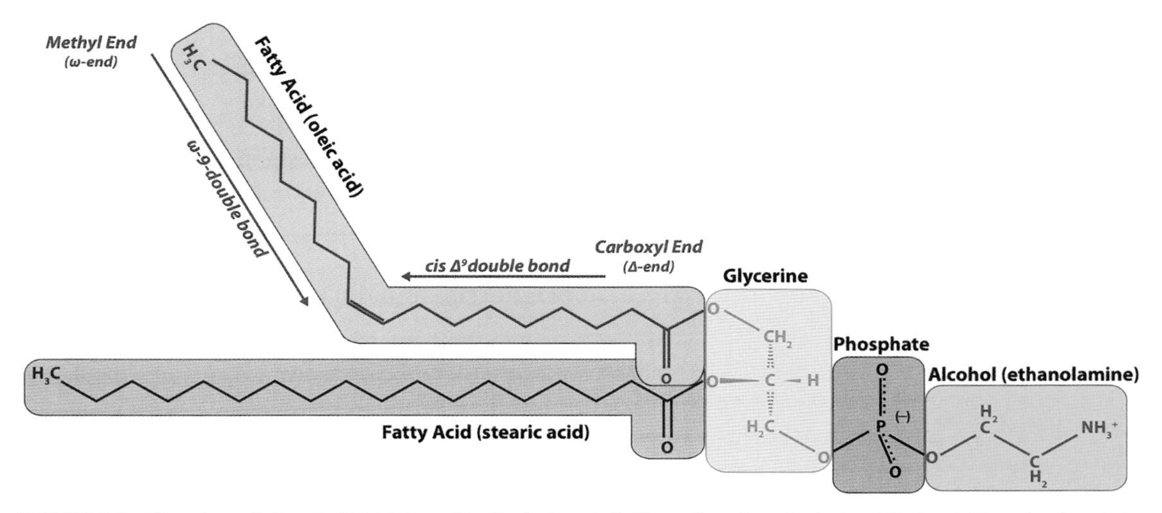

FIGURE 7.2 Structure of phospholipid fatty acids. Each phospholipid consists of a polar hydrophilic head (glycerin, phosphate, alcohol) and two hydrophobic fatty acid tails. The fatty acid structure affects the bilayer structure of the cell membrane. Unsaturated fatty acids (green) are bent, while saturated fatty acids (blue) are not bent, making it possible for them to be packed into the cell membrane more tightly than unsaturated fatty acids.

plant tissues in various stages of decomposition, it is difficult to draw conclusions about changes in soil microbial community structure from these patterns (Drenovsky et al., 2004). However, FAME analysis is more rapid than PLFA analysis. Phospholipids are found in the membranes of all living cells, but not in their storage products. Due to their taxonomic specificity and their rapid degradation after cell death, they are frequently used as signal molecules for microbial abundance and microbial community composition under different environmental conditions (Frostegård et al., 2011; Ruess and Chamberlain, 2010). PLFAs are extracted by single-phase solvent extractions and can be analyzed by: (1) colorimetric analysis of the phosphate after hydrolysis; (2) colorimetric analysis or GC after esterification; (3) capillary GC; and (4) GC-mass spectrometry or triple-quadruple mass spectrometry. The mass profile from mass spectrometry yields information on the present phospholipid classes and their relative intensities.

Characteristic fatty acids in phospholipid and neutral lipid fractions of bacteria, fungi, plants, and animals in soils are shown in Table 7.3. The presence of indicator PLFAs unique to certain taxa is inferred from pure culture studies. There is still debate as to whether or not some biomarkers common to a certain group of microorganisms might also be found in smaller amounts in other organisms. The following biomarkers, however, are widely accepted: (1) branched-chain fatty acids (iso, anteiso) as indicators for Gr^{+} bacteria; (2) cyclopropyl fatty acids for Gr^{-} bacteria; and (3) 18:2ω6,9 for fungi, which is usually well-correlated to ergosterol, an alternative fungal biomarker. Different fungal phyla contribute at different extents to the overall amount of 18:2ω6,9. Ascomycetes comprise 36 to 61 mol% and Basidiomycetes 45 to 57 mol% of linoleic acid (18:2ω6,9), whereas Zygomycetes contribute only 12 to 22 mol % of total PLFAs (Klamer and Bååth, 2004). The co-occurrence of the PLFA 16:1ω5 in both arbuscular mycorrhizae (AM) and bacteria makes it possible to use this indicator for AM only under environmental conditions of low bacterial abundance. An alternative procedure is to rely on neutral lipid fatty acids (NLFA) for certain groups of organisms (Table 7.3). The amount of neutral lipids (for energy storage) is usually higher than that of phospholipids (membrane constituents) in AM fungi, since these fungi store a large proportion of their energy C as neutral lipids (Olsson and Johansen, 2000). Straight-chain fatty acids, such as palmitic (16:0), stearic (18:0), palmitoleic (16:1ω7), or oleic acid (18:1ω9), frequently occur in lipid samples. Even though of limited taxonomic value, they can be used as indicators of microbial biomass.

7.5.3 Phospholipid etherlipids

Analyses of archaeal membrane lipids (e.g., phospholipid etherlipids [PLEL]) are increasingly included in ecological studies as a comparatively unbiased complement to gene-based microbiological approaches (Gattinger et al., 2003; Pitcher et al., 2010). The PLEL-derived isoprenoid side chains are measured by gas chromatography spectrometry/mass spectrometry (GC/MS) and provide a broad picture of the archaeal community in a soil extract, because lipids identified in isolates belonging to the sub-kingdom Eury- and Crenarchaeota are covered. Monomethyl-branched alkanes are dominant and account for 43% of the total identified ether-linked hydrocarbons, followed by straight chain (unbranched) and isoprenoid hydrocarbons, which account for 34.6% and 15.5%, respectively. The ether lipid crenarchaeol has been postulated as a specific biomarker for all archaea carrying out ammonia oxidation (AOA) (Pitcher et al., 2010). Since not many cultivated representatives outside the AOA lineages of the Group I Crenarchaeota are available, it is not yet clear how widespread crenarchaeol synthesis is among all archaea.

TABLE 7.3 Characteristic ester-linked fatty acids in the lipids of common soil biota.

Fatty acid type	Frequently found	Lipid fraction	Predominant origin
Saturated			
>C20, straight	22:0, 24:0	PLFA, NLFA	Plants
Iso/anteiso methyl-branched	i, a in C14-C18	PLFA	Gram-positive bacteria
10-methyl-branched	10ME in C15-C18	PLFA	Sulfate-reducing bacteria
Cyclopropyl ring	cy17:0, cy19:0	PLFA	Gram-negative bacteria
Hydroxy-substituted	OH in C10-C18	PLFA	Gram-negative bacteria, actinomycetes
Monounsaturated			
Double-bond C5	16:1ω5	PLFA	AM fungi, bacteria
		NLFA	AM fungi
Double-bond C7	16:1ω7	PLFA	Bacteria widespread
	18:1ω7	PLFA	Bacteria, AM fungi
		NLFA	AM fungi
Double-bond C8	18:1ω8	PLFA	Methane-oxidizing bacteria
Double-bond C9	18:1ω9	PLFA	Fungi
		PLFA	Gram-positive bacteria
		NLFA	Nematodes
	20:1ω9	PLFA	AM fungi (*Gigaspora*)
Polyunsaturated			
ω6 family	18:2ω6,9	PLFA	Fungi (saprophytic, EM)
		NLFA	Animals
	18:3ω6,9,12	PLFA	Zygomycetes
	20:4ω6,9,12,15	PLFA, NLFA	Animals widespread
ω3 family	18:3ω3,9,12	PLFA	Higher fungi
	20:5ω3,6,9,12,15	PLFA	Algae
		PLFA, NLFA	Collembola

NLFA, neutral lipid fatty acid; PLFA, phospholipid fatty acid.
Modified from Ruess and Chamberlain (2010).

7.5.4 Respiratory quinones

The quinone profile, which is represented as a molar fraction of each quinone type in a soil, is a simple and useful tool for analyzing population dynamics in soils and of biotic biomass (Fujie et al., 1998). These are essential components of the electron transport systems of most organisms that are present in membranes of mitochondria or chloroplasts. Isoprenoid quinones are chemically composed of benzoquinone (or naphthoquinone) and an isoprenoid side chain. There are two major groups of quinones in soils: (1) ubiquinones (1-methyl-2-isoprenyl-3,4-dimethoxyparabenzoquinone) and (2) menaquinones (1-isoprenyl-2-methyl-naphthoquinone). Most microorganisms contain only one major species of quinone, which remains unchanged by physiological conditions. The half-lives of those quinones released by dead soil microorganisms are very short (in the range of several days). The diversity of quinone species can be interpreted directly as an indicator of microbial diversity.

7.5.5 Ergosterol as a measure of fungal biomass

Sterols in fungi typically exist as a mixture of several sterols, with one comprising more than 50% of the total sterol composition (Weete et al., 2010). Ergosterol is the predominant sterol of many fungi (ascomycetes and basidiomycetes), yet it does not occur in plants (Fig. 7.3). In addition to ergosterol, there appear to be at least five taxon-specific end-products of sterol biosynthesis that occur as dominant sterols in certain lineages, including some subclades of ascomycetes and basidiomycetes. Glomeromycota, arbuscular mycorrhizal fungi (AMF), contain 24-ethylcholesterol and members of the fungal family Mortierallaceae contain desmosterol (Weete et al., 2010).

Ergosterol is extracted by methanol and detected using an HPLC with an ultraviolet detector (HPLC-UV) (Müller et al., 2016). Since chromatographic co-elution might be a problem, reversed-phase liquid chromatography with positive-ion atmospheric pressure chemical ionization tandem mass spectrometry (LC/APCI/MS/MS) can be used for full quantification and confirmation of ergosterol (Verma et al., 2002). The ergosterol content varies from 0.75 to 12.9 $\mu g\ g^{-1}$ soil in arable, grassland, and forest soils (Djajakirana et al., 1996). These values correspond to 5 to 31 mg ergosterol g^{-1} fungal dry weight depending on species and growth conditions. The ratio of ergosterol to microbial biomass C is used as an index of fungal biomass to total soil microbial biomass. Shifts in microbial community structure due to soil contamination or change in vegetation can be detected using the ergosterol to microbial biomass C ratio. The content of coprostanol, which is a sterol present after sewage sludge disposal and contamination by municipal wastes, is a useful marker of human fecal matter contamination of soils.

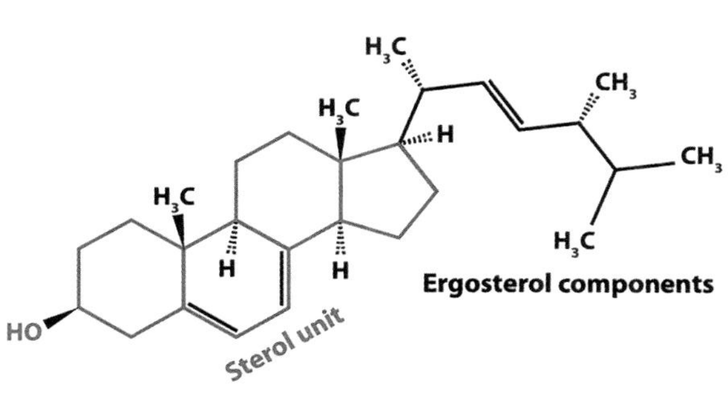

FIGURE 7.3 Chemical structure of ergosterol. The sterol unit, which is common to all sterols, is marked in blue.

7.5.6 Gene abundance as a measure of biomass of specific groups of soil microorganisms

Quantitative polymerase chain reaction (Q-PCR or real-time PCR) is increasingly applied to quantify the abundance and expression of taxonomic and functional gene markers within soils (Thiele-Bruhn et al., 2020). Q-PCR-based analyses combine end-point detection PCR with fluorescent detection to measure the accumulation of amplicons in "real time" during each cycle of the PCR amplification (Smith and Osborn, 2009). Q-PCR approaches targeting highly conserved regions of the 16S rRNA gene are currently used to quantify bacterial and/or archaeal copy numbers. The use of taxa-specific sequences within hypervariable regions of genes makes it possible to target specific groups of soil organisms at higher taxonomic resolution (e.g., β-proteobacteria, acidobacteria, actinobacteria) (Lennon et al., 2012; Philippot et al., 2011). In addition, Q-PCR is increasingly used to quantify functional genes that encode key enzymes in different biogeochemical cycles (e.g., ammonia oxidation, denitrification, methanogenesis, methane oxidation, and degradation of pesticides). Molecular biology methods estimating the relative abundance of functional guilds or taxa also give new insight into the ecology of microorganisms (Müller et al., 2020; Thiele-Bruhn et al., 2020). A description of different deoxyribonucleic acid (DNA) extraction procedures and PCR reactions, as well as application of these molecular techniques, can be found in Chapter 6.

7.5.7 Component specific analyses of microbial products

Several lipids and N-containing compounds (e.g., lipopolysaccharides, glycoproteins, amino sugars) can be used to track the residues of different fungi and bacteria (Amelung et al., 2008). The outer cell membrane of Gr^- bacteria contains unique lipopolysaccharide (LPS) polymers. The peptidoglycan of bacterial cell walls contains N-acetylmuramic acid and diaminopimelic acid. Chitin, a polymer of N-acetylglucosamine, is found in many fungi, but is also present in the exoskeleton of invertebrates. Three amino sugars that are mainly derived from dead microbial biomass can be used to trace the fate of C and N within residues of bacteria and fungi. These amino sugars in soils comprise about 1% to 5% of total SOC. Muramic acid and galactosamine primarily originate from bacterial peptidoglycan, whereas glucosamine is mainly derived from fungal cell walls. The enantiomers of amino acids have been used as markers for both bacterial residues and cell aging. D-amino acids are produced by racemization from their respective L-enantiomers during cell aging. Since aged and dead cells do not have any D-amino acid oxidase to decompose D-amino acids, these compounds may accumulate in soils. D-lysine is an especially promising age marker.

AMF belonging to the phylum Glomeromycota produce a recalcitrant, AM-specific glycoprotein, glomalin. Glomalin can be considered a biochemical marker of soil aggregate formation and stability (Rillig and Mummey, 2006). Glomalin, a putative homolog of plant heat shock protein 60, is a hydrophobic protein resistant to proteolysis, temperature, pH, and detergent denaturation. It is extracted from soil by applying several cycles of autoclaving, quantified using a Bradford assay, and evaluated for immuno-reactivity using a monoclonal antibody. Nevertheless, Reyna and Wall (2013) found that the use of the bicinchoninic acid (BCA) assay may be more appropriate than the Bradford assay to quantify easily extracted glomalin-related soil proteins (EE-GRSP) because it showed higher precision and reproducibility for determining the concentration values, with greater stability over time.

Water-insoluble polyester lipids, e.g., polyhydroxybutyrate (PHB), are produced by bacteria and some archaea as storage compounds under conditions of energy surplus. In continuous culture with glucose, the accumulation of these compounds can contribute up to 48% of bacterial dry cell mass. PHB in soil is primarily present in lipid inclusion bodies of living bacterial biomass. A new technique for its determination in soil samples was presented by Mason-Jones et al. (2019) which includes an extraction step with methanol and dichloromethane, followed by derivatization (ethanolysis) and quantification (GC/MS). Nevertheless, studies on the variation of, for example, PHB under natural environments, are rare. Mason-Jones et al. (2019) gave values for agricultural and forest soils from 1.2 to 14.4 μg PHB-C g^{-1}. Since glucose addition leads to high enrichment of PHB, this suggests that PHB content can be used as an indicator of bacterial nutritional status and unbalanced growth (e.g., N limitation). No universal fungal storage marker has been found. Nevertheless, the amount of triacylglycerol (TAG) and the neutral fatty acid 16:1ω5 can be used as storage markers for AMF (Calonne et al., 2014).

Component-specific analyses also include the extraction and characterization of proteins expressed by soil microorganisms within an ecosystem at a specific time (Becher et al., 2013; Keiblinger et al., 2016). Metaproteomic analyses offer functional information, especially at the intracellular level (Bastida et al., 2012). Metaproteomomic analyses start with the extraction, purification, and concentration of proteins. Major pitfalls are possible during protein extraction and processing of soil protein extracts. The first decision to make is whether the entire proteome of microorganisms consisting of extracellular, cytoplasmic, and membrane proteins should be extracted or whether only a certain fraction (e.g., extracellular enzymes) should be targeted. The indirect extraction protocols aim at enriching cells before extraction of proteins. This procedure reduces the possible interference by substances, such as humic substances and minerals. Nevertheless, this procedure reduces the extraction efficiency of proteins to a large extent. Much higher protein recovery rates are obtained by direct extraction procedures. This procedure involves a direct cellular lysis step (e.g., physical/mechanical lysis including heat and pressure), chemical lysis (e.g., stabilization agents and detergents), or enzymatic lysis. The next step after extraction of the protein fraction is the concentration of proteins by precipitation, followed by solubilization and MS analysis. Analytics of soil protein extracts involve, in most cases, tandem mass spectrometry of peptides after enzymatic protein digestions. Resulting spectra are compared with those of theoretical peptides from different protein databases. Data should be made publicly available in online repositories, such as Tranche or PRIDE. Keiblinger et al. (2016) suggest an innovative procedure (Voroni Treemaps) that can visualize highly complex hierarchically organized metaproteomic data in a space-optimized manner (Fig. 7.4). The top-level categories are represented by the different polygons which are constructed from a Voroni diagram. The top-level areas are then subdivided into subcategories, which is repeated down to the level of individual proteins. Thus, additional dimensions are introduced by area and/or color-coding.

7.6 Isotopic composition of microbial biomass and signal molecules

Radioactive and stable isotope tracers are used to follow the flow of C and nutrients (N, P, and S) into microbial biomass, specific groups of microorganisms, or into components of the soil microbial community. Fig. 7.5 gives the procedure and biomolecules used for stable isotope probing (SIP) of soil microorganisms using a ^{13}C-labeled substrate. This procedure also allows for tracing the ^{13}C label into other soil organisms (e.g., protozoa and nematodes). Studies have increasingly gained interest due to improved sensitivity of ^{13}C measurements and nonradioactive nature (in contrast to ^{14}C studies). Carbon use by the microbial community as a whole, or specific groups of microorganisms, are followed by labeling techniques that differ in their actual ^{13}C enrichment: natural abundance, near natural abundance (50−500‰), or highly enriched (>500‰) (Amelung et al., 2008). Natural abundance ^{13}C labeling tracer approaches are based on the physiological difference during the photosynthetic fixation of CO_2 between

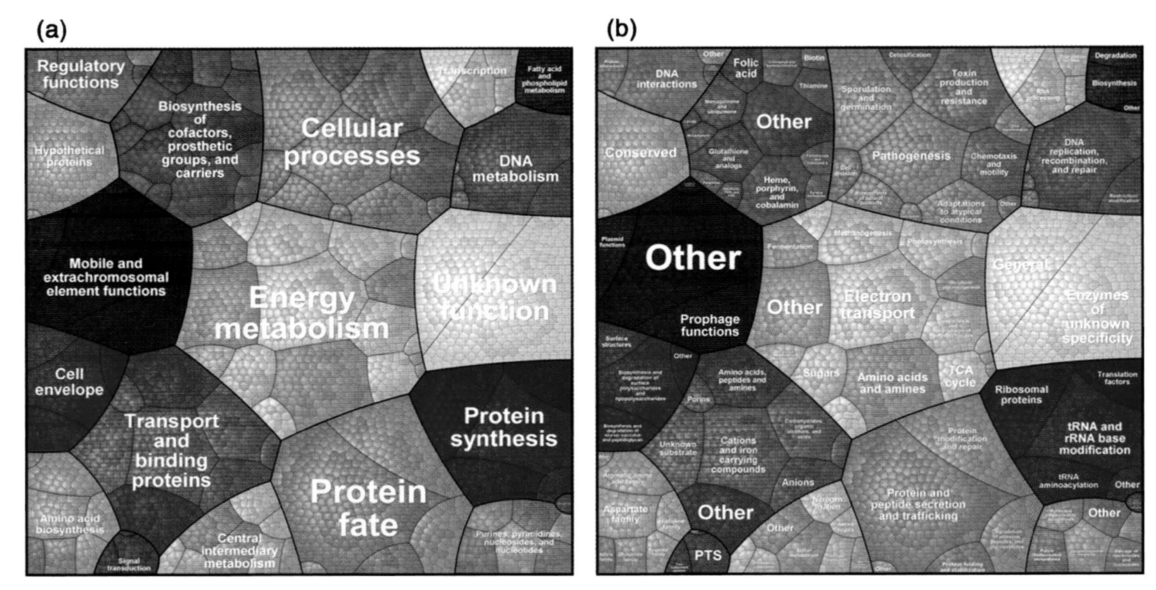

FIGURE 7.4 Voronoi treemaps can visualize highly complex hierarchically organized data in a space-optimized manner. Voronoi maps are applied to sequence data of proteins involved in the physiology of microbial cells. Coarse (a) and fine hierarchical resolution (b). *(Keiblinger et al., 2016.)*

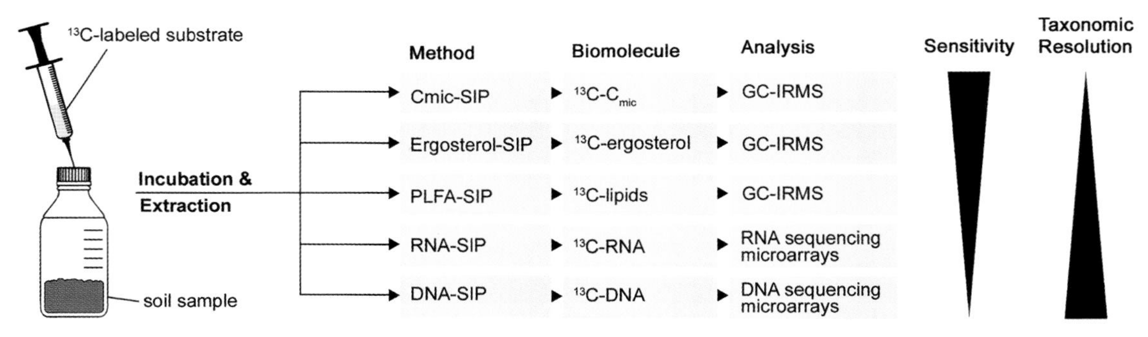

FIGURE 7.5 Schema of the stable isotope probing (SIP) procedure and biomolecules used for SIP of soil microorganisms. GC-IRMS, gas chromatography-ion ratio mass spectrometry. *(Pepper et al. 2014, redrawn by L. Brandt.)*

C3 and C4 plants. Since C3 plants (e.g., wheat) and C4 plants (e.g., maize) differ in their natural abundance of ^{13}C, these materials provide a natural label that can be used for decomposition studies in both microcosm and field experiments. If the use of natural abundance is not possible, ^{13}C pulse or continuous labeling introduces a distinct C signal into the soil system as well as to the soil microbial community (Kramer et al., 2012). The combined application of ^{13}C plant labeling and ^{13}C natural abundance allows linking aboveground C fixation by photosynthesis to belowground C dynamics of rhizosphere microorganisms.

7.6.1 Isotopic composition of microbial biomass

Determining the isotopic composition of microbial biomass C is an important tool for the study of soil ecology and the decomposition and transformation of soil organic C. CFE can be used for a large range of isotopes (i.e., ^{14}C, ^{13}C, ^{15}N, ^{32}P, and ^{35}S). The use of the CFI method is restricted to isotopes ^{14}C, ^{13}C, and ^{15}N due to the complex interaction of P and S with the soil matrix during incubation. Determinations of $^{13}C/^{12}C$ (reported in parts per thousand [per mil, ‰]) of the microbial C fraction in soil are performed with an off-line sample preparation technique combined with isotope analysis by a dual-inlet isotope ratio mass spectrometer (IRMS) or an on-line analysis using an element analyzer connected to an isotope ratio mass spectrometer (EA-IRMS). An alternative method is based on the UV-catalyzed liquid oxidation of fumigated and nonfumigated soil extracts combined with trapping the released CO_2 in liquid N; $\delta^{13}CO_2$-C is subsequently determined with a gas chromatograph connected to an IRMS (Potthoff et al., 2003) and is calculated as follows:

$$\delta^{13}C = \left(\frac{\left(\frac{^{13}C}{^{12}C}\right)_{\text{sample}}}{\left(\frac{^{13}C}{^{12}C}\right)_{\text{standard}}} - 1 \right) \times 1000‰$$

The ^{13}C analysis can also be performed by using an automated continuous-flow IRMS. In addition, isotopic analysis of $\delta^{15}N$ and $\delta^{13}C$ of DNA extracted from soil can determine the natural abundance of this important fraction of soil microbial biomass (Schwartz et al., 2007). This is a powerful tool for elucidating C and N cycling processes through soil microorganisms.

7.6.2 Stable-isotope probing (SIP) of fatty acids, ergosterol, and nucleic acids

SIP techniques introduce a stable isotope-labeled substrate into a microbial community, follow the fate of the substrate by extracting signal molecules, such as PLFAs or nucleic acids, and determine which specific molecules have incorporated the isotope (Coyotzi et al., 2016; Kreuzer-Martin, 2007). These techniques allow for the study of substrate assimilation in minimally disturbed communities and provide a tool for linking functional activity to specific members of microbial communities. PLFA-SIP techniques are characterized by a higher sensitivity than RNA-SIP or DNA-SIP techniques. Due to the high sensitivity of the PLFA-SIP method, incubations can be carried out at near in situ concentrations of labeled substrates. After incubation of the soil with labeled substrates, all lipids are extracted. The isotopic compositions of fatty acids are analyzed by gas chromatography-combustion-isotopic ratio mass spectrometry (GC-IRMS). The enrichment of ^{13}C in single biomarkers (Table 7.3) is used to identify the groups of microorganisms utilizing the added substrate. In contrast to other SIP-based methods (e.g., RNA-SIP), it is not necessary to purify unlabeled from labeled biomarkers. If two C sources with different isotopic signatures are available (e.g., ^{13}C litter and soil organic C), PLFA-SIP allows the quantification of the relative assimilation of each of these food resources. In contrast to the DNA-SIP technique which requires highly labeled substrates, the PLFA-SIP, NLFA-SIP, and ergosterol-SIP can be applied in studies using the natural abundance of substrates (e.g., C4 litter material) or substrates with low ^{13}C enrichment. For example, C flow from a C4 maize plant into rhizosphere microorganisms can be followed by growing maize in a soil which has an isotopic signature of C3 plants (Müller et al., 2016). Both PLFA and NLFA-

SIP techniques can also be used to study the trophic transfer of C from primary decomposers to higher levels of the soil food web (e.g., nematodes, collembolans, and earthworms) (Hünninghaus et al., 2019; Ruess and Chamberlain, 2010). For example, Murase et al. (2011) studied the strain-specific incorporation of methanotrophic biomass into eukaryotic grazers in a rice field soil using the PLFA-SIP technique.

Carbon flow from a given substrate into fungi can be followed by compound-specific analysis of ergosterol, combined with a stable isotopic probing technique. Soils are incubated with a substrate (e.g., ^{13}C labeled litter or litter of a C4 plant) for a period of days to months. After the incubation, ergosterol is extracted, purified by a preparative HPLC, followed by the determination of ^{13}C ergosterol using a GC-IRMS or tandem mass spectrometry (GC-MS-MS) (Kramer et al., 2012).

The DNA and RNA-SIP techniques involve labeling of soils with a substrate (e.g., ^{13}C glucose, ^{13}C cellulose, maize-derived litter) for a period of hours up to several weeks, extraction of DNA or RNA, and buoyant density gradient centrifugation (Neufeld et al., 2007). The nucleic acids of microorganisms having assimilated the ^{13}C have a higher buoyant density than the nucleic acids of organisms that have not. The ^{13}C-containing nucleic acid is separated from unlabeled nucleic acid via centrifugation, and labeled nucleic acid is concentrated in heavier gradient fractions. Nucleic acid fractions at various positions in the gradient are collected and amplified by the PCR or reverse transcription (RT) PCR using primers complementary to 16S rDNA or rRNA. Alternatively, specific functional genes are used as amplification targets (see Chapter 6).

7.6.3 Growth rates from signature molecules and leucine/thymidine incorporation

Growth and turnover rates can be determined by incorporating tracers into precursors of cytoplasmic constituents, membranes, or cell wall components. The increase in the tracer over short time periods yields estimates of growth. For example, a technique to estimate relative fungal growth rates is based on the addition of ^{14}C-acetate to a soil slurry and measurement of the subsequent uptake and incorporation by fungi of the labeled acetate into ergosterol (Newell and Fallon, 1991). The specificity of this incorporation was shown by using fungal and bacterial inhibitors. Incorporation rates were linear up to 18 hours after the acetate addition, but absolute growth rates cannot be calculated due to the uncertainty of conversion factors and problems associated with the incorporation of the added acetate. Similar techniques have also been developed for C isotopes in microbial lipids. In combination with their role as biomarkers, fatty acids can also be used to monitor the C flux in a bacterial community by measuring the ratios of their C isotopes. Before fatty acids can be used as chemotaxonomic markers and indicators of substrate usage in microbial communities, calibration studies on the degree and strain specificity of isotopic ^{13}C fractionation with regard to the growth substrate are necessary.

Bacterial and fungal growth rates can also be separately determined by using leucine/thymidine incorporation to estimate bacterial growth and acetate incorporation into ergosterol to estimate fungal growth (Rousk and Bååth, 2011). A parameter to quantify how C is partitioned between growth and respiration is carbon use efficiency (CUE), defined as the ratio of the amount of C employed in new biomass (excluding C excreted in the form of metabolites and enzymes) relative to the amount of C that has been consumed (Manzoni et al., 2012). High CUE values indicate efficient growth and relatively low release of C into the atmosphere. ^{13}C tracing can be used to estimate CUE based on substrate uptake and growth dynamics, whereas the ^{18}O stable isotope method is based on the uptake of labeled water and does not need any addition of a C resource. The ^{18}O stable isotope method has been improved by sequencing a marker gene from fractions retrieved from ultracentrifugation to produce taxon density curves to estimate the percent isotope composition of each microbial taxon's genome (Schwartz et al., 2016). Geyer et al. (2019) discuss in detail the advantages and disadvantages of different methods to estimate CUE (see also Chapters 6 and 9).

7.7 Physiological analyses

A wide array of physiologically based soil processes is known; e.g., decomposition, ammonification, nitrification, denitrification, N_2 fixation, P mineralization, and S transformations. These are described in detail in later chapters. Here we primarily cover culture studies, respiration techniques, N-mineralization, and enzyme measurements.

7.7.1 Culture studies, isolation, and characterization of specific organisms

Whereas culturing, isolation, and characterization of single colonies were performed in the past to study colony numbers, growth forms, and/or biochemical capabilities of soil microorganisms, these techniques are currently used as a prerequisite to understand the physiology, genomics, and evolution of individual taxa. Culturing a soil organism involves transferring its propagules to a nutrient medium conducive to growth. These techniques allow isolation of specific soil organisms from a wide range of soils. Nevertheless, culture techniques are selective and do not cover the complexity of the soil microbial community. Plate counts of bacteria in soils, waters, and sediments usually represent 1% to 5% of the number determined by direct microscopy. Plate count methods are not suitable for enumerating fungal populations or densities because spores and fragments of mycelia develop as single colonies that are counted. The high percentage of unculturability of bacteria is explained by: (1) slow-growing bacteria that do not form visible colonies within several days of incubation; (2) bacteria competing for specific nutrients or inhibited by bacteriocins and antibacterial substances; and (3) bacteria needing specific signal compounds released only under natural environmental conditions. Nevertheless, there are new techniques under development that will improve the culturing of the unculturable. These approaches include the use of low-nutrient media, co-cultivation with helper strains to enhance beneficial interactions, the addition of signaling molecules, and the encapsulation of individual cells in gel microdroplets. In addition, infrared lasers are used as optical tweezers to manipulate single cells and transfer them to specific media.

7.7.2 Isolation and characterization of specific organisms

Specific organisms are typically isolated from soil using liquid or solid substrates. These can involve general substrates from which individual colonies are picked for further characterization. Media containing chloramphenicol, tetracycline, and streptomycin inhibiting protein synthesis or nalidixic acid inhibiting DNA synthesis of gram-negative bacteria, are used. Specific growth environments include high or low pH, anaerobiosis, or high salinity. Plants are used to identify and enrich specific soil populations. Growth of compatible legumes makes it possible to isolate and identify rhizobia. Root or leaf pathogens and mycorrhizas are often similarly enriched and identified. Isolation can be performed without growth media by microscopic examination combined with micromanipulators or optical laser tweezers, which can separate single cells, spores, or hyphal fragments.

Genetic markers that are incorporated into isolated organisms before soil re-inoculation include antibiotic resistance, the lux operon for light emission, and enzyme-activity genes. Resistance to most presently used antibiotics is typically carried on a plasmid. The use of an appropriate vector makes it possible to transfer this plasmid to soil isolates. These isolates can then be reintroduced into soil and recovered by growth on the appropriate medium containing that antibiotic. Only organisms carrying the antibiotic resistance gene(s) will be capable of growth.

Automated, multi-substrate approaches have been used as a metabolic fingerprinting system. The "Biolog System" makes it possible to test growth reactions of many isolates on a broad range of media and to analyze community composition. This, combined with the use of a broad range of known organisms and mathematical clustering techniques, allows characterization of unknowns by sorting them into affinity types occurring in that habitat. A community-level physiological approach is based on the C utilization profiles of soil dilutions inoculated onto the redox-based plates used in multi-substrate systems such as Biolog. This bypasses the need to work with isolated culturable organisms, but still requires growth (reduction) in the appropriate substrate. Because of their generally oxidative and mycelial nature, fungi are not amenable to approaches that utilize substrate clustering analyses. Multiple substrate fingerprinting also has limitations (i.e., measuring the enzyme activity of a growing microbial population or the selectivity of the microbial response).

7.7.3 Microbial growth and respiration

Microbial communities oxidize naturally occurring organic material, such as carbohydrates, as:

$$CH_2O + O_2 \rightarrow CO_2 + H_2O + \text{intermediates} + \text{cellular material} + \text{energy}. \qquad \text{(Eq. 7.5)}$$

Under anaerobic conditions, the most common heterotrophic reaction is that of fermentation, which in its simplest form is:

$$C_6H_{12}O_6 \rightarrow 2CH_3CH_2OH + 2CO_2 + \text{intermediates}$$
$$+ \text{cellular material} + \text{energy}. \qquad \text{(Eq. 7.6)}$$

Measurement of microbial activity is complex under anaerobic conditions. Fermentation products and CH_4 produced within anaerobic microsites can diffuse to aerobic areas where oxidation to CO_2 and H_2O can occur. Microbial and faunal respiration is determined by measuring either the release of CO_2 or the uptake of O_2. In general, the faunal contribution to total soil respiration is roughly 10% (Foissner, 1992; Schaefer, 1990; see Chapter 5). Due to the present approximately 0.04% atmospheric CO_2 concentration versus 20% for O_2, measurements of CO_2 production are more sensitive than those for O_2. The application of an automated IRGA system is the preferred technique for quantifying CO_2. IRGAs are sensitive to CO_2 and can be used for both static and flow systems in the laboratory and in the field after H_2O (which absorbs in the same general wavelength) has been removed. An alternative simple and cheap method of CO_2 measurement involves aeration trains. Here, NaOH is used to trap evolved CO_2 in an air stream from which CO_2 is removed before the air is exposed to the soil sample. The reaction occurs as follows:

$$2NaOH + CO_2 \rightarrow Na_2CO_3 + H_2O. \qquad \text{(Eq. 7.7)}$$

Before titration, $BaCl_2$ or $SrCl_2$ is added to precipitate the CO_3^{2-} as $BaCO_3$, and excess NaOH is back titrated with acid. The use of carbonic anhydrase and a double endpoint titration provides greater accuracy when CO_2 concentrations are low. In the laboratory, NaOH containers placed in sealed jars are convenient and effective for CO_2 absorption. The jars must be opened at intervals so that the O_2 concentration does not drop below 10%. GC, with thermal conductivity detectors, can be used to measure the CO_2 concentration after CO_2 is separated from other constituents on column materials, such as Poropak Q. Computer-operated valves in conjunction with GC allow time-sequence studies to be automatically conducted. The results are expressed either per unit of soil dry weight ($\mu g\ CO_2\text{-}C\ g^{-1}$ soil h^{-1}) or per unit of microbial biomass ($mg\ CO_2\text{-}C\ g^{-1}\ C_{mic}\ h^{-1}$). The ratio of respiration to microbial biomass is termed the metabolic quotient (qCO_2) and is in the range of 0.5 to 3 $mg\ CO_2\text{-}C\ g^{-1}\ C_{mic}\ h^{-1}$. The metabolic quotient is particularly useful in differentiating the responses of soil biota to soil management techniques. For example, stress, heavy metal pollution, and nutrient deficiency increase qCO_2 due to a decrease in microbial biomass and an increase in respiration.

The measurement of CO_2 can be augmented by incorporating ^{14}C or ^{13}C into chosen substrates. The tracer may be molecules, such as glucose, cellulose, amino acids, and herbicides, or complex materials, such as microbial cells or plant residues. Substituting $SrCl_2$ for $BaCl_2$ allows the precipitate to be measured in mass spectrometers. The measurement of $^{14}CO_2$ or $^{13}CO_2$ makes it possible to calculate the decomposition rate of SOM, and establish a balance of the C used in growth relative to substrate decomposition and microbial byproducts. Mass spectrometers capable of directly analyzing a gaseous sample for $^{13}CO_2$ are preferable to the precipitation procedure.

A variety of methods exist to measure soil CO_2 efflux under field conditions (Sullivan et al., 2010). Static or dynamic chambers placed on the soil surface allow for measurement of soil CO_2 efflux by quantifying the increase in gas concentration within the chamber headspace over a certain time. This procedure does not alter soil structure, and therefore, field respiration rates of the indigenous microbial population are reliably measured. Gas samples are taken from the static field chamber using a gastight syringe and are then injected into a gas chromatograph. Measurements of other gases, such as N_2O and CH_4 from the same samples, are possible. A dynamic chamber is coupled with an IRGA in the field. This procedure allows for instantaneous soil CO_2 efflux measurements. An alternative option is to model CO_2 efflux from soil CO_2 gradient profiles. Solid-state IRGA probes (e.g., GMM 222; Vaisala Inc., Helsinki, Finland) are buried vertically at different soil depths (e.g., 2, 10, and 20 cm). The probes are connected to a data logger and multiplexer. The diffusivity of CO_2 in soil can be calculated by models (Sullivan et al., 2010).

The eddy covariance technique ascertains the net ecosystem exchange (NEE) rate of CO_2 (the NEE represents the balance of gross primary productivity and respiration of an ecosystem). The covariance between fluctuations in vertical wind velocity and CO_2 mixing ratio across the interface between the atmosphere and a plant canopy is measured at a flux tower (Baldocchi, 2003). Although flux tower data represent point measurements with a footprint of typically 1 km^2, they are especially useful for validating models and to spatialize biospheric fluxes at the regional scale (Papale and Valentini, 2003).

Litter decomposition in the field can be examined by following the decay process (i.e., weight loss) of added litter. Site-specific or standard litter (e.g., wheat straw, pine wood, or other standardized organic substances) is placed in nylon mesh bags and exposed on or just below the soil surface. Keuskamp et al. (2013) used an innovative, cost-effective, well-standardized method to gather data on decomposition rate and litter stabilization using commercially available tea bags (green and Rooibos tea). The Tea Bag Index (TBI) consists of two parameters describing decomposition rate (k) and litter stabilization factor (S).

Organic matter decomposition can also be followed by the minicontainer system (Eisenbeis et al. 1999) (Fig. 7.6a–e). The system consists of polyvinylchloride (PVC) bars as carriers and minicontainers enclosing organic material (Fig. 7.6a–b). Minicontainers can be exposed to the top layers of agricultural soils (Ap = plowed A horizon) and to forest soils for several weeks up to years. Vertical insertion of the bars will give information about gradients of decomposition within a soil profile; horizontal exposure of the bars explains spatial variation of decomposition within one horizon. The minicontainers are filled with 150 to 300 mg of organic substrate (e.g., straw, litter, cellulose, needles) and closed by nylon mesh of variable sizes (20 μm, 250 μm, 500 μm, or 2 mm) to exclude or include the faunal contribution to organic matter decomposition (Fig. 7.6e). After an exposure time of several weeks to months, minicontainers are removed (Fig. 7.6c–d), remaining organic materials are dried, and decomposition is calculated based on the weight loss of the oven-dried material taking into consideration the ash content of the substrate. A time series analysis allows for an investigation of the dynamics of decomposition processes.

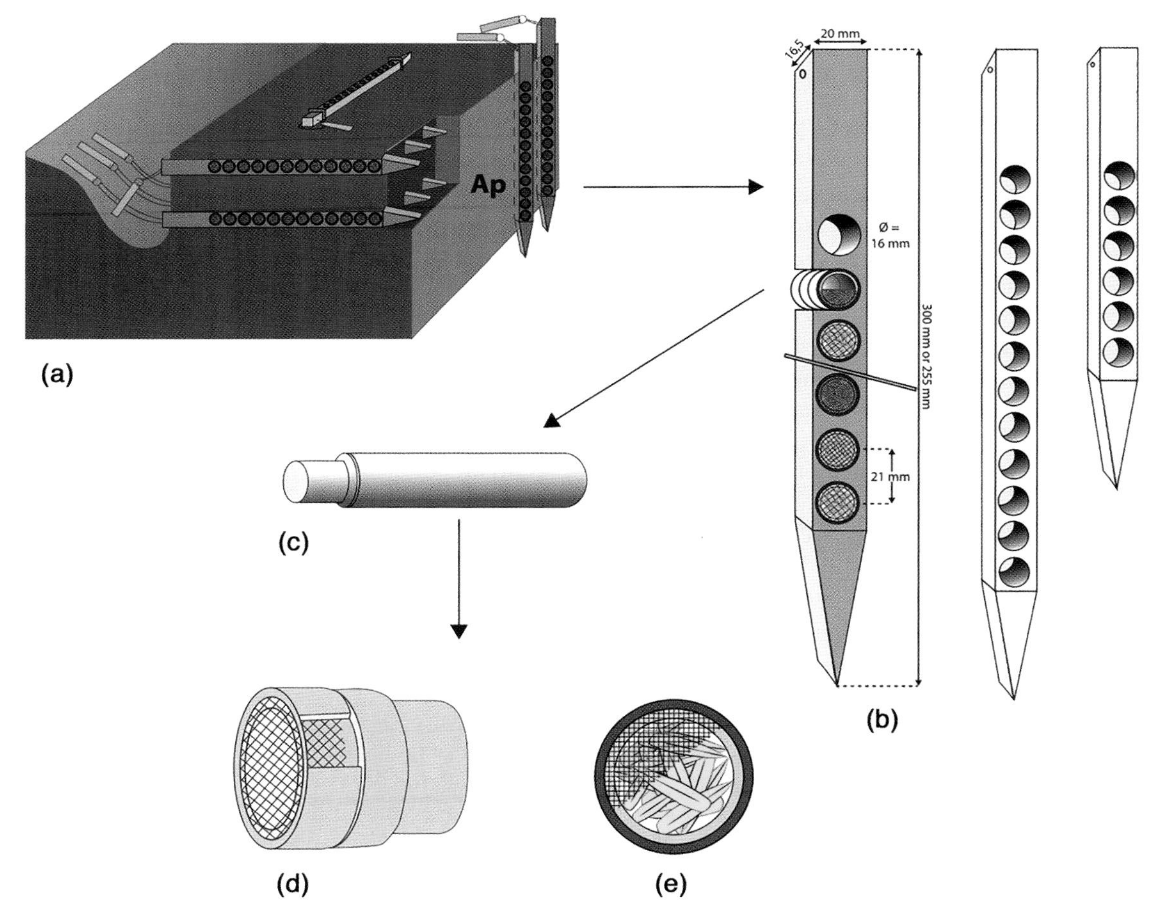

FIGURE 7.6 The decomposition of organic substrates can be estimated by exposing minicontainers filled with site-specific or standard material (e.g., maize straw) to particular conditions for a certain period of time (from weeks to several months). Detailed description can be found in the text. *(Eisenbeis et al., 1999.)*

7.7.4 Nitrogen mineralization

Nitrogen mineralization is estimated in field and laboratory experiments as the release of inorganic N from soil. Specific steps of the N mineralization process can be estimated (e.g., arginine deiminase, urease, ammonia monooxygenase). Nitrogen availability is measured using aerobic and anaerobic incubation tests, in addition to soil inorganic N measurements. The recommended methods differ in incubation time and temperature, moisture content, and extraction of ammonium and nitrate. Soils are frequently incubated under aerobic conditions and analyzed for ammonium, nitrite, and nitrate. Since mineral N is partly immobilized into microbial biomass during incubation, these incubation methods yield the net production of ammonium and nitrate. Isotope pool dilution techniques enable gross rates of nitrification (or mineralization) to be determined by monitoring the decline in the ^{15}N abundance in a nitrate or ammonium pool, labeled at $t = 0$ and receiving unlabeled N via nitrification or mineralization (Murphy et al., 2003). Labeled N can be applied as $^{15}NH_4^+$ solution or injected as $^{15}NH_3$ gas into soil. The ^{15}N pool dilution and enrichment can also be used to separate the heterotrophic and autotrophic pathways of nitrification. An isotopic dilution experiment using $^{14}NH_4^{15}NO_3$ yields rates of nitrification by the combined autotrophic and heterotrophic paths. A parallel isotope dilution experiment with $^{15}NH_4^{15}NO_3$ provides the gross mineralization rate and the size and ^{15}N abundance of the nitrate pool at different time intervals. Spatial variability of the tracer addition and extraction must be considered when interpreting such data.

Taylor et al. (2010) published an approach to estimate the archaeal and bacterial contributions to the nitrification potential of different soils. Ammonia monooxygenase of all soil microorganisms is irreversibly inactivated by acetylene; in a second step, acetylene is removed. Archaeal and bacterial contribution to the ammonia-oxidizing potential can be measured from the recovery of nitrification potential (RNP) in the presence and absence of kanamycin, an inhibitor of bacterial protein synthesis (Taylor et al., 2012). The inhibitor prevents resynthesis of ammonia monooxygenase by bacteria. Any RNP that recovers in the presence of a bacterial protein synthesis inhibitor is likely to be contributed by archaea.

7.8 Enzyme activities

Enzymes are specialized proteins that combine with a specific substrate and catalyze a biochemical reaction. Enzyme activities in soils are essential for energy transformation and nutrient cycling. The enzymes commonly extracted from soil, and their range of activities, are given in Table 7.4. Some enzymes (e.g., urease) are constitutive and routinely produced by cells. Others, such as cellulase, are adaptive or induced, and formed only in the presence of a susceptible substrate, some other initiator, or in the absence of an inhibitor. Enzymes associated with proliferating cells occur in the cytoplasm, the periplasmic membrane, and the cell membrane. Fig. 7.7 illustrates that soil enzymes are not only associated with proliferating cells, but also with humic colloids and clay minerals as extracellular enzymes. Since many of the enzymes that are frequently measured can be intracellular, extracellular, bound, and/or stabilized within their microhabitat, most assays determine enzymatic potential, but not necessarily the activity of proliferating microorganisms. The approach of Levakov et al. (2021) allows the determination of free extracellular enzyme activities in soil pore water in the soil and unsaturated zone in a simple, nondestructive, continuous manner.

TABLE 7.4 Enzymes extracted from soils, the reaction they catalyze, and their range of activities.

Enzyme	Reaction	Range of activity
Cellulase	Hydrolysis of the 1,4-β-D-glycosidic linkages in cellulose releasing oligosaccharides	0.4–80.0 μM glucose g^{-1} 24 h^{-1}
Invertase	Hydrolysis of sucrose into fructose and glucose	0.61–130 μM glucose g^{-1} h^{-1}
β-Glucosidase	Hydrolysis of oligosaccharides releasing β-D-glucose	0.09–405 μM *p*-nitrophenol g^{-1} h^{-1}
Proteinase	Hydrolysis of proteins releasing peptides and amino acids	0.5–2.7 μM tyrosine g^{-1} h^{-1}
Urease	Hydrolysis of urea to CO_2 and NH_4^+	0.14–14.3 μM N-NH_3 g^{-1} h^{-1}
Alkaline phosphatase	Hydrolysis of phosphomono- and phosphodiesters releasing an alcohol and orthophosphate under high pH conditions	6.76–27.3 μM *p*-nitrophenol g^{-1} h^{-1}
Acid phosphatase	Hydrolysis of phosphomono- and phosphodiesters to an alcohol and orthophosphate under low pH conditions	0.05–86.3 μM *p*-nitrophenol g^{-1} h^{-1}
Arylsulfatase	Hydrolysis of sulfate ester to sulfate and an alcohol	0.01–42.5 μM *p*-nitrophenol g^{-1} h^{-1}
Catalase	$2H_2O_2 \rightarrow O_2 + 2H_2O$	61.2–73.9 μM O_2 g^{-1} 24 h^{-1}

Adapted from Nannipieri et al. (2002).

Methods for analysis of a broad range of enzymes are described by Dick (2011), Schinner et al. (1996), and Tabatabai (1994). A general introduction to enzymes in the environment, their activity, ecology, and applications is given by Burns and Dick (2002). This section will primarily focus on colorimetric and fluorometric techniques to measure enzyme activities in soils and on some approaches for visualizing the locations of soil enzymes. Methods for assessing potential soil enzyme activities using spectroscopic and fluorometric approaches are available in bench scale and microplate format (Deng et al., 2013; Dick, 2011). Use of the microplate format offers the advantage of simultaneous analysis of multiple enzymes using a small amount of soil (Vepsäläinen et al., 2001). A microplate reader can simultaneously measure the absorbance or fluorescence of up to 96 wells.

7.8.1 Spectrophotometric methods

Many substrates and products of enzymatic reactions absorb light either in the visible or in the ultraviolet region of the spectrum, or they can be measured by a color reaction. Due to their higher sensitivity, methods based on the analysis of the released product are more frequently used than methods based on the analysis of substrate depletion. Analyzing enzyme activities involves incubating soils with the respective substrate at a fixed temperature, pH, and time, subsequently extracting the product and colorimetrically determining the product.

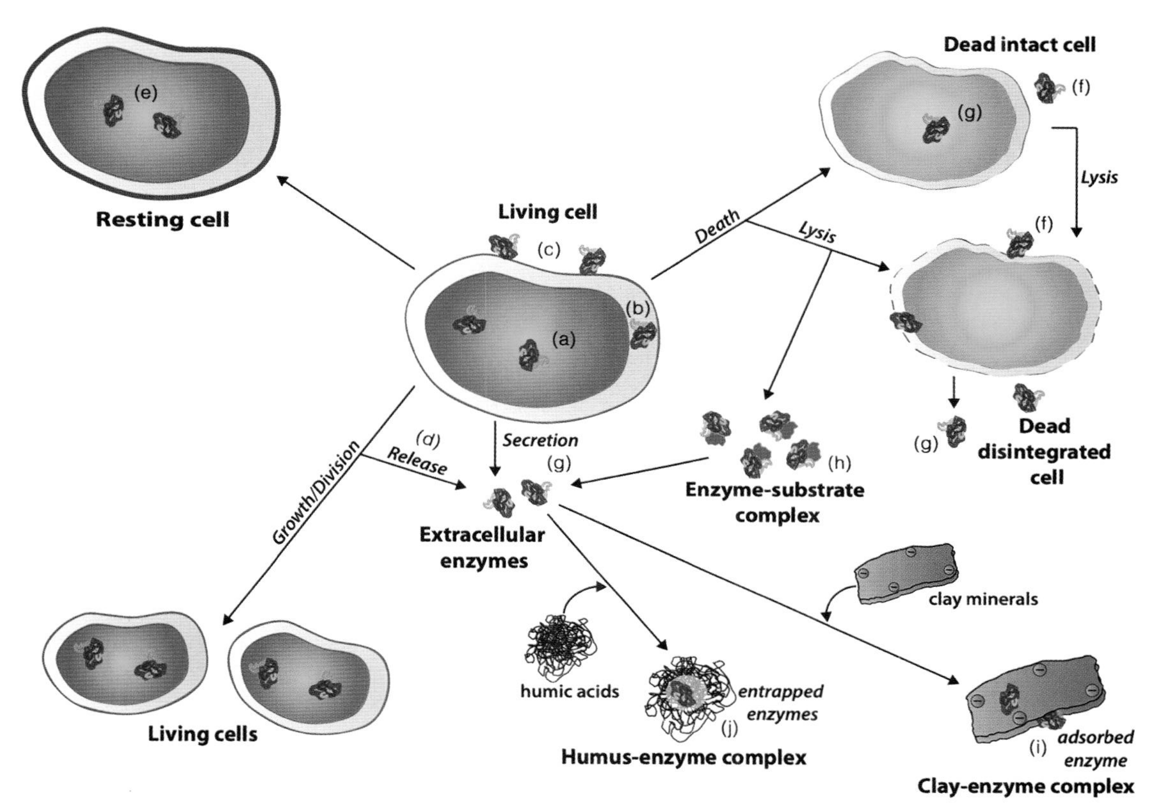

FIGURE 7.7 **Locations of enzymes within and outside cells.** (a) Intracellular; (b) periplasmatic; (c) attached to outer surface of cell membranes; (d) those released during cell growth and division; (e) within nonproliferating cells (spores, cysts, seeds, endospores); (f) attached to dead cells and cell debris; (g) leaking from intact cells or released from lysed cells; (h) temporarily associated in enzyme-substrate complexes; (i) adsorbed to surfaces of clay minerals; and (j) complexed with humic colloids. *(Burns, 1982 adapted by Klose, 2003 in Benckiser and Schnell, 2006, redrawn by R.S. Boeddinghaus.)*

Enzymes involved in C-cycling (e.g., xylanase, cellulase, invertase, and trehalase) are measured based on the release of sugars after soils are incubated with a buffered solution (pH 5.5) containing their corresponding substrates (xylan, carboxymethyl-cellulose, sucrose, or trehalose, respectively). The incubation period depends on the substrate used: high molecular weight substrates are incubated for 24 hours, whereas low molecular weight substrates are incubated for 1 to 3 hours. Reducing sugars released during the incubation period cause the reduction of potassium hexacyanoferrate(III) in an alkaline solution. Reduced potassium hexacyanoferrate(II) reacts with ferric ammonium sulfate in an acid solution to form a complex of ferric hexacyanoferrate(II) (Prussian blue), which is determined colorimetrically.

Enzymes involved in P-cycling (phosphomonoesterase, phosphodiesterase, phosphotriesterase) are preferably determined after the addition of a substrate analog. Phosphomonoesterase hydrolyzes the phosphate-ester-bond of the *p*-nitrophenylphosphate, and the *p*-nitrophenol released is measured as a yellow compound under alkaline conditions. For determining arylsulfatase, which catalyzes the hydrolysis of organic sulfate ester, *p*-nitrophenyl sulfate is used as a substrate analog. Different enzyme assays use *p*-nitrophenyl substrates because the released product, *p*-nitrophenol, can be quantitatively extracted. Urease is an important extracellular enzyme that hydrolyzes urea into CO_2 and NH_3. The urease assay involves measuring the released ammonia either by a colorimetric procedure or by steam distillation followed by a titration assay of ammonia. It is nearly impossible to link potential enzyme activities directly to process rates under field conditions or to results of transcriptomic and metabolomic studies. This missing interlinkage between potential enzyme activities and other physiological properties is caused by the varying contribution of intracellular, extracellular, as well as adsorbed and stabilized enzyme activities, to the overall activity (Fig. 7.7).

Whereas the described potential enzyme activities are based on the addition of an excess amount of suitable substrate and subsequent determination of product release, Bünemann (2008) suggests reversing the approach by adding excess of an enzyme. In this case, substrate availability becomes rate-limiting, and the maximum release of product indicates the availability of a given substrate. The method is applied to characterize the potential bioavailability of organic P compounds after adding different phosphatases to environmental samples (soil, manure, and sediment extracts) (Bünemann, 2008; Jarosch et al., 2019).

7.8.2 Fluorescence methods

Fluorogenic substrates are used to assay extracellular enzymes in aquatic and terrestrial environments. These substrates contain an artificial fluorescent molecule and one or more natural molecules (e.g., glucose, amino acids), and are linked by a specific type of binding (e.g., peptide or ester). The substrates used are conjugates of the highly fluorescent compounds 4-methylumbelliferone (MUB) and 7-amino-4-methyl coumarin (AMC). Marx et al. (2001) used this method to measure the activities of enzymes involved in C-cycling (β-D-glucosidase, β-D-galactosidase, β-cellobiase, β-xylosidase), N-cycling (leucine aminopeptidase, alanine amino peptidase, lysine-alanine aminopeptidase), P-cycling (acid phosphatase), and S-cycling (arylsulfatase). Fluorogenic model substrates are not toxic and are supplied to soil suspensions in high or increasing quantities to measure the maximum velocity of hydrolysis (v_{max}). Fluorescence is observed after enzymatic splitting of the complex molecules (Fig. 7.8). All calibrations are done with soil suspensions due to the interference in the detection of soil particles and quenching for fluorescence (Dick, 2011).

The increasing interest of soil microbiologists in the use of fluorogenic substrates to measure soil enzyme activities is primarily due to their high sensitivity. A comparative study between a fluorometric and a standard colorimetric enzyme assay based on *p*-nitrophenyl substrates generated similar values for the maximum rate of phosphatase and β-glucosidase (v_{max}), but the affinity for their respective substrates (as indicated by K_m values, Michaelis-Menten constant) was up to two orders of magnitude greater for the 4-methylumbelliferyl substrates compared to the *p*-nitrophenyl substrates (Marx et al., 2001). This high sensitivity of fluorometric enzyme assays provides an opportunity to detect enzyme activities in small samples (e.g., microaggregates and rhizosphere samples) and/or low-activity samples (subsoil, peat, and soil solutions). A wider applicability of measurement of soil enzymes comes from the interest of modelers to incorporate potential enzyme activities, temperature sensitivity, and other kinetic properties into their biogeochemical models (Ali et al. 2018). Considering temperature and moisture sensitivity of β-glucosidase activity, Ali et al. (2015) calculated modeled in situ enzyme activities as proxy for the seasonal variation of CO_2 production under field conditions.

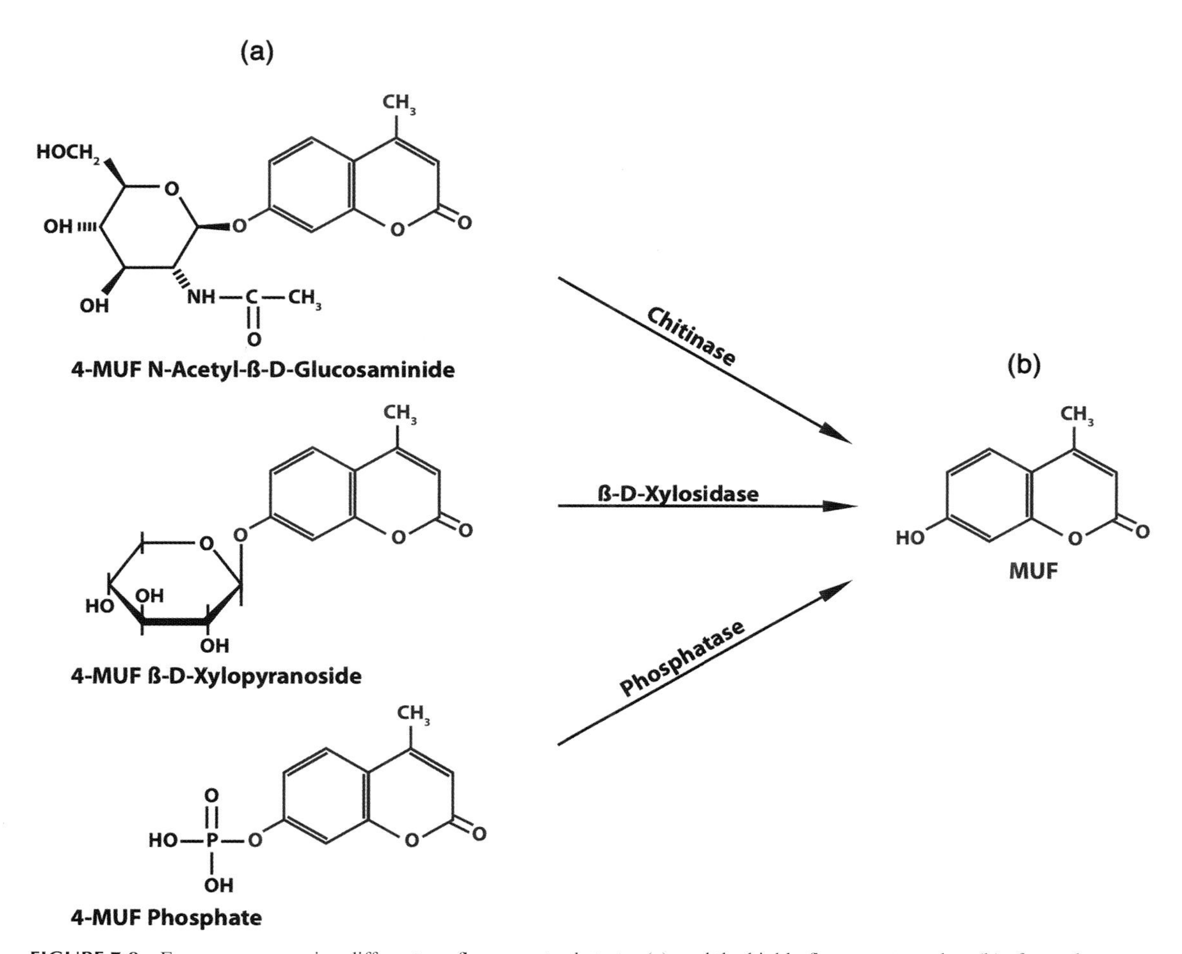

FIGURE 7.8 Enzyme assays using different nonfluorescent substrates (a), and the highly fluorescent product (b) after a short-term incubation during which the fluorescent product is cleaved from the original substrates.

7.9 Imaging microbial activities

Thin-section techniques combined with histochemical and imaging techniques can visualize the location of enzyme proteins and their activities. Foster et al. (1983), using either transmission electron microscopy (TEM) or scanning electron microscopy (SEM), showed peroxidase, succinic dehydrogenase, and acid phosphatase bound to root rhizosphere, bacterial cell walls, and organic matter in soil. Spohn and Kuzyakov (2013) used soil zymography with fluorescent substrates as an in situ method for the analysis of the two-dimensional distribution of enzyme activity at the soil–root interface. Soil zymography can be combined with ^{14}C imaging, a technique that gives insights into the distribution of photosynthates after labelling plants with ^{14}C. Whereas zymography was previously used to track enzyme activities at the root–soil interface, this method is now applied to larger, undisturbed soil surfaces (Fig. 7.9). Krueger

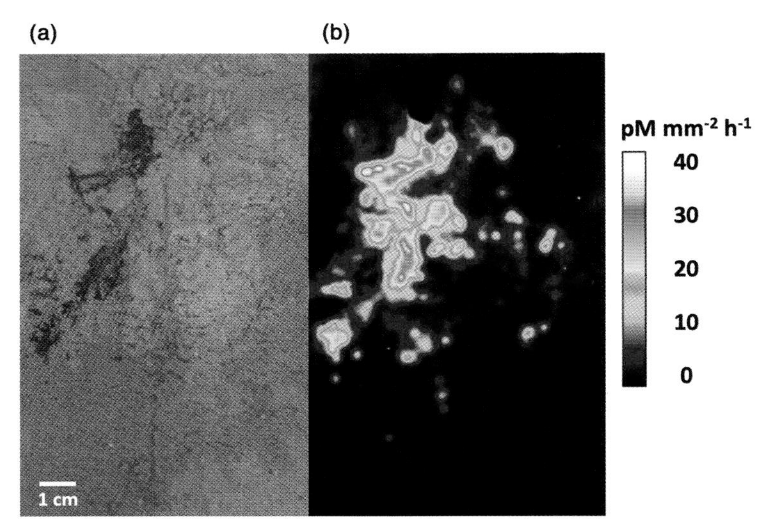

FIGURE 7.9 Spatial distribution of β-glucosidase in the subsoil of a beech forest. The surface of the undisturbed sample derives from a flow cell (size: 20 × 7.5 cm; Krueger et al., 2017) which allows the application of soil zymography. The comparison of a picture of the soil surface (a) with the β-glucosidase activity (b) shows that hotspots of β-glucosidase activity correspond to the dark brown color representing zones enriched in organic matter. *(Fig. 7.9a: S. Preusser; Fig. 7.9b: A. Lattacher.)*

et al. (2017) described a method for yielding undisturbed soil columns for flow cell (length 200 mm, width 75 mm) experiments. This approach provides a simultaneous application of biophysical and microbiological methods (e.g., pore size distribution, transport of labeled compounds, microbial C retention, and spatial distribution of enzyme activities).

Imaging and enumeration of specific soil microorganisms can be performed by fluorescence in situ hybridization (FISH) techniques. However, the detection of FISH-stained cells is often affected by strong autofluorescence of the background. If the FISH-approach is coupled with catalyzed reporter deposition (CARD), CARD-FISH-stained cells are suitable for automated counting using digital image analysis (Neufeld et al., 2007). In addition, very sensitive mass spectrometric techniques (SIMS) have facilitated the visualization of single-cell ecophysiology of microbes in environmental samples (Wagner, 2009). The SIMS uses an energetic primary ion beam to produce and expel secondary particles (primarily atoms or molecules) from the soil surface under high vacuum. The elemental, isotopic, and molecular compositions of these particles are determined by mass spectrometry. The combined approach of incubating environmental samples with stable isotope-labeled substrates, followed by single-cell analyses through high-resolution secondary ion mass spectrometry (NanoSIMS), gives insights into the in situ function of microorganisms (Eichhorst et al., 2015). Whereas the SIMS technique can provide results at the single-cell level, the technique by Stiehl-Braun et al. (2011) makes the study of the spatial distribution of active microorganisms in soil at larger scales (cm scale) possible. When undisturbed soil cores were exposed to $^{14}CH_4$, methanotrophs incorporated ^{14}C into their cells. The spatial distribution of ^{14}C incorporation was characterized by a combined micromorphological and autoradiographical technique (Stiehl-Braun et al., 2011).

7.10 Functional diversity

Different aspects of functional diversity have advanced our understanding of the significance of biodiversity to biogeochemical cycling. This includes several levels of resolution: (1) the importance of biodiversity to specific biogenic transformations; (2) the complexity and specificity of biotic interactions in soils that regulate biogeochemical cycling; and (3) how biodiversity may operate at different hierarchically arranged spatial and temporal scales to influence ecosystem structure and function (Beare et al., 1995). Most methods for measuring functional diversity consider only the biodiversity of those groups that regulate biogeochemical cycling. Several approaches enable functional diversity to be measured in situations where taxonomic information is poor. These include the use of binary biochemical and physiological descriptors to characterize isolates, evaluation of enzymatic capabilities for utilization of particular substrates, extraction of DNA and RNA from the soil, and the application of gene probes that code for functional enzymes. Recent advances in genomic analysis and SIP are the first steps toward a better understanding of linkages between structure and function in microbial communities.

Commercially available BIOLOG bacterial identification system plates or community-level physiological profiles (CLPPs) assess the functional roles of microorganisms based on utilization patterns of a wide range (up to 128) of individual C sources. Members of the microbial community that exhibit fast growth rates typical for r-strategists primarily contribute to CLPP analysis. The original CLPP method had severe limitations and has been replaced by alternative approaches based on CO_2 or O_2 detection. The CO_2 in the headspace of multiple flasks is measured by infrared detection or with sealed microtiter plates containing pH-sensitive dye. The microtiter system, based on an O_2-sensitive fluorophore, is used to detect basal soil respiration and substrate-induced responses at amendment levels 10 to 100-fold less than that used with any other CLPP approach (Garland et al., 2010).

An alternative approach (Kandeler et al., 1996) uses the prognostic potential of 16 soil microbial properties (microbial biomass, respiration, N-mineralization) and 13 soil enzymes involved in cycling of C, N, P, and S. Multivariate statistical analysis is used to calculate the functional diversity from measured soil microbial properties. The approach is based on the following assumptions: (1) the composition of the microbial species assemblage (taxonomic diversity) determines the community's potential for enzyme synthesis; (2) the actual rate of enzyme production and the fate of produced enzymes are modified by environmental effects as well as by ecological interactions; and (3) the spectrum and quantity of active enzymes are responsible for the functional capability of a microbial community, irrespective of whether they are active inside or outside the cell. According to the diversity concept in ecology, presence or absence of a certain function, as well as quantification of the potential of the community to realize this function, are considered. This approach may permit evaluation of the status of a changed ecosystem (e.g., by soil pollution, soil management, global change) while providing insight into the functional diversity of the soil microbial community of the undisturbed habitat.

Physiological methods are applied to better understand the physiology of single cells, soil biological communities, food webs, and biogeochemical cycling in terrestrial ecosystems. Small-scale studies explain biological reactions in aggregates, the rhizosphere, or at the soil—litter interface. Combining physiological and molecular methods helps us to understand gene expression, protein synthesis, and enzyme activities at both micro- and nanoscales. Linking these methods can also help explain whether the abundance and/or function of organisms are affected by soil management, global change, and soil pollution.

At the field scale, researchers use biochemical and physiological methods to investigate the functional response of soil organisms to the manipulation or preservation of soils. These applications include microbe-plant interactions and control of plant pathogens, as well as better characterization of organic matter decomposition and its impact on local and global C and N cycling. Soil biologists investigate the effect of soil management or disturbance (tillage, fertilizer, pesticides, crop rotation) on the function of soil organisms. In many cases, soil microbial biomass and/or soil microbial processes can be early predictors of the effects of soil management on soil quality and can also indicate the expected rapidity of these changes. Monitoring of soil microbial properties is also necessary in environmental studies that test the use of soil microorganisms in bioremediation and composting. Future challenges in functional soil microbiology can only be met by increasing our present knowledge for scaling up these data to regional and global scales.

7.11 Ecological approaches to understand and manipulate community composition

An important aim in soil ecology and biogeochemistry is to understand the functional role of different groups of soil organisms. Some complex approaches are used to separate different groups of organisms (e.g., meshes, exclosures, and microfluidic devices) to provide insight into plant—microbe, animal—microbe, fungal—fungal, and fungal—bacterial interactions. These procedures are also used to separate different microhabitats (e.g., rhizosphere, detritusphere, and mineralosphere) from the surrounding bulk soil. An overarching goal is to study the transfer of C or nutrients between different groups of organisms or to disentangle the functional role of competing groups of microorganisms (e.g., different functional guilds like ectomycorrhizal and saprotrophic fungi).

Animal-microbe interactions: Litterbags were used that selectively excluded macrofauna to assess their relative importance in relation to microbial, micro- and mesofaunal decomposition. Seeber et al. (2006) built plastic tubes (15 cm high, 15 cm diameter) that were covered by mesh (200 μm) at the top and bottom under lab conditions. The tubes were filled with moistened soil and the surface was covered with grass and dwarf shrub litter. The lab experiment was set up in a balanced factorial design with the factors as primary decomposers (*Lumbricus rubellus*, *Dendrobaena octaedra*, and *Cylindroiulus fulviceps*) and secondary decomposers (*Octolasion lacteum* and *Enantiulus nanus*). After 12 weeks, the study demonstrated that macrodecomposers are an important factor for litter decomposition. Whereas lab studies are frequently used, complex approaches under field conditions are rare (Kampichler et al., 2001). The highest complexity under field conditions was reached by using mesocosms; enclosed outdoor systems that are partially permeable to their surroundings. They are monoliths cut out of soil as whole "blocks," partly defaunated by deep freezing, and replanted into their natural environment. The use of different mesh sizes around the monoliths allows the recolonization by soil mesofauna (35 μm) or meso- and macrofauna (1 mm) (Bruckner et al., 1995). After several months of exposure, abundance and diversity of microorganisms (i.e., meso- and macrofauna) and different biochemical processes (e.g., C-mineralization, N mineralization, ^{13}C flow from substrates into the food web) can be estimated in different treatments. These results give insight into the importance of different size classes of soil animals on the biogeochemistry of soil systems.

Plant—microbe interactions: Separating roots from the bulk soil environment is a frequently used procedure to study organisms in the rhizosphere. In general, mesh of different sizes (e.g., <35 μm or <50μm) builds a barrier between roots and the surrounding soil. Small-scale soil samplings along the mesh are used to identify plant growth-promoting bacteria, microbial pathogens, as well as symbiotic microorganisms in the close vicinity of roots. These studies are often combined with ^{13}C and ^{15}N labeling techniques to follow the C flow from plants into rhizosphere microorganisms or to study N uptake and transport by microorganisms into plants. The application of "split-root boxes" (sensu Gorka et al., 2019) is a complex approach to separate the hyphosphere from the rhizosphere by dividing their root systems into two disconnected soil compartments. The hyphosphere is defined as a sphere where symbiotic and nonsymbiotic fungi are present in the soil environment and is characterized by intense fungal—bacterial—soil interactions (Kohler et al., 2017). Each of these compartments is separated from a litter compartment by a mesh (35 μm) penetrable for fungal hyphae but not by roots (Gorka et al., 2019). The authors showed a rapid transfer of recent photosynthates of beech trees (*Fagus sylvatica*) via ectomycorrhizal hyphae to bacteria in root-distant soil areas.

Fungal—fungal interactions: Complex microcosms in the lab or under field conditions are used to disentangle the role of different fungal guilds (Müller et al., 2020; Wallander et al., 2001). To separate the contributions of ectomycorrhizal (EM) and saprotrophic (SAP) fungi in situ, two different types of soil ingrowth tubes (IGTs) are used: (1) open IGTs with micromesh windows to allow all fungi to regrow into the tubes; and (2) closed IGTs (without mesh windows) in which only SAP fungi are allowed to grow and EM hyphae are unable to recolonize the IGTs (Fig. 7.10a). Tops and bottoms of the tubes had 2 cm mesh borders (50 mm mesh size) filled with sand to avoid fine root accession and hyphal migration of saprotrophic fungi (Wallander et al., 2001; modified by Müller et al., 2020). A comparison of microbial community composition and function after an exposure of IGTs for several months to years allows for an analysis on the importance of different fungal guilds for biogeochemical cycling (e.g., P cycling; Müller et al., 2020).

FIGURE 7.10 (a) In situ "open" and "closed" ingrowth tubes (IGTs, 10 cm height, 5 cm diameter) to separate ectomycorrhizal (EM) and saprotrophic fungal communities. Please see the text for more detailed information. (b) Exposure of mineral containers to study colonization and function of microorganisms in the mineralosphere. This picture shows the sampling of a goethite container which was exposed underneath the litter layer of a forest site for 5 years. *(Fig. 7.10a: P. Nassal; Fig. 7.10b: E. Kandeler.)*

Mineral-associated microorganisms: Mineral containers filled, with, for example, a mixture of illite, goethite, and quartz loaded with labile or recalcitrant organic matter are used to study the colonization and C use of microorganisms in the mineralosphere under field conditions (Kandeler et al., 2019; Fig. 7.10b). This approach closes the gap between complex lab experiments and experiments in the natural environment.

Microbial interactions: Microfluidic techniques (e.g., microfluidic chips) provide visual access to in situ soil ecology (Mafla-Endara et al., 2021). Microfluidic chips are engineered habitats for organisms and consist of channels, chambers, and other structures (porous walls, slits, etc.), creating artificial microscopic ecosystems (Fig. 7.11). Individual cells and populations in the microfluidic chip (Burmeister and Grünberger, 2020) can be tracked by using a microscope. This technique can also be combined with other molecular methods such as cell-isolation, DNA extractions, sequencing, etc. Since the geometry of these structures can be controlled with subcellular precision, habitats of various degrees of fragmentation or connectivity can be created. A further advantage of microfluidic chips is the ability to control flow and diffusion conditions (Aleklett et al., 2018). Under lab conditions, microfluid chips clarified hyphal behavior, fungal habitat colonization, their foraging strategies, and niche partitioning (Aleklett et al., 2021). The microfluidic chips can also be buried under field conditions and can sense microbial colonization of this defined habitat. Mafla-Endara et al. (2021) showed that both soil microbes and minerals enter the chips under field conditions. Consequently, diverse community interdependences (i.e., inter-kingdom and food-web interactions) and feedbacks between microbes and the pore space microstructures can be investigated. Soil chips hold a large potential for studying in situ microbial interactions and soil functions, and interconnecting field microbial ecology with laboratory experiments.

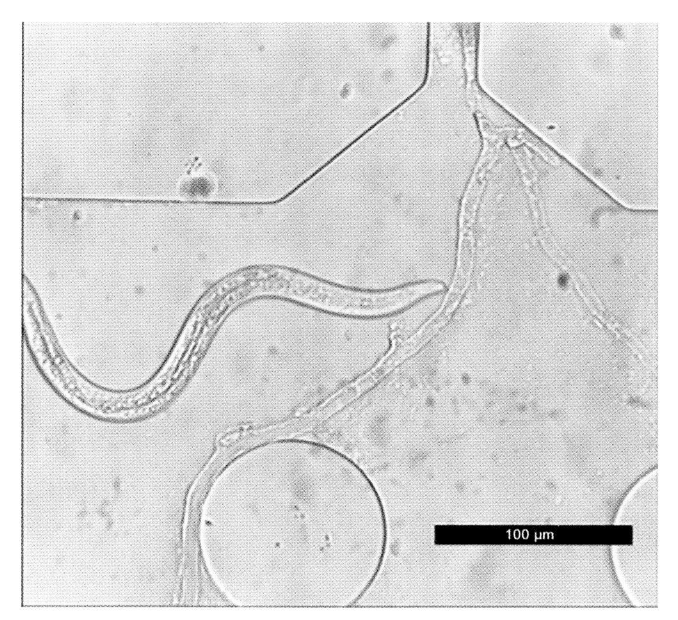

FIGURE 7.11 Application of microfluidic techniques in soil microbial ecology. A nematode migrates into a chip channel from a natural soil inoculum and feeds on hyphae of *Mycetinis scorodonius* growing in a pillar system with 100 μm wide pillars. *(Figs 7.11a–d: E. Hammer and F. Klinghammer.)*

References and Suggested Reading

Aleklett, K., Kiers, E.T., Ohlsson, P., Shimizu, T.S., Caldas, V.E., Hammer, E.C., 2018. Build your own soil: exploring microfluidics to create microbial habitat structures (Review). ISME J. 12, 312–319.

Aleklett, K., Ohlsson, P., Bengtsson, M., Hammer, E.C., 2021. Fungal foraging behaviour and hyphal space exploration in micro-structured soil chips. ISME J. 15, 1782–1793.

Ali, R.S., Ingwersen, J., Demyan, M.S., Funkuin, Y.N., Wizemann, H.D., Kandeler, E., et al., 2015. Modelling in situ activities of enzymes as a tool to predict seasonal variation of soil respiration from agro-ecosystems. Soil Biol. Biochem. 81, 291–303.

Ali, R.S., Kandeler, E., Marhan, S., Demyan, M., Ingwersen, J., Mirzaeitalarposhti, R., et al., 2018. Controls on microbially regulated soil organic carbon decomposition at the regional scale. Soil Biol. Biochem. 118, 59–68.

Amelung, W., Brodowski, S., Sandhage-Hofmann, A., Bol, R., 2008. Combining biomarker with stable isotope analyses for assessing the transformation and turnover of soil organic matter. Adv. Agron. 100, 155–250.

Anderson, J.P.E., Domsch, K.H., 1978. A physiological method for the quantitative measurement of microbial biomass in soils. Soil Biol. Biochem. 10, 215–221.

Bååth, E., Anderson, T.H., 2003. Comparison of soil fungal/bacterial ratios in a pH gradient using physiological and PLFA-based techniques. Soil Biol. Biochem. 35, 955–963.

Baldocchi, D.D., 2003. Assessing the eddy covariance technique for evaluating carbon dioxide exchange rates of ecosystems: past, present and future. Glob. Change Biol. 9, 479–492.

Barberán, A., Bates, S.T., Casamayor, E.O., Fierer, N., 2012. Using network analysis to explore co-occurrence patterns in soil microbial communities. ISME J. 6, 343–351.

Bastida, F., Algora, C., Hernandez, T., Garcia, C., 2012. Feasibility of a cell separation-proteomic based method for soils with different edaphic properties and microbial biomass. Soil Biol. Biochem. 45, 136–138.

Beare, M.H., Coleman, D.C., Crossley, D.A., Hendrix, P.F., Odum, E.P., 1995. A hierarchical approach to evaluating the significance of soil biodiversity to biochemical cycling. Plant Soil 170, 5–22.

Becher, D., Bernhardt, J., Fuchs, S., Riedel, K., 2013. Metaproteomics to unravel major microbial players in leaf litter and soil environments: challenges and perspectives. Proteomics 13, 2895–2909.

Beck, T., Joergensen, R.G., Kandeler, E., Makeschin, F., Nuss, E., Oberholzer, H.R., 1997. An inter-laboratory comparison of ten different ways of measuring soil microbial biomass C. Soil Biol. Biochem. 29, 1023–1032.

Bernhardt, E.S., Blaszczak, J.R., Ficken, C.D., Fork, M.L., Kaiser, K.E., Seybold, E.C., 2017. Control points in ecosystems: moving beyond the hot spot hot moment concept. Ecosystems 20, 665–682.

Bölscher, T., Ågren, A.G., Herrmann, A.M., 2020. Land-use alters the temperature response of microbial carbon-use efficiency in soils – a consumption-based approach. Soil Biol. Biochem. 140, 107639.

Boutton, T.W., Yamasaki, S.-I., 1996. Mass Spectrometry of Soils. Marcel Dekker, New York.

Brookes, P.C., Landman, A., Puden, G., Jenkinson, D.S., 1985. Chloroform fumigation and the release of soil nitrogen: a rapid direct extraction method to measure microbial biomass nitrogen in soil. Soil Biol. Biochem. 17, 837–842.

Bruckner, A., Wright, J., Kampichler, C., Bauer, R., Kandeler, E., 1995. A method to prepare mesocosms for assessing complex biotic processes in soils. Biol. Fertil. Soils 19, 257–262.

Bünemann, E.K., 2008. Enzyme additions as a tool to assess the potential bioavailability of organically bound nutrients. Soil Biol. Biochem. 40, 2116–2129.

Burmeister, A., Grünberger, A., 2020. Microfluidic cultivation and analysis tools for interaction studies of microbial co-cultures. Curr. Opin. Biotechnol. 62, 106–115.

Burns, R.G., Dick, R.P., 2002. Enzymes in the Environment – Activity, Ecology, and Applications. Marcel Dekker, New York.

Calonne, M., Fontaine, J., Debiane, D., Laruelle, F., Grandmougin-Ferjani, A., Lounès-Hadj Sahraoui, A., 2014. The arbuscular mycorrhizal *Rhizophagus irregularis* activates storage lipid biosynthesis to cope with the benzo[a]pyrene oxidative stress. Phytochemistry 97, 30–37.

Clode, P.L., Kilburn, M.R., Jones, D.L., Stockdale, E.A., Cliff III, J.B., Herrmann, A.M., 2009. In situ mapping of nutrient uptake in the rhizosphere using nanoscale secondary ion mass spectrometry. Plant. Physiol. 151, 1751–1757.

Coleman, D.C., Fry, B. (Eds.), 1991. Carbon Isotope Techniques. Academic Press, San Diego. California.

Coyotzi, S., Pratscher, J., Murrell, J.C., Neufeld, J.D., 2016. Targeted metagenomics of active microbial populations with stable-isotope probing. Curr. Opin. Biotechnol. 41, 1–8.

Daims, H., Wagner, M., 2007. Quantification of uncultured microorganisms by fluorescence microscopy and digital image analysis. Appl. Microbiol. Biotechnol. 75, 237–248.

Deng, S., Popova, I.E., Dick, L., Dick, R., 2013. Bench scale and microplate format assay of soil enzyme activities using spectroscopic and fluorometric approaches. Appl. Soil. Ecol. 64, 84–90.

Dick, R. (Ed.), 2011. Methods of Soil Enzymology. Soil Science of America Book Series 9. SSAJ, Inc., Madison, Wisconsin.

Djajakirana, G., Joergensen, R.G., Meyer, B., 1996. Ergosterol and microbial biomass relationship in soil. Biol. Fertil. Soils. 22, 299–304.

Drenovsky, R.E., Elliott, G.N., Graham, K.J., Scow, K.M., 2004. Comparison of phospholipid fatty acid (PLFA) and total soil fatty acid methyl esters (TSFAME) for characterizing soil microbial communities. Soil Biol. Biochem. 36, 1793–1800.

Eichhorst, S.A., Strasser, F., Woyke, T., Schintlmeister, A., Wagner, M., Woebken, D., 2015. Advancements in the application of NanoSIMS and Raman microspectroscopy to investigate the activity of microbial cells in soils. FEMS Microbiol. Ecol. 91, 2015, fiv106.

Eisenbeis, G., Lenz, R., Heiber, T., 1999. Organic residue decomposition: the Minicontainer-system – a multifunctional tool in decomposition studies. Environ. Sci. Pollut. R. 6, 220–224.

Foissner, W., 1992. Comparative studies on the soil life in ecofarmed and conventionally farmed fields and grasslands of Austria. Agric. Ecosyst. Environ. 40, 207–218.

Foster, R.C., Rovira, A.D., Cock, T.W., 1983. Ultrastructure of the Root-Soil Interface. American Phytopathological Society, St. Paul, Minnesota.

Frankland, J.C., Dighton, J., Boddy, L., 1991. Fungi in soil and forest litter. In: Grigorova, R., Norris, J.R. (Eds.), Methods in Microbiology, Vol. 22. Academic Press, London, pp. 344–404.

Frostegård, Å., Tunlid, A., Bååth, E., 2011. Use and misuse of PLFA measurements in soils. Soil Biol. Biochem. 43, 1621–1625.

Fujie, K., Hu, H.K., Tanaka, H., Urano, K., Saitou, K., Katayama, A., 1998. Analysis of respiratory quinones in soil for characterization of microbiota. J. Soil. Sci. Plant. Nutr. 44, 393–404.

Garland, J.L., Mackowiak, C.L., Zaboloy, M.C., 2010. Organic waste amendment effects on soil microbial activity in a corn-rye rotation: application of a new approach to community-level physiological profiling. Appl. Soil. Ecol. 44, 262–269.

Gattinger, A., Gunther, A., Schloter, M., Munch, J.C., 2003. Characterization of Archaea by polar lipid analysis. Acta. Biotechnol. 23, 21–28.

Geyer, K.M., Dijkstry, P., Sinsabaugh, R., Frey, S.D., 2019. Clarifying the interpretation of carbon use efficiency in soil through methods comparison. Soil Biol. Biochem. 128, 79–88.

Gorka, S., Dietrich, M., Mayerhofer, W., Gabriel, R., Wiesenbauer, J., Martin, V., et al., 2019. Rapid transfer of plant photosynthates to soil bacteria via ectomycorrhizal hyphae and its interaction with nitrogen availability. Front. Microbiol. 10. Article number 168.

Herrmann, A.M., Coucheney, E., Nunan, N., 2014. Isothermal microcalorimetry provides new insight into terrestrial carbon cycling. Environ. Sci. Technol. 48, 4344–4352.

Hünninghaus, M., Dibbern, D., Kramer, S., Koller, R., Pausch, J., Urich, T., et al., 2019. Disentangling carbon flow across microbial kingdoms in the rhizosphere of maize. Soil Biol. Biochem. 134, 122–130.

Jarosch, K.A., Kandeler, E., Frossard, E., Bünemann, E.K., 2019. Is the enzymatic hydrolysis of soil organic phosphorus compounds limited by enzyme or substrate availability? Soil Biol. Biochem. 139, 107628.

Jenkinson, D.S., Powlson, D.S., 1976. The effects of biocidal treatments on metabolism in soil. V. A method for measuring soil biomass. Soil Biol. Biochem. 8, 209–213.

Joergensen, R.G., 1996. The fumigation-extraction method to estimate soil microbial biomass: calibration of the $K_{(EC)}$ value. Soil Biol. Biochem. 28, 25–31.

Joergensen, R.G., Wu, J., Brookes, P.C., 2011. Measuring soil microbial biomass using an automated procedure. Soil Biol. Biochem. 43, 873–876.

Kampichler, C., Bruckner, A., Kandeler, E., 2001. The use of enclosed model ecosystems in soil ecology: a bias towards laboratory research. Soil Biol. Biochem. 33, 269–275.

Kandeler, E., Kampichler, C., Horak, O., 1996. Influence of heavy metals on the functional diversity of soil microbial communities. Biol. Fertil. Soils 23, 299–306.

Kandeler, E., Gebala, A., Boeddinghaus, R.S., Müller, K., Rennert, T., Soares, M., et al., 2019. The mineralosphere – abundance and carbon partitioning of bacteria and fungi colonizing mineral surfaces in grassland soils under different land use intensity. Soil Biol. Biochem. 136, 107534.

Keiblinger, K.M., Fuchs, S., Zechmeister-Boltenstern, S., Riedel, K., 2016. Soil and leaf metaproteomics – a brief guideline from sampling to understanding. FEMS Microbiol. Ecol. 92, 22016.

Keuskamp, J.A., Dingemans, B.J.J., Lehtinen, T., Sarneel, J.M., Hefting, M.M., 2013. Tea Bag Index: a novel approach to collect uniform decomposition data across ecosystems. Methods Ecol. Evol. 4, 1070–1075.

Klamer, A.J., Bååth, E., 2004. Estimation of conversion factors for fungal biomass determination in compost using ergosterol and PLFA 18:2ω6,9. Soil Biol. Biochem. 36, 57–65.

Knowles, R., Blackburn, T.H. (Eds.), 1993. Nitrogen Isotope Techniques. Academic Press, San Diego. California.

Kohler, J., Roldán, A., Campoy, M., Caravaca, F., 2017. Unraveling the role of hyphal networks from arbuscular mycorrhizal fungi in aggregate stabilization of semiarid soils with different textures and carbonate contents. Plant Soil 410, 273–281.

Kramer, S., Marhan, S., Ruess, L., Armbruster, W., Butenschoen, O., Haslwimmer, H., et al., 2012. Carbon flow into microbial and fungal biomass as a basis for the belowground food web of agroecosystems. Pedobiologia 55, 111–119.

Kreuzer-Martin, H.W., 2007. Stable isotope probing: linking functional activity to specific members of microbial communities. SSSAJ 71, 611–619.

Krueger, J., Heitkötter, J., Jeue, M., Schlüter, S., Vogel, H.J., Marschner, B., et al., 2017. Coupling of interfacial soil properties and bio-hydrological processes: the flow cell concept. Ecohydrology 11, e2024.

Levakov, L., Ronen, Z., Siebner, H., Dahan, O., 2021. Continuous in-situ measurement of free extracellular enzyme activity as direct indicator for soil biological activity. Soil Biol. Biochem. 163, 108448.

Lennon, J.T., Aanderud, Z.T., Lehmkuhl, B.K., Schoolmaster Jr., D.R., 2012. Mapping the niche space of soil microorganisms using taxonomy and traits. Ecology 93, 1867–1879.

Mafla-Endara, P.M., Arellano-Caicedo, C., Aleklett, K., Pucetaite, M., Ohlsson, P., Hammer, E.C., 2021. Microfluidic chips provide visual access to in situ soil ecology. Commun. Biol. 4 (1), 889.

Manzoni, S., Taylor, P., Richter, A., Porporato, A., Agren, G.I., 2012. Environmental and stoichiometric controls on microbial carbon-use efficiency in soils. New. Phytol. 196, 79–91.

Marx, M.C., Wood, M., Jarvis, S.C., 2001. A microplate fluorometric assay for the study of enzyme diversity in soils. Soil Biol. Biochem. 33, 1633–1640.

Mason-Jones, K., Banfield, C.C., Dippold, M.A., 2019. Compound-specific ^{13}C stable isotope probing confirms synthesis of polyhydroxybutyrate by soil bacteria. Rapid Commun. Mass Spectrom. 33, 795–802.

Müller, K., Kramer, S., Haslwimmer, H., Marhan, S., Scheunemann, N., Butenschön, N., et al., 2016. Carbon transfer from maize roots and litter into bacteria and fungi depends on soil depth and time. Soil Biol. Biochem. 93, 79–89.

Müller, K., Kubsch, N., Marhan, S., Mayer-Gruner, P., Nassal, P., Schneider, D., et al., 2020. Saprotrophic and ectomycorrhizal fungi contribute differentially to organic P mobilization in beech-dominated forest ecosystems. Front. For. Glob. Change 3 (47), 1–16.

Murase, J., Kees Hordijk, K., Tayasu, I., Bodelier, P.L.E., 2011. Strain-specific incorporation of methanotrophic biomass into eukaryotic grazers in a rice field soil revealed by PLFA-SIP. FEMS Microbiol. Ecol. 75, 284–290.

Murphy, D.V., Recous, S., Stockdale, E.A., Fillery, I.R.P., Jensen, L.S., Hatch, D.J., et al., 2003. Gross nitrogen fluxes in soil: theory, measurement and application of N-15 pool dilution techniques. Adv. Agron. 79, 69–118.

Nannipieri, P., Kandeler, E., Ruggiero, P., 2002. Enzyme activities and microbiological and biochemical processes in soil. In: Burns, R.G., Dick, R.P. (Eds.), Enzymes in the Environment – Activity, Ecology and Applications. Marcel Dekker, New York, pp. 1–34.

Neufeld, J.D., Wagner, M., Murrell, J.C., 2007. Who eats what, where and when? Isotope-labeling experiments are coming of age. ISME J. 1, 103–110.

Newell, S.Y., Fallon, R.D., 1991. Toward a method for measuring instantaneous fungal growth rates in field samples. Ecology 72, 1547–1559.

Oliverio, A., Geisen, S., Delgado-Baquerizo, M., Maestre, F.T., Turner, B.L., Fierer, N., 2020. The global-scale distributions of soil protists and their contributions to belowground systems. Sci. Adv. 6, eaax8787.

Olsson, P.A., Johansen, A., 2000. Lipid and fatty acid composition of hyphae and spores of arbuscular mycorrhizal fungi at different growth stages. Mycol. Res. 104, 429–434.

Papale, D., Valentini, R., 2003. A new assessment of European forests carbon exchange by eddy fluxes and artificial neural network spatialization. Glob. Change Biol. 9, 525–535.

Pepper, I., Gerba, C., Gentry, T., 2014. Environmental Microbiology, Third ed. Academic Press, Waltham, Massachusetts.

Philippot, L., Tscherko, D., Bru, D., Kandeler, E., 2011. Distribution of high bacterial taxa across the chronosequence of two alpine glacier forelands. Microb. Ecol. 61, 303–312.

Pinheiro, M., Pagel, H., Poll, C., Ditterich, F., Garnier, P., Streck, T., et al., 2018. Water flow drives small scale biogeography of pesticides and bacterial pesticide degraders – a microcosm study using 2,4-D as a model compound. Soil Biol. Biochem. 127, 137–147.

Pitcher, A., Rychlik, N., Hopmans, E.C., Spieck, E., Rijpstra, W.I.C., Ossebaar, J., et al., 2010. Crenarchaeol dominates the membrane lipids of *Candidatus Nitrososphaera gargensis*, a thermophilic Group I.1b Archaeon. ISME. J. 4, 542–552.

Poll, C., Pagel, H., Devers, M., Martin-Laurent, F., Ingwersen, J., Streck, T., et al., 2010. Regulation of bacterial and fungal MCPA degradation at the soil-litter interface. Soil Biol. Biochem. 42, 1879–1887.

Potthoff, M., Loftfield, N., Buegger, F., Wick, B., Jahn, B., Joergensen, R.G., et al., 2003. The determination of $\delta^{13}C$ in soil microbial biomass using fumigation-extraction. Soil Biol. Biochem. 35, 947–954.

Reyna, D.L., Wall, L.G., 2013. Revision of two colorimetric methods to quantify glomalin-related compounds in soils subjected to different managements. Biol. Fertil. Soils 50, 395–400.

Rillig, M.C., Mummey, D.L., 2006. Mycorrhizas and soil structure. New. Phytol. 171, 41–53.

Robertson, G.P., Coleman, D.C., Bledsoe, C.S., Sollins, P., 1999. Standard Soil Methods for Long-Term Ecological Research. Oxford University Press, New York.

Rousk, J., Bååth, E., 2011. Growth of saprotrophic fungi and bacteria in soil. FEMS Microbiol. Ecol. 78, 17–30.

Ruess, L., Chamberlain, P.M., 2010. The fat that matters: soil food web analysis using fatty acids and their carbon stable isotope signature. Soil Biol. Biochem. 42, 1898–1910.

Schaefer, M., 1990. The soil fauna of a beech forest on limestone: trophic structure and energy budget. Oecologia 82, 128–136.

Schimel, D.S., 1993. Theory and Application of Tracers. Academic Press, San Diego, California.

Schinner, F., Öhlinger, R., Kandeler, E., Margesin, R. (Eds.), 1996. Methods in Soil Biology. Springer, Berlin.

Schwartz, E., Blazewicz, S., Doucett, R., Hungate, B.A., Hart, S.C., Dijkstra, P., 2007. Natural abundance $d^{15}N$ and $d^{13}C$ of DNA extracted from soil. Soil Biol. Biochem. 39, 3101–3107.

Schwartz, E., Hayer, M., Hungate, B.A., Koch, B.J., McHugh, T.A., Mercurio, W., et al., 2016. Stable isotope probing with ^{18}O-water to investigate microbial growth and death in environmental samples. Curr. Opin. Biotechnol. 41, 14–18.

Seeber, J., Scheu, S., Meyer, E., 2006. Effects of macro-decomposers on litter decomposition and soil properties in alpine pastureland: a mesocosm experiment. Appl. Soil Ecol. 34, 168–175.

Setia, R., Verma, S.L., Marschner, P., 2012. Measuring microbial biomass carbon by direct extraction – comparison with chloroform fumigation-extraction. Europ. J. Soil Biol. 53, 103–106.

Sinsabaugh, R.L., Carreiro, M.M., Alvarez, S., 2002. Enzymes and microbial dynamics of litter decomposition. In: Burns, R.G., Dick, R.P. (Eds.), Enzymes in the Environment – Activity, Ecology, and Application. Decker, New York, pp. 249–266.

Slaets, J.I.F., Boeddinghaus, R.S., Piepho, H.P., 2020. Linear mixed models and geostatistics for designed experiments in soil science: two entirely different methods or two sides of the same coin? Eur. J. Soil. Sci. https://doi.org/10.1111/ejss.12976.

Smith, C.J., Osborn, A.M., 2009. Advantages and limitations of quantitative PCR (Q-PCR)-based approaches in microbial ecology. FEMS Microbiol. Ecol. 67, 6–20.

Spohn, M., Kuzyakov, Y., 2013. Distribution of microbial- and root-derived phosphatase activities in the rhizosphere depending on P availability and C allocation – coupling soil zymography with ^{14}C imaging. Soil Biol. Biochem. 67, 106–113.

Stiehl-Braun, P.A., Hartmann, A., Kandeler, E., Buchmann, N., Pascal, N., 2011. Interactive effects of drought and N fertilization on the spatial distribution of methane assimilation in grassland soils. Global. Change Biol. 17, 2629–2639.

Sullivan, B.W., Dore, S., Kolb, T.E., Hart, S.C., Montes-Helu, M.C., 2010. Evaluation of methods for estimating soil carbon dioxide efflux across a gradient of forest disturbance. Global Change Biol. 16, 2449–2460.

Tabatabai, M.A., 1994. Soil enzymes. In: Weaver, R.W., Angel, J.S., Bottomley, P.S. (Eds.), Methods of Soil Analysis. Part 2. Microbiological and Biochemical Properties. Soil Science Society of America, Inc., Madison, Wisconsin, pp. 775–833.

Tabatabai, M., Fung, M., 1992. Extraction of enzymes from soil. In: Stotzky, B., Bollag, J.-M. (Eds.), Soil Biochemistry, Vol. 7. Dekker, New York, pp. 197–227.

Taylor, A.E., Zeglin, L.H., Dooley, S., Myrold, D.D., Bottomley, P.J., 2010. Evidence of different contributions of archaea and bacteria to the ammonia-oxidizing potential of diverse Oregon soils. Appl. Environ. Microbiol. 76, 7691–7698.

Taylor, A.E., Zeglin, L.H., Wanzek, T.A., Myrold, D.D., Bottomley, P., 2012. Dynamics of ammonia-oxidizing archaea and bacteria populations and contributions to soil nitrification potentials. ISME J. 6, 2024–2032.

Thiele-Bruhn, S., Schloter, M., Wilke, B.M., Beaudette, L.A., Martin-Laurent, F., Cheviron, N., et al., 2020. Identification of new microbial functional standards for soil quality assessment. SOIL 6, 17–34.

Tscherko, D., Kandeler, E., Bárdossy, A., 2007. Fuzzy classification of soil microbial biomass and enzyme activity in grassland soils. Soil Biol. Biochem. 39, 1799–1808.

Van Elsas, J.D., Trevors, J.T., Rosado, A.S., Nannipieri, P., 2019. Modern Soil Microbiology, Third ed. Taylor & Francis Group, Boca Raton.

Vepsäläinen, M., Kukkonen, S., Vestberg, M., Sirviö, H., Niemi, R.M., 2001. Application of soil enzyme activity test kit in a field experiment. Soil Biol. Biochem. 33, 1665–1672.

Verma, B., Robarts, R.D., Headley, J.V., Peru, K.M., Christofi, N., 2002. Extraction efficiencies and determination of ergosterol in a variety of environmental matrices. Commun. Soil Sci. Plant Anal. 33, 3261–3275.

Wagner, M., 2009. Single-cell ecophysiology of microbes as revealed by Raman microspectroscopy or secondary ion mass spectrometry imaging. Annu. Rev. Microbiol. 63, 411–429.

Waldrop, M.P., Holloway, J.M., Smith, D.B., Goldhaber, M.B., Drenvsky, R.E., Scow, K.M., et al., 2017. The interacting roles of climate, soils, and plant production on soil microbial communities at a continental scale. Ecology 98, 1957–1967.

Wallander, H., Nilsson, L.O., Hagerberg, D., Bååth, E., 2001. Estimation of the biomass and seasonal growth of external mycelium of ectomycorrhizal fungi in the field. New. Phytol. 151, 753–760.

Weete, J.D., Abril, M., Blackwell, M., 2010. Phylogenetic distribution of fungal sterols. PLoS One 5, e10899.

Wu, H., Chen, W., Rong, X., Cai, P., Dai, K., Huang, Q., 2014. Soil colloids and minerals modulate metabolic activity of *Pseudomonas putida* measured using microcalorimetry. Geomicrobiol. J. 31, 590–596.

Chapter 8

The spatial distribution of soil biota and their functions

Serita D. Frey*

*Center for Soil Biogeochemistry and Microbial Ecology, Department of Natural Resources & the Environment, University of New Hampshire, Durham, NH, USA

Chapter outline

8.1 Introduction

Biota that occur in soils represent a large proportion of Earth's biodiversity, contributing to many critical ecosystem services, including food and fiber production, nutrient cycling, water storage and quality, and climate regulation. Terrestrial ecosystem multifunctionality (i.e., ability of ecosystems to simultaneously provide multiple services) is highly dependent on belowground organism activity (Soliveres et al., 2016). Soil organism abundances, diversity, and activities are not randomly distributed in soil but vary in a patchy fashion both horizontally across a landscape, vertically through the soil profile, and at the microscopic scale within the soil matrix. Different groups of organisms exhibit different spatial patterns because they each react to soil conditions and plant-soil interactions in different ways. This spatial heterogeneity in soil organism abundance, diversity, and function has been shown to correlate with gradients in site and soil characteristics, including abiotic factors (bulk density, texture, moisture, oxygen concentration, pH, soil organic matter [SOM] content, inorganic N availability, soil aggregation), biotic factors (food web interactions, vegetation dynamics), management factors (tillage, cropping system), and global change factors (soil warming, altered precipitation regimes, land-use change, atmospheric N deposition, invasive species). Some of these factors are important at the micro scale, whereas others act over larger distances. While there are still large gaps in our understanding of how soil organisms are distributed, there has been a dramatic increase in information obtained in this area over the past decade. This chapter summarizes what is known about the distribution of soil biota, from geographic differences at the landscape, regional and global scales, down to variability in microbial populations at the aggregate and pore level.

Soil Microbiology, Ecology, and Biochemistry. https://doi.org/10.1016/B978-0-12-822941-5.00008-9

8.2 The biogeography of soil biota and their functions

Biogeography, the science of documenting spatial patterns of biological diversity from local to continental scales, examines variation in genetic, phenotypic, and physiological characteristics at different scales (e.g., distantly located sampling sites or along environmental gradients). It emphasizes understanding the factors that generate and maintain organism distributions, with information on biogeographic patterns providing insight into the processes that determine and maintain biodiversity. Questions addressed include why global biodiversity is so great, why organisms live where they do, how many taxa can coexist, and how organisms will respond to environmental change. Biogeography is a relatively mature science, with the study of plant and animal diversity patterns going back more than two centuries to early plant and animal surveys. Thus, the global distributions of most of the world's flora and fauna are generally known. From this work, it is well established that most macroscopic plant and animal species are not globally distributed but instead have restricted geographic distributions because of climate sensitivity and natural barriers to migration. This isolation has, over geological time, led to the evolution of new species and the development of geographically distinct plant and animal communities.

Until recently, there has been less emphasis on understanding and mapping the biogeography of microscopic organisms, due primarily to methodological constraints. Advances in molecular sequencing technologies have revealed that microbial diversity far exceeds that of macroscopic organisms. It is now possible, by combining molecular sequencing and environmental data with advanced analytical approaches (e.g., neural networks, random forest modeling), to generate predictive models and maps of soil organism diversity with which to build a better understanding of their biogeography and ultimately, the relationships between diversity, community composition, and ecosystem function (Fig. 8.1).

A historically common perception was that microorganisms are cosmopolitan in their distribution, being capable of growth in many different places worldwide. This idea goes back more than a century to Baas Becking, a Dutch scientist, who suggested that "*everything is everywhere, but the environment selects*" (de Wit and Bouvier, 2006). Soil organisms, due to their small size and large numbers, are continuously being moved around, often across continental-scale distances. Dispersal mechanisms include water transport via rivers, groundwater, and ocean currents; airborne transport in association with dust particles and aerosols, especially during extreme weather events such as hurricanes and dust storms; transport on or in the intestinal tract of migratory birds, insects, and aquatic organisms; and human transport through air travel and shipping. Microbes are abundant in the atmosphere, where thousands of distinct bacterial taxa representing a wide range of identified phyla have been identified (Smith et al., 2013), along with large amounts of fungal spores (Chaudhary et al., 2020). Even isolated environments, like those found in the Antarctic, show a wide range of microbial species that appear to have been introduced from other places, often by human visitors. Despite the ability of soil organisms to be widely dispersed, not all survive their journey and the degree to which they become established in their new environment depends on resource availability, physicochemical conditions (temperature, salinity, pH), and ecological constraints (competition with native taxa). As a result, meaningful biogeographic patterns do exist for soil organisms, suggesting that they are not exempt from the fundamental evolutionary processes (i.e., geographic isolation and natural selection) that shape plant and animal communities. Thus, like plants and animals, not all soil organisms are found everywhere. Some can tolerate a broad range of environmental conditions, while others only persist under a specific set of environmental constraints and have more restricted ranges.

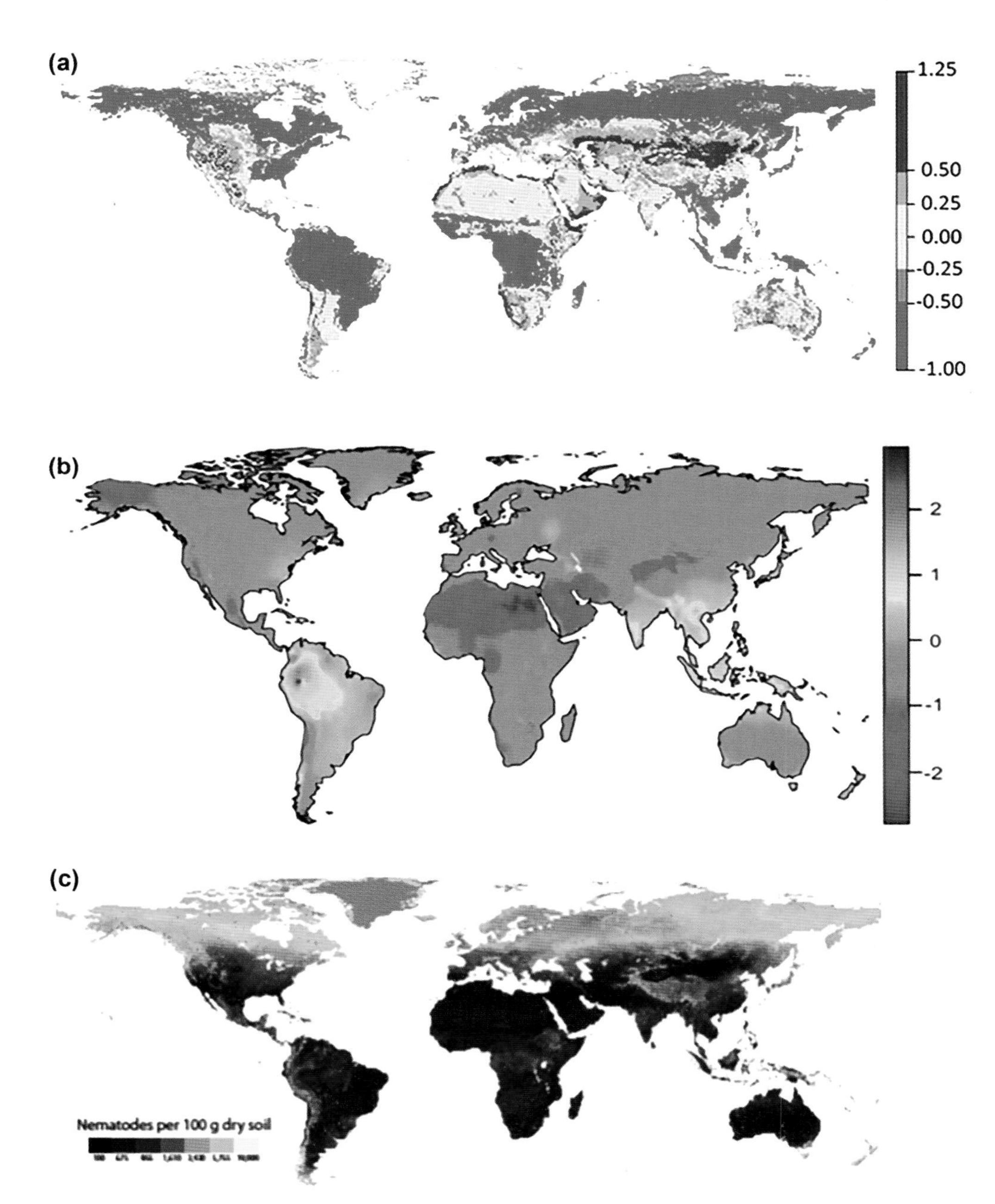

FIGURE 8.1 Global maps of: (a) dominant soil bacteria with preference for high pH (*warm* versus *cool colors* indicate higher versus lower relative abundance), (b) taxon richness of soil fungi (*warm colors* indicate more and *cold colors* indicate fewer taxa), and (c) soil nematode density. *(Adapted from (a) Delgado-Baquerizo et al., 2018; (b) Tedersoo et al., 2014; (c) van den Hoogen et al., 2019; used with permission.)*

Geographic patterns have been reported for several soil organism groups, including bacteria (Bahram et al., 2018; Delgado-Baquerizo et al., 2018), archaea (Karimi et al., 2018), fungi (Bahram et al., 2018; Tedersoo et al., 2014), protozoa (Bates et al., 2013), rotifers (Robeson et al., 2011), nematodes (van den Hoogen et al., 2019), and collembola (Baird et al., 2019). Despite the overwhelming diversity of soil organism communities, relatively few taxa are abundant in soils globally, with only a small fraction of taxa shared between individual soil samples. Delgado-Baquerizo et al. (2018) analyzed soils from over 200 locations, finding that only 2% of bacterial phylotypes accounted for nearly half of the soil bacteria observed worldwide. These dominant bacterial phylotypes were highly abundant in individual soil samples across the dataset, which represented six continents and multiple ecosystem types and climatic regions. It was also determined that the dominant bacterial taxa were clustered into predictable ecological groups that share similar habitat preferences linked to soil pH, moisture regime, and net primary productivity (Delgado-Baquerizo et al., 2018). Soil pH appears to be a particularly strong driver of bacterial diversity, with bacterial taxon richness tending to be highest in neutral soils and lowest in acidic soils. Particular bacterial groups (e.g., Acidobacteria, Actinobacteria, and Bacteroides) are more strongly affected by pH than others (Lauber et al., 2009). Bacterial diversity has also been linked to heterogeneity in soil texture, with certain taxa co-occurring with clay and fine silt–sized particles (Seaton et al., 2020). It has also been observed that certain genomic traits are potentially more useful than phylogeny in explaining bacterial distributions and habitat preferences (Barberán et al., 2014). Bacteria with larger genomes and more metabolic versatility appear more likely to have larger geographic and environmental distributions, perhaps explaining why some bacterial taxa are more widespread than others.

The global distribution of soil fungi has also been mapped (Fig. 8.1b), showing that fungal taxonomic diversity tends to increase at low latitudes, in general agreement with the biogeographic patterns of plants and animals (Bahram et al., 2018; Tedersoo et al., 2014). However, specific taxonomic and functional groups depart from this general trend. The diversity of saprotrophic fungi, parasites, and pathogens has been observed to increase toward the equator, while ectomycorrhizal fungi peak in temperate forests and Mediterranean ecosystems at mid-latitudes. As for bacteria, soil properties (especially soil C-to-N ratio) rather than climatic variables are strong predictors of fungal biomass and relative abundance (Bahram et al., 2018). The evolutionary histories of soil fungi also appear to have influenced their global distributions (Treseder et al., 2014).

An analysis of soil protistan communities, at sites representing a broad geographic range and variety of biome types, revealed a high level of diversity and distribution patterns that suggest that most taxa in this microbial group are not globally distributed (Bates et al., 2013). More than 1000 taxa representing 13 phyla were identified. A high number of rare taxa were observed, along with several taxa rarely found in soil environments. Only one taxon was found in more than 75% of the samples, and most taxa (84%) were found at five or fewer sites. Similar results have been observed for soil rotifers, a ubiquitously distributed group of microbial eukaryotes that are important contributors to biogeochemical cycling (Robeson et al., 2011). Rotifer diversity was found to be high, with global sampling revealing close to 800 taxa. Distant communities were different from one another even in similar environments, and there was almost no overlap in taxon occurrence between soil communities sampled at distances of ~150 meters apart. Many novel phylotypes were also observed at each sampling location, suggesting a high degree of endemism (where taxa are unique to a particular site). Global soil nematode abundance has also been recently mapped and shown to vary latitudinally and regionally (Fig. 8.1c), with the highest densities in tundra, boreal forest, and temperate forests regions (van den Hoogen et al., 2019). The lowest densities were observed in Mediterranean forests, the Antarctic, and hot deserts. While climatic factors (i.e., temperature and precipitation) played a role, soil characteristics (e.g., texture, soil organic C, pH, cation exchange capacity) were the most important drivers of global nematode abundance patterns. Soil organic C content and cation-exchange capacity were positively correlated with total nematode density, while pH had a negative effect.

It is also useful to consider regional and global patterns in soil microbial biomass, which represents a considerable fraction of living biomass on Earth, with estimates between 1 and 3% of total terrestrial soil C. A meta-analysis of belowground plant, microbial, and faunal biomass indicates that microbial biomass ranges from <5 to ~ 800 g C m^{-2} across biomes, with the highest concentrations occurring in temperate coniferous and tropical forests (Fierer et al., 2009). Climate, vegetation, soil characteristics, and land-use patterns all interact to influence soil biotic abundance and biomass at a given location. Microbial biomass tends to follow global patterns of plant biomass and productivity, but soil organic C is the best predictor of microbial biomass levels at a given location. Microbial biomass is generally positively related to SOM contents in most ecosystems, with peat and organic soils being an exception. Levels of microbial biomass are also typically correlated with soil clay content which is an important controller of SOM levels. Clay minerals promote microbial growth by maintaining the pH in an optimal range, buffering nutrient supply, adsorbing metabolites that are inhibitory to microbial growth, and providing protection from desiccation and grazing through increased aggregation. Total amounts of microbial biomass are also impacted by land use, with lower levels typically observed in arable compared to undisturbed forest and grassland soils due to cultivation-induced losses of organic matter. Microbial biomass is also correlated with latitude, with microbial biomass tending to be lower but highly variable at high latitudes.

Biogeographical information on the taxonomy, diversity, and abundance of soil biota is important for climate modeling, environmental decision making, and land management and conservation. Recent advances in understanding the regional and global biogeography of soil organisms have yielded new and useful insights into their distribution patterns. However, linking this biogeographical information to soil functioning continues to be a challenge. Most soil taxa do not match those found in existing genomic databases, and the majority cannot successfully be isolated and grown in vitro. Thus we lack a predictive understanding of the ecology of most soil organisms, with their traits, functional capacities, and environmental preferences still largely unknown, as well as their vulnerabilities to global change.

In a recent analysis, Guerra et al. (2020) identified gaps in soil organism research by compiling data on the global distribution of sampling sites where soil taxa or soil functions have been measured (Fig. 8.2). A fundamental mismatch in available observations on soil biodiversity and ecosystem functioning was observed, with only 0.3% of all sampling sites having an overlap in soil biodiversity and function data. The availability of comparable data on soil taxa and relevant ecosystem functions are particularly underrepresented for tropical and subtropical regions, dryland, and tundra ecosystems, and those areas with high climate variability. Additionally, only a fraction of soil types and environmental conditions are well represented, and most of the data represent a single sampling time without repeated measures over multiple years; thus information on temporal variability is lacking. In summary, understanding the linkages between soil biodiversity and ecosystem function, along with their responses to global change drivers, remains limited due to the absence of globally distributed and temporally explicit observational data. As a result, there has been a call for a global soil monitoring system that overcomes these limitations (Guerra et al., 2020).

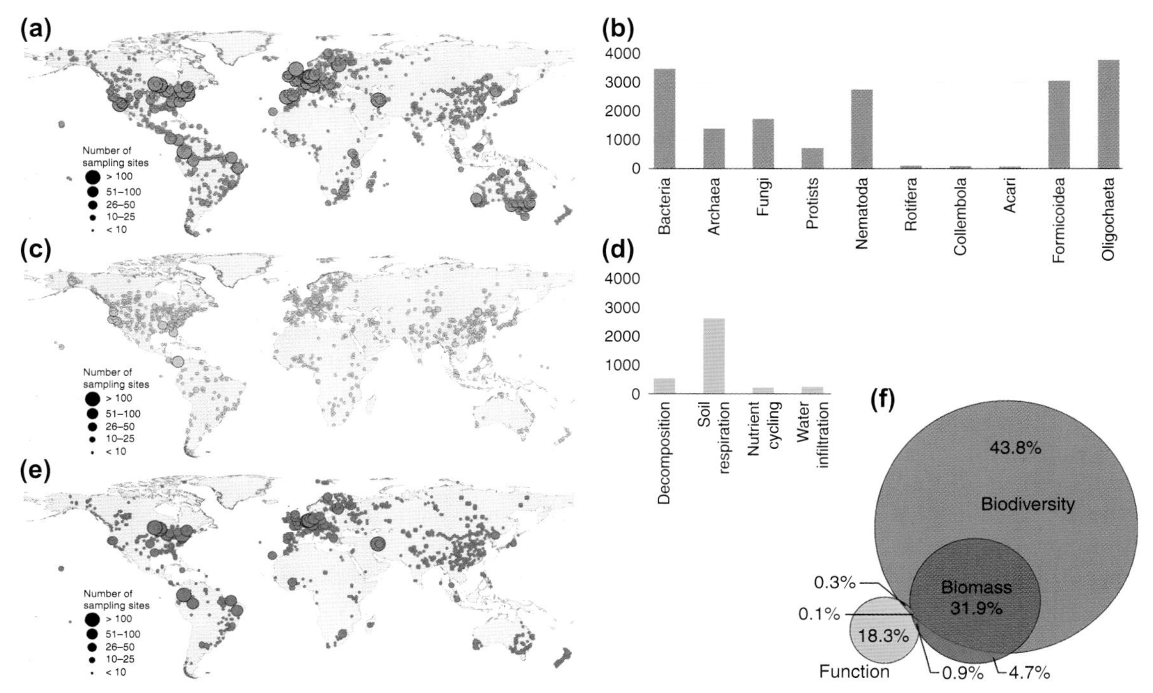

FIGURE 8.2 Distribution of sampling locations (a, c) for different soil biota groups (b) ecosystem functions (d), and sample locations with biomass data (e). The Venn diagram (f) shows that <0.3% of available data overlap for soil biodiversity (*green*), functions (*yellow*), and biomass (*blue*). The size of the circle corresponds to the number of sampling sites. *(From Guerra et al., 2020; used without changes and with permission under the Creative Commons Attribution 4.0 International License [https://creativecommons.org/licenses/by/4.0/].)*

8.3 Vertical distributions within the soil profile

The abundance, biomass, and activity of most soil organisms is highest in the top 10 to 20 cm of soil and declines with depth in parallel with SOM content and prey availability. Approximately 65% of the total microbial biomass is found in the top 25 cm of the soil profile. Below that depth, microbial densities typically decline by 1 to 3 orders of magnitude (Fig. 8.3). Potential enzyme activities, an index of microbial activity, also decline significantly with depth, tracking declines in soil nutrients and SOM (Stone et al., 2014). Abundances of G^- bacteria, fungi, and protozoa are typically highest at the soil surface, while G^+ bacteria, actinobacteria, and archaea tend to increase in proportional abundance with increasing depth (Fierer et al., 2003). In forest soils, the litter layer is dominated by fungi, with fungal biomass being up to three times higher than that of bacterial biomass (Baldrian et al., 2012). The relative importance of bacteria increases with depth. Hyphal density of, and root colonization by, mycorrhizal fungi decrease substantially below 20 cm. Microbial grazers (e.g., protozoa, collembola) also decrease with depth, often more rapidly than either their bacterial or fungal prey. Collembolan numbers peak at 1 to 5 cm below the soil surface, with almost none observed below 10 cm.

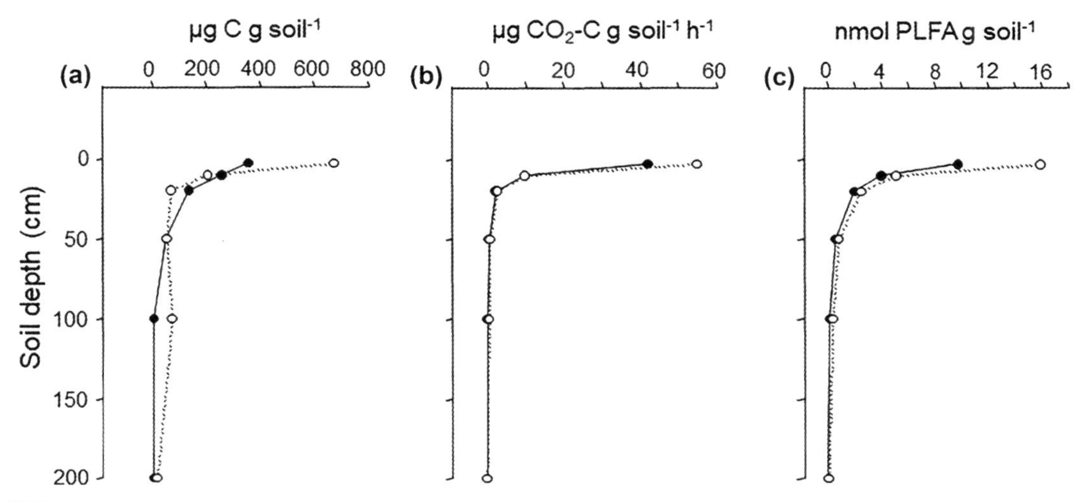

FIGURE 8.3 Microbial biomass with depth for a valley (*open circles*) and terrace (*closed circles*) soil profile as determined by three methods: (a) chloroform fumigation extraction, (b) substrate-induced respiration, and (c) phospholipid fatty acid (PLFA) analysis. *(Adapted from Fierer et al., 2003.)*

While generally low, the numbers and activities of soil organisms at depth vary spatially depending on gradients in texture, pH, temperature, water availability, and SOM content. Interfaces between soil layers often generate localized regions of greater water availability where microorganisms may exhibit increased numbers or activity due to improved access to water and nutrients. There are active cells in deeper soil horizons and on a depth-weighted basis, up to 30 to 40% of the microbial biomass in the soil profile is found below 25 cm. While potential enzyme activities tend to decline with depth, when normalized to microbial biomass, specific enzyme activities in the subsurface are either similar to or higher than those at the soil surface, suggesting a small but active microbial community at depth (Stone et al., 2014).

Microbial diversity and community composition are also vertically stratified. As with microbial biomass, microbial diversity is typically highest in the top 10 cm of soil, declining with depth. Bacterial diversity dropped by 20 to 40% from surface soil to deeper horizons in a montane forest in Colorado, USA (Eilers et al., 2012). There are distinct differences in the microbial community colonizing the litter layer versus more highly decomposed organic (Oe/Oa; L/H) layers and mineral horizons of forest soils. The litter layer (Oi/L) in a spruce forest in Central Europe was enriched in Acidobacteria and Firmicutes, while Actinobacteria were more abundant in the highly decomposed Oe/Oa (F/H) organic horizon layers (Baldrian et al., 2012). Surface mineral soils harbor a different microbial community than that found at depth, with the transition typically occurring between 10 and 25 cm (Eilers et al., 2012). The depth distribution of the bacterial community in a montane forest was driven primarily by a decline in the relative abundance of Bacteroidetes and a peak in the relative abundance of Verrucomicrobia between 10 and 50 cm (Fig. 8.4). The Actinobacteria were not as strongly structured by depth, showing similar relative abundances down to 150 cm. The depth distribution pattern of other bacterial groups (e.g., Acidobacteria, Alphaproteobacteria) was inconsistent and differed depending on sampling location.

Fungal community composition differs substantially with depth, with up to 40% of the most abundant taxa found in the organic horizon (Baldrian et al., 2012). There is a general shift from saprotrophic to mycorrhizal fungal dominance with increasing soil depth (Fig. 8.5) which coincides with vertical changes in

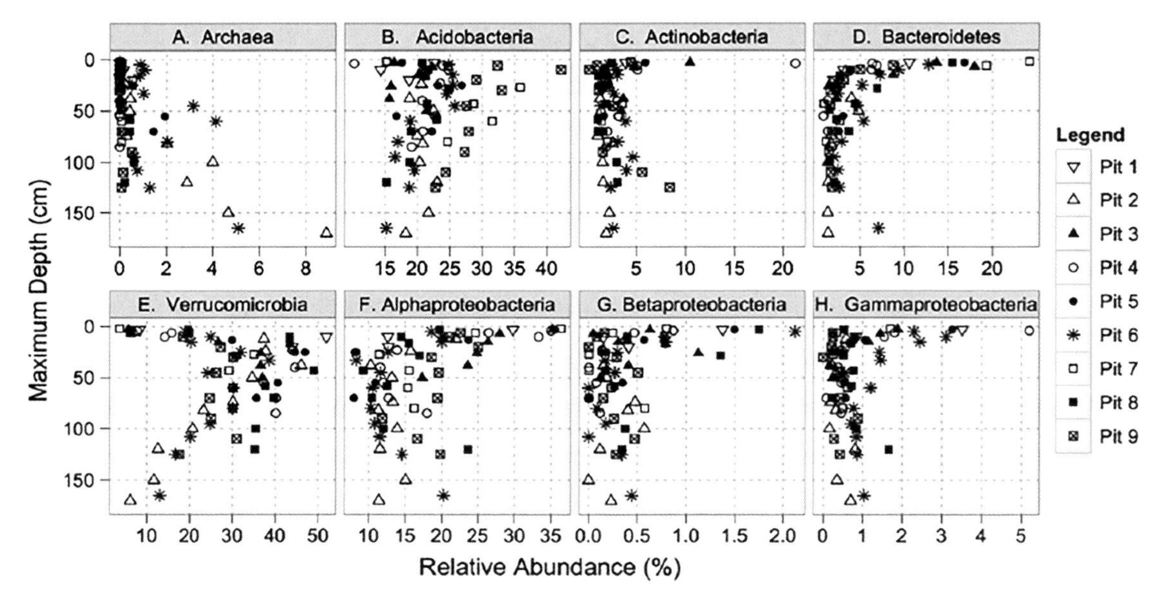

FIGURE 8.4 Relative abundance of archaea and bacterial taxa with depth at nine locations in a montane forest in Colorado, USA. *(From Eilers et al., 2012; used with permission.)*

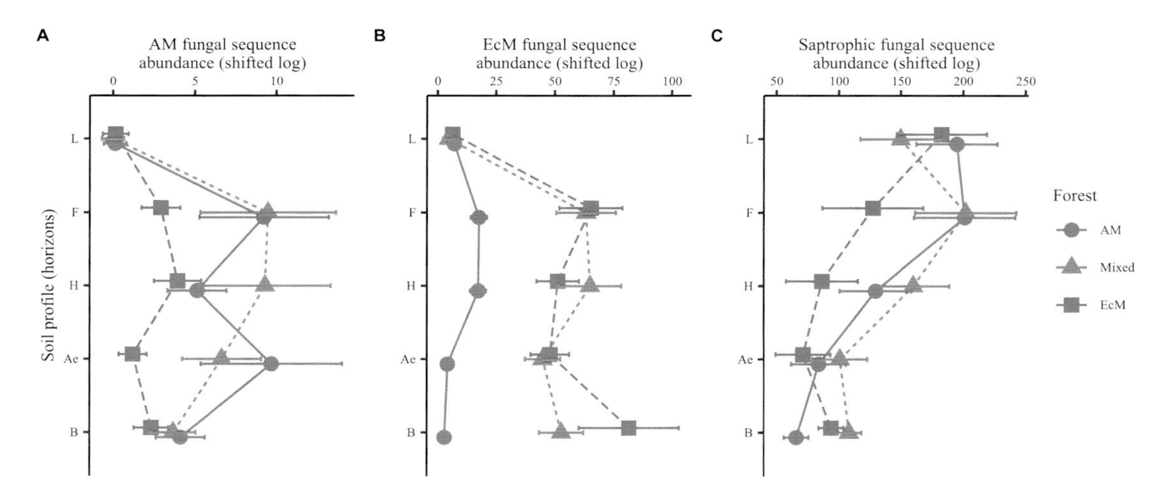

FIGURE 8.5 Depth distribution of (a) arbuscular (AM), (b) ectomycorrhizal (EcM), and (c) saprotrophic fungi from organic (L, F, H) to mineral (Ae, B) soil horizons across AM-dominated, EcM-dominated, and mixed forests. *(Modified from Carteron et al., 2021; used with permission.)*

soil chemistry (Carteron et al., 2021). The litter layer (Oi/L) in forest soils typically exhibits a high abundance of saprotrophic fungi, while mycorrhizal species dominate deeper organic horizon layers (Oe/Oa; F/H) and surface mineral soil. Observed differences in fungal taxa between soil horizons are likely related to their ecological functions in the system. Baldrian et al. (2012) analyzed forest soil fungi that harbor the *cbhI* gene that encodes for cellobiohydrolase, an extracellular enzyme that mediates the degradation of cellulose. The diversity of cellulolytic fungi was higher in the litter layer compared to deeper organic horizon layers. The fungal taxa mediating cellulose decomposition were distinct between these two organic horizon components. Some of the most abundant *cbhI* sequences were transcribed by fungi with low relative abundance, indicating that some species, while not highly abundant, may nonetheless make an important contribution to cellulose decomposition.

Soil invertebrate communities also vary with depth, particularly in the organic and upper mineral soil horizons. There is a succession from litter-dwelling to mineral soil-dwelling species of collembola at the interface between organic and mineral soil layers. Gut content analysis has shown that collembola near the top of the O-horizon feed preferentially on pollen grains, while those at the bottom of this layer feed mainly on fungal material and highly decomposed organic matter (Ponge, 2000).

Soil biota are widely distributed as biological soil crusts (also called cryptogamic, microbiotic, microphytic crusts) which form a protective soil covering in the interspaces between the patchy plant cover in arid and semiarid regions (Fig. 8.6). Soil crusts are globally distributed and play an important role in stabilizing soil and cycling water and nutrients in arid regions (Belnap and Lange, 2001). They also play a critical role in dryland water capture and retention by reducing sediment production (i.e., soil erosion) and increasing water storage (Eldridge et al., 2020). Soil crusts consist of a specialized community of archaea, bacteria, fungi, algae, mosses, and lichens. Sticky substances produced by crust organisms bind surface soil particles together, often forming a continuous layer that can reach 10 cm in thickness. In arid surface soils cyanobacteria vertically migrate to and from the soil surface in response to changing moisture conditions. This ability to follow water is likely an important mechanism for long-term survival of desert soil communities that are exposed to challenging conditions, including water stress, nutrient limitation, and intense solar radiation. An analysis of the fungal assemblages associated with soil crusts indicates a high level of diversity, which increases as crusts mature successionally (Bates et al., 2012). A few dominant fungal taxa were found to be widespread, with many additional taxa found to be site specific. Soil crusts are damaged by trampling disturbances caused by livestock grazing, tourist activities (e.g., hiking, biking), and off-road vehicle traffic. Such disturbances lead to reduced diversity of crust organisms and increased soil erosion. Crust recovery can take decades.

The deep subsurface environment, that region below the top few meters of soil, as well as deep aquifers, caves, bedrock, and unconsolidated sediments, was once considered hostile to and devoid of living organisms. It is now known that microorganisms reside in the "deep biosphere," often to depths of tens to hundreds of meters (Engel, 2019). Subsurface microbiology emerged as a discipline largely in response to groundwater quality issues. Modern drilling and coring techniques facilitated the retrieval and characterization of undisturbed, uncontaminated samples collected from deep environments. Scientists from many disciplines (e.g., geology, hydrology, geochemistry, and environmental engineering) are now interested in understanding the role of microorganisms in soil genesis, contaminant degradation, maintenance of groundwater quality, and the evolution of geological formations (e.g., caves). Many novel microorganisms with unique biochemical and genetic traits have been isolated from subsurface environments and may have important industrial and pharmaceutical uses.

FIGURE 8.6 Biological soil crusts. (a) Landscape with healthy soil crusts, (b) close-up of mature crusts on the Colorado Plateau, USA, and (c) cyanobacteria adhered to sand grains. *(Courtesy of Jayne Belnap, USGS Canyonlands Field Station, Moab, UT, USA; used with permission.)*

Estimates suggest that there are 3×10^{29} microbial cells residing in the deep biosphere globally (Hoehlert and Jorgensen, 2013). Deep subsurface microbial populations are dominated by bacteria and archaea. In addition, nearly all the major taxonomic and physiological groups have been found. Bacterial densities range from less than 10^1 up to 10^8 cells g^{-1} material, depending on the method of enumeration and the depth of sample collection. Groundwater sampled from aquifers or unconsolidated sediments have bacterial concentrations ranging from 10^3 to 10^6 cells mL^{-1}. Microbial biomass in the deep subsurface is typically several to many orders of magnitude lower than that observed for surface soils, where populations often reach 10^9 bacteria and archaea g^{-1} soil. The turnover of the subsurface biomass is typically on the timescale of centuries to millennia.

Subsurface microbial communities have a high proportion of inactive, nonviable, and nonculturable cells. Most subsurface microbes have no cultivated relatives, making it difficult to determine the physiological characteristics of even the dominant taxa (Hoehlert and Jorgensen, 2013). It is known that metabolic rates are very slow compared to terrestrial surface environments, with microbial activity at these depths being limited by temperature and the availability of water and energy sources. It is estimated that the energy flux available to the deep biosphere is $<1\%$ of the C fixed photosynthetically at the Earth's surface (Hoehlert and Jorgensen, 2013). Energy sources include low-concentration organic substrates or reduced inorganic substrates, such as H_2, CH_4, or S_2^-. As in surface soils, the size and interconnectivity of pores is an important factor regulating microbial growth. In the subsurface environment pores exist in unconsolidated materials or as fractures or fissures in consolidated rock. Microbial activity occurs in pores 0.2 to 15 μm in size, whereas little to no activity has been observed in pores with openings of <0.2 μm.

While bacteria and archaea dominate most subsurface microbial communities, protozoa and fungi have been observed at depth under certain conditions. Protozoan populations, typically at or below the level of detection at unpolluted sites, have been observed in the subsurface of contaminated sites. Protozoan presence may thus represent a useful indicator of environmental contamination. Protozoa have been shown to stimulate subsurface nitrification and bacterial degradation of dissolved organic C. Both ectomycorrhizal and arbuscular mycorrhizal hyphae have also been recovered in fractured, granitic bedrock at depths greater than 2 m (Egerton-Warburton et al., 2003), and root tips colonized by ectomycorrhizal fungi were observed at depths of up to 4 m in a tropical eucalypt plantation (Robin et al., 2019). The ability of mycorrhiza to grow and function in subsurface environments may enable their host plants to survive drought and/or low nutrient conditions by enhancing water and nutrient uptake from deep soil or bedrock sources.

The study of subsurface microorganisms is transforming our ideas regarding the extent of the biosphere and the role that microorganisms have and continue to play in the evolution of the subsurface environment (Engel, 2019). As one example, scientists have found evidence suggesting that caves and aquifers in Karst environments are formed, in part, by microbial activity rather than exclusively by abiotic, geochemical reactions as was previously thought. Sulfur-oxidizing bacteria colonize carbonate surfaces in the cave or aquifer, use H_2S as an energy source, and produce sulfuric acid as a by-product (Engel and Randall, 2011). This acid facilitates the conversion of limestone to gypsum, which is more easily dissolved by water.

8.4 Microscale heterogeneity of soil organisms

Soil is a complex, three-dimensional space comprised of mineral particles, plant roots, organic materials, pore spaces, and organisms. The shape and arrangement of soil minerals and organic particles is such that a network of pores of various shapes and sizes exists, resulting in a highly uneven distribution of water, oxygen, and solutes. Between 45 and 60% of the total soil volume is comprised of pores that are either air- or water-filled depending on moisture conditions. These pores may be open and connected to adjoining pores or closed and isolated from the surrounding soil. Pores of different shapes, sizes, and degrees of continuity provide a mosaic of organism habitats with very different physical, chemical, and biological characteristics, resulting in an uneven distribution of soil organisms (Fig. 8.7). Since soil

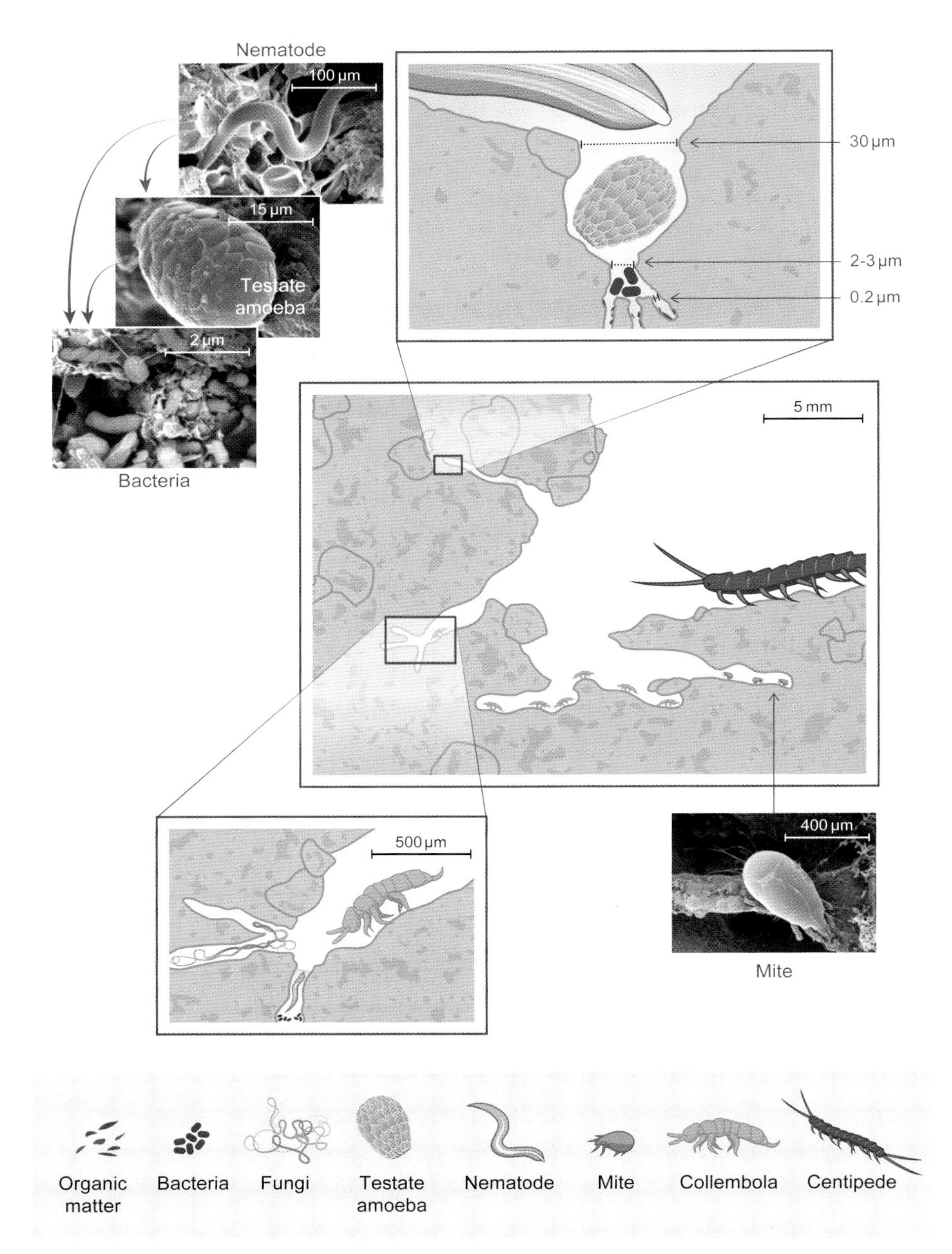

FIGURE 8.7 Size segregation of soil organisms within the soil pore network, where small pores serve as refugia for prey protecting them from larger consumers. *(From Erktan et al., 2020; used without changes and with permission under the Creative Commons Attribution BY-NC-ND 4.0 International License [https://creativecommons.org/licenses/by-nc-nd/4.0/].)*

organisms themselves vary in size, structural heterogeneity at this scale determines where a particular organism can reside, the degree to which its movement is restricted, its ability to sense and access food resources, and its interactions with other organisms. Soil heterogeneity fundamentally influences trophic dynamics, stabilizing predator-prey interactions and facilitating the coexistence of consumers and their prey (Hol et al., 2016; Petrenko et al., 2020).

The heterogeneous nature of the soil pore network plays a fundamental role in determining microbial abundance, diversity, activity, and community composition by affecting the relative proportion of air versus water-filled pores, which in turn regulates water and nutrient availability, gas diffusion, and biotic interactions such as competition and predation. Microbial activity is dependent on the location of organic matter substrates within the pore network and on optimum concentrations of water and oxygen. Microbial activity is maximized when about 60% of the total soil pore space is water filled. As soil moisture declines below this level, pores become poorly interconnected, water circulation becomes restricted, and dissolved nutrients carried by the soil solution become less available for microbial utilization. At the other extreme, when most or all pores are filled with water, oxygen becomes limiting since diffusion rates are significantly greater in air than through water. Gas diffusion into micropores is particularly slow, since small pores often retain water, even under dry conditions. Restricted oxygen diffusion into micropores, combined with biological oxygen consumption during the decomposition of organic matter, can lead to the rapid development and persistence of anaerobic conditions. Thus survival of soil biota residing in small pores depends on their ability to carry out anaerobic respiration (e.g., denitrification), replacing oxygen with an alternative electron acceptor (e.g., NO_3^-).

This chapter has focused thus far on the spatial distribution of biota at the scale of soil profiles, sites, or continents, but we might also ask whether there is a microbiogeography at the soil aggregate or pore scale. Recent evidence suggests that microbial diversity and community composition vary significantly across a gradient of soil mineral, aggregate, and pore size classes. Bach et al. (2018) observed that both bacterial and fungal taxonomic richness declined across a gradient of soil microaggregates (<250 μm) to small (250–1000 μm), medium (1000–2000 μm), and large (>2000 μm) macroaggregates, with microaggregates being more diverse and harboring microbial communities that were compositionally different from those of larger aggregate size classes. Bacteria can occupy both large and small soil pores; however, 80% of bacteria are thought to reside in small pores, and the relative abundance of many bacterial groups was found to be higher in microaggregates compared to macroaggregates. The maximum diameter of pores most frequently colonized by bacteria is estimated to range from 2.5 μm for fine to 9 μm in coarse-textured soils. Few bacteria have been observed in pores <0.8 μm in diameter, which means that 20 to 50% of the total soil pore volume, depending on soil texture and the pore size distribution, cannot be accessed and utilized by the microbial community.

Electron microscopy has revealed that bacteria often occur as isolated cells or small colonies (<10 cells) associated with decaying organic matter; however, larger colonies of several hundred cells have been observed on the surface of aggregates (Fig. 8.8). Bacterial cells are often embedded in mucilage, a sticky substance of bacterial origin to which clay particles attach. Clay encapsulation and residence in small pores may provide bacteria with protection against desiccation, predation, bacteriophage attack, digestion during travel through an earthworm gut, and the deleterious effects of introduced gases such as ethylene bromide, a soil fumigant.

Fungi with mycelial growth forms are generally found on aggregate surfaces and in large pores; however, they can use substrates in smaller pores if their hyphae can penetrate them (Ruamps et al, 2011). Fungi with a yeast growth form were found to be indicator taxa in soil microaggregates (Bach et al. 2018).

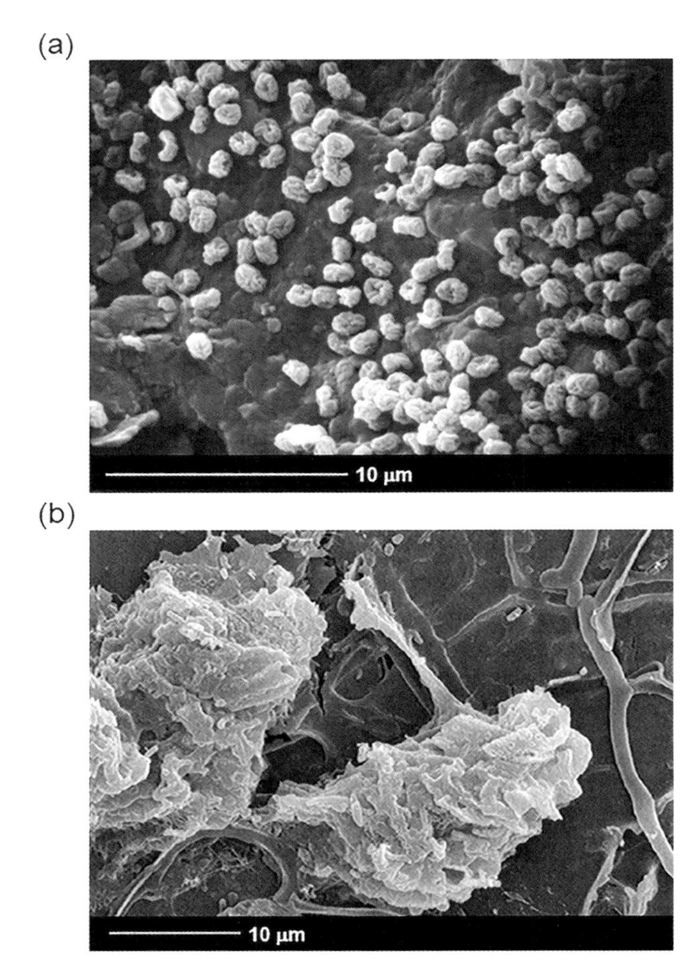

FIGURE 8.8 (a) Bacteria on the surface of an aggregate isolated from a grassland soil. (b) An amoeba with its extended pseudopodia engulfing bacteria. *(Photos courtesy of V.V.S.R. Gupta, CSIRO Land and Water, Glen Osmond, South Australia, Australia; used with permission.)*

Like bacteria, fungal hyphae are often sheathed in extracellular mucilage, which not only serves as protection against predation and desiccation but is also a gluing agent in the soil aggregation process. Extensive hyphal networks grow through the soil and over aggregate surfaces, binding soil particles together and thereby playing an important role in the formation and stabilization of aggregates.

The distribution of soil fauna is highly dependent on the size and arrangement of soil aggregates and pore spaces. Nematode abundance and community composition were found to be more highly influenced by aggregate size than by soil management practice (Briar et al., 2011). Nematode density was consistently lower in small macroaggregates (250–1000 μm) compared to large macroaggregates (>1000 μm) and interaggregate spaces. A higher juvenile-to-adult ratio of nematodes was observed in small macroaggregates, suggesting that juveniles, because of their smaller size, can more readily access nutrient resources (e.g., bacterial prey) in smaller pores. Smaller aggregates may also provide refuge for juvenile nematodes, protecting them from predation by larger soil fauna.

Soil heterogeneity influences nutrient cycling dynamics by restricting organism movement and thereby modifying the interactions between organisms. For example, small pores influence trophic relationships and nutrient mineralization by providing refuges and protection for smaller organisms, particularly bacteria, against attack from larger predators (e.g., protozoa, nematodes) that are typically unable to enter smaller pores. The location of bacteria within the pore network is a key factor in their growth and activity. Bacterial populations are consistently high in small pores but highly variable in large pores, where they are vulnerable to being consumed. This may explain why introduced bacteria (e.g., *Rhizobium* and biocontrol organisms) often exhibit poor survival relative to indigenous bacteria. When they are introduced in such a way as to be transported by water movement into small, protected pores, their ability to persist is enhanced.

Complex interactions and feedbacks occurring at the plant-soil interface and plant-mediated heterogeneity in soil physical and chemical characteristics also play a critical role in regulating the composition and diversity of soil microbial communities. Microbial abundance, diversity, community composition, and activity often reflect the plant species present in a given soil. In addition, plant-mediated differences in the microbial community are potentially attributable to specific variations between plant species in the quality and quantity of organic matter inputs to the soil. Plant resource quantity and quality can also be altered by disturbances such as herbivore grazing and global change (e.g., elevated atmospheric CO_2, N deposition). Plant herbivory has been shown to increase C allocation to roots, root exudation, fine root turnover, soil-dissolved organic C, microbial biomass and activity, and faunal activity. These changes in turn alter soil N availability, plant N acquisition, photosynthetic rates, and, ultimately, plant productivity. Soil biota−plant interactions are discussed in more detail in Chapter 11.

8.5 Summary

The spatial patterns of soil organisms and the factors that drive this heterogeneity have only begun to be explored in detail. Recent developments in molecular sequencing technologies, geostatistics, and other analytical approaches have resulted in an exponential increase in the number of studies published on the topic of microbial biogeography. Our understanding of how soil microbial communities are spatially organized across scales ranging from the soil pore to the globe has fundamentally changed over the past few years. It is becoming increasingly evident that soil organisms are not cosmopolitan in their distribution but exhibit meaningful biogeographic patterns. These distribution patterns often appear to be shaped by forces other than those typically observed for plant and animal taxa and likely shape ecosystem function in important ways.

References

Bach, E.M., Williams, R.J., Hargreaves, S.K., Yang, F., Hofmockel, K.S., 2018. Greatest soil microbial diversity found in micro-habitats. Soil Biol. Biochem. 118, 217−226.

Bahram, M., Hildebrand, F., Forslund, S.K., Anderson, J.L., Soudzilovskaia, N.A., Godegom, P.M., et al., 2018. Structure and function of the global topsoil microbiome. Nature 560, 233−237.

Baird, H.P., Janion-Scheepers, C., Stevens, M.I., Leihy, R.I., Chown, S.L., 2019. The ecological biogeography of indigenous and introduced Antarctic springtails. J. Biogeogr. 46, 1959−1973.

Baldrian, P., Kolařík, M., Štursová, M., Kopecký, J., Valášková, V., Větrovský, T., et al., 2012. Active and total microbial communities in forest soil are largely different and highly stratified during decomposition. ISME J. 6, 248–258.

Barberán, A., Ramirez, K.S., Leff, J.W., Bradford, M.A., Wall, D.H., Fierer, N., 2014. Why are some microbes more ubiquitous than others? Predicting the habitat breadth of soil bacteria. Ecol. Lett. 17, 794–802.

Bates, S.T., Nash, T.H., Garcia-Pichel, F., 2012. Patterns of diversity for fungal assemblages of biological soil crusts from the southwestern United States. Mycologia 104, 353–361.

Bates, S.T., Clemente, J.C., Flores, G.E., Walters, W.A., Parfrey, L.W., Knight, R., Fierer, N., 2013. Global biogeography of highly diverse protistan communities in soil. ISME J. 7, 652–659.

Belnap, J., Lange, O.L. (Eds.), 2001. Biological Soil Crusts: Structure, Function and Management. Ecological Studies. Springer-Verlag, Berlin.

Briar, S.S., Fonte, S.J., Park, I., Six, J., Scow, K., Ferris, H., 2011. The distribution of nematodes and soil microbial communities across soil aggregate fractions and farm management systems. Soil Biol. Biochem. 43, 905–914.

Carteron, A., Beigas, M., Joly, S., Turner, B.L., Laliberté, E., 2021. Temperate forests dominated by arbuscular or ectomycorrhizal fungi are characterized by strong shifts from saprotrophic to mycorrhizal fungi with increasing soil depth. Microb. Ecol. 82, 377–390. https://doi.org/10.1007/s00248-020-01540-7.

Chaudhary, V.B., Nolimal, S., Sosa-Hernández, M.A., Egan, C., Kastens, J., 2020. Trait-based aerial dispersal of arbuscular mycorrhizal fungi. New Phytol. 228, 238–252.

Delgado-Baquerizo, M., Reith, F., Dennis, P.G., Hamonts, K., Powell, J.R., Young, A., et al., 2018. Ecological drivers of soil microbial diversity and soil biological networks in the southern hemisphere. Ecol. 99, 583–596.

De Wit, R., Bouvier, T., 2006. *Everything is everywhere*, but, *the environment selects*'; what did Baas Becking and Beijerinck really say? Environ. Microbiol. 8, 755–758.

Egerton-Warburton, L.M., Graham, R.C., Hubbert, K.R., 2003. Spatial variability in mycorrhizal hyphae and nutrient and water availability in a soil-weathered bedrock profile. Plant Soil 249, 331–342.

Eilers, K.G., Debenport, S., Anderson, S., Fierer, N., 2012. Digging deeper to find unique microbial communities: the strong effect of depth on the structure of bacterial and archaeal communities in soil. Soil Biol. Biochem. 50, 58–65.

Eldridge, D.J., Reed, S., Travers, S.K., Bowker, M.A., Maestre, F.T., Ding, J.Y., et al., 2020. The pervasive and multifaceted influence of biocrusts on water in the world's drylands. Glob. Change Biol. 26, 6003–6014.

Engel, A.S., 2019. Microbes. In: White, W.B., Culver, D.C., Pipan, T. (Eds.), Encyclopedia of Caves, 3rd ed. Academic Press, Amsterdam, pp. 691–698.

Engel, A.S., Randall, K.W., 2011. Experimental evidence for microbially mediated carbonate dissolution from the saline water zone of the Edwards Aquifer, Central Texas. Geomicrobiol. J. 28, 313–327.

Erktan, A., Or, D., Scheu, S., 2020. The physical structure of soil: determinant and consequence of trophic interactions. Soil Biol. Biochem. 148, 107876.

Fierer, N., Schimel, J.P., Holden, P.A., 2003. Variations in microbial community composition through two soil depth profiles. Soil Biol. Biochem. 35, 167–176.

Fierer, N., Strickland, M.S., Liptzin, D., Bradford, M.A., Cleveland, C.C., 2009. Global patterns in belowground communities. Ecol. Lett. 12, 1238–1249.

Guerra, C.A., Heintz-Buschart, A., Sikorski, J., Chatzinotas, A., Guerrero-Ramírez, N., Cesarz, S., et al., 2020. Blind spots in global soil biodiversity and ecosystem function research. Nat. Commun. 11, 3870.

Hoehlert, T.M., Jorgensen, B.B., 2013. Microbial life under extreme energy limitation. Nat. Rev. Microbiol. 11, 83–94.

Hol, F.J.H., Rotem, O., Jurkevitch, E., Dekker, C., Koster, D.A., 2016. Bacterial predator–prey dynamics in microscale patchy landscapes. Proc. R. Soc. B 283, 20152154.

Karimi, B., Terrat, S., Dequiedt, S., Saby, N.P.A., Horrigue, W., Lelièvre, M., 2018. Biogeography of soil bacteria and archaea across France. Sci. Adv. 4 eaat1808.

Lauber, C.L., Hamady, M., Knight, R., Fierer, N., 2009. Pyrosequencing-based assessment of soil pH as a predictor of soil bacterial community structure at the continental scale. Appl. Environ. Microbiol. 75, 5111–5120.

Petrenko, M., Friedman, S.P., Fluss, R., Pasternak, Z., Huppert, A., Jurkevitch, E., 2020. Spatial heterogeneity stabilizes predator-prey interactions at the microscale while patch connectivity controls their outcome. Environ. Microbiol. 22, 694–704.

Ponge, J., 2000. Vertical distribution of collembola (Hexapdoa) and their food resources in organic horizons of beech forests. Biol. Fert. Soils 32, 508–522.

Robeson, M.S., King, A.J., Freeman, K.R., Birky, C.W., Martin, A.P., Schmidt, S.K., 2011. Soil rotifer communities are extremely diverse globally but spatially autocorrelated locally. Proc. Natl. Acad. Sci. USA 108, 4406–4410.

Robin, A., Pradier, C., Sanguin, H., Mahé, F., Lambais, G.R., de Araujo Pereira, A.P., et al., 2019. How deep can ectomycorrhizas go? A case study on *Pisolithus* down to 4 meters in a Brazilian eucalypt plantation. Mycorrhiza 29, 637–648.

Ruamps, L.S., Nunan, N., Chenu, C., 2011. Microbial biogeography at the soil pore scale. Soil Biol. Biochem. 43, 280–286.

Seaton, F.M., George, P.B.L., Lebron, I., Jones, D.L., Creer, S., Robinson, D.A., 2020. Soil textural heterogeneity impacts bacterial but not fungal diversity. Soil Biol. Biochem. 144, 107766.

Smith, D.J., Timonen, H.J., Jaffe, D.A., Griffin, D.W., Birmele, M.N., Perry, K.D., et al., 2013. Intercontinental dispersal of bacteria and archaea in transpacific winds. Appl. Environ. Microbiol. https://doi.org/10.1128/AEM.03029-12.

Soliveres, S., van der Plas, F., Manning, P., Prati, D., Gossner, M.M., Renner, S.C., et al., 2016. Biodiversity at multiple trophic levels is needed for ecosystem multifunctionality. Nature 536, 456–459.

Stone, M.M., DeForest, J.L., Plante, A.F., 2014. Changes in extracellular enzyme activity and microbial community structure with soil depth at the Luquillo Critical Zone Observatory. Soil Biol. Biochem. 75, 237–247.

Tedersoo, L., Bahram, M., Põlme, S., Kõljalg, U., Yorou, N.S., Wijesundera, R., et al., 2014. Global diversity and geography of soil fungi. Science 346 (6213), 1256688.

Treseder, K.K., Maltz, M.R., Hawkins, B.A., Fierer, N., Stajich, J.E., McGuire, K.L., 2014. Evolutionary histories of soil fungi are reflected in their large-scale biogeography. Ecol. Lett. 17, 1086–1093.

Van den Hoogen, J., Geisen, S., Routh, D., Ferris, H., Traunspurger, W., Wardle, D.A., et al., 2019. Soil nematode abundance and functional group composition at a global scale. Nature 572, 194–198.

Chapter 9

Biotic metabolism in soil

Alain F. Plante*, Maura Slocum*, Kevin Geyer[†], and William B. McGill[‡]
*University of Pennsylvania, Philadelphia, PA, USA; [†]Young Harris College, Young Harris, GA, USA; [‡]University of Northern British Columbia, Prince George, BC, Canada

Chapter outline

9.1 Introduction

The soil habitat (Chapter 2) is a highly complex environment at the intersection between the atmosphere, hydrosphere, lithosphere, and biosphere, with variable properties across space and time. This habitat variability generates a large diversity of selective pressures and available resources. Coupled with the evolutionary and metabolic flexibility of the soil biota, the result is the large phylogenetic and metabolic diversity found in soil biotic communities. Soil organisms can be classified based on phylogenetic markers in their genes (Chapters 3 and 4), dividing them into archaea, bacteria, and eukarya. Soil microorganisms comprise the two former groups, while fungi, protists, and the higher organisms comprising the soil fauna are in the eukarya group. Recent advances in high-throughput molecular tools have revealed soil to be the most taxonomically diverse habitat on Earth, with best estimates suggesting that there may be 10,000 species of microorganisms per cm^3 of soil (Fierer and Lennon, 2011). A different way to organize the soil biota is based on their metabolic capacities, which are directly linked to soil ecosystem processes. In this aspect, soils are also one of the most metabolically diverse habitats on Earth. There is some overlap between taxonomic/phylogenetic and metabolic classification systems. For example, eukaryotes have two basic metabolic strategies: autotrophy used by plants and aerobic heterotrophy used by fungi and the soil fauna. Archaea and bacteria exhibit these and many other unique metabolic strategies not found in eukaryotes. Groups of soil organisms defined by metabolic strategies or ecosystem

Soil Microbiology, Ecology, and Biochemistry. https://doi.org/10.1016/B978-0-12-822941-5.00009-0

functionality may include organisms that are only distantly related by phylogenetics, thus making the relationship between structure (i.e., phylogenetic classification) and function (i.e., metabolic classification) challenging to delineate. The metabolic and physiologic diversity of the soil microbial population is overlain by the diversity of soil fauna, which is largely limited metabolically to aerobic heterotrophy. This layering creates a hierarchical way of examining soil biotic physiology in terms of the metabolic products (e.g., CO_2) at the organismal, population, or community levels, with implications for the efficiency by which carbon (C) and nutrients are assimilated or respired.

It is important to remember what all soil organisms need to survive and reproduce. They require simple compounds to build and repair their cells, a source of energy from which adenosine triphosphate (ATP) can be synthesized, and a means of transferring electrons from one compound to another. The acquisition of resources and transfers of energy performed by soil microorganisms is subject to limitations imposed by chemical thermodynamics, soil environmental conditions, and the soil faunal population that feeds on them. This chapter examines the metabolic diversity of the soil biota and provides examples of several metabolic strategies relevant to soil function and biogeochemistry, concluding with an examination of how this diversity is expressed through the physiology of C utilization. Additional information about the basics of bioenergetics (including references to texts and links to online videos) and a glossary are provided in the online supplemental.

9.2 Foundations of soil biotic metabolism

9.2.1 Stoichiometry

The elemental composition or stoichiometry of soil microorganisms is relatively constrained though substantial variability exists, particularly among fungal tissues. The C:N:P ratio for soil microbes has been found to range between 60:7:1 (Cleveland and Liptzin, 2007) and 42:6:1 (Xu et al., 2013). The relatively consistent stoichiometry of the soil microbial biomass suggests that physiological and metabolic requirements drive the elemental composition of microorganisms. Conversely, the C to nutrient ratios of the substrates on which soil microorganisms grow is much larger and more variable. For instance, Cleveland and Liptzin (2007) reported a mean C:N:P ratio of 186:13:1 for total soil, while plant litter C:N ratios can vary from 30 to several hundreds, while C:P ratios can vary from 300 to several thousands (Manzoni et al., 2010). This dissimilarity in elemental ratios between microbes and their substrates has profound implications for understanding controls on microbial metabolism (Manzoni et al., 2010; Sinsabaugh et al., 2008). Microorganisms maintain their biomass elemental ratios in response to this stoichiometric imbalance by adjusting rates of element acquisition processes to assimilate missing elements, and by adjusting element partitioning and turnover times of elements in the microbial biomass (Spohn, 2016). At an ecosystem scale, these microbial processes influence rates of organic matter breakdown, nitrogen (N) mineralization/immobilization, microbial growth rates and C use efficiency, microbial community structure, and ultimately, the rate of energy and matter movement within soil food webs.

9.2.2 Redox reactions

As stated previously, all organisms require a source of energy to generate ATP. To understand the vast diversity of microbial energy-acquisition strategies, it is important to start with several basic concepts of thermodynamics, the branch of chemistry that predicts which reactions are energetically favorable and which are not. Metabolism is, in essence, a suite of oxidation-reduction (redox) reactions. Oxidation refers to a loss of electrons by a substance that leads to a new oxidized substance of lower potential energy. Conversely, reduction refers to the gain of electrons by a substance with greater potential energy. Soil microorganisms take advantage of the energy released from this transfer and use it in metabolic processes. A key concept is that oxidation and reduction reactions are always coupled; the oxidation of one compound necessarily leads to the reduction of another. Redox reactions lead to changes in the valence of elements and their oxidation state. The oxidation state of elements in a given reaction determines what is being oxidized and what is being reduced. Several general rules can be applied for determining the oxidation state of an element:

1. Elements in the free state have an oxidation number of zero.
2. Elements present as monoatomic ions have an oxidation state equal to the ionic charge (e.g., ferrous iron, or Fe^{2+}, has an oxidation state of +2).
3. In combination with other elements, hydrogen typically has an oxidation state of +1 and oxygen typically has an oxidation state of −2. An exception to this rule is that hydrides for hydrogen and peroxides for oxygen both have oxidation states of −1.
4. The sum of oxidation states in the atoms comprising a compound must equal the overall net charge of the compound.

The first step in nitrification (a set of oxidation reactions performed by soil microorganisms that transform ammonium to nitrate) involves the oxidation of ammonium to nitrite. The oxidation states of the constituent elements in this reaction are written above the elements in the following reaction:

$$\overset{-3\ +1}{NH_4^+} + 1.5\overset{0}{O_2} \rightarrow \overset{+3\ -2}{NO_2^-} + \overset{+1}{H^+} + \overset{+1\ -2}{H_2O}. \quad \text{(Eq. 9.1)}$$

Notice that the sum of the oxidation states of the elements in each compound equals the overall net charge. For example, a net charge of +1 can be determined for ammonium as follows:

$$(-3[N]) + (+1[H] \times 4) = +1 \quad \text{(Eq. 9.2)}$$

What is the likelihood of a redox reaction occurring in nature? Given that all redox reactions are composed of an oxidation and a reduction reaction, overall reactions can be written as two separate half reactions. The relative tendency of each half reaction to occur allows us to determine the likelihood of the overall reaction to proceed in a particular direction. For example, the reduction of CO_2 to methane, a common transformation in anaerobic soil environments, is accomplished by a class of anaerobic archaea, known as methanogens, and is composed of two half reactions:

$$\text{an oxidation}: 4H_2 \rightarrow 8H^+ + 8e^- \quad \text{(Eq. 9.3)}$$

$$\text{and a reduction}: CO_2 + 8H^+ + 8e^- \rightarrow CH_4 + 2H_2O \quad \text{(Eq. 9.4)}$$

$$\text{resulting in the overall reaction}: CO_2 + 4H_2 \rightarrow CH_4 + 2H_2O \quad \text{(Eq. 9.5)}$$

The tendency of the reaction in Eq. 9.5 to proceed as written will be determined by the tendencies of the reactions in Eqs. 9.3 and Eq. 9.4 to proceed in the direction indicated, from left (CO_2) to right (CH_4), as in the example above. This can be determined by measuring the electrical potential (E_o) of each half reaction relative to a reference substance (H_2) under conditions of standard temperature, acidity, and pressure (25 °C, pH = 0, 1 atm). This potential is then corrected to pH = 7 to reflect the fact that most metabolic reactions occur at near-neutral conditions in the intracellular environment. To facilitate comparison among reactions, all half reactions are written as reductions, and the sign of the corresponding electrical potential is adjusted accordingly. These standard values are known as reduction potentials (E_o). Low (more negative) reduction potentials indicate a low tendency for the reaction to occur, or a high tendency for the opposite oxidation reaction to occur. Summing the reduction potentials for two half reactions allows us to determine which direction a redox reaction tends to proceed. For example, the formation of water from hydrogen and oxygen can be written as two half reactions with the following reduction potentials:

$$2H^+ + 2e^- \rightarrow H_2 \; E_o = -0.42 \text{ V} \quad \text{(Eq. 9.6)}$$

$$\tfrac{1}{2} O_2 + 2H^+ + 2e^- \rightarrow H_2O \; E_o = +0.82 \text{ V} \quad \text{(Eq. 9.7)}$$

Notice how both half reactions are written as reductions, even though the reaction in Eq. 9.6 would need to be written in reverse for water to form such that diatomic hydrogen would be oxidized. The reaction in Eq. 9.6 has a more negative reduction potential than the reaction in Eq. 9.7, indicating that, under standard conditions, the reaction in Eq. 9.7 is more likely to proceed as written. For the reaction to proceed the other way, energy would have to be supplied from an external source. Reactions which require energy inputs to proceed are not as favorable to microorganisms because energy is a precious metabolic commodity. Microorganisms only exert energy to shift a naturally occurring reaction if it is necessary for metabolic function, usually in extreme environments.

9.2.3 Energetics

All compounds contain intrinsic energy, known as free energy or Gibbs free energy (G). An important property of a metabolic reaction is the corresponding change in free energy (ΔG) that the reaction produces. The ΔG value indicates the amount of energy released or required by a given reaction. A negative ΔG value indicates that a reaction is exergonic, or energy-releasing, and will thus proceed spontaneously, while a positive ΔG value indicates an endergonic reaction that requires external energy inputs to proceed. Equilibrium conditions are established at $\Delta G = 0$. Soil organisms obtain cellular energy through catabolism. Catabolic reactions serve as a destructive metabolism, breaking down complex molecules and releasing energy (i.e., exergonic) that microbes utilize for cellular functions. Anabolic reactions serve as a constructive metabolism, building complex molecules from smaller molecules, simultaneously storing and consuming energy (i.e., endergonic) (Fig. 9.1).

The following equation can be used to calculate the ΔG at standard conditions (ΔG^0) for a given redox reaction:

$$\Delta G^0 = -n\text{F}\Delta E_o \quad \text{(Eq. 9.8)}$$

where n is the number of electrons transferred, F is Faraday's constant (96.5 kJ V^{-1} mol^{-1} or 23.1 kcal V^{-1} mol^{-1}), and ΔE_o is the difference between the reduction potentials for two half reactions comprising a complete redox reaction when both half reactions are oriented to reflect the natural direction of the full reaction. For example, the first step in denitrification involves the reduction of nitrate to nitrite:

$$NO_3^- + 2H^+ + 2e^- \rightarrow NO_2^- + H_2O \quad \text{(Eq. 9.9)}$$

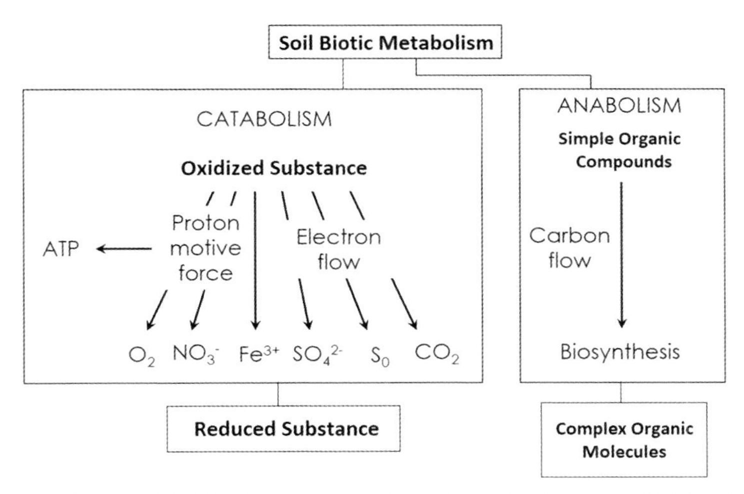

FIGURE 9.1 Biotic metabolism in soil is driven by two groups of reactions: (a) catabolism, which breaks down large molecules into smaller compounds using a set of redox reactions, terminal electron acceptors, and the proton motive force to store the released energy in ATP; and (b) anabolism, which consumes the energy stored in ATP to build complex molecules and biomass from simpler compounds.

ΔE_o for this reaction is the difference between the two reduction potentials for the half reactions $NO_3^- \rightarrow NO_2^-$ (+0.43V) and $2H^+ \rightarrow H_2O$ (-0.37V):

$$\Delta E_o = +0.43 - (-0.37) = 0.8\ V \quad \text{(Eq. 9.10)}$$

$$\Delta G^0 = -(1)(96.5)(0.8) = -77.2\ \text{kJ mol}^{-1} \text{ of } NO_2 \text{ produced} \quad \text{(Eq. 9.11)}$$

Eq. 9.9 shows that redox pairs with higher reduction potentials will produce more negative ΔG^0 values. The more negative the ΔG^0 for a reaction, the more energy is released that can then be used to generate ATP and other cellular energy-carrying molecules. In other words, the higher the ΔE_o value for a redox pair, the more energetically favorable that redox reaction is.

Fig. 9.2 depicts the relationship between oxidized and reduced substrates as a vertically arranged hierarchy of oxidation-reduction half reactions. Compounds on the left half of the diagram are in an oxidized state and therefore accept electrons, while compounds on the right half of the diagram are in a reduced state and donate electrons. Redox conditions are highly oxidizing at the top of the hierarchy and reduce more progressively, moving down the hierarchy. Reduction potentials for the various oxidant-reductant couples are listed on the right side of the figure. Thermodynamically favorable reactions are determined by linking pairs of half reactions such that the lower reaction in the pair proceeds leftward, therefore producing electrons, and the upper reaction proceeds rightward, accepting electrons. The order of redox pairs listed in Fig. 9.2 reflects a hierarchical order of energy yield, which in turn dictates the metabolic strategies of soil microorganisms. Key catabolic reactions in microbial metabolism that drive the soil C cycle pair the oxidation of organic C in the lower right section of the figure to a series of terminal electron acceptors in the upper left. That a large range of redox couples are employed by soil microorganisms underscores the complexity and variability of the soil environment.

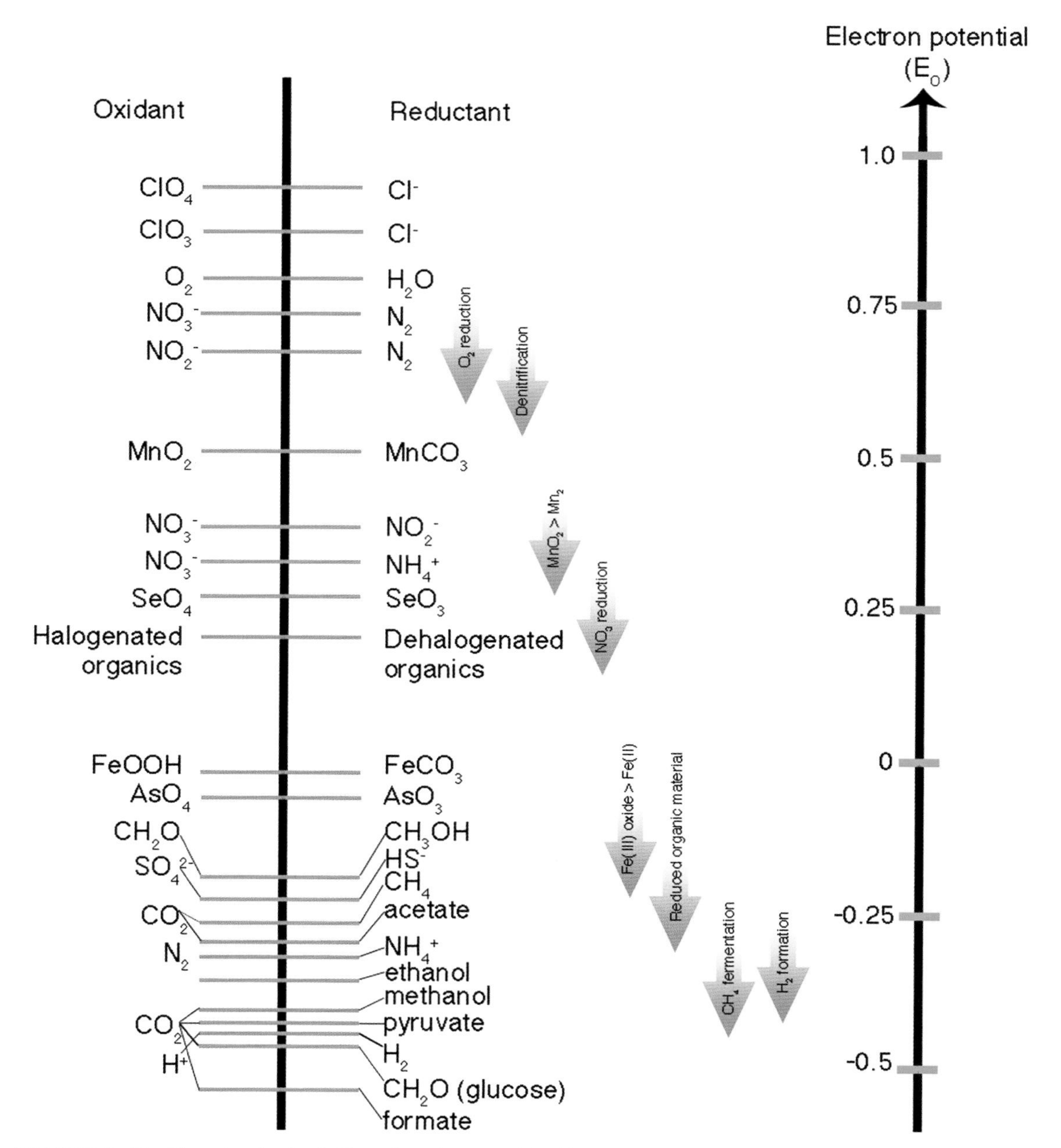

FIGURE 9.2 The hierarchy of redox half reactions that control the biogeochemical reactions carried out by soil microorganisms. *(Reproduced with permission from Madsen, 2015.)*

9.2.4 Role of enzymes in metabolism

Many reactions in nature do not proceed spontaneously at rates that are compatible with cellular life. Some reactions occur too quickly, releasing energy that cannot be harvested for cellular processes. Other reactions, even reactions with very negative ΔG^0 values, may occur too slowly in nature to be of any use to organisms. Reactions often occur slowly because the existing chemical bonds require an initial energy input to be broken. This energy input is called the activation energy, or the energy input required for a reaction to proceed spontaneously. Thus, for a reaction to proceed spontaneously, an external energy source is often required or the activation energy must be reduced.

A catalyst is a substance that promotes a reaction by reducing the activation energy required for that reaction to proceed, without being altered in the process. In biological systems, enzymes are specialized proteins that act as catalysts for a range of catabolic and anabolic reactions, thus allowing these reactions to occur at biologically useful rates. Enzymes function by temporarily binding reactants at an "active site" where they are physically oriented in a manner that facilitates the formation of the desired product. The result of enzymatic catalysis is that the metabolic reactions of life can proceed at biologically controlled rates with minimum energy inputs (i.e., reduced activation energies). Fig. 9.3 illustrates this by depicting the activation energy required for an enzyme-catalyzed versus a noncatalyzed reaction.

While most enzymes in multicellular organisms function intracellularly, it is often advantageous for soil microorganisms to release enzymes into the environment. Such enzymes are known as extracellular enzymes (or exoenzymes). In soils, extracellular enzymes play a key role in decomposing large, complex organic polymers into oligomeric and monomeric substances that are small enough to be transported inside the cell (Sinsabaugh and Follstad Shah, 2012). Extracellular enzymes can be broadly divided into two classes: hydrolases and oxidases. Hydrolases are substrate-specific; their structure enables them to catalyze reactions that cleave specific bonds (such as C-O and C-N) in organic matter. Oxidases use either oxygen (oxygenases) or hydrogen peroxide (peroxidases) to oxidize a broad suite of molecules that share

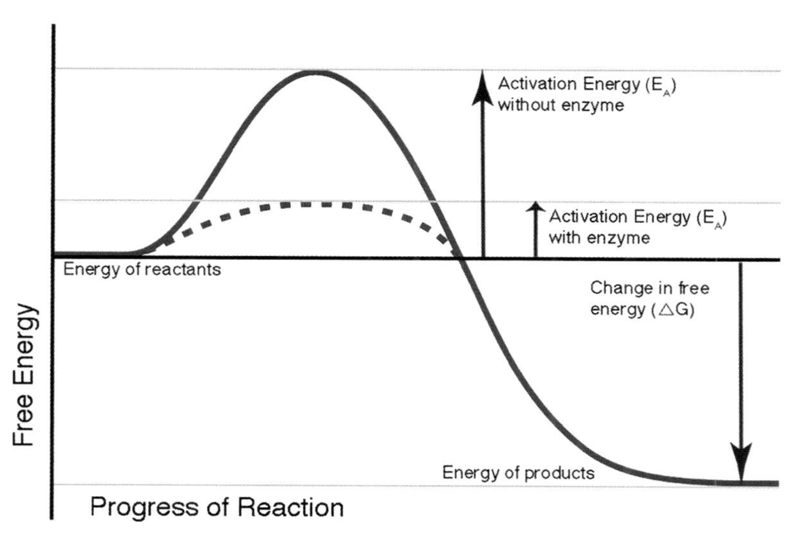

FIGURE 9.3 Conceptual illustration of the role of enzymes in mediating the chemical thermodynamics of reactions.

TABLE 9.1 Commonly measured extracellular enzymes, their functions in soils, and classification.

Enzyme	Enzyme function	Enzyme class
Acid phosphatase	Mineralizes organic P into phosphate by hydrolyzing phosphoric (mono) ester bonds under acidic conditions.	Hydrolase
α-glucosidase	Principally a starch-degrading enzyme that catalyzes the hydrolysis of terminal, non-reducing 1,4-linked α -D-glucose residues, releasing α-D-glucose.	Hydrolase
β-glucosidase	Catalyzes the hydrolysis of terminal 1,4 linked β-D-glucose residues from β-D-glucosides, including short-chain cellulose oligomers.	Hydrolase
β-xylosidase	Degrades xylooligomers (short xylan chains) into xylose.	Hydrolase
Cellobiohydrolase	Catalyzes the hydrolysis of 1,4-β-D-glucosidic linkages in cellulose and cellotetraose, releasing cellobiose.	Hydrolase
N-acetyl glucosaminidase	Catalyzes the hydrolysis of terminal 1,4 linked N-acetyl-beta-D-glucosaminide residues in chitooligosaccharides (chitin-derived oligomers).	Hydrolase
Leucine aminopeptidase	Catalyzes the hydrolysis of leucine and other amino acid residues from the N-terminus of peptides. Amino acid amides and methyl esters are also readily hydrolyzed by this enzyme.	Hydrolase
Urease	Catalyzes the hydrolysis of urea into ammonia and carbon dioxide.	Hydrolase
Phenol oxidase	Also known as polyphenol oxidase or laccase. Oxidizes benzenediols to semiquinones with O_2.	Oxidase
Peroxidase	Catalyzes oxidation reactions via the reduction of H_2O_2. It is considered to be used by soil microorganisms as a lignolytic enzyme because it can degrade molecules without a precisely repeated structure.	Oxidase

(Modified with permission from German et al., 2011)

similar bonds (such as C-C or C-O). Table 9.1 lists commonly measured extracellular enzymes in soils and their functions. These enzymes have been demonstrated to play an important role in the cycling of at least one of the three most biotically important nutrients: C, N, and P.

9.3 Metabolic classification of soil organisms

Classification of soil organisms in terms of how they acquire energy, the raw materials to build biomass (particularly C), and terminal electron acceptors for redox reactions provide a means to divide virtually all soil organisms (Fig. 9.4). Energy sources are divided into two main categories: chemicals and light. Organisms that derive energy by converting light energy into chemical energy are called phototrophs, while organisms that derive energy from the breakdown of energy-rich chemicals are called chemotrophs. The chemical sources of energy are further broken down into organic sources (e.g., compounds

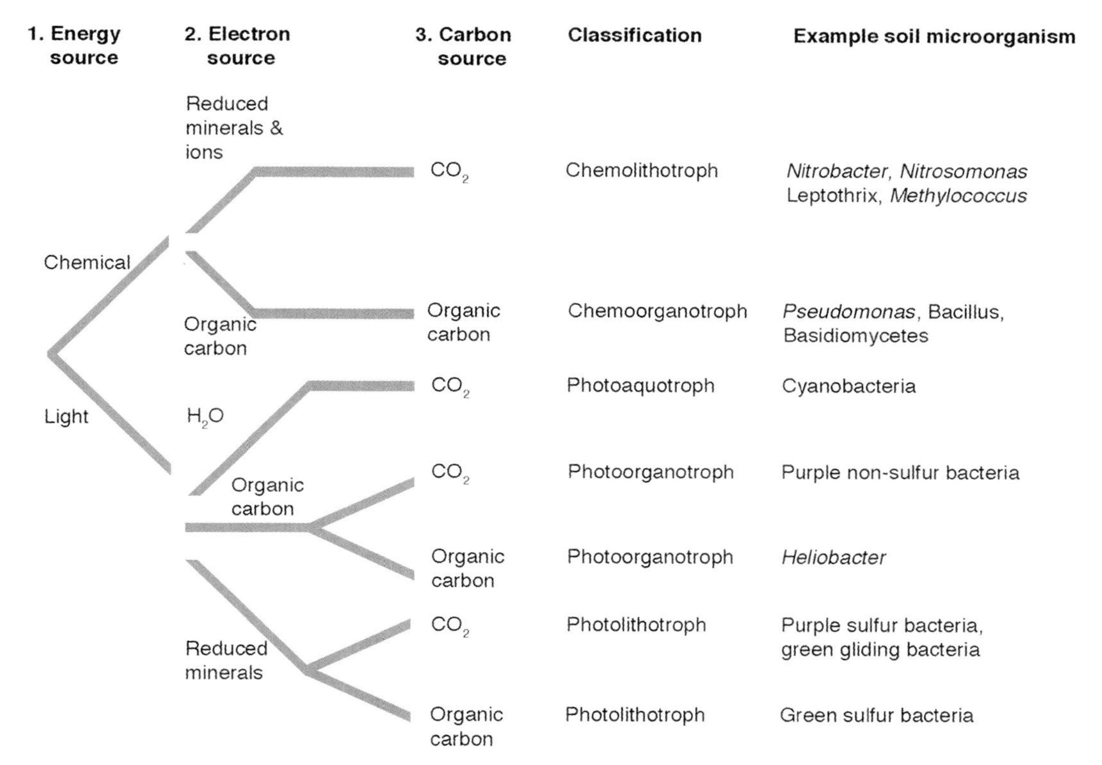

FIGURE 9.4 Classification of soil microorganisms by metabolic physiology based on sources of energy, carbon, and terminal electron acceptor (i.e., reducing equivalents).

containing C-H bonds) and inorganic sources (e.g., ammonia, hydrogen gas, or elemental sulfur). The C sources used by soil microbes as building blocks for biosynthesis are divided into two categories: organic (or fixed) compounds and inorganic (or gaseous) compounds. Organisms that use an organic form of C are referred to as heterotrophs, while organisms that fix inorganic C (i.e., CO_2) are referred to as autotrophs. Finally, the electron source can be either organic (organotrophs), inorganic (lithotrophs), or water (aquatrophs).

According to this classification, plants and algae could be called photosynthetic autotrophs or photoaquatrophs. Soil fungi or other soil fauna that consume plant-derived organic substrates could be called chemosynthetic organoheterotrophs or heterotrophic chemoorganotrophs. Microorganisms that oxidize inorganic compounds for energy and use CO_2 as their C source, such as nitrifying bacteria, could be called chemosynthetic lithoautotrophs. These metabolic designations to describe the metabolic strategy of an organism are often abbreviated to "autotrophs," "heterotrophs," and "lithotrophs" because certain combinations of metabolic strategies typically occur together. The boundaries between some of these classifications can be unclear because some microorganisms may change their metabolic strategies in response to changes in environmental conditions. Some soil microorganisms can switch between chemoorganotrophy and chemolithotrophy depending on substrate availability.

A key distinction between autotrophs and heterotrophs is that autotrophs make their own biosynthetic molecules directly via C fixation, while heterotrophs must degrade pre existing organic compounds via catabolism to synthesize new biomolecules. In soils, the use of reduced C as a C, energy, or electron source, or some combination of the three, drives the soil C cycle.

9.4 Cellular energy transformations

Recall from above that microbial metabolism consists of two complementary pathways (Fig. 9.1). The ultimate result of catabolism is the production of the energy storage molecule ATP. The energy released from the reaction ATP $\rightarrow$ ADP + P_i is used to fuel many intracellular reactions. Formation of ATP is accomplished by two pathways: substrate-level phosphorylation and electron transport phosphorylation, ATP, which is also known as oxidative phosphorylation.

Substrate-level phosphorylation can be a significant source of ATP in anaerobic organisms but is generally insignificant in aerobic and facultatively anaerobic organisms when compared with the ATP production from oxidative phosphorylation. Substrate-level phosphorylation involves the direct production of ATP during the enzymatic oxidation of another substance. Hydrolysis of the phosphoryl group of intermediates produced after the oxidation of an organic molecule releases enough energy to form ATP. Yields of ATP are low, and consequently biomass production is low. Organisms that rely on this pathway tend to grow slowly. The production of fermentation intermediates is high, and they frequently become available as substrates for other microbes.

Oxidative phosphorylation is a more generalized means of ATP production because electrons from any number of oxidation reactions can be processed through the same electron transport chain. In the initial energy-yielding oxidation, performed by chemoorganotrophs (i.e., heterotrophs), reduced organic C serves as the primary electron donor. These electrons are transferred to the electron carriers NAD^+ (nicotinamide-adenine-dinucleotide) and FAD (flavin adenine dinucleotide), reducing these molecules to NADH and FADH. NADH and FADH then shuttle electrons to an electron transport chain composed of a series of membrane-localized electron acceptors, which ultimately pass their electrons to a terminal acceptor. In aerobic respiration, this terminal acceptor is O_2 because oxygen has a high reduction potential, but any electron acceptor from Fig. 9.2 can be used, provided its reduction potential is greater than that of the original, oxidized substance. Reduction of some of the membrane-localized electron acceptors draws in an H^+ ion from the cytoplasm. As it flips to a reduced state, the H^+ is passed through the membrane, thereby increasing the H^+ concentration within the cell. As this process continues, a gradient of H^+ develops across the membrane. A proton-motive force (ΔP) is consequently developed to drive H^+ back across the membrane, and a portion of the energy released from this process is used to produce ATP from ADP.

Four variations for generating ATP and reducing equivalents in autotrophic and aerobic chemotrophic organisms are illustrated in Fig. 9.5. In this figure, electron transport through the respiratory chain to B_{ox}, generates the ΔP (proton-motive force) needed to both synthesize ATP and to drive reverse electron transport (e.g., chemolithotrophs, such as *Nitrobacter* [Proteobacteria]). Reducing equivalents (NADH) are produced by oxidation of substrates within the cytoplasm. In Fig. 9.5b, reverse electron transport is used to generate NADH, which is used to reduce CO_2 in the Calvin cycle.

Photosynthesis is the ultimate source of energy to allow soil organisms to work. It requires no oxygen (although it may partially cycle O_2) and may or may not generate O_2. If photosynthesis uses H_2O as an

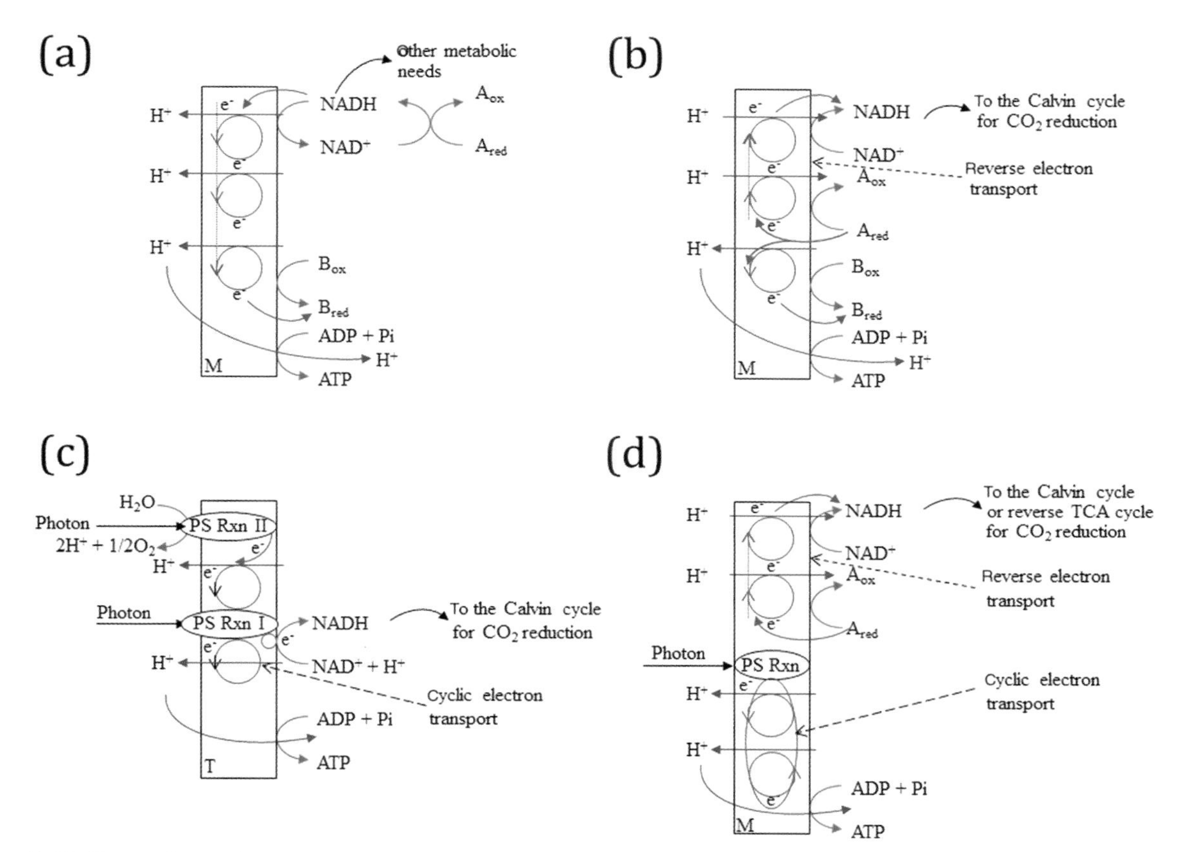

FIGURE 9.5 Generation of ATP and reducing equivalents (NADH) in: (a) chemoorganotrophs where substrates are oxidized within the cytoplasm (e.g., *Pseudomonas* spp.), (b) chemo-lithotrophs in which substrates are oxidized at the membrane (e.g., *Nitrobacter* spp.), (c) oxygenic phototrophic cyanobacteria, or (d) anoxygenic phototrophic bacteria, such as Rhodospirillaceae (purple non-sulfur bacteria), Chromatiaceae (purple sulfur bacteria), Chlorobiaceae (green sulfur bacteria), and Chloroflexaceae (green gliding bacteria) when oxidizing substrates, such as H_2. Small circles within the plasma membrane (M) or thylakoid membrane (T; an intracytoplasmic membrane) represent redox complexes. PS Rxn is the photosynthesis reaction site. Red arrows indicate transformations of energy carriers occurring at the membrane. Blue arrows indicate the motion of protons, while green arrows indicate the motion of electrons.

electron donor, as it does for photoaquatrophs, then it generates O_2 and is called oxygenic photosynthesis (Table 9.2, Reaction 1). In oxygenic photosynthesis, ATP is synthesized during e^- transfer from H_2O to $NADP^+$. The reducing equivalents ($NADPH + H^+$) are used in the Calvin cycle for reduction of C. In Fig. 9.5c, the electrons released from the splitting of water are passed linearly to NAD^+ along a series of redox couples, during which ΔP is generated for ATP synthesis. Two photosynthesis reaction (PS Rxn) sites are involved. This linear transfer of electrons may not generate enough ΔP to produce enough ATP for CO_2 reduction. Cyclic electron transfer can make up for the shortfall. This is the process used by plants and algae.

TABLE 9.2 Anoxygenic and oxygenic photosynthesis.

Oxygenic Photosynthesis, e.g., *Panicum* (grass) *Nostoc* (cyanobacteria)			
$h\nu + 2\ HOH + 3\ (ADP + P_i)$	→	$4\ e^- + 4\ H^+ + O_2 + 3\ ATP$	
$3\ ATP + CO_2 + 4\ e^- + 4\ H^+$	→	$CH_2O + HOH + 3\ (ADP + P_i)$	
Overall: $h\nu + CO_2 + HOH$	→	$CH_2O + O_2$	[1]
Anoxygenic Photosynthesis, e.g., *Chromatium* (purple S bacteria)			
$2\ H_2S$	→	$2\ S^0 + 4\ e^- + 4\ H^+$	
$h\nu$ + Pigment	→	[Pigment$^+$ + e^-][a]	
$3\ (ADP + P_i)$ + [Pigment$^+$ + e^-][a]	→	3 ATP + Pigment	
$3\ ATP + CO_2 + 4\ e^- + 4\ H^+$	→	$CH_2O + HOH + 3\ (ADP + P_i)$	
Overall: $h\nu + CO_2 + 2\ H_2S$	→	$CH_2O + 2\ S^0 + HOH$	[2]
Or Overall: $h\nu + 3\ CO_2 + 2\ S^0 + 5\ HOH$	→	$2\ SO_4^{-2} + 3\ CH_2O + 4\ H^+$	[3]

[a]*Complex including several components of the photosynthetic system and the respiratory chain.*
(Reproduced from McGill, 1996, with permission from Sociedade Brasileira de Ciência do Solo [SBCS] and Sociedade Latino-Americana de Ciência do Solo [SLCS].)

In evolutionary terms, photosynthesis predates current levels of oxygen in the atmosphere and thus has been proposed to be the primary source of oxygen for atmospheric accumulation over geologic time during the Great Oxidation event 2–3 billion years ago. In anoxygenic photosynthesis, photolithotrophs use photosynthesis with a reduced mineral, such as H_2S as electron donor to produce S^0 (Table 9.2, Reaction 2) or SO_4^{-2} (Table 9.2, Reaction 3). All anoxygenic photosynthetic organisms represented in Fig. 9.5d can use H_2 as H-donor. In addition, the purple non-sulfur bacteria Rhodospirillaceae (Proteobacteria) and the green sulfur bacteria Chloroflexaceae (Chloroflexi) can use organic substrates. The purple-sulfur bacteria Chromatiaceae (Proteobacteria) can use H_2S and organic substrates, and the green sulfur bacteria Chlorobiaceae (Chlorobi) can use H_2S, but not organic substrates.

9.5 Examples of biotic metabolism in soil

The biogeochemical fluxes and transformations of C, N, and other nutrient elements observed at the soil aggregate, pedon, and landscape scales can be reduced to the various metabolic strategies used by soil organisms to acquire cellular building blocks, energy, and reducing equivalents. We highlight below some examples of soil biotic transformations that illustrate important metabolic strategies.

9.5.1 Organotrophy

The physiologic variation among soil fauna species is vast in terms of both body shape and size. The large number of species results in large, complicated food webs as larger predators consume smaller species, which in turn feed on primary producers and consumers. Although food webs connecting soil organisms are complex, the metabolic approach of soil fauna is straightforward. As organoheterotrophs, all fauna consumes organic C by feeding on other organisms or organic residues and uses the consumed C as the primary electron donor in their metabolism, releasing CO_2 as a by-product. This common metabolic approach serves an important role in cycling organic matter in the soil environment. Due to their ability to metabolize organic C, soil fauna represents a higher energetic (i.e., trophic) level in the soil food web. The large energy differential in organoheterotrophy is what makes the growth of large, complicated body structures possible. Soil fauna use the energy from oxidizing organic C to grow large body forms and for greater motility. These larger soil organisms play an important role in restructuring the physical soil environment and driving the biogeochemical cycling of nutrients, specifically soil respiration and N mineralization, which play an important role in the global C and N cycles. Despite the potential for soil fauna to increase soil respiration through litter consumption, it is suggested that soil fauna may alternatively suppress microbial respiration in soil through grazing or alterations to leachate chemistry. Increases in stoichiometric C:N ratio of leachates in the presence of soil fauna were observed to create a downward pressure on soil microbial populations and activity, as the leachate N is presumed to be more readily available for assimilation compared to the N derived from decomposing litter (Frouz et al., 2020). Further investigation is needed to better predict and generalize how soil faunal interactions with microbial populations might affect soil respiration.

Soil fauna may dominate in body structure and size diversity, but what soil microorganisms lack there, they make up for in metabolic diversity. Organoheterotrophs can only exist where organic C is available for consumption. A recent global meta-analysis revealed that out of numerous soil properties examined, C-availability was the best predictor of phyla-level bacterial abundances (Fierer et al., 2007). It is therefore not surprising that the most dominant microorganisms in soil are those that use reduced organic compounds as their source of C, energy, and electrons. These heterotrophic chemoorganotrophs are responsible for the decomposition of organic matter, such as plant residues and microbial necromass. For example, lignin degradation is accomplished primarily by fungi in the phyla Basidiomycota and Ascomycota, which can be either saprotrophic (free-living) or root associated (Osono, 2007). Cellulose degradation is primarily accomplished by specialized saprotrophic fungi (e.g., *Trichoderma* [Ascomycota]) and common bacterial genera, including *Bacillus* (Firmicutes), *Pseudomonas* (Proteobacteria), and *Clostridium* (Firmicutes). In its simplest form, the aerobic oxidation of organic matter (frequently represented by $C_6H_{12}O_6$ or CH_2O) to CO_2 is the reverse of the overall oxygenic photosynthesis reaction (Table 9.3). In addition to CO_2, the oxidation of nutrient element-containing organic compounds releases mineral forms of N (NH_3), phosphorous (PO_4^{3-}), and sulfur (SO_4^{2-}), which subsequently become available for plant uptake during primary production or for microbial assimilation. The oxidation of one mole of C in the form of CH_2O generates 4 electrons, whereas the oxidation of one mole of C as L-glutamic acid

TABLE 9.3 Oxidation of reduced C (represented as CH_2O) using O_2 as terminal electron acceptor.

$CH_2O + HOH$	→	$CO_2 + 4\ e^- + 4\ H^+$
$O_2 + 4\ e^- + 4\ H^+$	→	2 HOH + energy
Overall: $CH_2O + O_2$	→	CO_2 + HOH + energy

(Reproduced from McGill, 1996, with permission from Sociedade Brasileira de Ciência do Solo [SBCS] and Sociedade Latino-Americana de Ciência do Solo [SLCS].)

TABLE 9.4 Oxidation of N-containing organic molecules to CO_2 as a way to obtain energy. L-glutamic acid oxidation via α-keto glutaric acid is used as an example.

$2\ (C_5NH_8O_4) + 2\ HOH$	→	$2\ NH_3 + 2\ (C_5H_4O_5) + 6\ H^+ + 6\ e^-$
$2\ (C_5H_4O_5) + 10\ HOH$	→	$10\ CO_2 + 28\ H^+ + 28\ e^-$
$34\ H^+ + 34\ e^- + 8.5\ O_2$	→	$17\ HOH$ + energy
Overall: $2\ (C_5NH_8O_4) + 8.5\ O_2$	→	$2\ NH_3 + 10\ CO_2 + 5\ HOH$ + energy

(Reproduced from McGill, 1996, with permission from Sociedade Brasileira de Ciência do Solo [SBCS] and Sociedade Latino-Americana de Ciência do Solo [SLCS].)

($C_5NH_8O_4$) generates only 3.4 electrons (Table 9.4). The amount of ATP synthesized (and therefore of energy gained) from the oxidation of one mole of C from L-glutamate is therefore less than the oxidation of one mole of C from glucose. N-containing energy sources are therefore less energetically favorable than pure carbohydrates or lipids. When the supply of N exceeds the stoichiometric demands of the microbial biomass, the search for energy through the continued catabolism of N-containing organic molecules is the basis for much of the mineral N supply to plants.

Some soil microorganisms can substitute other oxidized inorganic compounds for oxygen as the terminal electron acceptor during anaerobic respiration. Oxygen is the most energetically favorable electron acceptor, but other common alternative electron acceptors in soil are nitrate, sulfate, and carbon dioxide (Ehrlich, 1993). Even under well-aerated conditions, soils provide microsites within aggregates where O_2 is excluded or in which it is consumed faster than it can diffuse to microorganisms.

Soil microorganisms, almost all of which are bacteria, that can use nitrate (NO_3^-) as an alternative terminal electron acceptor are facultative anaerobes (e.g., *Pseudomonas denitrificans*), meaning they can perform aerobic respiration in the presence of O_2 or switch to nitrate in its absence (Table 9.5, Reaction 9.3). The reduction of NO_3- during oxidation of reduced C is called dissimilatory NO_3- reduction. More specifically, nitrate reduction to N_2 or N_2O is called denitrification, and to NH_3, it is called nitrate

TABLE 9.5 Carbon oxidation using NO_3^- or SO_4^{-2} as terminal electron acceptors.

NO_3^- accepts e^-			
$5\ CH_2O + 5\ HOH$	→	$5\ CO_2 + 20\ H^+ + 20\ e^-$	
$4\ NO_3^- + 20\ H^+ + 20\ e^- + 4\ H^+$	→	$2\ N_2 + 12\ HOH$ + energy	
Overall: $5\ CH_2O + 4\ NO_3^- + 4\ H^+$	→	$5\ CO_2 + 2\ N_2 + 7\ HOH$ + energy	[1]
SO_4^{-2} accepts e^-			
$2\ CH_2O + 2\ HOH$	→	$2\ CO_2 + 8\ H^+ + 8\ e^-$	
$SO_4^{-2} + 8\ H^+ + 8\ e^- + 2\ H^+$	→	$H_2S + 4\ HOH$ + energy	
Overall: $2\ CH_2O + SO_4^{-2} + 2\ H^+$	→	$2\ CO_2 + H_2S + 2\ HOH$ + energy	[2]

(Reproduced from McGill, 1996, with permission from SBCS and SLCS.)

respiration. Dissimilatory nitrate reduction is an important biogeochemical process because: (1) it returns biologically available N into the atmosphere in a process that is effectively the reverse of N fixation, (2) it represents a loss of soil fertility, and (3) it is a source of a potent greenhouse gas (e.g., N_2O).

In an analogous process, a diverse group of soil bacteria and archaea can use sulfate (SO_4^{2-}) as an alternative terminal electron acceptor in a process called dissimilatory SO_4^{2-} reduction (Table 9.5, Reaction 2). These organisms are strict or obligate anaerobes (e.g., *Desulfovibrio desulfuricans*). The free energy change (ΔG) and ATP/2e$^-$ ratio is greater for NO_3^- respiration than for SO_4^{2-} respiration (Gottschalk, 1986). These metabolic distinctions between NO_3^- respiration and dissimilatory SO_4^{2-} reduction suggest that the latter would not occur in a soil well-supplied with NO_3^- or oxygen.

The use of CO_2 as a terminal electron acceptor to produce methane (CH_4) is called methanogenesis. Methanogens, all of which are archaea, are the strictest anaerobes normally found in nature and use a limited array of substrates: $H_2 + CO_2$, formate, methanol, methylamines, and acetate. These substrates are formed during fermentation or converted from fermentation products in other anaerobic systems. Two groups of organisms produce methane, one of which is chemoorganotrophic. These produce CH_4 from substrates such as methanol, acetate, or methylamines, which contain methyl groups. For methanol fermentation to CH_4, the overall reaction is: $4\ CH_3OH \rightarrow 3\ CH_4 + CO_2 + 2\ HOH$. Acetate is the most common and important, and its conversion to methane is written simply as: $C_2H_4O_2 \rightarrow CH_4 + CO_2$. In the presence of SO_4^{2-}, acetate is oxidized to CO_2 rather than being split into CH_4 and CO_2. Here we see the preference of SO_4^{2-} as a terminal electron acceptor over methane fermentation. The other group are strictly chemolithotrophic organisms that grow on H_2 and CO_2. They are fascinating in their ability to produce all their needs for energy and C from H_2 and CO_2 alone.

This sequence of metabolic strategies, using alternative terminal electron acceptors during the oxidation of reduced organic compounds, represents a "redox ladder" where the selection of the terminal electron acceptor is a function of oxygen or redox status and results in decreasing yields of free energy change (Table 9.6; Fig. 9.1).

TABLE 9.6 **The "redox ladder" of half reactions for various terminal electron acceptors and Gibb's free energy change associated with important metabolic reactions carried out by organotrophs during carbon oxidation.**

Process	Half reaction	Heterotrophic reaction	ΔG^0 (kJ eq^{-1})
Aerobic respiration	$^1/_4\ O_2 + H^+ + e^- \rightarrow {}^1/_2\ H_2O$	$CH_2O + O_2 \rightarrow CO_2 + H_2O$	-125
Denitrification	$^1/_5\ NO_3^- + {}^6/_5\ H^+ + e^- \rightarrow {}^1/_{10}\ N_2 + {}^3/_5\ H_2O$	$5CH_2O + 4NO_3^- + 4H^+ \rightarrow 5CO_2 + 2N_2 + 7H_2O$	-119
Iron reduction	$FeOOH + HCO_3^- + 2H^+ + e^- \rightarrow FeCO_3 + 2H_2O$	$CH_2O + 4FeOOH + 8H^+ \rightarrow CO_2 + 4Fe^{2+} + 7H_2O$	-42
Sulfate reduction	$^1/_8\ SO_4^{2-} + {}^9/_8\ H^+ + e^- \rightarrow {}^1/_8\ H_2S + {}^1/_2\ H_2O$	$2CH_2O + SO_4 + 2H^+ \rightarrow 2CO_2 + H_2S + 2H_2O$	-25
Methanogenesis	$^1/_8\ CO_2 + H^+ + e^- \rightarrow {}^1/_8\ CH_4 + {}^1/_4\ H_2O$	$2CH_2O \rightarrow CO_2 + CH_4$	-23

9.5.2 Lithotrophy

Chemolithotrophic organisms are relatively restricted in soils compared to chemoorganotrophs because the oxidation of inorganic compounds for energy typically yields much less energy than the oxidation of reduced C (Table 9.7). Balanced against the low energy yields of the oxidation reactions catalyzed by chemolithotrophs is the ability to use energy sources unavailable to other microorganisms. This has the advantage of both reducing resource competition and allowing chemolithotrophs to grow in a range of environments that would be hostile to other organisms. Typically, chemolithotrophs obtain C from CO_2 and are therefore often called chemoautotrophs or chemolithoautotrophs. Most chemolithotrophs are aerobic; however, the diversity of chemolithotrophic strategies is an area of active research, and new chemolithotrophic processes, such as the anaerobic oxidation of ammonia (ANAMOX), are still being discovered.

Methanotrophs are bacterial and archaeal organisms that gain energy from the oxidation of CH_4 to CO_2. Methanotrophy occurs most frequently in soils that are often submerged or water saturated, where a high level of methanogenesis occurs. Methanotrophs have also been found to oxidize atmospheric methane in upland soils. This restricted but ecologically important microbial process serves to offset some of the methane emissions produced by methanogens in periodically saturated upland and wetland soils.

Nitrifiers are a group of chemolithoautotrophs that use energy from oxidation of ammonium or nitrite to fix CO_2. The process of nitrification is accomplished by two separate groups of bacteria; those that oxidize ammonium to nitrite (e.g., *Nitrosomonas* spp., Table 9.8) and those that convert nitrite to nitrate (e.g., *Nitrobacter* spp., Table 9.9). Nitrifiers are agriculturally and environmentally important because the transformation of ammonium to nitrate is an important step in the soil N cycle that counters the processes of N fixation and denitrification. Some systems have shown an imbalance in these processes (i.e., N saturation in forests), which can lead to soil acidification, nitrate leaching, and ultimately reduced primary production and pollution.

Current research suggests that the ability to oxidize reduced sulfur is evolutionarily ancient, possibly representing the earliest form of self-sustaining metabolism (Ghosh and Dam, 2009). Various species of

TABLE 9.7 Comparison of Gibb's free energy change from glucose oxidation versus other important oxidation reactions carried out by chemolithotrophs.

Reaction	ΔG^0 (kJ mol^{-1})
$CH_2O + O_2 \rightarrow CO_2 + H_2O$	-2870
$CH_4 + 2O_2 \rightarrow CO_2 + 2H_2O$	-871
$S + \frac{1}{2} O_2 + H_2O \rightarrow SO_4^{2-} + 2H^+$	-587
$NH_4^+ + 1\frac{1}{2} O_2 \rightarrow NO_2^- + 2H^+ + H_2O$	-275
$H_2 + \frac{1}{2} O_2 \rightarrow H_2O$	-237
$HS^- + H^+ + \frac{1}{2} O_2 \rightarrow S + H_2O$	-209
$NO_2^- + \frac{1}{2} O_2 \rightarrow NO_3^-$	-76
$2Fe^{2+} + 2H^+ + \frac{1}{2} O_2 \rightarrow 2Fe^{3+} + 2H_2O$	-31

TABLE 9.8 The first step in nitrification where NH_4^+ is oxidized for energy and to reduce CO_2 for biomass formation by autotrophic chemolithotrophs, such as *Nitrosomonas* spp.

Energy: $NH_4^+ + 2\ HOH$	→	$NO_2^- + 6\ e^- + 8\ H^+$
$6\ e^- + 6\ H^+ + 1.5\ O_2$	→	$3\ HOH - 270.7$ kJ
$NH_4^+ + 1.5\ O_2$	→	$NO_2^- + HOH + 2\ H^+ - 270.7$ kJ
C Reduction: $4\ NH_4^+ + 8\ HOH$	→	$4\ NO_2^- + 24\ e^- + 32\ H^+$
$6\ CO_2 + 24\ e^- + 24\ H^+$	→	$6\ CH_2O + 6\ HOH$
$4\ NH_4^+ + 6\ CO_2 + 2\ HOH$	→	$4\ NO_2^- + 6\ CH_2O + 8\ H^+$

(Reproduced from McGill, 1996, with permission from SBCS and SLCS.)

TABLE 9.9 The second step in nitrification where NO_2^- (nitrite) is oxidized for energy and to reduce CO_2 for biomass formation by autotrophic chemolithotrophs, such as *Nitrobacter* spp.

Energy: $NO_2^- + HOH$	→	$NO_3^- + 2\ e^- + 2\ H^+$
$2\ e^- + 2\ H^+ + 0.5\ O_2$	→	$HOH - 77.4$ kJ
$NO_2^- + 0.5\ O_2$	→	$NO_3^- - 77.4$ kJ
C Reduction $2\ NO_2^- + 2\ HOH$	→	$2\ NO_3^- + 4\ e^- + 4\ H^+$
$ATP + CO_2 + 4\ e^- + 4\ H^+$	→	$CH_2O + HOH$
$ATP + 2\ NO_2^- + CO_2 + HOH$	→	$2\ NO_3^- + CH_2O$
$3\ NO_2^- + CO_2 + 0.5\ O_2 + HOH$	→	$3\ NO_3^- + CH_2O$

(Reproduced from McGill (1996), with permission from SBCS and SLCS.)

reduced S compounds, such as sulfide (H_2S), inorganic S (S_0), and thiosulfate ($S_2O_3^{2-}$), are used as an energy source by S oxidizing bacteria and archaea (e.g., *Beggiatoa*, *Paracoccus*, *Thiobacillus* [Proteobacteri]). Sulfur oxidation generally occurs in stages by different groups of microorganisms, and the reducing power generated is used to form NADH via reverse electron flow, which is used for CO_2 fixation in the Calvin cycle (Table 9.10).

Ferrous iron (Fe^{2+}) is a soluble form of iron that can be oxidized by several groups of iron-oxidizing bacteria and archaea. The first group (acidophiles, e.g., *Acidithiobacillus ferrooxidans*) oxidizes iron in very low pH environments and is important in acid mine drainage. The second group (microaerophiles) oxidizes iron at near-neutral pH and lives at the oxic-anoxic interface (e.g., *Leptothrix ochracea*). The third group of iron-oxidizing microbes is the anaerobic photosynthetic bacteria, such as Rhodopseudomonas. Aerobic iron oxidation is an energetically poor process that requires large amounts of iron to be oxidized to generate ATP. Similar to sulfur oxidation, reverse electron flow is used to form NADH used for CO_2 fixation in the Calvin cycle.

TABLE 9.10 **Aerobic sulfur oxidation for energy and to reduce CO_2 for biomass formation by autotrophic chemolithotrophs, such as *Thiobacillus* spp.**

Energy: $S^0 + 4\ HOH$	→	$SO_4^{-2} + 6\ e^- + 8\ H^+$
$6\ e^- + 6\ H^+ + 1.5\ O_2$	→	$3\ HOH - 584.9$ kJ
$S^0 + HOH + 1.5\ O_2$	→	$SO_4^{-2} + 2\ H^+ - 584.9$ kJ
C Reduction: $4S^0 + 16\ HOH$	→	$4\ SO_4^{-2} + 24\ e^- + 32\ H^+$
$6\ CO_2 + 24\ e^- + 24\ H^+$	→	$6\ CH_2O + 6\ HOH$
$4\ S^0 + 6\ CO_2 + 10\ HOH$	→	$4\ SO_4^{-2} + 6\ CH_2O + 8\ H^+$

(Reproduced from McGill, 1996, with permission from SBCS and SLCS.)

9.5.3 Phototrophy

While organotrophy and lithotrophy refer specifically to strategies used by organisms to acquire electrons (either reduced C or inorganic minerals), phototrophy refers to a strategy used to acquire energy. In contrast to chemotrophs (which can be chemoorganotrophs or chemolithotrophs, referring to either of the two electron sources discussed above), phototrophs use light energy. Most commonly in nature, phototrophs use light energy to fix inorganic CO_2 into reduced C compounds for their own cellular metabolism in a process known as photosynthesis. Photosynthesis produces the vast majority of the reduced C required by all forms of life for biosynthesis.

All higher plants are photoaquatrophs, meaning they use water as an electron source to drive CO_2 fixation, producing oxygen as a by-product. Archaea and bacteria, however, employ a much broader range of phototrophic strategies, in some cases using inorganic minerals as an electron source (photolithotrophs) and in other cases, using reduced organic compounds as an electron source (photoorganotrophs). Additionally, many phototrophic microorganisms are metabolically flexible and can use a range of organic and inorganic electron sources. The C source employed by phototrophic bacteria is also variable, while photoaquatrophs are exclusively autotrophic (i.e., CO_2-fixers), and photolithotrophs and photoorganotrophs can either be autotrophic or heterotrophic.

Phototrophy is far more phylogenetically restricted among soil microorganisms than chemotrophic strategies because the light requirement restricts this strategy to the uppermost part of the soil. For example, the phyla Cyanobacteria are the only known group of photoaquatrophs in soils. Several prominent groups of anoxyphototrophic bacteria (or photolithotrophs) use a reduced form of S, such as sulfide or thiosulfate, as an electron source. These include the autotrophic purple sulfur bacteria (Chromatiaceae) and green gliding bacteria (Chloroflexaceae), as well as the predominantly heterotrophic green sulfur bacteria (Chlorobiaceae). Photosynthetic members of Chloroflexaceae are noteworthy for exhibiting what may be the earliest form of photosynthetic reaction centers and CO_2 fixation mechanisms. While cyanobacteria and purple and green phototrophic bacteria are indeed found in soils, they occur primarily in aquatic habitats. By contrast, Heliobacteriaceae (Proteobacteria) are a phylogenetically and ecologically distinct group of photoheterotrophs that have been found only in soils.

Nitrogen fixation is often found in conjunction with phototrophy. The cyanobacteria are considered one of the most important and widespread groups of N-fixers. Members of Rhodospirillaceae, a family of

TABLE 9.11 Nitrogen fixation accompanying oxygenic and anoxygenic photosynthesis.

Oxygenic Photosynthesis and N_2 Fixation (e.g., *Nostoc*)		
hυ + 6 HOH + 19 (ADP + P_i)	→	3 O_2 + 12 e^- + 12 H^+ 19 ATP
3 ATP + CO_2 + 4 e^- + 4 H^+	→	CH_2O + HOH + 3 (ADP + P_i)
12 ATP + N_2 + 6 e^- + 6 H^+	→	2 NH_3 + 12 (ADP + P_i)
4 ATP + 2 e^- + 2 H^+	→	H_2 + 4 (ADP + P_i)
Overall: hυ + 5 HOH + CO_2 + N_2	→	3 O_2 + CH_2O + 2 NH_3 + H_2
Anoxygenic Photosynthesis and N_2 Fixation (e.g., *Chromatium* [purple S bacteria])		
6 H_2S	→	6 S^0 + 12 e^- + 12 H^+
hυ + Pigment[a]	→	[Pigment$^+$ + e^-][a]
19 (ADP + P_i) + [Pigment+ + e^-][a]	→	10 ATP + Pigment[a]
3 ATP + CO_2 + 4 e^- + 4 H^+	→	CH_2O + HOH + 3 (ADP + P_i)
4 ATP + 2 e^- + 2 H^+	→	H_2 + 4 (ADP + P_i)
12 ATP + N_2 + 6 e^- + 6 H^+	→	2 NH_3 + 12 (ADP + P_i)
Overall: hυ + 6 H_2S + CO_2 + N_2	→	6 S^0 + CH_2O +2 NH_3 + H_2 + HOH

[a]*Complex including several components of the photosynthetic system and the respiratory chain.*
(Reproduced from McGill, 1996, with permission from SBCS and SLCS.)

purple nonsulfur bacteria, fix N by coupling light energy with electrons from reduced C or H_2. Some methanogenic archaea are also N-fixers (e.g., *Methanococcus* [Euryarchaeota]). All described species of heliobacteria are also avid N-fixers and are thought to be important in the fertility of rice paddy soils, demonstrating that both oxygenic and anoxygenic forms of photosynthesis are associated with the ability to fix N_2 and to reduce C for release to subsequent organisms along the food chain (Table 9.11).

9.6 Unifying views of biotic metabolism in soil

The biogeochemical transformations mediated by soil organisms result from their search for energy and C. Energy is obtained by passing electrons (e^-) from donors to acceptors in multiple interconnected oxidation-reduction couples, which lead to cycles through which electrons flow. Similarly, the efficiency with which microorganisms allocate C toward biomass synthesis is a controlling step in the cycling of C between abiotic and biotic pools. These conceptualizations are simple and robust ways to unite the myriad details about metabolic transformations mediated by soil organisms with broader biogeochemical principles.

9.6.1 Interconnected cycles of electrons

The soil system of flowing e^- is like an electrical system with two cycles: an anoxygenic (not O_2-producing) cycle and an oxygenic (O_2-producing) cycle, and four circuits—two phototrophic, one chemoorganotrophic, and one chemolithotrophic hooked in parallel (Fig. 9.6). The microbial component consists of four groups of organisms described using the metabolic classification in Fig. 9.4. Three groups of organisms are responsible for C addition to soil and only one (chemoorganotrophs) for its removal. The two cycles are distinguished on the basis of the photosynthetic mechanisms: anoxygenic or oxygenic. The chemoorganotrophic circuit unites the anoxygenic cycle with the oxygenic cycle.

Anoxygenic photosynthesis uses energy from sunlight to couple the reduction of C in CO_2 to the anaerobic oxidation of S in S^0 or H_2S. If one is accustomed to thinking of S oxidation in a strictly aerobic sense, then anaerobic S oxidation appears contradictory. Anoxygenic photosynthesis would have been compatible with the anoxic (O_2-free) conditions of Earth's primordial atmosphere. It could have been mediated by anaerobic organisms like present-day photosynthetic sulfur bacteria and is believed to have preceded oxygenic photosynthesis (Blankenship, 2010). Dominance of anoxygenic photosynthesis would have favored anaerobic respiration or fermentative pathways for obtaining energy from the products of photosynthesis. Consequently, one finds a wide representation of anaerobic microorganisms in soil environments. Many elements can cycle under entirely anaerobic conditions due to the syntropic relationships among photolithotrophs and anaerobic chemoorganotrophs. One can thus summarize the anoxygenic cycle as a combination of photosynthesis powered by electromagnetic radiation by

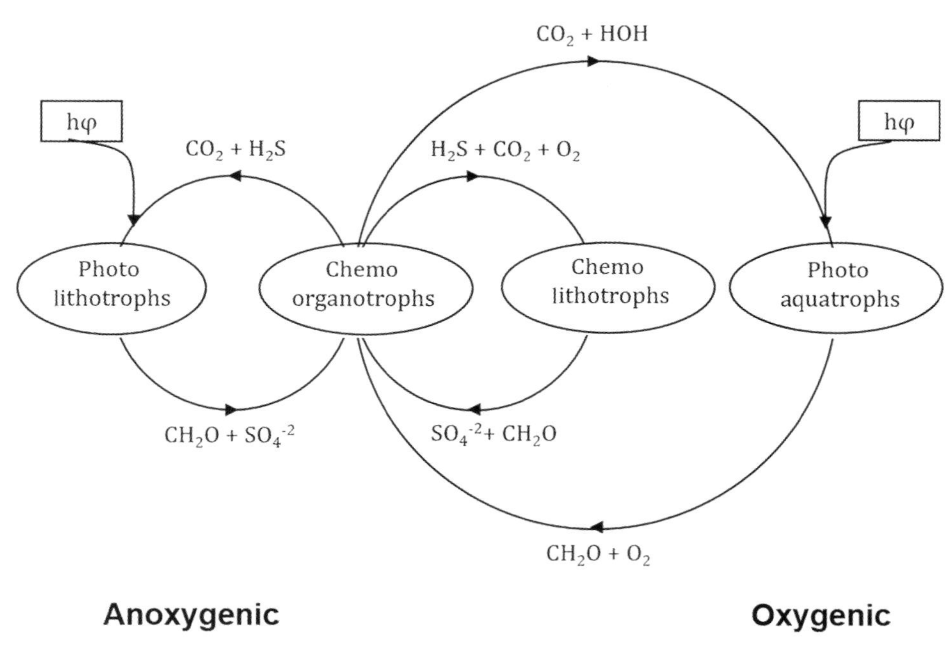

FIGURE 9.6 Schematic outline of electron flow through cycles starting with anoxygenic (not oxygen-producing) and oxygenic photosynthesis (oxygen-producing). Anoxygenic photosynthesis generates oxidized S species from reduced S under anaerobic conditions, thereby maintaining anaerobic conditions; oxygenic photosynthesis generates oxygen from water, thereby producing aerobic conditions. Both yield CH_2O for biomass production.

photolithotrophs while they reduce C and oxidize elements such as S, combined with decomposition by anaerobic chemoorganotrophs to re-oxidize C and re-reduce S. The cycle can be represented as:

$$4\ CO_2 + 2\ H_2S + 4\ HOH \leftrightarrow 2\ SO_4^{-2} + 4\ CH_2O + 4\ H^+ \qquad \text{(Eq. 9.12)}$$

Energy trapped as organic compounds during anoxygenic photosynthesis could be released by oxidation of the reduced C (e-donor), coupled with reduction of the oxidized minerals (e-acceptor), and formed during anoxygenic photosynthesis.

Oxygenic photosynthesis can be carried out by many eukaryotes but only by the cyanobacteria among the prokaryotes. Consequently, anoxygenic photosynthesis is dominantly prokaryotic and oxygenic photosynthesis dominated by eukaryotes. The oxygenic cycle has two circuits. First, a photo-chemo circuit consisting of photoaquatrophs to reduce C coupled with chemoorganotrophs, which may be either aerobic or anaerobic to re-oxidize it. Second, a chemo-chemo circuit consisting of aerobic chemolithotrophs to oxidize minerals coupled with anaerobic (or facultative) chemoorganotrophs to oxidize C and re-reduce minerals (Fig. 9.6). Electrons flow among the cycles, thereby connecting them. For example, an electron from water may be passed to CH_2O during photosynthesis, proceed through the photo-chemo circuit, and then be returned to water through aerobic oxidation. From there it may again be passed to CO_2 (photosynthesis) to form CH_2O, proceed to the chemo-chemo circuit, and under anaerobic conditions, be used to reduce SO_4^{2-} to H_2S. In the presence of O_2, it proceeds through the aerobic part of the chemo-chemo circuit and the electron is again used to reduce CO_2 to CH_2O concurrently with oxidation of (loss of electrons from) S^{2-}. As it continues its journey, under anaerobic conditions, the electron may again be transferred to H_2O, which brings it to the intersection again with the photo-chemo circuit. Hence the oxidation and reduction of many elements, although not conducted by photosynthetic organisms, is tied to photosynthesis by transfers of electrons among organisms through the reduced C and O_2 produced by photosynthesis. The oxidation of elements is made possible by O_2 from photosynthesis and their reduction by the reduced C from photosynthesis. This alternating oxidation—reduction system involving chemolithotrophs requires that O_2 and CH_2O from photosynthesis travel separately and that there be habitats from which the O_2 is excluded. Soils are uniquely suited to providing such habitats.

The two chemotrophic circuits of Fig. 9.6 can be expanded to distinguish aerobic, facultative, and anaerobic domains based on sensitivity to O_2 (Fig. 9.7). The aerobic domain is set at Eh > 300 mV comprising aerobic chemoorganotrophs and photoaquatrophs as a syntropic system. The facultative domain, between Eh 100 and 300 mV, comprises chemoorganotrophs, which may be either aerobic or anaerobic in syntropic associations with chemolithotrophs, which are aerobic. The anaerobic domain consists of strictly anaerobic chemoorganotrophs in association with photoaquatrophs. Anaerobic chemoorganotrophs in the anaerobic domain reduce oxidized minerals generated by aerobic chemolithotrophs in the aerobic domain. Some energy is dissipated through heat loss, etc., and must be made up by subsequent photosynthesis.

O_2 is an overall control because it inhibits anaerobic processes. Consequently, the balance among the three domains in Fig. 9.7 is a function of O_2 availability in local environments or micro-sites. N_2 fixation is interesting in that it is inhibited by O_2 and mediated by three of the four groups of organisms, with only chemolithotrophs excluded. In addition, the chemoorganotrophs are responsible for both removing N from the pedosphere by reducing NO_3^- to N_2 and for returning it by further reducing N_2 to NH_3.

Fig. 9.7 further shows that as the oxidation-reduction potential (Eh) becomes increasingly negative, oxidants become decreasingly effective. Given that the energy available through a redox reaction is

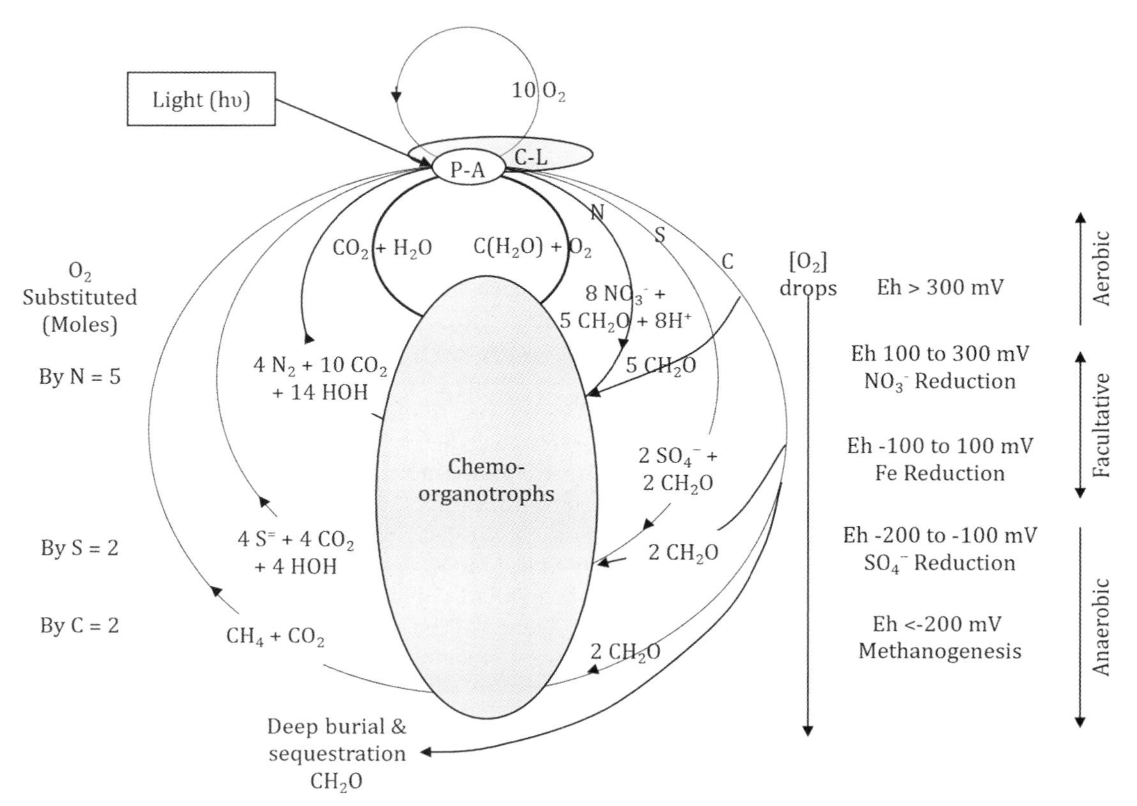

FIGURE 9.7 The oxygenic cycles in soil entail cyclic oxidations and reductions of C, N, and S (among other elements), which are driven by solar radiation and controlled by the availability of O_2. Lowering O_2 availability is reflected in lowering Eh values. Oxygenic photoaquatrophs (P-A) produce CH_2O and release O_2; aerobic chemo-lithotrophs (C-L) oxidize N to NO_3^- or S to SO_4^{-2} using O_2 from photosynthesis and reduce CO_2 to CH_2O autotrophically. Aerobic chemoorganotrophs oxidize CH_2O using O_2 through aerobic respiration (heavy circle); and facultative anaerobic chemoorganotrophs oxidize CH_2O using NO_3^- (nitrate respiration). Anaerobic chemoorganotrophs oxidize CH_2O by reducing SO_4^{-2} or under extremely anaerobic conditions by reducing a portion of the CH_2O itself, thereby splitting it into CO_2 (oxidized) and CH_4 (more reduced). For the stoichiometry represented here the moles of O_2 substituted by N, S, or C are designated at the left. Dinitrogen fixation using reduced C ultimately from P-A is required to reduce N_2 to NH_3 prior to its oxidation to NO_3^-. Deep burial of CH_2O opens the oxygenic cycles, thereby allowing O_2 to accumulate.

directly proportional to the change in oxidation potential between the two couples, less and less energy is released as one moves from aerobic to strictly anaerobic metabolism. Soil organisms have evolved in an energy-limited environment; thus communities of soil organisms use the most energetically favorable energy sources available to them. Such a strategy favors use of O_2 as an electron acceptor followed by NO_3^-, SO_4^{2-}, and eventually the use of a portion of the C in CH_2O during methanogenesis. In other words, CH_2O is allocated first to reduction of O_2, then NO_3^-, followed by SO_4^{2-}, and finally methanogenesis. From a practical perspective then, NO_3^- might be a useful electron acceptor for metabolism of organic contaminants under anaerobic conditions. Indeed, NO_3^- reducing and denitrifying populations have been proposed for removal of organic contaminants as C substrates under anaerobic conditions, such as those in riparian soils, to prevent or reduce stream contamination.

These coupled cycles of electron flow operate at the metabolic or molecular scale of soil microorganisms but have important implications in the global biogeochemical cycling of several elements (see Falkowski et al., 2008). For instance, the O_2 produced during photosynthesis is consumed stoichiometrically during complete decomposition of photosynthate. It is not possible for O_2 to accumulate in the atmosphere from photosynthesis unless large quantities of CH_2O are not decomposed. We are not awash in non-decomposed plant litter, so how could residual O_2 have accumulated in the atmosphere? Disrupting the cycle of photosynthesis and decomposition would retain O_2 in the atmosphere rather than consuming it in oxidation of photosynthate. Disruption occurs by deep burial and sequestration of CH_2O as represented in Fig. 9.7. Processes yielding soil humus, peat, shale, coal, or petroleum and deep ocean organic sediments, among others over geologic time, all remove CH_2O that microbes would otherwise use to reduce O_2 to water (see Logan et al., 1995). Given that C storage could allow O_2 accumulation, what might be the control on maximum O_2 accumulation? Mineral weathering is a sink for O_2 and participates in regulating atmospheric O_2 concentration. In addition, oxidation-reduction reactions, over geological time, control Fe solubility, and due to the reactivity of Fe species with P, control P concentration. Fe and P concentrations in solution, especially in marine environments, often limit photosynthesis and N_2 fixation. Regulation of Fe and P solubility by oxidation state of Fe provides a feedback to atmospheric O_2 concentration (Van Cappellen and Ingall, 1996).

9.6.2 Carbon use efficiency

The flow of C and nutrients through soil food webs is controlled by properties of the soil biota and their environment. An important concept used to synthesize many of these properties, and thus better understand soil biochemistry, is C use efficiency (CUE). Only a portion of the substrate C taken up (U) by biota is retained as biomass (C_B) (Fig. 9.8). CUE is therefore defined as the proportion of C resources a microorganism (or potentially any organism) uses for building biomass (i.e., CUE $= C_B/U$). So-called "efficient" uses of substrate are thus anabolic, biomass-yielding processes, such as cell mass production, cell division, and replacement of intracellular molecules. Conversely, "inefficient" uses of substrate are catabolic, energy-yielding activities, such as respiration to generate the energy needed for anabolism (i.e., growth respiration; R_G) or to regulate intracellular pH and osmotic balance (i.e., maintenance respiration; R_M). CUE estimates range from 0 to $\sim$0.8, with the upper limit reflecting that even under the most favorable conditions for biosynthesis, perfect conversion of C from substrate to biomass is not possible. Other terms, such as "microbial growth efficiency," have been replaced by CUE as even non-growing, dormant organisms undergo C metabolism. In other words, organisms need to consume C to grow but do not have to grow to consume C.

Interest in CUE has expanded in recent decades from what was a purely microbiological field to include ecological and Earth systems perspectives. Understanding when, how, and why CUE fluctuates provides clues to the origins and turnover of soil organic matter (SOM), emissions of greenhouse gases from soil, and ultimately, soil-climate feedbacks. Here we explore concepts across scales that reflect these varied interests and which make CUE a unifying concept across soil science disciplines.

CUE is foremost a metabolic attribute of a single cell, population, or community of microorganisms. Metabolism is fueled by extracellular substrates, although recycling of internal polysaccharides, lipids, and proteins will occur during starvation as endogenous metabolism. Substrate will undergo dissimilation,

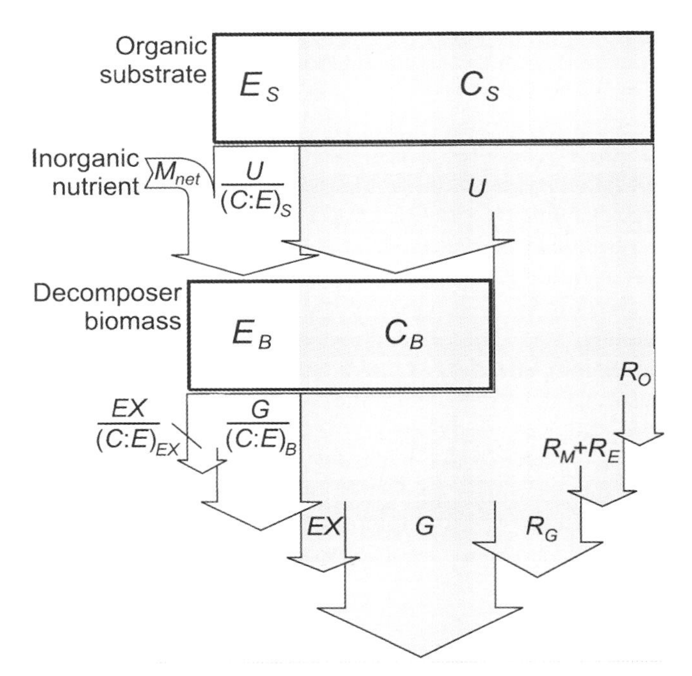

FIGURE 9.8 Mass balance of carbon (C; shaded portions) and nutrient elements (E; unshaded portions) during uptake and metabolism by microorganisms as regulated by the stoichiometric differences between microbial biomass C (subscript B) and substrate (subscript S). U, G, EX, indicate C uptake, microbial growth, and C exudation; RG, RM, RE, and RO represent respiration for growth, maintenance, enzyme production, and C overflow, respectively. In this example, net immobilization (Mnet) of the inorganic nutrient and overflow respiration (RO) occur. *(Reproduced with permission from Manzoni et al., 2012.)*

the complete catabolic breakdown of bonds to liberate energy and generate ATP, or assimilation, the amphibolic process wherein catabolism liberates both energy and precursor compounds (e.g., pyruvate, glucose-6-phosphate) that are shuttled toward anabolic biosynthesis. Assimilation and dissimilation both occur within the linked steps of central metabolism, differing only in whether all (dissimilation) or only some (assimilation) substrate bonds are broken to yield energy.

Whether a substrate undergoes dissimilation or assimilation depends in part on the biochemical nature of the substrate. Substrate of limited quantity, or which is relatively oxidized and of low energy, will be dissimilated/respired to prioritize the harvest of energy, resulting in a low CUE. As substrate becomes more abundant or reduced, surplus energy beyond maintenance needs can support assimilation. CUE increases with increasing energy availability until an asymptote is reached when the substrate matches the degree of reduction of microbial biomass ($\gamma S \sim 4.2$ moles of electrons per mass of C) (Fig. 9.9). A transition occurs beyond this point from energy-limited to C-limited growth (Manzoni et al., 2012). A peak CUE of ~ 0.8 is reached during C-limited growth because substrate undergoing complete assimilation still incurs some catabolic cost alongside anabolism, and thermodynamic inefficiency accompanies any chemical reaction. Most substrates that a soil microbe utilizes are believed to be energy-limited relative to microbial biomass (i.e., $\gamma S < 4.2$), making energy-limited growth and sub-maximum CUE the norm.

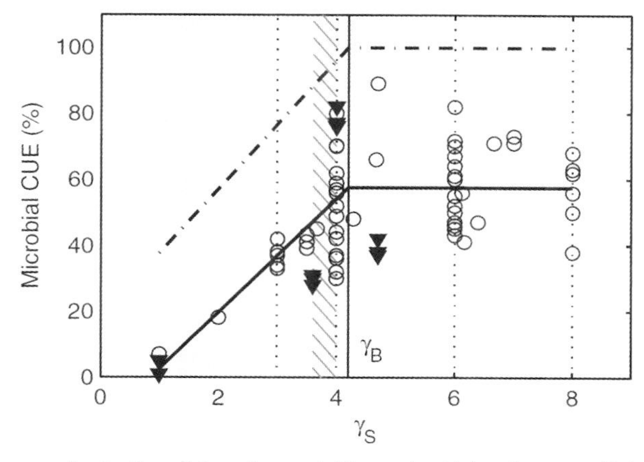

FIGURE 9.9 Effect of the degree of reduction of the substrate (ɣS) on microbial carbon use efficiency (CUE; open symbols, data from culture studies; closed symbols, data from soil incubations). A transition from energy-limitation (shaded area for ɣS < ɣB, where ɣB ~ 4.2 is the approximate degree of reduction of microbial biomass) to C-limitation (ɣS > ɣB) is depicted. The dot-dashed line illustrates the theoretical maximum assimilation efficiency. The degree of reduction of malate (ɣS = 3), glucose (ɣS = 4), ethanol (ɣS = 6), and methane (ɣS = 8) are indicated for reference with vertical dotted lines. The hatched area highlights the range of ɣS for typical soil solutes. *(Reproduced with permission from Manzoni et al., 2012.)*

Substrate quality has been identified among the more influential drivers of CUE and may underlie differences in mean CUE observed at ecosystem scales, such as among forests (0.41), grasslands (0.65), and shrublands (0.73) (Qiao et al., 2019). It should be noted that the substrate-dependence of CUE unfortunately hinders comparison of CUE estimates gathered by isotope tracing experiments wherein labeled glucose, acetate, and many other substrates of differing quality have been used.

A host of other factors interact to regulate CUE. These can be summarized as taxonomic and environmental in origin. One result of efforts to simplify the enormous taxonomic diversity of microbial communities has been the emergence of discrete functional groupings. One example is the hypothesized tradeoff between growth rate maximization (r-selected species) versus CUE (K-selected species). This tradeoff is phylogenetically deep. Clades, such as the bacterial phylum Proteobacteria and the fungal subphylum Mucormycotina, share fast-growing and rapidly reproducing life histories that maintain a "metabolic alertness" to best compete for resource pulses. Being primed for growth, however, incurs a C cost and thus reduces CUE. K-selected organisms, such as the bacterial phylum Acidobacteria and fungal Basidiomycota (i.e., wood-rot fungi), on the other hand, achieve slow but consistent growth by efficiently exploiting less energy-rich, but often more available, resources. Further division of the r/K dichotomy can be made by assessing traits, such as potential for C decomposition and stress tolerance, both of which require investment into compounds like exoenzymes and exopolysaccharides that may not be "efficient" uses of C resources if exuded from the cell. These results are broadly supported by traditional culturing in prepared media, in situ stable isotope probing, and full genome sequencing. One of the more pervasive assumptions is that the phylogenetic division between bacteria and fungi also divides copiotrophic r-selected from oligotrophic K-selected groups, respectively. Although the higher biomass C:N of many fungi (~8–15) compared to bacteria (~6–10) may require more efficient carbon use, the predictability of CUE is likely greater by functional grouping (e.g., r/K) than phylogenetics since relatively copiotrophic fungi and oligotrophic bacteria can be found throughout soils.

The net effect of changing substrate availability on CUE is dependent on the scale of observation. A positive relationship generally exists between substrate quality, microbial growth rate, and CUE for any single species. This is attributed to diminishing maintenance costs as the rate of growth increases such that relatively greater proportions of C are directed to biomass formation. On the other hand, the availability of substrate generally reduces CUE at a community scale because of proliferation by copiotrophic r-selected taxa. Labile rhizosphere root exudates have been shown to reduce CUE by up to ~35% compared to nearby bulk soils. This nuance warrants consideration of population vs. community scale CUE (Geyer et al., 2016).

Finally, CUE is strongly affected by the environment in which a microorganism operates. In general, stressful conditions (e.g., warming, nutrient scarcity, drought, interspecific competition) result in reduced CUE as investments in cell maintenance increase. Climate change and other anthropogenic activities continue pushing these environmental conditions to new extremes. Warming, for example, accelerates respiration rate faster than growth rate for microbial communities, resulting in declining CUE that may underlie the reduced microbial biomass, more rapid SOM breakdown, and depleted SOM stocks of 10−15% both predicted and observed for warming soils (Frey et al., 2013). The negative relationship between soil warming and CUE results in enhanced loss of C from soil as CO_2, which may further exacerbate global climatic warming in a positive feedback. This threat is all the more real given that the largest soil C stocks reside in boreal systems where rapid warming is exposing once-frozen organic matter to microbial decomposition. This constitutes a new source of greenhouse emission from soil regardless of the CUE that occurs upon melting.

Nutrient availability affects microbial CUE by imposing an imbalance between the stoichiometry of microbial biomass and the organic substrates being metabolized (Fig. 9.9). For example, nutrient limitation will result from decomposition of recalcitrant plant litter (possibly orders of magnitude greater in C:nutrients than microbial biomass) if no other source of nutrients such as N exist. Microorganisms will mine additional organic matter for limiting N, mineralizing the excess C as overflow respiration, and lowering their CUE. Elevated substrate C:nutrients may also induce exoenzyme production by decomposers or energy-intensive N fixation by diazotrophs, both of which may also reduce CUE. Nitrogen availability will alleviate the limitations to growth and increase CUE until C becomes limiting (e.g., the threshold element ratio). An increase in CUE is expected in the succession from heterotrophic microorganisms colonizing fresh, nutrient-limited plant litter to the breakdown of relatively nutrient-rich, microbially processed SOM. Latitudinal patterns in CUE reflect the complex interaction between temperature and substrate availability that can occur. In general, CUE increases at lower latitudes in response to more favorable temperatures for growth and higher C and N inputs to soil as a proportion of net primary production (Sinsabaugh et al., 2017).

A number of important questions about CUE in soils remain to be answered. Early definitions of efficiency classified the production of exudates, such as exoenzymes, extracellular polysaccharides, and antibiotics, as insignificant compared to growth. Recent work has revealed that soil microbes may invest 30−40% of C resources into exudates; proper accounting of these C expenses is necessary for accurate CUE estimation (Geyer et al., 2020). Furthermore, understanding when and why exudate synthesis occurs is key to understanding the role that microorganisms play in generating precursor compounds which stabilize as SOM. An integrative knowledge of what results in high CUE (e.g., which microbial species, metabolizing which substrates, under which environmental conditions) further holds many promises for managing soil systems. Reduced tillage, for example, can increase fungal:bacterial ratios and thereby enhance soil C concentrations as a result of increasing CUE (Sauvadet et al., 2018). Practices like this help to achieve goals of retaining and/or restoring soil C, improving soil fertility, and combating climate change.

Supplemental reading

The following references are textbooks that provide more in-depth coverage of the basics of microbial physiology and metabolism. While most do not focus exclusively on soil ecosystems and cover a much broader scope of topics, each contains one or several chapters on microbial energy acquisition, physiology, and metabolism from an organismal perspective.

Kim, B.H., Gadd, G.M., 2008. Bacterial Physiology and Metabolism. Cambridge University Press.

White, D., Drummond, J., Fuqua, C., 2011. The Physiology and Biochemistry of Prokaryotes. Fourth Edition. Oxford University Press.

The following textbook focuses more closely on the connections between bacterial metabolism and environmental biogeochemistry.

Fenchel, T., King, G.M., Blackburn, T.H., 2012. Bacterial Biogeochemistry: The Ecophysiology of Mineral Cycling. third ed. Academic Press.

References

Blankenship, R.E., 2010. Early evolution of photosynthesis. Plant Physiol. 154, 434–438.

Cleveland, C.C., Liptzin, D., 2007. C:N:P stoichiometry in soil: is there a "Redfield ratio" for the microbial biomass? Biogeochemistry 85, 235–252.

Ehrlich, H.L., 1993. Bacterial mineralization of organic carbon under anaerobic conditions. In: Bollag, J.-M., Stotzky, G. (Eds.), Soil Biochemistry. Dekker, New York, NY, pp. 219–247.

Falkowski, P.G., Fenchel, T., Delong, E.F., 2008. The microbial engines that drive Earth's biogeochemical cycles. Science 320, 1034–1039.

Fierer, N., Lennon, J.T., 2011. The generation and maintenance of diversity in microbial communities. Am. J. Bot. 98, 439–448.

Fierer, N., Bradford, M.A., Jackson, R.B., 2007. Toward an ecological classification of soil bacteria. Ecology 88, 1354–1364.

Frey, S.D., Lee, J., Melillo, J.M., Six, J., 2013. The temperature response of soil microbial efficiency and its feedback to climate. Nat. Clim. Change 3, 395–398.

Frouz, J., Novotná, K., Čermáková, L., Pivokonský, M., 2020. Soil fauna reduce soil respiration by supporting N leaching from litter. Appl. Soil Ecol. 153, 103585.

German, D.P., Weintraub, M.N., Grandy, A.S., Lauber, C.L., Rinkes, Z.L., Allison, S.D., 2011. Optimization of hydrolytic and oxidative enzyme methods for ecosystem studies. Soil Biol. Biochem. 43, 1387–1397.

Geyer, K.M., Kyker-Snowman, E., Grandy, A.S., Frey, S.D., 2016. Microbial carbon use efficiency: accounting for population, community, and ecosystem-scale controls over the fate of metabolized organic matter. Biogeochemistry 127, 173–188.

Geyer, K., Schnecker, J., Grandy, A.S., Richter, A., Frey, S., 2020. Assessing microbial residues in soil as a potential carbon sink and moderator of carbon use efficiency. Biogeochemistry 151, 237–249.

Ghosh, W., Dam, B., 2009. Biochemistry and molecular biology of lithotrophic sulfur oxidation by taxonomically and ecologically diverse bacteria and archaea. FEMS Microbiol. Rev. 33, 999–1043.

Gottschalk, G., 1986. Bacterial Metabolism, second ed. Springer-Verlag, New York, NY.

Logan, G.A., Hayes, J.M., Hieshima, G.B., Summons, R.E., 1995. Terminal proterozoic reorganization of biogeochemical cycles. Nature 376, 53–56.

Madsen, E.L., 2015. Environmental Microbiology: From Genomes to Biogeochemistry, second ed. Wiley-Blackwell Publishing.

Manzoni, S., Trofymow, J.A., Jackson, R.B., Porporato, A., 2010. Stoichiometric controls dynamics on carbon, nitrogen, and phosphorus in decomposing litter. Ecol. Monogr. 80, 89–106.

Manzoni, S., Taylor, P., Richter, A., Porporato, A., Ågren, G.I., 2012. Environmental and stoichiometric controls on microbial carbon-use efficiency in soils. New Phytol. 196, 79–91.

McGill, W.B., 1996. Soil sustainability: Microorganisms and electrons. In: Solo Suelo 96 Conference 2, CD Version. Sociedade Brasileira de Ciência do Solo (SBCS) and Sociedade Latino-Americana de Ciência do Solo (SLCS), Viçosa/MG.

Osono, T., 2007. Ecology of ligninolytic fungi associated with leaf litter decomposition. Ecol. Res. 22, 955–974.

Qiao, Y., Wang, J., Liang, G., Du, Z., Zhou, J., Zhu, C., et al., 2019. Global variation of soil microbial carbon-use efficiency in relation to growth temperature and substrate supply. Sci. Rep. 9, 5621.

Sauvadet, M., Lashermes, G., Alavoine, G., Recous, S., Chauvat, M., Maron, P.-A., et al., 2018. High carbon use efficiency and low priming effect promote soil C stabilization under reduced tillage. Soil Biol. Biochem. 123, 64–73.

Sinsabaugh, R.L., Follstad Shah, J.J., 2012. Ecoenzymatic stoichiometry and ecological theory. Ann. Rev. Ecol. Evol. Syst. 43, 313–343.

Sinsabaugh, R.L., Lauber, C.L., Weintraub, M.N., Ahmed, B., Allison, S.D., Crenshaw, C., et al., 2008. Stoichiometry of soil enzyme activity at global scale. Ecol. Lett. 11, 1252–1264.

Sinsabaugh, R.L., Moorhead, D.L., Xu, X., Litvak, M.E., 2017. Plant, microbial and ecosystem carbon use efficiencies interact to stabilize microbial growth as a fraction of gross primary production. New Phytol. 214, 1518–1526.

Spohn, M., 2016. Element cycling as driven by stoichiometric homeostasis of soil microorganisms. Basic Appl. Ecol. 17, 471–478.

Van Cappellen, P., Ingall, E.D., 1996. Redox stabilization of the atmosphere and oceans by phosphorus-limited marine productivity. Science 271, 493–496.

Xu, X., Thornton, P.E., Post, W.M., 2013. A global analysis of soil microbial biomass carbon, nitrogen and phosphorus in terrestrial ecosystems. Glob. Ecol. Biogeogr. 22, 737–749.

Chapter 10

The ecology of soil biota and their function

Sherri J. Morris* and Christopher B. Blackwood[†]
*Department of Biology, Bradley University, Peoria, IL, USA; [†]Department of Biological Sciences, Kent State University, Kent, OH, USA

Chapter outline

10.1 Introduction

Ecology is the study of the interactions between organisms and their environment. The term *ecology* was initially based on the Greek word "oikos," which describes the family household; thus "ecology" describes the environment and organisms as a household (Kingsland, 1991). The ecological field of study allowed scientists to construct functional mechanisms to understand the patterns of organisms identified in the field of natural history. Because the origins in the field of ecology were based on understanding mechanisms, not just patterns, ecology has always been heavily dependent on the principles of evolution, and ecological relationships are understood to be based on long-term evolutionary history.

Most ecological research is predicated on the existence of a set of unique species that can be detected across a range of environments. While this works well for macroflora and fauna, it is challenging to achieve for soil organisms. Evolutionarily divergent and poorly studied phylogenetic lineages of soil biota continue to be discovered using new technologies (Hug et al., 2016; James et al., 2020). While our

Soil Microbiology, Ecology, and Biochemistry. https://doi.org/10.1016/B978-0-12-822941-5.00010-7

understanding of well-studied lineages continues to expand, the periodic dramatic expansions to the scope of the tree of life create new opportunities to ask ecological questions. The ever-expanding molecular and genetic methodologies for determining characteristics of soil organisms, especially genomes, have allowed for a more rapid progress in this area than would have been allowed 20 years ago.

Difficulties with defining a microbial species are a lurking problem for the conceptual foundation of microbial ecology, but recent genomic evidence seems to indicate that these issues may be resolvable. The "biological species concept," adopted by many plant and animal biologists, defines a species as an interbreeding group of organisms that is reproductively isolated from other organisms, stopping gene flow between species except in rare cases. Bacteria, archaea, and many microbial eukaryotes seem to act differently in two ways: (1) reproducing asexually and (2) acquiring genes through processes that are independent of cell division (i.e., horizontal gene transfer). This makes application of the biological species concept problematic for microorganisms. Recent examination of tens of thousands of bacterial and archaeal whole genome sequences showed genetically distinct clusters of organisms, with rampant gene flow within clusters but not between clusters. This is consistent with the current understanding of prokaryotic species (Bobay and Ochman, 2017; Parks et al., 2020). Acquisition and loss of genes may be part of the strategy of some bacterial species, allowing for adaptation to a wide range of environmental conditions. For example, the bacterial group *Pseudomonas fluorescens* (class Gammaproteobacteria) is widely distributed in soil, including strains that interact with a large diversity of other organisms, such as plants, bacteria, archaea, fungi, oomycetes, nematodes, and insects. This phenotypic diversity may be related to a highly variable genome, with only 45–52% of genes in a particular genome also being found in all other *P. fluorescens* strains (Loper et al., 2012). Recombination is less common between distantly related lineages because of their lack of homologous genes and promoters and the limited host ranges of plasmids and viruses that mediate gene transfer (Thomas and Nielsen, 2005). If the biological species concept is refocused on identifying cohesive networks of genetic exchange, rather than sexual reproduction per se as the mechanism maintaining genetic cohesion, most microbial species can be accommodated (Bobay, 2020). Methods focusing on genetic markers for microbial phylogeny, coupled with analysis of key genes and physiological traits, are allowing us to make dramatic progress in understanding the ecology of soil organisms despite an evolutionary framework that is still developing.

In this chapter we focus on the processes that drive community structure (number and types of species) and the resultant impacts of their function in ecosystems (processes of energy transformations and nutrient turnover). As soil biota are essential components of every ecosystem, developing an understanding of the ecology of these organisms is essential to understanding terrestrial systems. The need to understand and accurately predict the impacts of global climate change and to develop sustainable practices in agriculture, forestry, ecosystem restoration, and natural resource management provides additional impetus to emphasize the ecology of soil organisms and its importance in the 21st century. This understanding is necessary to protect these systems from becoming degraded over the long term. This is especially important at a time when disturbance to ecosystems by humans is high. Examining soils as complex, integrated systems has great potential to influence ecological concepts and the science of ecology in general, with soils representing a critical ecological frontier.

10.2 Mechanisms that drive community structure

A large part of ecology is the study of how organisms become distributed in the environment, which is related to questions such as: Why are some species found in one area and not in others? How predictable are the groups of species found together? The paradigm that dominated scientific understanding of the distribution of microorganisms throughout the 20th century was articulated by Lourens G.M. Baas Becking based on the work of Bejerinck: "Everything is everywhere, but the environment selects" (O'Malley, 2007). The implication of this statement is that the distribution of microorganisms is strictly determined by environmental conditions. However, this has been vigorously challenged since molecular methods have allowed the study of microbial biogeographic patterns to be unified with general ecological theories (Nemergut et al., 2013).

The area an organism lives in is called its habitat, and biotic and abiotic conditions of the habitat affect the physiological ability of each organism to survive and reproduce. Functional traits are properties of an organism that affect how well the organism performs under a certain set of conditions (McGill et al., 2006). Organisms that live in one habitat make up the community, and the numbers and kinds of organisms present are referred to as community structure. Communities are composed of populations or sub-populations of various species. A population is a collection of all organisms belonging to a single species with potential for interaction. This makes the spatial scale of a population dependent on the mobility of the species. A study may encompass only part of a true population (e.g., migrating species) or encompass multiple, isolated populations (e.g., soil bacteria). Given the degree to which species are differentially mobile, it is normal for both situations to arise in the same study.

10.2.1 Physiological limitations to survival

Ecologists often need quantitative information on the suitability of habitats for a particular species to predict population dynamics under changing environmental conditions in new areas or with changes in community structure. Species' functional traits include limitations on the conditions under which individuals, and therefore populations, can grow and reproduce. Shelford's law of tolerance states that there is a maximum and minimum value for each environmental factor, beyond which a given species cannot survive. This law is usually discussed with respect to environmental characteristics, known as modulators, such as temperature, pH, or salinity. Modulators impact the physiology of organisms by altering the conformation of proteins and cell membranes and the thermodynamic and kinetic favorability of biochemical reactions (see Chapter 9). Species also have an optimal range for each environmental modulator where maximum population growth occurs. Tolerance to modulators can be interactive; for example, in some fungi, tolerance to cold temperatures depends on water potential (Hoshino et al., 2009). Normally the geographical range of a species coincides with areas where environmental conditions are within the optimal ranges for the species, with the most optimal conditions at the center of the geographical range. The effects of species being in habitats with modulators outside their tolerance levels are listed in Table 10.1, along with biochemical strategies used by microorganisms that exist under these "extreme" conditions.

The response of a species to environmental conditions depends on its genetic makeup. Limits and optima are determined through natural selection and other mechanisms that affect the genome. Some species are adapted to varying conditions through high levels of genomic diversity among strains, as described for *P. fluorescens* above. In addition, all organisms have some degree of phenotypic or trait plasticity, the ability to alter their form or physiology to adjust to the environment. In microorganisms a change in an environmental condition can induce expression of alternative phenotypes that can allow them to acclimate to the new conditions, which broadens the range of conditions acceptable for the species. The cost associated with carrying genes allowing for a large degree of plasticity is extra genetic

TABLE 10.1 The effects of physical stresses (modulators) on microorganisms and the biochemical adaptations they induce.

Modulator	Effect on cell	Biochemical adaptation	Organisms that have required adaptation
Temperature	Denaturation of enzyme; change in membrane fluidity	Production of proteases and ATP-dependent chaperones (Derré et al., 1999); production of cold-tolerant enzymes by amino acid substitution (Lönn et al., 2002); increases in intracellular trehalose and polyol concentrations and unsaturated membrane lipids, secretion of antifreeze proteins and enzymes active at low temperatures (Robinson, 2001)	Thermophiles; psychrophiles
Water deficit or salt stress	Dehydration and inhibition of enzyme activity	Changes in composition of polysaccharides produced (Coutinho et al., 1999); maintaining salt in cytoplasm, and uptake or synthesis of compatible solutes (Roeßler and Müller, 2001)	Osmophiles; xerophiles; halophiles
pH	Protein denaturation; enzyme inhibition	Increased intrasubunit stability in proteins afforded by increased hydrogen bonds and stronger salt bridges (Settembre et al., 2004); organisms that can secrete a surplus of protons or block extracellular protons from the cytoplasm by blocking membrane composition (de Jonge et al., 2003); stress regulator genes (de Vries et al., 2001)	Acidophiles
Aeration stress	Oxygen radicals damage membrane lipids, proteins, and DNA	Detoxification of oxygen radicals by catalase and superoxide dismutase (Wu and Conrad, 2001)	Obligate anaerobes; methanogens; sulfur and N users

Modified from Paul and Clark (1996).

material that must be duplicated with each cell division, resulting in lower efficiency of resource use. This strategy can be particularly efficient in fluctuating environments, such as the soil surface.

Resources are physical components of the environment that are captured by organisms for their use, such as nutrients, energy, or territory. Shelford's law can be applied to most resources, but the responses to different resources are highly interactive. This has been partially described in Liebig's law of the minimum, which states that the resource in lowest supply relative to organismal needs will limit growth. At very low levels of a resource, the organism is unable to accumulate that resource in adequate quantities for metabolism. The ratio of resources that an organism needs for metabolism and growth (e.g., the C:N:P ratio of biomass) is an important functional trait determining how species performance varies with resource availability, as well as excretion of waste nutrients that are consumed in excess of need. Liebig's law, developed in 1840 to help understand agricultural plant production, is now recognized to apply to species and communities in natural systems that may have positive growth responses to an increased supply of alternate nutrients (C, N, or P). This situation is known as multiple nutrient limitation or colimitation. Multiple nutrient limitation may also be important for soil microorganisms, particularly in the tropics (Camenzind et al., 2018; Soong et al., 2018). Individual organisms experience multiple nutrient limitation if trait plasticity can balance the use of some resources for acquisition of other resources, optimizing overall growth. For example, in the presence of only poorly available, polymeric C, microbial growth may be enhanced by the influx of readily available soluble C. However, an influx of N may also enhance microbial growth if the additional N can be allocated toward the production of N-rich extracellular enzymes that will enhance the availability of the polymeric C (Allison et al., 2010). As a second example, N-fixing, mycorrhizal plants allocate a great deal of photosynthate C to microbial root symbionts that enhance supply of N and P. Increasing the supply of either mineral N or P allows the plant to reduce photosynthate investment in one of the root symbioses, freeing photosynthate to be allocated to the other symbiosis and aboveground growth.

Multiple nutrient limitation can also occur through shifts in biomass nutrient ratios themselves, altering the resource needs to maximize growth. Although soil microbial biomass converges on a fairly consistent global average C:N:P ratio (Cleveland and Liptzin, 2007), these ratios can shift locally in response to environmental resource supply (e.g., Soong et al., 2018). It is possible that shifts in biomass nutrient ratios occur through individual species trait plasticity, although it is probably more likely driven by changes in species composition of the soil microbial community. The impact of changes in microbial resource requirements can in turn impact function of the microbial community. Differences in the C:N ratio of fungal and bacterial biomass (as well as other trait differences between fungi and bacteria, such as growth efficiency) have significant consequences for soil C and N cycling (Waring et al., 2013).

10.2.2 Intraspecific competition

As organisms grow and reproduce, resources are consumed, reducing resource availability for other organisms with the same requirements. Resource limitations reduce activity, growth, and reproduction rates yet also increase death rates. This reduction in the performance of one organism by another organism that consumes the same resources is called competition. One of Darwin's key observations was that because members of the same species have nearly identical resource needs, competition for resources among members of the same species (intraspecific competition) is fiercer than among members of different species (interspecific competition) (Kingsland, 1991). The logistic growth equation is a mathematical model that describes the effect of intraspecific competition on the change in population size over time (population dynamics). The probability of an individual reproducing minus the probability of death per unit time is equal to the overall amount by which a population grows or shrinks in that period. This is represented in the logistic growth equation as the population-specific growth rate, μ, and is a direct manifestation of the suitability of the habitat. The differential equation for logistic growth is:

$$\frac{\mathrm{d}N}{\mathrm{d}t} = \mu \cdot N \qquad \text{Eq. 10.1}$$

where

$$\mu = r \cdot \left(\frac{K - N}{K}\right) \qquad \text{Eq. 10.2}$$

and where N is the number of individuals in the population, t is time, and $\mathrm{d}N/\mathrm{d}t$ is the change in N over time. The first important trait represented in this equation is the intrinsic growth rate of the population (r), which is the value μ approaches when resources are not limiting growth and there is no intraspecific competition. The number of individuals that the resources of a habitat can support (K) is referred to as the carrying capacity and models intraspecific competition with a constant level of resource supply. As a population grows, the resources must be shared among many individuals, decreasing the reproduction rate and increasing the rate of death. In the equation N approaches K, causing μ to approach zero. If the population is above the carrying capacity, it cannot be supported by the resources present, μ becomes negative, and the population declines. The relationship where population growth rate is sensitive to population size is known as density-dependent population regulation.

Lifetime patterns of growth and reproduction, including timing of reproductive and dormant stages, are known as a species' life history. Life history strategies have important consequences for population dynamics. The logistic growth equation led to MacArthur and Wilson's (1967) concept of *r*- and *K*-selection, which is used to generalize species' life histories. *K*-selected species (with high *K* values and low *r* values) have traits that favor the persistence of individuals under conditions of scarce resources and high intraspecific competition (e.g., elephants, humans, nut-bearing trees, Actinobacteria). These conditions occur when populations remain near their carrying capacity (*K*). In contrast, *r*-selected species have the opposite characteristics, with relatively high efficiency in converting resources to offspring (e.g., rodents, weeds, Gammaproteobacteria). The *K*-selected strategy is an adaptation to environments where conditions are relatively stable, resulting in density-dependent mechanisms of population regulation, while the *r*-selected strategy is an adaptation to a variable environment with high resource levels. Pianka (1970) made further predictions about a variety of traits that could be associated with *r*- versus *K*-selected species (e.g., *r*-selected species should have more variable population size, weak competitive interactions, rapid maturation, short lifespan, and high productivity). However, it was soon realized that selection pressure and reproductive value at different ages can lead to opposite traits (Reznick et al., 2002). Both the logistic growth equation and the *r*-/*K*-selection model are now used as conceptual tools. A more detailed understanding of population dynamics is reflected by explicitly modeling resource consumption using the Monod model (Panikov, 1995) or, similarly, using structured demographic models to better understand the effects of life history and age-related variation (Reznick et al., 2002).

The classification of organisms as *r*- or *K*-selected is common in soil biology. Microbial colonies are typically classified based on the amount of time it takes for them to appear in laboratory isolation media. These designations must be made with reference to a particular environment. Laboratory isolation conditions represent a small range of the conditions encountered in the environment. The environment of a batch culture changes continuously as nutrients are not replenished, and wastes are not removed. Hence *r*-selected species can be positively identified (with respect to the isolation conditions), but *K*-selected species cannot. Soil microbiologists have also created other classifications similar to the *r*-/*K*-selection dichotomy that are more focused on species' preferred resources than on intraspecific competition. In 1925 S. Winogradsky used the term "autochthonous" to describe organisms that grow steadily on resistant organic matter with a constant presence in the environment, and "zymogenous" for organisms that proliferate on fresh organic matter (Panikov, 1995). In another scheme oligotrophs grow only at low nutrient levels, while copiotrophs grow quickly at high nutrient levels.

The above discussion is affected in the soil habitat by the very large number of resting and protected bodies that allow population maintenance during adverse environments.

10.2.3 Dispersal in space and time

As resources become depleted, organisms may move, or disperse, to new areas to avoid the negative effects of competition. Such movement can be short range to exploit a nearby resource patch, or long range, resulting in the establishment of a new population. Active dispersal involves the expenditure of energy by the organism. Passive dispersal occurs due to the movement of material the organism is attached to or caught in (e.g., wind or water). Passive dispersal can be truly passive with no energy expended, or an organism may prepare morphologically and physiologically by entering a new life stage. Stages for dispersal are typically more resistant, dormant, or mobile than growth stages. The fruiting bodies and spores of fungi and myxococci are examples of elaborate life stages for passive dispersal.

Passive dispersal of bacteria with water flow can occur if the bacteria are not adsorbed onto immobile soil particles or protected by soil structure. Cell size limits the passive movement of organisms with water in soil due to the sieving effect of soil particles. Larger-celled organisms, such as yeasts and protozoans, do not experience passive dispersal to as great a degree as bacteria and viruses; however, nondormant protozoans are typically engaged in active dispersal to obtain food. Plant roots, seeds, fungal spores, and chemical substrates found within several centimeters of particular soil bacteria have been shown to induce chemotactic responses (active dispersal) that may be important for responses, such as rhizosphere colonization (Compant et al., 2010).

For hyphal fungi growing in soil, passive dispersal is normally restricted to spores. Spores may be produced in the soil, such as by arbuscular mycorrhizal fungi or in sporocarps (fruiting bodies) above the soil surface. Sporulation is often induced by environmental cues, such as moisture. Spores are also dispersed by animal activity both above- and belowground. Vegetative growth of fungal hyphae can also be considered a form of active dispersal since new areas of the soil are being explored. Fungi often have distinct forms of hyphal growth for nutrient acquisition versus dispersal. Hyphae for dispersal, such as rhizomorphs, grow more rapidly, are thicker and tougher, and may be formed by anastomosis (cellular fusion) of multiple smaller hyphae (Rayner et al., 1999). The strategy behind rhizomorphs is to invest little and maintain impermeable surfaces during exploration of a resource-poor environment until a resource-rich patch is encountered. When such a patch is encountered, there is proliferation of thinner, more permeable hyphae with a higher surface-to-volume ratio.

The ability of an organism to enter a dormant phase can also be seen as a dispersal mechanism, but through time rather than space. This form of dispersal is one version of the storage effect, where reproductive potential is stored across time, resulting in higher reproduction rates under favorable environmental conditions. Entrance of individuals into a dormant life stage may be a developmentally programmed event for some species or may be induced to avoid density-dependent competition. For many organisms, life stages that facilitate passive dispersal in space are also optimal for dispersal in time. This is true of plant seeds and fungal spores.

Species are often distributed in multiple populations that are linked by dispersal and dormancy, together called a metapopulation, which can create significant feedback between populations and contribute to species persistence (Harrison, 1991). This may be particularly important in soil, where there is a large degree of spatial and temporal heterogeneity (Ettema and Wardle, 2002). Many bacteria proliferate in spatially and temporally discrete habitats with increased availability of resources, including near roots, on particulate organic matter, and within worm casts. However, soil microorganisms appear to be generally dormant, as indicated by the increase in numbers and metabolic activity when soil is amended with water or nutrients. Bacterial cells entering a dormant stage are known to undergo a suite of biochemical and morphological changes, including reduction in size. "Dwarf" cells ($<0.07\ \mu m^3$ biovolume or $<0.3\ \mu m$ diameter) make up most bacterial cells in soil (Kieft, 2000). These dormant, passively dispersed cells provide the population base for colonization of new patches of resource-rich habitats (Berg and Smalla, 2009; Poll et al., 2010).

10.2.4 Interspecific competition

We have discussed in the previous sections how a species' geographic range and habitat occupation is limited by the species' traits resulting in overall adaptation to a set of environmental conditions. The effects of abiotic factors on the survival or growth rate of a population could be plotted with each axis corresponding to one factor. If we imagine many axes, each defining one dimension of an *n*-dimensional space, the region of this space suitable for growth of a species is what Hutchinson (1957) envisioned as the species' fundamental niche. The fundamental niche of a species describes all combinations of environmental conditions that are acceptable for the persistence of a population. Because of predation and competition with other species, populations of a species will not be present in all habitats that satisfy the species' fundamental niche. The reduced hypervolume corresponding to the conditions that a species is able to occupy is called its realized niche. Interactions between species are fundamental processes in determining the species that will be present in a given location.

Interspecific competition can operate according to the same mechanism as intraspecific competition, except that the individuals competing are from different species. A finite pool of resources is available at any given time, yet if they are consumed faster than they are replenished, growth rates decline. The strongest competitors are able to maintain the highest growth rates despite low levels of resources, thus driving these resources to even lower levels. This form of interspecific competition is known as resource-based or exploitative competition. Given this description of interspecific competition, how do similar species coexist? One common explanation is that spatial and temporal resource variability leads to limited growth in different times and places. Because a species trait can only be optimal under a certain set of conditions, species that are superior competitors for one resource are typically not as competitive for others. Tilman (1982) suggested that the number of similar species that can coexist in a habitat should be equal to the number of potentially limiting resources used in that habitat. It is often assumed that the high heterogeneity of soil conditions supports the enormous microbial diversity.

Species coexistence may also be supported by fluctuating rates of mortality. An event that causes the sudden mortality of an otherwise competitively dominant species or group of species is known as a disturbance. Soil tillage is an example of a disturbance that is well known to be detrimental to fungi, favoring bacteria. Species that are *r*-selected may be out-competed under normal conditions but flourish when mortality spikes for the dominant species. Fugitive species avoid competition by dispersing into habitat patches where the dominant species have become locally extinct. Mortality rates can also be altered through interference competition, where one competing species impacts another through direct aggressive action rather than resource use. This type of competition would be likely in systems containing antibiotic-producing organisms. Davelos et al. (2004) confirmed Waksman's earlier observations, where a wide variety of antibiotic producing and antibiotic resistant phenotypes were found to be present in soil streptomycetes (phylum Actinobacteria) at one location, suggesting many organisms are capable of competitive interference. It also suggests that organisms have developed mechanisms to avoid this type of interference.

Similar species could also evolve to use different subtypes of the same resource, or their traits and niches can shift in other ways. This is known as resource partitioning and was taken as some of the first evidence of competition and natural selection. The concept that no two species having identical niches can coexist is the competitive exclusion principle. Extinctions due to competition have been documented, yet evolution can allow an environment to be partitioned into an astonishing array of niches. For example, it has been shown that pure cultures of *Escherichia coli* (class Gammaproteobacteria) evolve into distinct, coexisting subtypes because of physiological trait differences (Marx, 2013).

In soil competition has been exploited as a mechanism for biocontrol but has also been blamed for the failure of many soil inoculation programs. Fluorescent pseudomonads have been shown to suppress a variety of plant pathogens by secretion of antibiotics (interference competition) and siderophores which sequester iron (resource competition). Nonpathogenic strains of fungi can be superior competitors for carbon (C) and root colonization sites. Organisms introduced to sterilized soil often survive, while populations decline rapidly in nonsterile soil. Promising newer approaches to manage microbial competition with plant pathogens include inoculation of a greater diversity of biocontrol agents (Hu et al., 2016) or use of organic amendments (Bonanomi et al., 2018), both of which also stimulate the complex community in the plant microbiome. Besides biocontrol agents, a relatively short half-life of introduced populations has been observed for a variety of other types of microbes introduced to soil, including rhizobia, fecal organisms, and genetically modified microbes. This has been attributed to competition but could also be a result of trophic interactions.

10.2.5 Direct effects of exploitation

Biological interactions that affect assembly of microbial communities include exploitation and mutualism, as well as intra- and interspecific competition. Exploitation has many forms in the microbial world, including predation, herbivory, parasitism, and pathogenesis. These are trophic interactions, where energy or nutrients are transferred from one organism (prey) to another (the consumer). While studying energy flow through an ecosystem, it can be useful to categorize consumers into trophic levels (e.g., herbivores and carnivores). To understand the impact of exploitation on population dynamics, it is important to know if the consumer is a predator, causing the immediate death of prey, or a parasite, obtaining only a portion of the prey's resources without killing it. Another important distinction is between a generalist that consumes many different types of prey species and a specialist that consumes a very narrow, specifically defined range of prey species. As with all categories in ecology, there is a gradient of lifestyles that fall between the extremes. Exploitation in soil biota is widespread. Most soil animals, including protozoa, nematodes, collembola, mites, and earthworms, obtain their resources through exploitation of bacteria, fungi, or plant roots. Invertebrates that ingest plant detritus normally get most of their energy and nutrients from microorganisms residing on the detritus (their "prey") rather than directly from the detritus. Many species of fungi have been shown to attack bacterial colonies and other fungi; there are also fungi that attack soil animals, such as nematodes. *Bdellovibrio* (class Deltaproteobacteria) is a bacterial predator that attacks other bacteria. All organisms also appear to serve as a habitat for an assemblage of smaller organisms, many of which are parasitic. An extreme example of this is the presence of secondary, smaller viral genomes within larger viral genomes, which are dependent on both their bacterial and viral hosts (Swanson et al., 2012).

Predators and parasites will aggregate in patches of high prey density and in areas where prey populations are growing. High numbers of bacteria near roots and decomposing plant litter are thought to increase nematode and protozoan populations, which in turn increases microbial N mineralization (Irshad et al., 2011; Koller et al., 2013). High growth rates of bacteria can increase production of viruses in soil (Srinivasiah et al., 2008). Predatory pressure is also a factor in habitat quality for the prey. Predator-free patches serve as refugia for prey populations and can significantly impact metapopulation dynamics. In soil there is a hierarchy of pore sizes that can serve as refuges for progressively smaller soil fauna and microorganisms while also regulating volatile signals and the ability for organisms to move toward resources (Erktan et al., 2020). Elliott et al. (1980) found that a finer-textured soil contained more bacteria protected from predation by nematodes compared to a coarse-textured soil. The finer-textured soil contained a larger proportion of pores too small for the nematodes to utilize, serving as a refuge for bacteria from nematodes. Amoebae were able to use these pores, consume bacteria, and emerge as food for nematodes. There was a greater growth of nematodes when amoebae were added to a fine- compared to a coarse-textured soil.

The effect of predation on prey population dynamics is an increased death rate. Parasitism is a considerably more complicated phenomenon to model than predation because prey are weakened by parasites, which impacts reproduction and death rates. Parasitism can decrease the accumulation of biomass or rate of development. Parasitism also increases the death rate, either through prolonged exposure to the parasite or by making the prey more sensitive to other causes of mortality.

The details of a parasite's transmission route between hosts are critical to understanding how the parasite is spread. Some parasites can colonize new hosts from dead tissue. The plant root pathogens in the genera *Gaeumannomyces*, *Rhizoctonia*, and *Pythium* (phyla Ascomycota, Basidiomycota, and Oomycota, respectively) can live saprotrophically within plant residue and then colonize new roots from these habitats. High-quality habitat patches allow pathogenic hyphae to grow further through the soil (to at least 15 cm) to colonize new roots. The probability distribution of colonization of a root from a particular inoculum source also depends on a variety of other factors, such as the species involved, as well as soil temperature, moisture, and texture. Planting crops at wider distances (i.e., reducing host density) is known to reduce the spread of root diseases. Some parasites are transferred by other species or other components of the environment (vectors), with their spread tightly linked to the dynamics of these factors.

10.2.6 Indirect effects of exploitation

The genetic makeup of prey species will respond to exploitation through evolution resulting in defensive adaptations. Defenses from exploitation can take a variety of forms, including behavioral, morphological, or biochemical. Evolution can also result in the development of new attack strategies in consumers, resulting in a continual coevolutionary arms race between consumers and their prey. Exploitation can have a large influence on the outcome of competitive interactions between prey species. It can contribute to the coexistence of competing prey species by reducing the population size of the superior competitor. This results in increased resource abundance and ameliorates competition. Most predatory soil fungi, protozoa, nematodes, and collembola consume multiple prey species. However, they either show feeding preferences for or receive enhanced benefits from particular prey species, which in turn affects microbial community composition and activity (Bray et al., 2019; Lucas et al., 2020). These trophic relationships can therefore impact soil biogeochemical processes and competition among plant species by altering nutrient availability and mycorrhizal fungal effectiveness (Crowther et al., 2012). In addition, the indirect effects of soil pathogens exploiting plants are a dominant driver of plant community composition and diversity (Bever et al., 2015).

Exploitation pressure at lower trophic levels can be regulated by exploitation at higher trophic levels in a process called a trophic cascade. If the population size of fungal-feeding, invertebrate grazers is limited through heavy predation, then the pressure on fungi from the grazers will be low. Although trophic cascades may be common in other environments, a meta-analysis by Sackett et al. (2010) found that they were not widespread in soils when considering entire trophic levels and effects on nutrient cycling. However, exploitation can still regulate species-specific interactions and are often the basis of biocontrol strategies of plant pests.

The many species-specific trophic relationships between organisms in ecosystems result in a complex web of interactions (a food web). When microorganisms are included in soil food webs, the increase in complexity at the species level has been viewed as overwhelming. Thus microbes are typically represented by a single trophic level (i.e., an undifferentiated pool of microbial biomass) or are divided into very broad groups (e.g., bacteria and fungi). This is understandable because of the enormous diversity of soil microorganisms, the unknown role of each taxon in a food web, and the fact that the focus of soil food web studies has typically been on biogeochemical processes, not community structure. However, this approach masks unique features of food webs arising when microbial species are included explicitly. There are no "top predators" in food webs containing microorganisms because all organisms are exploited by parasites of varying lethality. Also, the presence of "three-species loops" has been the subject of controversy in food webs of macroscopic organisms and may be possible only when there is differential predation on species due to developmental stage. In microbial systems this food web structure has not been explicitly investigated, but since many predators within the system are generalists, it seems likely that such loops can frequently occur due to random encounters. Soil food webs must also account for the presence of decomposer organisms, the resources they utilize, dead organisms, and their by-products (detritus). This decomposition is critical to the recycling of nutrients for use in primary production. Decomposer organisms affect population dynamics of primary producers by supplying nutrients and often by competing with primary producers for the same resources (immobilization). In addition, including detritus, nutrients, and decomposers in food webs creates a variety of indirect pathways for organisms to interact (Moore et al., 2004; Moore and de Ruiter, 2012).

10.2.7 Mutualistic interactions

Mutualisms, interspecific relationships beneficial to both organisms involved, are also of great ecological significance in ecosystem dynamics. A diverse array of cross-kingdom partnerships has existed throughout evolutionary history. Organisms in soil collaborate with a wide variety of plants to perform nutrient acquisition services in exchange for plant-derived carbohydrates. While the relationships were originally perceived as bacteria in symbiotic relationships for acquiring N and fungi acquiring P, studies have indicated that fungi are involved with the acquisition of almost any limiting nutrient in soil depending on partnering species (Allen, 1991; Smith and Read, 1997).

Mycorrhizae, the relationship between a plant root and fungus, is one of the most ubiquitous soil mutualisms and may be one of the oldest plant-microbe relationships (Brundrett, 2002; Stubblefield and Taylor, 1988). There is evidence that this relationship evolved and was lost multiple times in different divisions in the kingdom Fungi and in different groups of plants. While the original understanding of mycorrhizae suggested the interaction was simple, there have been many studies that have revealed a more complex system that goes beyond one plant and one fungus to whole ecosystem influences that are complex (Delavaux et al., 2017). Some mycorrhizal fungi acquire nutrients directly from decomposing litter (Leake and Read, 1997) and can increase the rate of decomposition by adding C for free-living decomposers or decrease the rate of decomposition by competing with other soil organisms for nutrients (Frey, 2019; Lang et al., 2021). Some mycorrhizal fungi can also get nutrients from live animals, such as springtails (Klironomos and Hart, 2001), and act both as mutualists and predators. Beyond the acquisition of nutrients, studies have found that mycorrhizal fungi can ameliorate stress in plants. It has long been known that mycorrhizae influence plant water relations (Allen, 1991); however, in more recent studies it has been found that they can increase the productivity of plants growing under drought conditions (Jayne and Quigley, 2014) and in highly saline soils (Chandrasekaran et al. 2019).

Mycorrhizal relationships alter the aboveground community both directly, by changing rates of reproduction and death of participant species and indirectly by altering competition among plant species. In terms of species specificity there is not a great deal of consensus on the degree to which partnerships require or benefit from species-specific fungal-plant pairings. There is evidence that there is a wide array of generalists. However, the discovery of what is now called common mycorrhizal networks has drawn attention to the fact that while a single fungus may be a mutualist for one plant, it may also be taking resources from another, suggesting that the ecological relationships are more complex than previously thought (Simard et al., 2012).

While mycorrhizae are some of the most terrestrially important mutualists, other soil mutualists are also essential. Much of the N available in soil systems is present because of N-fixing bacteria, such as rhizobia. Studies over the last 100 years focused on understanding this relationship were often designed with single plant and rhizobial species, with an emphasis on plant growth impacts. More recent analyses of these data and new experiments that include a more realistic inclusive approach have found that, as with mycorrhizal fungi, there are factors beyond the one plant-one rhizobia interaction that impact plant growth characteristics (Friesen, 2012; Friesen and Heath, 2013). Rhizobia are essential in newly establishing plant communities during primary succession following catastrophic disturbances to soil. Since plants require N for survival, those that are early colonists on highly disturbed sites have often formed symbioses with bacterial N fixers to acquire this necessary nutrient. There is evidence that rhizobia transferred genes to plants deep in their evolutionary history, supporting the evolutionary importance of this relationship for plant survival (Lacroix and Citovsky, 2016).

Early soil formation from rock is impacted by lichen, mutualistic associations of fungi, algae, and bacteria. Lichens stimulate breakdown of rock through hydrolysis, the production of organic forms of N through fixation, and perform a wide range of other activities that influence ecosystem characteristics (Asplund and Wardle, 2017). These associations are also important as macrobiotic crusts necessary for soil stabilization in easily erodible soils (see Chapter 8).

10.2.8 Community impacts on abiotic factors

Interactions between organisms have been discussed in terms of resource use. However, environmental modulators can also be affected by organisms (Table 10.1). The activities of both nitrifying bacteria and plant roots decrease soil pH, and soil temperature is affected by plant and litter cover. There are also a variety of differences that develop under arbuscular mycorrhizal and ectomycorrhizal trees, including in soil pH and C and N chemistry (Tedersoo and Bahram, 2019). This can have positive or negative impacts on the growth of another species, depending on the species' niche requirements.

Some organisms alter the spatial arrangement of components of the environment or serve as new habitats themselves. These organisms, often referred to as ecosystem engineers, have widespread effects on an ecosystem beyond their own resource use (Jones et al., 1994). Large, competitively dominant organisms, such as trees, are obvious examples of ecosystem engineers. In soils earthworms are important ecosystem engineers because they bury plant litter and create macropores, with large impacts on water infiltration, microbial community structure, and C and nutrient cycling (Eisenhauer et al., 2011; Stromberger et al., 2012).

10.2.9 Community variation among soil habitats

The ecological processes described thus far in this chapter (physiological constraints, competition for resources, exploitation, and mutualism) act locally to determine the combination of species that can persist while interacting within a habitat patch with particular environmental conditions. Together, these processes result in "selection" of organisms based on species niches, which is one of four forces shaping community composition described by Vellend (2010) and adopted by many microbial ecologists (Nemergut et al., 2013). Because soil conditions are highly variable across a broad range of spatial and temporal scales, there is enormous potential for variation in soil microbial communities to be generated based on selection alone. The implication is that community composition would be predictable from soil habitat type or environmental variables, and indeed, many studies have identified soil conditions with a strong selection effect on microbial community composition. In this section we explore these factors across multiple spatial scales. Note, however, that the predictability in community composition due to selection is counteracted by the other three forces that Vellend (2010) identified as affecting communities (dispersal, drift, and diversification), which we will return to in Section 10.2.11 below.

Starting at the scale at which single-celled microorganisms interact with the soil environment, the most active microorganisms are in habitat patches characterized by available labile organic matter, moisture, and nutrients ("microbial hot spots;" Kuzyakov and Blagodatskaya, 2015). The rhizosphere, fecal matter, and decomposing plant tissue are examples of this type of habitat. These are areas of increased biogeochemical activity and interactions with plant roots. These habitats harbor increased microbial biomass and food web activity, and there is strong selection for species able to compete under these conditions (Kuzyakov and Blagodatskaya, 2015). Microbial hot spots tend to be transient, and most bacterial and archaeal cells in soil are not found in these active patches but reside in a very different microhabitat: on mineral surfaces in pore spaces, which are inside soil aggregates (Erktan et al., 2020). Most microorganisms in soil pores are isolated from each other, root exudates, and other resources by complex pore networks, small pore neck sizes, and air-filled pores (Tecon and Or, 2017). Because of the lack of labile organic matter or other resources, these cells are thought to be mostly dormant. Many dormant microbes in soil will rapidly become active if local conditions become favorable, and the microorganisms that proliferate in response to hot spots are often recruited from the surrounding soil. This soil habitat can be further divided based on pore size classes (micropores, mesopores, and macropores), aggregate size classes (microaggregates and macroaggregates), and aggregate types (fungal, root, invertebrate, or abiotically generated), reflecting further variability in habitat conditions that affect microbial activity and select for different microbial species (Gupta and Germida, 2015; Lavelle et al., 2020).

Soil organisms with hyphal growth forms (i.e., mostly fungi) are able to span the soil environment on a larger spatial scale than individual aggregates, rhizospheres, or decomposing organic matter particles. This growth form allows these organisms to grow across pore spaces that cannot be crossed by single-celled organisms and to integrate over multiple patches of microhabitats (Tecon and Or, 2017). Hyphal organisms can transport resources from one area to another, effectively correcting nutrient imbalances across habitats within the soil, a function that is critical for many ecosystem processes in the context of spatial variation in soil (Collins et al., 2008; Watkinson et al., 2006).

At the somewhat larger spatial scale of an individual plant or research soil pit, biotic communities are usually differentiated by soil depth, which is an indicator for many other soil variables. Increasing soil depth corresponds with reductions in the following: root activity, deposition of senesced plant tissue, disturbance, soil mineral weathering, hydrology, and atmospheric gas exchange. These factors result in reduced biotic diversity and biomass at increasing depths, with soil depth typically causing greater differentiation in community composition than other factors investigated (Too et al., 2018; Upton et al., 2020; see Chapter 8).

At larger spatial scales (tens of meters to thousands of kilometers), there are many environmental factors that have been investigated for effects on community composition, biomass, and activity within soil habitats. Fierer (2017) arranged commonly investigated factors into a hierarchy based on their importance in determining soil bacterial communities in a generic setting, with soil pH, organic C quality and quantity, and soil redox status as the most important factors. However, the ability to predict microbial community composition, or even predict the most important drivers of community composition, is highly context dependent (e.g., is pH varying by <1 or >3 units?). Investigations over the strongest environmental gradients and at the broadest taxonomic levels tend to result in the clearest patterns of species selection shaping community composition (Rousk et al., 2010; Vasco-Palacios et al., 2019). For example, the effects of land use and management practices on soil biotic communities will be strongest in areas with the most dramatic shifts in soil chemical and physical properties and vegetation (Barnett et al., 2020; Wang et al., 2017). However, there are many confounding factors among soil types, ecosystem types, and land uses, with a complex combination of disturbance and soil factors likely involved in differentiating biotic communities at the regional scale.

An important case of context dependence involves the expectation that growth of different plant species will select for divergence in the composition of soil microbial communities. Beyond altering soil physicochemical properties, plant species also affect microbial communities by interacting directly with microbial mutualists and pathogens. They release different suites of compounds into the rhizosphere, from simple organic acids to complex secondary metabolites, through root exudation or the decomposition of various plant tissues. It is well established that plant species select for particular pathogens, mycorrhizal fungi, and endophytic and rhizosphere organisms (collectively called the "plant microbiome;" Trivedi et al., 2020). This species selection within the plant microbiome has strong implications for plant health and success. Although plants are adapted to favor microbes that provide them the greatest benefit, the microbes that flourish the most in the rhizosphere are typically those best able to exploit that environment (Bever et al., 2015). The result is the common phenomenon of negative plant-soil feedback, which can regulate plant population size (Hovatter et al., 2013), community composition (Bever et al., 2015), and ecosystem processes (Mommer et al., 2018).

10.2.10 Changes in community structure through time

A landscape is comprised of a variety of habitats created by environmental heterogeneity and disturbance events, whether we are considering a biotic landscape within the soil or a vegetation landscape over hundreds of km (Wiens, 1997). Communities also change over time, including over very short to very long timescales. Succession is the replacement of populations in a habitat through time due to ecological interactions. Because creation of habitat patches and community succession is often predictably repeated over time, the habitats and communities in a landscape can be thought of as a shifting mosaic in a dynamic, steady state (Wu and Loucks, 1995). After a habitat patch is created by introduction of environmental heterogeneity or a disturbance event, the initial species that colonize the habitat patch are typically *r*-selected species with dispersal strategies to increase the chances that they are the first organisms to colonize newly created habitats. These pioneer or "fugitive" species make opportunistic use of available resources or have mechanisms to increase rates of nutrient cycling, such as N fixation. Fugitive species are replaced over time by more competitive species; for example, plant species more tolerant of shade or low soil nutrients, or soil microbes capable of producing antibiotics or utilizing more recalcitrant organic matter.

Many small-scale microbial hotspots in soil, such as decomposing plant litter or fecal particles, are habitat patches defined by a limited pool of labile resources. In addition to colonization and competition within the microbial community, succession in these habitats is driven by a constant change in environmental conditions as resources are used up and the environment is restructured (Kuzyakov and Blagodatskaya, 2015). The resulting turnover in major groups of soil biota is somewhat predictable, although also determined by the type of hotspot (e.g., the biochemical composition of plant tissue) as well as the surrounding habitat (e.g., surface soil versus buried, decomposing wood; Frankland, 1998; Kohout et al., 2021). Soil communities vary over longer timescales of soil development (thousands to millions of years), which is likely to be primarily associated with changes in vegetation and soil physicochemical properties (Freedman and Zak, 2015; Turner et al., 2019).

10.2.11 Diversification, drift, selection, and dispersal

According to the Baas Becking paradigm (O'Malley, 2007) quoted at the beginning of this section, with perfect knowledge of organism traits and environmental conditions, one should be able to use the theories described above to predict species abundance and community composition in any habitat patch. However, there is almost always more variation in community composition than one might predict, with communities in similar habitats including different species and varying abundances. It has become clear that there are ecological and evolutionary processes that do not allow populations and communities to reach true stable equilibria predicted by deterministic theories alone. Vellend (2010) described three categories of processes that can counteract the deterministic aspect of niche-based species selection, modified for microbial ecology by Nemergut et al. (2013). Diversification involves genetic mutations and evolution of species, ongoing processes that generate variation by introducing new entities at particular locations. Drift describes the unpredictable effects of stochastic variation in population birth and death rates on community composition. Such variation among populations of the same species could be the result of genetic differences, chance spatial arrangements, or historical events. If either drift or diversification rates are faster than processes involved in species selection (i.e., physiological, competitive, trophic, and mutualistic interactions), environmental conditions will not fully determine the presence/absence and abundance of species within a particular area. Diversification and drift create heterogeneity among communities, while species selection reduces heterogeneity among areas with similar environmental conditions and maintains heterogeneity among dissimilar environments.

The final category in Vellend's (2010) synthesis is dispersal, or the movement of individuals to new areas. This is essential for species to colonize new habitats. Dispersal rates that are either very low or very high can counteract species selection, particularly in the context of simultaneous drift and diversification (Langenheder and Lindström, 2019). The best adapted species may be unable to occupy a habitat patch if there are physical or other barriers preventing dispersal to the area, resulting in dispersal limitation. Similarly, dispersal limitation resulting in stochasticity in the order and timing of colonization by different species can amplify the effects of drift during succession. At the other end of the spectrum, poorly adapted species may be found in a habitat due to frequent colonization from nearby source populations, a pattern known as mass effects. The potential importance of dispersal is highlighted by the concept of a metacommunity, which is a group of communities linked by dispersal across a landscape, resulting in emergent, regionally driven community dynamics (Leibold et al., 2004).

Given the importance of dispersal in the balance among the other three processes, the arrangement of habitats in the landscape can have a dramatic effect. The characteristics of a patch's edges and surrounding habitat and ease of dispersal across other elements of the landscape may all be important determinants of local patch population dynamics. The environment can also change, resulting in evolutionarily novel conditions and nonequilibrium dynamics. Thus, at the microhabitat scale within soil, there is likely a significant variation in all four processes over time and among locations. Many microbial communities experience long periods of dormancy and short dispersal distances through a complex pore space (Tecon and Or, 2017), punctuated by population explosions in microbial hotspots, such as those due to root exudation and invertebrate activity (Kuzyakov and Blagodatskaya, 2015). Because the timing and placement of microbial hotspots relative to soil aggregates is at least partly stochastic, sudden microbial growth would result in drift, followed by a period of species selection as a food web develops and resources shift and decline. Increased populations in microbial hotspots also result in longer-range dispersal, and perhaps more rapid diversification, followed by some fraction of the altered microbial community becoming dormant again. With a diversity of microhabitat types appearing and declining over time, an extremely heterogeneous network of pore and hotspot communities likely develops.

The processes of selection, dispersal, drift, and diversification are difficult to measure, but the effects of stochastic processes counteracting selection are evident from spatial heterogeneity at small scales (Blackwood et al., 2006; O'Brien et al., 2016). Decomposing leaves on the forest floor represent an example situation, with spatial structure in the fungal community suggesting that each fresh leaf represents a defined habitat patch, appearing in a highly heterogenous mixture, with drift, selection, and dispersal shaping the local fungal community (Feinstein and Blackwood, 2013).

A shifting mosaic of habitats and process rates can also be envisioned at larger spatial scales due to a wide variety of changes in the environment, including shifting plant species, human activity, geological activity, or climate change. At scales from tens of meters to thousands of kilometers, microbial communities are affected by niche-based species selection as described above but also exhibit a substantial amount of unexplained variation and independent spatial structure. However, the specific influence of drift, diversification, and dispersal on larger-scale biogeography of soil microbial communities remains ambiguous (Fierer, 2017; Langenheder and Lindström, 2019; Nemergut et al., 2013). These processes are even harder to study at large spatial scales than within the soil matrix. They are difficult to disentangle due to complex and confounded spatial patterns in environmental drivers and the extreme diversity of taxa and ecological niches (Bissett et al., 2010; Daws et al., 2020; Tedersoo et al., 2020).

The importance of diversification coupled with dispersal limitation may be most evident in cases where barriers to dispersal have broken down, resulting in migration of exotic species that establish populations outside their native range. Human activity has greatly increased the transport of materials around the globe. Earthworms from Europe were introduced to the Atlantic coast of North America and have been steadily colonizing new soils each year. The root pathogen *Phytophthora infestans* (phylum Oomycota) was introduced from Mexico to the United States, then transferred to Europe (causing the

Irish potato famine), and from there to the rest of the world (Goodwin et al., 1994). Transport of soil is now the subject of international law and regulations. The difference in effects of the introduced species in these two examples is interesting, given the questions raised above. *P. infestans* in agroecosystems has a substantial impact because it is involved in aggressive exploitation of crop plants (an important ecosystem engineer) and causes system reorganization. Earthworms play the role of a detritivore involved in comminution of plant tissue, with large effects on water infiltration, aggregation, and the maintenance of surface plant litter. There is great interest in elucidating the traits that lead some introduced species to become invasive because of the deleterious effects that many invasives have on native species and ecosystems (van der Putten et al., 2007).

10.3 Consequences of microbial community structure for ecosystem function

The structure of microbial communities and the unique contributions of their genetic makeup determine the range of activities that we describe as ecosystem functions. Soil microorganisms are richly diverse in the fundamental niches they can occupy. They span from strict aerobes to anaerobes, from consumers of simple inorganic to complex organic substrates, and from autotrophs to heterotrophs, with some organisms capable of multiple strategies. As a result of their physiologies and complex life history strategies, soil organisms leave a footprint on the systems in which they live. The characteristics of that footprint are a result of the metaphenome or the genes expressed by a specific community under specific in situ environmental constraints (Jansson and Hofmockel, 2018). Microbial functioning drives many aspects of the biogeochemical cycles required for terrestrial life, and the specifics concerning rate, process, product, etc., are driven by the metaphenome.

Ecosystems are systems defined by a specific composition of interacting populations and the environment within which these organisms interact. They are spatially defined by the interactions of the organisms and their relationship to physical space as an integrated system (Fig 10.1). The specific components of an ecosystem and the controls over their characteristics were best described as soil-

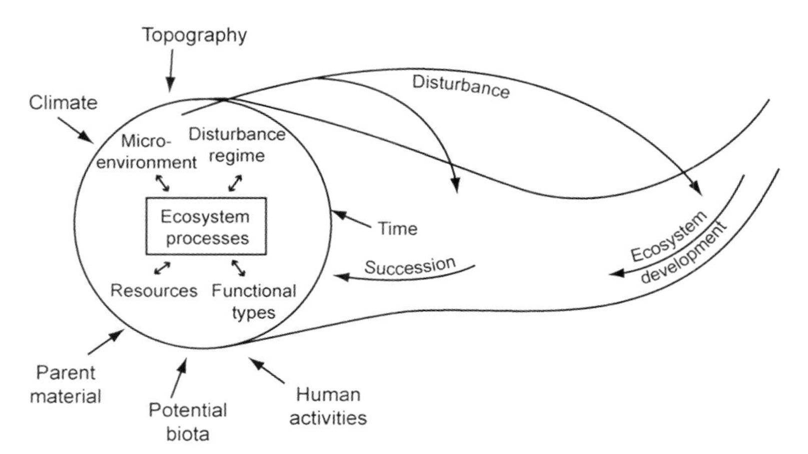

FIGURE 10.1 The relationships among state factors, the interactive controls (ecosystem development, disturbance, succession), and ecosystem processes (inside the box) that drive the specific and unique characteristics of ecosystems. *(Chapin et al., 2011; reprinted by permission from Principles of Ecosystem Ecology, 2nd ed., Springer.)*

forming factors by Dokuchaev (Jenny, 1961), and by ecosystem processes and interactive controls (Fig. 10.1). State factors include climate, parent material, biota, and topography, all interacting over time. Climate determines the rate at which biological and abiotic processes occur. The parent material determines soil nutrient and water-holding capacity, along with the types of biota that occupy that soil. Topography, slope, and aspect determine availability of water, movement of materials, soil depth, degree of weathering of parent material, and the total annual energy budget. As the severity of a slope increases, the sediment type can change as a result of erosion, with the top of the slope losing organic matter and the bottom of the slope gaining it. Aspect, the direction that the slope faces, has a large impact on moisture and plant productivity due to differences in moisture and the amount of radiant energy received. The time needed for soil development is related interactively with climate as environmentally harsh conditions require a great deal more time for soil development, while warmer environments with ample moisture and moderate temperatures require considerably less.

Potential biota include all organisms that can or have existed in an area. Deep-rooting grasses will contribute materials at a depth that will be converted into organic matter with slow turnover. Rooting depth, C:N ratio of materials added to soils, and the density and diversity of plants, animals, microbes, etc., all contribute differentially to the soil produced. Community function alters soil chemistry and soil development through processes that increase nutrient availability (e.g., N fixation, P solubilization) and/or alter decomposition rates. It also alters soil physical structure as a result of aggregation, which has a direct bearing on water infiltration, erosion resistance, and nutrient availability.

The high diversity of organic compounds available on the terrestrial surface has resulted in a wide range of organisms with a broad spectrum of enzymes. It has been suggested that the microbial community is functionally redundant, i.e., many species have duplicate enzyme systems and roles in decomposition processes so that alterations to microbial community composition do not alter function. This assumption has been challenged as we develop more sophisticated approaches for understanding the unique roles that individual taxa play in the ecosystem in which they are found. Strickland et al. (2009) used a microcosm approach to detect differences in community-level C mineralization rates across different communities, suggesting that each combination provided a unique set of metabolic physiologies resulting in different process rates. These findings are supported by metagenomic analyses of metabolic gene diversity (Röling et al., 2010). One recent hypothesis focuses on the phylogenetic niche conservatism theory, which states that bacterial taxa, given their small genome sizes, will all have the same basic core set of enzymes exhibiting small differences in the way each species crafts a niche, thus resulting in overlap in their metabolic processes (Jia and Whalen, 2020). In addition, there is empirical evidence that ecosystems with high functional redundancy are more resilient compared to those ecosystems with less redundancy or less biodiversity (Biggs et al., 2020).

10.3.1 Energy flow

The flow of energy in ecosystems begins with the energy source, flowing to autotroph, and then to heterotroph. For nearly all systems, the energy source is the sun and the autotrophs are plants, though there are some autotrophic bacteria. The term used to describe this flow is net primary productivity (NPP), which is the total energy uptake by plants that is subsequently available for use by other trophic levels. The amount of NPP in an ecosystem can most easily be predicted by climate (i.e., moisture and temperature). Biomes are easily plotted along moisture and temperature gradients (Fig. 10.2), showing corresponding increases in NPP. At shorter time scales, such as across seasons, NPP is controlled by leaf

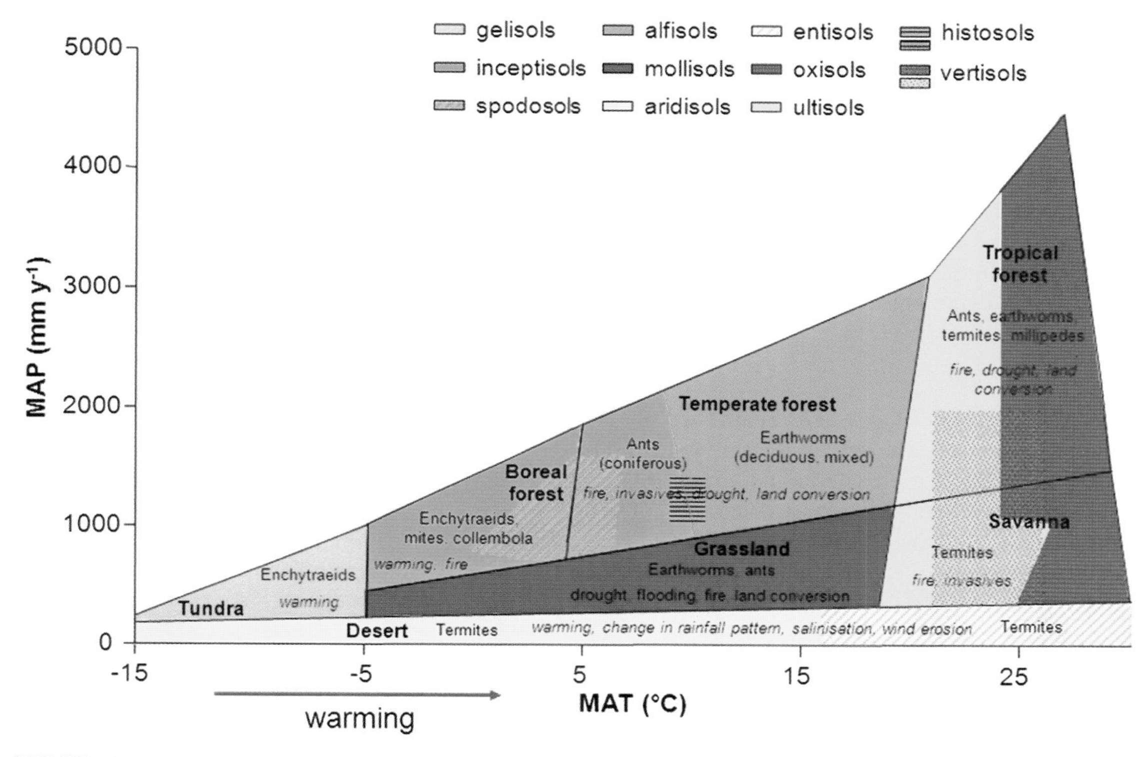

FIGURE 10.2 Dominant soil types and characteristic soil forming invertebrates across biomes (major global change threats are shown in italics). *MAT*, mean annual temperature; *MAP*, mean annual precipitation. For data and biome sources, see Brussaard et al. (2012). *(From Filser et al. (2016) with CC Attribution 3.0 License, which was modified from the original in Brussaard et al. (2012) and is reproduced with permission of Oxford University Press through PLSclear.)*

area, N content, season length, temperature, light, and CO_2 (Chapin et al., 2011). While autotrophs create a stable energy source using energy from the sun and CO_2 from the atmosphere, they require nutrients, water, and support from the soil. This requirement makes the contributions by soil biota to key ecosystem processes, such as decomposition and N fixation, every bit as essential to life on this planet as photosynthesis, allowing the addition of soils and organisms to the traditional Whittaker Biome Classification (Fig. 10.2). Conceptualizing these relationships together emphasizes that biome dynamics are driven by state factors. Energy transfers in ecosystems from the autotrophs to heterotrophs through consumption and decomposition are all essential for the development of the complex and diverse systems found on the terrestrial surface of the planet.

While photosynthesis and decomposition are key ecosystem properties, there are other essential trophic interactions. Plants and plant products influence the structure of microbial food webs. Studies following plant community change, such as that which occurs through restoration or the introduction of invasive species, have documented that individual plant characteristics have the greatest impact on microbial community composition (Bezemer et al., 2010), suggesting that it might be more appropriate to examine soil food webs under individual plant species rather than at the plant community level. However,

N turnover rates were influenced by the higher-order characteristics of the plant community, such as plant diversity; hence plant community structure overall does matter to soil food web dynamics. Other studies have found that plant species can alter microbial community composition (Schmid et al., 2021). Many studies have also found suppression of mycorrhizal fungi following colonization of soils by invasive species, such as garlic mustard (Rout and Callaway, 2012; Wolfe et al., 2008). This alteration to the microbial community decreases nutrient flow to plants and energy flow to microbial communities through alteration to community structure and, specifically, microbial food web dynamics. In the case of obligate symbionts trees requiring mycorrhizal fungi for growth and survival will decrease in number impacting diversity. Disturbances, such as global climate change and N enrichment, also have potential to alter belowground communities in ways that will impact plant productivity (Anthony et al., 2020).

10.3.2 Nutrient cycles

The term *biogeochemical cycles* emphasizes the intertwined roles of biotic and abiotic components for providing necessary molecules for the growth and reproduction of living organisms. The long evolutionary histories of the bacteria and archaea have allowed them to develop the machinery necessary to recycle nearly all naturally constructed, complex molecules to the atomic building blocks necessary for the nutrition of higher organisms. While the process of decomposition results in the simultaneous recycling of many different elements, the C, N, and P cycles provide examples of the activities of microbes in these cycles and the consequences of their actions for terrestrial, aquatic, and atmospheric element pools.

Flux in the global C cycle from terrestrial systems is determined by the relative rates of photosynthesis (C uptake) versus autotrophic and heterotrophic respiration as a consequence of decomposition and other microbial processes (C release) (Chapin et al., 2011). For example, the establishment of a mycorrhizal symbiosis increases photosynthetic rates, especially under conditions of stress including nutrient or water limitations (Augé et al., 2014). These relationships also alter photosynthetic rates indirectly as they mediate competition for resources and alter incidence of root pathogens. The rate at which material is decomposed is impacted, for example, by competition among decomposers for food resources, predation on decomposers by soil animals or other nonsaprophytes, and alterations to abiotic conditions in which the organisms live. The interactions among these systems are more complex than previously detailed (Tedersoo et al., 2020).

Human-induced alteration to global fluxes in the C cycle is probably the most important ecological experiment of all time. The global C cycle has been greatly altered by the large flux of CO_2 to the atmosphere created by fossil fuel use and land use change (IPCC, 2014). The increase in global temperature and increasingly variable weather patterns have altered the relative contribution of microbes to C emissions relative to C storage belowground. Net C flux to the atmosphere is predicted to increase over time under most scenarios as consumption of organic matter by bacteria and fungi increases with increased temperature and moisture, especially in arctic systems. Several studies have examined the relative contributions of different belowground organism groups to C flux (illustrated in Fig 10.3) and the impacts of elevated CO_2 and climate change on these fluxes (Manzoni et al., 2018).

Soil organisms directly and indirectly control C flux through decomposition and indirectly through impacts on plant productivity. They thus have an overwhelmingly large impact on biogeochemistry (Crowther et al., 2019). They have substantial control over the production and consumption of methane and soil N_2O emissions, which, as greenhouse gases, directly influence global temperatures. Most of the

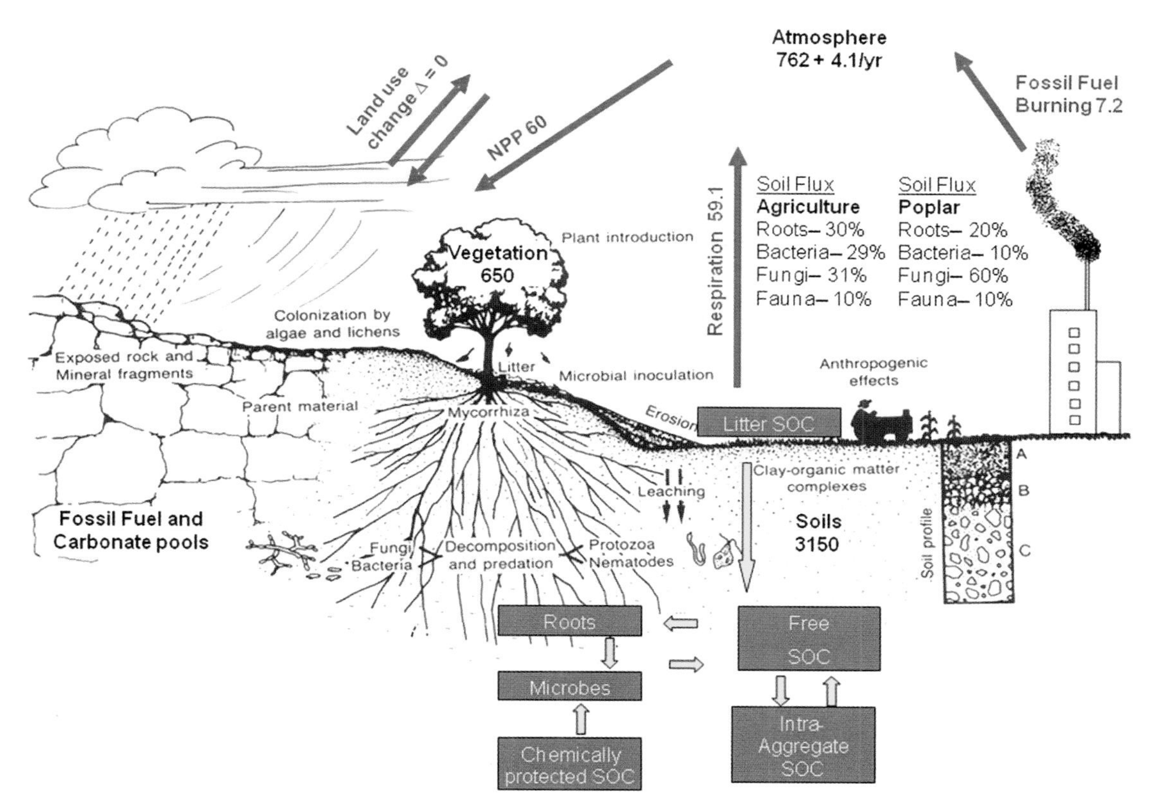

FIGURE 10.3 The terrestrial global carbon cycle, with pool and flux values in units of Pg C and Pg C yr^{-1}, respectively. Soil fluxes are driven largely by microorganisms. The global C cycle currently includes net flux rates of −2.2 from oceanic systems providing for the balance between surface flux rate (terrestrial + ocean = −3.1 per year), C release to atmosphere from fossil fuels (7.2), and current atmospheric accrual (4.1). While there have been many data sets published since the one these data are drawn from, this diagram illustrates the ecosystem characteristics necessary to understand controls on C flux within ecosystems. *(Modified from Paul and Clark, 1989; 1996, with pool and flux values from Denman et al., 2007.)*

true decomposers are heterotrophic osmotrophs (Richards and Talbot, 2018) that release enzymes to break down organic materials into simple compounds that can be used to make necessary macromolecules and energy sources. Soil microorganisms also include lithotrophs capable of using materials, such as ammonium and some sulfur compounds, as energy sources. Many soil organisms also use nitrate or sulfate as the ultimate electron acceptors rather than O_2 in O_2-limited areas, allowing them to live anaerobically.

Fluxes within the N cycle are primarily driven by N fixation (conversion from atmospheric N to soil N either by natural or anthropogenic processes), mineralization (ammonification + nitrification) of organic N from sources, such as leaf litter and other plant or animal materials, and gaseous loses, such as through denitrification and ammonia volatilization (Chapter 14). Mutualistic relationships formed by plants with soil organisms, such as *Rhizobium* (class Alphaproteobacteria) and *Frankia* (phylum Actinobacteria), are known for their importance in agriculture; however, these relationships are essential in early successional communities where N may not yet be available in soil. As plant production increases, the N available for plant growth in these community transitions from atmospheric N fixed by microorganisms to N that is available through decomposition of plant materials.

The rate at which N is returned to the ecosystem following plant uptake is primarily determined by plant form, nutrient use efficiency, and, ultimately, the rate of mineralization once the organic material becomes available to decomposer organisms. Nitrogen mineralization, the conversion from organic to inorganic forms, has been studied extensively, and the process is largely governed by climate and soil physical and chemical properties. The need to understand the impact of global climate change on the N cycle has resulted in new global studies of N mineralization that suggest there are still some uncertainties that should be clarified so that N models are more robust (Li et al., 2018). Nitrogen that becomes available for uptake following release from litter or soil organic matter can either be taken up by plants or by soil microorganisms through immobilization, where available N becomes part of microbial tissues rather than plant tissues. When N is limited in soils, plant-microbe competition for N is most often "won" by microbes setting up N limitation for plants, reducing NPP and litter quality (van der Heijden et al., 2008).

Historically, the amount of N that occurred naturally in ecosystems was a consequence of either N fixation by symbiotic or free-living microbes or recycling of organic materials by microbes. This is no longer the case, as anthropogenic generation and application of fertilizer N and pollutant dispersal has resulted in a doubling of the amount of N currently available for plants in many ecosystems (Chapter 14). This has changed N availability for plant uptake, decomposition rates, plant species germination and competition, and microbial control of N availability in terrestrial ecosystems. Additions of N have been found to alter respiration, microbial biomass, and enzyme activities across a broad range of soils, suggesting that these additions have had significant impacts on the functioning of soil microorganisms (Ramirez et al., 2012).

A great deal of the early literature on the plant-fungal mutualisms focused exclusively on P. The need for P in plants is so great and competition so intense that plants with mutualistic relationships can often grow larger than plants without fungal partners, even though the plants must provide C for fungal growth. As was described for C and N, relationships (e.g., competition, predation, and exploitation) between mycorrhizal fungi and soil organisms determine the rate at which nutrients and water become available for plants and determine competition within plant communities. The examination of the relationship between N and P availability has, to date, provided no robust model for understanding the impact of the mutual limitation on plant growth (Deng et al., 2017). There is evidence that P has a greater impact than N in limiting the rate of succession following disturbances such as glacial retreat (Darcy et al., 2018). Disturbances, such as fires, that release nutrients into soil can decrease reliance on biogeochemical cycles by hastening the rate at which nutrients become available.

According to Liebig's law of the minimum, low quantities of any essential nutrient can cause stress and decrease productivity, so the cycling of all nutrients is important for understanding ecosystem dynamics. In most ecosystems the nutrients that plants depend on for growth are those that are returned through recycling as a result of microbial action, also called internal cycling, rather than through fresh nutrient inputs as a result of biotic or abiotic weathering processes. The importance of internal cycling on ecosystem characteristics was documented at the Hubbard Brook Forest (Likens et al., 1977). The literature suggests that, in general, decomposition of plant litter by soil biota can return up to 100% of nutrients required for plant growth (van der Heijden et al., 2008).

10.4 Conclusion

Integrating across scales and disciplines is a challenge that defines ecological study. Scientists that study soil microbiology, ecology, and biochemistry can contribute to and benefit from approaching soil organisms from an ecological perspective. Society has placed a great burden on scientists by damaging systems before understanding how they operate. Science, especially the field of soil ecology, is now charged with developing an understanding of these systems and finding ways to mitigate the damage. This can only be accomplished by integration across scientific fields to develop a more comprehensive understanding of the structure and function of soil microbial communities and their influences on ecosystems and the globe.

References

Allen, M.F., 1991. Ecology of Mycorrhizae. Cambridge University Press, Cambridge.

Allison, S.D., Weintraub, M.N., Gartner, T.B., Waldrop, M.P., 2010. Evolutionary-economic principles as regulators of soil enzyme production and ecosystem function. In: Shukla, G., Varma, A. (Eds.), Soil Enzymology. Springer, Berlin, pp. 229–243.

Anthony, K.R.N., Helmstedt, K.J., Bay, L.K., Fidelman, P., Hussey, K.E., Lundgren, P., et al., 2020. Interventions to help coral reefs under global change – a complex decision challenge. PLoS One 15 (8), e0236399.

Asplund, J., Wardle, D.A., 2017. How lichens impact on terrestrial community and ecosystem properties. Biol. Rev. 92, 1720–1738.

Augé, R.M., Toler, H.D., Saxton, A.M., 2014. Arbuscular mycorrhizal symbiosis alters stomatal conductance of host plants more under drought than under amply watered conditions: a meta-analysis. Mycorrhiza 25, 13–24.

Barnett, S.E., Youngblut, N.D., Buckley, D.H., 2020. Soil characteristics and land-use drive bacterial community assembly patterns. FEMS Microbiol. Ecol. 96, fiz194.

Berg, G., Smalla, K., 2009. Plant species and soil type cooperatively shape the structure and function of microbial communities in the rhizosphere. FEMS Microbiol. Ecol. 68, 1–13.

Bever, J.D., Mangan, S.A., Alexander, H.M., 2015. Maintenance of plant species diversity by pathogens. Annu. Rev. Ecol. Evol. Syst. 46, 305–325.

Bezemer, T.M., Fountain, M.T., Barea, J.M., Christensen, S., Bekker, S.C., Duyts, H., et al., 2010. Divergent composition but similar function of soil food webs of individual plants: plant species and community effects. Ecology 91, 3027–3036.

Biggs, C.R., Yeager, L.A., Bolser, D.G., Bonsell, C., Dichiera, A.M., Hou, Z., et al., 2020. Does functional redundancy affect ecological stability and resilience? A review and meta-analysis. Ecosphere 11 (7), e03184.

Bissett, A., Richardson, A.E., Baker, G., Wakelin, S., Thrall, P.H., 2010. Life history determines biogeographical patterns of soil bacterial communities over multiple spatial scales. Mol. Ecol. 19, 4315–4327.

Blackwood, C.B., Dell, C.J., Smucker, A.J., Paul, E.A., 2006. Eubacterial communities in different soil macroaggregate environments and cropping systems. Soil Biol. Biochem. 38, 720–728.

Bobay, L.M., 2020. The prokaryotic species concept and challenges. In: Tettelin, H., Medini, D. (Eds.), The Pangenome: Diversity, Dynamics and Evolution of Genomes. Springer, Switzerland, pp. 21–49.

Bobay, L.M., Ochman, H., 2017. Biological species are universal across life's domains. Genome. Biol. Evol. 9, 491–501.

Bonanomi, G., Lorito, M., Vinale, F., Woo, S.L., 2018. Organic amendments, beneficial microbes, and soil microbiota: toward a unified framework for disease suppression. Annu. Rev. Phytopathol. 56, 1–20.

Bray, N., Kao-Kniffin, J., Frey, S.D., Fahey, T., Wickings, K., 2019. Soil macroinvertebrate presence alters microbial community composition and activity in the rhizosphere. Front. Microbiol. 10, 256.

Brundrett, M.C., 2002. Coevolution of roots and mycorrhizas of land plants. New Phytol. 154, 275–304.

Brussaard, L., Aanen, D.K., Briones, M.J.I., Decaëns, T., De Deyn, G.B., Fayle, T.M., et al., 2012. Biogeography and phylogenetic community structure of soil invertebrate ecosystem engineers: global to local patterns, implications for ecosystem functioning and services and global environmental change impacts. In: Wall, D.H., Bardgett, R.D., Behan-Pelletier, V., Herrick, J.E., Jones, T.H., Ritz, K., et al. (Eds.), Soil Ecology and Ecosystem Services. Oxford University Press, Oxford, pp. 201–232.

Camenzind, T., Hättenschwiler, S., Treseder, K.K., Lehmann, A., Rillig, M.C., 2018. Nutrient limitation of soil microbial processes in tropical forests. Ecol. Monogr. 88, 4–21.

Chandrasekaran, M., Chanratana, M., Kim, K., Seshadri, S., Sa, T., 2019. Impact of arbuscular mycorrhizal fungi on photosynthesis, water status, and gas exchange of plants under salt stress – a meta-analysis. Front. Plant Sci. 10, 457. https://doi.org/10.3389/fpls.2019.00457.

Chapin, F.S., Matson, P.A., Vitousek, P.M., 2011. Principles of Terrestrial Ecosystem Ecology, second ed. Springer, New York, NY.

Cleveland, C.C., Liptzin, D., 2007. C:N:P stoichiometry in soil: is there a "Redfield ratio" for the microbial biomass? Biogeochemistry 85, 235–252.

Collins, S.L., Sinsabaugh, R.L., Crenshaw, C., Green, L., Porras-Alfaro, A., Stursova, M., et al., 2008. Pulse dynamics and microbial processes in arid land ecosystems. J. Ecol. 96, 413–420.

Compant, S., Clément, C., Sessitsch, A., 2010. Plant growth-promoting bacteria in the rhizo- and endosphere of plants: their role, colonization, mechanisms involved and prospects for utilization. Soil Biol. Biochem. 42, 669–678.

Coutinho, H.L.C., Kay, H.E., Manfio, G.P., Neves, M.C.P., Ribeiro, J.R.A., Rumjanek, N.G., et al., 1999. Molecular evidence for shifts in polysaccharide composition associated with adaptation of soybean *Bradyrhizobium* strains to the Brazilian Cerrado soils. Environ. Microbiol. 1, 401–408.

Crowther, T.W., Boddy, L., Jones, T.H., 2012. Functional and ecological consequences of saprotrophic fungus–grazer interactions. ISME J. 6, 1992–2001.

Crowther, T.W., van den Hoogen, J., Wan, J., Mayes, M.A., Keiser, A.D., Averill, M.C., et al., 2019. The global soil community and its influence on biogeochemistry. Science 365, 772.

Darcy, J.L., Schmidt, S.K., Knelman, J.E., Cleveland, C.C., Castle, S.C., Nemergut, D.R., 2018. Phosphorus, not nitrogen, limits plants and microbial primary producers following glacial retreat. Sci. Adv. 4, eaaq0942.

Davelos, A.L., Kinkel, L.L., Samac, D.A., 2004. Spatial variation in frequency and intensity of antibiotic interactions among streptomycetes from prairie soil. Appl. Environ. Microbiol. 70, 1051–1058.

Daws, S.C., Cline, L.A., Rotenberry, J., Sadowsky, M.J., Staley, C., Dalzell, B., et al., 2020. Do shared traits create the same fates? Examining the link between morphological type and the biogeography of fungal and bacterial communities. Fungal. Ecol. 46, 100948.

de Jonge, R., Takumi, K., Ritmeester, W.S., van Leusden, F.M., 2003. The adaptive response of *Escherichia coli* O157 in an environment with changing pH. J. Appl. Microbiol. 94, 555–560.

de Vries, N., Kuipers, E.J., Kramer, N.E., van Vliet, A.H.M., Bijlsma, J.J.E., Kist, M., et al., 2001. Identification of environmental stress-regulated genes in *Helicobacter pylori* by a *lacZ* reporter gene fusion system. Helicobacter 6, 300–309.

Delavaux, C.S., Smith-Ramesh, L.M., Kuebbing, S.E., 2017. Beyond nutrients: a meta-analysis of the diverse effects of arbuscular mycorrhizal fungi on plants and soils. Ecology 98 (8), 2111–2119.

Deng, Q., Hui, D., Dennis, S., Reddy, K.C., 2017. Responses of terrestrial ecosystem phosphorus cycling to nitrogen addition: a meta-analysis. Global Ecol. Biogeogr. 26, 713–728.

Denman, K.L., Brasseur, G., Chidthaisong, A., Ciais, P., Cox, P.M., Dickinson, R.E., et al., 2007. Couplings between changes in the climate system and biogeochemistry. In: Solomon, S., Qin, D., Manning, M., Chen, Z., Marquis, M., Averyt, K.B., et al. (Eds.), Climate Change 2007: The Physical Science Basis. Contribution of Working Group I to the Fourth Assessment Report of the Intergovernmental Panel on Climate Change. Cambridge University Press, Cambridge, United Kingdom and New York, NY, USA, pp. 499–587.

Derré, I., Rapoport, G., Msadek, T., 1999. CtsR, a novel regulator of stress and heat shock response, controls clp and molecular chaperone gene expression in gram-positive bacteria. Mol. Microbiol. 31, 117–131.

Eisenhauer, N., Schlaghamerský, J., Reich, P.B., Frelich, L.E., 2011. The wave towards a new steady state: effects of earthworm invasion on soil microbial function. Biol. Invasions 13, 2191–2196.

Elliott, E.T., Anderson, R.V., Coleman, D.C., Cole, C.V., 1980. Habitable pore space and microbial trophic interactions. Oikos 35, 327–335.

Erktan, A., Or, D., Scheu, S., 2020. The physical structure of soil: determinant and consequence of trophic interactions. Soil Biol. Biochem. 148, 107876.

Ettema, C.H., Wardle, D.A., 2002. Spatial soil ecology. Trends Ecol. Evol. 17, 177–183.

Feinstein, L.M., Blackwood, C.B., 2013. The spatial scaling of saprotrophic fungal beta diversity in decomposing leaves. Mol. Ecol. 22, 1171–1184.

Fierer, N., 2017. Embracing the unknown: disentangling the complexities of the soil microbiome. Nat. Rev. Microbiol. 15, 579–590.

Filser, J., Faber, J.H., Tiunov, A.V., Brussaard, L., Frouz, J., De Deyn, G., et al., 2016. Soil fauna: key to new carbon models. SOIL 2, 565–582.

Frankland, J.C., 1998. Fungal succession – unravelling the unpredictable. Mycol. Res. 102, 1–15.

Freedman, Z., Zak, D.R., 2015. Soil bacterial communities are shaped by temporal and environmental filtering: evidence from a long-term chronosequence. Environ. Microbiol. 17, 3208–3218.

Frey, S.D., 2019. Mycorrhizal fungi as mediators of soil organic matter dynamics. Annu. Rev. Ecol. Evol. Syst. 50, 237–259.

Friesen, M.L., 2012. Widespread fitness alignment in the legume-rhizobium symbiosis. New Phytol. 194, 1096–1111.

Friesen, M.L., Heath, K.D., 2013. One hundred years of solitude: integrating single-strain inoculations with community perspectives in the legume-rhizobium symbiosis. New Phytol. 198, 7–9.

Goodwin, S.B., Cohen, B.A., Fry, W.E., 1994. Panglobal distribution of a single clonal lineage of the Irish potato famine fungus. Proc. Natl. Acad. Sci. USA 91, 11591–11595.

Gupta, V.V., Germida, J.J., 2015. Soil aggregation: influence on microbial biomass and implications for biological processes. Soil Biol. Biochem. 80, A3–A9.

Harrison, S., 1991. Local extinction in a metapopulation context: an empirical evaluation. In: Gilpin, M., Hanski, I. (Eds.), Metapopulation Dynamics: Empirical and Theoretical Investigations. Academic Press, London, pp. 73–88.

Hoshino, T., Xiao, N., Tkachenko, O.B., 2009. Cold adaptation in the phytopathogenic fungi causing snow molds. Mycoscience 50, 26–38.

Hovatter, S., Blackwood, C.B., Case, A.L., 2013. Conspecific plant–soil feedback scales with population size in *Lobelia siphilitica* (Lobeliaceae). Oecologia 173, 1295–1307.

Hu, J., Wei, Z., Friman, V.P., Gu, S.H., Wang, X.F., Eisenhauer, N., et al., 2016. Probiotic diversity enhances rhizosphere microbiome function and plant disease suppression. MBio. 7, e01790–16.

Hug, L.A., Baker, B.J., Anantharaman, K., Brown, C.T., Probst, A.J., Castelle, C.J., et al., 2016. A new view of the tree of life. Nat. Microbiol. 1, 1–6.

Hutchinson, G.E., 1957. Concluding remarks. Cold Spring Harb. Symp. Quant. Biol. 22, 415–427.

IPCC, 2014. Climate change 2014: Synthesis Report. Contribution of Working Groups I, II, and III to the fifth assessment report of the intergovernmental panel on climate change. In: Pachauri, R.K., Meyer, L.A. (Eds.), Core Writing Team. IPCC, Geneva, Switzerland, p. 151.

Irshad, U., Villenave, C., Brauman, A., Plassard, C., 2011. Grazing by nematodes on rhizosphere bacteria enhances nitrate and phosphorus availability to *Pinus pinaster* seedlings. Soil Biol. Biochem. 43, 2121–2126.

James, T.Y., Stajich, J.E., Hittinger, C.T., Rokas, A., 2020. Toward a fully resolved fungal tree of life. Annu. Rev. Microbiol. 74, 291–313.

Jansson, J.K., Hofmockel, K.S., 2018. The soil microbiome – from metagenomics to metaphenomics. Curr. Opin. Microbiol. 43, 162–168.

Jayne, B., Quigley, M., 2014. Influence of arbuscular mycorrhiza on growth and reproductive response of plants under water deficit: a meta-analysis. Mycorrhiza 24, 109–119.

Jenny, H., 1961. Derivation of state factor equations of soils and ecosystems. Soil Sci. Soc. Am. Proc. 25, 385–388.

Jia, Y., Whalen, J.K., 2020. A new perspective on functional redundancy and phylogenetic niche conservatism in soil microbial communities. Pedosphere 30 (1), 18–24.

Jones, C.G., Lawton, J.H., Shachak, M., 1994. Organisms as ecosystem engineers. Oikos 69, 373–386.

Kieft, T.L., 2000. Size matters: dwarf cells in soil and subsurface terrestrial environments. In: Colwell, R.R., Grimes, D.J. (Eds.), Nonculturable Microorganisms in the Environment. American Society for Microbiology Press, Washington, DC, pp. 19–46.

Kingsland, S.E., 1991. Defining ecology as a science. In: Real, L.A., Brown, J.H. (Eds.), Foundations of Ecology: Classic Papers with Commentaries. University of Chicago Press, Chicago, pp. 1–12.

Klironomos, J.N., Hart, M.M., 2001. Food-web dynamics – animal nitrogen swap for plant carbon. Nature 410, 651–652.

Kohout, P., Sudová, R., Brabcová, V., Vosolsobě, S., Baldrian, P., Albrechtová, J., 2021. Forest microhabitat affects succession of fungal communities on decomposing fine tree roots. Front. Microbiol. 12, 541583.

Koller, R., Scheu, S., Bonkowski, M., Robin, C., 2013. Protozoa stimulate N uptake and growth of arbuscular mycorrhizal plants. Soil Biol. Biochem. 65, 204–210.

Kuzyakov, Y., Blagodatskaya, E., 2015. Microbial hotspots and hot moments in soil: concept & review. Soil Biol. Biochem. 83, 184–199.

Lacroix, B., Citovsky, V., 2016. A functional bacterium-to-plant DNA transfer machinery of *Rhizobium etli*. PLoS Pathog. 12 (3), e1005502.

Lang, A.K., Jevon, F.V., Vietorisz, C.R., Ayres, M.P., Matthes, J.H., 2021. Fine roots and mycorrhizal fungi accelerate leaf litter decomposition in a northern hardwood forest regardless of dominant tree mycorrhizal associations. New Phytol. 230, 316–326.

Langenheder, S., Lindström, E.S., 2019. Factors influencing aquatic and terrestrial bacterial community assembly. Environ. Microbiol. Rep. 11, 306–315.

Lavelle, P., Spain, A., Fonte, S., Bedano, J.C., Blanchart, E., Galindo, V., et al., 2020. Soil aggregation, ecosystem engineers and the C cycle. Acta Oecol. 105, 103561.

Leake, J.R., Read, D.J., 1997. Mycorrhizal fungi in terrestrial habitats. In: Wicklow, D.T., Söderström, B. (Eds.), The Mycota IV. Environmental and Microbial Relationships. Springer-Verlag, Berlin, pp. 281–301.

Leibold, M.A., Holyoak, M., Mouquet, N., Amarasekare, P., Chase, J.M., Hoopes, M.F., et al., 2004. The metacommunity concept: a framework for multi-scale community ecology. Ecol. Lett. 7, 601–613.

Li, Z., Tian, D., Wang, B., Wang, J., Wang, S., Chen, H.Y.H., 2018. Microbes drive global soil nitrogen mineralization and availability. Glob. Chang. Biol. 25, 1078–1088.

Likens, G.E., Bormann, F.H., Pierce, R.S., Eaton, K.S., Johnson, N.M., 1977. Biogeochemistry of a Forested Ecosystem. Springer Verlag, Heidelberg, Berlin, New York.

Lönn, A., Gárdonyi, M., van Zyl, W., Hahn-Hägerdal, B., Otero, R.C., 2002. Cold adaptation of xylose isomerase from *Thermus thermophilus* through random PCR mutagenesis gene cloning and protein characterization. Eur. J. Biochem. 269, 157–163.

Loper, J.E., Hassan, K.A., Mavrodi, D.V., Davis II, E.W., Kim, C.K., Shaffer, B.T., et al., 2012. Comparative genomics of plant-associated Pseudomonas spp.: insights into diversity and inheritance of traits involved in multitrophic interaction. PLoS Genet. 8, e1002784.

Lucas, J.M., McBride, S.G., Strickland, M.S., 2020. Trophic level mediates soil microbial community composition and function. Soil Biol. Biochem. 143, 107756.

MacArthur, R.H., Wilson, E.O., 1967. The Theory of Island Biogeography. Princeton University Press, Princeton, NJ.

Manzoni, S., Čapek, P., Porada, P., Thurner, M., Winterdahl, M., Beer, C., 2018. Reviews and syntheses: carbon use efficiency from organisms to ecosystems- definitions, theories, and empirical evidence. Biogeosciences 15, 5929–5949.

Marx, C.J., 2013. Can you sequence ecology? Metagenomics of adaptive diversification. PLoS Biol. 11, e1001487.

McGill, B.J., Enquist, B.J., Weiher, E., Westoby, M., 2006. Rebuilding community ecology from functional traits. Trends Ecol. Evol. 21, 178–184.

Mommer, L., Cotton, T.A., Raaijmakers, J.M., Termorshuizen, A.J., van Ruijven, J., Hendriks, M., et al., 2018. Lost in diversity: the interactions between soil-borne fungi, biodiversity and plant productivity. New Phytol. 218, 542–553.

Moore, J., de Ruiter, P., 2012. Energetic Food Webs. Oxford University Press, UK.

Moore, J.C., Berlow, E.L., Coleman, D.C., de Ruiter, P.C., Dong, Q., Hastings, A., et al., 2004. Detritus, trophic dynamics and biodiversity. Ecol. Lett. 7, 584–600.

Nemergut, D.R., Schmidt, S.K., Fukami, T., O'Neill, S.P., Bilinski, T.M., Stanish, L.F., et al., 2013. Patterns and processes of microbial community assembly. Microbiol. Mol. Biol. Rev. 77, 342–356.

O'Brien, S.L., Gibbons, S.M., Owens, S.M., Hampton-Marcell, J., Johnston, E.R., Jastrow, J.D., et al., 2016. Spatial scale drives patterns in soil bacterial diversity. Environ. Microbiol. 18, 2039–2051.

O'Malley, M.A., 2007. The nineteenth century roots of "everything is everywhere". Nat. Rev. Microbiol. 5, 647–651.

Panikov, N.S., 1995. Microbial Growth Kinetics. Chapman & Hall, London, UK.

Paul, E.A., Clark, F.E., 1989. Soil Microbiology and Biochemistry. Academic Press, San Diego, p. 272.

Paul, E.A., Clark, F.E., 1996. Soil Microbiology and Biochemistry, second ed. Academic Press, London.

Parks, D.H., Chuvochina, M., Chaumeil, P.A., Rinke, C., Mussig, A.J., Hugenholtz, P., 2020. A complete domain-to-species taxonomy for Bacteria and Archaea. Nat. Biotechnol. 38, 1079–1086.

Pianka, E.R., 1970. On r- and K-selection. Am. Nat. 104, 592–597.

Poll, C., Brune, T., Begerow, D., Kandeler, E., 2010. Small-scale diversity and succession of fungi in the detritusphere of rye residues. Microb. Ecol. 59, 130–140.

Ramirez, K.S., Craine, J.M., Fierer, N., 2012. Consistent effects of nitrogen amendments on soil microbial communities and processes across biomes. Glob. Chang. Biol. 18, 1918–1927.

Rayner, A.D.M., Beeching, J.R., Crowe, J.D., Watkins, Z.R., 1999. Defining individual fungal boundaries. In: Worrall, J.J. (Ed.), Structure and Dynamics of Fungal Populations. Kluwer Academic Publishers, Dordrecht, pp. 19–41.

Reznick, D., Bryant, M.J., Bashey, F., 2002. r- and K-selection revisited: the role of population regulation in life-history evolution. Ecology 83, 1509–1520.

Richards, T.A., Talbot, N.J., 2018. Osmotrophy. Curr. Biol. 28, R1171–R1189.

Robinson, C.H., 2001. Cold adaptation in arctic and antarctic fungi. New Phytol. 151, 341–353.

Roeßler, M., Müller, V., 2001. Osmoadaptation in bacteria and archaea: common principles and differences. Environ. Microbiol. 3, 743–754.

Röling, W.F., Ferrer, M., Golyshin, P.N., 2010. Systems approaches to microbial communities and their functioning. Curr. Opin. Biotechnol. 21, 532–538.

Rousk, J., Bååth, E., Brookes, P.C., Lauber, C.L., Lozupone, C., Caporaso, J.G., et al., 2010. Soil bacterial and fungal communities across a pH gradient in an arable soil. ISME J. 10, 1340–1351.

Rout, M.E., Callaway, R.M., 2012. Interactions between exotic invasive plants and soil microbes in the rhizosphere suggest that "everything is not everywhere". Ann. Bot. 110, 213–222.

Sackett, T.E., Classen, A.T., Sanders, N.J., 2010. Linking soil food web structure to above- and belowground ecosystem processes: a meta-analysis. Oikos 119, 1984–1992.

Schmid, M.W., van Moorsel, S.J., Hahl, T., De Luca, E., De Deyn, G.B., Wagg, C., et al., 2021. Effects of plant community history, soil legacy and plant diversity on soil microbial communities. J. Ecol. https://doi.org/10.1111/1365-2745.13714.

Settembre, E.C., Chittuluru, J.R., Mill, C.P., Kappock, T.J., Ealick, S.E., 2004. Acidophilic adaptations in the structure of *Acetobacter aceti* N5-carboxyaminoimidazole ribonucleotide mutase (PurE). Acta Crystallogr. D 60, 1753–1760.

Simard, S.W., Beiler, K.J., Bingham, M.A., Deslippe, J.R., Philip, L.J., Teste, F.P., 2012. Mycorrhizal networks: mechanisms, ecology, and modeling. Fungal. Biol. Rev. 26, 39–60.

Smith, S.E., Read, D., 1997. Mycorrhizal Symbiosis, second ed. Academic Press, London, UK.

Soong, J.L., Marañon-Jimenez, S., Cotrufo, M.F., Boeckx, P., Bodé, S., Guenet, B., et al., 2018. Soil microbial CNP and respiration responses to organic matter and nutrient additions: evidence from a tropical soil incubation. Soil Biol. Biochem. 122, 141–149.

Srinivasiah, S., Bhavsar, J., Thapar, K., Liles, M., Schoenfeld, T., Wommack, K.E., 2008. Phages across the biosphere: contrasts of viruses in soil and aquatic environments. Res. Microbiol. 159, 349–357.

Strickland, M.S., Lauber, C., Fierer, N., Bradford, M.A., 2009. Testing the functional significance of microbial community composition. Ecology 90, 441–451.

Stromberger, M.E., Keith, A.M., Schmidt, O., 2012. Distinct microbial and faunal communities and translocated carbon in *Lumbricus terrestris* drilospheres. Soil Biol. Biochem. 46, 155–162.

Stubblefield, S.P., Taylor, T.N., 1988. Recent advances in palaeomycology. New Phytol. 108, 3–25.

Swanson, M.M., Reavy, B., Makarova, K.S., Cock, P.J., Hopkins, D.W., Torrance, L., et al., 2012. Novel bacteriophages containing a genome of another bacteriophage within their genomes. PLoS One 7, e40683.

Tecon, R., Or, D., 2017. Biophysical processes supporting the diversity of microbial life in soil. FEMS Microbiol. Rev. 41, 599–623.

Tedersoo, L., Bahram, M., 2019. Mycorrhizal types differ in ecophysiology and alter plant nutrition and soil processes. Biol. Rev. 94, 1857–1880.

Tedersoo, L., Anslan, S., Bahram, M., Drenkhan, R., Pritsch, K., Buegger, F., et al., 2020. Regional-scale in-depth analysis of soil fungal diversity reveals strong pH and plant species effects in Northern Europe. Front. Microbiol. 11, 1953.

Thomas, C.M., Nielsen, K.M., 2005. Mechanisms of, and barriers to, horizontal gene transfer between bacteria. Nat. Rev. Microbiol. 3, 711–721.

Tilman, D., 1982. Resource Competition and Community Structure. Princeton University Press, Princeton, NJ.

Too, C.C., Keller, A., Sickel, W., Lee, S.M., Yule, C.M., 2018. Microbial community structure in a Malaysian tropical peat swamp forest: the influence of tree species and depth. Front. Microbiol. 9, 2859.

Trivedi, P., Leach, J.E., Tringe, S.G., Sa, T., Singh, B.K., 2020. Plant—microbiome interactions: from community assembly to plant health. Nat. Rev. Microbiol. 18, 607—621.

Turner, B.L., Zemunik, G., Laliberté, E., Drake, J.J., Jones, F.A., Saltonstall, K., 2019. Contrasting patterns of plant and microbial diversity during long-term ecosystem development. J. Ecol. 107, 606—621.

Upton, R.N., Checinska Sielaff, A., Hofmockel, K.S., Xu, X., Polley, H.W., Wilsey, B.J., 2020. Soil depth and grassland origin cooperatively shape microbial community co-occurrence and function. Ecosphere 11, e02973.

van der Heijden, M.G., Bargett, R.D., Van Straalen, N.M., 2008. The unseen majority: soil microbes as drivers of plant diversity and productivity in terrestrial ecosystems. Ecol. Lett. 11, 296—310.

van der Putten, W.H., Klironomos, J.N., Wardle, D.A., 2007. Microbial ecology of biological invasions. ISME J. 1, 28—37.

Vasco-Palacios, A.M., Bahram, M., Boekhout, T., Tedersoo, L., 2019. Carbon content and pH as important drivers of fungal community structure in three Amazon forests. Plant Soil 450, 111—131.

Vellend, M., 2010. Conceptual synthesis in community ecology. Q. Rev. Biol. 85, 183—206.

Wang, H., Marshall, C.W., Cheng, M., Xu, H., Li, H., Yang, X., et al., 2017. Changes in land use driven by urbanization impact nitrogen cycling and the microbial community composition in soils. Sci. Rep. 7, 44049.

Waring, B.G., Averill, C., Hawkes, C.V., 2013. Differences in fungal and bacterial physiology alter soil carbon and nitrogen cycling insights from meta-analysis and theoretical models. Ecol. Lett. 16, 887—894.

Watkinson, S., Bebber, D., Darrah, P., Fricker, M., Tlalka, M., Boddy, L., 2006. The role of wood decay fungi in the carbon and nitrogen dynamics of the forest floor. In: Gadd, G.M. (Ed.), Fungi in Biogeochemical Cycles. Cambridge University Press, Cambridge, pp. 151—181.

Wiens, J.A., 1997. Metapopulation dynamics and landscape ecology. In: Hanski, I., Gilpin, M.E. (Eds.), Metapopulation Biology: Ecology, Genetics, and Evolution. Academic Press, San Diego, pp. 43—61.

Wolfe, B.E., Rodgers, V.L., Stinson, K.A., Pringle, A., 2008. The invasive plant *Alliaria petiolata* (garlic mustard) inhibits ecto-mycorrhizal fungi in its introduced range. J. Ecol. 96, 777—783.

Wu, X.L., Conrad, R., 2001. Functional and structural response of a cellulose-degrading methanogenic microbial community to multiple aeration stress at two different temperatures. Environ. Microbiol. 3, 355—362.

Wu, J., Loucks, O.L., 1995. From balance of nature to hierarchical patch dynamics: a paradigm shift in ecology. Q. Rev. Biol. 70, 439—466.

Chapter 11

Plant–soil biota interactions

R. Balestrini, V. Bianciotto, S. Ghignone, E. Lumini, A. Mello, F. Sillo, and E. Zampieri
National Research Council, Institute for Sustainable Plant Protection (CNR-IPSP), Torino, Italy

Chapter outline

11.1 Soil biota: definitions and functions

Because of the exceptional diversity of soil biota, it has been stated that "a 'biological universe' exists in a gram of soil" (Fortuna, 2012; Jeffery et al., 2010). Biotic biodiversity differs from soil to soil (e.g., grassland, arable land, forest) because these soils are inhabited by different plant species (Wardle et al., 2004) and vary in soil type. The activity of soil biota across land, air, and water interfaces (Madsen, 2008) affects plants through: (1) soil organic matter (SOM) formation and turnover of carbon (C); (2) nutrient cycling of elements such as nitrogen (N), sulfur (S), and phosphorus (P); (3) plant disease transmission and control; (4) pollutant degradation; and (5) soil structure formation and modification. The soil quality concept is based on traditional indicators, such as biological, chemical, and physical properties and processes, as well as molecular characterization using advanced genomics tools. The latter provides a useful approach that allows plant–soil interactions to be characterized at a fine resolution (Schloter et al., 2018), with a detailed description of organismal diversity, community composition, and functional capacity at genus or subgenus levels which are now possible (Baldrian, 2019). Genome annotations can now be linked with functional annotations, allowing inference of specific functions (e.g., N and C cycling) using data from soil metagenomics and metatranscriptomics (Deng et al., 2019; Djemiel et al., 2017; Mackelprang et al., 2018). Use of multi-omics approaches offers information about the involvement in soil processes of the soil biota as either individual taxa or whole communities, leading to a better understanding of the plant root–soil interface.

Soil Microbiology, Ecology, and Biochemistry. https://doi.org/10.1016/B978-0-12-822941-5.00011-9

11.2 The rhizosphere and its inhabitants

The term "rhizosphere" was coined in 1904 by the German agronomist and plant physiologist Lorenz Hiltner (Hartmann et al., 2008; Hiltner, 1904). It describes the root–soil interface, i.e., the zone of soil around roots inhabited by a diverse community of organisms that play critical roles in plant growth and reproduction. This zone is under the direct influence of plant root exudates that transform the physico-chemical properties of this soil compartment and helps drive community composition through signaling and chemo-attractant molecules (Sylvia, 2005). It is possible to distinguish three root-associated compartments: (1) the endorhizosphere (the root interior), including cortex and endodermis; (2) the rhizoplane (the root surface), which is the medial region adjacent to the root, including the root epidermis and mucilage; and (3) the ectorhizosphere (the soil close to the root surface) (McNear, 2013). The rhizosphere is thus a complex system and is considered a hot spot of biotic activity and diversity (Jones and Hinsinger, 2008), where chemical, biological, and physical properties and processes dynamically change both radially and longitudinally along the root.

It has been estimated that the rhizosphere extends to about 0.5 to 4 mm away from the root surface. Despite this rather limited extent, it determines many processes, dynamics, and cycling of C, nutrients, and water in terrestrial ecosystems because of the presence of root microbes (Kuzyakov and Razavi, 2019). The rhizosphere may contain up to 10^{11} microbial cells per gram of roots (Berendsen et al., 2012). Plants and their microbiomes are thus considered as a "holobiont," an assemblage of a host and the many other species living in or around it (Gordon et al., 2013; Hassani et al., 2018).

The release of organic C compounds from plant roots into the adjacent soil is termed rhizodeposition (Koranda et al., 2011) and accounts for up to 11% of photosynthetically fixed C (Hinsinger et al., 2012). Rhizodeposit composition and amount are dependent on factors such as plant type, climatic conditions, insect herbivory, nutrient deficiency or toxicity, and the chemical, physical, and biological properties of the surrounding soil (Dennis et al., 2010; McNear, 2013). Root exudates mediate interactions with other nearby plants (Fragasso et al., 2013; Pangesti et al., 2013), influencing physiological processes such as photosynthesis, respiration, protein biosynthesis, and cell division and elongation (Field et al., 2007). Iannucci et al. (2017) demonstrated that soil type, plant domestication, and breeding can dramatically affect root exudate composition. Root exudates are involved in plant–soil microbe interactions (Bais et al., 2006), affecting plant health and growth and nutrient uptake (Pérez-Jaramillo et al., 2016).

Rhizosphere biota include bacteria, archaea, fungi, viruses, protists, and nematodes, all of which can potentially affect host biology (Orozco-Mosqueda et al., 2018). Root-associated microbiota, such as root symbionts and growth-promoting rhizobacteria that live together with endophytes, saprotrophic microbes, and pathogens, have been demonstrated to be an important element in disease resistance (Chialva et al., 2018). Biotic factors elicit specific responses and shape plant molecular profiles. For example, tomato genes involved in defense responses, such as those related to oxidative stress, phenol biosynthesis, lignin deposition, and innate immunity, were enhanced in native soils suppressive to *Fusarium oxysporum* f. sp. *lycopersici*. Genes associated with phenylpropanoid pathways were also elicited in the presence of the arbuscular mycorrhizal fungal species *Funneliformis mosseae* (Chialva et al., 2018). The control of wilt caused by *F. oxysporum* is an example of a specific suppressive soil with a high concentration of one or more specific microbial species (e.g., nonpathogenic *Fusarium* spp., *Pseudomonas fluorescens*; Mousa and Raizada, 2016).

Different rhizosphere-associated microorganisms are used as biofertilizers, biostimulants, and biocontrol agents. However, applying these beneficial microorganisms to soils may have limited success, since the new microorganisms may be excluded by the more resilient, and already adapted, native microbial communities that have been shaped through complex interactions with the environment (Ke et al., 2020). Even if plant growth-promoting bacteria or rhizobacteria (PGPB or PGPR) have been isolated and their potential positive effects verified in the presence of defined environmental and soil conditions, most results are not reproducible under other conditions. The precise action mode remains unclear, raising doubts about the wide-scale application of these microorganisms (Bulgarelli et al., 2013; dos Santos et al., 2020). Discovering new microbial strains/isolates able to support plant growth, nutrition, fitness, disease control, and productivity requires the development of novel strategies to manage and optimize the plant-biota biome. An emerging field of research, mainly for bacteria, is so-called microbiome engineering, aimed at developing approaches useful to "control" microbial communities associated with plants (Ke et al., 2020; Mueller and Sachs, 2015). Bioengineering can be carried out through host-mediated and multigenerational microbiome selection through inoculation into bulk soil or the rhizosphere, atomization into tissues such as stems, leaves, and flowers, or direct injection into tissues or wounds (Ke et al., 2020; Orozco-Mosqueda et al., 2018). Bacterial and fungal taxa predominantly contribute to plant-associated microbiome interactions; however, there is a knowledge gap concerning the role of other organisms (such as viruses, archaea, protists, and nematodes) that influence and shape bacterial and fungal communities (Trivedi et al., 2020).

11.2.1 Bacteria

Microbial communities able to perform beneficial activities, such as potential phytopathogen biocontrol and plant growth promotion, include PGPB that primarily live in the rhizosphere and/or are associated with roots as endophytic bacteria (inside of root tissues) (Orozco-Mosqueda et al., 2018). Rhizosphere PGPB can improve N, P, and iron (Fe) nutrition (Calvo et al., 2017) and can synthesize hormones (Bhattacharyya et al., 2015; Pérez-Flores et al., 2017). They increase ethylene concentrations through 1-aminocyclopropane-1-carboxylate (ACC) deaminase activity (Glick, 2012; Ravanbakhsh et al., 2018). They can also reduce plant injury after infection by fungal and bacterial pathogens, inhibiting them by secretion of antibiotics, proteases, chitinases, bacteriocins, siderophores, lipopeptides, and volatile organic compounds (VOCs) (Orozco-Mosqueda et al., 2018). These positive functions are not carried out by the total microbiota, but only by a few microbial species (Timm et al., 2016), or even by a single species (Finkel et al., 2020). There is great interest in using these to enhance plant defense systems, increase yield, or in bioremediation protocols (Mueller and Sachs, 2015; Yuan et al., 2016). Consequently, it is important to conduct research aimed at understanding the interaction and/or the effects that PGPB exert on the plant microbiome (Orozco-Mosqueda et al., 2018). The archaea are also represented in soil suppressive to *Rhizoctonia* damping-off disease (Mendes et al., 2013). These organisms are also important actors in the N cycle in which they oxidize ammonia under unfavorable environmental conditions (e.g., limited nutrient availability, extreme pH, or a low, dissolved oxygen content) (Kuppardt et al., 2018; Zheng et al., 2017).

Bacterial pathogens present in the soil infect plants via roots, with the most dangerous as: *Agrobacterium tumefaciens*, *Ralstonia solanacearum*, *Dickeya dadanthi* and *D. solani*, and *Pectobacterium carotovorum* and *P. atrosepticum* (Mansfield et al., 2012).

11.2.2 Fungi

Fungi develop interactions with plants in the rhizosphere, ranging from antagonistic to mutualistic. Fungi inhabiting plant roots without the formation of typical mycorrhizal associations are called endophytes (Stone and Bacon, 2000). They are commensals and can affect plant performance by improving mineral element uptake and/or potentially enhancing resistance to drought, insect damage, and other stress (He et al., 2019). However, the effects of endophytic fungi on plant performance are still open to debate (Addy et al., 2005; Jumpponen, 2001; Wu et al., 2020). The same population of endophytes can be both weak parasites or mutualists, depending on the host plant species, ecotype, and environmental conditions (Mandyam and Jumpponen, 2015). Endophytes display a range of lifestyles along the saprotrophy-biotrophy continuum based on the growth and performance benefits conferred to the host plant, production of secondary metabolites, and extent of colonization (Zuccaro et al., 2011). *Piriformospora indica* (Sebacinales, Basidiomycota), discovered in 1998 in the Indian Thar desert (Varma et al., 1999), is an interesting endophytic fungus. It shows a remarkable positive impact on plant fitness, improving plant growth and yield as well as resistance/tolerance to biotic and abiotic stresses (Daneshkhah et al., 2013; Xu et al., 2017). *P. indica*, which has been reported to have a biphasic lifestyle, can colonize the roots of a broad range of plants, including members of the Brassicaceae (such as *Arabidopsis*) that are typically nonmycorrhizal (Zuccaro et al., 2011). Comparison of the *P. indica* genome with other fungi with diverse lifestyles allowed the identification of features typically associated with both biotrophism and saprotrophism. The fungal transcriptional responses associated with the colonization of living and dead barley roots have been analyzed (Zuccaro et al., 2011), and the capacity of this fungus to antagonize *Meloidogyne incognita*, a plant-parasitic nematode, has been verified (Atia et al., 2020).

Plant-associated saprophytic fungi, such as *Trichoderma* spp. (Macías-Rodríguez et al., 2020) are involved in cellulose and lignin decomposition, and their hyphae redistribute C, N, and P in soil (Crowther et al., 2012; Pepper, 2019). These fungi, belonging to the Ascomycota (Order Hypocreales, Family Hypocreaceae), can reduce the inoculum of phytopathogenic microorganisms by releasing gliovirin and siderophores (Macías-Rodríguez et al., 2020).

Soil-borne plant pathogenic fungi comprise fungal-like organisms such as species belonging to the oomycetes group, i.e., eukaryotic organisms that superficially resemble filamentous fungi, but are phylogenetically related to diatoms and brown algae in the stramenopiles (Kamoun et al., 2015). Within oomycetes, the most important genera are *Pythium* and *Phytophthora*, while the major species among "true" fungi belong to the *Fusarium*, *Phoma*, *Sclerotinia*, and *Verticillium* genera within the Ascomycota, and to *Armillaria* and *Rhizoctonia* within the Basidiomycota (Katan, 2017). Soil-borne plant pathogenic fungi are historically grouped into two functional categories: soil inhabitants and soil invaders (Lockwood, 1986). The first category generally includes unspecialized microbes that infect seedlings and young root tissues, while the second comprises disease agents that show a relevant degree of host specificity (Lockwood, 1986). Seed decay, damping-off, and root rots of seedlings are the most common diseases caused by soil-borne fungi. However, some species (e.g., *Fusarium oxysporum*) colonize plant xylem, resulting in vascular diseases. In the absence of a host or in the presence of an unfavorable environment, these fungi can produce a wide variety of survival structures which may germinate later under favorable conditions. These structures may be as simple as thick-walled cells called chlamydospores, which are commonly produced by a large number of pathogens or may be more complex like the sclerotia typical of some fungi (e.g., *Sclerotium*, *Sclerotinia*, *Botrytis*). In addition to providing a survival function, root-like aggregation of hyphae called rhizomorphs, typical of some root-rot fungi (e.g., *Armillaria* spp.), may play a crucial role for fungal spread, host infection, and disease transmission. Survival may also depend on the production of resting fruiting bodies (several Ascomycota) or propagules (oospores of Oomycota) (Katan, 2017).

11.2.3 Viruses

Another group of plant pathogens is represented by viruses that infect plant roots by using vectors such as zoosporic fungi or nematodes (Campbell, 1996; Macfarlane, 2003). It has been estimated that 97% of viruses in the world may be found in soil and sediment (Cobián Güemes et al., 2016). Soil viral diversity is still largely unexplored. At least 17 genera in 8 plant single-stranded RNA (ssRNA) virus families are known to be transmitted by soil-inhabiting organisms, e.g., nematodes (Williamson et al., 2017). Some of the most remarkable examples include the beet necrotic yellow vein virus (BNYVV; genus *Benyvirus*) causing rhizomania disease (a massive proliferation of lateral roots and rootlets) in sugar beets (Tamada, 2016), and the potato mop-top virus (PMTV; genus *Pomovirus*) that causes brown rings in potato tuber flesh (Harrison and Reavy, 2002). Recent research has primarily focused on soil-borne, virus-infecting bacteria (Emerson, 2019). Rhizobiophages (phages that infect rhizobia) seem to be ubiquitous (Williamson et al., 2017) and are known to be lytic viruses that can play an important role in rhizobial ecology due to their effects on the legume-rhizobia symbiosis which may in turn have consequences on N cycling (Williamson et al., 2017).

11.2.4 Nematodes and control agents

Nematodes play a considerable role in the soil food web and are considered by some to be keystone species (Neher et al., 2005). Only a few nematode species are pest organisms that can cause severe damage to crops (e.g., soybean, tomato, rice), with most playing a critical role in the release of plant-available nutrients by feeding on bacteria and fungi (Neher, 2010). Others may be of interest for the biocontrol of insect pests. Root-knot nematodes (RKN) are plant-parasitic nematodes primarily belonging to the genus *Meloidogyne* (Wesemael et al., 2011). Specific relationships exist between bacterial and nematode populations (Castillo et al., 2017), with some bacteria, such as *Bacillus* and *Pseudomonas*, serving as potentially promising biological agents against RKN in crops (Siddiqui et al., 2009). Zhou et al. (2019) showed that soils from noninfested areas in fields with high RKN pressure have more microbial diversity than those from infested areas. A reduction in RKN root-gall number (Zhou et al., 2019) was elicited when microbiota from noninfested soils were used to inoculate tomato roots. It may therefore be possible to control RKN by enrichment of the abundance and diversity of specific microbial groups (Zhou et al., 2019). Diverse nematode antagonists have been described as belonging to several taxonomic groups (Stirling, 2014) with different modes of action: (1) by directly parasitizing diverse or specific RKN development stages as shown for *Pasteuria penetrans* (Davies et al., 2011); (2) by indirectly repelling, immobilizing, and/or killing the nematode through metabolite production and/or inducing a plant defense response, such as observed for *Fusarium oxysporum* strain Fo162 (Dababat and Sikora, 2007a,b); or (3) both directly and indirectly as reported for *Trichoderma* spp. (de Medeiros et al., 2017; Martínez-Medina et al., 2017). *Pochonia chlamydosporia* is a root endophyte fungus antagonistic to RKN and cyst nematodes which can act both directly by parasitizing eggs and indirectly. This fungal species colonizes the roots of diverse plant species, such as barley (Maciá-Vicente et al., 2009), tomato (Bordallo et al., 2002), potato (Manzanilla-López et al., 2011), and *Arabidopsis* (Zavala-Gonzalez et al., 2017). It can induce plant defenses (Ghahremani et al., 2019) or activate genes related to salicylic and jasmonic acid pathways that play a role in plant defense (Larriba et al., 2015; Zavala-Gonzalez et al., 2017). Although *P. chlamydosporia* has been extensively studied for its exploitation as a nematophagous fungus, the suppression mechanisms of root infection, development, and/or reproduction of RKN have still to be fully elucidated (Escudero and Lopez-Llorca, 2012; Ghahremani et al., 2019).

Arbuscular mycorrhizal fungi (AMF) are known to reduce the numbers of parasitic nematodes (e.g., *Meloidogyne*, *Heterodera*, *Globodera*, and *Tylenchorhynchus* spp.). Gough et al. (2020) studied the effects of AMF on the root-lesion nematode *Pratylenchus* spp., revealing that their interaction is influenced by crop species, crop cultivar, AMF order and genus. This finding suggests that the AMF biocontrol effect may likely be due to a combination of host tolerance, competition between organisms, and systemic resistance (Gough et al., 2020).

11.3 Mycorrhizal fungi

Mycorrhizal fungi are root symbionts that establish associations with a great variety of plants (Balestrini and Lumini, 2018; Bonfante, 2001). Mycorrhizal associations occur in roots of seed plants and underground gametophytes of many bryophytes and pteridophytes. They also occur in the sporophytes of most pteridophytes and specialized structures such as orchid protocorms (Bonfante and Genre, 2008). However, about 29% of all vascular plants, including the model plant *Arabidopsis thaliana* and major crops, such as sugar beet and rapeseed (canola), are thought to be nonhosts. Species belonging to the Brassicaceae, Proteaceae, Chenopodiaceae, and Caryophyllaceae are also considered nonhost plants, although under specific conditions, roots of presumed nonhosts show various degrees of fungal colonization (Cosme et al., 2018).

Mycorrhizal fungi give nutritional benefits to their host by absorbing phosphate, N, and other macronutrients, microelements, and water from the soil, providing these resources to the host plant (Smith and Read, 2008). The fungus, in turn, receives C compounds such as carbohydrates and lipids (Balestrini and Lumini, 2018). Mycorrhizal interactions achieve full symbiotic functionality through the development of an extensive contact surface between the plant and fungal cells where signals and nutrients are exchanged (Balestrini and Bonfante, 2014; Smith and Smith, 1990). Endomycorrhizae form structures inside plant roots, while ectomycorrhizal structures are exterior to root surfaces. Sequencing of mycorrhizal fungal genomes of taxa with different lifestyles (e.g., ecto- and endomycorrhizal fungi, endophytes, saprotrophs, pathogens) has helped to reveal how evolution has played a crucial role in fashioning fungal genomes to suit their own specialization. Recent evidence indicates that mycorrhizal fungi drive plant population biology and community ecology by affecting dispersal and establishment and regulating plant coexistence (Tedersoo et al., 2020). Mycorrhizal networks physically connect that conspecific and heterospecific plant individuals in soil, play an important role in C and nutrient transfer, and transmit signal molecules among plants (Alaux et al., 2021; Tedersoo et al., 2020). Some processes, i.e., interplant nutrient transfer through mycelial networks and the C-to-nutrient exchange at the mycorrhizal interface, are still to be fully elucidated (Tedersoo et al., 2020). The correlation of these processes with soil processes and plant communities will allow us to develop predictions about the impacts of global change and pollution and to elaborate technologies to improve production in agriculture and forestry. Mycorrhizal fungal mycelium also transports plant-derived C in soil as sugars, amino acids, and polyols to sustain the mycorrhizosphere microbiome (Tarkka et al., 2018).

The most common mycorrhizal association is formed by the AMF that interact with the roots of most terrestrial plants, including several important crop species (e.g., maize, wheat, soybean, rice, tomato) (Balestrini and Lumini, 2018; Berruti et al., 2016; Chialva et al., 2020; Ontivero et al., 2020). AMF belong to the subphylum Glomeromycotina (Spatafora et al., 2016) and are obligate biotrophs. In the AMF symbiosis, the host plasma membrane invaginates and proliferates around the developing intracellular fungal structures, and cell wall material is laid down between this membrane and the fungal cell surface, leading to the creation of an interface space where nutrients and signals are exchanged (Balestrini and Bonfante, 2005, 2014). The AMF receive plant-fixed C and can gradually release this C to the microbial community present in the rhizosphere, thus pointing out the prominent role of AMF in the transfer of C between plants and soil (Drigo et al., 2010; Fitter et al., 2005; Frey, 2019). Kiers et al. (2011) demonstrated that in the AMF symbiosis, the host plant allocates C to cooperative fungal partners, increasing nutrient (phosphate) transfer to the roots.

AM fungal communities have been extensively studied in diverse ecosystems (Opik et al., 2006). Effects of soil depth on AM fungal community structure have also been studied (Oehl et al., 2005) in deep soil layers where their spores differ from those in the topsoil, suggesting that deep soil layers should be included in studies to obtain a complete picture of AM fungal diversity.

Ectomycorrhizal (ECM) fungi belong to the Ascomycota and Basidiomycota (Smith and Read, 2008), including species able to produce visible mushrooms such as truffles, boletes, amanitas, and chanterelles (Anderson and Cairney, 2007). ECM fungi interact with some woody plants. During the symbiotic phase, they form a fungal sheath (the mantle) outside the root that is linked to extramatrical hyphae which are involved in water uptake, substrate exploration, and mineral nutrition. Some hyphae enter between the epidermal and outer cortical cells, from the inner layers of the mantle, to build an intercellular hyphal network inside the root tissues, i.e., the Hartig net. In ECM, plant and fungal cell walls are in direct contact forming the interface between the two partners (Hacquard et al., 2013). While many ECM fungi have been studied from the ecological point of view, only four have been used as models in in vitro associations: *Laccaria bicolor* with poplars, *Hebeloma cylindrosporum* with pine, *Tuber melanosporum* with hazelnut, and *Pisolithus* spp. with eucalyptus.

Ericoid mycorrhizal (ERM) fungi form endomycorrhizae with the Ericaceae and include some soil fungi in the Leotiomycetes (Martino et al., 2018). ERM plants survive in harsh habitats thanks to the nutrient mobilization from SOM by their fungal partners. The genomes of the ERM fungi *Meliniomyces bicolor*, *M. variabilis*, *Oidiodendron maius*, and *Rhizoscyphus ericae* show a capacity for a dual saprotrophic and biotrophic lifestyle (Martino et al., 2018).

Recent information on the mechanisms involved in orchid mycorrhizal (ORM) establishment and functioning has been obtained (Balestrini et al., 2014; Fochi et al., 2017a,b; Ghirardo et al., 2020; Perotto et al., 2014), mainly in the orchid model system *Serapias vomeracea/Tulasnella calospora*. Orchids require a fungal association to support seed germination and to assist the adult plant during growth (Selosse and Roy, 2009; Waterman and Bidartondo, 2008). During initial colonization of the young, nonphotosynthetic host, the fungus provides organic C and other essential nutrients to the plant (Dearnaley and Cameron, 2017). Proteomics and transcriptomics have identified changes in the steady-state levels of proteins and transcripts in the roots of the green, terrestrial orchid *Oeceoclades maculate* (Valadares et al., 2020).

A mycorrhizal root, and protocorms in the case of ORM, represents a heterogeneous environment formed by a mixture of cell types and tissues (Fiorilli et al., 2019). Examination of the process at the molecular level suffers from the use of whole organs since cell type-specific differences might be masked. In the last decade, laser microdissection, which permits rapid collection of selected cell-type populations from a section of heterogeneous tissues, has been widely used to study cell-specificity in gene expression profiles in mycorrhizal interactions (Table 11.1). Most of these studies (see Table S11.1 in Supplementary Material) have focused on verifying the expression of specific genes in colonized cells with neighboring noncolonized cells (Balestrini and Fiorilli, 2020).

TABLE 11.1 List of original published papers applying laser microdissection technology to study gene expression in ECM and ORM mycorrhizal symbioses.

ECM; Hacquard et al. (2013)	The two compartments forming an ectomycorrhiza (i.e., the mantle and the Hartig net) were micro-dissected from ectomycorrhizal sections of *Tuber melanosporum*/*Corylus avellana* and distinct genetic programs associated with each compartment were identified. While N and water acquisition from soil, synthesis of secondary metabolites, and detoxification mechanisms seemed to be important processes in the fungal mantle, transport activity was enhanced in the Hartig net, playing a key role in the reciprocal transfer of resources between the symbiotic partners.
ORM; Balestrini et al. (2014)	*SvNod1*, a *Serapias vomeracea* gene coding for a nodulin-like protein containing a plastocyanin-like domain, was expressed only in protocorm cells containing intracellular fungal hyphae.
ORM; Fochi et al. (2017a)	Fungal pelotons are thought to be key structures for nutrient exchange in the symbiotic orchid protocorms. The cell-specific expression pattern of genes involved in N uptake and metabolism was studied in *Serapias vomeracea*/*Tulasnella calospora* mycorrhizal protocorms. Specific cell-type expression of the fungal *AMT* genes was analyzed and transcripts corresponding to *TcAMT1* and *TcAMT2* were identified in microdissected protocorm cells containing both younger (i.e., occupying the whole plant cell) and more condensed fungal pelotons.
ORM; Fochi et al. (2017b)	Localization of specific plant gene transcripts in different cell-type populations collected from mycorrhizal protocorms and roots of the Mediterranean orchid *Serapias vomeracea* colonized by *Tulasnella calospora*. Genes potentially involved in N uptake and transport and previously identified as up-regulated in symbiotic protocorms were evaluated. Results clearly showed that some plant N transporters are differentially expressed in cells containing fungal coils at different developmental stages, as well as in noncolonized cells, allowing the identification of new functional markers associated with coil-containing cells.
ORM; Balestrini et al. (2018b)	The expression pattern for fungal genes was reported for *Serapias vomeracea*/*Tulasnella calospora* mycorrhizal roots.

11.4 Threats to soil biota—plant interactions

One of the major challenges in a changing environment is to understand how complex plant-soil biotic communities recover from disturbances such as climate extremes or human impacts, which are predicted to increase in frequency and intensity with climate change (de Vries and Shade, 2013). Many studies have demonstrated that climate extremes (e.g., extended drought, precipitation intensity, temperature fluctuations) can considerably affect soil biotic communities, often with consequences for ecosystem processes and plant community dynamics (de Vries et al., 2018). Fungi are generally more resistant, but less resilient, with differences in the recovery of fungi and bacteria as governed by plant physiological responses (de Vries et al., 2018). Changes in vegetation resulting from climate extremes could have long-lasting effects on soil communities, potentially influencing plant community composition and the ability of above- and belowground communities to withstand future disturbances. Other important threats to soil

TABLE S11.1 List of original published papers applying the laser microdissection technology to study gene expression in arbuscular mycorrhizal symbioses.

Reference	Major findings
Balestrini et al. (2007)	Five tomato phosphate transporter (PT) transcripts have been contemporaneously localized in arbusculated cells together with transcripts corresponding to one fungal PT gene.
Guether et al. (2009a)	Identification of several arbuscule marker genes whose transcripts were detected exclusively in arbusculated cells. Among them, genes coding for plasma membrane proteins (*LjMlo2*; *LjCesA*; *LjCel1*), putatively involved in fungal accommodation, a phosphate (*LjPT4*) and a peptide transporter (*LjPTR*), a MYB transcription factor (*LjMYB*), and a gene coding for a lipid-transfer protein (*LjLTP*).
Guether et al. (2009b)	Localization of transcripts corresponding to a mycorrhizal-induced ammonium transporter (*LjAMT*) in AMF-colonized roots of *Lotus japonicus*.
Fiorilli et al. (2009)	Among the genes regulated in mycorrhizal tomato roots in a cDNA array experiment, five genes showing similarity to a β-xylosidase α-l-arabinosidase, a putative kinesin-like protein, a bifunctional nuclease, a UDP-glucoronosyl UDP-glucosyl transferase protein, and a cytochrome P450 (putative CYP707 A3) were exclusively expressed in arbuscule-containing cells. Transcripts for a gene encoding a putative indole-3-acetic acid amido synthetase were also observed to be expressed more in arbuscule-containing cells with respect to the noncolonized neighboring cells.
Gomez et al. (2009)	Identification of novel *Medicago truncatula* and *Rhizophagus intraradices* genes expressed in colonized cortical cells and in arbuscules. Expression of genes associated with the urea cycle, amino acid biosynthesis, and cellular autophagy was detected in arbusculated cells. Analysis of gene expression in colonized cortical cell revealed up-regulation of a lysine motif (LysM) receptor-like kinase (TF GRAS family) and a symbiosis-specific ammonium transporter, which is a likely candidate for mediating ammonium transport in the AM symbiosis.
Gómez-Ariza et al. (2009)	Collection of cortical, epidermal, and central cylinder cells with the aim of analyzing the expression of the five tomato PT genes (Balestrini et al., 2007) in different root tissues. Using RT-PCR analysis, *LePT1* and *LePT2* transcripts were detected in epidermal and cortical cells, while *LePT3*, *LePT4*, and *LePT5* transcripts were confirmed to be exclusive to arbuscule-containing cells.
Guether et al. (2011)	Transcripts corresponding to a *Lotus japonicus* amino acid transporter (AAT) gene (*LjLHT1.2*) have been demonstrated to be located above all in arbusculated cells.
Campos-Soriano et al. (2011)	Expression of a rice calcium-dependent protein kinase (*OsCPK18*) was found in cortical cells and not in epidermal cells of *Rhizophagus intraradices* inoculated rice roots, suggesting a preferential role of this gene in the root cortex.
Hogekamp et al. (2011)	Identification of 25 novel arbuscule-specific genes and 37 genes expressed both in the arbuscule-containing and the adjacent cortical cells colonized by fungal hyphae.
Pérez-Tienda et al. (2011)	Fungal ammonium transporter (*GintAMT1* and *GintAMT2*) genes were localized in the arbuscules formed by *Rhizophagus intraradices* in *Medicago* mycorrhizal roots.

Continued

TABLE S11.1 List of original published papers applying the laser microdissection technology to study gene expression in arbuscular mycorrhizal symbioses.—cont'd

Reference	Major findings
Gaude et al. (2012)	Comparative analyses of different cell types were performed using laser microdissection combined with microarray hybridization. A high proportion of transcripts regulated in arbuscule-containing cells and nonarbusculated cells encodes proteins involved in transport processes, transcriptional regulation, and lipid metabolism, indicating that reprogramming of these processes is of particular importance for AM symbiosis.
Giovannetti et al. (2012)	Two *Lotus* aquaporin genes, *LjXIP1* and *LjNIP1*, observed to be regulated during a microarray experiment. One gene (*LjNIP1*), capable of transferring water when expressed in yeast protoplasts, is expressed exclusively in arbuscule-containing cells (ARB). *LjXIP1*, which was the first XIP family aquaporin shown to be transcriptionally regulated during symbiosis, was expressed in all three cell-type populations considered (ARB, MNM from the colonized roots, and cortical cells from nonmycorrhizal roots). Through one-step RT-qPCR on RNA extracted from microdissected cells, it has been revealed that the gene was significantly overexpressed in arbusculated cells compared to other analyzed cell types, thus confirming an additional or enhanced role within the arbuscule-containing cells.
Tisserant et al. (2012)	Identification of several fungal transcripts expressed in arbuscules.
Hogekamp and Küster (2013)	Early developmental stages of the AM symbiosis were analyzed by monitoring gene expression in appressorial and nonappressorial areas from roots harboring infection units at 5—6 days after inoculation. To cover the late stages of AM formation, arbusculated cells, cortical cells colonized by intraradical hyphae, and epidermal cells from mature mycorrhizal roots were studied at 21 days. Overall, the cell-specific expression patterns of 18,014 genes were revealed, including 1392 genes whose transcription was influenced by mycorrhizal colonization at different stages.
Fiorilli et al. (2013)	The *Rhizophagus irregularis RiPEIP1* expression profile (preferentially expressed *in planta*) was evaluated during the intraradical stage showing that *RiPEIP1* was expressed in the ARB population.
Belmondo et al. (2014)	Transcripts corresponding to a dipeptide transporter from the arbuscular mycorrhizal fungus *Rhizophagus irregularis* (*RiPTR2*) were detected in arbuscules.
Giovannetti et al. (2014)	Expression corresponding to a gene (*LjSultr1;2*) coding for a group 1 sulfate transporter, induced by both S starvation and mycorrhizal formation, was located in arbuscule-containing cells.
López-Ráez et al. (2015)	Apocarotenoids are a class of compounds that play important roles in nature. A prominent role for these compounds in AM symbiosis was shown. They are derived from carotenoids by the action of the carotenoid cleavage dioxygenase (CCD) enzyme family. Induction of *SlCCD7* and *SlCCD1* expression in arbusculated cells promoted the production of C^{13}/C^{14} apocarotenoid derivatives, leading to a reduction in the biosynthesis of the apocarotenoid strigolactones and maintaining a functional symbiosis.

TABLE S11.1 List of original published papers applying the laser microdissection technology to study gene expression in arbuscular mycorrhizal symbioses.—cont'd

Reference	Major findings
Belmondo et al. (2016)	The expression pattern of five NADPH oxidase (also called RBOH) encoding genes in *Medicago truncatula* showed that only one of them (*MtRbohE*) is specifically upregulated in arbuscule-containing cells.
Zeng et al. (2018)	Combined LMD of mycorrhizal root sections with RNAseq analysis to obtain new information into stage-specific fungal gene expression, focusing the attention on fungal-secreted protein genes. The LMD-RNA-seq data also reflected the expression pattern of studied plant marker genes known from the literature.

biota are massive anthropogenic activities, such as soil disturbance due to intensive agriculture (Yang et al., 2020) and urbanization (Guilland et al., 2018), as well as climate warming and biological invasions (van der Putten et al., 2007). Biological invasions occur when species are introduced outside of their native range and successfully spread in their new environment (Valéry et al., 2008). Invasive species are considered one of the most important threats to biodiversity, negatively influencing ecosystems and reducing species richness (Paini et al., 2016).

Invasive plant species often experience more positive effects from soil biota compared to their native range, resulting in an increased level of invasiveness (Maron et al., 2014; Reinhart and Callaway, 2006). A lack of compatible mutualists may cause the failure of invasive plant establishment (Zenni and Nuñez, 2013). Co-introduced root symbionts may enhance plant invasiveness, as observed during the co-invasion of pines with their ectomycorrhizal symbionts in the southern hemisphere (Dickie et al., 2010; Hayward et al., 2015). Almost two million hectares are covered by pines in New Zealand. Co-introduction of mutualist rhizosphere fungi associated with pines played a crucial role in the invasive spread of these trees (Nuñez et al., 2017; Richardson et al., 2000). An increase in the competitive ability of invading plants due to the presence of associated invasive fungi has also been observed in grasses. For example, *Centaurea maculosa* benefitted indirectly from mycorrhizal fungi through enhanced competitive ability against native grasses (Marler et al., 1999). Divergent hypotheses have been proposed, suggesting that plant invasions either enhance or degrade the mutualisms between plants and AMF, but the relative support for these hypotheses remains unknown (Bunn et al., 2015; Hayward et al., 2015). While some AMF may facilitate invasiveness of some alien plants, such plants may also potentially impact mycorrhizal community structure and functions in invaded habitats (Reinhart and Callaway, 2006). If relatively few individual plants are introduced, the bottleneck can create strong founder effects in associated fungal symbiont populations. This loss of symbiont diversity may lead to mutualism reduction or missed mutualists useful for beneficial symbioses (Catford et al., 2009).

The alteration of mycorrhizal soil environments and the deprivation of local mycorrhizal fungi by invasive plants have been documented for a variety of species, leading to the "Mycorrhizal Degradation Hypothesis" (van der Putten et al., 2007; Vogelsang et al., 2004). Networks of native symbiotic soil microorganisms directly affect the successful naturalization of invasive plants as well as their ability to outcompete native plant species. Microbiome shifts between the native and invasive ranges facilitate the invasion process and the subsequent establishment of invasive plants (Coats and Rumpho, 2014). Since specific plant–microbe interactions can determine the suitability of habitats with competitive advantage for invasive plants, high-quality predictions for susceptible habitats will be one of the best strategies to prevent invasive plant introductions (Coats and Rumpho, 2014). The detrimental effect of plant invasion on belowground biodiversity is still debated. While a decrease in aboveground diversity due to shifts in ecosystem structure and function has been well-documented, the effects on microbial diversity from invasion processes appear to be less significant (Custer and van Diepen, 2020).

Another important threat to soil biodiversity is represented by emerging plant diseases. Invasive fungal pathogens have been reported to rapidly adapt to their new environment, with the ability to interact with novel hosts (i.e., "spillover"). They may display a plasticity of their genomes that lead to genetic changes either through natural selection or via hybridization (Hulme, 2014; Jarić et al., 2019; Jeschke et al., 2013; Ricciardi et al., 2017; Stukenbrock et al., 2012). Fungal plant pathogens can trigger a defense response in plants, and the release of polyphenol oxidases may inhibit both the belowground formation of root nodules produced by rhizobia and mycorrhizal colonization (Ballhorn et al., 2014; Sillo et al., 2015). For example, invasive and native fungal pathogens affected the expression of genes related to cell-wall hydrolytic enzymes and hydrophobins of an ectomycorrhizal fungus (i.e., *Suillus luteus*). Zampieri et al. (2017) highlighted that fungal symbionts (i.e., *Tuber borchi* in association with *Pinus pinea*) responded more to native fungal pathogens (*Heterobasidion annosum*) compared to invasive ones (*H. irregulare*), with increased regulation for genes related to symbiosis functioning. This result suggests that the symbiont recognizes a self/native and a nonself/nonnative pathogen during interaction with the host plant (Zampieri et al., 2017). ECM fungi can act antagonistically versus fungal pathogens in saprotrophic conditions by displaying antibiotic activity (Mucha et al., 2006, 2009). In addition, mycorrhizal fungi can improve resistance against pathogen-induced stresses, reducing the disease impact (Pfabel et al., 2012; Whipps, 2004). Gonthier et al. (2019) demonstrated that *Pinus sylvestris* inoculated with the ECM fungus *S. luteus* was significantly less susceptible to *H. annosum* than nonmycorrhizal plants; however, the protective role of the symbiont was not effective against the invasive *H. irregulare*, thus highlighting that native mycorrhizal fungi may be differently affected by native or invasive pathogens belonging to the same genus.

11.5 Insights into plant–soil biota through -*omics* approaches

A better understanding of soil biodiversity (Girvan et al., 2005; Philippot et al., 2013; Wagg et al., 2014) should help in the manipulation of plant–soil interactions, including resistance to environmental threats (Nannipieri et al., 2014). *Omics* approaches can be distinguished by their molecular targets, i.e., DNA, RNA, proteins, or metabolites (Fig. 11.1).

The availability of DNA sequencing methods (454 pyrosequencing) and second- and third-generation sequencing platforms, such as Illumina MiSeq, HiSeq, NovaSeq, Ion Torrent PGM GeneStudio, PacBio RSII Sequel, Oxford Nanopore MinION, GridION, and PrometION (Nilsson et al., 2019) have facilitated the exploration of microbial community complexity in a wide range of environments. Two main approaches used for such studies are metabarcoding and shotgun metagenomics (Escobar-Zepeda et al., 2015). In metabarcoding, a single marker gene is amplified from environmental DNA (e.g., 16S rRNA gene for prokaryotes and intergenic transcribed spacer (ITS) or the large ribosomal subunit (LSU) for eukaryotes). Metagenomics is defined as the genomic analysis of organisms in environmental samples, such as bulk or rhizosphere soil, as well as plant roots. This approach provides information about taxonomic composition and relative abundance (Singer et al., 2016; Vogel et al., 2009). DNA barcode sequences provide species identification (Xu, 2016) and enhanced knowledge on microbiota communities

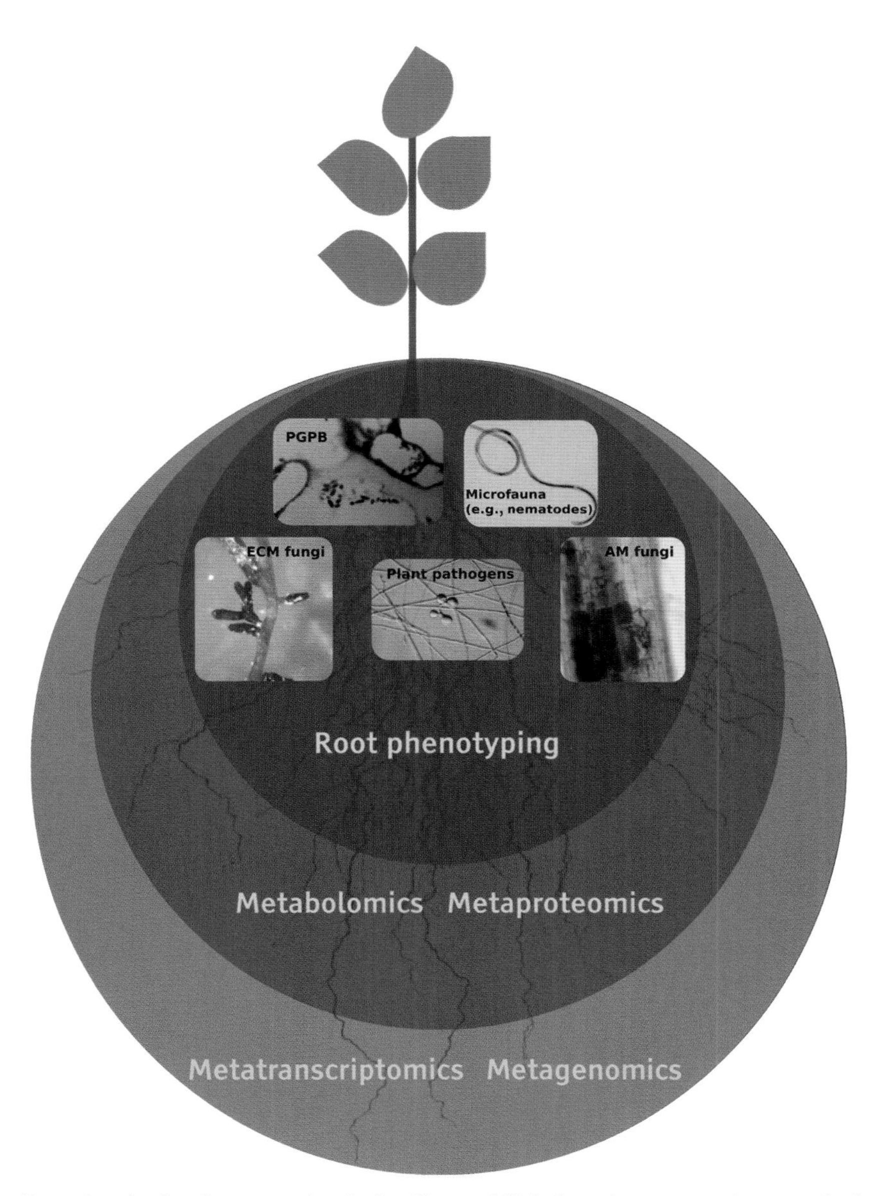

FIGURE 11.1 Examples of soil and root-associated microbiota and high throughput approaches to study their biodiversity and functions. From left to right, above: plant growth promoting bacteria (PGPB) associated with roots; nematode, *Xiphinema index* (provided with permission by F. De Luca and A. Troccoli); below: *Pinus sylvestris* inoculated with the ECM fungus *Suillus luteus*; the plant pathogen fungus *Phytopthora cinnamoni* (in vitro culture; provided with permission by L. Giordano); tomato roots colonized by an AM fungus.

interacting with plant roots. In addition to classical approaches to analyze metagenomic data, such as Quantitative Insights into Microbial Ecology (QIIME) (Bolyen et al., 2019; Caporaso et al., 2010), functional profiles of bacterial communities identified through the bacterial 16S rRNA gene sequence dataset can be predicted with Phylogenetic Investigation of Communities by Reconstruction of Unobserved States (PICRUSt) (Langille et al., 2013). FUNGuild (Nguyen et al., 2016) can be used to partition fungal sequences into key functional groups (e.g., saprotrophs, pathotrophs, ectomycorrhizae).

The major constraint in both DNA-based approaches, i.e., metabarcoding and metagenomics, is that neither approach can differentiate between expressed and nonexpressed genes, and/or between living or dead organisms (Mello and Zampieri, 2017; Zampieri et al., 2016). For this reason, metagenomic studies are often supported by metatranscriptomic and metaproteomic approaches that can provide functional data to decipher the mechanisms involved in plant—biota interactions. Metatranscriptomics target expressed genes, thus functional aspects of the active fungal community (Nilsson et al., 2019). Fungal transcript analysis demonstrates that fungi play a dominant role in plant litter decomposition (Žifčáková et al., 2017). Metatranscriptomic of mycorrhizal communities provides a deeper understanding of the functional roles of these fungi in nature through the identification of genes associated with symbiosis and ecosystem function (Liao et al., 2014). In *Salix* spp., a metatranscriptomic analysis showed that root and fungal gene expression involved transcripts encoding carbohydrate/amino acid (C/N) dialogue, whereas bacterial gene expression included the apparatus necessary for biofilm interaction and direct reduction of contamination stress (Gonzalez et al., 2018). A limitation that characterizes both metagenomics and metatranscriptomics is the low number of available fungal genomes, which does not allow taxonomic classification of fungal transcripts much beyond the phylum level (Nilsson et al., 2019).

Metaproteomic is the study of all the proteins expressed in an environment (Bastida et al., 2014; Wilmes and Bond, 2004), thus targeting active metabolic processes (Jansson and Hofmockel, 2018; Liu et al., 2017; Mello and Zampieri, 2017; Qian and Hettich, 2017; Siggins et al., 2012). Applying metagenomic and metaproteomic together connects microbial community composition to ecological processes (Martinez-Alonso et al., 2019; Zampieri et al., 2016). The application of metaproteomics techniques to soils is challenged by technical and computational limitations, such as humic acids, contaminants, and the absence of a comprehensive protein database (Chiapello et al., 2020).

The monitoring of the volatilome (volatilomics) is the study of VOCs that are characterized by high chemical diversity, low molecular weight, high vapor pressure, and polarity (Lee et al., 2019). VOCs play an important role in the formation and regulation of symbiotic interactions (Mhlongo et al., 2018) and can also be involved in the plant defense response, having antimicrobial as well as allopathic effects (Brilli et al., 2019a). Guo et al. (2020) studied the volatile metabolic fingerprints of four *Trichoderma* spp. over a 48-hour growth period, revealing how the metabolic profile changes in relation to the fungal species and development status.

High-throughput approaches can be used in plant phenomics, the study of plant growth, performance, and composition under different environmental conditions (Furbank and Tester, 2011). A sub-branch is represented by shovelomics, which is the high-throughput phenotyping of root systems, based on the measure of root numbers, angles, densities, and lengths (Lynch and Brown, 2012). This approach allows for the potential identification of species/genotypes/individuals more efficient in water, and P and N uptake (Burridge et al., 2016). It could also be used to verify the impact of rhizosphere microorganisms on root system architecture. A further effort will integrate results from different omics techniques in a complex network that provides a complete picture of the environment and the organisms living within it.

11.6 How soil microbiota improve plant responses to abiotic stress factors

Important changes in the global climate, primarily due to the increase in atmospheric CO_2 concentrations, will lead to an increase of the global surface temperatures during the 21st century (Kannojia et al., 2019). Above- and belowground terrestrial ecosystems will be impacted by climate change, including soil biota that play a pivotal role in the mitigation of the negative effects of environmental stresses (Salwan et al., 2019). Currently, several beneficial soil microorganisms are exploited in agriculture against abiotic stress, such as drought, high salinity, extreme temperatures, flooding, or heavy metal pollution. For example, AMF can improve the growth performance and abiotic stress tolerance of host plants by mediating different physiological processes (Evelin et al., 2009). Soil salinity is also associated with climate change caused by coastal flooding and seawater intrusion (Daliakopoulos et al., 2016). AMF can mitigate the damage from salt stress in both halophytes and glycophytes (Kosová et al., 2013). Pan et al. (2020) demonstrated that glycophytes are more dependent on AM symbiosis than halophytes in the presence of salt stress. Through a combined approach, including eco-physiological, morphometric, biochemical, and molecular analyses, Chitarra et al. (2016) and Volpe et al. (2018) demonstrated that AM symbiosis positively impacts the tolerance to water stress in tomatoes dependent on considered traits and fungal species. Rivero et al. (2018) studied the impact of three AMF of different genera on tomato tolerance to drought or salt stress. Some responses were common to all tested AMF, while others were specifically related to single isolates (Berruti et al., 2016).

PGPB or PGPR can also ameliorate water deficit (Rubin et al., 2017). Armada et al. (2018) demonstrated that under drought stress, inoculation with *Bacillus thuringiensis* strain IAM 12077 impacted plant nutrition and AM fungal root colonization in *Santolina chamaecyparissus*. It also promoted *Lavandula dentate* shoot growth, with no effect on the rhizosphere microbial community. Brilli et al. (2019b) demonstrated that tomato plants inoculated with *Pseudomonas chlororaphis* subsp. *aureofaciens* strain M71 show improved antioxidant activity under mild water stress by increasing stress tolerance. The presence of the M71 strain modulated stomatal closure, improving water use efficiency and biomass in water-stressed plants, with an increase in the abscisic acid level in leaves (Brilli et al., 2019b). The term "induced systemic tolerance" has been proposed for the physical and chemical changes that improve tolerance to abiotic stresses and that are induced by PGPB (Yang et al., 2009). Mechanisms related to salt stress or drought modulation, mediated by soil microorganisms, are not yet completely deciphered (de Vries et al., 2020). Improved tolerance to abiotic stresses and secondary metabolite production has been noted (Alagna et al., 2020; Balestrini et al., 2018a; Irankhah et al., 2020).

11.7 Summary

Plant—soil biota interactions affect plant health and productivity (Song et al., 2020; Trivedi et al., 2020; Yang et al., 2020). This includes the role of root exudates in the complex dialog between the soil biota and plants (Bais et al., 2006). Plant VOCs, such as terpenes, phenylpropanoids, and benzenoid, are involved in structuring plant—microbe interactions (Dudareva et al., 2006; Huang et al., 2012). The organisms involved in these interactions and how they dialogue are still unresolved issues; however, it is known that PGPB or PGP fungi may induce resistance/tolerance to stress factors (Alagna et al., 2020). *Omics* approaches have allowed us to obtain more details to answer questions related to what soil biota are present and what they are doing. López-Angulo et al. (2020) explained the variations in soil microbial diversity, the root biomass/composition that influences microbial diversity, and the community composition in a semiarid Mediterranean scrubland. Aboveground plant composition affected fungal community structure. These findings suggest that it is crucial to study the plant-soil biota system in its totality (Gilbert et al., 2014; López-Angulo et al., 2020). This is particularly important for microbial activities related to plant productivity and crop quality. For example, changes in soil microbial communities of grape berries involved in fermentation influence the wine organoleptic properties correlated with *terroir*.

Advances in bioinformatics have helped decipher and interconnect the data from various *-omics* approaches (Berg et al., 2020). Further development of these approaches will be useful to establish the linkage between structure and function of the soil biotic community and lead to new insights on the ecological processes in the environment with special emphasis on plant–microbe interactions (Biswas and Sarkar, 2018).

Acknowledgments

The authors of this chapter were funded by the Italian National Research Council (CNR) project FOE-2019 DBA.AD003.139.

References

Addy, H.D., Piercey, M.M., Currah, R.S., 2005. Microfungal endophytes in roots. Can. J. Bot. 83, 1–13.

Alagna, F., Balestrini, R., Chitarra, W., Marsico, A.D., Nerva, L., 2020. Getting ready with the priming: innovative weapons against biotic and abiotic crop enemies in a global changing scenario. In: Hossain, M.A., Liu, F., Burritt, D.J., Fujita, M., Huang, B. (Eds.), Priming-Mediated Stress and Cross-Stress Tolerance in Crop Plants. Academic Press, Elsevier, London, pp. 35–56.

Alaux, P.L., Zhang, Y., Gilbert, L., Johnson, D., 2021. Can common mycorrhizal fungal networks be managed to enhance ecosystem functionality? New Phytol. https://doi.org/10.1002/ppp3.10178.

Anderson, I.C., Cairney, J.W.G., 2007. Ectomycorrhizal fungi: exploring the mycelial frontier. FEMS Microbiol. Rev. 31, 388–406.

Armada, E., Leite, M.F.A., Medina, A., Azcón, R., Kuramae, E.E., 2018. Native bacteria promote plant growth under drought stress condition without impacting the rhizomicrobiome. FEMS Microbiol. Ecol. 94, 10.

Atia, M.A.M., Abdeldaym, E.A., Abdelsattar, M., Ibrahim, D.S.S., Saleh, I., Abd Elwahab, M., et al., 2020. *Piriformospora indica* promotes cucumber tolerance against Root-knot nematode by modulating photosynthesis and innate responsive genes. Saudi J. Biol. Sci. 27, 279–287.

Bais, H.P., Weir, T.L., Perry, L.G., Gilroy, S., Vivanco, J.M., 2006. The role of root exudates in rhizosphere interactions with plants and other organisms. Annu. Rev. Plant Biol. 57, 233–266.

Baldrian, P., 2019. The known and the unknown in soil microbial ecology. FEMS Microbiol. Ecol. 95. https://doi.org/10.1093/femsec/fiz005.

Balestrini, R., Bonfante, P., 2005. The interface compartment in arbuscular mycorrhizae: a special type of plant cell wall? Plant Biosyst. 139, 8–15.

Balestrini, R., Bonfante, P., 2014. Cell wall remodeling in mycorrhizal symbiosis: a way towards biotrophism. Front. Plant Sci. 5, 237.

Balestrini, R., Lumini, E., 2018. Focus on mycorrhizal symbioses. Appl. Soil Ecol. 123, 299–304.

Balestrini, R., Fiorilli, V., 2020. Laser microdissection as a useful tool to study gene expression in plant and fungal partners in AM symbiosis. Methods Mol. Biol. 2146, 171–184.

Balestrini, R., Nerva, L., Sillo, F., Girlanda, M., Perotto, S., 2014. Plant and fungal gene expression in mycorrhizal protocorms of the orchid *Serapias vomeracea* colonized by *Tulasnella calospora*. Plant Signal. Behav. 9, e977707.

Balestrini, R., Chitarra, W., Antoniou, C., Ruocco, M., Fotopoulos, V., 2018a. Improvement of plant performance under water deficit with the employment of biological and chemical priming agents. J. Agric. Sci. 156, 680–688.

Balestrini, R., Fochi, V., Lopa, A., Perotto, S., 2018b. The use of laser microdissection to investigate cell-specific gene expression in orchid tissues. In: Lee, Y.I., Yeung, E.T. (Eds.), Orchid Propagation: From Laboratories to Greenhouses. Springer Protocols Handbooks. Humana Press, New York, NY.

Ballhorn, D.J., Younginger, B.S., Kautz, S., 2014. An aboveground pathogen inhibits belowground rhizobia and arbuscular mycorrhizal fungi in *Phaseolus vulgaris*. BMC Plant Biol. 14, 321.

Bastida, F., Hernández, T., García, C., 2014. Metaproteomics of soils from semiarid environment: functional and phylogenetic information obtained with different protein extraction methods. J. Proteomics. 101, 31—42.

Berendsen, R.L., Pieterse, C.M.J., Bakker, P.A.H.M., 2012. The rhizosphere microbiome and plant health. Trends Plant Sci. 17, 478—486.

Berg, G., Rybakova, D., Fischer, D., Cernava, T., Vergès, M.C., Charles, T., et al., 2020. Microbiome definition re-visited: old concepts and new challenges. Microbiome 8, 103.

Berruti, A., Lumini, E., Balestrini, R., Bianciotto, V., 2016. Arbuscular mycorrhizal fungi as natural biofertilizers: let's benefit from past successes. Front. Microbiol. 6, 1559.

Bhattacharyya, D., Garladinne, M., Lee, Y.H., 2015. Volatile indole produced by rhizobacterium *Proteus vulgaris* JBLS202 stimulates growth of *Arabidopsis thaliana* through auxin, cytokinin, and brassinosteroid pathways. J. Plant Growth Regul. 34, 158—168.

Biswas, R., Sarkar, A., 2018. "Omics" tools in soil microbiology: the state of the art. In: Adhya, T., Lal, B., Mohapatra, B., Paul, D., Das, S. (Eds.), Advances in Soil Microbiology: Recent Trends and Future Prospects. Microorganisms for Sustainability, Vol 3. Springer, Singapore.

Bolyen, E., Rideout, J.R., Dillon, M.R., Bokulich, N.A., Abnet, C.C., Al-Ghalith, G.A., et al., 2019. Reproducible, interactive, scalable and extensible microbiome data science using QIIME 2. Nat. Biotechnol. 37, 852—857.

Bonfante, P., 2001. At the interface between mycorrhizal fungi and plants: the structural organization of cell call, plasma membrane and cytoskeleton. In: Hock, B. (Ed.), Fungal Associations. The Mycota, Vol 9. Springer, Berlin.

Bonfante, P., Genre, A., 2008. Plants and arbuscular mycorrhizal fungi: an evolutionary-developmental perspective. Trends Plant Sci. 13, 492—498.

Bordallo, J.J., Lopez-Llorca, L.V., Jansson, H.B., Salinas, J., Persmark, L., Asensio, L., 2002. Colonization of plant roots by egg-parasitic and nematode-trapping fungi. New Phytol. 154, 491—499.

Brilli, F., Loreto, F., Baccelli, I., 2019a. Exploiting plant volatile organic compounds (VOCs) in agriculture to improve sustainable defense strategies and productivity of crops. Front. Plant Sci. 10, 264.

Brilli, F., Pollastri, S., Raio, A., Baraldi, R., Neri, L., Bartolini, P., et al., 2019b. Root colonization by *Pseudomonas chlororaphis* primes tomato (*Lycopersicum esculentum*) plants for enhanced tolerance to water stress. J. Plant Physiol. 232, 82—93.

Bulgarelli, D., Schlaeppi, K., Spaepen, S., Ver Loren van Themaat, E., Schulze-Lefert, P., 2013. Structure and functions of the bacterial microbiota of plants. Annu. Rev. Plant Biol. 64, 807—838.

Bunn, R.A., Ramsey, P.W., Lekberg, Y., 2015. Do native and invasive plants differ in their interactions with arbuscular mycorrhizal fungi? A meta-analysis. J. Ecol. 103, 1547—1556.

Burridge, J., Jochua, C.N., Bucksch, A., Lynch, J.P., 2016. Legume shovelomics: high-throughput phenotyping of common bean (*Phaseolus vulgaris* L.) and cowpea (*Vigna unguiculata* subsp, *unguiculata*) root architecture in the field. Field Crops Res. 192, 21—32.

Calvo, P., Watts, D.B., Kloepper, J.W., Torbert, H.A., 2017. Effect of microbial-based inoculants on nutrient concentrations and early root morphology of corn (*Zea mays*). J. Plant Nutr. Soil Sci. 180, 56—70.

Campbell, R.N., 1996. Fungal transmission of plant viruses. Annu. Rev. Phytopathol. 34, 87—108.

Caporaso, J.G., Kuczynski, J., Stombaugh, J., Bittinger, K., Bushman, F.D., Costello, E.K., et al., 2010. QIIME allows analysis of high-throughput community sequencing data. Nat. Methods 7, 335—336.

Castillo, J.D., Vivanco, J.M., Manter, D.K., 2017. Bacterial microbiome and nematode occurrence in different potato agricultural soils. Microb. Ecol. 74, 888—900.

Catford, J., Jansson, R., Nilsson, C., 2009. Reducing redundancy in invasion ecology by integrating hypotheses into a single theoretical framework. Divers. Distrib. 15, 22—40.

Chialva, M., di Fossalunga, A.S., Daghino, S., Ghignone, S., Bagnaresi, P., Chiapello, M., et al., 2018. Native soils with their microbiotas elicit a state of alert in tomato plants. New Phytol. 220, 1296–1308.

Chialva, M., Ghignone, S., Cozzi, P., Lazzari, B., Bonfante, P., Abbruscato, P., et al., 2020. Water management and phenology influence the root-associated rice field microbiota. FEMS Microbiol. Ecol. 96, fiaa146.

Chiapello, M., Zampieri, E., Mello, A., 2020. A small effort for researchers, a big gain for soil metaproteomics. Front. Microbiol. 11, 88.

Chitarra, W., Pagliarani, C., Maserti, B., Lumini, E., Siciliano, I., Cascone, P., et al., 2016. Insights on the impact of arbuscular mycorrhizal symbiosis on tomato tolerance to water stress. Plant Physiol. 171, 1009–1023.

Coats, V.C., Rumpho, M.E., 2014. The rhizosphere microbiota of plant invaders: an overview of recent advances in the microbiomics of invasive plants. Front. Microbiol. 5, 368.

Cobián Güemes, A.G., Youle, M., Cantú, V.A., Felts, B., Nulton, J., Rohwer, F., 2016. Viruses as winners in the game of life. Annu. Rev. Virol. 3, 197–214.

Cosme, M., Fernández, I., Van der Heijden, M.G., Pieterse, C.M., 2018. Non-mycorrhizal plants: the exceptions that prove the rule. Trends Plant Sci. 23, 577–587.

Crowther, T.W., Boddy, L., Hefin Jones, T., 2012. Functional and ecological consequences of saprotrophic fungus-grazer interactions. ISME J. 6, 1992–2001.

Custer, G.F., van Diepen, L.T., 2020. Plant invasion has limited impact on soil microbial α-diversity: a meta-analysis. Diversity 12, 112.

Dababat, A.E.F.A., Sikora, R.A., 2007a. Influence of the mutualistic endophyte *Fusarium oxysporum* 162 on *Meloidogyne incognita* attraction and invasion. Nematology 9, 771–776.

Dababat, A.A., Sikora, R.A., 2007b. Induced resistance by the mutualistic endophyte, *Fusarium oxysporum* strain 162, toward *Meloidogyne incognita* on tomato. Biocontrol Sci. Technol. 17, 969–975.

Daliakopoulos, I.N., Tsanis, I.K., Koutroulis, A., Kourgialas, N.N., Varouchakis, A.E., Karatzas, G.P., et al., 2016. The threat of soil salinity: a European scale review. Sci. Total Environ. 573, 727–739.

Daneshkhah, R., Cabello, S., Rozanska, E., Sobczak, M., Grundler, F., Wieczorek, K., et al., 2013. *Piriformospora indica* antagonizes cyst nematode infection and development in *Arabidopsis* roots. J. Exp. Bot. 64, 3763–3774.

Davies, K.G., Rowe, J., Manzanilla-López, R., Opperman, C.H., 2011. Re-evaluation of the life-cycle of the nematode-parasitic bacterium *Pasteuria penetrans* in root-knot nematodes, *Meloidogyne* spp. Nematology 13, 825–835.

de Medeiros, H.A., de Araújo Filho, J.V., de Freitas, L.G., Castillo, P., Rubio, M.B., Hermosa, R., et al., 2017. Tomato progeny inherit resistance to the nematode *Meloidogyne javanica* linked to plant growth induced by the biocontrol fungus *Trichoderma atroviride*. Sci. Rep. 7, 40216.

de Vries, F.T., Shade, A., 2013. Controls on soil microbial community stability under climate change. Front. Microbiol. 4, 265.

de Vries, F.T., Griffiths, R.I., Bailey, M., Craig, H., Girlanda, M., Gweon, H.S., et al., 2018. Soil bacterial networks are less stable under drought than fungal networks. Nat. Commun. 9, 3033.

de Vries, F.T., Griffiths, R.I., Knight, C.G., Nicolitch, O., Williams, A., 2020. Harnessing rhizosphere microbiomes for drought-resilient crop production. Science 368, 270–274.

Dearnaley, J.D., Cameron, D.D., 2017. Nitrogen transport in the orchid mycorrhizal symbiosis – further evidence for a mutualistic association. New Phytol. 213, 10–12.

Deng, J., Zhang, Y., Yin, Y., Zhu, X., Zhu, W., Zhou, Y., 2019. Comparison of soil bacterial community and functional characteristics following afforestation in the semi-arid areas. PeerJ 7, e7141.

Dennis, P.G., Miller, A.J., Hirsch, P.R., 2010. Are root exudates more important than other sources of rhizodeposits in structuring rhizosphere bacterial communities? FEMS Microbiol. Ecol. 72, 313–327.

Dickie, I.A., Bolstridge, N., Cooper, J.A., Peltzer, D.A., 2010. Co-invasion by *Pinus* and its mycorrhizal fungi. New Phytol. 187, 475–484.

Djemiel, C., Grec, S., Hawkins, S., 2017. Characterization of bacterial and fungal community dynamics by high-throughput sequencing (HTS) metabarcoding during flax dew-retting. Front. Microbiol. 8, 2052.

dos Santos, R.M., Diaz, P.A.E., Lobo, L.L.B., Rigobelo, E.C., 2020. Use of plant growth-promoting rhizobacteria in maize and sugarcane: characteristics and applications. Front. Sustain. Food Syst. 4, 136.

Drigo, B., Pijl, A.S., Duyts, H., Kielak, A.M., Gamper, H.A., Houtekamer, M.J., et al., 2010. Shifting carbon flow from roots into associated microbial communities in response to elevated atmospheric CO2. Proc. Natl. Acad. Sci. USA 107, 10938–10942.

Dudareva, N., Negre, F., Nagegowda, D.A., Orlova, I., 2006. Plant volatiles: recent advances and future perspectives. Crit. Rev. Plant Sci. 25, 417–440.

Emerson, J.B., 2019. Soil viruses: a new hope. mSystems 4, e00120-19.

Escobar-Zepeda, A., Vera-Ponce de León, A., Sanchez-Flores, A., 2015. The road to metagenomics: from microbiology to DNA sequencing technologies and bioinformatics. Front. Genet. 6, 348.

Escudero, N., Lopez-Llorca, L.V., 2012. Effects on plant growth and root-knot nematode infection of an endophytic GFP transformant of the nematophagous fungus *Pochonia chlamydosporia*. Symbiosis 57, 33–42.

Evelin, H., Kapoor, R., Giri, B., 2009. Arbuscular mycorrhizal fungi in alleviation of salt stress: a review. Ann. Bot. 104, 1263–1280.

Field, B., Jordan, F., Osbourn, A., 2007. First encounters – deployment of defense-related natural products by plants. New Phytol. 172, 193–207.

Finkel, O.M., Salas-González, I., Castrillo, G., Conway, J.M., Law, T.F., Teixeira, P.J.P.L., et al., 2020. A single bacterial genus maintains root growth in a complex microbiome. Nature 587, 103–108. https://doi.org/10.1038/s41586-020-2778-7.

Fiorilli, V., Volpe, V., Balestrini, R., 2019. Microscopic techniques coupled to molecular and genetic approaches to highlight cell-type specific differences in mycorrhizal symbiosis. In: Reinhardt, D., Sharma, A. (Eds.), Methods in Rhizosphere Biology Research. Rhizosphere Biology. Springer, Singapore.

Fitter, A.H., Gilligan, C.A., Hollingworth, K., Kleczkowski, A., Twyman, R.M., Pitchford, J.W., et al., 2005. Biodiversity and ecosystem function in soil. Funct. Ecol. 19, 369–377.

Fochi, V., Chitarra, W., Kohler, A., Voyron, S., Singan, V.R., Lindquist, E.A., et al., 2017a. Fungal and plant gene expression in the *Tulasnella calospora-Serapias vomeracea* symbiosis provides clues about nitrogen pathways in orchid mycorrhizas. New Phytol. 213, 365–379.

Fochi, V., Falla, N., Girlanda, M., Perotto, S., Balestrini, R., 2017b. Cell-specific expression of plant nutrient transporter genes in orchid mycorrhizae. Plant Sci. 263, 39–45.

Fortuna, A., 2012. The soil biota. Nat. Edu. Know 3, 1.

Fragasso, M., Iannucci, A., Papa, R., 2013. Durum wheat and allelopathy: toward wheat breeding for natural weed management. Front. Plant Sci. 4, 375.

Frey, S.D., 2019. Mycorrhizal fungi as mediators of soil organic matter dynamics. Ann. Rev. Ecol. Evol. S. 50, 237–259.

Furbank, R.T., Tester, M., 2011. Phenomics – technologies to relieve the phenotyping bottleneck. Trends Plant Sci. 16, 635–644.

Ghahremani, Z., Escudero, N., Saus, E., Gabaldón, T., Sorribas, F.J., 2019. *Pochonia chlamydosporia* induces plant-dependent systemic resistance to *Meloidogyne incognita*. Front. Plant Sci. 10, 945.

Ghirardo, A., Fochi, V., Lange, B., Witting, M., Schnitzler, J.P., Perotto, S., et al., 2020. Metabolomic Adjustments in the Orchid Mycorrhizal Fungus *Tulasnella calospora* during Symbiosis with *Serapias vomeracea*. New Phytol. 10. https://doi.org/10.1111/nph.16812.

Gilbert, J.A., van der Lelie, D., Zarraonaindia, I., 2014. Microbial terroir for wine grapes. Proc. Natl. Acad. Sci. USA 111, 5–6.

Girvan, M.S., Campbell, C.D., Kilham, K., Prosser, J.I., Glover, L.A., 2005. Bacterial diversity promotes community stability and functional resilience after perturbation. Environ. Microbiol. 7, 301–313.

Glick, B.R., 2012. Plant growth-promoting bacteria: mechanisms and applications. Scientifica 963401.

Gonthier, P., Giordano, L., Zampieri, E., Lione, G., Vizzini, A., Colpaert, J.V., et al., 2019. An ectomycorrhizal symbiosis differently affects host susceptibility to two congeneric fungal pathogens. Fungal Ecol. 39, 250–256.

Gonzalez, E., Pitre, F.E., Pagé, A.P., Marleau, J., Nissim, W.G., St-Arnaud, M., et al., 2018. Trees, fungi and bacteria: tripartite metatranscriptomics of a root microbiome responding to soil contamination. Microbiome 6, 53.

Gordon, J., Knowlton, N., Relman, D.A., Rohwer, F., Youle, M., 2013. Superorganisms and holobionts. Microbe 8, 152–153.

Gough, E.C., Owen, K.J., Zwart, R.S., Thompson, J.P., 2020. A systematic review of the effects of arbuscular mycorrhizal fungi on root-lesion nematodes, *Pratylenchus* spp. Front. Plant Sci. 11, 923.

Guilland, C., Maron, P., Damas, O., Ranjard, L., 2018. Biodiversity of urban soils for sustainable cities. Environ. Chem. Lett. 16, 1267–1282.

Guo, Y., Jud, W., Ghirardo, A., Antritter, F., Benz, J.P., Schnitzler, J.P., et al., 2020. Sniffing fungi – phenotyping of volatile chemical diversity in *Trichoderma* species. New Phytol. 227, 244–259.

Hacquard, S., Tisserant, E., Brun, A., Legué, V., Martin, F., Kohler, A., 2013. Laser microdissection and microarray analysis of *Tuber melanosporum* ectomycorrhizas reveal functional heterogeneity between mantle and Hartig net compartments. Environ. Microbiol. 15, 1853–1869.

Harrison, B.D., Reavy, B., 2002. Potato Mop-Top Virus. AAB Descriptions of Plant Viruses No. 389. DPV Database.

Hartmann, A., Rothballer, M., Schmid, M., 2008. Lorenz Hiltner, a pioneer in rhizosphere microbial ecology and soil bacteriology research. Plant Soil 312, 7–14.

Hassani, M.A., Durán, P., Hacquard, S., 2018. Microbial interactions within the plant holobiont. Microbiome 6, 58.

Hayward, J., Horton, T.R., Pauchard, A., Nuñnez, M.A., 2015. A single ectomycorrhizal fungal species can enable a *Pinus* invasion. Ecology 96, 1438–1444.

He, C., Wang, W.Q., Hou, J.L., 2019. Characterization of dark septate endophytic fungi and improve the performance of liquorice under organic residue treatment. Front. Microbiol. 10, 1364.

Hiltner, L., 1904. Ueber neuere Erfahrungen und Probleme auf dem Gebiete der Bodenbakteriologie und unter besonderer Ber-Ucksichtigung der Grundungung und Brache. Arb. Deut. Landw. Gesell. 98, 59–78.

Hinsinger, P., Jones, D.L., Marschner, P., 2012. Biogeochemical, biophysical, and biological processes in the rhizosphere. In: Huang, P.M., Li, Y., Sumner, M.E. (Eds.), Handbook of Soil Sciences Resource Management and Environmental Impacts. CRC Press, Boca Raton, Florida, pp. 1–30.

Huang, M., Sanchez-Moreiras, A.M., Abel, C., Sohrabi, R., Lee, S., Gershenzon, J., et al., 2012. The major volatile organic compound emitted from *Arabidopsis thaliana* flowers, the sesquiterpene (E)-β-caryophyllene, is a defense against a bacterial pathogen. New Phytol. 193, 997–1008.

Hulme, P.E., 2014. Invasive species challenge the global response to emerging diseases. Trends Parasitol. 30, 267–270.

Iannucci, A., Fragasso, M., Beleggia, R., Nigro, F., Papa, R., 2017. Evolution of the crop rhizosphere: impact of domestication on root exudates in tetraploid wheat (*Triticum turgidum* L.). Front. Plant Sci. 8, 2124.

Irankhah, S., Chitarra, W., Nerva, L., Antoniou, C., Lumini, E., Volpe, V., et al., 2020. Impact of an arbuscular mycorrhizal fungal inoculum and exogenous MeJA on fenugreek secondary metabolite production under water deficit. Environ. Exp. Bot. 176, 104096.

Jansson, J.K., Hofmockel, K.S., 2018. The soil microbiome-from metagenomics to metaphenomics. Curr. Opin. Microbiol. 43, 162–168.

Jarić, I., Heger, T., Castro Monzon, F., Jeschke, J.M., Kowarik, I., McConkey, K.R., et al., 2019. Crypticity in biological invasions. Trends Ecol. Evol. 34, 291–302.

Jeffery, S., Gardi, C., Jones, A., Montanarella, L., Marmo, L., Miko, L., et al., 2010. European Atlas of Soil Biodiversity. European Commission, Publications Office of the European Union, Luxembourg.

Jeschke, J.M., Keesing, F., Ostfeld, R.S., 2013. Novel organisms: comparing invasive species, GMOs, and emerging pathogens. Ambio 42, 541–548.

Jones, D.L., Hinsinger, P., 2008. The rhizosphere: complex by design. Plant Soil 312, 1–6.

Jumpponen, A., 2001. Dark septate endophytes – are they mycorrhizal? Mycorrhiza 11, 207–211.

Kamoun, S., Furzer, O., Jones, J.D., Judelson, H.S., Ali, G.S., Dalio, R.J., et al., 2015. The top 10 oomycete pathogens in molecular plant pathology. Mol. Plant Pathol. 16, 413–434.

Kannojia, P., Sharma, P.K., Sharma, K., 2019. Climate change and soil dynamics: effects on soil microbes and fertility of soil. In: Choudhary, K.K., Kumar, A., Singh, A.K. (Eds.), Climate Change and Agricultural Ecosystems. Woodhead Publishing, Sawston, UK, pp. 43–64.

Katan, J., 2017. Diseases caused by soilborne pathogens: biology, management and challenges. J. Plant Pathol. 9, 305–315.

Ke, J., Wang, B., Yoshikuni, Y., 2020. Microbiome engineering: synthetic biology of plant-associated microbiomes in sustainable agriculture. Trends Biotechnol. 30203–302011. S0167-7799.

Kiers, E.T., Duhamel, M., Beesetty, Y., Mensah, J.A., Franken, O., Verbruggen, E., et al., 2011. Reciprocal rewards stabilize cooperation in the mycorrhizal symbiosis. Science 333, 880–882.

Koranda, M., Schnecker, J., Kaiser, C., Fuchslueger, L., Kitzler, B., Stange, C.F., et al., 2011. Microbial processes and community composition in the rhizosphere of European beech: the influence of plant C exudates. Soil Biol. Biochem. 43, 551–558.

Kosová, K., Vítámvás, P., Urban, M.O., Prášil, I.T., 2013. Plant proteome responses to salinity stress – comparison of glycophytes and halophytes. Funct. Plant Biol. 40, 775–786.

Kuppardt, A., Fester, T., Härtig, C., Chatzinotas, A., 2018. Rhizosphere protists change metabolite profiles in *Zea mays*. Front. Microbiol. 9, 857.

Kuzyakov, Y., Razavi, B.S., 2019. Rhizosphere size and shape: temporal dynamics and spatial stationarity. Soil Biol. Biochem. 135, 343–360.

Langille, M.G.I., Zaneveld, J., Caporaso, J.G., McDonald, D., Knights, D., Reyes, J., et al., 2013. Predictive functional profiling of microbial communities using 16S rRNA marker gene sequences. Nat. Biotechnol. 31, 814–821.

Larriba, E., Jaime, M.D.L.A., Nislow, C., Martín-Nieto, J., Lopez-Llorca, L.V., 2015. Endophytic colonization of barley (*Hordeum vulgare*) roots by the nematophagous fungus *Pochonia chlamydosporia* reveals plant growth promotion and a general defense and stress transcriptomic response. J. Plant Res. 128, 665–678.

Lee, S., Behringer, G., Hung, R., Bennett, J., 2019. Effects of fungal volatile organic compounds on *Arabidopsis thaliana* growth and gene expression. Fungal Ecol 37, 1–9.

Liao, H.L., Chen, Y., Bruns, T.D., Peay, K.G., Taylor, J.W., Branco, S., et al., 2014. Metatranscriptomic analysis of ectomycorrhizal roots reveals genes associated with *Piloderma–Pinus* symbiosis: improved methodologies for assessing gene expression in situ. Environ. Microbiol. 16, 3730–3742.

Liu, D., Keiblinger, K.M., Schindlbacher, A., Wegner, U., Sun, H., Fuchs, S., et al., 2017. Microbial functionality as affected by experimental warming of a temperate mountain forest soil: a metaproteomics survey. Appl. Soil Ecol. 117, 196–202.

Lockwood, J.L., 1986. Soilborne plant pathogens: concepts and connections. Phytopathol. 76, 20–27.

López-Angulo, J., de la Cruz, M., Chacón-Labella, J., Illuminati, A., Matesanz, S., Pescador, D.S., et al., 2020. The role of root community attributes in predicting soil fungal and bacterial community patterns. New Phytol. https://doi.org/10.1111/nph.16754.

Lynch, J.P., Brown, K.M., 2012. New roots for agriculture: exploiting the root phenome. Philos. Trans. R. Soc. Lond. B Biol. Sci. 367, 1598–1604.

Macfarlane, S.A., 2003. Molecular determinants of the transmission of plant viruses by nematodes. Mol. Plant Pathol. 4, 211–215.

Maciá-Vicente, J.G., Rosso, L.C., Ciancio, A., Jansson, H.B., Lopez-Llorca, L.V., 2009. Colonization of barley roots by endophytic *Fusarium equiseti* and *Pochonia chlamydosporia*: effects on plant growth and disease. Ann. Appl. Biol. 155, 391–401.

Macías-Rodríguez, L., Contreras-Cornejo, H.A., Adame-Garnica, S.G., Del-Val, E., Larsen, J., 2020. The interactions of *Trichoderma* at multiple trophic levels: inter-kingdom communication Microbiol. Res. 240, 126552.

Mackelprang, R., Grube, A.B., Lamendella, R., Jesus, E.D., Copeland, A., Liang, C., et al., 2018. Microbial community structure and functional potential in cultivated and native tallgrass prairie soils of the Midwestern United States. Front. Microbiol. 9, 1775.

Madsen, E.L., 2008. Microbial biogeochemistry: a grand synthesis. In: Environmental Microbiology: From Genomes to Biogeochemistry. Wiley-Blackwell Publishing, Malden, Massachusetts, pp. 281–299.

Mandyam, K.G., Jumpponen, A., 2015. Mutualism-parasitism paradigm synthesized from results of root-endophyte models. Front. Microbiol. 5, 776.

Mansfield, J., Genin, S., Magori, S., Citovsky, V., Sriariyanum, M., Ronald, P., et al., 2012. Top 10 plant pathogenic bacteria in molecular plant pathology. Mol. Plant Pathol. 13, 614–629.

Manzanilla-López, R.H., Esteves, I., Powers, S.J., Kerry, B.R., 2011. Effects of crop plants on abundance of *Pochonia chlamydosporia* and other fungal parasites of root-knot and potato cyst nematode. Ann. Appl. Biol. 159, 118–129.

Marler, M.J., Zabinski, C.A., Callaway, R.M., 1999. Mycorrhizae indirectly enhance competitive effects of an invasive forb on a native bunchgrass. Ecology 80, 1180–1186.

Maron, J., Klironomos, J., Waller, L., Callaway, R.M., 2014. Invasive plants escape from suppressive soil biota at regional scales. J. Ecol. 102, 19–27.

Martinez-Alonso, E., Pena-Perez, S., Serrano, S., Garcia-Lopez, E., Alcazar, A., Cid, C., 2019. Taxonomic and functional characterization of a microbial community from a volcanic englacial ecosystem in Deception Island, Antarctica. Sci. Rep. 9, 12158.

Martínez-Medina, A., Fernandez, I., Lok, G.B., Pozo, M.J., Pieterse, C.M., Van Wees, S.C., 2017. Shifting from priming of salicylic acid- to jasmonic acid-regulated defenses by *Trichoderma* protects tomato against the root knot nematode *Meloidogyne incognita*. New Phytol. 213, 1363–1377.

Martino, E., Morin, E., Grelet, G.A., Kuo, A., Kohler, A., Daghino, S., et al., 2018. Comparative genomics and transcriptomics depict ericoid mycorrhizal fungi as versatile saprotrophs and plant mutualists. New Phytol. 217, 1213–1229.

McNear Jr., D.H., 2013. The rhizosphere-roots, soil and everything in between. Nat. Edu. Know 4, 1.

Mello, A., Zampieri, E., 2017. Who is out there? What are they doing? Application of metagenomics and metaproteomics to reveal soil functioning. Italian J. Mycol. 46, 1–7.

Mendes, R., Garbeva, P., Raaijmakers, J.O., 2013. The rhizosphere microbiome: significance of plant beneficial, plant pathogenic, and human pathogenic microorganisms. FEMS Microbiol. Rev. 37, 634–663.

Mhlongo, M.I., Piater, L.A., Madala, N.E., Labuschagne, N., Dubery, I.A., 2018. The chemistry of plant–microbe interactions in the rhizosphere and the potential for metabolomics to reveal signaling related to defense priming and induced systemic resistance. Front. Plant Sci. 9, 112.

Mousa, W.K., Raizada, M.N., 2016. Natural disease control in cereal grains. In: Reference Module in Food Science. Elsevier, pp. 1–7.

Mucha, J., Dahm, H., Strzelczyk, E., Werner, A., 2006. Synthesis of enzymes connected with mycoparasitism by ectomycorrhizal fungi. Arch. Microbiol. 185, 69–77.

Mucha, J., Zadworny, M., Werner, A., 2009. Cytoskeleton and mitochondrial morphology of saprotrophs and the pathogen *Heterobasidion annosum* in the presence of *Suillus bovinus* metabolites. Mycol. Res. 113, 981–990.

Mueller, U.G., Sachs, J.L., 2015. Engineering microbiomes to improve plant and animal health. Trends Microbiol. 23, 606–617.

Nannipieri, P., Pietramellara, G., Renella, G., 2014. Soil as a biological system. In: Nannipieri, P., Pietramellara, G., Renella, G. (Eds.), Omics in Soil Science. Caster Academic Press, Norfolk, UK, pp. 1–7.

Neher, D.A., 2010. Ecology of plant and free-living nematodes in natural and agricultural soil. Ann. Rev. Phytopathol. 48, 371–394.

Neher, D.A., Wu, J., Barbercheck, M.E., Anas, O., 2005. Ecosystem type affects interpretation of soil nematode community measures. Appl. Soil Ecol. 30, 47–64.

Nguyen, N.H., Song, Z., Bates, S.T., Branco, S., Tedersoo, L., Menke, J., et al., 2016. FUNGuild: an open annotation tool for parsing fungal community datasets by ecological guild. Fungal Ecol. 20, 241–248.

Nilsson, R.H., Anslan, S., Bahram, M., Wurzbacher, C., Baldrian, P., Tedersoo, L., 2019. Mycobiome diversity: high-throughput sequencing and identification of fungi. Nat. Rev. Microbiol. 17, 95–109.

Nuñez, M.A., Chiuffo, M.C., Torres, A., Paul, T., Dimarco, R.D., Raal, P., et al., 2017. Ecology and management of invasive Pinaceae around the world: progress and challenges. Biol. Invasions 19, 3099–3120.

Oehl, F., Sieverding, E., Ineichen, K., Ris, E.A., Boller, T., Wiemken, A., 2005. Community structure of arbuscular mycorrhizal fungi at different soil depths in extensively and intensively managed agroecosystems. New Phytol. 165, 273–283.

Ontivero, E.R., Voyron, S., Risio Allione, L.V., Bianco, P., Bianciotto, V., Iriarte, H.J., et al., 2020. Impact of land use history on the arbuscular mycorrhizal fungal diversity in arid soils of Argentinean farming fields. FEMS Microbiol. Lett. 367, 14.

Opik, M., Moora, M., Liira, J., Zobel, M., 2006. Composition of root-colonizing arbuscular mycorrhizal fungal communities in different ecosystems around the globe. J. Ecol. 94, 778–790.

Orozco-Mosqueda, M.D.C., Rocha-Granados, M.D.C., Glick, B.R., Santoyo, G., 2018. Microbiome engineering to improve biocontrol and plant growth-promoting mechanisms. Microbiol. Res. 208, 25–31.

Paini, D.R., Sheppard, A.W., Cook, D.C., De Barro, P.J., Worner, S.P., Thomas, M.B., 2016. Global threat to agriculture from invasive species. Proc. Natl. Acad. Sci. USA 113, 7575–7579.

Pan, J., Peng, F., Tedeschi, A., Xue, X., Wang, T., Liao, J., et al., 2020. Do halophytes and glycophytes differ in their interactions with arbuscular mycorrhizal fungi under salt stress? A meta-analysis. Bot. Stud. 61, 13.

Pangesti, N., Pineda, A., Pieterse, C.M., Dicke, M., van Loon, J.J., 2013. Two-way plant mediated interactions between root-associated microbes and insects: from ecology to mechanisms. Front. Plant Sci. 4, 414.

Pepper, I.L., 2019. Chapter 5 – biotic characteristics of the environment. In: Brusseau, M.L., Pepper, I.L., Gerba, C.P. (Eds.), Environmental and Pollution Science, third ed. Academic Press, Cambridge, USA, pp. 61–87.

Pérez-Flores, P., Valencia-Cantero, E., Altamirano-Hernández, J., Pelagio-Flores, R., López-Bucio, J., García-Juárez, P., et al., 2017. *Bacillus methylotrophicus* M4-96 isolated from maize (*Zea mays*) rhizoplane increases growth and auxin content in *Arabidopsis thaliana* via emission of volatiles. Protoplasma 254, 2201–2213.

Pérez-Jaramillo, J.E., Mendes, R., Raaijmakers, J.M., 2016. Impact of plant domestication on rhizosphere microbiome assembly and functions. Plant Mol. Biol. 90, 635–644.

Perotto, S., Rodda, M., Benetti, A., Sillo, F., Ercole, E., Rodda, M., et al., 2014. Gene expression in mycorrhizal orchid protocorms suggests a friendly plant-fungus relationship. Planta 239, 1337–1349.

Pfabel, C., Eckhardt, K.U., Baum, C., Struck, C., Frey, P., Weih, M., 2012. Impact of ectomycorrhizal colonization and rust infection on the secondary metabolism of poplar (*Populus trichocarpa deltoides*). Tree Physiol. 32, 1357–1364.

Philippot, L., Spor, A., Hénault, C., Bru, D., Bizouard, F., Jones, C.M., et al., 2013. Loss in microbial diversity affects nitrogen cycling in soil. ISME J. 7, 1609–1619.

Qian, C., Hettich, R.L., 2017. Optimized extraction method to remove humic acid interferences from soil samples prior to microbial proteome measurements. J. Proteome Res. 16, 2537–2546.

Ravanbakhsh, M., Sasidharan, R., Voesenek, L.A.C.J., Kowalchuk, G.A., Jousset, A., 2018. Microbial modulation of plant ethylene signaling: ecological and evolutionary consequences. Microbiome 6, 52.

Reinhart, K.O., Callaway, R.M., 2006. Soil biota and invasive plants. New Phytol. 170, 445–457.

Ricciardi, A., Blackburn, T.M., Carlton, J.T., Dick, J.T.A., Hulme, P.E., Iacarella, J.C., et al., 2017. Invasion science: a horizon scan of emerging challenges and opportunities. Trends Ecol. Evol. 32, 464–474.

Richardson, D.M., Pyšek, P., Rejmánek, M., Barbour, M.G., Panetta, D.F., West, C.J., 2000. Naturalization and invasion of alien plants – concepts and definitions. Divers. Distrib. 6, 93–107.

Rivero, J., Álvarez, D., Flors, V., Azcón-Aguilar, C., Pozo, M.J., 2018. Root metabolic plasticity underlies functional diversity in mycorrhiza-enhanced stress tolerance in tomato. New Phytol. 220, 1322–1336.

Rubin, R.L., van Groenigen, K.J., Hungate, B.A., 2017. Plant growth promoting rhizobacteria are more effective under drought: a meta-analysis. Plant Soil 416, 309–323.

Salwan, R., Sharma, A., Sharma, V., 2019. Microbes mediated plant stress tolerance in saline agricultural ecosystem. Plant Soil 442, 1–22.

Schloter, M., Nannipieri, P., Sørensen, S.J., van Elsas, J.D., 2018. Microbial indicators for soil quality. Biol. Fertil. Soils 54, 1–10.

Selosse, M.A., Roy, M., 2009. Green plants that feed on fungi: facts and questions about mixotrophy. Trends Plant Sci. 14, 64–70.

Siddiqui, Z.A., Qureshi, A., Akhtar, M.S., 2009. Biocontrol of root-knot nematode *Meloidogyne incognita* by *Pseudomonas* and *Bacillus* isolates on *Pisum sativum*. Arch. Phytopathol. Plant Protect. 42, 1154–1164.

Siggins, A., Gunnigle, E., Abram, F., 2012. Exploring mixed microbial community functioning: recent advances in metaproteomics. FEMS Microbiol. Ecol. 80, 265–280.

Sillo, F., Zampieri, E., Giordano, L., Lione, G., Colpaert, J.V., Balestrini, R., et al., 2015. Identification of genes differentially expressed during the interaction between the plant symbiont *Suillus luteus* and two plant pathogenic allopatric *Heterobasidion* species. Mycol. Progr. 14, 106.

Singer, E., Bushnell, B., Coleman-Derr, D., Bowman, B., Bowers, R.M., Levy, A., et al., 2016. High-resolution phylogenetic microbial community profiling. ISME J. 10, 2020–2032.

Smith, S.E., Read, D.J. (Eds.), 2008. Mycorrhizal Symbiosis. Academic Press, London.

Smith, S.E., Smith, F.A., 1990. Structure and function of the interfaces in biotrophic symbioses as they relate to nutrient transport. New Phytol. 114, 1–38.

Song, C., Zhu, F., Carrión, V.J., Cordovez, V., 2020. Beyond plant microbiome composition: exploiting microbial functions and plant traits via integrated approaches. Front. Bioeng. Biotechnol. 8, 896.

Spatafora, J.W., Chang, Y., Benny, G.L., Lazarus, K., Smith, M.E., Berbee, M.L., et al., 2016. A phylum-level phylogenetic classification of zygomycete fungi based on genome-scale data. Mycologia 108, 1028–1046.

Stirling, G. (Ed.), 2014. Biological Control of Plant-Parasitic Nematodes: Soil Ecosystem Management in Sustainable Agriculture, second ed. CABI International, Wallingford, UK.

Stone, J.K., Bacon, C.W., 2000. An overview of endophytic microbes: endophytism defined. In: Bacon, C.W., White Jr., J.F. (Eds.), Microbial Endophytes. Marcel Dekker, New York, pp. 29–33.

Stukenbrock, E.H., Christiansen, F.B., Hansen, T.T., Dutheil, J.Y., Schierup, M.H., 2012. Fusion of two divergent fungal individuals led to the recent emergence of a unique widespread pathogen species. Proc. Natl. Acad. Sci. USA 109, 10954–10959.

Sylvia, D.M. (Ed.), 2005. Biological control of soilborne plant pathogens and nematodes. Principles and Applications of Soil Microbiology. Pearson Prentice Hall, Upper Saddle River, New Jersey.

Tamada, T., 2016. General features of beet necrotic yellow vein virus. In: Biancardi, E., Tamada, T. (Eds.), Rhizomania. Springer, Cham, Switzerland, pp. 55–83.

Tarkka, M.T., Drigo, B., Deveau, A., 2018. Mycorrhizal microbiomes. Mycorrhiza 28, 403–409.

Tedersoo, L., Bahram, M., Zobel, M., 2020. How mycorrhizal associations drive plant population and community biology. Science 367, eaba1223.

Timm, C.M., Pelletier, D.A., Jawdy, S.S., Gunter, L.E., Henning, J.A., Engle, N., et al., 2016. Two poplar-associated bacterial isolates induce additive favorable responses in a constructed plant-microbiome system. Front. Plant Sci. 7, 497.

Trivedi, P., Leach, J.E., Tringe, S.G., Sa, T., Singh, B.K., 2020. Plant-microbiome interactions: from community assembly to plant health. Nat. Rev. Microbiol. 18, 607–621.

Valadares, R.B.S., Perotto, S., Lucheta, A.R., Santos, E.C., Oliveira, R.M., Lambais, M.R., 2020. Proteomic and transcriptomic analyses indicate metabolic changes and reduced defense responses in mycorrhizal roots of *Oeceoclades maculata* (Orchidaceae) collected in nature. J. Fungi (Basel) 6, E148.

Valéry, L., Fritz, H., Lefeuvre, J.C., Simberloff, D., 2008. In search of a real definition of the biological invasion phenomenon itself. Biol. Inv. 10, 1345–1351.

Van der Putten, W.H., Klironomos, J.N., Wardle, D.A., 2007. Microbial ecology of biological invasions. ISME J. 1, 28–37.

Varma, A., Verma, S., Sahay, N.S., Butehorn, B., Franken, P., 1999. *Piriformospora indica*, a cultivable plant growth promoting root endophyte. Appl. Environ. Microbiol. 65, 2741–2744.

Vogel, T.M., Hirsch, P.R., Simonet, P., Jansson, J.K., Tiedje, J.M., van Elsas, J.D., et al., 2009. Advantages of the metagenomic approach for soil exploration: reply from Vogel et al. Nat. Rev. Microbiol. 7, 756–757.

Vogelsang, K.M., Bever, J.D., Griswold, M., Schulz, P.A., 2004. The Use of Mycorrhizal Fungi in Erosion Control Applications. Final Report for Caltrans. California Department of Transportation Contract no. 65A0070, Sacramento, California.

Volpe, V., Chitarra, W., Cascone, P., Volpe, M.G., Bartolini, P., Moneti, G., et al., 2018. The association with two different arbuscular mycorrhizal fungi differently affects water stress tolerance in tomato. Front. Plant Sci. 9, 1480.

Wagg, C., Bender, S.F., Widmer, F., van der Heijden, M.G.A., 2014. Soil biodiversity and soil community composition determine ecosystem multifunctionality. Proc. Natl. Acad. Sci. USA 111, 5266–5270.

Wardle, D.A., Bardgett, R.D., Klironomos, J.N., Setälä, H., van der Putten, W.H., Wall, D.H., 2004. Ecological linkages between aboveground and belowground biota. Science 304, 1629–1633.

Waterman, R.J., Bidartondo, M.I., 2008. Deception above, deception below: linking pollination and mycorrhizal biology of orchids. J. Exp. Bot. 59, 1085–1096.

Wesemael, W., Viaene, N., Moens, M., 2011. Root-knot nematodes (*Meloidogyne* spp.) in Europe. Nematology 13, 3–16.

Whipps, J.M., 2004. Prospects and limitations for mycorrhizas in biocontrol of root pathogens. Can. J. Bot. 82, 1198–1227.

Williamson, K., Fuhrmann, J., Wommack, K.E., Radosevich, M., 2017. Viruses in soil ecosystems: an unknown quantity within an unexplored territory. Annu. Rev. Virol. 4, 201–219.

Wilmes, P., Bond, P.L., 2004. The application of two-dimensional polyacrylamide gel electrophoresis and downstream analyses to a mixed community of prokaryotic microorganisms. Environ. Microbiol. 6, 911–920.

Wu, F.L., Li, Y., Tian, W., Sun, Y., Chen, F., Zhang, Y., et al., 2020. A novel dark septate fungal endophyte positively affected blueberry growth and changed the expression of plant genes involved in phytohormone and flavonoid biosynthesis. Tree Physiol. 40, 1080–1094.

Xu, J., 2016. Fungal DNA barcoding. Genome 59, 913–932.

Xu, L., Wang, A., Wang, J., Wei, Q., Zhang, W., 2017. *Piriformospora indica* confers drought tolerance on *Zea mays* L. through enhancement of antioxidant activity and expression of drought-related genes. Crop. J. 5, 251–258.

Yang, J., Kloepper, J.W., Ryu, C.M., 2009. Rhizosphere bacteria help plants tolerate abiotic stress. Trends Plant Sci. 14, 1–4.

Yang, T., Siddique, K.H.M., Liu, K., 2020. Cropping systems in agriculture and their impact on soil health, A review. Glob. Ecol. Conserv. 23, e01118.

Yuan, Z., Druzhinina, I.S., Labbé, J., Redman, R., Qin, Y., Rodriguez, R., et al., 2016. Specialized microbiome of a halophyte and its role in helping non-host plants to withstand salinity. Sci. Rep. 6, 32467.

Zampieri, E., Chiapello, M., Daghino, S., Bonfante, P., Mello, A., 2016. Soil metaproteomics reveals an inter-kingdom stress response to the presence of black truffles. Sci. Rep. 6, 25773.

Zampieri, E., Giordano, L., Lione, G., Vizzini, A., Sillo, F., Balestrini, R., et al., 2017. A non-native and a native fungal plant pathogen similarly stimulate ectomycorrhizal development but are perceived differently by a fungal symbiont. New Phytol. 213, 1836–1849.

Zavala-Gonzalez, E.A., Rodríguez-Cazorla, E., Escudero, N., Aranda-Martinez, A., Martínez-Laborda, A., Ramírez-Lepe, M., et al., 2017. *Arabidopsis thaliana* root colonization by the nematophagous fungus *Pochonia chlamydosporia* is modulated by jasmonate signaling and leads to accelerated flowering and improved yield. New Phytol. 213, 351–364.

Zenni, R.D., Nuñez, M., 2013. The elephant in the room: the role of failed invasions in understanding invasion biology. Oikos 122, 801–815.

Zheng, L., Zhao, X., Zhu, G., Yang, W., Xia, C., Xu, T., 2017. Occurrence and abundance of ammonia-oxidizing archaea and bacteria from the surface to below the water table, in deep soil, and their contributions to nitrification. Microbiologyopen 6, e00488.

Zhou, D., Feng, H., Schuelke, T., De Santiago, A., Zhang, Q., Zhang, J., et al., 2019. Rhizosphere microbiomes from root knot nematode non-infested plants suppress nematode infection. Microb. Ecol. 78, 470–481.

Žifčáková, L., Větrovský, T., Lombard, V., Henrissat, B., Howe, A., Baldrian, P., 2017. Feed in summer, rest in winter: microbial carbon utilization in forest topsoil. Microbiome 5, 122.

Zuccaro, A., Lahrmann, U., Güldener, U., Langen, G., Pfiffi, S., Biedenkopf, D., et al., 2011. Endophytic life strategies decoded by genome and transcriptome analyses of the mutualistic root symbiont *Piriformospora indica*. PLoS Pathog. 7, e1002290.

Supplementary Material

References

Balestrini, R., Gómez-Ariza, J., Lanfranco, L., Bonfante, P., 2007. Laser microdissection reveals that transcripts for five plant and one fungal phosphate transporter genes are contemporaneously present in arbusculated cells. Mol. Plant Microbe Interact. 20, 1055–1062.

Belmondo, S., Calcagno, C., Genre, A., Puppo, A., Pauly, N., Lanfranco, L., 2016. The *Medicago truncatula MtRbohE* gene is activated in arbusculated cells and is involved in root cortex colonization. Planta 243, 251–262.

Belmondo, S., Fiorilli, V., Pérez-Tienda, J., Ferrol, N., Marmeisse, R., Lanfranco, L., 2014. A dipeptide transporter from the arbuscular mycorrhizal fungus *Rhizophagus irregularis* is upregulated in the intraradical phase. Front. Plant Sci. 5, 436.

Campos-Soriano, L., Gómez-Ariza, J., Bonfante, P., San Segundo, B., 2011. A rice calcium-dependent protein kinase is expressed in cortical root cells during the presymbiotic phase of the arbuscular mycorrhizal symbiosis. BMC Plant Biol. 11, 90.

Fiorilli, V., Catoni, M., Miozzi, L., Novero, M., Accotto, G.P., Lanfranco, L., 2009. Global and cell-type gene expression profiles in tomato plants colonized by an arbuscular mycorrhizal fungus. New Phytol. 184, 975–987.

Fiorilli, V., Lanfranco, L., Bonfante, P., 2013. The expression of *GintPT*, the phosphate transporter of *Rhizophagus irregularis*, depends on the symbiotic status and phosphate availability. Planta 237, 1267–1277.

Gaude, N., Bortfeld, S., Duensing, N., Lohse, M., Krajinski, F., 2012. Arbuscule-containing and non-colonized cortical cells of mycorrhizal roots undergo extensive and specific reprogramming during arbuscular mycorrhizal development. Plant J. 69, 510–528.

Giovannetti, M., Balestrini, R., Volpe, V., Guether, M., Straub, D., Costa, A., et al., 2012. Two putative-aquaporin genes are differentially expressed during arbuscular mycorrhizal symbiosis in *Lotus japonicus*. BMC Plant Biol. 12, 186.

Giovannetti, M., Tolosano, M., Volpe, V., Kopriva, S., Bonfante, P., 2014. Identification and functional characterization of a sulfate transporter induced by both sulfur starvation and mycorrhiza formation in *Lotus japonicus*. New Phytol. 204, 609–619.

Gomez, S.K., Javot, H., Deewatthanawong, P., Torres-Jerez, I., Tang, Y., Blancaflor, E.B., et al., 2009. *Medicago truncatula* and *Glomus intraradices* gene expression in cortical cells harboring arbuscules in the arbuscular mycorrhizal symbiosis. BMC Plant Biol. 9, 10.

Gómez-Ariza, J., Balestrini, R., Novero, M., Bonfante, P., 2009. Cell-specific gene expression of phosphate transporters in mycorrhizal tomato roots. Biol. Fertil. Soils 45, 845–853.

Guether, M., Balestrini, R., Hannah, M., He, J., Udvardi, M.K., Bonfante, P., 2009a. Genome-wide reprogramming of regulatory networks, transport, cell wall and membrane biogenesis during arbuscular mycorrhizal symbiosis in *Lotus japonicus*. New Phytol. 182, 200–212.

Guether, M., Neuhäuser, B., Balestrini, R., Dynowski, M., Ludewig, U., Bonfante, P., 2009b. A mycorrhizal-specific ammonium transporter from *Lotus japonicus* acquires nitrogen released by arbuscular mycorrhizal fungi. Plant. Physiol. 150, 73–83.

Guether, M., Volpe, V., Balestrini, R., Requena, N., Wipf, D., Bonfante, P., 2011. LjLHT1.2-a mycorrhiza-inducible plant amino acid transporter from *Lotus japonicus*. Biol. Fertil. Soils 47, 925.

Hogekamp, C., Arndt, D., Pereira, P.A., Becker, J.D., Hohnjec, N., Küster, H., 2011. Laser microdissection unravels cell-type-specific transcription in arbuscular mycorrhizal roots, including CAAT-box transcription factor gene expression correlating with fungal contact and spread. Plant. Physiol. 157, 2023–2043.

Hogekamp, C., Küster, H., 2013. A roadmap of cell-type specific gene expression during sequential stages of the arbuscular mycorrhiza symbiosis. BMC Genom. 14, 306.

López-Ráez, J.A., Fernández, I., García, J.M., Berrio, E., Bonfante, P., Walter, M.H., et al., 2015. Differential spatio-temporal expression of carotenoid cleavage dioxygenases regulates apocarotenoid fluxes during AM symbiosis. Plant Sci. 230, 59–69.

Pérez-Tienda, J., Testillano, P.S., Balestrini, R., Fiorilli, V., Azcón-Aguilar, C., Ferrol, N., 2011. GintAMT2, a new member of the ammonium transporter family in the arbuscular mycorrhizal fungus *Glomus intraradices*. Fungal Genet. Biol. 48, 1044–1055.

Tisserant, E., Kohler, A., Dozolme-Seddas, P., Balestrini, R., Benabdellah, K., Colard, et al., 2012. The transcriptome of the arbuscular mycorrhizal fungus *Glomus intraradices* (DAOM 197198) reveals functional tradeoffs in an obligate symbiont. New Phytol. 193, 755–769. https://doi.org/10.1111/j.1469-8137.2011.03948.

Zeng, T., Holmer, R., Hontelez, J., te Lintel-Hekkert, B., Marufu, L., de Zeeuw, T., et al., 2018. Host- and stage-dependent secretome of the arbuscular mycorrhizal fungus *Rhizophagus irregularis*. Plant J. 94, 411–425.

Chapter 12

Soil carbon formation and persistence

William R. Horwath
Department of Land, Air, and Water Resources, University of California, Davis, CA, USA

Chapter outline

Soil Microbiology, Ecology, and Biochemistry. https://doi.org/10.1016/B978-0-12-822941-5.00012-0

12.1 Introduction

The sequestration and persistence of soil organic carbon (SOC) make it a significant C pool within the global C cycle. Soil C, including organic and inorganic forms, represents more than twice the C found in the atmosphere and terrestrial vegetation combined. In addition to being a C reservoir, SOC contains essential plant nutrients of nitrogen (N), phosphorus (P), sulfur (S), and micronutrients that are important for sustaining terrestrial net primary production (NPP). A critical function of SOC is to promote soil structure through aggregation of mineral particles which in turn promotes water regulation and SOC persistence. For this reason, SOC is described as an intrinsic soil property that supports plant, microbial, and faunal life, providing important ecosystem services related to air and water quality (Jackson et al., 2017; Schmidt et al., 2011).

12.2 Short- and long-term C cycles

The global C cycle consists of short- and long-term cycles. Both C cycles interact in space and time to influence Earth's climate by controlling atmospheric trace gas composition. Carbon dioxide (CO_2) is an important gas within both cycles and is also the main greenhouse gas (GHG) affecting Earth's climate. The short-term C cycle is dominated by biological processes of photosynthesis and decomposition and operates at days to millennial time scales (Fig. 12.1).

The long-term C cycle is dominated by rock weathering and storage of fossil C with its processes unfolding on the order of millennial to millions of years. The long-term C cycle is considered Earth's major thermostat through its key influence on atmospheric CO_2 concentrations (Fig. 12.2). Rock weathering by atmospheric CO_2, dissolved in precipitation, forms carbonates (CO_3) that dissolve calcium (Ca) in silicate rocks to form $CaCO_3$. The $CaCO_3$ moves in surface and groundwater flow into oceans and

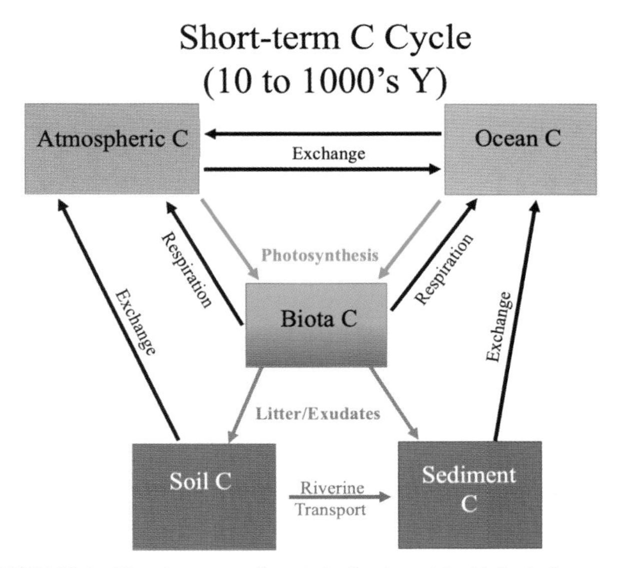

FIGURE 12.1 The short-term C cycle is dominated by biological processes.

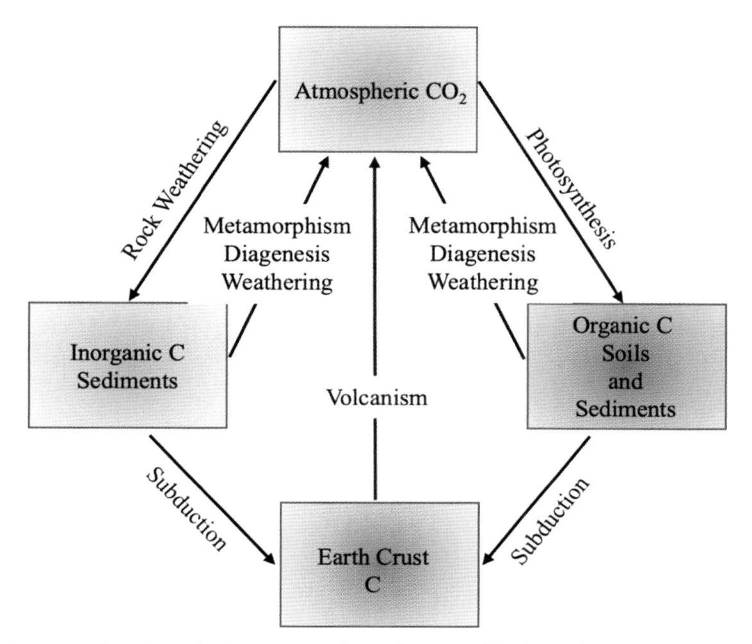

FIGURE 12.2 The long-term C cycle is dominated by rock weathering and is the major thermostat regulating Earth's climate.

other large water bodies, and precipitates to form sedimentary rocks, an extremely stable and long-lived C pool. It also dominates the fate of major C pools, including crustal, terrestrial, ocean, and fossil (Brenner, 2003).

12.2.1 Soil inorganic carbon

The products of rock weathering produce secondary minerals such as clay, iron (Fe) and aluminum (Al) oxides, and other reactive minerals that represent nascent soil substrates where the process of SOC sequestration begins. Some of the $CaCO_3$ is deposited during soil development and is called soil inorganic C (SIC). The majority of SIC is derived from a combination of atmospheric CO_2 and respiratory CO_2 from plant roots and soil microbial and faunal biomass and is found primarily in soils developed in temperate regions, especially in arid areas (Table 12.1). In desert soils, SIC exceeds SOC due to high organic C turnover rates and low NPP (Table 12.2). Soil inorganic C formation rates in arid environments average 10 to 35 kg ha^{-1} y^{-1} (Schlesinger, 2017). Its mean residence time is about 85,000 years (Marion, 1989). Globally, soils contain less than half as much SIC as SOC at about 940 Pg C (Batjes, 1997; Eswaran et al., 2000; Wilding et al., 2017). In undisturbed soils, the gain and loss of SOC reach quasi-steady state within 100 to 10,000 years depending on soil development and regolith weathering (Jenny, 1980). However, the formation and leaching of SIC continue to contribute to the longer-term C cycle (Olson et al., 1985). As soils age, less C is diverted from NPP/respiration into the long-term C cycle implying that highly weathered soils, such as Ultisols and Oxisols, can potentially lose SIC through acidification and thus play a smaller role in long-term C sequestration. In contrast, less weathered soils

TABLE 12.1 Soil inorganic C stocks according to ecological regions.

Biome	Soil inorganic C (Pg)	% Global	Soil inorganic C/TC % Global*
Tundra	18	1.9	0.8
Boreal	256	27.2	11.4
Temperate	518	55.1	23.1
Tropical	149	15.9	6.6
Total	940	100	

[a]Total global soil C from Table 12.2.
Adapted from Wilding et al., 2017.

TABLE 12.2 Global SOC stocks by depth and associated area.

				Global SOC (Pg C)	
	NPP	Area		Soil depth	
Biome	Pg C y^{-1}	$km^2 \times 10^6$	0–0.3 m	0–1 m	0–2 m
FOREST					
Tropical	20.1	37	190	375	537
Temperate	7.4	30	196	373	539
Boreal	2.4	13	165	354	623
GRASSLANDS					
Tropical	13.7	18	56	110	154
Temperate	5.1	8	37	70	95
Montane/shrubland	1.3	2	10	19	25
		108	654	1301	1973
OTHER					
Cropland	7.9	21			248
Tundra	0.5	4			1700
Peatlands and Wetlands	4.3	3			500
Deserts	3.2	8	15	31	44
		36	15	31	2492
	Total	144	669	1332	4465

From Amundson, 2001; Friedlingstein et al., 2020; Jackson et al., 2017; Jobbágy and Jackson, 2000; Tarnocai et al., 2009.

remove atmospheric C at a higher rate through intense weathering and CO_3 leaching (Chadwick et al., 1994). Soil inorganic C is susceptible to climate change (i.e., changes in precipitation patterns and amount) and or disturbance (i.e., irrigated agriculture), and can release C back to the atmosphere through the dissolution of SIC (Schlesinger, 2017).

12.2.2 Global carbon cycle budget

The evolution of photosynthetic organisms was the beginning of large organic C inputs into the environment spanning both short- and long-term C cycles. Terrestrial NPP reaches 50 to 70 and marine production from 35 to 50 Gt of C (billion tons of C) y^{-1} (Webb, 2020). Both terrestrial and ocean C sequestration are sinks of about 2 to 3 Gt C (Friedlingstein et al., 2020; Fig. 12.3). Global fluxes of C from the terrestrial biosphere to rivers into the ocean is estimated at 1.7 Gt C y^{-1} (Ciais et al., 2013), representing about 1% of the SOC sequestered by terrestrial systems. The oxidation and flux of fossil C through anthropogenic activities is estimated at about 9.4 Gt C y^{-1}, and the release of soil C to the atmosphere from land use change is estimated to be 1.6 Gt C y^{-1} (Friedlingstein et al., 2020; Fig. 12.3). The total net flux attributable to land use change since 1850 is estimated to have been 145 ± 16 Gt C (Houghton and Nassikas, 2017). Since 1750, approximately 337 Gt C have been released to the atmosphere including fossil C, with half of these emissions occurring since the mid-1970s (Boden et al., 2010). The release of soil and fossil C back to the atmosphere represents a significant perturbation to the global C cycle where the long-and short-term C cycles C cannot remove the additional atmospheric C loading resulting in a continuous risk to the climate system.

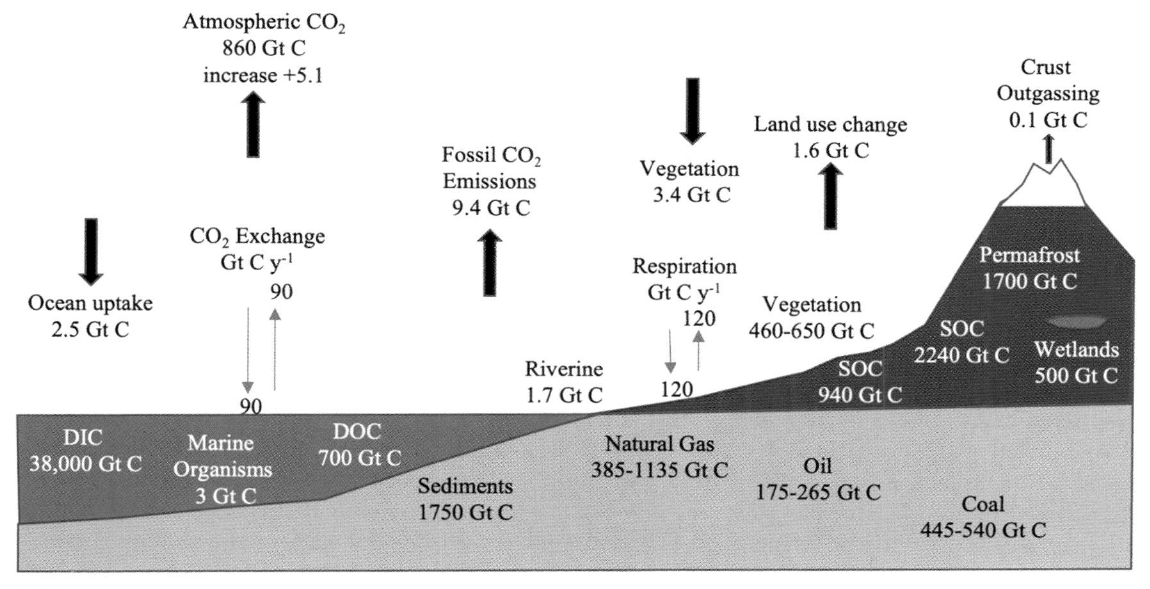

FIGURE 12.3 **The global C budget (2010–2019) showing major C fluxes among terrestrial and ocean C pools in Gt C (equivalent to a billion metric tons).** Soil, permafrost, peatlands, and wetlands contain more than two times the amount of C than found in the atmosphere showing the importance of the soil C pool in the global C cycle.

12.3 Soil organic carbon

Soil organic C formation and persistence are influenced by the five soil-forming factors of parent material, topography, biota (i.e., plants and soil organisms), climate, and time (Jenny, 1980). It represents a continuum of atmospheric to soil C where SOC in a typical soil profile is concentrated at the surface; however, a significant amount is located below 30 cm in depth in typical soils (Balesdent et al., 2018). Deeper SOC is a result of root inputs and leaching of dissolved organic C (DOC) from higher in the soil profile. From an ecosystem perspective, SOC directly influences food security and fiber production with increasing amounts corresponding to increased NPP (Lal, 2004). Ecosystem services, such as water and air quality, are directly impacted by SOC which is sensitive to soil and crop management, disturbance, land use change, and climate change. All these factors raise important questions concerning SOC's persistence and role within the global C cycle (Balesdent et al., 2018). A historical perspective of SOC is important to ensure continued research to understand its nature, formation, and persistence (Feller et al., 2000).

12.3.1 History of SOC research before 1950

The scientific pursuit to uncover the nature and function of SOC began more than two centuries ago. Humic acids, an operationally defined fraction of soil humus was first extracted from peat bogs in Germany by Achard (1786). Following this initial discovery, humic acids were extracted from decomposing plant matter. Theodore de Saussure shocked the scientific world in 1804 by demonstrating that plants obtain their C from CO_2 in the air and that they were the source of SOC which he termed "humus." The importance of SOC was recognized by Albrecht Thaër in 1805 who related it to crop production and the economics of agriculture (Waksman, 1942). de Saussure (1804) and Döbereiner (1822) began crude alkali extraction studies of humic acids to further understand the nature of SOC. Braconnot (1819) added acids to starch and sucrose forming a dark precipitate, thus inferred that humic acids were formed from simple carbohydrates. Glucose was demonstrated to form synthetic humic acids suggesting an important role for sugars in SOC formation (Malguti, 1835). Mulder (1839) synthesized humic acids from cellulose, further supporting the idea that carbohydrates are involved in SOC formation. By the mid-1800s, chemical formulas were routinely used to describe humic acids and humus (Hermann, 1842).

The role of carbohydrates in SOC formation continued to be supported by findings of furan structures, a cyclic sugar in coal (Gortner, 1916; Marcusson, 1920). Fischer and Schrader (1921) questioned the role of carbohydrates in humus formation arguing that carbohydrates and polysaccharides should be rapidly consumed by soil microorganisms and therefore were unlikely sources of SOC. They argued that lignin correlated with the production of humic acids. The role of lignin in humus formation was promoted by coal researchers. However, humus also contained N indicating a role for proteins. In 1912, Maillard demonstrated that the accumulation of N in humic substances occurred through the interaction of reducing sugars and amino acids to form dark brown polymers, calling it the "Browning" reaction (Stevenson, 1994). The Browning reaction gained recognition because microorganisms readily produce sugars and amino compounds in addition to that from plant sources. This mechanism could also explain the formation of humic substances in aquatic environments, such as the ocean, where no lignin is formed in situ, but rather is transported from the terrestrial environment. The mechanism involves rearrangement and fragmentation of compounds to form three intermediates: (1) aldehydes and ketones, (2) reductones, and (3) furfurals (Stevenson, 1994). All these compounds react readily with amino compounds to form dark-colored products through the formation of double bonds.

Waksman and Iyer (1932) hypothesized that the microbial alteration of lignin and its interaction with proteins led to the formation of humic acids, explaining its N content. They postulated that the formation of tannin-protein or lignin-protein complexes could explain the resistance to microbial attack and called it the "Protein-Lignin Theory" (Waksman, 1936). The term "proximate analysis" was introduced to describe various analytical steps, including such procedures as boiling, autoclaving, acid digests, and alkali extractions in formulating the Protein-Lignin Theory. Despite fierce debate on the origin and formation of SOC, there was a consensus that its chemical structure has an approximate stoichiometry of C/N/P/S of 100/10/1/1 (Horwath, 2014).

12.3.2 History of SOC research after 1950

The belief that SOC had a significant aromatic component was likely related to the hypothesized role of lignin in its formation. Bremner (1951) criticized the use of nonisolated methods, such as proximate analyses, because they created chemical artifacts that were likely not representative of the original material analytical methods, such as oxidative procedures, were suspected of producing aromatic compounds (Bone et al., 1930; Stevenson, 1994; Wright and Schnitzer, 1961). Despite evolving evidence that artifacts were a serious issue connected to proximal and oxidative analyses, the Protein-Lignin Theory continued to get support into the 1960s (Kononova, 1966). The lignin origin theory was supported by Flaig et al. (1975) who coined the term "Polyphenol Theory," suggesting that quinone structures from lignin polymerize with amino acids to form high-molecular-weight nitrogenous substances.

The debate on the aromaticity as an intrinsic chemical structure of SOC was addressed by Shapiro (1957) using chromatography, solution-phase IR, and organic solvents showing that a major portion of humic acids is aliphatic, not aromatic. Research using modified chemical oxidation and reduction methods to degrade humic substances confirmed that amphiphilic and alkane compounds were dominant features of SOC (Hatcher et al., 1981; Schnitzer, 1978). Preston and Schnitzer (1984) identified chemical structures and side chains using nuclear magnetic resonance (NMR) techniques showing both aliphatic and N-containing substances. Farmer and Pisaniello (1985) argued against a dominant aromatic component of fulvic acids, concluding that oxidative methods also produced artifacts of aromaticity in this fraction.

Haider and Schulten (1985) used advanced analytical mass spectrometry which bolstered the argument for the significant aliphatic content of soil organic matter (SOM). Their results showed aromatic structures were cross-linked by longer chain aliphatic compounds with carboxyl and amine groups. A molecular SOC model based on emerging evidence for the role of aliphatic compounds and described soil N, was proposed by Schulten and Schnizter (1993). Schulten et al. (1997) used pyrolysis gas chromatography/mass spectrometry (Py-GC/MS) to confirm that alkyl-aryl compounds consisted of aromatic rings covalently bonded to aliphatic chains. They also showed that the aromatic fraction contained significant heterocyclic N (pyrroles and pyridines), N-derivatives of benzene, and long-chain nitriles. Whether these structures impart inherent recalcitrance versus stabilization by the mineral phase was still up for debate. In addition, the role of lignin and other aromatic components in SOC formation and sequestration was still being debated into the 1990s despite the overwhelming evidence of an aliphatic component. For example, Shevchenko and Bailey (1996) argued that the tentative structural formula of soil humic acids should include more lignin-related linkages and structures.

Horwath and Elliott (1996) studied ryegrass decomposition using enzymatic degradation and elemental stoichiometry analyses and concluded that lignin was highly altered during initial phases of decomposition and that the resulting acid-insoluble fraction was likely microbial in origin. Wedin et al. (1995) showed that the ^{13}C signature of lignin phenols was not a significant part of SOC; lignin is highly depleted in ^{13}C ($-2‰$ to $-6‰$) compared to other plant residue components (~ -25 to $-28‰$) such as cellulose (Benner et al., 1987). Rather, the ^{13}C signature of SOC resembles the ^{13}C signature of bulk plant material and exudates (Ehleringer et al., 2000). Hall et al. (2020a) documented that in nonacidic soils, lignin degradation is rapid. Both bacteria and fungi can incorporate lignin C into DNA showing the potential for catabolic activity; however, the amounts are low compared to other sources of litter C (Brink et al., 2019; Wilhelm et al., 2019). From a microbial ecology perspective, eliminating lignin from plant residues to free sugars in polymeric polysaccharides, such as cellulose, is a key strategy to access a labile C source in bulk plant residues.

12.4 Inputs for soil organic carbon formation

12.4.1 Belowground versus aboveground inputs as sources of SOC

Photosynthetically fixed C is the major source of chemical inputs to soil. Some studies suggest that up to half of net assimilated C is allocated belowground to roots (Hartman et al., 2020; Jackson et al., 2017; Jones et al., 2009; Lambers, 1987). It is estimated that up to 20% of net assimilated C is released as root exudates (Mendez-Millan et al., 2010; Shen et al., 2020). However, $^{14}CO_2$ labeling studies showed less than 20% of net assimilated C being allocated to roots of both trees and grasses including exudates and soil respiration (Milchunas et al., 1985; Warembourg and Paul, 1973). Determining C allocation to belowground inputs is experimentally challenging and is the reason for the continued debate on which source, above- or belowground, contributes more to SOC formation and persistence.

Roots and associated rhizosphere processes contribute C to soil via turnover, exudation, and mycorrhizal inputs (Gross and Harrison, 2019). Rhizodeposition was described by Whipps and Lynch (1985) as organic material lost from plant roots, including exudates, lysates from microbial processing, microbial turnover, and root senescence. Ecto- and arbuscular mycorrhizal fungi produce significant hyphal biomass that is an important C input in both the rhizosphere and bulk soil (Moore et al., 2015). Root exudates and microbial products can sorb directly to minerals and represent a direct pathway to stable SOC (Pett-Ridge and Firestone, 2017). A meta-analysis conducted by Mathieu et al. (2015) showed that belowground C inputs in deeper soil are controlled by mineralogy, such as clay content and type, with climate playing a secondary role. Finer textured soils can have greater contributions from root derived and microbial inputs to SOC compared to coarser soil textures (Creamer et al., 2016). The extensive microbial processing of rhizodeposition results in decomposition products having similar chemical composition to that of bulk SOC, supporting the role of root exudates and microbial biomass turnover in influencing belowground C processes (Angst et al., 2018).

Rhizosphere C mean residence time is estimated at about twice or more than that of litter inputs (Rasse et al., 2005), with some studies estimating up to 10 times higher amounts of C contributing to SOC formation than from aboveground C sources (Angst et al., 2018; Balesdent and Balabane, 1996; Villarino et al., 2021). In agricultural systems, root-derived C can contribute up to six times as much C to SOC compared to aboveground inputs (Jackson et al., 2017). Root exudates influence SOC formation because they are utilized more efficiently than above- or belowground litter residues (Rasse et al., 2005; Sokol and

Bradford, 2019). More efficient utilization of root exudates by the microbial community leads to higher biomass production and the potential to add C sources directly to SOC formation and maintenance.

Litter and root manipulation experiments provide corroborating evidence that root and rhizosphere processes are important determinants of SOC formation and maintenance (Xu et al., 2013). Aboveground forest litter manipulation studies show minimal changes in SOC, confirming the important role of rhizodeposition as an important source of SOC (Bowden et al., 2014). Modifying aboveground inputs in forest and prairie systems produced variable and inconsistent results on SOC changes in both particulate and mineral-associated fractions (Lajtha et al., 2014). Long-term (decadal) litter exclusion experiments consistently show that rhizosphere processes play a major role in maintaining SOC (Wang et al., 2017).

12.4.2 Soil biotic biomass is the source of SOC

The decomposition and conversion of above- and belowground inputs into microbial biomass represents an important source of SOC (Angst et al., 2018; Godbold et al., 2006; Kindler et al., 2006; Miltner et al., 2012; Simpson et al., 2007). In a study by Voroney et al. (1989), a field experiment documented that glucose-derived C was more persistent than C derived from bulk wheat residues. The results are analogous to the utilization of labile root exudates where labile C sources (e.g., glucose) lead to a more efficient and greater production of decomposer biomass compared to litter residues. Creamer et al. (2016) showed that increasing clay content stabilized more microbial C derived from glucose showing the role of soil mineralogy as an additive effect of stabilizing SOC.

The predominance of microbial-derived aliphatic compounds in SOC from cell walls provides further evidence that they play a major role in contributing to SOC (Schurig et al., 2013). Fungi and bacteria comprise approximately 90% of total soil biotic biomass (Rinnan and Bååth, 2009), and their turnover to nonliving biomass (e.g., necromass) is estimated to contribute as much as 80% to the maintenance and accumulation of SOC (Liang and Balser, 2010). Fungal hyphae, with more complex cell walls and pigmentation, were shown to contribute more C to SOC than other microbial groups (Martin and Haider, 1979).

Fungi are hypothesized to have higher C use efficiency (CUE) than bacteria (Rinnan and Bååth, 2009). Kallenbach et al. (2016) showed increased substrate use efficiency in soils with higher fungal abundances. Bölscher et al. (2016) showed forest systems dominated by fungi were more efficient at utilizing added model substrates supporting the role of fungi in contributing to SOC. However, Six et al. (2006) questioned the idea that fungi have a higher C utilization efficiency compared to bacteria due to overlapping ranges of CUEs. In field studies of temperate and tropical forest soils experiment, Throckmorton et al. (2012) found no difference in either the turnover rate or contribution to SOC of different microbial groups that included fungi, gram-negative bacteria, gram-positive bacteria, and actinomycetes. They also found that C inputs from different microbial groups had no effect on the amount of mineral-associated organic C (MAOC; Throckmorton et al., 2015). The lack of consensus on the contribution to SOC from diverse microbial groups requires additional research to determine the role of microbial diversity on SOC formation and persistence.

12.4.3 The role of dissolved organic C inputs

Dissolved OC represents the product of turnover and decomposition processes from both plant and microbial sources. It is present throughout the soil and is found in the highest concentration within the

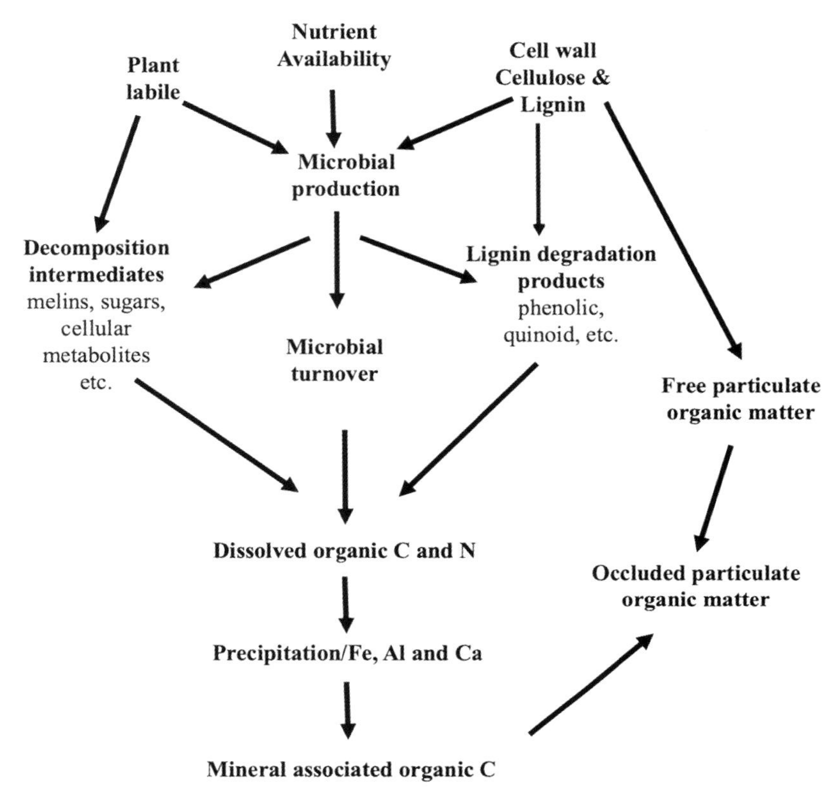

FIGURE 12.4 Conceptual pathway for degradation of plant inputs to dissolved organic C (DOC) and particulate organic matter (POM) and conversion to mineral-associated organic C (MAOC) and occluded POM (oPOM). *(Adapted from Horwath and Elliott, 1996.)*

rhizosphere. Dissolved OC is ionically charged, is soluble, and can interact with minerals and oxides (Fig. 12.4). It also equilibrates between MAOC and soil solution (Toosi et al., 2012). The concentration of DOC ranges up to 100 mg C per liter of soil solution with highest levels found in peat soils or below animal feedlots (Kaiser and Kalbitz, 2012). Annual fluxes of DOC from surface litter to mineral soil can be as high as 400 kg C ha^{-1} y^{-1} (Michalzik et al., 2001). However, the fluxes of DOC through all soil horizons are generally less than 100 kg C ha^{-1} y^{-1} (Ahrens et al., 2015).

Surface soils above 10 cm in depth contain the highest concentrations of DOC with concentration decreasing deeper in the soil profile (Gross and Harrison, 2019). The higher DOC concentrations coincide with the distribution of roots and microbial biomass (Hagedorn et al., 2012; Neff and Asner, 2001). The DOC derived from leaf litter at the soil surface is highly processed with less entering the soil at deeper depths compared to root-derived C (Xu et al., 2013). DOC below 20 to 30 cm in soil depth, is suggested to be primarily derived from microbial activity, showing a strong relationship to community composition (Gabor et al., 2014). Throughout the soil profile, DOC is more related to rhizosphere C chemistry (Sanderman et al., 2008). The transport of DOC past the C horizon is generally an order of magnitude less than transport within the A to C horizons.

The chemical nature of DOC is highly oxidized and contains a significant component of carboxyl groups (Stevenson, 1994). Carboxyl groups are important to impart chemical reactivity needed for reactions leading to SOC formation (Lehmann and Kleber, 2015). Studies that reviewed or modeled DOC dynamics in forest ecosystems estimated that DOC-derived SOC represented ~20% to 89% of total SOC showing its importance as a first step in its formation

12.5 Developing a stabilization framework for soil organic carbon

The processes leading to the persistence of SOC in soils and sediments remain highly debated (Eglinton et al., 2021; Lavallee et al., 2019; Schmidt et al., 2011). A persistent view of SOC stabilization has centered on the molecular composition of inputs, such as lignin content, and/or bulk chemical composition, such as C:N ratio. Other important factors include soil texture, mineralogy, and climate. One limitation to understanding SOC persistence is determining its chemical nature and structure. Great strides in analytical capability have been made to identify the chemical constituents of SOC, yet the debate on its actual structure continues.

12.5.1 Conceptualizing SOC pools (early research)

The initial analytical limitations for uncovering the exact nature and structure of SOC led to the development of conceptual models to describe its formation and persistence. By using a combination of operationally defined fractions, such as alkali extractions, C:N ratios, acid-insoluble fractions, litter bag decomposition results, and advanced isotope methods such as ^{14}C dating and stable isotopes (^{13}C and ^{15}N), an understanding of accumulation and turnover of SOC was developed (Nieder and Benbi, 2008; Paul and van Veen, 1978). Litter bag mass loss studies are still routinely used, but their results are often questioned since they exclude macro-decomposers and alter hydrology, resulting in the potential underestimation of decomposition (Moore et al., 2017).

Initial models used simple, additive first-order rates to describe SOC transfer among three primary compartments: (1) labile (active fraction), (2) resistant (slow), and (3) recalcitrant (passive) to describe SOC maintenance and persistence (Fig. 12.5). The labile SOC pool is composed of recent inputs and microbial biomass with turnover rates measured from days to years. The resistant pool was considered a continuum between labile and recalcitrant pools with turnover rates of years to decades. Carbon-14 dating of the operationally defined fractions, such as acid-insoluble and alkali-insoluble humin, were used to define recalcitrant SOC with mean residence times of decades to 1000 years (Paul and van Veen, 1978).

Isotopically labeled, plant residue studies are experimentally robust and provide realistic C turnover rates assuming plant residues are uniformly isotopically labeled (all components of the plant tissue contain the same concentration of an isotope). An early experiment done by Jenkinson (1977) followed the decomposition of ^{14}C-uniformly labeled ryegrass residue in temperate and tropical soils and found soil temperature and moisture influenced the rate of decomposition only within the first year or two. After 5 years, the amount of labeled residue remaining was independent of climate and soil. Isotopically labeled plant residue decomposition experiments underpin mechanistic models that often contain the conceptual framework described above to predict SOC turnover and accumulation (Paustian et al., 2019). Additional factors, such as plant residue inputs, soil texture, climate, C:N ratio of inputs, lignin to N ratio of inputs, management (no-till), etc., are often included to increase predictive capability of mechanistic models.

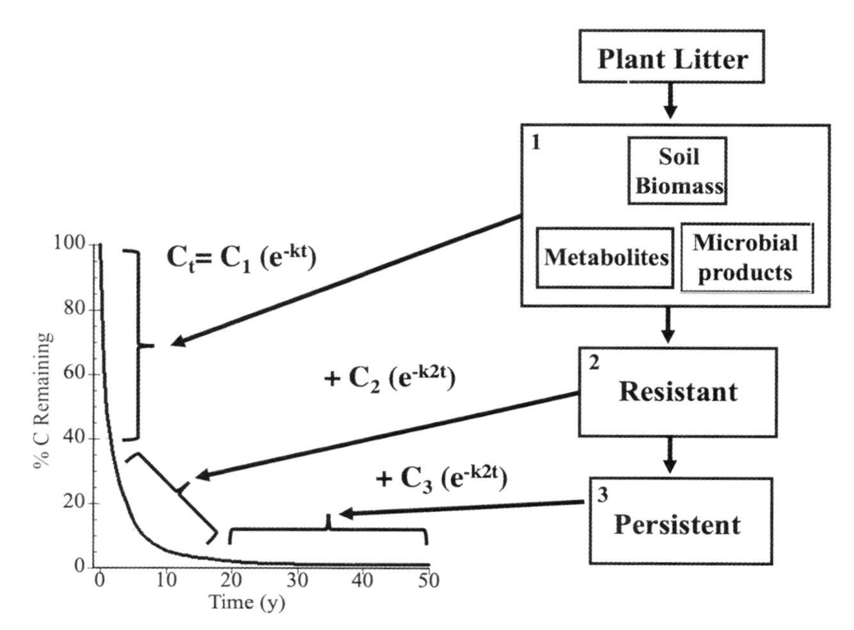

FIGURE 12.5 A simple conceptual model showing C pools with various turnover rates. Simple additive first-order equations can reasonably predict the fate of SOC based on plant litter studies where the amount of plant litter remaining over time can be used to determine decomposition rate constants (e).

One of the major limitations of the conceptual model approach is relegating microbial processes to a "Black Box" using fixed variables of substrate use efficiency and turnover to constrain and fit observational data of C movement among SOC pools (Cotrufo et al., 2013, Marshland III et al., 2020). In addition, it's assumed that the microbial substrate use efficiency is constant with depth despite possible changes in SOC concentration, variations in temperature and moisture, and changes in mineralogy.

12.5.2 Insights of stable isotope studies on SOC

The age and ^{13}C content of SOC increase progressively with soil depth suggesting C mean residence time and C isotope fractionation can provide insights into SOC dynamics and persistence (Ehleringer et al., 2000; Paul and van Veen, 1978). Root litter residues are naturally enriched in ^{13}C compared to aboveground inputs and can be a source of deep SOC ^{13}C enrichment, yet likely not a significant source for changing bulk soil ^{13}C isotope signatures (Powers and Schlesinger, 2002). The labile components of plant litter and root exudates, such as organic acids and sugars, are enriched in ^{13}C compared to complex components like lignin and fats, and their contribution to deeper soil SOC can be significant (Ehleringer et al., 2000). Soils also act as a chromatograph to sorb ^{13}C depleted lipids and leaf waxes resulting in a depleted ^{13}C signature in surface soil. In contrast to fungi, bacterial metabolism can enrich biomass with ^{13}C, with bacteria more abundant as soil depth increases. All these factors result in observed increases in SOC age and ^{13}C content with depth. In addition, protected SOC pools, such as soil aggregate-protected, particulate organic matter (POM) referred to as occluded (oPOM; discussed later), result in increased age

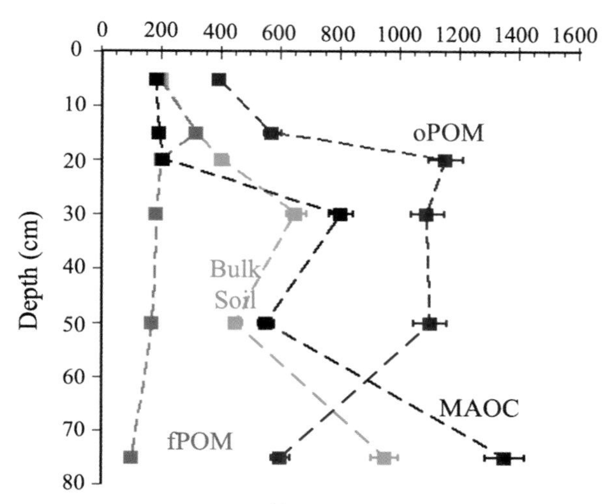

FIGURE 12.6 Mean residence time of SOC determined by ^{14}C-dating in different fractions of soil and by depth. *(From Heckman et al., 2014.)*

and ^{13}C content of SOC with soil depth (Heckman et al., 2014; Fig. 12.6). Though not well integrated into mechanistic models, natural abundance isotope studies provide information on sources and fate of SOC throughout the soil profile.

12.5.3 Mineral-associated organic C

SOC's association with secondary minerals, such as clay and amorphous oxides (notably iron and aluminum oxides) was recognized decades ago (Allison et al., 1949; Hayes and Clapp, 2001; Theng, 1982; Fig. 12.4). The incorporation of MAOC into soil aggregates was recognized as a physical-spatial isolation protection mechanism (Tisdall and Oades, 1982). Protected microsites can experience nutrient and oxygen limitations and can exclude and limit decomposer activity. Hassink (1995, 1997) showed that regardless of soil texture, the heavy fraction of soil was dominated by MAOC and small aggregate (<20 μm) fractions having turnover rates of more than 14 years.

12.5.3.1 Properties of MAOC associations

Research progress and conceptual thinking coalesced around the hypothesis that MAOC was partly protected in part by micellar properties, analogous to biological membranes, in which the interior of the structures is hydrophobic, and the exteriors are hydrophilic. These organo-mineral interactions occur through physical forces (i.e., van der Waal Forces) and hydrophobic and electrostatic interactions (i.e., pi bonding, dipole interactions, ionic bonds, and ligand exchange) (Wershaw, 1986). A fluorescence study showed salts destabilized humic structures in solution and decreased the sorption of the hydrocarbon pyrene demonstrating the destruction of micellar hydrophobic properties (Von Wandruszka, 1998).

Micellar protection was hypothesized to promote the macromolecular structure of humic substances through self-aggregation onto mineral and oxide surfaces (Piccolo, 2001; Swift, 1999).

The Sutton and Sposito (2005) review of NMR spectroscopy, X-ray absorption near-edge structure spectroscopy, electrospray ionization, and pyrolysis mass spectrometry studies promoted the theory of dynamic associations that are stabilized by hydrophobic interactions and hydrogen bonding capable of organizing into micellar structures in aqueous environments. Von Lützow et al. (2006) summarized important mechanisms controlling the persistence of SOC as: (1) formation of structures such as micelles; (2) metal bridging to oxides and clay minerals; and (3) physical isolation within aggregates, both micellar and POM. It was the beginning of the understanding that SOC stabilization was an assemblage of compounds that of themselves were likely easily degraded, but when combined into MAOC associations, it promoted persistence through interactions and specific structure.

The concept of micellar properties and metal bridging that promote MAOC associations was included in a zonal model (Kleber et al., 2007). It assumes SOC is stabilized through a self-organization of a heterogeneous mixture of compounds that display amphiphilic and hydrophobic properties (Fig. 12.7). The interactions in the zonal model are characterized by nonionic, covalent, and H-bonding to the soil mineral matrix (Chien et al., 1997; Lehmann and Kleber, 2015). Multivalent metals, such as Fe, Al, and Ca, covalently bridge to minerals attracting amphiphilic compounds like proteins and long-chain aliphatics. The amphiphilic portions of the compounds interact with water and cationic metals to sorb additional OC. The near-mineral zone of MAOC associations is enriched in N and likely contains proteins or other N compounds such as heterocyclic N adhering to the mineral surface. This explains the low C:N ratio of isolated MAOC (Hassink, 1997; Rasmussen et al., 2008). Heckman et al. (2014) show that MAOC generally contains the most N of soil fractions in a forest system (Fig. 12.8).

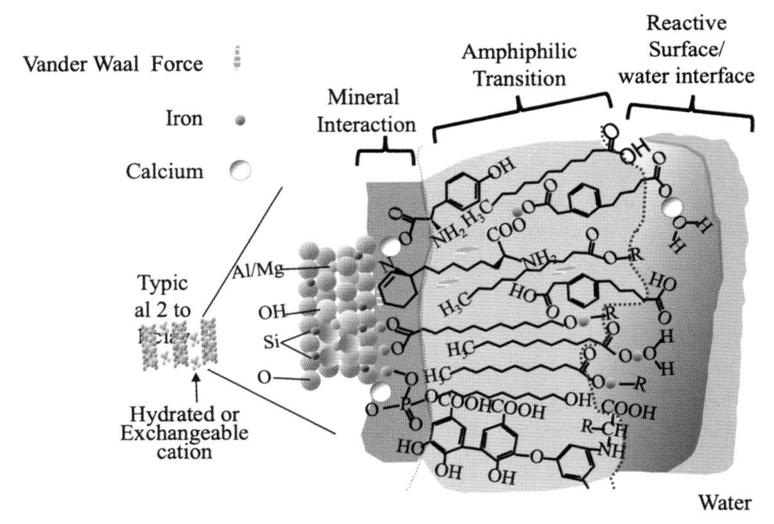

FIGURE 12.7 The zonal or layered model of SOC in soil showing mineral interactions, amphiphilic transition, and reactive interface moving from the mineral surface to the aqueous phase. Various stabilizing mechanisms including Vander Waal forces and metal bridging are indicated.

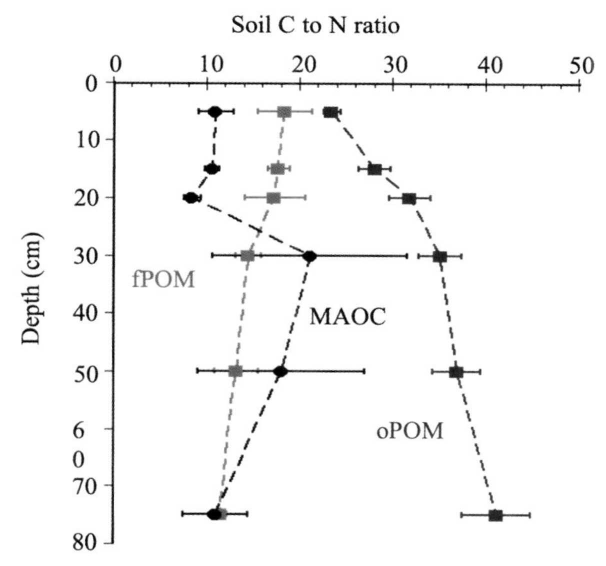

FIGURE 12.8 The C:N ratio of soil fractions by depth. *(From Heckman et al., 2014.)*

The properties of the zonal model agree with earlier studies showing an often-limitless partitioning of hydrophobic and amphiphilic compounds (Chiou et al., 1983; Wershaw, 1986). The partitioning mechanism is analogous to a simple hexane/water mixture where hydrophobic and amphiphilic compounds, such as pollutants and pesticides, readily partition out of the aqueous phase to the nonpolar phase. The entropic heat of a hydrophobic or amphiphilic compound partitioning into a nonpolar phase is less than staying in the solution, thus providing a favorable free energy reaction driving the phase partitioning process. At the boundary or surface of the hydrophobic layer, the conglomerated structure interacts with water (soil solution) containing amphiphilic molecules and polar compounds. This active outer boundary is thought to have high exchange rates with the soil solution and maintains a DOC equilibrium with it (Kleber et al., 2007; Lehmann and Kleber, 2015).

Toosi et al. (2012) showed that soil solution DOC is at equilibrium with MAOC in the presence and/or absence of fresh OC inputs. In the presence of added DOC, they observed a rapid 80% exchange with existing MAOC fractions. The rapid exchange is supported in fertilizer ^{15}N studies where a significant portion of the added ^{15}N fertilizer is found in MAOC within the first year after application (Bird et al., 2002). Daly et al. (2021) proposed a conceptual model to explain N availability, postulating that when N-associated MAOC is equivalent to N in POM fractions, there can be significant exchange of N from mineral surfaces. There is a dearth of information on the fate of N in MAOC fractions, yet additional study is required to understand N dynamics in SOC fractions, and also to address N use efficiency of agroecosystems and its loss to the environment.

12.5.3.2 Redox processes affecting MAOC associations

The incorporation of redox reactions into process-based biogeochemical models would improve observed C and N exchange in MAOC associations. Iron has been shown to be a major transition element responsible for bridging SOC to mineral surfaces and in bridging other compounds within the zonal model away from the mineral surface. The crystallization of oxidized Fe reduces its capacity to form precipitates with DOC. Preserving the oxidized state of Fe^{3+} is important for stabilizing MAOC associations. Reducing Fe^{3+} to Fe^{2+} in alkali extractions containing the reductant sodium dithionate was shown to release additional SOC and organic N. Boudot et al. (1989) observed that the degradation of Fe^{3+} and Al coprecipitates of model compounds during a 44-day incubation progressively decreased with increasing molar ratio of metal to C, demonstrating the stabilization effect of metals. Mikutta et al. (2011) demonstrated that coprecipitation of microbial, extracellular polysaccharides with Al reduced their mineralization rates relative to their free forms. Henneberry et al. (2012) showed that coprecipitated Fe^{3+} and DOC formed a noncrystalized precipitate, but in the absence of DOC, the crystalline mineral goethite (a common Fe mineral found in soil and other low-temperature sediment environments) was formed. Through scanning transmission X-ray microscopy (STXM) and near edge X-ray absorption fine structure (NEXAFS) analysis, it was shown that carboxyl groups were the primary functional group adhering to Fe^{3+}. Up to 70% of bonds within SOC structures can be carboxyl-forming H bonds, showing their importance in promoting stability (Stevenson, 1994). In many of the studies mentioned above, the coprecipitated interacts with aromatics and compounds, such as amphiphilic, aliphatic, and simple sugars, adhering to the nascent micellar structures (Henneberry et al., 2010). These observations support the zonal model of SOC protection in MAOC associations.

These studies show a strong influence of iron on SOC stability. Increasing the C to Fe^{3+} ratio is a key aspect affecting SOC stability. Adhikari et al. (2017) found that a 25.1% reduction of Fe^{3+} to Fe^{2+} within ferrihydrite (a common soil mineral), with a C to Fe ratio of 3.7, released 54.7% of the associated SOC. They found that aromatic C was retained in the remaining coprecipitate. Ye and Horwath (2017) observed similar effects in rice paddy soils where the majority of DOC released during the growing season was associated with the reduction of Fe^{3+} to Fe^{2+}. The abundance of metal-OM precipitates is also a prominent process in upland soils (Kleber et al., 2015). There is a need for further research into the factors that affect metal oxide reduction in soils to better predict the stability of SOC and the sources of available N in soils.

12.5.3.3 Microbial processes affecting MAOC associations

The largest fraction of the soil microbial biomass is found in the smallest size soil fraction (<2 μm) representing MAOC (Bolton et al., 1993; Fig. 12.9). Huysman and Verstraete (1993) found that the hydrophobicity of bacterial cells was the major factor influencing adhesion to soil particles. They also showed that multivalent cations play a significant role in the ability of bacteria to adhere to surfaces. Nkoh et al. (2020) found that bacterial extracellular polysaccharides, cation exchange capacity, and ferric oxide (Fe_2O_3) were determinants of bacterial adhesion to soil surfaces. The ability of Fe-reducing bacteria to regulate membrane proteins in response to Fe oxide surfaces was shown to promote their attachment (Lower, 2005). Cell surface metal binding in the periplasm, cytoplasm, or on EPS, is a detoxication strategy that microbes use to promote the extracellular precipitation of metals as oxides, sulfides, or metal-protein aggregates (Learman et al., 2011).

This colocation of microorganisms with the finest soil fraction, such as MAOC associations, suggests an intense competition for substrates and energy. Microorganisms, such as Geobacter metallireducens, can reduce Fe^{3+} to Fe^{2+} to release OC as a strategy to access substrates resulting in the destabilization of SOC. The microbial reduction of Fe^{3+} could explain the exchange of MAOC-associated OC and N to bioavailable

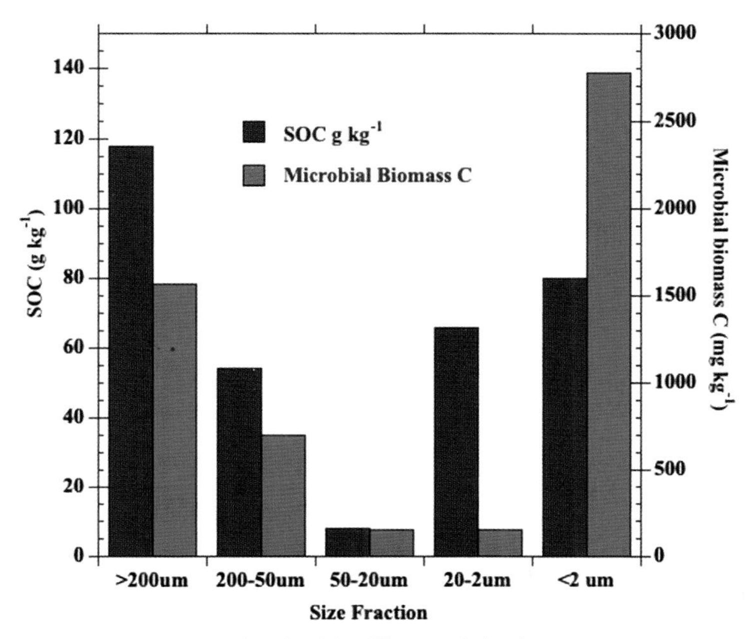

FIGURE 12.9 The distribution of soil microbial C in different soil size fractions. *(Data from Bolton et al., 1993.)*

forms as presented in the framework of integrating plant–microbe–mineral–POM interactions (Daly et al., 2021). The result is confirmed by a study by Wenjuan and Hall (2017) showing that mineralization of older C was promoted by the reduction Fe^{3+} to Fe^{2+}.

The influence of soil moisture regimes on the accumulation MAOC associations suggests they are a key aspect of SOC persistence regardless of input quality and climatic influences. Despite these redox-associated destabilizing processes, MAOC fractions tend to make up the majority of SOC, especially deeper in the soil profile (Heckman et al., 2014; Fig. 12.10). These results demonstrate that soil moisture regimes will influence the redox status of Fe, an important metal bonding or bridging OC in MAOC associations. The ability of microbes to oxidize reduced metals no doubt plays a role in affecting precipitation and stability of OC within MAOC and aggregates.

12.5.4 Particulate organic matter and SOC protection in soil aggregates

POM exists in a state of recent to highly decomposed plant litter inputs, both shoot and root residues. It is found in the free state (fPOM) or as occluded within aggregates (oPOM). The POM fraction represents 5% to 15% of the total C of cultivated soils and up to 25% or more in surface layers of grassland and forest soil horizons (Haile-Mariam et al., 2008). POM is identified as a significant determinant of nutrient cycling. The fPOM fraction is composed of available C substrates that support microbial activity and influences the fate of nutrient cycling. This fraction represents the large fraction of soil >200 μm and often contains elevated microbial biomass (Fig. 12.9). The oPOM fraction, physically protected within soil aggregates, contributes less to actively cycling SOC.

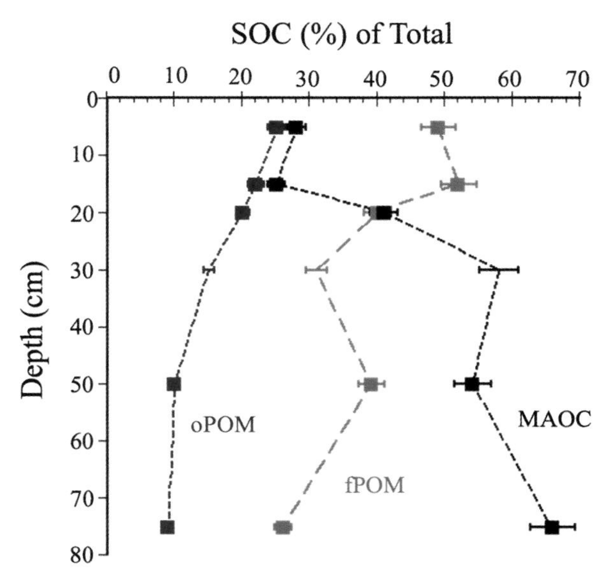

FIGURE 12.10 Soil organic C (%) in different soil fractions by depth. *(From Heckman et al., 2014.)*

Jastrow et al. (2007) proposed that soil aggregation was a mechanism to protect SOC that would otherwise readily be decomposed. Carbon-14 dating shows long mean residence times (Heckman et al., 2014; Fig. 12.6). The interaction of individual MAOC units is likely the first step in the formation of soil aggregates (Fig. 12.11). The process is likely iterative if sufficient soil moisture and C inputs and microbial growth and turnover support precipitating DOC on metal (i.e., ferrihydrite and other transition metals), mineral, and oxide surfaces. Six et al. (2000) proposed an aggregate formation scheme where small aggregates are building blocks of larger aggregates based on C:N ratio and stable isotope data. As more mineral and metal oxide surfaces interact with DOC, they likely associate through overlapping hydrophobic domains to create larger aggregates, while in the process of entrapping oPOM. In a study on opencast mine restoration, adding biosolids to exposed regolith resulted in increased fPOM, MAOC (silt and clay; <53 μm), and (microaggregate; 53–250 μm) formation within 3 to 6 years, followed by an increase in macroaggregate (>250 μm) formation (Silva et al., 2013). It was shown through stable isotope analysis that more than 65% of the new sequestered SOC was found in the above-mentioned fractions, demonstrating the importance of MAOC and aggregates in stabilizing SOC. The results support the aggregate formation theory of small to large aggregates proposed by Six et al. (2000).

The protection of oPOM indicates aggregates are persistent despite the observed exchange of OC in MAOC associations that bind them together. In an elevated CO_2 study in a forage production system, up to 15% of the total soil C was replaced in 3 years, with the majority of new C found in most aggregate size fractions >53 μm (Horwath et al., 2001; Six et al., 2000). The accumulation of POM is an early indicator of long-term C sequestration because of changes in soil management, such as reduced tillage and/or greater OC inputs (Six et al., 2002). Inputs that support an increase in fPOM can result from increased NPP including cover crops or addition of organic amendments such as manure, compost, biosolids, and food processing waste. As indicated in Fig. 12.9, the microbial biomass is concentrated in certain size

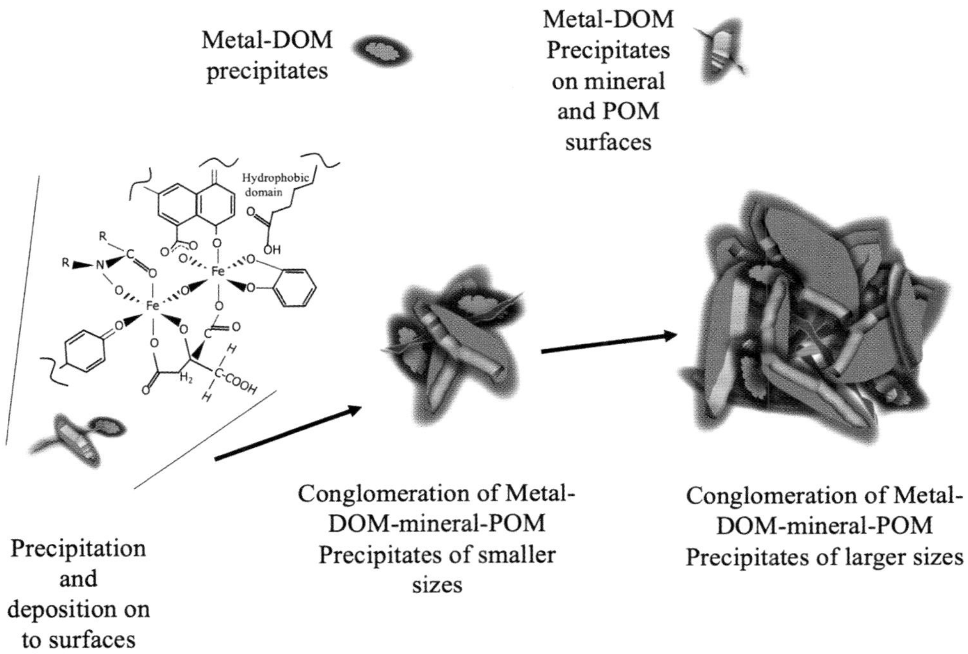

FIGURE 12.11 Precipitation of DOC with metals is the first step in soil aggregate formation. Conglomeration of precipitates forms small aggregates that can further assemble into larger aggregates. The entrapment of POM into aggregates is a key SOC protection mechanism.

fractions. Vos et al. (2013) and Rillig et al. (2017) provide additional information on micro-scale determinants and the role of aggregates on microbial diversity in soils.

12.5.5 Other SOC stability considerations

The processes leading to the formation and persistence of SOC are a complicated intermingling of NPP potential and NSP outcomes. One enduring question revolves around the importance of molecular complexity of OC inputs and resulting SOC, which are not often linked. Nitrogen is an integral element of SOC but is rarely mentioned when discussing the factors affecting SOC persistence. The priming effect (PE), where OC inputs can lead to increased or decreased decomposition of existing SOC, is well-documented, but typically not included in conceptual or mechanistic models describing SOC persistence.

12.5.5.1 Molecular diversity of inputs and resulting SOC

The importance of molecular diversity (hydrophobic and amphiphilic compounds) in influencing SOC stability was previously discussed. The range of identifiable compounds, such as aliphatic amides, saccharide ethers, hydrolyzable fractions, and heterocyclic N compounds, may differ yet can be grouped by chemical functionality, such as degree of hydrophobicity and/or polarity (Drosos et al., 2018; Galicia-Andrés et al., 2021; Saiz-Jimenez and De Leeuw, 1985; Tfaily et al., 2015). Using X-ray spectroscopy, Lehmann et al. (2008) showed that SOC at nanoscale resolution is identifiable as distinct plant and microbial polymers; however, at larger scales, OC was similar in bulk soils. The molecular diversity is likely a function of microbial diversity and expression of functional genes (Deng et al., 2019). Hall et al. (2020b) showed that soils from across North America had over 90% variability in types of carbohydrate, lipid, protein, lignin, and biochar; however, their abundance was similar across soils. These results show that molecular diversity can be large, but redox reactions, interaction with metals, and propensity to form micellar attributes result in similar outcomes that influence SOC persistence.

12.5.5.2 Nitrogen and SOC

The occurrence for N in the structure of SOC (8−12 C:N ratio) is an important factor affecting its sequestration and persistence. Soils with a large POM fraction, such as grasslands and forests, can have a higher C:N ratio due to the higher C:N ratio in the fPOM fraction. The proportion of OC between MAOC and POM fractions and their associated C:N ratio can mediate the effects of other soil variables (Cotrufo et al., 2019). The results indicate that the physical distribution of OC in MAOC versus POM can inform land management opportunities for N-efficient C sequestration. This has practical consequences affecting policy initiatives that promote SOC sequestration, such as the 4 per Thousand (4PT) Initiative, which suggests sequestration rates of 2 to 3 Gt C per year to offset about one-third of current annual fossil C emissions (Minasny et al., 2017). The implications of increased SOC sequestration are that additional N inputs would be needed to sequester SOC to achieve the 4PT goal. Van Groenigen et al. (2017) estimate this would require an additional 100 Tg of N inputs to soil per year. Approximately one-third of the N could come from crop residues following harvest or cover crops. However, agricultural systems have low N-use efficiency. Therefore, the promotion of SOC sequestration should be carefully weighed against the potential to overload soils with N causing losses in the form of nitrous oxide (a potent GHG) and leaching of nitrates to surface and groundwater resulting in water and air quality degradation.

12.5.5.3 Priming effects on SOC

Studies have shown a close relationship between SOC and the content of microbial N (Fig. 12.12). Bastida et al. (2021) support the role of including a microbial component in SOC models by showing that their diversity-to-biomass ratio can explain SOC content across global biomes. The observation confirms our understanding that the majority of SOC is derived from the turnover of the decomposer, microbial community through NSP. However, current SOC models that predict its persistence often omit or have limited microbial processes such as the priming effect (PE) (Bingeman et al., 1953). Jenkinson (1966) further defined the PE concept as being either positive (increasing SOC mineralization) or negative (decreasing SOC mineralization) in the presence of organic inputs. Overall, the addition of organic materials generally causes a positive PE (Jenkinson et al., 1985). The mineralization of CO_2 from SOC during the PE leads to the release of N (Kuzyakov et al., 2000). The addition of crop residues to soils, C-rich root exudates, and other organic materials such as manure, cause a positive PE. Some easily decomposable plant materials with low C:N ratios (i.e., living roots), toxic substances (pollutants, pesticides, etc.), and mineral N fertilizers can cause a negative PE. Mechanisms of a PE are complex and may result from the competition for energy and nutrients among specialized microorganisms feeding on simple and complex organic substances (Fontaine et al., 2003). The PE under future climate change is uncertain; however, recent evidence suggests that the temperature sensitivity of SOC will affect stable C pools more intensively as a result of the labile C pool being quickly utilized under warmer soil conditions. Soil moisture will dictate the magnitude of the response by controlling microbial growth and activity.

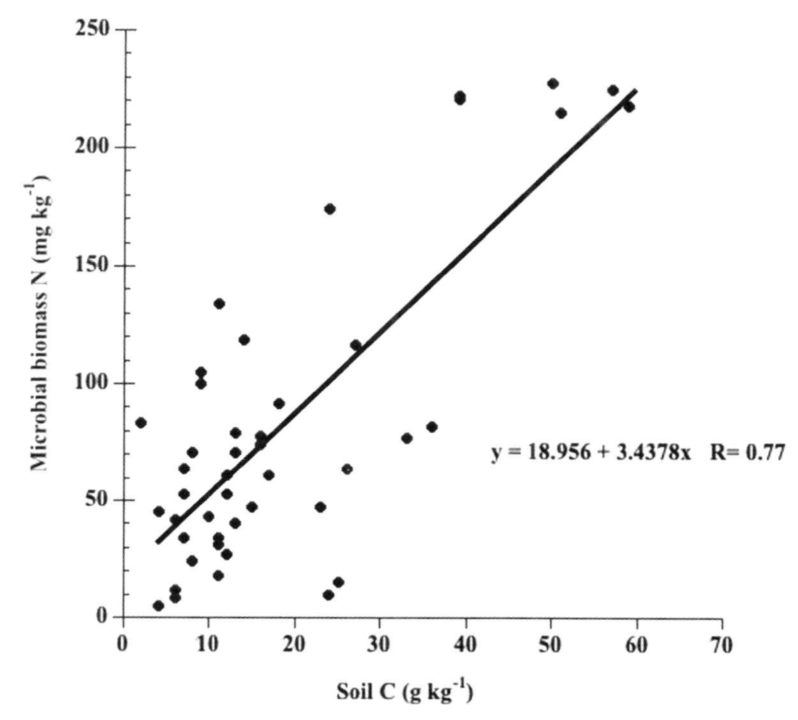

FIGURE 12.12 The relationship of SOC to microbial biomass N. *(Adapted from Horwath, 2017.)*

Regardless of the SOC stabilizing mechanism proposed, it is evident that the strong interaction between climate, NPP, microbial diversity, and soil mineralogy influences SOC persistence. To date, all proposed models use hypotheses that are constrained by observations of bulk SOC sequestration and persistence (Schaumann and Thiele-Bruhn, 2011). Future research must integrate climate effects, microbial processes, and MAOC chemistry and structure into a unified framework.

12.6 Climate impacts on soil organic carbon persistence

Climate change is highly likely to impact SOC sequestration and persistence. The main reasons for impacts on SOC will be changes in the quality and amounts of NPP, N availability, and soil temperature and moisture. The quality changes in NPP inputs may be reflected in changes in the C:N ratio or changes, such as accumulation of tannins. These changes will influence decomposer and metabolic efficiency outcomes, and potentially affect soil biomass production, activity, and turnover.

12.6.1 Surface and subsurface SOC

The interaction of mean annual temperature (MAT) and mean annual precipitation (MAP) has a strong influence on SOC (Fig. 12.13). The MAT by MAP interaction is a strong selector of biome type and

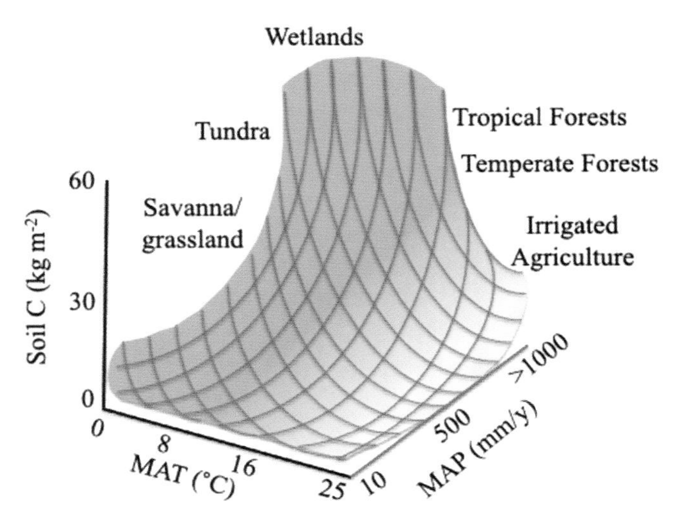

FIGURE 12.13 The effect of mean annual temperature (MAT) and mean annual precipitation (MAP) on SOC accumulation. Soils and biomes receiving increased MAP and mesic temperatures show the most SOC accumulation.

corresponding NPP potential, which directly affects SOC levels and persistence (Tables 12.2 and 12.3). Tropical biomes have the highest NPP inputs and fastest SOC turnover rates, while tundra have the lowest NPP and centennial SOC turnover rates. Peatlands and wetlands approach millennial SOC turnover rates due to reduced decomposition in water inundated environments.

The turnover of SOC differs by depth with subsoils often having about an order of magnitude longer turnover rates (Table 12.3). Soil temperatures are expected to increase by 2.3°C to 4.5°C and to a depth of 100 cm with greater warming in higher latitudes with the expectation this will affect the persistence of SOC, including deep SOC (Soong et al., 2020). Surface soil temperatures where SOC is highly concentrated will closely track a rapid increase in air temperature (Chen et al., 2021). As a result, both root and soil biomass respiration will increase (Hicks Pries et al., 2017; Jarvi and Burton, 2020). In surface soils, fPOM, oPOM, and MAOC could decrease, but the magnitude will be dependent on changes in NPP and the amount and frequency of MAP (Gross and Harrison, 2019). Many soil warming studies show that the sensitivity of soil respiration will remain unchanged despite elevating specific root and microbial biomass respiration rates with increasing temperature (Carey et al., 2016; Hicks Pries et al., 2017; Schindlbacher et al., 2015). Subsoils will respond to a lesser degree than surface soils but will experience similar SOC loss and temperature-related sensitivity responses. Soong et al. (2021) found the loss of subsoil SOC was primarily from the fPOM fraction, with less loss from oPOM and MAOC.

Subsurface soils appear more susceptible to a proportionately greater SOC loss under rising temperatures, when interacting effects of increased rhizosphere priming are considered. Warming soils will likely experience increased evapotranspiration causing soil drying and force roots to explore larger volumes and deeper soil volumes for moisture. Increased root exploration and activity will likely promote root turnover and exudation. As root exploration increases, the role of root exudates in SOC formation, turnover, and maintenance may contribute to SOC as previously discussed, but destabilizing effects

TABLE 12.3 Turnover rates of soil C inputs in various soil depths.

	Soil depth (cm)			
	0–20 cm		0–300 cm	
	K	Turnover	K	Turnover
Biome	(y^{-1})	(y)	(y^{-1})	(y)
Tropical forests	0.187	5.3	0.045	22.2
Tropical savannas	0.162	6.2	0.033	29.9
Temperate forests	0.105	9.5	0.039	25.9
Temperate grasslands	0.063	15.9	0.016	63.3
Deserts	0.039	25.4	0.007	144.2
Mediterranean	0.060	16.7	0.014	69.7
Boreal forests	0.099	10.1	0.037	27.4
Tundra	0.019	52.3	0.006	164.8
Croplands	0.159	6.3	0.041	24.4
Wetlands			0.001	945.4

Data from Amundson, 2001; Jackson, et al., 2017; Jobbágy and Jackson, 2000; Tarnocai et al., 2009.

through the PE can also occur (Villarino et al., 2021). These contrasting outcomes become more complicated when adding in elevated CO_2 effects, making the prediction of SOC persistence a challenge to incorporate into biogeochemical models as climate changes. The results suggest increased SOC loss with increasing MAT, combined with the predicted change in the amount and frequency of MAP (Bradford et al., 2019; Sulman et al., 2018).

In permafrost soils, temperatures at all soil depths will likely increase faster than in lower-latitude soils (Chen et al., 2021). The lack of MAOC protection in these soils will likely increase total ecosystem C flux and lead to SOC loss (Hicks Pries et al., 2016). These authors found that subsoil SOC contribution to soil respiration can be as high as 50% under increasing temperature. Overall, increasing temperature effects on subsoil SOC is understudied, with its effect on total soil profile SOC losses likely underestimated.

12.6.2 Changes in DOC flux under climate warming and elevated CO_2

DOC fluxes could potentially increase under predicted climate change and lead to SOC losses. Two contrasting questions concerning whether an increase in DOC production under climate warming can be summarized include: (1) will PE effects on older SOC increase DOC? (Hicks Pries et al., 2016) and (2) will the increased DOC stabilize or destabilize MAOC interactions in the subsoil? (Evans et al., 2007; Sokol and Bradford, 2019). The increased transport of DOC to subsoil could enhance MAOC stabilizing interactions (Evans et al., 2007). These potential outcomes could occur concurrently and result in increased DOC production throughout the soil profile under rising temperatures and atmospheric CO_2. Shahzad et al. (2015) found that under these conditions, root exploration of deeper soil stimulated the mineralization of old SOC. The potential for rhizosphere PEs can increase both SOC (27−245%) and N (36−62%) mineralization (Zhu et al., 2014). They also found increased microbial biomass and ß-glucosidase and oxidative enzyme activity in the rhizosphere under higher soil temperature.

The overall effect observed in field-elevated CO_2 studies is an initial increase in NPP. This initial effect is often associated with a positive interaction between root production and exudation. In most field elevated CO_2 studies, most notably Free Atmospheric Carbon Exchange (FACE) experiments, after a brief increase in NPP, there is a strong acclimation of plants returning to previous growth rates within a few years. The results of FACE experiments show little to no change in SOC under elevated CO_2 in a broad range of ecosystems.

Some FACE experiments observed losses in SOC when NPP inputs increased suggesting PEs on SOC (Ainsworth and Long, 2005; Phillips et al., 2012). After analyzing numerous FACE results, Terrer et al. (2021) found evidence that when plants don't respond to elevated CO_2, SOC can increase, with the opposite effect with plant response. This is attributed to increased nutrient needs of stimulated biomass under elevated CO_2 leading to increased mineralization and loss of SOC. In forest systems, loss of SOC was greater than in grasslands despite the latter having overall lower NPP inputs. These results have great implications for the ability to predict SOC changes under elevated CO_2 and require more research (Bradford et al., 2016).

12.7 Land use and management impacts on surface and subsurface soil organic carbon

The conversion of forests and grasslands to cropland is the foremost land use change that has occurred globally. The need for increased food production is the primary reason more arable land is needed to support increasing human and animal populations (FAO, 2015). The conversion to agriculture commonly results in SOC loss in both surface and subsurface soil. The losses result for a variety of reasons including changes in the mass and quality of NPP input and soil disturbance from tillage and accompanying soil loss from erosion (Fig. 12.14). Tillage, monoculture, and reduced OC inputs (e.g., fewer manure and cover crops) are among the top reasons for the decline in SOC as agriculture intensified (Chambers et al., 2016).

12.7.1 SOC losses from grasslands and forests following conversion to agriculture

The loss of C from agriculture activities is responsible for 21% to 37% of global annual GHG emissions which include SOC loss and fossil C oxidation (Guo and Gifford, 2002; Mbow et al., 2019). The loss of subsoil SOC in forest and grassland soils converted to agriculture occurs within decades as confirmed by radiocarbon studies (Hobley et al., 2017). Conversion of forests to agriculture consistently reduces SOC (Mayer et al., 2020). Forest harvesting activities alone cause an average decline of 8% in SOC (Nave et al., 2010). When reforestation occurs, SOC levels often return to previous levels unless harvest pressure is high or monocultures are used, such as for bioenergy production (Mayer et al., 2020). The intensity of forest harvesting practices can also reduce deep SOC >60 cm by 18% (James and Harrison, 2016).

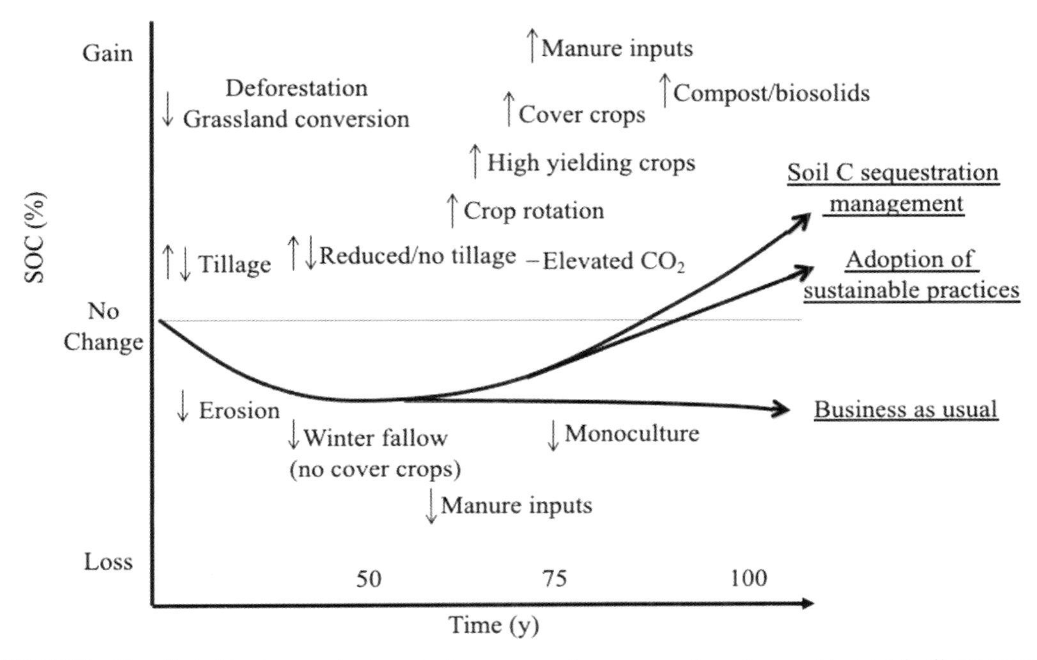

FIGURE 12.14 The effect of land conversion to agriculture, agronomic management practices, and improved soil management on the loss and gain of SOC.

Deeper forest soils are typically sensitive to all forest harvesting practices, including thinning (Mobley et al., 2015).

12.7.2 Agronomic management effects on SOC

The adoption of no-till soil management, a conservation tillage practice, shows significant SOC gains in near-surface soil (Reicosky and Lindstrom, 1993). A recent meta-analysis concluded that SOC increased in the 0 to 30 cm surface soil under no-till by about 0.8 to 8.4 Mg Ha^{-1} when practiced for 10 or more years (Haddaway et al., 2017). However, they also found that when comparing no-till or till management beyond 30 cm up to 1 m in depth, no significant differences were found in SOC mass. Often tilled soils contain deeper SOC in the soil profile. The results indicate that no-till management shifts SOC from deeper soil to the surface.

The use of conservation tillage and cover crops to sequester SOC depends on initial SOC levels, soil type, and region, with some soils increasing and others decreasing in fPOM and MAOC (Jilling et al., 2020). Leichty et al. (2020) found that SOC production in irrigated maize was less efficient under no-till compared to till. They concluded that soil contact with plant residues was important to efficiently form and sequester SOC. The results have important implications on the potential of no-till management to sequester SOC in protocols designed for soil C sequestration in C markets.

Optimized harvest index (defined as the ratio of grain to total aboveground biomass) resulting from crop improvement since the Green Revolution is approaching greater than 50%. The increased grain yields are accompanied by comparable crop residue return to soil and can result in SOC increases. In a study on maize yields and SOC levels in USA Mollisols since the 1980s, a net gain in SOC was observed even when maize was practiced in monoculture regardless of tillage management (Clay et al., 2012). In California's hot, intensively managed irrigated agriculture, yields of cropping systems are higher than the former dryland grasslands resulting in gains in SOC across San Joaquin Valley (DeClerck and Singer, 2003). The increased crop residue and root exudate inputs may lead to SOC increases, but the effects of climate, crop diversity, and soil properties will determine the extent.

The use of cover crops and/or management to increase plant diversity, such as the use of diverse crop rotations, often results in additional crop residue and root exudate inputs to soil. In a meta-analysis, consistent use of cover crops resulted in an increase in SOC averaging 1.11 Mg C ha^{-1} (McClelland et al., 2021). A meta-analysis on plant diversity effects showed an average increase of 5% to 8% in SOC in mixed species versus monoculture systems (Chen et al., 2020). The downside of implementing SOC building management is the burden of increased costs, especially in low-value grain crops (Chahal et al., 2020). A carbon price of $50 per Mg soil C (currently about $20 per Mg C) would be required to maintain profitable grain cropping systems, the most critical crops needed for future global food security.

12.7.3 SOC gains following restoration of grasslands and forests from agriculture

The SOC in surface and subsoils of restored grasslands and forests on former agricultural land can initially decrease during restoration (Mobley et al., 2015; Richter et al., 1999). Some losses can be attributed to PEs from changes in OC quality and quantity inputs. Reintroduced, deep-rooted trees and grasses will provide inputs that can both increase and promote decomposition of older SOC (Mobley et al., 2015). Nitrogen fertilization of forests and grasslands has the potential to promote SOC PEs in deeper soil (Tian et al., 2016). The loss of SOC following reforestation is attributed to reduced inputs initially, higher soil temperatures during tree establishment, potential soil erosion after harvest and before canopy closure, and nutrient losses from runoff and leaching (Guo and Gifford, 2002). Once restoration activities are established for more than 5 years, the conversion of agricultural land to secondary forest can result in a 53% gain in SOC at 30 cm soil depth. Estimates for SOC sequestration following grassland restoration from former agricultural land vary widely. Restoring grasslands through improved management, such as fertilization, use of legumes, improved grasses, and irrigation, can result in SOC sequestration of 0.1 to 1.1 Mg C ha^{-1} y^{-1} (Conant et al., 2017).

As more grassland and forests are restored, increased microbial biomass and activity, combined with increasing plant residue and root exudate inputs, have the potential to sequester SOC. Proper selection of plant species and restoration management will be required for optimal results (Gross and Harrison, 2019). Restoration activities can provide opportunities to reverse SOC losses from land use change.

12.8 Climate sensitive ecosystems and soil organic carbon persistence

Tundra, peatlands, and wetlands are sensitive to climate change making them vulnerable to changes in SOC stocks compared to temperate and tropical biomes. The land area associated with these sensitive biomes is immense with the impact of climate on SOC processes important to understand. These soils are poorly mapped, and their SOC distribution can vary considerably (Gross and Harrison, 2019). As climate change impacts increase, the magnitude of SOC loss due to warming and PEs will likely be exacerbated (Bradford et al., 2016).

12.8.1 Tundra and permafrost

Permafrost systems are characterized by low temperature and decomposition rates resulting in the storage of large amounts of POM and SOC. Soils in the northern permafrost area are estimated to contain approximately 1700 Pg C ha^{-1} to a depth of 300 cm (Scharlemann et al., 2014). The stability of permafrost SOC is highly dependent on microbial activity and less so on protective MAOC or oPOM fractions. The main driver controlling SOC stability in these soils is hydrology (Berendse et al., 2001). As these soils warm, the permanent frozen state is altered allowing both drainage and temperature to increase and along with it, microbial activity. The thawing of permafrost soils allows for decomposition of old POM and SOC to be released back into the atmosphere. With predicted warming, SOC reductions of 37% to 81% can be expected (Collins et al., 2013).

The continued defrosting of permafrost soils may lead to some soil development and SOC sequestration through developing MAOC associations, but the risk of erosional losses will likely negate this outcome. Future management to mitigate SOC losses will be temperature dependent and likely need NPP producers with large inputs to offset SOC losses (Gross and Harrison, 2019). Mineral and aggregate protection will be minor within a century timeframe leading to additional SOC losses from this climate-sensitive biome that took millennia to form. The potential for GHG emissions, particularly methane and nitrous oxide, will increase as labile C and N are mineralized complicating the assessment of climate impacts associated with these systems (Limpens et al., 2008).

12.8.2 Wetlands and peatlands

Peatland and wetlands contain about 500 Pg C globally (Berendse et al., 2001; Hobley et al., 2017). Like permafrost soils, mineral interactions to stabilize SOC are lower in wetlands and peatlands. Ephemeral wetlands with upland water inputs may have more MAOC protection capacity due to fluvial soil deposition processes. In addition to the impact of climate change on these fragile climate systems, cultivation of drained peatland and wetland soils has resulted in large losses of SOC in addition to methane and nitrous oxide emissions (Moomaw et al., 2018; Salimia et al., 2021). Northern peatlands converted to croplands are estimated to have emitted 72 Pg C over the last 1000 years (Qiu et al., 2021). In relation to concerns associated with permafrost soils, expected decreases in moisture and higher temperatures could dry out these systems and promote decomposition. As for permafrost soils, additional SOC sequestration in vegetation and a reduction in methane emissions may offset SOC losses; however, as these areas dry due to rising temperatures, decomposition will increase the loss of century-old SOC. Additional threats from wildfire losses of both vegetation and SOC complicate the assessment of future climate impacts on these systems (Moomaw et al., 2018).

12.9 Synthesis and outlook for soil organic carbon research

The processes affecting SOC persistence across both temporal and spatial scales have been a scientific pursuit for centuries. This chapter emphasized the short-term C cycle outcomes, but it is clear soils have a role in the long-term C cycle with both cycles affecting atmospheric composition and climate.

Climate change and soil management and disturbance from land use change can impact all aforementioned factors and can also change critical processes related to microbial diversity and activity, PEs, and overall biogeochemical processes that affect the stability and persistence of SOC. One of the major limitations in understanding SOC processes and outcomes is that most of our knowledge concerns near surface soil that represents the smallest proportion of total SOC globally. Pools of SOC beyond 15 to 30 cm in depth often comprise 70% of total soil C (Angst et al., 2018; Hobley et al., 2017; Jandl et al., 2014). Deep SOC is older than surface SOC which consequentially affects how we understand and manage soil as a resource. The stability of SOC is dependent on changes in the quality and quantity of inputs (particularly root exudates), PEs (rhizosphere processes such as root exudation), and changes in hydrology and temperature. The influence of biological processes, both microbial and rhizosphere, extend to SOC greater than 100 cm in depth (Sierra et al., 2017). In examining protection mechanisms, it's clear, despite numerous proposed models including mineral and microbial centric, that scientific consensus is still lacking.

The metal-mineral/protection line of thought explains a fundamental property of sorption of organic pollutants with hydrophobic properties of SOC. This behavior can be modeled with a simple hexane/water system showing SOC stability is partly explained by its hydrophobic characteristics. Microbial substrate use efficiency, combined with the quality and quantity of NPP inputs, is an extremely important concept since it controls the amount of precursor formation such as DOC for the SOC synthesis. Additional important concepts emphasize the role of substrate dissolution and transport that can explain stability factors associated with PEs, provoking the need for a more microbial-centric approach. Kleber et al. (2015) suggest that MAOC is strongly affected by the intensity of pedogenesis which, depending on the state of weathering, provides different amounts of reactive surfaces for the precipitation of DOC.

Concentrating on specific attachment or bonding mechanisms or molecular complexity to predict the stability of SOC is not predictive by itself. Secondary reactions, such as coprecipitation of DOC with metals and minerals/oxides, create spatially and temporally distinct hydrophobic domains that scale to the aggregate level and express chemical and physical protection of SOC. Crucial drawbacks of extraction and molecular identification approaches are that there is little reference to the original state leading to a conceptual exercise to fit observed outcomes of turnover and interaction of pollutants. For example, how important is the hydrophobic isolation of SOC in preventing enzymatic attack? What is the importance of metal reduction and oxidation in both the synthesis and persistence of SOC? Understanding the physical and chemical structure of MAOC and its surface chemistry would increase current knowledge of processes such as microbial adhesion to SOC dissolution potential to flux/equilibrium outcomes.
The following are important considerations for future SOC research:

(1) Impact of soil management and disturbance on SOC stability.

Much is understood on this topic; however, many interacting factors can confound interpretation of results. For example, primary forests are declining in productivity which decreases NPP inputs and therefore, SOC maintenance and accrual. Terrer et al. (2021) determined plants not responding to elevated CO_2 gained SOC while plants experiencing lower and PP increased SOC. This discrepancy must be resolved. As climate change manifests itself, these and other issues will need to be addressed.

(2) Factors affecting the fate of deep SOC need resolution.

Some studies show increasing PEs can occur under increasing soil temperatures affecting the stability of SOC. Other studies show greater MAOM interactions from increased DOM and root exudate inputs. Since deep SOC represents the largest terrestrial C pool, more research is needed to understand its stability.

(3) Lack of understanding of the molecular structure of MAOM inhibits our understanding of factors affecting the persistence of SOC.

Current models of SOC can predict long-term C flux, interaction with pollutants, and an understanding of microbial processes and turnover. Much effort into the understanding of the chemical makeup of SOC has been completed. The next step is a morphological model to discern microbial and chemical interactions involving molecular configuration and bonding and substrate flux and equilibrium outcomes that control microbial processes leading to SOC production and stabilization.

(4) What are the climatic influences, such as increasing temperature and variable hydrology, on stabilization and destabilization of surface and deep SOC?

Climate change will impact SOC dynamics. Changes in NPP quality and inputs will change. How will this affect metabolic efficiency outcomes of the soil microbial community? Microbial models with minimal protection schemes show that long-term SOC persistence can be explained (Lehmann et al., 2020). Is the persistence of SOC caused by functional complexity through the interplay between spatial and temporal variation of SOC molecular diversity and composition? How does this explain important properties such as partitioning into hydrophobic domains?

(5) The case for a microbial-centric approach to describe SOC persistence.

The incorporation of microbial, redox reactions within the spatial constraints of hydrophobic domains provides an explanation for the observed C and N exchange in MAOC associations. It also has relevance to the persistence of SOC in MAOC associations. Iron and other redox-sensitive elements (i.e., manganese) have been shown as major transition elements responsible for bridging SOC to the mineral surface and bridging other compounds further from the mineral or metal oxide surface. Preserving the oxidized-redox state of Fe^{3+} is important for stabilizing SOC within organo-mineral complexes. What controls the reduction of transition metals? A microbial/physiochemical structural model would likely address many current unknowns.

(6) What controls the flux of C and N in and out of MAOC associations?

There is a dearth of information on the fate of N in MAOC fractions requiring more study to understand both N dynamics in SOC models and to address N use efficiency of crops, loss from fertilized and disturbed soils, and N-deposition effects. The microbial reduction of Fe^{3+} could help explain the exchange of MAOC-associated N to bioavailable N forms as presented in the framework of integrating plant-microbe-mineral interactions (Daly et al., 2021). Factors that affect metal and oxide reduction need more research to better predict the stability of SOC and the sources of available N in soils.

(7) What is the magnitude of PEs under changes in the quality and quantity of inputs from NPP, rhizosphere exudates, and organic amendments such as food processing waste, biosolids, etc. that affect surface and deep SOC persistence?

Deeper soils are more susceptible to SOC loss than surface soils under rising temperatures when interacting effects of increased rhizosphere priming are considered.

Future soil scientists must form collaborations within interdisciplinary teams to better understand the effects of management, disturbance, and climate change on the spectrum of areas from plant and microbial diversity, NPP constraints, and biogeochemistry to provide new insights into SOC formation and persistence. SOC is an intrinsic characteristic of ecosystems and will remain a crucial research area to ensure food security for humans and animals and also beneficial environmental outcomes. There is an urgency to increase the knowledge that exists and to provide insight and new science needed to better understand soil resources.

References

Achard, F.K., 1786. Chemische Utersuchung des Torfs. Crell's Chem. Ann. 2, 391−403.

Adhikari, D., Zhao, Q., Da, K., Mejia, J., Huang, R., Wan, S., 2017. Dynamics of ferrihydrite-bound organic carbon during microbial Fe reduction. Geochem. Cosmochim. Acta. 212, 221−233.

Ahrens, B., Braakhekke, M.C., Guggenberger, G., Schrumpf, M., Reichstein, M., 2015. Contribution of sorption, DOC transport and microbial interactions to the ^{14}C age of a soil organic carbon profile: insights from a calibrated process model. Soil Biol. Biochem. 88, 390−402.

Ainsworth, E.A., Long, S.P., 2005. What have we learned from 15 years of free-air CO_2 enrichment (FACE)? A meta-analytic review of the responses of photosynthesis, canopy. New Phytol. 165, 351−371.

Allison, F.E., Sherman, M.S., Pinck, L.A., 1949. Maintenance of soil organic matter: I. Inorganic soil colloid as a factor in retention of carbon during formation of humus. Soil Sci. 68, 463−478.

Amundson, R., 2001. The carbon budget in soils. Annu. Rev. Earth Planet Sci. 2, 535−562.

Angst, G., Messinger, J., Greiner, M., Häusler, W.R., Hertel, D., Kirfel, K., et al., 2018. Soil organic carbon stocks in topsoil and subsoil controlled by parent material, carbon input in the rhizosphere, and microbial-derived compound. Soil Biol. Biochem. 122, 19−30.

Balesdent, J., Balabane, M., 1996. Major contribution of roots to soil carbon storage inferred from maize cultivated soils. Soil Biol. Biochem. 28, 1261−1263.

Balesdent, J., Basile-Doelsch, I., Chadoeuf, J., Cornu, S., Derrien, D., Fekiacova1, Z., 2018. Atmosphere-soil carbon transfer as a function of soil depth. Nature 559, 599−602.

Bastida, F., Eldridge, D.J., García, C., Kenny Png, G., Bardgett, R.D., Delgado-Baquerizo, M., 2021. Soil microbial diversity−biomass relationships are driven by soil carbon content across global biomes. ISME J. 15, 2081−2091. https://doi.org/10.1038/s41396-021-00906-0.

Batjes, N.H., 1997. Total carbon and nitrogen in soils of the world. Eur. J. Soil Sci. 47, 151−163.

Benner, R., Fogel, M.L., Sprague, E.K., Hodson, R.E., 1987. Depletion of ^{13}C in lignin and its implications for stable carbon isotope studies. Nature 329, 708−710.

Berendse, F., Van Breemen, N., Rydin, H., Buttler, A., Heijmans, M., Hoosbeek, M.R., 2001. Raised atmospheric CO_2 levels and increased N deposition cause shifts in plant species composition and production in Sphagnum bogs. Glob. Change Biol. 7, 591−598.

Bingeman, C.W., Varner, J.E., Martin, W.P., 1953. The effect of the addition of organic materials on the decomposition of an organic soil. Soil Sci. Soc. Am. J. 17, 3−38.

Bird, J.A., van Kessel, C., Horwath, W.R., 2002. Nitrogen dynamics in humic fractions under alternative straw management in temperate rice. Soil Sci. Soc. Am. J. 66, 478−488.

Boden, T.A., Marland, G., Andres, R.J., 2010. Global, Regional, and National Fossil-Fuel CO_2 Emissions. Carbon Dioxide Information Analysis Center. Oak Ridge National Laboratory, U.S. Department of Energy, Oak Ridge, TN. https://doi.org/10.3334/CDIAC/00001_V2010.

Bölscher, T., Wadsö, L., Börjesson, G., Herrmann, A.M., 2016. Differences in substrate use efficiency: impacts of microbial community composition, land use management, and substrate complexity. Biol Fert. Soils 52, 547−559.

Bolton Jr., H., Smith, J.L., Link, S.O., 1993. Soil microbial biomass and activity of a disturbed and undisturbed shrub-steppe ecosystem. Soil Biol. Biochem. 25, 545−552.

Bone, W.A., Horton, L., Ward, S.G., 1930. Researches on the chemistry of coal IV. Its benzenoid constitution as shown by its oxidation with alkaline permanganate. Proc. Roy. Soc. London, Series A. 127, 480−510.

Boudot, J.P., Brahim, A.B.H., Steiman, R., Seiglemurandi, F., 1989. Biodegradation of synthetic organo-metallic complexes of iron and aluminum with selected metal to carbon ratios. Soil Biol. Biochem. 21, 961–966.

Bowden, R.D., Deem, L., Plante, A.F., Peltre, F.C., Nadelhoffer, K., Lajtha, K., 2014. Litter input controls on soil carbon in a temperate deciduous forest. Soil Sci. Soc. Am. J. 78, S66.

Braconnot, H., 1819. Memoir on the conversion of wood particles in rubber, in sugar, and in a special natural acid, by means of sulfuric acid; conversion of the same woody substance in ulmin by potash (Translated from French). Ann. Chim. Phys 12, 172–195.

Bradford, M.A., Wieder, W.R., Bonan, G.B., Fierer, N., Raymond, P.A., Crowther, T.W., 2016. Managing uncertainty in soil carbon feedbacks to climate change. Nat. Clim. Change 6, 751–758.

Bradford, M.A., Carey, C.J., Atwood, L., Bossio, D., Fenichel, E.P., Gennet, S., 2019. Soil carbon science for policy and practice. Nat. Sustain. 2, 1070–1072.

Bremner, J.M., 1951. A review of recent work on soil organic matter. Eur. J. Soil Sci. 2, 67–82.

Brenner, B.A., 2003. The long-term carbon cycle, fossil fuels and atmospheric composition. Nature 426, 323–326.

Brink, D.P., Ravi, K., Lidén, G., Gorwa-Grauslund, M.F., 2019. Mapping the diversity of microbial lignin catabolism: experiences from the eLignin database. Appl. Microbiol. Biotechnol. 103, 3979–4002. https://doi.org/10.1007/s00253-019-09692-4.

Carey, J.C., Tanga, J., Templer, P.H., Kroeger, K.D., Crowther, T.W., Burton, A.F., 2016. Temperature response of soil respiration largely unaltered with experimental warming. Proc. Natl. Acad. Sci. 113, 13797–13802.

Chadwick, O.A., Kelly, E.F., Merritts, D.M., Amundson, R.G., 1994. Carbon dioxide consumption during soil development. Biogeochemistry 24, 115–127.

Chahal, I., Vyn, R.J., Mayers, D., Van Eerd, L.L., 2020. Cumulative impact of cover crops on soil carbon sequestration and profitability in a temperate humid climate. Sci. Rep. 10, 13381. https://doi.org/10.1038/s41598-020-70224-6.

Chambers, A., Lal, R., Paustian, K., 2016. Soil carbon sequestration potential of US croplands and grasslands: implementing the 4 per Thousand Initiative. J. Soil Water Conserv. 71, 68A–74A. https://doi.org/10.2489/jswc.71.3.68A.

Chen, L., Aalto, J., Luoto, M., 2021. Significant shallow–depth soil warming over Russia during the past 40 years. Glob. Planet. Change 197, 103394.

Chen, X., Chen, H.Y.H., Chen, C., Ma, Z., Searle, E.B., Yu, Z., et al., 2020. Effects of plant diversity on soil carbon in diverse ecosystems: a global meta-analysis. Biol. Rev. 95, 167–183.

Chien, Y.Y., Kim, E.G., Bleam, W.F., 1997. Paramagnetic relaxation of atrazine solubilized by humic micellar solutions. Environ. Sci. Technol. 31, 3204–3208.

Chiou, C.T., Porter, P.E., Schmedding, D.W., 1983. Partition equilibria of nonionic organic compounds between soil organic matter and water. Environ. Sci. Technol. 17, 227–231.

Ciais, P., Sabine, C., Bala, G., Bopp, L., Brovkin, V., Canadell, J.G., et al., 2013. Carbon and other biogeochemical cycles. In: Stocker, T.F., Qin, D., Plattner, G., Tignor, M., Allen, S.K., Boschung, J., et al. (Eds.), Climate Change 2013: The Physical Science Basis. Contribution of Working Group 1 to the Fifth Assessment Report of the Intergovernmental Panel on Climate Change. Cambridge University Press, Cambridge.

Clay, D.E., Chang, J., Clay, S.A., Stone, J., Gelderman, R.H., Carlson, G.C., et al., 2012. Corn yields and no-tillage affects carbon sequestration and carbon footprints. Agron. J. 104, 763–770.

Collins, M., Knutti, R., Arblaster, J., Dufresne, J.L., Fichefet, T., Friedlingstein, P., et al., 2013. Long-term climate change: projections, commitments and irreversibility. In: Stocker, T.F., Qin, D., Plattner, G.-K., Tignor, M., Allen, S.K., Boschung, J., et al. (Eds.), Climate Change 2013: The Physical Science Basis. Contribution of Working Group I to the Fifth Assessment Report of the Intergovernmental Panel on Climate Change. Cambridge University Press, New York, pp. 1029–1136. ISBN 9781107415324.

Conant, R.T., Cerri, C.E.P., Osborne, B.B., Paustian, K., 2017. Grassland management impacts on soil carbon stocks: a new synthesis. Ecol. Appl. 27, 662–668.

Cotrufo, M.F., Wallenstein, M.D., Boot, C.M., Denef, K., Paul, E.A., 2013. The Microbial Efficiency-Matrix Stabilization (MEMS) framework integrates plant litter decomposition with soil organic matter stabilization: do labile plant inputs form stable soil organic matter? Glob. Chang. Biol. 19, 988–995. https://doi.org/10.1111/gcb.12113.

Cotrufo, M.F., Ranalli, M.G., Haddix, M.L., Six, J., Lugato, E., 2019. Soil carbon storage informed by particulate and mineral-associated organic matter. Nat. Geosci. 12, 989–994.

Creamer, C.A., Jones, D.L., Baldock, J.A., Rui, Y., Murphy, D.V., Hoyle, F.C., et al., 2016. Is the fate of glucose-derived carbon more strongly driven by nutrient availability, soil texture, or microbial biomass size? Soil Biol. Biochem. 103, 201–212.

Daly, A.B., Jilling, A., Bowles, T.M., Buchkowski, R.W., Frey, S.D., Kallenbach, C.M., et al., 2021. A holistic framework integrating plant-microbe-mineral regulation of soil bioavailable nitrogen. Biogeochemistry 154, 211–229. https://doi.org/10.1007/s10533-021-00793-9.

de Saussure, N.T., 1804. Recherches chimiques sur la vegetation. Chez la Ve. Nyon, Paris.

DeClerck, F., Singer, M.J., 2003. Looking back 60 years, California soils maintain overall chemical quality. Calif. Agric. 57, 38–41. https://doi.org/10.3733/ca.v057n02p38.

Deng, J., Zhu, W., Zhou, Y., Yin, Y., 2019. Soil organic carbon chemical functional groups under different revegetation types are coupled with changes in the microbial community composition and the functional genes. Forests 10, 240. https://doi.org/10.3390/f10030240.

Döbereiner, J.W., 1822. Zur Pneumatischen Chemie. III. Zur Pneumatischen. Phytochemie. Jena, Germany, pp. 64–74.

Drosos, M., Savy, D., Spiteller, M., Piccolo, A., 2018. Structural characterization of carbon and nitrogen molecules in the Humeome of two different grassland soils. Chem. Biol. Technol. Agric. 5, 14. https://doi.org/10.1186/s40538-018-0127-y.

Eglinton, T.L., Galy, V.V., Hemingway, J.D., Feng, X., Bao, H., Blattmann, T.M., et al., 2021. Climate control on terrestrial biospheric carbon turnover. Proc. Natl. Acad. Sci. 118 (8) e2011585118.

Ehleringer, J.R., Buchmann, N., Flannagen, L.B., 2000. Carbon isotope ratios in belowground carbon cycle processes. Ecol. Appl. 10, 412–422.

Eswaran, H., Reich, P.F., Kimble, J.M., Beinroth, F.H., Padmanabhan, E., Moncharoen, P., 2000. Global carbon stocks. In: Lal, R., Kimble, J.M., Eswaran, H., Stewart, B.A. (Eds.), Global Climate Change and Pedogenic Carbonates. Lewis Publishers, Boca Raton, pp. 15–25.

Evans, C.D., Freeman, C., Cork, L.G., Thomas, D.N., Reynolds, B., Billett, M.F., et al., 2007. Evidence against recent climate-induced destabilization of soil carbon from ^{14}C analysis of riverine dissolved organic matter. Geophys. Res. Lett. 34, 1–5.

FAO, 2015. FAO STAT Statistical Database, 2013. Food and Agriculture Organization of the United Nations, Rome, Italy.

Farmer, V.C., Pisaniello, D.L., 1985. Against an aromatic structure for soil fulvic acid. Nature 313, 474–475.

Feller, C., Balesdent, J., Nicolardot, B., Cerri, C., 2000. Approaching functional soil organic matter pools through particle-size fractionation examples for tropical soils. In: Lal, R., Kimble, J.M., Follett, R.F., Stewart, B.A. (Eds.), Assessment Methods for Soil Carbon, Advances in Soil Science. CRC Press, Boca Raton, pp. 53–68.

Fischer, F., Schrader, H., 1921. The origin and chemical structure of coal. Brennstoff-Chem. 2, 37–45.

Flaig, W., Beutelspacher, H., Rietz, E., 1975. Chemical composition and physical properties of humic substances. In: Gieseking, J.E. (Ed.), Soil Components. Springer-Verlag, Berlin.

Fontaine, S., Mariotti, A., Abbadie, L., 2003. The priming effect of organic matter: a question of microbial competition? Soil Biol. Biochem. 35, 837–843.

Friedlingstein, P., O'Sullivan, M., Jones, M.W., Andrew, R.M., Hauck, J., Olsen, A., et al., 2020. Global carbon budget 2020. Earth Syst. Sci. Data 12, 3269–3340. https://doi.org/10.5194/essd-12-3269-2020.

Gabor, R.S., Eilers, K., McKnight, D.M., Fierer, N., Anderson, S.P., 2014. From the litter layer to the saprolite: chemical changes in water-soluble soil organic matter and their correlation to microbial community composition. Soil Biol. Biochem. 68, 166–176.

Galicia-Andrés, E., Escalona, Y., Grančič, P., Oostenbrink, C., Tunega, D., Gerzabek, M.H., 2021. A molecular Dynamic Study of Soil Organic Matter Stabilization Mechanisms, EGU General Assembly 2021 online, 19–30 April 2021. https://doi.org/10.5194/egusphere-egu21-2486.

Godbold, D.L., Hoosbeek, M.R., Lukac, M., Cotrufo, M.F., Janssens, I.A., Ceulemans, R., 2006. Mycorrhizal hyphal turnover as a dominant process for carbon input into soil organic matter. Plant Soil 281, 15–24. https://doi.org/10.1007/s11104-005-3701-6.

Gortner, R.A., 1916. The origin of humin formed by the acid hydrolysis of proteins: II. Hydrolysis in the presence of carbohydrates and of aldehydes. J. Biol. Chem. 26, 177–204.

Gross, C.D., Harrison, R.B., 2019. The case for digging deeper: soil organic carbon storage, dynamics, and controls in our changing world. Soil Syst. 3, 1–28.

Guo, L.B., Gifford, R.M., 2002. Soil carbon stocks and land use change: a metaanalysis. Glob. Chang. Biol. 8, 345–360.

Haddaway, N.R., Hedlund, K., Jackson, L.E., Kätterer, T., Lugato, E., Thomsen, I.K., 2017. How does tillage intensity affect soil organic carbon? A systematic review. Environ. Evid. 6, 30. https://doi.org/10.1186/s13750-017-0108-9.

Hagedorn, F.A., Kammer, M.W., Schmidt, I., Goodale, C.I., 2012. Nitrogen addition alters mineralization dynamics of ^{13}C-depleted leaf and twig litter and reduces leaching of older DOC from mineral soil. Glob. Chang. Biol. 18, 1412–1427.

Haider, K., Schulten, H.R., 1985. Pyrolysis field ionization mass spectrometry of lignins, soil humic compounds and whole soil. J. Anal. Appl. Pyrolysis 8, 317–331.

Haile-Mariam, S., Collins, H.P., Wright, S., Paul, E.A., 2008. Fractionation and long-term laboratory incubation to measure soil organic matter dynamics. Soil Sci. Soc. Am. J. 72, 370–378.

Hall, S.J., Huang, W., Timokhin, V.I., Hammel, K.E., 2020a. Lignin lags, leads, or limits the decomposition of litter and soil organic carbon. Ecology 101, e03113.

Hall, S.J., Ye, C., Weintraub, R.B., Hockaday, W.C., 2020b. Molecular trade-offs in soil organic carbon composition at continental scale. Nat. Geosci. 13, 687–692.

Hartmann, H., Bahn, M., Carbone, M., Richardson, A.D., 2020. Plant carbon allocation in a changing world – challenges and progress: introduction to a virtual issue on carbon allocation. New Phytol. 227, 981–988.

Hassink, J., 1995. Decomposition rate constants of size and density fractions of soil organic matter. Soil Sci. Soc. Am. J. 59, 1631–1635.

Hassink, J., 1997. The capacity of soils to preserve organic C and N by their association with clay and silt particles. Plant Soil 191, 77–87.

Hatcher, P.G., Maciel, G.E., Dennis, L.W., 1981. Aliphatic structure of humic acids; a clue to their origin. Org. Geochem. 3, 43–48.

Hayes, M.H.B., Clapp, C.E., 2001. Humic substances: considerations of compositions, aspects of structure, and environmental influences. Soil Sci. 166, 723–737.

Heckman, K., Throckmorton, H., Clingensmith, C., Vila, F.J.G., Horwath, W.R., Rasmussen, C., 2014. Factors affecting the molecular structure and mean residence time of occluded organics in a lithosequence of soils under ponderosa pine. Soil Biol. Biochem. 77, 1–11.

Henneberry, Y.K., Kraus, T.E.C., Fleck, J.A., Krabbenhoft, D.P., Bachand, P.M., Horwath, W.R., 2010. Removal of inorganic mercury and methylmercury from surface waters following coagulation of dissolved organic matter with metal-based salts. Sci. Total Environ. 409, 631–637.

Henneberry, Y.K., Kraus, T.E.C., Peter, N., Horwath, W.R., 2012. Structural stability of coprecipitated natural organic matter and ferric iron under reducing conditions. Org. Geochem. 48, 81–89.

Hermann, R., 1842. Untersuchungen über den Moder. J. Prakt. Chem. 27, 165–172.

Hicks Pries, C.E., Schuur, E.A.G., Natali, S.M., Crummer, K.G., 2016. Old soil carbon losses increase with ecosystem respiration in experimentally thawed tundra. Nat. Clim. Chang. 6, 214–218.

Hicks Pries, C.E., Castanha, C., Porras, R., Torn, M.S., 2017. The whole-soil carbon flux in response to warming. Science 355, 1420–1423.

Hobley, E., Baldock, J., Hua, Q., Wilson, B., 2017. Land-use contrasts reveal instability of subsoil organic carbon. Glob. Chang. Biol. 23, 955–965.

Horwath, W.R., 2014. Carbon cycling: the dynamics and formation of organic matter. In: Paul, E.A. (Ed.), Soil Microbiology, Ecology and Biochemistry, fourth ed. Elsevier, New York, pp. 339–382.

Horwath, W.R., 2017. The role of the soil microbial biomass in cycling nutrients. In: Tate, K.R. (Ed.), Microbial Biomass: A Paradigm Shift in Terrestrial Biogeochemistry. World Scientific, Hackensack, NJ, pp. 41–66.

Horwath, W.R., Elliott, L.F., 1996. Ryegrass straw component decomposition during mesophilic and thermophilic incubations. Biol. Fertil. Soils 21, 227–232.

Horwath, W.R., van Kessel, C., Hartwig, U., Harris, D., 2001. Use of ^{13}C isotopes to determine net carbon sequestration in soil under ambient and elevated CO_2. In: Lal, R., Kimble, J.M., Follett, R.F., Stewart, B.A. (Eds.), Assessment Methods for Soil Carbon. Lewis Publishers, Boca Raton, Florida, pp. 221–232.

Houghton, R.A., Nassikas, A.A., 2017. Global and regional fluxes of carbon from land use and land cover change 1850–2015. Glob. Biogeochem. Cy. 31, 456–472. https://doi.org/10.1002/2016GB005546.

Huysmann, F., Verstraete, W., 1993. Effect of cell surface characteristics on the adhesion of bacteria to soil particles. Biol. Fert. Soils 16, 21–26.

Jackson, R.B., Lajtha, K., Crow, S.E., Hugelius, G., Kramer, M.G., Piñeiro, G., 2017. The ecology of soil carbon: pools, vulnerabilities, and biotic and abiotic controls. Annu. Rev. Ecol. Evol. Syst. 48, 419–445.

James, J., Harrison, R., 2016. The effect of harvest on forest soil carbon: a meta-analysis. Forests 7, 308.

Jandl, R., Rodeghiero, M., Martinez, C., Cotrufo, M.F., Bampa, F., van Wesemael, B., et al., 2014. Current status, uncertainty and future needs in soil organic carbon monitoring. Sci. Total Environ. 468–469, 376–383.

Jarvi, M.P., Burton, J., 2020. Root respiration and biomass responses to experimental soil warming vary with root diameter and soil depth. Plant Soil 451, 435–446. https://doi.org/10.1007/s11104-020-04540-1.

Jastrow, J.D., Amonette, J.E., Bailey, V.L., 2007. Mechanisms controlling soil carbon turnover and their potential application for enhancing carbon sequestration. Clim. Change 80, 5–23. https://doi.org/10.1007/s10584-006-9178-3.

Jenkinson, D.S., 1966. The priming action. In: The Use of Isotopes in Soil Organic Matter Studies. Report of Food and Agriculture Organization and International Atomic Energy Agency. Braunschweig-VoE lkenrode, Technical Meeting, pp. 199–207, 1963.

Jenkinson, D.S., 1977. Studies on the decomposition of plant material in soil. V. The effects of plant cover and soil type on the loss of carbon from C-14 labelled ryegrass decomposing under field conditions. J. Soil Sci. 28, 424–434. https://doi.org/10.1111/j.1365-2389.1977.tb02250.x.

Jenkinson, D.S., Fox, R.H., Rayner, J.H., 1985. Interactions between fertilizer nitrogen and soil nitrogen, Ð. the so-called "priming" effect. J. Soil Sci. 36, 425–444.

Jenny, H., 1980. The Soil Resource, Origin and Behavior. Springer-Verlag, New York.

Jilling, A., Kane, D., Williams, A., Yannarell, A.C., Davis, A., Jordane, N.R., et al., 2020. Rapid and distinct responses of particulate and mineral-associated organic nitrogen to conservation tillage and cover crops. Geoderma 359, 114001.

Jobbágy, E.G., Jackson, R.B., 2000. The vertical distribution of soil organic carbon and its relation to climate and vegetation. Ecol. Appl. 10, 423–436.

Jones, D.L., Nguyen, C., Finlay, R.D., 2009. Carbon flow in the rhizosphere: carbon trading at the soil-root interface. Plant Soil 321, 5–33.

Kaiser, K., Kalbitz, K., 2012. Cycling downwards – dissolved organic matter in soils. Soil Biol. Biochem. 52, 29–32.

Kallenbach, C.M., Frey, S.A., Grandy, A.S., 2016. Direct evidence for microbial-derived soil organic matter formation and its ecophysiological controls. Nat. Commun. 7, 136330. https://doi.org/10.1038/ncomms13630.

Kindler, R., Miltner, A., Richnow, H., Kastner, M., 2006. Fate of gram-negative bacterial biomass in soil – mineralization and contribution to SOM. Soil Biol. Biochem. 38, 2860–2870.

Kleber, M., Eusterhues, K., Keiluweitk, M., Mikutta, C., Mikutta, R., Nico, P.S., 2015. Mineral–organic associations: formation, properties, and relevance in soil environments. Adv. Agron. 130, 1–140.

Kleber, M., Sollins, P., Sutton, R., 2007. A conceptual model of organo-mineral interactions in soils: self-assembly of organic molecular fragments into zonal structures on mineral surfaces. Biogeochemistry 85, 9–24.

Kononova, M.M., 1966. Soil Organic Matter: Its Nature, its Role in Soil Formation and in Soil Fertility. Pergamon Press Ltd., Oxford, pp. 45–49.

Kuzyakov, Y., Friedel, J.K., Stahr, K., 2000. Review of mechanisms and quantification of priming effects. Soil Biol. Biochem. 32, 1485–1498.

Lajtha, K., Townsend, K.L., Kramer, M.G., Swanston, C., Bowden, R.D., Nadelhoffer, K., 2014. Changes to particulate versus mineral-associated soil carbon after 50 years of litter manipulation in forest and prairie experimental ecosystems. Biogeochemistry 119, 341–360. https://doi.org/10.1007/s10533-014-9970-5.

Lal, R., 2004. Soil carbon sequestration impacts on global climate change and food security. Science 304, 1623–1627.

Lambers, H., 1987. Growth, respiration, exudation and symbiotic associations: the fate of carbon translocated to the roots. In: Gregory, P.J., Lake, J.V., Rose, D.A. (Eds.), Root Development and Function – Effects of the Physical Environment. Cambridge University Press, Cambridge, pp. 125–145.

Lavallee, J.M., Soong, J.L., Cotrufo, M.F., 2019. Conceptualizing soil organic matter into particulate and mineral-associated forms to address global change in the 21st century. Glob. Change Biol. 26, 261–273.

Learman, D.R., Voelker, B.M., Vazquez-Rodriguez, A.I., Hansel, C.M., 2011. Formation of manganese oxides by bacterially generated superoxide. Nat. Geosci. 4, 95−98.

Lehmann, J., Kleber, M., 2015. The contentious nature of soil organic matter. Nature 528, 60−68.

Lehmann, J., Solomon, D., Kinyangi, J., Wirick, S., Jacobsen, C., 2008. Spatial complexity of soil organic matter forms at nanometer scales. Nat. Geosci. 1, 238−242.

Lehmann, J., Hansel, C.M., Kaiser, C., Kleber, M., Maher, K., Manzoni, S., et al., 2020. Persistence of soil organic carbon caused by functional complexity. Nat. Geosci. 13, 529−534.

Leichty, S., Cotrufo, S.M., Stewart, C.E., 2020. Less efficient residue-derived soil organic carbon formation under no-till irrigated corn. Soil Sci. Soc. Am. J. 84, 1928−1942.

Liang, C., Balser, T.C., 2010. Microbial production of recalcitrant organic matter in global soils: implications for productivity and climate policy. Nat. Rev. Microbiol. 9, 75. https://doi.org/10.1038/nrmicro2386.cl.

Limpens, J., Berendse, F., Blodau, C., Canadell, J.G., Freeman, C., Holden, J., et al., 2008. Peatlands and the carbon cycle: from local processes to global implications − a synthesis. Biogeosciences 5, 1475−1491.

Lower, S.K., 2005. Directed natural forces of affinity between a bacterium and mineral. J. Sci. 305, 752−765.

Malguti, M., 1835. Action des Acides etendus sur les Sucre. Ann. Chim. Phys. 59, 407−423.

Marcusson, J., 1920. Origin of asphalt and coal. Chem. Ztg. 44, 43−44.

Marion, G.M., 1989. Correlation between long-term pedogenic $CaCO_3$ formation rate and modern precipitation in deserts of the American southwest. Quaternary Res. 32, 291−295.

Marshland III, R., Cui, W., Mehta, P., 2020. A minimal model for microbial biodiversity can reproduce experimentally observed ecological patterns. Sci. Rep. 10, 3308.

Martin, J.P., Haider, K., 1979. Biodegradation of ^{14}C-labeled model and cornstalk lignins, phenols, model phenolase humic polymers, and fungal melanins as influenced by a readily available carbon source and soil. Appl. Environ. Microbiol. 38, 91−110.

Mathieu, J.A., Hatte, C., Balesdent, J., Parent, E., 2015. Deep soil carbon dynamics are driven more by soil type than by climate: a worldwide meta-analysis of radiocarbon profiles. Glob. Chang. Biol. 221, 4278−4292.

Mayer, M., Prescott, C.E., Wafa Abaker, E.A., Augusto, L., Cécillon, L., Ferreira, G.W.D., et al., 2020. Tamm review: influence of forest management activities on soil organic carbon stocks: a knowledge synthesis. For. Ecol. Manage. 466, 118127.

Mbow, C.C., Rosenzweig, L.G., Barioni, T.G., Benton, M., Herrero, M., Krishnapillai, E., et al., 2019. Food security. In: P.R. Shukla, J., Skea, E., Calvo Buendia, V., Masson-Delmotte, H.-O., Pörtner, D.C., Roberts, P., et al. (Eds.), Climate Change and Land: An IPCC Special Report on Climate Change, Desertification, Land Degradation, Sustainable Land Management, Food Security, and Greenhouse Gas Fluxes in Terrestrial Ecosystems. Intergovernmental Panel on Climate Change.

McClelland, S.C., Paustin, K., Schipanski, M.E., 2021. Management of cover crops in temperate climates influences soil organic carbon stocks: a meta-analysis. Ecol. Appl. 31, e02278.

Mendez-Millan, M., Dignac, M.-F., Rumpel, C., Rasse, D.P., Derenne, S., 2010. Molecular dynamics of shoot vs. root biomarkers in an agricultural soil estimated by natural abundance ^{13}C labelling. Soil Biol. Biochem. 42, 169−177.

Michalzik, B., Kalbitz, K., Park, J., Solinger, S., Matzner, E., 2001. Fluxes and concentrations of dissolved organic carbon and nitrogen − a synthesis for temperate forests. Biogeochemistry 52, 173−205.

Mikutta, C., 2011. X-ray absorption spectroscopy study on the effect of hydroxybenzoic acids on the formation and structure of ferrihydrite. Geochim. Cosmochim. Acta 75, 5122−5139.

Milchunas, D.G., Lauenroth, W.K., Singh, J.S., Cole, C.V., Hunt, H.W., 1985. Root turnover and production by 14C dilution − implications for C partitioning in plants. Plant Soil 88, 353−365.

Miltner, A., Bombach, P., Schmidt-Brücken, B., Kästner, M., 2012. SOM genesis: microbial biomass as a significant source. Biogeochemistry 111, 41−55. https://doi.org/10.1007/s10533-011-9658-z.

Minasny, B., Malone, B.P., McBratney, A.B., Angers, D.A., Arrouays, D., Chambers, A., et al., 2017. Soil carbon 4 per mille. Geoderma 292, 59−86.

Mobley, M.L., Lajtha, K., Kramer, M.G., Bacon, A.R., Heine, P.R., Richter, D.D., 2015. Surficial gains and subsoil losses of soil carbon and nitrogen during secondary forest development. Glob. Chang. Biol. 21, 986−996.

Moomaw, W.R., Chmura, G.L., Davies, G.T., Finlayson, C.M., Middleton, B.A., Natali, S.M., et al., 2018. Wetlands in a changing climate: science, policy, and management. Wetlands 38, 183−205. https://doi.org/10.1007/s13157-018-1023-8.

Moore, J.A.M., Jiang, J., Patterson, C.M., Mayes, M.A., Wang, G., Classen, A.T., 2015. Interactions among roots, mycorrhizas and free-living microbial communities differentially impact soil carbon processes. J. Ecol. 103, 1442–1453.

Moore, T.R., Trofymow, J.A., Prescott, C.E., Titus, B.D., 2017. Can short-term litter-bag measurements predict long-term decomposition in northern forests? Plant Soil 416, 419–426.

Mulder, G.J., 1839. Ueber die Zusammensetzung einiger thierischen Substanzen. J. prakt. Chem. 16, 495–497.

Nave, L.E., Vance, E.D., Swanston, C.W., Curtis, P.S., 2010. Harvest impacts on soil carbon storage in temperate forests. For. Ecol. Manag. 259, 857–866.

Neff, J.C., Asner, G.P., 2001. Dissolved organic carbon in terrestrial ecosystems: synthesis and a model. Ecosystems 4, 29–48.

Nieder, R., Benbi, D.K., 2008. Carbon and Nitrogen in the Terrestrial Environment. Springer, New York.

Nkoh, N.J., Liu, Z., Yan, J., Cai, S., Hong, Z., Xu, R., 2020. The role of extracellular polymeric substances in bacterial adhesion onto variable charge soils. Arch. Agron Soil Sci. 66, 1780–1793. https://doi.org/10.1080/03650340.2019.1696016.

Olson, J.S., Garrels, R.M., Berner, R.A., Armentano, T.V., Dyer, M.I., Yaalon, D.H., 1985. The natural carbon cycle. In: Trabalka, J.R. (Ed.), Atmospheric Carbon Dioxide and the Global Carbon Cycle. US Department of Energy, Washington, DC, pp. 175–213.

Paul, E.A., van Veen, J.A., 1978. The use of tracers to determine the dynamic nature of organic matter. Trans. Int. Conf. Soil Sci. 61–102. Edmonton.

Paustian, K., Collier, S., Baldock, J., Burgess, R., Creque, J., DeLonge, M., et al., 2019. Quantifying carbon for agricultural soil management: from the current status toward a global soil information system. Carbon Manag. 10 (6), 567–587. https://doi.org/10.1080/17583004.2019.1633231.

Pett-Ridge, J., Firestone, M.K., 2017. Using stable isotopes to explore root-microbe-mineral interactions in soil. Rhizosphere 3, 244–253.

Phillips, R.P., Meier, I.C., Bernhardt, E.S., Grandy, A.S., Wickings, K., Finzi, A.C., 2012. Roots and fungi accelerate carbon and nitrogen cycling in forests exposed to elevated CO_2. Ecol. Lett. 15, 1042–1049.

Piccolo, A., 2001. The supramolecular structure of humic substances. Soil Sci. 166, 810–832.

Powers, J.S., Schlesinger, W.H., 2002. Relationships among soil carbon distributions and biophysical factors at nested spatial scales in rain forests of northeastern Costa Rica. Geoderma 109, 165–190. https://doi.org/10.1016/S0016-7061(02)00147-7.

Preston, C.M., Schnitzer, M., 1984. Effects of chemical modifications and, extractants on the carbon-13 NMR spectra of humic materials. Soil Sci. Soc. Am. J. 48, 305–311.

Qiu, C., Ciais, P., Zhu, D., Guenet, B., Peng, S., Petrescu, A.M.R., et al., 2021. Large historical carbon emissions from cultivated northern peatlands. Sci. Adv. 7, eabf1332.

Rasmussen, C., Southard, R.J., Horwath, W.R., 2008. Litter type and soil minerals control temperate forest soil carbon response to climate change. Glob. Chang. Biol. 14, 2064–2080.

Rasse, D.P., Rumpel, C., Dignac, M.F., 2005. Is soil carbon mostly root carbon? Mechanisms for a specific stabilization. Plant Soil 269, 341–356.

Reicosky, D.C., Lindstrom, M.J., 1993. Fall tillage method: effect on short-term carbon dioxide flux from soil. Agron. J. 85, 1237–1243.

Richter, D.D., Markewitz, D., Trumbore, S.E., Wells, C.G., 1999. Rapid accumulation and turnover of soil carbon in a re-establishing forest. Nature 400, 56–58.

Rillig, M.C., Mullerand, L.A.H., Lehmann, A., 2017. Soil aggregates as massively concurrent evolutionary incubators. ISME J. 11, 1943–1948. https://doi.org/10.1038/ismej.2017.56.

Rinnan, R., Bååth, E., 2009. Differential utilization of carbon substrates by bacteria and fungi in tundra soil. Appl. Environ. Microbiol. 75, 3611–3620.

Saiz-Jimenez, C., De Leeuw, J.W., 1985. Chemical characterization of soil organic matter fractions by analytical pyrolysis-gas chromatography-mass spectrometry. J. Anal. Appl. Pyrolysis 9, 99–119.

Salimia, S., Almuktar, S.A.A.N., Scholz, M., 2021. Impact of climate change on wetland ecosystems: a critical review of experimental wetlands. J. Environ. Manag. 286, 121160. https://doi.org/10.1016/j.jenvman.2021.112160.

Sanderman, J., Baldock, J.A., Amundson, R., 2008. Dissolved organic carbon chemistry and dynamics in contrasting forest and grassland soils. Biogeochemistry 89, 181–198.

Scharlemann, J.P., Tanner, E.V., Hiederer, R., Kapos, V., 2014. Global soil carbon: understanding and managing the largest terrestrial carbon pool. Carbon Manag. 5, 81–91.

Schaumann, G.E., Thiele-Bruhn, S., 2011. Molecular modeling of soil organic matter: squaring the circle? Geoderma 166, 1–14.

Schindlbacher, A., Schnecker, J., Takriti, M., Borken, W., Wanek, W., 2015. Microbial physiology and soil CO_2 efflux after 9 years of soil warming in a temperate forest – no indications for thermal adaptations. Glob. Chang. Biol. 21, 4265–4277. https://doi.org/10.1111/gcb.12996.

Schlesinger, W.H., 2017. Inorganic carbon: global carbon cycle. In: Rattan, L. (Ed.), Encyclopedia of Soil Science, third ed. Taylor & Francis, New York, pp. 1203–1205 https://doi.org/10.1081/E-ESS3-120042705.

Schmidt, M.W.I., Torn, M.S., Abiven, S., Dittmar, T., Guggenberger, G., Janssens, I.A., et al., 2011. Persistence of soil organic matter as an ecosystem property. Nature 478, 49–56.

Schnitzer, M., 1978. Humic substances: chemistry and reactions. In: Schnitzer, M., Khan, S.U. (Eds.), Soil Organic Matter, Vol. 8. Elsevier Scientific Publication Company, New York, pp. 1–64.

Schulten, H.-R., Schnitzer, M., 1993. A state-of-the-art structural concept for humic substances. Naturwissenschaften 80, 29–30.

Schulten, H.-R., Sorge-Lewin, C., Schnitzer, M., 1997. Structure of "unknown" soil nitrogen investigated by analytical pyrolysis. Biol. Fertil. Soils 24, 249–254.

Schurig, C., Smittenberg, R.R., Berger, G., Kraft, J., Woche, S., Goebel, M.O., et al., 2013. Microbial cell envelope fragments and the formation of soil organic matter: a case study from a glacier forefield. Biogeochemistry 113, 595–612.

Shahzad, T., Chenu, C., Genet, P., Barot, S., Perveen, N., Mougin, C., et al., 2015. Contribution of exudates, arbuscular mycorrhizal fungi and litter depositions to the rhizosphere priming effect induced by grassland species. Soil Biol. Biochem. 80, 146e155.

Shapiro, J., 1957. Chemical and biological studies on the yellow organic acids of lake water. Limnol. Oceanogr. 2, 161–179.

Shen, F., Yang, F., Xiao, C., Zhou, Y., 2020. Increased contribution of root exudates to soil carbon input during grassland degradation. Soil Biol. Biochem. 146, 107817.

Shevchenko, S.M., Bailey, G.W., 1996. Life after death: lignin-humic relationships reexamined. Crit. Rev. Environ. Sci. Technol. 26, 95–15.

Sierra, C.A., Müller, M., Metzler, H., Manzoni, S., Trumbore, S.E., 2017. The muddle of ages, turnover, transit, and residence times in the carbon cycle. Glob. Chang. Biol. 23, 1763–1773.

Silva, C.R.S., Corrêa, R.S., Doane, T.A., Pereira, E.I.P., Horwath, W.R., 2013. Unprecedented carbon accumulation in mined soils: the synergistic effect of resource input and plant species invasion. Ecol. Appl. 23, 1345–1356.

Simpson, A.J., Simpson, M.S., Smith, E., Kelleher, B.P., 2007. Microbially derived inputs to soil organic matter: are current estimates too low? Environ. Sci. Technol. 41, 8070–8076.

Six, J., Elliott, E.T., Paustian, K., 2000. Soil macroaggregate turnover and microaggregate formation: a mechanism for C sequestration under no-tillage agriculture. Soil Biol. Biochem. 32, 2099–2103.

Six, J., Conant, R.T., Paul, E.A., Paustian, K., 2002. Stabilization mechanisms of soil organic matter: implications for C-saturation of soils. Plant Soil 241, 155–176.

Six, J., Frey, S.D., Thiet, R.K., Batten, K.M., 2006. Bacterial and fungal contributions to carbon sequestration in agroecosystems. Soil Sci. Soc. Am. J. 70, 555–569.

Sokol, N.W., Bradford, M.A., 2019. Microbial formation of stable soil carbon is more efficient from belowground than aboveground input. Nat. Geosci. 12, 46–53.

Soong, J.L., Phillips, C.L., Ledna, C., Koven, C.D., Torn, M.S., 2020. CMIP5 models predict rapid and deep soil warming over the 21st century. Biogeosciences 125, e2019JG005266. https://doi.org/10.1029/2019JG005266.

Soong, J.L., Castanha, C., Hicks Pries, C.E., Ofiti, N., Porras, C., Riley, W.J., et al., 2021. Five years of whole-soil warming led to loss of subsoil carbon stocks and increased CO_2 efflux. Sci. Adv. 7 (21), eabd1343.

Stevenson, F.J., 1994. Humus Chemistry: Genesis, Composition, Reactions. John Wiley & Sons, Inc., New York.

Sulman, B.N., Moore, J.A.M., Abramoff, R., Averill, C., Kivlin, S., Georgiou, K., et al., 2018. Multiple models and experiments underscore large uncertainty in soil carbon dynamics. Biogeochemistry 141, 109–123.

Sutton, R., Sposito, G., 2005. Molecular structure in soil humic substances: the new view. Environ. Sci. Technol. 39, 9009–9015.

Swift, R.S., 1999. Macromolecular properties of soil humic substances: fact, fiction and opinion. Soil Sci. 164, 790–802.

Tarnocai, C., Canadell, J.G., Schuur, E.A.G., Kuhry, P., Mazhitova, G., Zimov, S., 2009. Soil organic carbon pools in the northern circumpolar permafrost region. Glob. Biogeochem. Cy. 23, GB2023. https://doi.org/10.1029/2008GB003327.

Terrer, C., Phillips, R.P., Hungate, B.A., Rosende, J., Pett-Ridge, J., Craig, M.E., et al., 2021. A trade-off between plant and soil carbon storage under elevated CO_2. Nature 591, 599–603.

Tfaily, M.M., Chu, R.K., Tolic, N., Roscioli, K.M., Anderton, C.R., LPasa-Tolic, L., et al., 2015. Advanced solvent based methods for molecular characterization of soil organic matter by high-resolution mass spectrometry. Anal. Chem. 87, 5206–5215.

Theng, B.K.G., 1982. Clay-polymer interactions: summary and perspectives. Clays Clay Miner. 30, 1–10.

Throckmorton, H.M., Bird, J.A., Dane, L., Firestone, M.K., Horwath, W.R., 2012. The source of microbial C has little impact on soil organic matter stabilization in forest ecosystems. Ecol. Lett. 15, 1257–1265.

Throckmorton, H.M., Bird, J.A., Monte, N., Doane, T., Firestone, M.K., Horwath, W.R., 2015. The soil matrix increases microbial C stabilization in temperate and tropical forest soils. Biogeochemistry 122, 35–45. https://doi.org/10.1007/s10533-014-0027-6.

Tian, Q., Yang, X., Wang, X., Liao, C., Li, Q., Wang, M., et al., 2016. Microbial community mediated response of organic carbon mineralization to labile carbon and nitrogen addition in topsoil and subsoil. Biogeochemistry 128, 125–139.

Tisdall, J.M., Oades, J.M., 1982. Organic matter and water-stable aggregates in soils. J. Soil Sci. 33, 141–163.

Toosi, E.R., Doane, T.A., Horwath, W.R., 2012. Abiotic solubilization of soil organic matter, a less-seen aspect of dissolved organic matter production. Soil Biol. Biochem. 5, 12–21.

Van Groenigen, J.W., van Kessel, C., Hungate, B.A., Oenema, O., Powlson, D.S., van Groenigen, K.J., 2017. Sequestering soil organic carbon: a nitrogen dilemma. Environ. Sci. Technol. 51, 4738–4739.

Villarino, S.H., Pinto, P., Jackson, R.B., Piñeiro, G., 2021. Plant rhizodeposition: a key factor for soil organic matter formation in stable fractions. Sci. Adv. 7, eabd3176.

Von Lützow, M., Kögel-Knaber, I., Ekschmitt, K., Matzner, E., Guggenberger, G., Marschner, B., et al., 2006. Stabilization of organic matter in temperate soils: mechanisms and their relevance under different soil conditions – a review. Eur. J. Soil Sci. 57, 426–445.

Von Wandruszka, R., 1998. The micellar model of humic acid: evidence from pyrene fluorescence measurements. Soil Sci. 163, 921–930.

Voroney, R.P., Paul, E.A., Anderson, D.W., 1989. Decomposition of wheat straw and stabilization of microbial products. Can. J. Soil Sci. 69, 63–77.

Vos, M., Wolf, A.B., Jennings, S.J., Kowalchuk, G.A., 2013. Micro-scale determinants of bacterial diversity in soil. FEMS Microbiol. Rev. 37, 936–954.

Waksman, S.A., 1936. Humus Origin, Chemical Composition, and Importance in Nature. Williams and Wilkins Co., Baltimore.

Waksman, S.A., 1942. Liebig-The humus theory and the role of humus in plant nutrition. In: Moulton, F.R. (Ed.), Liebig and after Liebig. Amer. Assn. Adv. Sci. Special Publ. 16, Washington, DC, pp. 56–63.

Waksman, S.A., Iyer, R.R.N., 1932. Contribution to our knowledge of the chemical nature and origin o humus: II. The influence of synthesized humus compounds and of natural humus upon soil microbiological processes. Soil Sci. 34, 71–79.

Wang, J.J., Pisani, O., Lin, L.H., Lun, O.O.Y., Bowden, R.D., Lajtha, K., et al., 2017. Long-term litter manipulation alters soil organic matter turnover in a temperate deciduous forest. Sci. Total Environ. 865–875, 607–608.

Warembourg, F.R., Paul, E.A., 1973. The use of $^{14}CO_2$ canopy techniques for measuring carbon transfer through the plant-soil system. Plant Soil 38, 331–345.

Webb, P., 2020. Introduction to Oceanography, tenth ed. Pressbooks, Book Oven, Inc., Montreal.

Wedin, D.A., Tieszen, L.L., Dewey, B., Pastor, J., 1995. Carbon isotope dynamics during grass decomposition and soil organic matter formation. Ecology 76, 1383–1392.

Wenjuan, H., Hall, S.J., 2017. Elevated moisture stimulates carbon loss from mineral soils by releasing protected organic matter. Nat. Commun. 8, 1774. https://doi.org/10.1038/s41467-017-01998-z.

Wershaw, R.L., 1986. A new model for humic materials and their interactions with hydrophobic organic chemicals in soil-water or sediment-water systems. J. Contam. Hydrol. 1, 29–45.

Whipps, J.M., Lynch, J.M., 1985. Energy losses by the plant in rhizodeposition. Ann. Proc. Phytochem. Soc. Eur. 26, 59–71.

Wilding, L.P., Nordt, L.C., Kimble, J.M., 2017. Inorganic carbon: global stocks. In: Lal, R. (Ed.), Encyclopedia of Soil Science, third ed. CRC Press, Taylor and Francis Group, Boca Raton, New York, pp. 1206–1209.

Wilhelm, R.C., Singh, R., Eltis, L.D., Mohn, W.W., 2019. Bacterial contributions to delignification and lignocellulose degradation in forest soils with metagenomic and quantitative stable isotope probing. ISME J. 13, 413–429. https://doi.org/10.1038/s41396-018-0279-6.

Wright, J.R., Schnitzer, M., 1961. An estimate of the aromaticity of the organic matter of a Podzol soil. Nature 190, 703–704.

Xu, S., Liu, L.L., Sayer, E.J., 2013. Variability of above-ground litter inputs alters soil physicochemical and biological processes: a meta-analysis of litterfall-manipulation experiments. Biogeosciences 10, 7423–7433.

Ye, R., Horwath, W.R., 2017. Influence of rice straw on priming of soil C for dissolved organic C and CH_4 production. Plant Soil 417, 231–241.

Zhu, B., Gutknecht, J.L.M., Herman, D.J., Keck, D.C., Firestone, M.K., Cheng, W., 2014. Rhizosphere priming effects on soil carbon and nitrogen mineralization. Soil Biol. Biochem. 76, 183–192.

Chapter 13

Methods for studying soil organic matter: nature, dynamics, spatial accessibility, and interactions with minerals

Claire Chenu*, Cornelia Rumpel[†], Charlotte Védère*, and Pierre Barré[‡]
*Université Paris-Saclay, INRAE, Agro Paris Tech, UMR Ecosys, Palaiseau, France; [†]CNRS, UMR IEES, Paris, France; [‡]CNRS, École Normale Supérieure, PSL Université, Laboratoire de Géologie, Paris, France

Chapter outline

Soil Microbiology, Ecology, and Biochemistry. https://doi.org/10.1016/B978-0-12-822941-5.00013-2

13.1 Introduction

Soil organic matter (SOM) can be defined as organic materials found in soil that are, or have been, part of living organisms. It is a continuum of materials at various stages of transformation due to biotic and abiotic processes. Despite its low concentration in soil, organic matter is of major qualitative importance. It contributes to soil chemical fertility, being a reserve of nutrients released by mineralization and retaining nutrient cations on its negative charges. It is also the basis of soil biological activity, being the source of carbon (C) and energy for heterotrophs. SOM also plays a major role in soil physical fertility, increasing water retention, aggregating mineral particles, and thus reducing soil erosion. Moreover, SOM has major impacts on the environment by retaining organic pollutants, heavy metals, and radionuclides, thus protecting water quality while at the same time producing nitrates and phosphates by mineralization that can reduce water quality. These dynamics lead to mineralization or storage of CO_2, methane, and nitrous oxide, all of which are major greenhouse gases (see Chapter 12).

Characterization of SOM fulfills different aims (Table 13.1): (a) providing an understanding of its origin and formation in relation to pedogenesis, (b) understanding and predicting soil organic C (SOC) stocks as a function of land use and of cropping and forestry practices in different climatic and pedogenic contexts, (c) understanding and predicting soil properties and functions influenced by SOM, (d) identifying critical SOM contents relative to these functions and identifying indicators of soil quality, and (e) predicting ecosystem response to disturbance.

TABLE 13.1 Rationale for methods to study soil organic matter.

Questions	Methods category	Examples of methods
What is SOM made of?	Characterization methods	Elemental analysis, biochemistry methods, FTIR, NMR, pyrolysis, NEXAFS
What are SOM chemical and physical properties?	Fractionation methods (physical and chemical)	Particle size fractionation, solubility in acids and bases
	Characterization methods	Measurement of charge, complexation properties, hydrophobicity
Where is SOM located in soil?	Physical fractionation methods	Aggregate size fractionation
	Visualization methods	Optical microscopy, scanning electron microscopy, nanoSIMS, STXM and NEXAFS, FTIR
Which processes control SOM dynamics and stabilization in soil?	Fractionation methods (physical and chemical)	Aggregate size fractionation, HF hydrolysis,
	Visualization methods	Optical microscopy, scanning electron microscopy, nanoSIMS, STXM and NEXAFS, FTIR
Is it possible to isolate-characterize SOM fractions corresponding to kinetic pools?	Fractionation methods (physical and chemical)	Particle size fractionation, chemical and thermal oxidation, long-term incubation
What is the turnover rate of SOM?	Isotopic methods	^{13}C natural labeling, ^{14}C dating, ^{14}C bomb labeling, tracer addition experiments

Given the variety of inputs (i.e., plant tissues, microorganisms, animals) and different stages of decomposition, SOM is extremely complex. Because of this complexity, which intrigued early chemists and soil scientists, various approaches have been used to subdivide SOM into fractions, each of which provided a different picture, resulting in increased misconceptions.

Soils are extremely heterogeneous materials and environments due to: (a) their diverse mineral and organic constituents and that of their living inhabitants, (b) their size spectrum from the nanometer to the meter scale (Fig. 13.1), and (c) the complex 3-D spatial arrangements of these constituents, which define a network of voids of various sizes, more or less filled with water and air. Soils are the juxtaposition of many microenvironments in which the constituents and the physical and physicochemical conditions (e.g., redox potential) may differ considerably between microenvironments and average soil characteristics. Therefore the traditional and pragmatic approach for predicting soil properties and functions from their average composition and characteristics is very limited. This particularly applies to SOM dynamics and functions. Two categories of methods are used to account for the structural heterogeneity of soil, even at nanometer spatial scales: (1) fractionation methods that attempt to separate SOM into meaningful subsets, and (2) visualization methods used to observe SOM in its environment. An integrated understanding of organic matter functions within a specific pedological context is best attained by the combination of these two approaches, along with detailed molecular analysis of SOM composition and measurement of its dynamics.

13.2 Quantifying soil organic matter

SOM quantification is generally performed on soil that has passed a 2 mm sieve, thus excluding plant roots, litter, coarse plant debris, or large fauna. It is rare and extremely difficult to measure SOM directly.

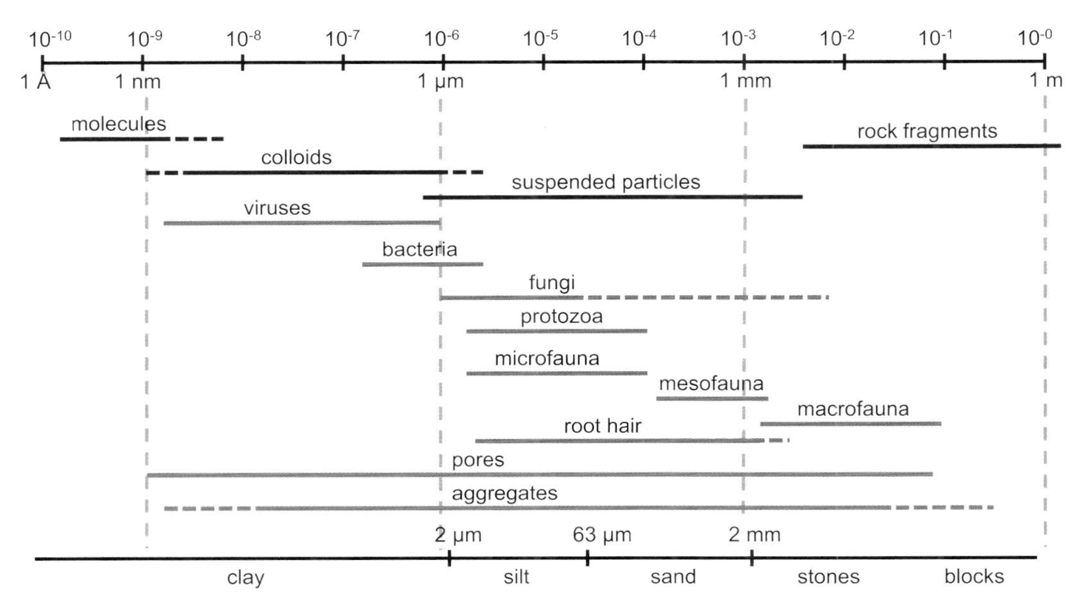

FIGURE 13.1 Size of soil constituents and structural features. *(Adapted from Totsche and Kogel-Knabner, 2004.)*

Its content can be roughly estimated by determining loss on ignition at temperatures between 300 and 600°C, at contrasting durations. Although it is a relatively easy method, no common protocol exists, and the result can be influenced by a variety of factors and is soil type dependent (Hoogsteen et al., 2015). Most methods measure its most abundant element, C, and multiply the SOC content by a factor of 1.72 to 2 to obtain the SOM content. These conversion factors correspond to the fact that the C content of SOM usually ranges from 50% to 58% for most organic materials. Organic C contents can be determined by wet digestion, using dichromate to oxidize organic molecules, as in the classical Walkley-Black method (Nelson and Sommers, 1996), or by dry combustion. Both methods convert all soil C to CO_2, which is then quantified by various methods, such as infrared detection or thermoconductivity. If the soil contains carbonates, either combustion may be performed after acid pretreatment to remove the carbonates (following adjustment of pH in acid solution or performing an acid fumigation) or the total C needs to be corrected for the inorganic C content measured separately.

13.3 Fractionation methods

The complexity of SOM has historically led to numerous attempts to separate SOM into subsets or fractions, homogeneous in their characteristics or in their dynamics. The aim of fractionation methods is to subdivide the complex continuum of SOM to identify and characterize its constituents and to isolate "functional pools." These correspond to the following:

1. Kinetic pools, i.e., subsets of SOM characteristically distinct from other pools, each characterized by specific turnover rates and pool size (Christensen, 2001);
2. Pools of SOM that are controlled by a specific stabilization mechanism (e.g., biochemical recalcitrance, spatial inaccessibility, and organomineral interactions) (von Lutzow et al., 2007);
3. Pools of SOM that have a major implication for a given soil function, either biological functions such as the recycling of C and N by biodegradation and mineralization, or physical and chemical functions such as cation exchange capacity or pollutant sorption (Feller et al., 2001).

SOM dynamics are usually described in models by several kinetic pools, three pools being the most frequent: (a) one pool, named either labile or active, is assigned turnover rates of weeks to months; (b) an intermediate or slow pool with turnover rates of years to decades; and (c) a passive to inert pool with turnover rates of centuries to millennia (Falloon et al., 1998). Much research has been devoted to identifying SOM fractions matching these conceptual pools, under the well-known headline of "Modeling the Measurable or Measuring the Modelable" (Christensen, 1996; Elliott et al., 1996). Achieving this would considerably increase the efficiency of the calibration and testing of models.

Fractionation methods separate the SOM continuum into fractions by either biological, chemical, or physical approaches. The separated fractions can be characterized by: (a) their size (soil mass, % of soil C, N, S, or P), (b) their elemental and biochemical composition, as well as their physicochemical or biological characteristics, and (c) their C and N turnover rates. In addition to being meaningful (i.e., able to separate functional pools or distinct organic components), fractionation methods should be: (a) adaptable to soils with different textures, mineralogy, pH, and water logging conditions; (b) sensitive to changes in management and environmental conditions; and (c) repeatable and standardizable.

13.3.1 Chemical and thermal fractionation

13.3.1.1 Extraction methods

The objective of extraction methods is to bring as much as possible of the target SOM fraction into solution for analysis (see Supplementary Table S13.1).

13.3.1.1.1 Dissolved organic matter

Dissolved organic matter (DOM) is typically defined as SOM smaller than 0.45 μm. Either it is obtained with different extractants, from cold water to saline solutions with different ionic strengths after mixing a given mass of soil with the extractant and centrifuging the slurry to recover DOM, or it is recovered directly in situ by using suction cups or lysimeters (Haynes, 2005; Zsolnay, 2003). DOM generally represents less than 1% of total organic C in agricultural soils and a few percent in forest soils. It is made of diverse molecules from low molecular weight to colloidal moieties and is heterogeneous in turnover rates, including labile components with SOM that has decadal turnover times (Haynes, 2005; Kalbitz et al., 2000; Zsolnay, 2003).

13.3.1.1.2 Solubility in bases and acids: humic substances and extracts

The first fractionation of SOM was introduced in the late 18th century by Achard (1786). Alkali solutions, such as NaOH or pyrophosphate ($Na_4P_2O_7$), solubilize large amounts of SOM (up to 80% of TOC with NaOH; Stevenson, 1982). Oxygen-containing functional groups (carboxyl, alcohols, phenols) are ionized at high pH, and thus molecules rich in such functional groups are rendered soluble. In contrast, nonpolar alkyl-C molecules are not extracted by alkali solutions. The enzymatic and oxidative depolymerization of plant polymers increases the abundance of phenolic and carboxylic functional groups; thus the solubility of plant material in alkali increases with decomposition. The organic molecules solubilized by alkali have been named *humic substances* but are, strictly speaking, an operational definition of the organic matter extracted using the method (Kleber and Johnson, 2010; Kleber and Lehmann, 2019; Lehmann and Kleber, 2015). Lowering the pH below 2 and using HCl precipitates organic molecules named *humic acid extracts*, whereas others, named *fulvic acid extracts*, remain soluble. The material insoluble in alkali is named *humin*. Extensive studies have been performed since the 19th century to characterize SOM using this approach. The relative abundance of humic and fulvic acid extracts varies according to soil type and was used to decipher and characterize pedogenesis (Duchaufour, 1976). Although these methods allow the extraction and thus analyses of large proportions of SOM, they suffer serious drawbacks: (a) formation of artifacts — the alkaline extraction products do not represent SOM as it is present in situ as extracted organic molecules have distinct characteristics from those existing in soil in an adsorbed state (Baldock and Nelson, 2000) or found using spectromicroscopy (Lehmann et al., 2008); (b) alkaline extractions cannot distinguish between molecules that are issued from secondary synthesis in soils and molecules inherited from biological tissues (nonhumic substances) by oxidative decomposition (Kleber and Lehmann, 2019); and (c) although it was thought that humic substance extracts consist of pedogenic macromolecules, it appears that both humic and fulvic acid fractions are a mixture of biomolecules (Kelleher and Simpson, 2006), including inherited macromolecules and supramolecular associations of small organic entities (Piccolo, 2002). This separation procedure has been extensively used to study the interaction between SOM and pollutants because the chemical reactivity of SOM constituents can be

TABLE S13.1 Chemical fractionation methods.

Procedure	Reagent	Fraction targeted	Extracted compounds	Abundance of fraction (% TOC)	Assumed turnover or stabilization process	Uniform residence time?	Assumption verified?
Extraction	Cold water or dilute saline solutions	Dissolved organic matter	Sugars, amino acids, phenols ...	0.2–2%	Extracted fraction is labile	no	no
	Alkali solutions (NaOH 0.1–0.5M)	Humic substance extracts	Mixture of biogenic compounds (carbohydrates, proteins, lipids ...)	30–80%	Solubilized fraction is stable	no	no
	Organic solvents	Lipids	Lipids	2–6%	–	no	–
Hydrolysis	Hot water		Carbohydrates, amino acids	1–5%	Extracted fraction is labile, management sensitive		yes
	Concentrated acids (6N HCl, 1M H_2SO_4)		Carbohydrates, proteins, fatty acids	30–80%	Resistant fraction is chemically recalcitrant	no	Yes, fraction is millennia older than SOM
Oxidation	$KMnO_4$			10–30%	Extracted fraction is labile, management sensitive	?	?
	H_2O_2		Wide range, except natural waxes and pyrogenic compounds	20–93%	Resistant fraction is chemically recalcitrant or protected in micropores or by adsorption	no	Yes, fraction is millennia older than SOM
	$Na_2S_2O_8$		Wide range, except long-chain aliphatics	16–99%	Resistant fraction is chemically recalcitrant or protected in micropores or by adsorption	?	?
	NaOCl		Wide range but alkyl C and pyrogenic compounds	26–96%	Resistant fraction is chemically recalcitrant or protected in micropores or by adsorption	Yes	Yes, fraction is millennia older than SOM

directly investigated; however, it has been recommended that these approaches be set aside as a method to understand the processes of SOM formation and persistence in soils (Lehmann and Kleber, 2015).

13.3.1.2 Hydrolysis methods

Hot water (i.e., 80−100°C) is used to isolate a fraction of SOM that is hypothesized to be readily available to microorganisms (Haynes, 2005; see Supplementary Table S13.1). Hot water extracts typically correspond to 1% to 5% of SOC and primarily contain carbohydrates and N-rich compounds. This fraction is sensitive to land use and agricultural practices and is thus used as an early indicator of changes in management and as a proxy for organic compounds promoting soil aggregation (Haynes, 2005). Water-soluble compounds are mainly composed of those with rapid turnover (Balesdent, 1996).

Concentrated solutions of acids, such as 6N HCl or 1M H_2SO_4, have been extensively used to extract and characterize soil C and N based on their reactivity (Paul et al., 2006). In the presence of acids and usually at high temperatures organic bonds are cleaved by dissociated water molecules. Fatty acids, proteins, and polysaccharides (including hemicellulose and cellulose) are thus susceptible to acid hydrolysis treatment, while long-chain alkyls, waxes, lignin, and other aromatics, including pyrogenic C, are resistant. Moreover, the composition of the HCl insoluble fraction could be affected by the formation of artifacts, the so-called melanoidins, by recombination of carbohydrates and proteins following hot acid treatment. Reflux acid hydrolysis with 6N HCl at 116°C typically solubilizes 65% to −80% of soil N (Bremner, 1967) and 30% to 87% of SOC (Paul et al., 2006). On average, acid hydrolysis−resistant SOC is 1200 to 1400 years older than bulk SOC (Paul et al., 2006). It was proposed as a proxy for the passive pool of SOC that would separate chemically recalcitrant organic C. The size of the acid-resistant organic fraction varies with soil management, such as afforestation or cultivation, showing that the amount of SOC remaining after acid hydrolysis is due not only to its intrinsic chemical properties but also to its interaction with minerals.

13.3.1.3 Oxidation methods

As biodegradation of SOM is an oxidative process, mild oxidants, such as $KMnO_4$ (0.33 or 0.03 M), have been proposed to mimic it (Blair et al., 1995). Potassium permanganate oxidizes 13% to 28% of SOC, the abundance of this fraction being sensitive to land management (Culman et al., 2012). While permanganate oxidizable soil C reflects a relatively processed pool of organic C, the turnover rate of organic C resistant to $KMnO_4$ has not been measured to date, and it is not clear if this method offers a proxy for a SOC kinetic pool.

Strong oxidants have traditionally been used to eliminate SOM prior to textural or mineralogical analyses. As not all SOM can be oxidized, these methods have been proposed to isolate a kinetically stable organic C pool. Hydrogen peroxide (H_2O_2, 10−50%), disodium peroxydisulfate ($Na_2S_2O_8$ 8%, buffered with $NaHCO_3$), and sodium persulfate (NaOCl, 6%) oxidize most SOM (see Supplementary Table S13.1) with variations depending on soil constituents and pH. Disodium peroxydisulfate and sodium persulfate are considered more efficient because of the additional dispersing action of sodium, which increases the accessibility of SOM to the oxidant in the soil matrix (Mikutta et al., 2005). Organic compounds resistant to these oxidants are essentially long-chain alkyl and pyrogenic C. Resistance to wet oxidation seems to also depend on soil microaggregation and the presence of sorbents such as clay minerals and oxides. Centennial-old SOM from a long-term bare fallow was not enriched in H_2O_2 or NaOCl-resistant organic matter (Lutfalla et al., 2014). However, organic matter resistant to these oxidants

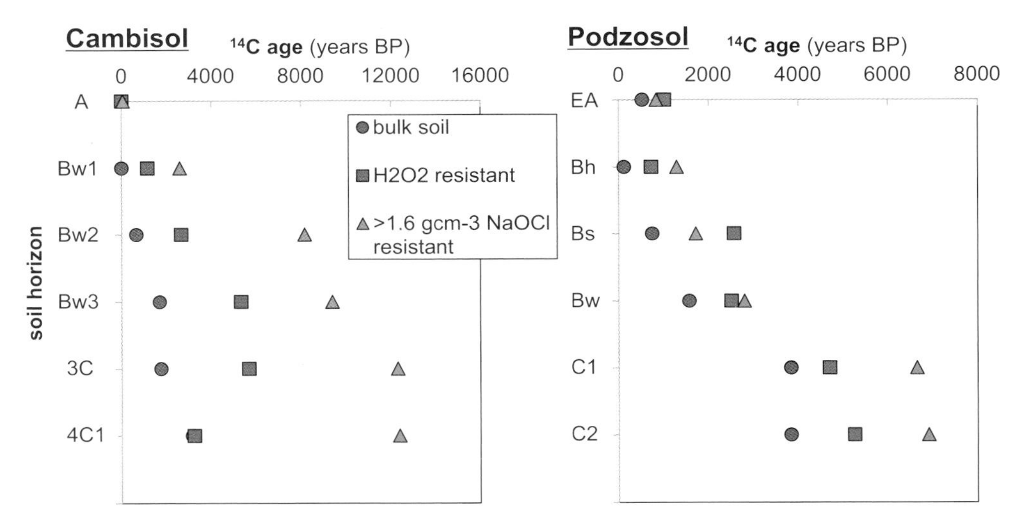

FIGURE 13.2 Age of bulk soil organic carbon and organic fractions resistant to oxidation in two temperate soils, a dystric Cambisol *(left)* and a Podzosol *(right)*. *(Data from Kögel-Knabner et al., 2008.)*

was found to be older than average SOM by 75 to 3000 years (Eusterhues et al., 2005; Kleber et al., 2005; Fig. 13.2). Strong oxidants therefore have the potential to isolate a kinetically stable fraction.

13.3.1.4 Destruction of the mineral phase

Hydrofluoric acid (HF) hydrolyzes silicates, carbonates, and most oxides but does not affect apatite, spinel, and zircon. Treatment of soil with HF is assumed to free the OM adsorbed to hydrolyzed minerals or coprecipitated with them. If the previously adsorbed or coprecipitated OM is solubilized, it may be lost with the HF solution. Loss of OC after HF treatment can be used to quantify nonparticulate SOM adsorbed to or coprecipitated with minerals. Application of HF treatment to subsoil horizons dissolved up to 92% of SOC, showing that most of the SOM in these horizons is adsorbed to minerals (Eusterhues et al., 2003). However, moderate losses of SOC (10–30%) and small alterations of the chemical composition are found in simple organic compounds and litter layer samples after HF treatment, showing that SOM can be affected by the treatment as well (von Lutzow et al., 2007). The HF soluble fraction is not always older than HF-resistant SOM; hence HF treatments do not isolate kinetic pools. This chemical fractionation was proposed for the quantification of organic matter that is stabilized by adsorption to minerals (Eusterhues et al., 2007). However, a recent study indicated that C losses could not be explained by soil properties. Sanderman et al. (2017) suggested that the results obtained with this method should be interpreted carefully, taking into consideration the specific physicochemical environment and stabilization mechanisms operating in the soil under study. Moreover, as the HF reagent is extremely toxic, this fractionation procedure should be adopted only with adequate safety considerations.

Sodium dithionite solutions are known to solubilize all iron and aluminum oxides, whereas sodium oxalate is used to solubilize poorly crystallized iron and aluminum oxides (Shang and Zelazny, 2008).

These reactants thus break the bonds between oxide minerals and adsorbed or coprecipitated organic compounds. Although these methods solubilize a fair amount of SOM in different soil types, the use of a bicarbonate buffer or of the oxalate itself adds C in solution and impedes any measurement of the turnover of the extracted or the resistant SOC by methods such as ^{13}C natural abundance or ^{14}C dating.

13.3.1.5 Thermal analysis methods

Thermal analysis methods consist of gradually heating a soil sample along a temperature gradient in the presence of oxygen or under an inert atmosphere (pyrolysis) and tracking the gases produced (evolved gas analysis methods) or the energy generated (calorimetric methods) during the combustion or cracking of SOM (Plante et al., 2009). Thermal analysis methods showed that labile organic matter is richer in energy compared to more stable organic matter (Barré et al., 2016; Plante et al., 2011). Several studies have also shown that thermally stable organic matter is on average older (Plante, 2013; Sanderman and Grandy, 2020). Recent results have reported that Rock-Eval® thermal analysis, in which the sample is successively subjected to pyrolysis and oxidation steps, allows quantitatively linking thermal analysis results to SOM biogeochemical stability. Barré et al. (2016) observed that several Rock-Eval® parameters (e.g., thermal stability under pyrolysis and oxidation conditions, amount of organic matter cracked during the pyrolysis stage, etc.) correlate with biogeochemical stability measured in situ using long-term bare fallow trials. Using these data, Cécillon et al. (2021) developed a machine-learning model ("PARTY$_{SOC}$") to quantitatively relate biogeochemical stability of SOM to Rock-Eval® thermal analysis results (Fig. 13.3). The performance of this machine-learning model is expected to improve in the coming years with the inclusion of additional data.

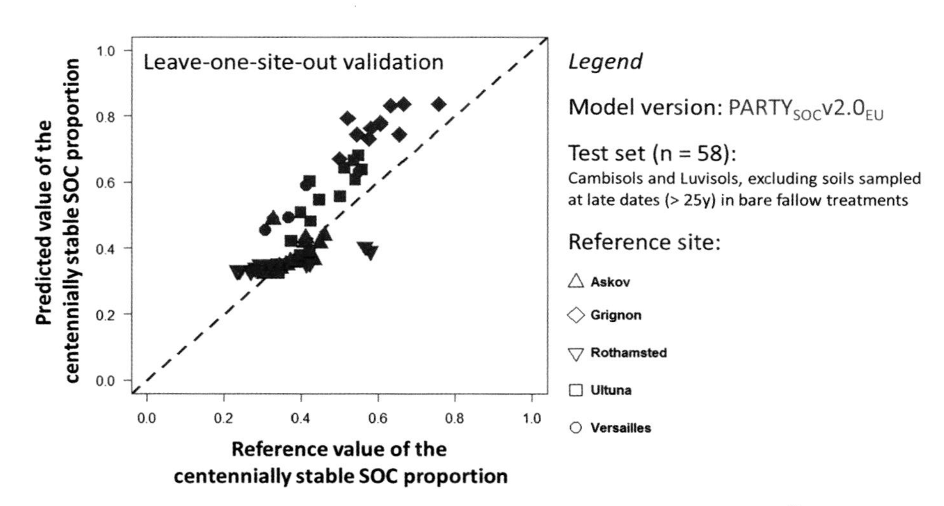

FIGURE 13.3 Performance of PARTY$_{SOC}$v2.0 machine-learning models based on Rock-Eval® thermal analysis for predicting the centennially stable organic carbon proportion in topsoils. Results of the leave-one-site-out procedure ($R^2 = 0.71$; RMSEP = 0.11) show that the model can accurately predict the proportion of centennially stable SOC proportion in surface layers of Cambisols and Luvisols. *(Adapted from Cécillon et al., 2021.)*

13.3.2 Physical fractionation

13.3.2.1 Introduction and definitions

Physical fractionation separates subsets of soil according to physical criteria such as size, density, or magnetic susceptibility. This is done after various degrees of dispersion are applied to soil to break bonds between elements of the soil structure. Physical fractionation procedures seek to minimize chemical alterations and aim to separate primary soil particles, primary organomineral, or secondary organomineral associations. Primary organomineral associations relate to the "primary structure of soils as defined by soil texture" (Christensen, 1992). They result from the association of SOM with primary mineral particles and are isolated after complete dispersion of soils. Secondary organomineral associations are aggregates separated after limited dispersion of soil. In many soils the complete dispersion of particles, especially clay-sized particles, is not possible if organic matter is not oxidized; hence true primary organomineral complexes cannot be isolated (Chenu and Plante, 2006). However, these terms will be used here for clarity and their limits discussed when needed.

Physical fractionation pursues various objectives, such as the separation of soil organic constituents in their original state, i.e., soils without chemical alteration. This is one of the reasons behind the separation of particulate organic matter (POM), which is essentially composed of weakly decomposed plant debris, root fragments, fungal hyphae, spores, seeds, fecal pellets, and faunal bodies. Another objective is to separate SOM from its physical surroundings under the assumption that the association of organic matter with minerals, and their location within the 3-D soil architecture, controls SOM dynamics through the processes of physical or physicochemical protection. Physical fractionation methods have revealed a considerable amount of information on SOM in the last decades leading to major advances in the understanding of soil C dynamics and stabilization processes.

13.3.2.2 Soil dispersion

Physical fractionation schemes represented in Fig. 13.4 imply some degree of soil dispersion prior to application of the separation criteria. A variety of dispersion techniques and protocols exist, complicated by a lack of standardization. Indeed, standardization is seldom possible using energy units (e.g., how to express the energy provided by agitating a certain mass of soil in water). Furthermore, a given objective of dispersion (e.g., to disperse all particles >50 μm diameter) will require different energy inputs depending on the soil type and land use.

Weak dispersion of soil is achieved by immersing it in water with no or little agitation. The least dispersion is obtained using field capacity moist soil as is done to measure aggregate stability (Angers and Mehuys, 1993; Arias et al., 1997). Strong dispersion of soil can be achieved by mechanical agitation of soil in water with the addition of glass or agate beads (e.g., Balesdent et al., 1998) or by the use of ultrasonic vibrations. The energy released by ultrasonic probes depends on the probes themselves, their diameter, depth of immersion, and volume of the vessel. Calorimetric measurements can be used to calibrate and standardize ultrasonic dispersion (Schmidt et al., 1999). Excessive energy might cause breakdown of small plant remnants, such as POM or mica-type mineral particles. For a range of acidic to mildly acidic temperate soils, sonication energies between 450 and 500 J mL^{-1} were shown to completely disperse the soil without abrasion artifacts (i.e., to yield the same particle size distribution as the textural analysis; Balesdent et al., 1991; Schmidt et al., 1999). To avoid POM abrasion, soil dispersion can also be performed by: (1) agitation with glass beads to allow for the separation of POM and a fraction <50 μm,

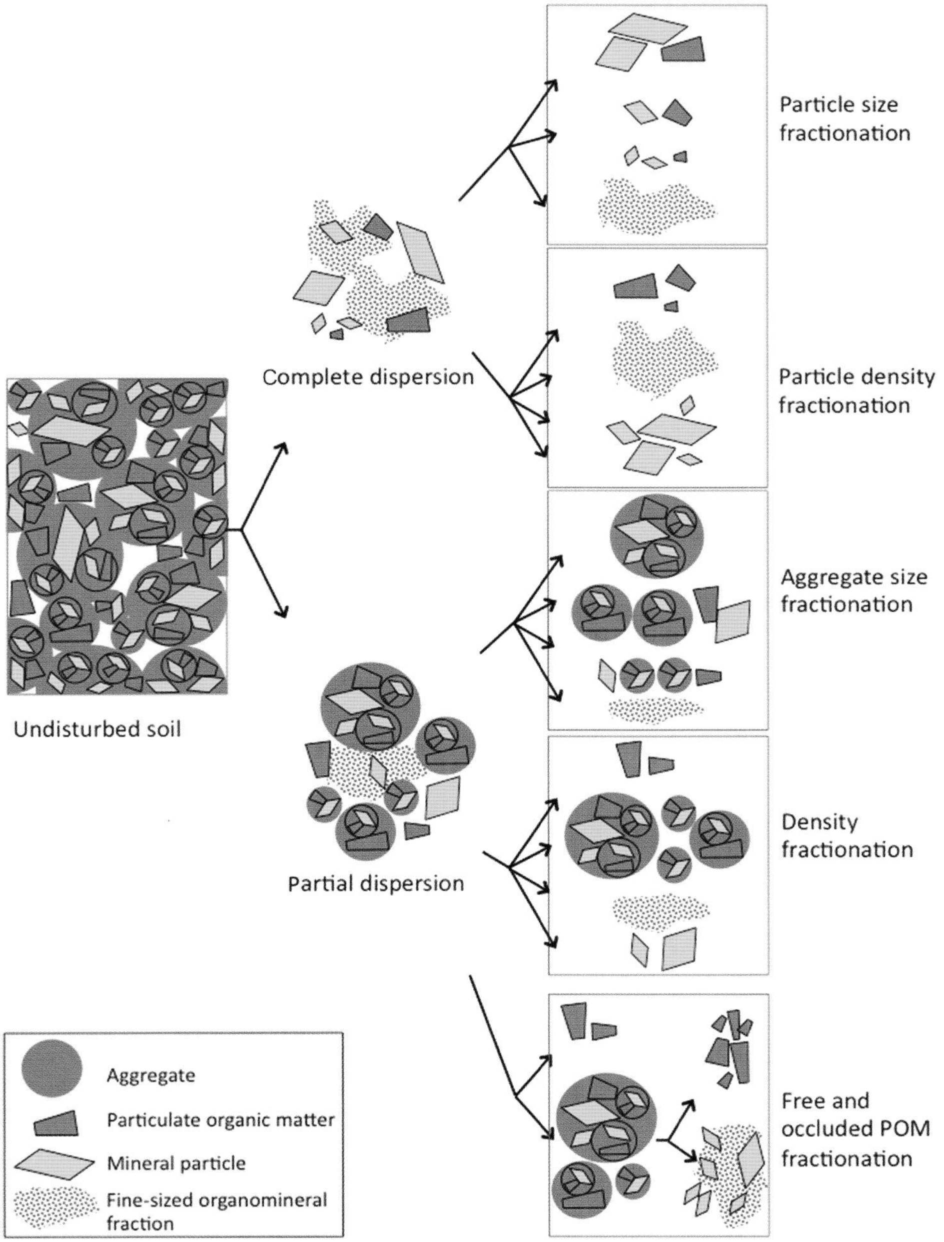

FIGURE 13.4 Schematic presentation of physical fractionation methods. Fine-sized organomineral fraction can be clay + silt fraction or clay size fraction depending on protocols.

and (2) an ultrasonic dispersion of the <50 μm fraction (Balesdent et al., 1998, 1991). Chemical dispersants, such as sodium hexametaphosphate and Na resins, have also been used to aid complete dispersion of soil (Feller et al., 1991). However, use of chemicals during physical fractionation should be cautiously considered and, if possible, avoided if composition and turnover of SOM fractions are to be characterized.

13.3.2.3 Particle fractionation

Once dispersed, soil particles may be separated by either size, density, or both (Fig. 13.4). For SOM, particle size fractionation relies on the idea that as decay induces fragmentation and as more processed organic matter associates with minerals, fine-sized OM should have longer residence times (Cambardella and Elliott, 1992; Christensen, 1987, 1992). This assumption has largely been confirmed, yet coarser fractions remain a mixture of POM and mineral sands and may also contain char, and clay size fractions are heterogeneous during turnover (Christensen, 1992). Even so, conceptualization and quantification of POM and mineral-associated organic matter (MaOM) fractions have since been developed (Lavallee et al., 2020).

Both particle size and density fractionation were introduced to separate POM from mineral particles of the same size, using either water as the flotation medium (Balesdent et al., 1988; 1998; Feller, 1979) or denser liquids. It is then possible to separate POM fractions in different decomposition stages (Balesdent, 1996; Fig. 13.5). Particle density fractionation assumes that the association of SOM with minerals, or absence thereof, determines its turnover time. Light fraction organic matter, which roughly corresponds to POM, is separated using organic liquids (bromoform), followed by dense mineral liquids such as NaI or $ZnCL_2$ (Bremer et al., 1995; Monnier et al., 1962; Strickland and Sollins, 1987). Sodium polytungstate ($Na_6[H_2W_{12}O_{40}]$) is now extensively used because it spans a wide range of densities (from 1.0 to 3.1 g

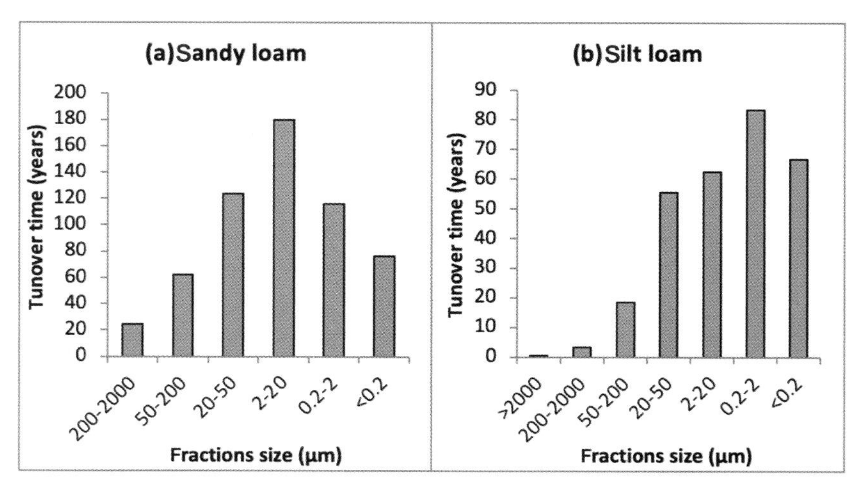

FIGURE 13.5 Turnover time of particle fractions in temperate cultivated soils. Particles were separated by size (fractions ≤50 μm) or by both size and density (POM, i.e., fractions >50 μm). Turnover time was estimated from ^{13}C natural abundance. *(From Balesdent, 1996; Balesdent et al., 1987.)*

cm^{-3}) with a low viscosity and little toxicity. Moreover, it can be recycled (Six et al., 1999). However, some solubilization of organic matter takes place, which is negligible when separating POM, but may amount to 20% of TOC when using it to fractionate fine-sized OM (Virto et al., 2008).

The nature of soil minerals affects their capacity to protect organic matter from decomposition. Therefore a second rationale for performing density fractionation is the separation of distinct mineral species with their associated organic matter (Sollins et al., 2006). Poorly crystallized aluminosilicates, feldspars, and iron oxides have been separated using solutions of sodium polytungstate based on their different densities (Basile-Doelsch et al., 2007). A limit to this approach is the frequent incomplete dispersion of particles in the protocols used, such that both aggregates and primary organomineral associations are being separated.

13.3.2.4 Aggregate fractionation

Aggregate size fractionation is based on the premise that: (1) organic matter binds mineral particles together such that aggregate size fractionation can then be used to separate the organic matter responsible for aggregation (e.g., Puget et al., 1995), and (2) aggregates of different sizes correspond to different microbial habitats and provide contrasting conditions for decomposition (Beare et al., 1994; Elliott, 1986). Aggregates can be separated after immersion in water of initially dry or wet soil and use of different intensities of agitation or slaking. Increasing the applied energies allows focus on differently sized aggregates since the smaller the aggregates, the higher the energy needed to break them. According to this concept, macroaggregates >250 μm (e.g., Elliott, 1986), sand-sized microaggregates (e.g., Besnard et al., 1996), or silt-sized microaggregates (e.g., Virto et al., 2008) can be separated (Fig. 13.4). A two-step procedure was introduced to separate sand-sized microaggregates located within macroaggregates (Six et al., 2000; Fig. 13.6). One limitation of aggregate fractionation is that the sized fractions contain true aggregates as well as free organic and mineral particles, and the contribution becomes larger as the size of the separated objects decreases.

The use of aggregate fractionation methods led to the development of an aggregate hierarchy in soils and the investigation of the complex interactions between soil structure and SOM dynamics (see review by Six et al. [2004]). Although SOM in aggregates of different sizes is characterized by different turnover rates (Puget et al., 1995; Six and Jastrow, 2002), this approach does not allow for the separation of distinct

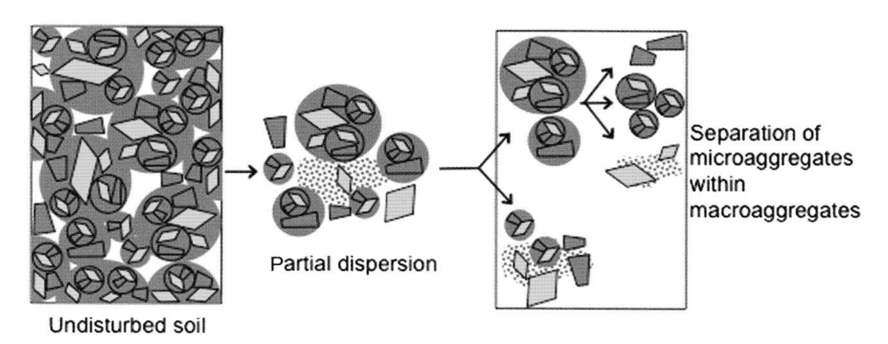

FIGURE 13.6 Schematic presentation of separation of microaggregates within macroaggregates. *(Method developed by Six et al., 2000.)*

SOC kinetic pools (Poeplau et al., 2018). It also does not allow for the isolation of SOM stabilized by aggregation as the sole stabilization mechanism because the SOM located within an aggregate can also be stabilized due to its biochemical recalcitrance and adsorption to minerals.

Organomineral fractions have a density intermediate between that of pure organic matter and pure mineral particles (Fig. 13.7; Chenu and Plante, 2006). Dense mineral solutions can thus be used to separate true aggregate fractions with intermediate densities from dispersed POM and mineral fractions (Fig. 13.4). This approach has shown that occluded POM is physically protected from biodegradation by virtue of its location within aggregates (Besnard et al., 1996; Six et al., 2000).

Several methods have been developed to peel aggregates with the idea that organic matter should be more accessible in the outer volume of aggregates and, consequently, has faster turnover rates (Bol et al., 2004; Wilcke et al., 1996). However, these methods only show gradients in soils where aggregates turn over very slowly.

A comparison of 20 different SOC fractionation methods was performed to assess their suitability to isolate SOM fractions with varying turnover rates (Poeplau et al., 2018). Particle-size fractionation significantly outperformed aggregate-size fractionation since coarser aggregates contain finer aggregates and OM particles of various sizes. Coarse light SOC separated by a combination of size and density fractionation approaches (Balesdent, 1996) had the highest turnover rate, while oxidation-resistant SOC left after extraction with NaOCl (Zimmermann et al., 2007) had the lowest turnover rate (Poeplau et al., 2018).

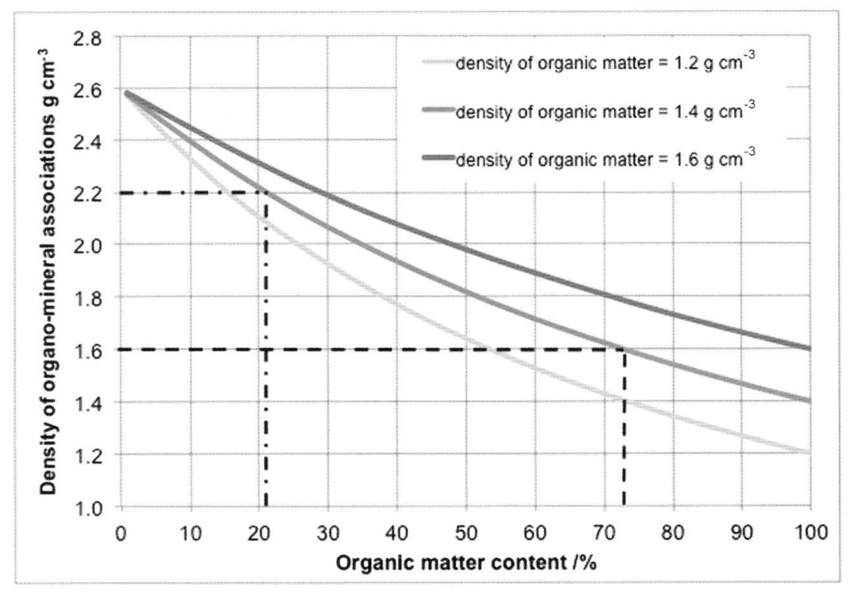

FIGURE 13.7 Calculated density of organomineral associations assuming a density of either 1.2, 1.4, or 1.6 g cm^{-3} for the organic matter and a density of 2.6 g cm^{-3} for the mineral phase. Dashed drop lines allow to visualize the OM content at the boundary when separations are performed at densities of 1.6 and 2.2 g cm^{-3}. *(Redrawn from Chenu and Plante, 2006.)*

13.4 Characterization methods

Molecular characterization of SOM composition is essential because most pedological processes operate at the molecular scale. The objective of the analysis of biochemical constituents, as well as molecular composition of SOM, is to gain a better understanding of their origin and transformation processes. Since the early 19th century, individual molecular compounds have been characterized after NaOH extraction (see above). Solubilization of SOM was necessary because of the technical limitations for the characterization of SOM in its solid state. The advent of technical developments allowing for the characterization of the composition of functional groups using solid-state nuclear magnetic resonance (NMR) spectroscopy and molecular composition using analytical pyrolysis, in addition to the findings of analytical bias with NaOH-extraction, has led to a significant reduction in the use of NaOH fractionation. In general, a combination of different methods is needed to obtain a good understanding of the origin and chemical composition of SOM. All methods presented in this section can be applied to bulk soils as well as to soil fractions, provided the sample amount is sufficient (generally several hundred milligrams: Fig. 13.8).

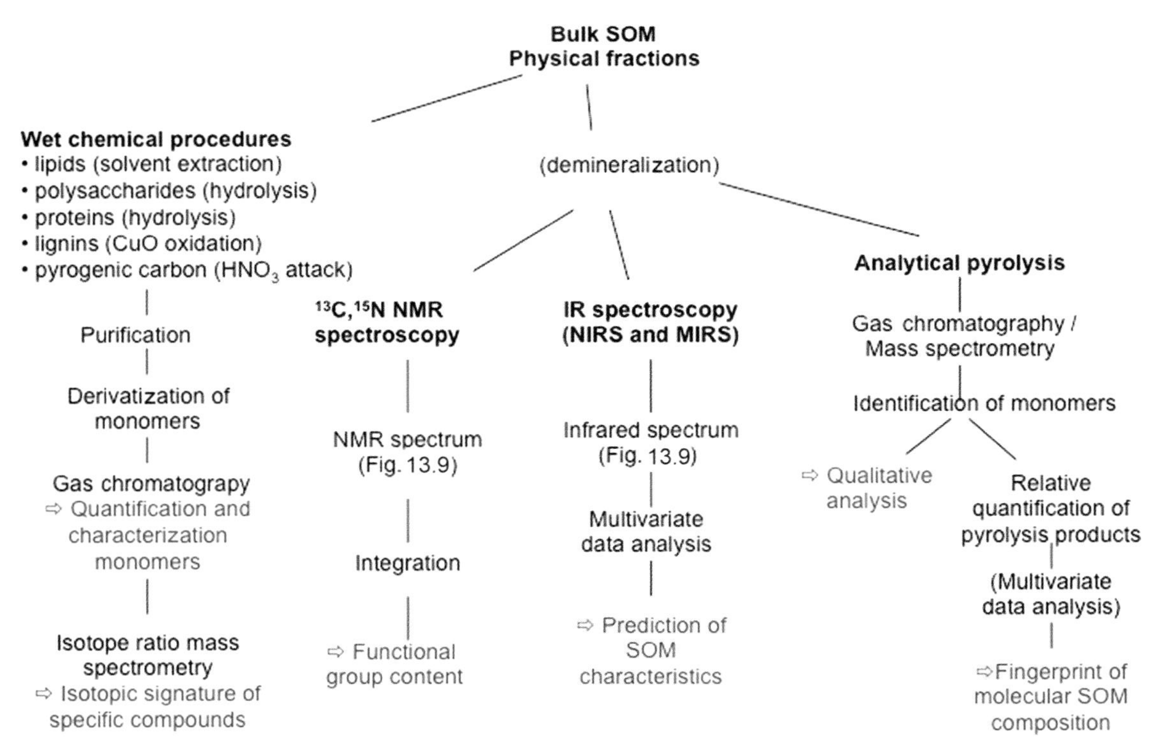

FIGURE 13.8 Analysis scheme and information to be obtained with biogeochemical methods commonly used for SOM analyses.

13.4.1 Wet chemical methods for analyses of biochemical SOM constituents

Wet chemical methods include solvent extraction for free lipids, as well as acid and base hydrolysis or chemical oxidation, to break macromolecules into their monomer constituents before analyses (Fig. 13.8). While the biochemical composition of pure organic matter in plant litter or organic soil horizons can be easily analyzed with wet chemical methods, the presence of mineral interactions hampers the wet chemical analysis of SOM compounds in mineral soil horizons. Less than 50% of SOM can usually be accounted for when using these methods (Hedges et al., 2000).

13.4.1.1 Carbohydrates

Carbohydrates are abundant biochemical constituents of SOM. They account for approximately 50% of plant litter entering the soil system (Kogel-Knabner, 2002). Although carbohydrates are labile compounds, which are usually rapidly metabolized by the microbial community, their degradation may be reduced by adsorption to the mineral matrix (Miltner and Zech, 1998), as well as by their incorporation into soil aggregates (Puget et al., 1999). To understand the dynamics and roles of carbohydrates in soil, it is important to know whether they are plant or microbial derived. Their monomers can be of specific origin (e.g., xylose and arabinose are of plant origin, whereas mannose and galactose are predominantly microbial) (Cheschire, 1979). Determination of carbohydrates in soil at the molecular level involves acid hydrolysis, purification, and derivatization of the monosaccharides, which are subsequently analyzed. Due to its stable crystalline structure, cellulose requires harsh hydrolysis conditions involving strong acids and high temperatures, such as hot H_2SO_4 (Cheschire, 1979). Cold H_2SO_4 (Cheschire, 1979), cold HCl (Kiem et al., 2000), or hot TFA (Amelung et al., 1996) are used for the hydrolysis of hemicelluloses and other amorphous polysaccharides. Determination of total monomer concentration after hydrolysis can be achieved by colorimetric methods, e.g., by using MBTH (3-methyl-2-benzothiazolinone hydrazone hydrochloride). Monomers can also be analyzed by liquid or gas chromatography after derivatization (Cheschire, 1979). Characterization of the carbohydrate composition of SOM in mineral-associated fractions showed a predominant microbial signature, indicating possible stabilization by interaction with the mineral phase (Derrien et al., 2006).

13.4.1.2 Aromatic compounds

Aromatic compounds in soil include plant-derived lignin and fire-derived pyrogenic C, which are degraded at a slower rate by soil microbes. The most frequently used method for soil lignin analysis is cupric oxide (CuO) oxidation, which includes chemical breakdown of the macromolecular structure by CuO in an alkaline medium and purification using solid phase extraction and subsequent gas chromatographic analyses of the released monomers (Hedges and Ertel, 1982). The CuO oxidation products include vanillyl, syringyl, and p-coumaryl units with their aldehyde, ketone, and acid side chains. The sum of the oxidation products is used as an indicator of soil lignin quantity, even if only a small part of the lignin is accessible with this procedure. The ratios between CuO products characterize lignin composition and its state of degradation (Kogel-Knabner, 2000). CuO oxidation can be combined with stable isotope analyses allowing for assessment of its dynamics (Dignac et al., 2005). The composition and fate of lignin was widely studied using this method. It is now generally accepted that soil lignin turnover occurs at a decadal timescale rather than at centennial timescales, as previously thought (Thévenot et al., 2010).

Several methods can be used to quantify pyrogenic C (i.e., condensed fire-derived C; Hammes et al., 2007). Wet-chemical analyses of pyrogenic C quantity and composition can be achieved by biomarker approaches such as the analyses of benzene carboxylic acids after digestion with TFA followed by nitric acid oxidation and silylation of the monomers before analysis by gas chromatography (Brodowski et al., 2005). Pyrogenic C consists of a continuum ranging from partly burned OM to completely graphitized materials. Therefore different methods target different sections of this pyrogenic C spectrum and have specific strengths and weaknesses that must be carefully evaluated before choosing one particular approach.

13.4.1.3 Aliphatic compounds

Aliphatic compounds in soil comprise free and ester-bound lipids. Free lipids are analyzed after solvent extraction using methanol/chloroform, methanol/dichloromethane mixtures, or an accelerated solvent extraction system at high temperature and pressure (Bligh and Dyer, 1959; Wiesenberg et al., 2004). Total free lipid content can be determined gravimetrically after solvent evaporation. Characterization of lipid composition is usually carried out by gas chromatography-mass spectrometry (GC-MS) after derivatization using silylation or transmethylation. GC-MS analyses identify alkanes, alkenes, alcohols, sterols, fatty acids, and aromatic hydrocarbons. Lipids are commonly used as biomarkers and have successfully been applied to assess SOM composition following land use change (Bull et al., 2000).

Phospholipid fatty acids are used as biomarkers for all living microorganisms in soil except archaea. They are fractionated after solvent extraction using silicic acid chromatography and analyzed by capillary gas chromatography (Zelles et al., 1992). Branched glycerol dialkyl glycerol tetraethers (GDGTs) may be used as indicators of past temperature and pH effects. They are separated by solid phase extraction and analyzed by high-pressure liquid chromatography/mass spectrometry (HPLC/MS) (Hopmans et al., 2000). Ester-bound lipids, which include cutin and suberin, are released by either saponification or transesterification, the former being more efficient (Mendez-Millan et al., 2010). Their relative abundance allows for the differentiation of root and shoot origin of SOM (e.g., Mendez-Millan et al., 2010).

13.4.1.4 Organic nitrogen

Organic N compounds make up most of the N in soils and are potentially important contributors to stabilized SOM. The most important N-containing compounds are, by order of decreasing abundance, proteinsamino acids, amino sugars, and DNA. Proteins and amino acids can be analyzed after acid hydrolysis using 6M HCl at high temperatures. Hydrolyzable proteins can be quantified calorimetrically by detection of the amino group of amino acids after their reaction with ninhydrin (Stevenson and Cheng, 1970). Quantification may also be carried out by gas chromatography after purification and derivatization. Amino acid enantiomers in soils may be used to assess the residence time of amino acids (Amelung and Zhang, 2001). The common hydrolysis procedure with 6M HCl leaves a major part of organic N unextracted, but the use of methanesulfonic acid (MSA) hydrolysis and nonderivatized amino acid and amino sugar quantification by ion chromatography with pulsed amperometric detection greatly increases recovered N (Martens and Loeffelmann, 2003). Amino sugars are only produced by microorganisms in soils, with muramic acid and glucosamine being used as biomarkers for bacterial and fungal residues, respectively (Joergensen, 2018). Amino sugars are quantified by HPLC or GC after acid hydrolysis using 6M HCl at high temperatures (Joergensen, 2018).

13.4.2 Spectroscopic methods to characterize functional groups of SOM

Spectroscopic methods may give a more complete picture of bulk SOM composition than wet-chemical methods. However, due to low sensitivity and signal interference by the mineral phase or paramagnetic compounds, demineralization is often recommended as a pretreatment. This is typically carried out with hydrofluoric acid (see Section 13.3.1), which was found to only slightly alter SOM composition while greatly improving spectral quality (e.g., Rumpel et al., 2006)

13.4.2.1 Infrared spectroscopy

Fourier-transformed infrared analyses are especially useful for elucidating degradation pathways of litter compounds (Wershaw et al., 1996). Near-infrared spectroscopy (NIRS) uses the near-infrared region of the electromagnetic spectrum (from about 0.75 to 2.5 μm wavelength), whereas midinfrared spectroscopy (MIRS) exploits the region between 3 and 8 μm. The interpretation of the spectra is difficult due to the overlap of inorganic soil components with signals with SOM absorption bands. This can be overcome by applying IR to SOM in chemical extracts or with demineralization prior to measurement (Rumpel et al., 2006; Fig. 13.9) or by subtracting IR spectra recorded before and after SOM removal from the samples (Skjemstad et al., 1996). Infrared peaks of soils and SOM are often too complex to provide quantitative data; thus multivariate data analysis is coupled with regression modeling to model relationships between two sets of measurements to predict certain parameters from IR spectra (Nduwamungu et al., 2009). Such predictions can reduce the time and costs of the analyses of the predicted parameter. MIRS or NIRS in combination with prediction modeling has been used to predict lignite C contribution to mine soils (Rumpel et al., 2001), pyrogenic C content (Cotrufo et al., 2016), and POM and resistant fractions of SOM (Baldock et al., 2013), as well as bulk SOM composition (Terhoeven-Urselmans et al., 2006).

13.4.2.2 Solid-state NMR spectroscopy

Solid-state NMR spectroscopy provides information on an atomic level that depends on the chemical environment of the observed nucleus. The technique gives an overview of the functional group composition of C- and N-containing compounds in SOM (Kogel-Knabner, 2000). Two NMR pulse programs may be used for solid samples: (1) cross-polarization magic angle spinning (CPMAS), which detects C via its interaction with H atoms, and (2) single pulse or Bloch decay experiments, which detect C directly. As NMR spectroscopy is an expensive, time-consuming method due to the low abundance of C compared to H, CPMAS is generally preferred because it requires less measurement time. However, at high magnetic fields and low spinning speeds, spinning sidebands can overlap with relevant signals and obscure the intensity distribution (Knicker, 2011). Aromatic C with low content of H-atoms, close to C, may not be seen by this method. Despite these drawbacks, ^{13}C and ^{15}N CPMAS NMR spectroscopy has been widely used in soil science, assuming that quantitative information about alkyl, O-alkyl, and aromatic and carboxylic groups present in SOM may be obtained with the right NMR parameters (Knicker, 2011; Fig. 13.9).

Conventional solid-state ^{13}C CPMAS NMR analyses are of great use for characterizing biodegradation of plant litter (Kogel-Knabner, 2002), the state of SOM biodegradation (Baldock et al., 1997), and the contribution of pyrogenic C in soil (Knicker, 2011) or for evaluating the changes in SOM composition with soil management (Audette et al., 2021). It can be used in combination with ^{13}C-labeled substances to

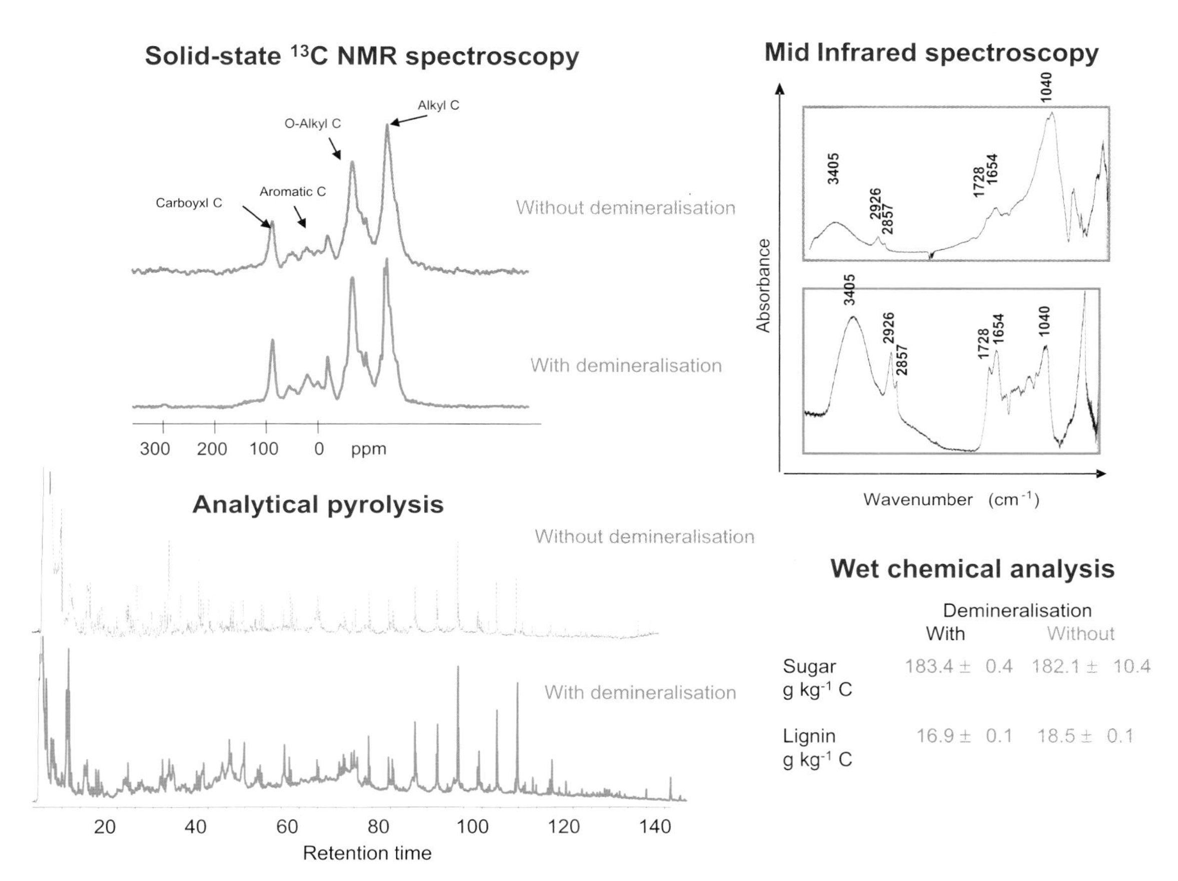

FIGURE 13.9 Characterization of SOM in topsoil (A horizon) of a Dystric Cambisol under forest with four methods before and after demineralization. *(Data from Rumpel et al., 2006.)*

study the fate of plant material in terrestrial environments (Kelleher et al., 2006). ^{15}N CPMAS NMR spectroscopy is not as widely used given that most organic N measured by this method occurs in soils in the form of amide. NMR spectroscopy is a nonsensitive method because NMR spectra consist of the accumulation of several hundred or thousands of signals (Knicker, 2011). ^{31}P NMR spectroscopy is applied only in liquid state due to the low soil P concentration. The technique allows for the identification of organic P forms, which comprise 35% to 65% of soil P (Cade-Menun, 2017). To overcome the limitations of conventional NMR, novel (high resolution) forms of NMR have been developed. One of these techniques, known as comprehensive multiphase (CMP) NMR, now allows analyzing soil in its natural swollen state. The technique may be used to study SOM composition at all its physical states allowing the analysis of liquid, semisolid, and solid components within intact and unaltered soil samples and has the potential to fundamentally improve our understanding of the SOM formation processes involving soil biota, minerals, and water (Masoom et al., 2016).

13.4.3 Molecular analysis of SOM

Molecular analyses of SOM in solid samples can be achieved by analytical pyrolysis or thermochemolysis. These two techniques are based on the heating of SOM with or without tetramethylammonium hydroxide (TMAH) under an N_2 atmosphere to thermally break down SOM. The pyrolysis products are analyzed by GC-MS or pyrolysis molecular beam mass spectrometry (py-MBMS). Carbohydrate-derived compounds can be distinguished from lignin-derived compounds, N-containing, aliphatic, and unspecific compounds. The interpretation of pyrolysis data requires a detailed knowledge of pyrolysis behavior of the compounds under study, as many pyrolysis products can originate from a SOM component and the presence of thermal secondary reactions. Choice of polarity of the chromatographic column is decisive: polar compounds (i.e., sugars, lignins, and proteins) should be analyzed on polar columns, whereas for polyaromatic and aliphatic compounds, an apolar column is necessary (Dignac et al., 2006). Py-MBMS is a more rapid analysis that does not result in losses on the GC columns and is especially useful when used in conjunction with multivariate analysis to exploit the wealth of data recorded. Major drawbacks of pyrolysis techniques include problems related to the presence of the mineral phase, such as retention of SOM compounds and catalytic reactions induced by clay minerals (Kogel-Knabner, 2000).

Progress concerning the complexity of SOM may be made following the application of ultra-high-resolution electrospray ionization Fourier transform ion cyclotron resonance—mass spectrometry (FTICR-MS) (Sleighter and Hatcher, 2007). This application, unlike conventional mass spectrometry, allows for molecular weight determination of large macromolecules provided they are soluble. It is also possible to obtain chemical formulas of organic matter compounds (Sleighter and Hatcher, 2007). The data-rich mass spectra obtained with this approach need to be analyzed by statistical and graphical tools. Water-soluble fractions of SOM have been characterized, and condensed aromatic structures have been detected in DOM, pore water, and river- and groundwater from a fire-impacted watershed, providing evidence of microbial dissolution of pyrogenic C (Hockaday et al., 2007).

13.5 Visualization methods

The objectives of visualization methods are twofold: (1) to gain morphological information on SOM to better understand its dynamics or roles (e.g., the state of decomposition of plant debris), and (2) to locate SOM within its natural environment and to study its relationships with mineral soil constituents, its accessibility to decomposers (Védère et al., 2022), or its contribution to soil structure. Visualization methods allow for the interpretation of the spatial heterogeneity and complexity of soils and can be distinguished on the basis of:

- their spatial resolution that determines which constituents of SOM can be observed;
- how SOM is distinguished from minerals or voids; and
- the information given on the elemental or chemical nature of SOM.

Methods for the visualization of SOM, its principles, resolution, and associated preparation are presented in Supplementary Table S13.2.

TABLE S13.2 Visualization methods for soil organic matter.

Category	Name	Principle and radiation used	Image	Resolution	Sample preparation	Visualization of organic matter	Additional information on organic matter
Light microscopy	Stereomicroscopy	Incident light	3-D	≈10 μm	None (fresh samples)	Shape of coarse organic objects or organisms, color	
	Light microscopy bright field	Transmitted light	2-D	<1 μm	Thin sample, either a deposit on glass slide or thin section of sample embedded in resin	Shape, color of organic matter	
	Epifluorescence microscopy	Transmitted light at a given wavelength that excites fluorescence in the sample	2-D	0.2 μm	Thin sample, either a deposit on glass slide or thin section of sample embedded in resin	Fluorescent stains: biological stains, immuno-stains	Autofluorescence of organic matter (and not minerals)
	Confocal (laser) microscopy	Transmitted light at a given wavelength that excites fluorescence in the sample	3-D	0.16 μm	Fresh sample (hydrated), but staining required	Fluorescent stains: biological stains, immuno-stains	Autofluorescence of organic matter (and not minerals)
Scanning probe microscopy	Atomic force microscopy (AFM)	A cantilever scans the surface of the sample and reacts to forces between surface atoms of the sample and its tip	3-D	1 nm	No preparation, can be hydrated, but flat surface needed	Shape of organic molecules deposited on mineral surfaces	Also gives information on surface properties such as hydrophobicity
Electron microscopy	Scanning electron microscopy (SEM)	Electrons reflected by surface of the sample allow a	3-D	10 nm	Fresh (environmental SEM), fresh and frozen (cryoSEM), air	Shape of organic matter	Backscattered electrons give information on atomic number contrast,

Continued

TABLE S13.2 Visualization methods for soil organic matter.—cont'd

Category	Name	Principle and radiation used	Image	Resolution	Sample preparation	Visualization of organic matter	Additional information on organic matter
		3-D image to be reconstructed			dried (conventional SEM) (under vacuum)		energy-dispersive X-ray spectroscopy on elemental composition
	Transmission electron microscopy (TEM)	Electrons that go through the sample. Image is issued from transmitted and diffracted electrons	2-D	0.2 nm	Thin deposit on a grid or ultrathin section of sample embedded in a resin (under vacuum)	Shape of organic matter, reaction to specific stains	Energy-dispersive X-ray spectroscopy gives information on elemental composition. Energy loss spectrum gives information on elemental composition and chemical bonding
X-ray spectromicroscopy	Soft X-ray spectromicroscopy in the water window (TXM, STXM)	X-rays transmitted through the sample	3-D	30 nm	Thin samples, deposits, or thin sections of samples embedded in resin or shock frozen	Shape of organic matter, elemental distribution maps	Information can be obtained on binding states of elements using X-ray absorption near-edge structures (XANES or NEXAFS)
Secondary ion mass spectrometry	NanoSIMS	Incident ions detach sample ions of opposite charge which are collected and analyzed	2-D	150 nm	Dry and flat sample (under vacuum)	Elemental and isotopic distribution maps	
X-ray computed tomography	X-ray μCT	X-rays crossing the samples are attenuated, depending on density and atomic number of materials	3-D	1 μm	Moist sample, no preparation	3-D image of solid particles and voids	Not yet imaged except for large particles (>100 μm)

13.5.1 Light microscopy

It is possible to observe the soil fabric using transmitted light after embedding soil in a resin and preparing thin sections. This allows identification of several organic matter features based on their shape and color: e.g., plant remnants, fecal pellets, and organic matter coatings on mineral grains. The origin of SOM (e.g., plant vs. leachates), its processing by soil fauna or flora, and its degree of association with minerals can be identified at scales of several micrometers to millimeters (Fig. 13.10a). Incident light at selected wavelengths can excite fluorescence within the sample either due to autofluorescence of organic molecules (such as chitin) or to stains bound to organic molecules, the latter, for example, targeting polysaccharides. Epifluorescence microscopy (0.2 μm) has a better spatial resolution than conventional light microscopy and can be used to visualize and enumerate bacteria and fungi within soil fabrics (Nunan et al., 2001;

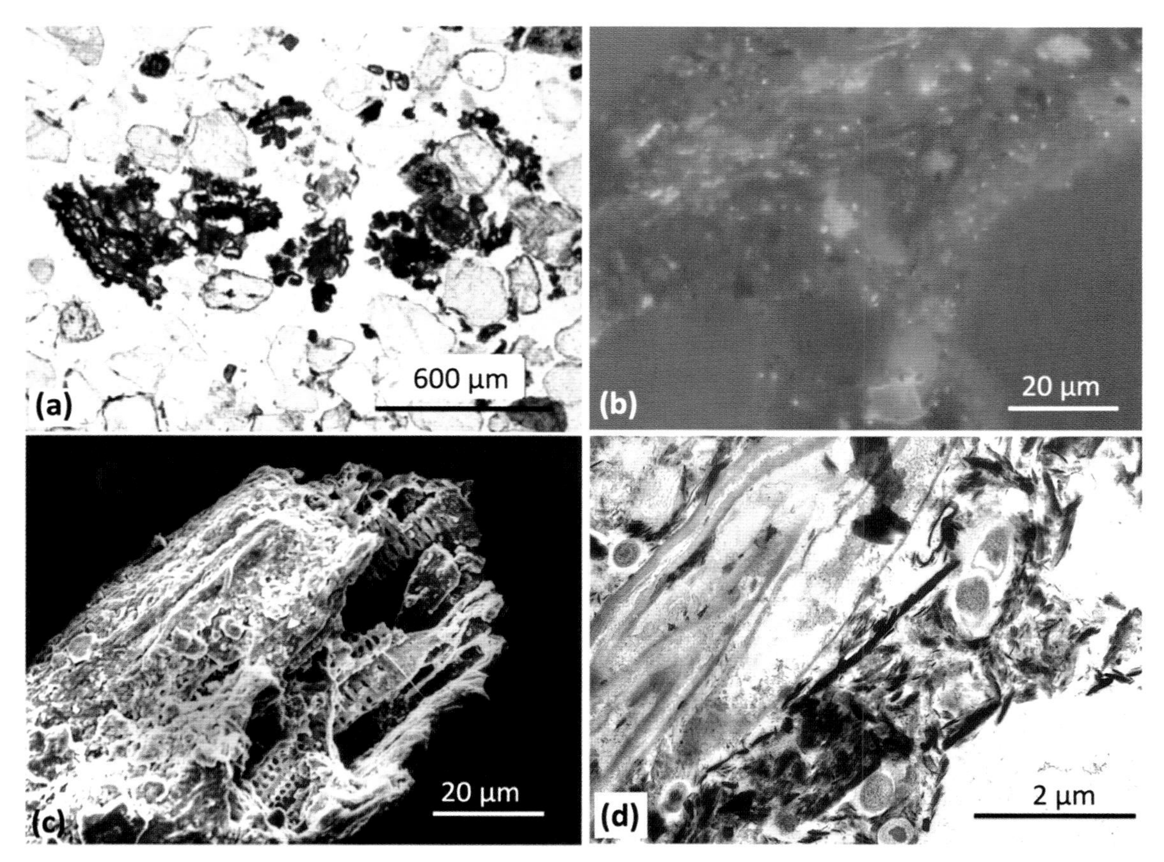

FIGURE 13.10 Visualization of organic matter in soil fabrics: (a) partly decomposed root debris in a Podzol observed with bright field light microscopy (Buurman et al., 2008); (b) bacteria located in a soil thin section using epifluorescence microscopy after the sample had been stained with Calcofluor white (Nunan et al., 2003); (c) plant vessel occluded by clay particles, SEM (Chenu, unpublished); and (d) bacteria in their soil habitat as visualized with TEM (UPb staining). Six bacteria can be observed in close vicinity to the plant cell wall remnant in the half-left section of the photograph (Chenu, unpublished).

Postma and Altemüller, 1990). It is not used to locate dead organic matter, because of the weak fluorescence recorded and the autofluorescence of minerals. The same limitation applies to confocal laser microscopy, which has seldom been applied to soils, although it can be used on fresh soils to locate bacteria previously stained (Gilbert et al., 1998; Minyard et al., 2012).

13.5.2 Scanning electron microscopy

Using scanning electron microscopy (SEM), the surface of a sample is scanned by an electron beam and secondary emitted electrons are recorded. This allows for reconstruction of a 3-D image of the topography of the sample with a resolution <1 μm. Specific preparation methods (low-temperature SEM or cryoSEM) or specific SEMs (environmental SEM) are necessary to preserve the fragile hydrated structures of organic matter, microorganisms, or clay minerals (Chenu and Tessier, 1995). In SEM organic matter is identified by its shape (Fig. 13.11c) and can be combined with energy-dispersive X-ray fluorescence, which gives access to the elemental composition of samples. In this case the presence of organic matter can be deduced from C and N signals. SEM has been very useful in studies of soil aggregation by organic matter (Chenu, 1993; Dorioz et al., 1993), identification of particulate organic matter in soil fractions (Besnard et al., 1996; Golchin, 1994), and localization of bacteria in their microhabitat (Chenu et al., 2001; Gaillard et al., 1999).

13.5.3 Transmission electron microscopy

When transmission electron microscopes send an electron beam through very thin samples, the transmitted electrons form an image in which thicker regions of the sample, or regions with a higher atomic number, appear darker. Diffraction of part of the incident electrons provides additional contrast and information, such as the visualization of clay lattices. With the resolution of TEM being 0.2 nm, it is well adapted to the study of fine soil particles. Sample preparation involves the deposition of a suspension of soil particles on a grid or the preparation of an ultrathin section of the sample after chemical fixation to preserve organic matter features, followed by dehydration and embedding in a resin (Elsass et al., 2007). Organic matter appears white or pale grey due to its low mean atomic number. It is thus usually stained with heavy metals to enhance its contrast either prior to embedment or directly on the thin section. Stains are contrasting agents with broader or narrower (e.g., polysaccharides) specificity. Excellent visualization of decaying plant material, microorganisms, and their association with minerals can be obtained (Fig. 13.11d). Energy-dispersive X-ray spectroscopy (EDS) additionally informs on the atomic composition of the sample. Another method, electron energy loss spectroscopy (EELS), analyzes the energy distribution of electrons that have crossed the sample. This carries information on the atomic composition and chemical bonding (Watteau et al., 2002). One advantage of EELS over EDS X-ray spectroscopy is its higher sensitivity to low-atomic-number elements (i.e., those being characteristic of organic matter) and the ability to obtain information about chemical bonds, not just atomic identification. EELS has been used to visualize polyphenols in decomposing litters (Watteau et al., 2002), whereas TEM provided early evidence of the heterogeneity of soils at the microscale, including the description of diverse bacterial habitats and sites where SOM could be inaccessible to microorganisms (Foster, 1988; Foster and Rovira, 1976; Kilbertus and Proth, 1979). TEM also demonstrated the patchy distribution of organic matter among clay-sized particles and the occurrence of clay-organic matter aggregates at submicrometric scales (Chenu and Plante, 2006).

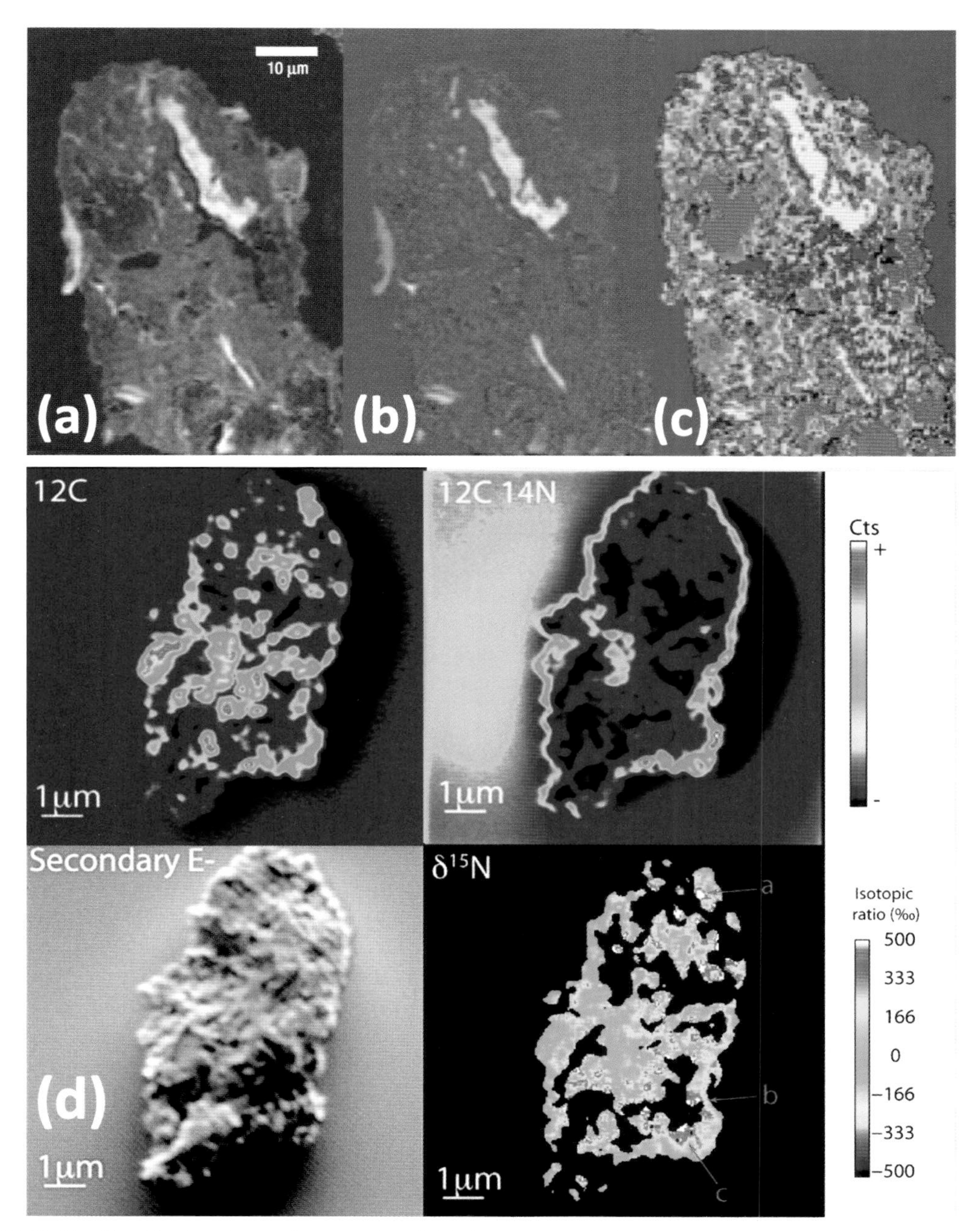

FIGURE 13.11 Visualization of organic matter in soil fabric: (a) to (c) distribution of organic matter in a microaggregate from an Oxisol using NEXAFS combined with STXM. (a) The map of total C concentration using a given portion of the NEXAFS spectra (subtraction of energy region at 280–282 eV from 290 to 292 eV); the brighter the zone, the higher the C concentration. (b) The map of aromatic C using another spectral region; again, the brighter zones are richer in aromatic compounds. (c) A cluster map of different C forms (takes into account aliphatic C, carboxylic C, and phenolic C, where different colors correspond to different types of spectra recorded in NEXAFS; Lehmann et al., 2008). (d) NanoSIMS images of a microaggregate from a soil to which ^{15}N labeled litter had been added showing the secondary electron image for morphology and the ^{12}C, ^{12}C-^{14}N, and $\delta^{15}N$ spatial distributions. Points a, b, and c have very high $\delta^{15}N$ showing sites of high assimilation of the labeled litter *(Remusat et al., 2012).*

13.5.4 Atomic force microscopy

In atomic force microscopy (AFM) the surface of the sample is scanned with a physical probe attached to a cantilever that records the forces arising between the probe and the surface of the sample. This process gives an image of the topography of the sample with unmatched lateral and vertical resolution and can be applied to wet samples, yet only if nearly flat. Thus far, AFM is seldom used in soil science, presumably because of the roughness of soil particles. It has, however, aided in the visualization of soil mineral and organic colloids and the identification of the patchy distribution of organic matter at the surface of clay minerals (Cheng et al., 2009; Citeau et al., 2006).

13.5.5 Infrared spectromicroscopy

Infrared spectromicroscopy allows for the detection of organic matter functional groups (e.g., aromatic C, aliphatic C, hydroxyl). Fourier transform infrared (FTIR) spectroscopy, combined with light microscopy (FTIR microscopy), allows for analysis of the vibrational molecular surface of a sample in a nondestructive way. When connected with synchrotron sources, it allows for 5 μm resolution to observe C speciation within microaggregates (Hernandez-Soriano et al., 2018). In addition, Raman microscopy using Raman spectroscopy can provide nondestructive, vibrational, and rotational molecular characterization of SOM. Raman microscopy requires no special preparation and can be used on hydrated samples (Xing et al., 2016). Visible and near-infrared (VNIR) hyperspectral imaging also allows for the characterization and quantification of SOM using hyperspectral cameras on undisturbed soil samples with almost no preparation. Hyperspectral image models can predict SOC and SON distributions on relatively large samples as they are acquired directly from a camera with a resolution of about 50 μm (Steffens et al., 2021).

13.5.6 X-ray spectromicroscopy

Scanning transmission X-ray microscopes (STXM) are synchrotron-based, soft X-ray instruments that permit visualization and chemical mapping of thin specimens. An image of the transmitted X-rays is formed with the same principles as with TEM with a spatial resolution of a few tens of nanometers. Element distribution is obtained using X-rays absorbed by the sample at energy levels characteristic of atoms or bonds of atoms (Lehmann et al., 2009; Thieme et al., 2010). STXM can be applied to hydrated samples provided they are very thin. It has been used on fine particle deposits and ultrathin sections of soil embedded in sulfur and of shock-frozen microaggregates (Kinyangi et al., 2006; Lehmann et al., 2008; Solomon et al., 2012; Wan et al., 2007). STXM can be coupled with near-edge X-ray absorption fine structure (NEXAFS) spectroscopy (for higher-z elements). This method, also called X-ray absorption near-edge structure (XANES), gives information on the functional groups within organic matter at the same time as spatial resolution. This extremely powerful set of methods and instruments has allowed visualization of the spatial distribution of organic matter at a submicrometric scale as discrete particles, as well as coatings among soil mineral particles. For example, a spatially distinct distribution of different chemical qualities of organic matter has been found between aromatic vs. carboxylic compounds (Kinyangi et al., 2006; Lehmann et al., 2007, 2008; Solomon et al., 2012; Wan et al., 2007; Fig. 13.12). Provided the energy of the X-ray beam is strong enough, elements such as Si, Ca, Fe, Al, and K can also be mapped (Solomon et al., 2012; Wan et al., 2007). This shows that the heterogeneity of the distribution

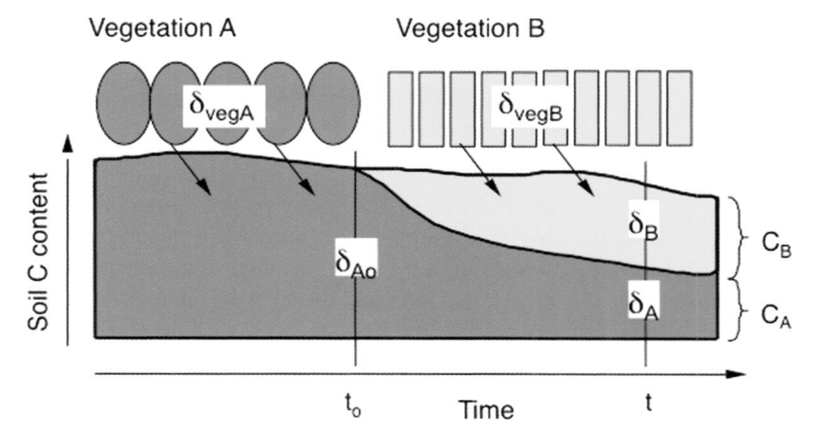

FIGURE 13.12 Principle of the ^{13}C natural abundance measurement of soil C turnover. At a given time *t*, the total C content of soil is: ($C = C_A + C_B$). The isotopic composition of SOM at this time is given by the mass balance equation: ($(C_A + C_B) = \delta_A.C_A + \delta_B.C_B$), from which the fraction of new carbon in sample F can be derived: ($F = C_B/C = (\delta - \delta_A)/(\delta_B - \delta_A)$). Under steady-state conditions, F is a direct expression of the turnover of soil. *(Reprinted from Balesdent and Mariotti, 1996.)*

of OM is also dependent on the adsorbent mineral phase (Lutfalla et al., 2019). Spatial distribution of S and P can be determined, albeit with slightly lower spatial resolution of about 80 nm (Brandes et al., 2007; Lehmann et al., 2009).

13.5.7 Mass spectrometry imaging

Secondary ion mass spectrometry is based on sputtering a few atomic layers from the surface of a sample using a primary ion beam and analyzing the emitted secondary ions, distinguished by their mass-to-charge ratio. Primary ions of cesium (Cs^+) or oxygen (O^-) are used to eject ions of opposite charge, e.g., $^{12}C^-$, $^{13}C^-$, $^{12}C^{14}N^-$, $^{12}C^{15}N^-$, and $^{28}Si^-$. Maps of elements and isotopes can be recorded with a spatial resolution of 1 μm with nanoscale secondary ion mass spectrometry (nanoSIMS). The operation must be performed under high vacuum, so samples should be dry and flat (nm scale rugosity). It is used on thin sections of previously embedded samples or on deposits of fine-sized particles of microaggregates (Herrmann et al., 2007a; Hoschen et al., 2015; Mueller et al., 2012; Remusat et al., 2012). This method has succeeded in localizing bacteria in a coarse-textured sand having incorporated ^{15}N ammonium sulfate (Herrmann et al., 2007b) and analyzing the spatial and temporal distribution of assimilated ^{13}C-^{15}N-labeled amino acids or ^{15}N-labeled litter at the scale of individual soil particles or microaggregates (Fig. 13.11d; Mueller et al., 2012; Remusat et al., 2012). The method helped to overcome the paradigm of C saturation of clay minerals by showing that new C may be attached to locations already covered with SOM (Vogel et al., 2014). In recent years nanoSIMS has been used to understand C and N fluxes in soil (Vidal et al., 2021; Witzgall et al., 2021). NanoSIMS has a very high mass resolution, but adequate standards are necessary to obtain quantitative analysis. To date, it is the only visualization method that provides dynamic information on biogeochemical processes affecting SOM given the possibility to locate isotopes of C and N. Matrix-assisted laser desorption ionization (MALDI) is a mass spectrometry imaging method requiring the application of a matrix on the sample which is hit by a laser, causing desorption and ionization. The evaporated fraction is then analyzed with a mass spectrometer. While it allows the targeting of molecules in situ, such as root exudates (Velickovic and Anderton, 2017), it has seldom been used for soils.

13.5.8 X-ray computed microtomography

Another methodology rapidly expanding in soil science is X-ray microcomputed tomography (μCT) that generates 3-D images of fresh, undisturbed soil samples. The sample is subjected to a beam of X-rays and X-ray radiographs are recorded at different angles during stepwise rotation of the sample around its vertical axis. The transmission of the X-rays is attenuated through the sample depending on the density and composition (average atomic number) of the object. After image processing, cross-sectional images of the attenuation coefficients allow for reconstruction of a 3-D image of the sample. Contrast in the images comes from the differences in the average atomic number. Mineral particles and voids can easily be distinguished, but visualization of organic matter, with the exception of large organic debris, poses problems that have not yet been resolved (Elyeznasni et al., 2012). The spatial resolution of μCT is currently several micrometers and depends on the size of the scanned sample. Using small samples and monochromatic X-ray beams issued from synchrotron lines improves the resolution to a micrometer. Accurate image processing, thresholding, and analysis are still an open issue in μCT (Peth et al., 2008; Sleutel et al., 2008). One possibility that allows for visualization of organic matter is to stain it with heavy elements, as done for TEM, to enhance its contrast with surrounding minerals (Arai et al., 2019; Peth et al., 2014; Rawlins et al., 2016).

13.5.9 Zymography

Zymography is a nondestructive method often used to map enzyme spatial distribution in rhizoboxes. The approach requires the application of a membrane or gel in which a substrate signals the presence of a specific enzyme under UV light, the signal being recorded by a camera (Razavi et al., 2019). Its resolution reaches tens of micrometers allowing the study of soil microbial hotspots (Heitkotter and Marschner, 2018).

13.6 Methods to measure the turnover rate of soil organic matter

The dynamics of SOM can be approached either by following the fate of organic material added to soil or by measuring the long-term turnover of whole soil C using lab incubations and modeling, ^{13}C natural abundance or labeling, and ^{14}C dating or labeling or by following an ^{15}N signal.

13.6.1 Soil incubation and modeling of CO_2 evolution

Incubation of soil samples under controlled conditions is used to measure the mineralization of SOM by capturing the CO_2 emitted or mineral N that accumulates. The mineralization curve is modeled using first-order kinetics to derive turnover rate constants (Paul et al., 2006). This approach can also be used to biologically fractionate SOM (i.e., separate kinetic pools of SOM-C based on the shape of the mineralization curve). Paul et al. (1999) used the results of long-term incubation, acid hydrolysis, and C dating to establish kinetically derived pools and the mean residence time (MRT) necessary to model field CO_2

evolution rates. The active SOM pool, representing 2% to 3% of the SOC of three cultivated field sites, was found to have a field-equivalent MRT of 36 to 66 days. The slow SOC pool of these sites accounted for 40% to 42% of the SOC, with MRTs of 9 to 13 years. Approximately 50% of the SOC resistant to acid hydrolysis, and thus calculated as the resistant pools, had ^{14}C MRTs of 1350 to 1450 years. Utilization of these values in the Century SOM model, together with measured field moisture and temperature measurements, produced daily field CO_2 predictions very similar to those measured. One of the limits of the incubation approach is that it concentrates on organic materials that have relatively short residence times in soils. The analysis of long-term bare fallow soil is useful for characterizing SOM with longer residence times (i.e., where field soil has been left bare of any vegetation for years to decades, allowing the organic matter initially present to progressively decompose over time without additional C inputs; Barré et al., 2010).

13.6.2 Decomposition of organic materials added to soil

The decomposition and mineralization of plant residues, organic wastes, or organic molecules in soil are classically measured by adding them to soil in ^{14}C-, ^{13}C-, or ^{15}N-labeled forms and incubating the whole. The evolved CO_2 from the added organic matter (^{14}C-CO_2, ^{13}C-CO_2) or mineral N (^{15}N) is then measured, or the remaining ^{14}C or ^{13}C in soil is monitored over time. Because of the high cost of organic matter enriched in isotopes, such studies are typically performed at laboratory scale or on small plots at field scale. Mineralization of the added material can also be estimated by measuring the difference of evolved CO_2 between the sample with the organic matter addition and a reference soil, provided the amounts of added C are sufficient and no priming takes place (i.e., enhanced mineralization of native SOM after addition of fresh C substrate).

The decomposition of litter is currently measured in situ using litterbags (i.e., small nylon or fiberglass mesh bags with known amount of plant litter) that are placed on top of or within soil. Decomposition rates, using first-order kinetics, are determined from the mass loss of the material within the litterbags. Litterbag experiments account for decomposition due to microbial decay and leaching, but depending on the mesh size, they may not account for litter fragmentation by soil fauna, which alters decomposition rates measured (Cotrufo et al., 2010).

13.6.3 ^{13}C natural abundance

The ^{13}C natural abundance method utilizes known vegetation changes from C_3 plants to C_4 plants with contrasting photosynthetic pathways. Plant tissues of these two photosynthetic types have contrasting ^{13}C to ^{12}C ratios (typically $\delta^{13}C \approx -27‰$ and $-12‰$ for C_3 and C_4 vegetation, respectively). Although microbial decomposition of these materials causes some isotopic fractionation, the resulting SOM still bears the isotopic signature of the parent vegetation. Hence, where a change of vegetation has occurred at some known date, measurement of the change in ^{13}C-SOC natural abundance allows determination of the rate of C loss derived from the initial vegetation and the rate of incorporation of C from the new vegetation using first-order kinetics (Fig. 13.12; Baldesdent, 1996).

Despite a narrow range of situations where it can be applied, this method has allowed major progress in measuring and understanding the dynamics of soil C, because the turnover rate of C is measured in situ and over decades and can be applied to bulk soil C as well as to any physical and chemical SOM fractions. Its coupling with gas chromatography-based separation methods permits the measurement of the turnover rate of molecular entities in situ (Amelung et al., 2008). However, this approach has several requirements and limitations. The isotopic signature of SOM is slightly different from that of the parent vegetation (δ_A and δ_B cannot be measured directly, Fig. 13.12). A reference situation with no vegetation change is therefore needed as its isotopic composition provides an estimate of δ_A. It also accounts for the variability of ^{13}C content between plant organs or among organic molecules; isotopic measurements of any considered fraction need to be performed on both the chronosequence samples as well as on the reference situation. The ^{13}C natural abundance method is well suited for estimation of C turnover within years to decades, and data from such measurements should specify the length of the tracer exchange.

13.6.4 ^{14}C dating

After the death of organisms, their ^{14}C content, initially in equilibrium with that of the atmosphere, decreases with first-order kinetics by radioactive decay, with a half-life of 5568 years. Soil C pools constantly receive new C inputs from plants and lose C through decomposition. The $^{14}C/^{12}C$ ratio of a given organic matter fraction reflects both the radioactive decay (time elapsed since the C was fixed by photosynthesis) and the rate of decomposition of the organic matter. Measuring the ^{14}C content in soil provides an estimate of the age of SOM within a time frame of 200 to 40,000 years; samples with age less than 200 years are considered modern. This method is suited to measure SOM turnover rates of centuries to millennia (Paul et al., 1997; Scharpenseel and Schiffmann, 1977; Torn et al., 1997).

Aboveground nuclear bomb testing released large amounts of ^{14}C into the atmosphere between 1950 and 1965, causing a sharp increase and then a slow decrease in the atmospheric ^{14}C concentration. This peak of bomb ^{14}C entered the vegetation and is progressively incorporated into SOM, creating a large-scale in situ labeling of SOM that allows us to measure turnover rates from $\sim$4 to 100 years (Trumbore, 2009).

13.6.5 Compound-specific assessment of SOM turnover

Compound-specific stable isotope analyses can be carried out at natural abundance or after labeling. Isotopic mass ratio spectrometers were first coupled to gas chromatographs, and methods for stable isotope measurements of lignins, sugars, and lipids were established. This includes stable isotope measurements of molecules separated by liquid chromatography (Bai et al., 2013; Malik et al., 2016). The advantage of this technique is that no derivatization, and therefore no correction for C additions, is required. This method has been used to measure the turnover rate of rhizodeposits of different molecular masses (Malik et al., 2016). Measurement of ^{14}C at the molecular level was performed using preparative gas chromatography to analyze the ^{14}C age of single molecules (Hatté et al., 2016; Kramer and Gleixner, 2006; van der Voort et al., 2017).

In summary, the different methods available for measuring turnover rates of bulk soil C or specific compounds allow quantification of SOM with a range of turnover times. The duration of laboratory incubations is insufficient to mobilize C having turnover rates of centuries to millennia, but instead, it addresses C pools with turnover times of days to a few decades. Available vegetation changes for ^{13}C natural abundance studies usually go on for a few decades, yet very seldom for a century because the tracer (the ^{13}C signature of the new vegetation) cannot label SOM with turnover rates of millennia. ^{13}C natural abundance and ^{14}C bomb spiking are appropriate methods for measuring turnover rates of years-to-decades. In turn, ^{14}C dating gives access to the organic matter that has the slowest turnover rate (i.e., centuries-to-millennia). Appropriate modeling is then necessary to account for the heterogeneous character of SOM turnover in soils.

A wealth of highly sophisticated methods is now available to study SOM with increased spatial, temporal, and molecular resolutions. These techniques allow the processes governing the nature, functions, and dynamics of SOM to unravel. However, they are often costly, time-consuming, and of limited accessibility. As a consequence, they have too often been applied to small samples raising the issue of upscaling and applied to only a few pedoclimatic conditions. However, some high throughput and lower cost methods such as infrared spectroscopies are able to capture information gained from other methods and can be used, once duly calibrated, to gain information from much wider datasets. Moreover, a priority in the study is to quantify SOM, both as concentrations and as stocks. Current efforts to disseminate standard protocols are particularly welcome in this field.

References

Achard, F.K., 1786. Chemische Untersuchung des Torfs. Chem. Ann. Freunde Naturlehre 2, 391−403.

Amelung, W., Brodowski, S., Sandhage-Hofmann, A., Bol, R., 2008. Combining biomarker with stable isotope analyses for assessing the transformation and turnover of soil organic matter. In: Sparks, D.L. (Ed.), Advances in Agronomy. Academic Press Inc., pp. 155−250

Amelung, W., Cheshire, M.V., Guggenberger, G., 1996. Determination of neutral and acidic sugars in soil by capillary gas-liquid chromatography after trifluoroacetic acid hydrolysis. Soil Biol. Biochem. 28, 1631−1639.

Amelung, W., Zhang, X.D., 2001. Determination of amino acid enantiomers in soils. Soil Biol. Biochem. 33, 553−562.

Angers, D.A., Mehuys, G.R., 1993. Aggregate stability to water. In: Carter, M.R. (Ed.), Soil Sampling and Methods of Analysis. CRC Press, Boca Raton, pp. 651−657.

Arai, M., Uramoto, G.I., Asano, M., Uematsu, K., Uesugi, K., Takeuchi, A., et al., 2019. An improved method to identify osmium-stained organic matter within soil aggregate structure by electron microscopy and synchrotron X-ray micro-computed tomography. Soil Tillage Res. 191, 275−281.

Arias, M., Barral, M.T., Diaz Fierros, F., 1997. Influence of dispersion procedures on measurements of particle size distribution and composition of the granular fraction of pasture and arable soils. [Spanish]. Agrochimica 41, 63−77.

Audette, Y., Congreves, K.A., Schneider, K., Zaro, G.C., Nunes, A.L.P., Zhang, H.J., et al., 2021. The effect of agroecosystem management on the distribution of C functional groups in soil organic matter: a review. Biol. Fertil. Soils 57 (7), 881−894.

Bai, Z., Bode, S., Huygens, D., Zhang, X.D., Boeckx, P., 2013. Kinetics of amino sugar formation from organic residues of different quality. Soil Biol. Biochem. 57, 814−821.

Baldock, J.A., Nelson, P.N., 2000. Soil organic matter. In: Sumner, M.E. (Ed.), Handbook of Soil Science. CRC Press, Boca Raton, pp. 825−884.

Baldock, J.A., Oades, J.M., Nelson, P.N., Skene, T.M., Golchin, A., Clarke, P., 1997. Assessing the extent of decomposition of natural organic materials using solid-state 13C NMR spectroscopy. Aust. J. Soil Res. 35, 1061−1083.

Baldock, J.A., Hawke, B., Sanderman, J., Macdonald, L.M., 2013. Predicting contents of carbon and its component fractions in Australian soils from diffuse reflectance mid-infrared spectra. Soil Res. 51 (7−8), 577−583.

Balesdent, J., 1996. The significance of organic separates to carbon dynamics and its modelling in some cultivated soils. Eur. J. Soil Sci. 47, 485−494.

Balesdent, J., Besnard, E., Arrouays, D., Chenu, C., 1998. The dynamics of carbon in particle-size fractions of soil in a forest-cultivation sequence. Plant Soil 201, 49−57.

Balesdent, J., Mariotti, A., 1996. Measurement of soil organic matter turnover using 13C natural abundance. In: Boutton, J.W., Yamakasi, S.I. (Eds.), Mass Spectrometry of Soils. Marcel Dekker, New York, pp. 83−111.

Balesdent, J., Mariotti, A., Guillet, B., 1987. Natural 13C abundance as a tracer for studies of soil organic matter dynamics. Soil Biol. Biochem. 19, 25−30.

Balesdent, J., Pétraud, J.P., Feller, C., 1991. Effet des ultrasons sur la distribution granulométrique des matières organiques des sols. Sci. Sol. 29, 95–106.

Balesdent, J., Wagner, G.H., Mariotti, A., 1988. Soil organic matter turnover in long-term field experiments as revealed by carbon-13 natural abundance. Soil Sci. Soc. Am. J. 52, 118–124.

Barré, P., Eglin, T., Christensen, B.T., Ciais, P., Houot, S., Kätterer, T., et al., 2010. Quantifying and isolating stable soil organic carbon through long-term bare fallow experiments. Biogeosciences 7, 3839–3850.

Barré, P., Plante, A.F., Cécillon, L., Lutfalla, S., Baudin, F., Bernard, S., et al., 2016. The energetic and chemical signatures of persistent soil organic matter. Biogeochemistry 130, 1–12.

Basile-Doelsch, I., Amundson, R., Stone, W.E.E., Borschneck, D., Bottero, J.Y., Moustier, S., Masin, F., Colin, F., 2007. Mineral control of carbon pools in a volcanic soil horizon. Geoderma 137 (3–4), 477.

Beare, M.H., Cabrera, M.L., Hendrix, P.F., Coleman, D.C., 1994. Aggregate-protected and unprotected organic matter pools in conventional-tillage and no-tillage soils. Soil Sci. Soc. Am. J. 58, 787–795.

Besnard, E., Chenu, C., Balesdent, J., Puget, P., Arrouays, D., 1996. Fate of particulate organic matter in soil aggregates during cultivation. Eur. J. Soil Sci. 47, 495–503.

Blair, G.J., Lefroy, R.D.B., Lise, L., 1995. Soil carbon fractions based on their degree of oxidation and the development of carbon management index for agricultural systems. Aust. J. Soil Res. 46 (7), 1459–1466.

Bligh, E.G., Dyer, W.J., 1959. A rapid method of total lipid extraction and purification. Can. J. Biochem. Physiol. 54, 911–917.

Bol, R., Amelung, W., Friedrich, C., 2004. Role of aggregate surface and core fraction in the sequestration of carbon from dung in a temperate grassland soil. Eur. J. Soil Sci. 55 (1), 71–77.

Brandes, J.A., Ingall, E., Paterson, D., 2007. Characterization of minerals and organic phosphorus species in marine sediments using soft X-ray fluorescence spectromicroscopy. Mar. Chem. 103, 250–265.

Bremer, E., Ellert, B.H., Janzen, H.H., 1995. Total and light-fraction carbon dynamics during four decades after cropping changes. Soil Sci. Soc. Am. J. 59 (5), 1398–1403.

Bremner, J.M., 1967. Nitrogenous compounds in soils. In: McLaren, A.D., Peterson, G.H. (Eds.), Soil Biochemistry. Dekker, New York, pp. 19–66.

Brodowski, S., Amelung, W., Haumaier, L., Abetz, C., Zech, W., 2005. Morphological and chemical properties of black carbon in physical soil fractions as revealed by scanning electron microscopy and energy-dispersive X-ray spectroscopy. Geoderma 128 (1–2), 116.

Bull, I.D., Nott, C.J., vanBergen, P.F., Poulton, P.R., Evershed, R.P., 2000. Organic geochemical studies of soils from the Rothamsted classical experiments – VI. The occurrence and source of organic acids in an experimental grassland soil. Soil Biol. Biochem. 32, 1367–1376.

Buurman, P., Jongmans, A.G., Nierop, K.G.J., 2008. Comparison of Michigan and Dutch podzolized soils: organic matter characterization by micromorphology and pyrolysis-GC/MS. Soil Sci. Soc. Am. J. 72 (5), 1344–1356.

Cade-Menun, B.J., 2017. Characterizing phosphorus forms in cropland soils with solution P-31-NMR: past studies and future research needs. Chem. Biol. Tech. Agric. 4, 19.

Cambardella, C.A., Elliott, E.T., 1992. Particulate organic matter across a grassland cultivation sequence. Soil Sci. Soc. Am. J. 56, 777–783.

Cécillon, L., Baudin, F., Chenu, C., Christensen, B.T., Franko, U., Houot, S., et al., 2021. Partitioning soil organic carbon into its centennially stable and active fractions with machine-learning models based on Rock-Eval (R) thermal analysis. Geosc. Model. Dev. 14 (6), 3879–3898.

Cheng, S.Y., Bryant, R., Doerr, S.H., Wright, C.J., Williams, P.R., 2009. Investigation of surface properties of soil particles and model materials with contrasting hydrophobicity using atomic force microscopy. Environ. Sci. Technol. 43 (17), 6500–6506.

Chenu, C., 1993. Clay- or sand-polysaccharides associations as models for the interface between microorganisms and soil: water-related properties and microstructure. Geoderma 56, 143–156.

Chenu, C., Hassink, J., Bloem, J., 2001. Short-term changes in the spatial distribution of microorganisms in soil aggregates as affected by glucose addition. Biol. Fertil. Soils 34, 349–356.

Chenu, C., Plante, A.F., 2006. Clay-sized organo-mineral complexes in a cultivation chronosequence: revisiting the concept of the "primary organo-mineral complex. Eur. J. Soil Sci. 56 (4), 596–607.

Chenu, C., Tessier, D., 1995. Low temperature scanning electron microscopy of clay and organic constituents and their relevance to soil microstructures. Scanning Microsc. 9, 989–1010.

Cheschire, M.V., 1979. Nature and Origin of Carbohydrates in Soils. Academic Press, London.

Christensen, B.T., 1987. Decomposability of organic matter in particle size fractions from field soils with straw incorporation. Soil Biol. Biochem. 19, 429–436.

Christensen, B.T., 1992. Physical fractionation of soil and organic matter in primary particle size and density separates. Adv. Soil Sci. 20, 1–90.

Christensen, B.T., 1996. Matching measurable soil organic matter fractions with conceptual pools in simulation models of carbon turnover: revision of model structures. In: Powlson, D., Smith, P., Smith, J. (Eds.), Evaluation of Soil Organic Matter Models. Springer Verlag, Berlin, p. 38.

Christensen, B.T., 2001. Physical fractionation of soil and structural and functional complexity in organic matter turnover. Eur. J. Soil Sci. 52 (3), 345–353.

Citeau, L., Gaboriaud, F., Elsass, F., Thomas, F., Lamy, I., 2006. Investigation of physico-chemical features of soil colloidal suspensions. Colloids Surf. A Physicochem. Eng. Asp. 287 (1–3), 94–105.

Cotrufo, M.F., Ngao, J., Marzaioli, F., Piermatteo, D., 2010. Inter-comparison of methods for quantifying above-ground leaf litter decomposition rates. Plant Soil 334 (1–2), 365–376.

Cotrufo, M.F., Boot, C., Abiven, S., Foster, E.J., Haddix, M., Reisser, M., et al., 2016. Quantification of pyrogenic carbon in the environment: an integration of analytical approaches. Org. Geochem. 100, 42–50.

Culman, S.W., Snapp, S.S., Freeman, M.A., Schipanski, M.E., Beniston, J., Lal, R., et al., 2012. Permanganate oxidizable carbon reflects a processed soil fraction that is sensitive to management. Soil Sci. Soc. Am. J. 76 (2), 494–504.

Derrien, D., Marol, C., Balabane, M., Balesdent, J., 2006. The turnover of carbohydrate carbon in a cultivated soil estimated by 13C natural abundances. Eur. J. Soil Sci. 57 (4), 547–557.

Dignac, M.-F., Bahri, H., Rumpel, C., Rasse, D.P., Bardoux, G., Balesdent, J., et al., 2005. Carbon-13 natural abundance as a tool to study the dynamics of lignin monomers in soil: an appraisal at the Closeaux experimental field (France). Geoderma 128, 1–17.

Dignac, M.F., Houot, S., Derenne, S., 2006. How the polarity of the separation column may influence the characterization of compost organic matter by pyrolysis-GC/MS. J. Anal. Appl. Pyrolysis 75, 128–139.

Dorioz, J.M., Robert, M., Chenu, C., 1993. The role of roots, fungi and bacteria on clay particle organization. An experimental approach. Geoderma 56, 179–194.

Duchaufour, P., 1976. Dynamics of organic matter in soils of temperate regions. Its action on pedogenesis. Geoderma 15 (1), 31–40.

Elliott, E.T., 1986. Aggregate structure and carbon, nitrogen and phosphorus in native and cultivated soils. Soil Sci. Soc. Am. J. 50, 627–633.

Elliott, E.T., Paustian, K., Frey, S.D., 1996. Modeling the measurable or measuring the modelable: a hierarchical approach to isolating meaningful soil organic matter fractionations. In: Powlson, D., Smith, P., Smith, J. (Eds.), Evaluation of Soil Organic Matter Models. Springer Verlag, Berlin, pp. 161–180.

Elsass, F., Chenu, C., Tessier, D., 2007. Transmission electron microscopy for soil samples: preparation methods and use. In: Drees, R., Kingery, B. (Eds.), Methods in Soil Analysis. Soil Science Society of America, Madison, pp. 235–268.

Elyeznasni, N., Sellami, F., Pot, V., Benoit, P., Vieuble-Gonod, L., Young, I., et al., 2012. Exploration of soil micromorphology to identify coarse-sized OM assemblages in X-ray CT images of undisturbed cultivated soil cores. Geoderma 179, 38–45.

Eusterhues, K., Rumpel, C., Kleber, M., Kogel-Knabner, I., 2003. Stabilisation of soil organic matter by interactions with minerals as revealed by mineral dissolution and oxidative degradation. Org. Geochem. 34 (12), 1591–1600.

Eusterhues, K., Rumpel, C., Kogel-Knabner, I., 2005. Organo-mineral associations in sandy acid forest soils: importance of specific surface area, iron oxides and micropores. Eur. J. Soil Sci. 56 (6), 753–763.

Eusterhues, K., Rumpel, C., Kogel-Knabner, I., 2007. Composition and radiocarbon age of HF-resistant soil organic matter in a Podzol and a Cambisol. Org. Geochem. 38 (8), 1356–1372.

Falloon, P., Smith, P., Coleman, K., Marshall, S., 1998. Estimating the size of the inert organic matter pool from total soil organic carbon content for use in the Rothamsted carbon model. Soil Biol. Biochem. 30 (8–9), 1207–1211.

Feller, C., 1979. Une méthode de fractionnement granulométrique de la matière organique des sols. Application aux sols tropicaux à textures grossières, très pauvres en humus. Cah. ORSTOM, sér. Pédologie XVII 339–345.

Feller, C., Balesdent, J., Nicolardot, B., Cerri, C., 2001. Approaching "functional" soil organic matter pools through particle-size fractionation: examples for tropical soils. In: Lal, R., Kimble, J.M., Stewart, B.A. (Eds.), Assessment Methods for Soil Carbon. CRC Press, Boca Raton, pp. 53–67.

Feller, C., Burtin, B., Gérard, B., Balesdent, J., 1991. Utilisation des résines sodiques et des ultrasons dans le fractionnement granulométrique de la matière organique des sols. Intérêts et limites. Sci. Sol. 29, 77–93.

Foster, R.C., 1988. Microenvironments of soil microorganisms. Biol. Fertil. Soils 6, 189–203.

Foster, R.C., Rovira, A.D., 1976. Ultrastructure of wheat rhizosphere. New Phytol. 76, 343–352.

Gaillard, V., Chenu, C., Recous, S., Richard, G., 1999. Carbon, nitrogen and microbial gradients induced by plant residues decomposing in soil. Eur. J. Soil Sci. 50, 567–578.

Gilbert, B., Assmus, B., Hartmann, A., Frenzel, P., 1998. In situ localization of two methanotrophic strains in the rhizosphere of rice plants. FEMS Microbiol. Ecol. 25 (2), 117–128.

Golchin, A., 1994. Soil structure and carbon cycling. Aust. J. Soil Res. 32, 1043–1068.

Hammes, K., Schmidt, M.W.I., Smernik, R.J., Currie, L.A., Ball, W.P., Nguyen, T.H., et al., 2007. Comparison of quantification methods to measure fire-derived (black/elemental) carbon in soils and sediments using reference materials from soil, water, sediment and the atmosphere. Glob. Biogeochem. Cycles. 21 (3), 1–18.

Hatté, C., Noury, C., Kheirbeik, L., Balesdent, J., 2016. Turnover of soil organic matter amino-acid fraction investigated by 13C and 14C signatures of carboxyl carbon. Radiocarbon 59 (2), 1–9.

Haynes, R.J., 2005. Labile organic matter fractions as central components of the quality of agricultural soils: an overview. In: Sparks, D.L. (Ed.), Advances in Agronomy. Academic Press, p. 221.

Hedges, J.I., Eglinton, G., Hatcher, P.G., Kirchman, D.L., Arnosti, C., Derenne, S., et al., 2000. The molecularly-uncharacterized component of nonliving organic matter in natural environments. Org. Geochem. 31, 945–958.

Hedges, J.I., Ertel, J.R., 1982. Characterization of lignin by gas capillary chromatography of cupric acid oxidation products. Anal. Chem. 54 (2), 174–178.

Heitkotter, J., Marschner, B., 2018. Soil zymography as a powerful tool for exploring hotspots and substrate limitation in undisturbed subsoil. Soil Biol. Biochem. 124, 210–217.

Hernandez-Soriano, M.C., Dalal, R.C., Warren, F.J., Wang, P., Green, K., Tobin, M.J., et al., 2018. Soil organic carbon stabilization: mapping carbon speciation from intact microaggregates. Environ. Sci. Technol. 52 (21), 12275–12284.

Herrmann, A.M., Ritz, K., Nunan, N., Clode, P.L., Pett-Ridge, J., Kilburn, M.R., et al., 2007a. Nano-scale secondary ion mass spectrometry – a new analytical tool in biogeochemistry and soil ecology: a review article. Soil Biol. Biochem. 39 (8), 1835–1850.

Herrmann, A.M., Clode, P.L., Fletcher, I.R., Nunan, N., Stockdale, E.A., O'Donnel, A.G., et al., 2007b. A novel method for the study of the biophysical interface in soils using nano-scale secondary ion mass spectrometry. Rapid Commun. Mass Spectrom. 21 (1), 29–34.

Hockaday, W.C., Grannas, A.M., Kim, S., Hatcher, P.G., 2007. The transformation and mobility of charcoal in a fire-impacted watershed. Geochim. Cosmochim. Acta. 71, 3432–3445.

Hoogsteen, M.J.J., Lantinga, E.A., Bakker, E.J., Groot, J.C.J., Tittonell, P.A., 2015. Estimating soil organic carbon through loss on ignition: effects of ignition conditions and structural water loss. Eur. J. Soil Sci. 66 (2), 320–328.

Hopmans, E.C., Schouten, S., Pancost, R.D., van der Meer, M.T.J., Damste, J.S.S., 2000. Analysis of intact tetraether lipids in archaeal cell material and sediments by high performance liquid chromatography/atmospheric pressure chemical ionization mass spectrometry. Rapid Commun. Mass Spectrom. 14 (7), 585–589.

Hoschen, C., Hoschen, T., Mueller, C.W., Lugmeier, J., Elgeti, S., Rennert, T., et al., 2015. Novel sample preparation technique to improve spectromicroscopic analyses of micrometer-sized particles. Environ. Sci. Technol. 49 (16), 9874–9880.

Joergensen, R.G., 2018. Amino sugars as specific indices for fungal and bacterial residues in soil. Biol. Fertil. Soils 54 (5), 559–568.

Kalbitz, K., Solinger, S., Park, J.H., Michalzik, B., Matzner, E., 2000. Controls on the dynamics of dissolved organic matter in soils: a review. Soil Sci. 165 (4), 277–304.

Kelleher, B.P., Simpson, A.J., 2006. Humic substances in soils: are they really chemically distinct? Environ. Sci. Technol. 40, 4605–4611.

Kelleher, B.P., Simpson, M.J., Simpson, A.J., 2006. Assessing the fate and transformation of plant residues in the terrestrial environment using HR-MAS NMR spectroscopy. Geochim. Cosmochim. Acta. 70 (16), 4080–4094.

Kiem, R., Knicker, H., Korschens, M., Kogel-Knabner, I., 2000. Refractory organic carbon in C-depleted arable soils, as studied by 13C NMR spectroscopy and carbohydrate analysis. Org. Geochem. 31 (7/8), 655–668.

Kilbertus, G., Proth, J., 1979. Observation d'un sol forestier (rendzine) en microscopie électronique. Can. J. Microbiol. 25, 943–946.

Kinyangi, J., Solomon, D., Liang, B.I., Lerotic, M., Wirick, S., Lehmann, J., 2006. Nanoscale biogeocomplexity of the organo-mineral assemblage in soil: application of STXM microscopy and C 1s-NEXAFS spectroscopy. Soil Sci. Soc. Am. J. 70 (5), 1708–1718.

Kleber, M., Johnson, M.G., 2010. Advances in understanding the molecular structure of soil organic matter: implications for interactions in the environment. Adv. Agron. 106, 78–143.

Kleber, M., Lehmann, J., 2019. Humic substances extracted by alkali are invalid proxies for the dynamics and functions of organic matter in terrestrial and aquatic ecosystems. J. Environ. Qual. 48 (2), 207–216.

Kleber, M., Mikutta, R., Torn, M.S., Jahn, R., 2005. Poorly crystalline mineral phases protect organic matter in acid subsoil horizons. Eur. J. Soil Sci. 56 (6), 717–725.

Knicker, H., 2011. Solid state CPMAS 13C and 15N NMR spectroscopy in organic geochemistry and how spin dynamics can either aggravate or improve spectra interpretation. Org. Geochem. 42, 867–889.

Kogel-Knabner, I., 2000. Analytical approaches for characterizing soil organic matter. Org. Geochem. 31 (7–8), 609–625.

Kogel-Knabner, I., 2002. The macromolecular organic composition of plant and microbial residues as inputs to soil organic matter. Soil Biol. Biochem. 34 (2), 139–162.

Kögel-Knabner, I., Guggenberger, G., Kleber, M., Kandeler, E., Kalbitz, K., Scheu, S., Eusterhues, K., Leinweber, P., 2008. Organo-mineral associations in temperate soils: integrating biology, mineralogy, and organic matter chemistry. J. Plant Nutr. Soil Sci. 171 (1), 61–82.

Kramer, C., Gleixner, G., 2006. Variable use of plant- and soil-derived carbon by microorganisms in agricultural soils. Soil Biol. Biochem. 38 (11), 3267–3278.

Lavallee, J.M., Soong, J.L., Cotrufo, M.F., 2020. Conceptualizing soil organic matter into particulate and mineral-associated forms to address global change in the 21st century. Glob. Chang. Biol. 26 (1), 261–273.

Lehmann, J., Brandes, J., Fleckenstein, H., Jacobsen, C., Solomon, D., Thieme, J., 2009. Synchrotron-based near-edge X-ray spectroscopy of natural organic matter in soils and sediments. In: Senesi, N., Xing, P., Huang, P.M. (Eds.), Biophysico-Chemical Processes Involving Natural Nonliving Organic Matter in Environmental Systems. IUPAC Series on Biophysico-Chemical Processes in Environmental Systems. Wiley, New Jersey, pp. 729–781.

Lehmann, J., Kinyangi, J., Solomon, D., 2007. Organic matter stabilization in soil microaggregates: implications from spatial heterogeneity of organic carbon contents and carbon forms. Biogeochemistry 85, 45–57.

Lehmann, J., Kleber, M., 2015. The contentious nature of soil organic matter. Nature 528 (7580), 60–68.

Lehmann, J., Solomon, D., Kinyangi, J., Dathe, L., Wirick, S., Jacobsen, C., 2008. Spatial complexity of soil organic matter forms at nanometre scales. Nat. Geosci. 1 (4), 238–242.

Lutfalla, S., Barré, P., Bernard, S., Guillou, C.L., Alléon, J., Chenu, C., 2019. Multidecadal persistence of organic matter in soils: investigations at the submicrometer scale. Biogeosciences 16, 1401–1410. https://doi.org/10.5194/bg-16-1401-2019.

Lutfalla, S., Chenu, C., Barre, P., 2014. Are chemical oxidation methods relevant to isolate a soil pool of centennial carbon? Biogeochemistry 118 (1–3), 135–139.

Malik, A.A., Roth, V.N., Hebert, M., Tremblay, L., Dittmar, T., Gleixner, G., 2016. Linking molecular size, composition and carbon turnover of extractable soil microbial compounds. Soil Biol. Biochem. 100, 66–73.

Martens, D.A., Loeffelmann, K.L., 2003. Soil amino acid composition quantified by acid hydrolysis and anion chromatography–pulsed amperometry. J. Agric. Food Chem. 51, 6521–6529.

Masoom, H., Courtier-Murias, D., Farooq, H., Soong, R., Kelleher, B.P., Zhang, C., et al., 2016. Soil organic matter in its native state: unravelling the most complex biomaterial on Earth. Environ. Sci. Technol. 50 (4), 1670–1680.

Mendez-Millan, M., Dignac, M.F., Rumpel, C., Rasse, D.P., Derenne, S., 2010. Molecular dynamics of shoot vs. root biomarkers in an agricultural soil estimated by natural abundance C-13 labelling. Soil Biol. Biochem. 42 (2), 169–177.

Mikutta, R., Kleber, M., Kaiser, K., Jahn, R., 2005. Review: organic matter removal from soils using hydrogen peroxide, sodium hypochlorite, and disodium peroxodisulfate. Soil Sci. Soc. Am. J. 69, 120–135.

Miltner, A., Zech, W., 1998. Carbohydrate decomposition in beech litter as influenced by aluminium, iron and manganese oxides. Soil Biol. Biochem. 30, 1–7.

Minyard, M.L., Bruns, M.A., Liermann, L.J., Buss, H.L., Brantley, S.L., 2012. Bacterial associations with weathering minerals at the regolith-bedrock interface, Luquillo Experimental Forest, Puerto Rico. Geomicrobiol. J. 29 (9), 792–803.

Monnier, G., Turc, L., Jeanson-Lusignang, C., 1962. Une méthode de fractionnement densimétrique par centrifugation des matières organiques des sols. Ann. Agron. 13, 55.

Mueller, C.W., Koelbl, A., Hoeschen, C., Hillion, F., Heister, K., Herrmann, A.M., et al., 2012. Submicron scale imaging of soil organic matter dynamics using NanoSIMS – from single particles to intact aggregates. Org. Geochem. 42 (12), 1476–1488.

Nduwamungu, C., Ziadi, N., Parent, L.E., Tremblay, G.F., Thuries, L., 2009. Opportunities for, and limitations of near infrared reflectance spectroscopy applications in soil analysis: a review. Can. J. Soil Sci. 89 (5), 531–541.

Nelson, D.W., Sommers, L.E., 1996. Total carbon, organic carbon and organic matter. In: Sparks, D.L., Page, A.L., Helmke, P.A., Loeppert, R.H., Soltanpour, P.N., Tabatabai, M.A., Johnston, C.T., Sumner, M.E. (Eds.), Methods of Soil Analysis. Soil Science Society of America Madison, pp. 961–1010.

Nunan, N., Ritz, K., Crabb, D., Harris, K., Wu, K., Crawford, J.W., et al., 2001. Quantification of the in situ distribution of soil bacteria by large-scale imaging of thin sections of undisturbed soil. FEMS Microbiol. Ecol. 37 (1), 67–77.

Nunan, N., Wu, K., Young, I.M., Crawford, J.W., Ritz, K., 2003. Spatial distribution of bacterial communities and their relationships with the micro-architecture of soil. FEMS Microbiol. Ecol. 44 (2), 203–215.

Paul, E.A., Follett, R.F., Leavitt, S.W., Halvorson, A., Peterson, G.A., Lyon, D.J., 1997. Radiocarbon dating for determination of soil organic matter pool sizes and dynamics. Soil Sci. Soc. Am. J. 61, 1058–1067.

Paul, E.A., Harris, D., Collins, H.P., Schulthess, U., Robertson, G.P., 1999. Evolution of CO_2 and soil carbon dynamics in biologically managed, row crop agroecosystems. Appl. Soil Ecol. 11, 53–65.

Paul, E.A., Morris, S.J., Conant, R.T., Plante, A.F., 2006. Does the acid hydrolysis-incubation method measure meaningful soil organic carbon pools? Soil Sci. Soc. Am. J. 70 (3), 1023–1035.

Peth, S., Chenu, C., Leblond, N., Mordhorst, A., Garnier, P., Nunan, N., et al., 2014. Localization of soil organic matter in soil aggregates using synchrotron-based X-ray microtomography. Soil Biol. Biochem. 78, 189–194.

Peth, S., Horn, R., Beckmann, F., Donath, T., Fischer, J., Smucker, A.J.M., 2008. Three-dimensional quantification of intra aggregate pore space features using synchrotron-radiation-based microtomography. Soil Sci. Soc. Am. J. 72, 897–907.

Piccolo, A., 2002. The supramolecular structure of humic substances: a novel understanding of humus chemistry and implications in soil science. In: Advances in Agronomy. Academic Press, San Diego, pp. 57–134.

Plante, A.F., 2013. Distribution of radiocarbon ages in soil organic matter by thermal fractionation. Radiocarbon 55, 1077–1083.

Plante, A.F., Fernandez, J.M., Haddix, M.L., Steinweg, J.M., Conant, R.T., 2011. Biological, chemical and thermal indices of soil organic matter stability in four grassland soils. Soil Biol. Biochem. 43 (5), 1051–1058.

Plante, A.F., Fernandez, J.M., Leifeld, J., 2009. Application of thermal analysis techniques in soil science. Geoderma 153 (1–2), 1–10.

Poeplau, C., Don, A., Six, J., Kaiser, M., Benbi, D., Chenu, C., et al., 2018. Isolating organic carbon fractions with varying turnover rates in temperate agricultural soils – a comprehensive method comparison. Soil Biol. Biochem. 125, 10–26.

Postma, J., Altemüller, H.J., 1990. Bacteria in thin soil sections stained with the fluorescent brightener calcofluor white M2R. Soil Biol. Biochem. 22, 89–96.

Puget, P., Angers, D.A., Chenu, C., 1999. Nature of carbohydrates associated with water-stable aggregates of two cultivated soils. Soil Biol. Biochem. 31, 55–63.

Puget, P., Chenu, C., Balesdent, J., 1995. Total and young organic carbon distributions in aggregates of silty cultivated soils. Eur. J. Soil Sci. 46, 449–459.

Rawlins, B.G., Wragg, J., Reinhard, C., Atwood, R.C., Houston, A., Lark, R.M., et al., 2016. Three-dimensional soil organic matter distribution, accessibility and microbial respiration in macro-aggregates using osmium staining and synchrotron X-ray CT. SOIL. Discussions. 0, 1–24.

Razavi, B.S., Zhang, X.C., Bilyera, N., Guber, A., Zarebanadkouki, M., 2019. Soil zymography: simple and reliable? Review of current knowledge and optimization of the method. Rhizosphere 11, 100161.

Remusat, L., Hatton, P.J., Nico, P.S., Zeller, B., Kleber, M., Derrien, D., 2012. NanoSIMS study of organic matter associated with soil aggregates: advantages, limitations, and combination with STXM. Environ. Sci. Technol. 46 (7), 3943−3949.

Rumpel, C., Janik, L.J., Skjemstad, J.O., Kogel-Knabner, I., 2001. Quantification of carbon derived from lignite in soils using mid-infrared spectroscopy and partial least squares. Org. Geochem. 32 (6), 831−839.

Rumpel, C., Rabia, N., Derenne, S., Quenea, K., Eusterhues, K., Kogel-Knabner, I., et al., 2006. Alteration of soil organic matter following treatment with hydrofluoric acid (HF). Org. Geochem. 37 (11), 1437−1451.

Sanderman, J., Grandy, A.S., 2020. Ramped thermal analysis for isolating biologically meaningful soil organic matter fractions with distinct residence times. Soil 6 (1), 131−144.

Sanderman, J., Farrell, M., Macreadie, P.I., Hayes, M., McGowan, J., Baldock, J., 2017. Is demineralization with dilute hydrofluoric acid a viable method for isolating mineral stabilized soil organic matter? Geoderma 304, 4−11.

Scharpenseel, H.W., Schiffmann, H., 1977. Radiocarbon dating of soils, a review. Zeitschrift. fur. Pflanzenernahrung. und. Bodenkunde 140 (2), 159−174.

Schmidt, M.W., Rumpel, C., Kögel-Knabner, I., 1999. Evaluation of an ultrasonic dispersion procedure to isolate primary organomineral complexes from soils. Eur. J. Soil Sci. 50, 87−94.

Shang, C., Zelazny, L.W., 2008. Selective dissolution techniques of mineral analysis of soils and sediments. In: Ulery, A.L., Drees, L.R. (Eds.), Methods of Soil Analysis. Part 5. Mineralogical Methods. Soil Science Society of America, Madison, pp. 33−80.

Six, J., Bossuyt, H., De Gryze, S., Denef, K., 2004. A history of research on the link between (micro)aggregates, soil biota, and soil organic matter dynamics. Soil Tillage Res. 79, 7−31.

Six, J., Elliott, E.T., Paustian, K., 2000. Soil macroaggregate turnover and microaggregate formation: a mechanism for C sequestration under no-tillage agriculture. Soil Biol. Biochem. 32 (14), 2099−2103.

Six, J., Jastrow, J., 2002. Organic matter turnover. In: Lal, R. (Ed.), Encyclopedia of Soil Science. Marcel Dekker Inc., New York, pp. 936−942.

Six, J., Schultz, P.A., Jastrow, J.D., Merckx, R., 1999. Recycling of sodium polytungstate used in soil organic matter studies. Soil Biol. Biochem. 31 (8), 1193−1196.

Skjemstad, J.O., Clarke, P., Taylor, J.A., Oades, J.M., McClure, S.G., 1996. The chemistry and nature of protected carbon in soil. Aust. J. Soil Res. 34, 251−271.

Sleighter, R.L., Hatcher, P.G., 2007. Molecular characterization of dissolved organic matter (DOM) along a river to ocean transect of the lower Chesapeake Bay by ultrahigh resolution electrospray ionization Fourier transform ion cyclotron resonance mass spectrometry. Mar. Chem. 110, 140−152.

Sleutel, S., Cnudde, V., Masschaele, B., Vlassenbroek, J., Dierick, M., Van Hoorebeke, L., et al., 2008. Comparison of different nano- and micro-focus X-ray computed tomography set-ups for the visualization of the soil microstructure and soil organic matter. Comput. Geosci. 34 (8), 931−938.

Sollins, P., Swanston, C., Kleber, M., Filley, T., Kramer, M., Crow, S., et al., 2006. Organic C and N stabilization in a forest soil: evidence from sequential density fractionation. Soil Biol. Biochem. 38 (11), 3313.

Solomon, D., Lehmann, J., Harden, J., Wang, J., Kinyangi, J., Heymann, K., et al., 2012. Micro- and nano-environments of carbon sequestration: multi-element STXM-NEXAFS spectromicroscopy assessment of microbial carbon and mineral associations. Chem. Geol. 329, 53−73.

Steffens, M., Zeh, L., Rogge, D.M., Buddenbaum, H., 2021. Quantitative mapping and spectroscopic characterization of particulate organic matter fractions in soil profiles with imaging VisNIR spectroscopy. Sci. Rep. 11 (1), 16725. https://doi.org/10.1038/s41598-021-95298-8.

Stevenson, F.J., 1982. Humus Chemistry: Genesis, Composition and Reactions. John Wiley, New York.

Stevenson, F.J., Cheng, C.N., 1970. Amino acids in sediments: recovery by acid hydrolysis and quantitative estimation by a colorimetric procedure. Geochim. Cosmochim. 34, 77−88.

Strickland, T.C., Sollins, P., 1987. Improved method for separating light and heavy fraction organic material from soil. Soil Sci. Soc. Am. J. 51, 1390−1393.

Terhoeven-Urselmans, T., Michel, K., Helfrich, M., Flessa, H., Ludwig, B., 2006. Near-infrared spectroscopy can predict the composition of organic matter in soil and litter. J. Plant Nutr. Soil Sci. 169 (2), 168–174.

Thévenot, M., Dignac, M.-F., Rumpel, C., 2010. Fate of lignins in soils : a review. Soil Biol. Biochem. 42, 1200–1211.

Thieme, J., Sedlmair, J., Gleber, S.C., Prietzel, J., Coates, J., Eusterhues, K., et al., 2010. X-ray spectromicroscopy in soil and environmental sciences. J. Synchrotron Radiat. 17, 149–157.

Torn, M.S., Trumbore, S.E., Chadwick, O.A., Vitousek, P.M., Hendricks, D.M., 1997. Mineral control of soil organic carbon storage and turnover. Nature 389 (6647), 170–173.

Totsche, K.U., Kogel-Knabner, I., 2004. Mobile organic sorbent affected by contaminant transport in soil: numerical case studies for enhanced and reduced mobility. Vadose Zone J. 3, 352–367.

Trumbore, S., 2009. Radiocarbon and soil carbon dynamics. Annu. Rev. Earth Planet Sci. 37 (1), 47–66.

van der Voort, T.S., Zell, C.I., Hagedorn, F., Feng, X., McIntyre, C.P., Haghipour, N., et al., 2017. Diverse soil carbon dynamics expressed at the molecular level. Geophys. Res. Lett. 44 (23), 11840–11850.

Védère, C., Vieublé Gonod, L., Nunan, N., Chenu, C., 2022. Opportunities and limits in imaging microorganisms and their activities in soil microhabitats. Soil Biol. Biochem. 174, 108807.

Velickovic, D., Anderton, C.R., 2017. Mass spectrometry imaging: towards mapping the elemental and molecular composition of the rhizosphere. Rhizosphere 3, 254–258.

Vidal, A., Kloffel, T., Guigue, J., Angst, G., Steffens, M., Hoeschen, C., et al., 2021. Visualizing the transfer of organic matter from decaying plant residues to soil mineral surfaces controlled by microorganisms. Soil Biol. Biochem. 160, 108347.

Virto, I., BarrÈ, P., Chenu, C., 2008. Microaggregation and organic matter storage at the silt-size scale. Geoderma 146 (1–2), 326.

Vogel, C., Babin, D., Pronk, G.J., Heister, K., Smalla, K., Kogel-Knabner, I., 2014. Establishment of macro-aggregates and organic matter turnover by microbial communities in long-term incubated artificial soils. Soil Biol. Biochem. 79, 57–67.

von Lutzow, M., Kogel-Knabner, I., Ekschmittb, K., Flessa, H., Guggenberger, G., Matzner, E., et al., 2007. SOM fractionation methods: relevance to functional pools and to stabilization mechanisms. Soil Biol. Biochem. 39 (9), 2183–2207.

Wan, J., Tyliszczak, T., Tokunaga, T.K., 2007. Organic carbon distribution, speciation, and elemental correlations within soil micro aggregates: applications of STXM and NEXAFS spectroscopy. Geochim. Cosmochim. Acta 71 (22), 5439–5449.

Watteau, F., Villemin, G., Ghanbaja, J., Genet, P., Pargney, J.C., 2002. In situ ageing of fine beech roots (*Fagus sylvatica*) assessed by transmission electron microscopy and electron energy loss spectroscopy: description of microsites and evolution of polyphenolic substances. Biol. Cell 94 (2), 55–63.

Wershaw, R.L., Leenheer, J.A., Kennedy, K.R., Noyes, T.I., 1996. Use of C-13 NMR and FTIR for elucidation of degradation pathways during natural litter decomposition and composting. 1. Early stage leaf degradation. Soil Sci. 161 (10), 667–679.

Wiesenberg, G.L.B., Schwark, L., Schmidt, M.W.I., 2004. Improved automated extraction and separation procedure for soil lipid analyses. Eur. J. Soil Sci. 55 (2), 349–356.

Wilcke, W., Baumler, R., Deschauer, H., Kaupenjohann, M., Zech, W., 1996. Small scale distribution of Al, heavy metals and PAHs in an aggregated Alpine Podzol. Geoderma 71, 19–30.

Witzgall, K., Vidal, A., Schubert, D.I., Hoschen, C., Schweizer, S.A., Buegger, F., et al., 2021. Particulate organic matter as a functional soil component for persistent soil organic carbon. Nat. Commun. 12 (1), 4115.

Xing, Z., Du, C.W., Tian, K., Ma, F., Shen, Y.Z., Zhou, J.M., 2016. Application of FTIR-PAS and Raman spectroscopies for the determination of organic matter in farmland soils. Talanta 158, 262–269.

Zelles, L., Bai, Q.Y., Beck, T., Beese, F., 1992. Signature fatty acids in phospholipids and lipopolysaccharides as indicators of microbial biomass and community structure in agricultural soils. Soil Biol. Biochem. 24, 317–323.

Zimmermann, M., Leifeld, J., Schmidt, M.W.I., Smith, P., Fuhrer, J., 2007. Measured soil organic matter fractions can be related to pools in the RothC model. Eur. J. Soil Sci. 58 (3), 658–667. https://doi.org/10.1111/j.1365-2389.2006.00855.x.

Zsolnay, A., 2003. Dissolved organic matter: artefacts, definitions, and functions. Geoderma 113 (3/4), 187–209.

Chapter 14

Nitrogen transformations

G.P. Robertson* and P.M. Groffman†

*Department of Plant, Soil, and Microbial Sciences and W.K. Kellogg Biological Station, Michigan State University, MI, USA; †Cary Institute of Ecosystem Studies, Millbrook, NY and Advanced Science Research Center, City University of New York, NY, USA

Chapter outline

14.1 Introduction

Nitrogen (N) is essential for life on Earth. Soil biota are responsible for its accumulation, persistence, and loss from ecosystems. Biotic N transformations in soil include its capture from the atmosphere, mineralization from soil organic matter (SOM), nitrification into forms more likely to be taken up by plants or lost, and denitrification back to atmospheric forms. Our understanding of N cycles in soil has been transformed in recent years with the discoveries of new microbial taxa via the application of modern genomic technologies, new processes via the application of new analytical approaches, and new insights into the functional importance of biotic biodiversity. Understanding N cycle transformations in soil is key to understanding the terrestrial and thus the global N cycle, including the cumulative environmental impact of reactive N – i.e., N that is environmentally active – as it accelerates plant productivity, contributes to climate change, and suppresses biodiversity in ecosystems worldwide.

The importance of N to ecosystem productivity is most evident in agriculture, where the amount of N fertilizer added to the biosphere each year to support crop growth (112 Tg N; FAO 2019) now exceeds the amount of N added naturally from other terrestrial sources (61 Tg; Vitousek et al., 2013). This has enormous implications for both human welfare – we are now feeding more than 7 billion people – and

Soil Microbiology, Ecology, and Biochemistry. https://doi.org/10.1016/B978-0-12-822941-5.00014-4

the environment. Consequences of more anthropogenic N in the biosphere are legion, ranging from degraded ground and surface water quality and increased greenhouse gas loading of the atmosphere to plant biodiversity loss and poor rural air quality.

What do soil biota have to do with this? Nitrogen exists in more forms than any other element essential for life on Earth (Table 14.1), all of which are affected by microbial activity. In fact, once N enters the biosphere in a form that is biologically available, microbes largely control its transformation from one form to another. Even with the advent of chemical fertilizers, microbes remain largely responsible for their entry into the biosphere. Understanding the cycling and fate of reactive N, whether at global or local scales, thus requires understanding the organisms responsible for driving each part of the cycle.

It is a daunting and exciting task. Of the 14 discrete N transformations known to be mediated by microbes, 4 have been discovered in only the past decade (Kuypers et al., 2018). Whole classes of microbes with newly recognized metabolic capacities have been revealed by recent genomic advances, yet we are only now learning how plants can affect — perhaps even control — their N environment by altering their microbiome. Soil microbiology thus plays another crucial role in life on Earth by regulating the form and availability of N to all terrestrial and many aquatic and marine organisms. Since N often limits plant productivity, it follows that soil microbes often regulate net primary production, the productive capacity of ecosystems, whether agricultural or natural. Understanding N transformations and the soil organisms that perform them is thus essential for understanding and managing ecosystem health and productivity.

Nitrogen takes nine different chemical forms in soil corresponding to different oxidative states (Table 14.1). Dinitrogen gas (N_2) comprises 79% of our atmosphere and is by far the most abundant form of N in the biosphere, but it is unusable by most organisms, including plants. Biological N_2 fixation (BNF), whereby N_2 is transformed by microbes into simple organic compounds, is the dominant natural process by which N enters soil biological pools. All other soil N transformations happen subsequently: (1) N mineralization, the conversion of organic N to inorganic forms; (2) N immobilization, the uptake or

TABLE 14.1 Main forms of nitrogen in the environment and their oxidation states.

Name	Chemical formula	Oxidation state
Nitrate	NO_3^-	+5
Nitrogen dioxide (g)[a,b]	NO_2	+4
Nitrite	NO_2^-	+3
Nitric oxide (g)[b]	NO	+2
Nitrous oxide (g)	N_2O	+1
Dinitrogen (g)	N_2	0
Ammonia (g)	NH_3	-3
Ammonium	NH_4^+	-3
Organic N	R	-3

[a]*Gases (g) occur free in both the soil atmosphere and dissolved in soil water.*
[b]*NO and NO_2 are collectively known as NO_x.*

assimilation of inorganic N into biomass by plants, microbes, and other soil organisms; (3) nitrification, the conversion of ammonium (NH_4^+) to nitrite (NO_2^-) and then nitrate (NO_3^-); and (4) denitrification, the conversion of NO_3^- to nitrous oxide (N_2O) and then N_2, closing the global cycle. Other forms of N (Table 14.1) are primarily involved in these conversions as intermediaries. During conversion, they can escape into the environment where they can participate in chemical reactions or be transported elsewhere for further transformation.

Löhnis (1913) first formulated the concept of the N cycle, which formalizes the notion that N is converted from one form to another in an orderly and predictable fashion (Fig. 14.1) with no global loss. That is, to maintain atmospheric equilibrium at the global scale, the same amount of N_2 that is fixed each year must either be permanently stored in geologic reservoirs or converted back to N_2 via denitrification.

Nitrogen fixation — both biological and industrial — now far outpaces historical rates of denitrification and is the principal reason reactive N has accumulated in the biosphere to become a major pollutant (Galloway et al., 2008). Making agricultural and other managed ecosystems more N-conservative, and removing excess N from soils, water bodies, and urban waste streams, are major environmental challenges that require a fundamental knowledge of microbial N transformations (Robertson and Vitousek, 2009).

The microbiology, physiology, and biochemistry of N cycle processes have been studied for over a century, and much of our understanding has been derived from molecular and organismal scale studies in the laboratory. Laboratory observations and experiments have been used to characterize the nature and regulation of N transformations, but their reductionist nature has caused us to sometimes overlook the surprising possibilities for microbial activity in nature, thus impairing our ability to understand the

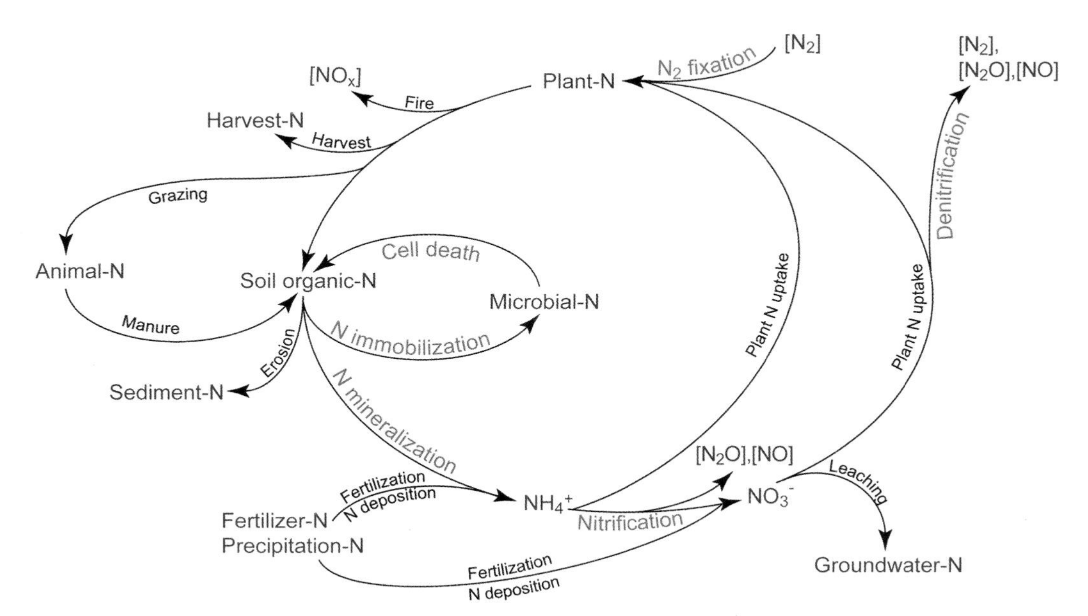

FIGURE 14.1 Schematic representation of the major elements of the terrestrial nitrogen cycle. Those processes mediated by soil microbes appear in red. *(Gases appear in brackets.)*

ecological significance of these processes. Genomic advances have allowed us to overcome some of these limitations by revealing the occurrence and activity of N cycle genes in surprising places; even so, we are still dependent on laboratory observations and experiments to define the functional significance of such genes. For example, theory and years of laboratory work suggested that denitrification ought to occur only in anaerobic wetland and muck soils, but when new field-based methods became available in the 1970s, and genomic methods in the 2000s, it became clear that almost all soils, including dry and even desert soils, support active denitrifiers.

The genomic revolution has also challenged our taxonomic understanding of soil N microbes. Whereas traditional taxonomy often classified microbial taxa on the basis of metabolic capabilities — nitrifiers, denitrifiers, N_2 fixers, and such — we now know of many crossover examples, such as N_2 fixers or nitrifiers that can also denitrify, or the occurrence of N_2 fixation genes in many taxa for which N_2 fixation has not been documented. Archaeal nitrifiers were unknown 20 years ago but are now known to numerically dominate nitrifier populations in almost all soils. Taxonomic classification based on N cycle processes is no longer useful or applicable.

In this chapter we will detail the major soil processes responsible for driving the terrestrial N cycle: N_2 fixation, N mineralization, nitrification, and denitrification. An understanding of these four processes forms the foundation for understanding N in the environment. That said, at the end of the chapter we also consider several other processes that can be important in specialized environments.

14.2 Nitrogen fixation

Nitrogen is rarely present in soil parent materials, creating the basis for widespread limitation of primary production by N. Although atmospheric N_2 is abundant, it is also extraordinarily stable, and large amounts of energy are required to convert it to a form useable by plants and other organisms. Thus the global N cycle depends on energy-intensive mechanisms, either natural or anthropogenic, to "fix" N from the atmosphere.

Biological nitrogen fixation (BNF) dominates natural inputs to the terrestrial biosphere (Table 14.2), but because of the rise of industrial fertilizer production, today it comprises approximately 35% of total global sources. The environmental cost of the shift from biological to synthetic N production is substantial — N fertilizers represent the principal source of greenhouse gas costs in most fertilized cropping systems and contribute to low system-wide N use efficiency (Robertson and Vitousek, 2009). There is thus substantial interest in moving fertilized agriculture toward systems that rely more on BNF to provide the N needed for high yields while increasing N use efficiency and lowering N losses from soil.

BNF in natural systems represents the primary process of N availability in plants. Atmospheric N deposition via precipitation is minor in most ecosystems (certainly in preindustrial times) and SOM (important in all ecosystems) represents the legacy of past BNF. In unfertilized systems an understanding of BNF and its consequences is crucial for understanding ecosystem function. This may become especially important as climate change solutions that involve sequestering C in ecosystems will necessarily also sequester N — creating a new demand for N that could be difficult to satisfy without additional BNF or fertilizer inputs (Hungate et al., 2003).

Biological nitrogen fixation, mediated exclusively by microbes with the enzymatic capacity to reduce atmospheric N_2 to ammonia (NH_3), is subsequently assimilated into amino acids or leaked into the soil

TABLE 14.2 Sources of nitrogen to the terrestrial biosphere.

Inputs	Tg N
Natural	
Biological N_2 fixation (BNF)	44
Lightning fixation	4
Rock N	10
N in aerosols from ocean	3
Total	61
Anthropogenic	
N fertilizer	112
Crop biological N_2 fixation	48
Fossil fuel combustion	25
Other industrial production	22
Total	207

From Vitousek et al. (2013) and (FAO 2019).

environment as NH_4^+. Nitrogen (N_2) fixers fall into three functional groups described by life history habits that evolved in response to the high energy requirements for converting N_2 to NH_3: (1) symbiotic N_2 fixers that live in close association with a host that provides their symbiont fixed C and a BNF-compatible microenvironment in exchange for assimilable N; (2) associative N_2 fixers that live in close proximity to plants, often in the plant rhizosphere (adjacent to roots), use C that is provided by the plants (purposefully or not) to fix N_2; and (3) free-living N_2 fixers that use C available to all heterotrophs, or in the case of phototrophic cyanobacteria, fix their own through photosynthesis.

14.2.1 The biochemistry of biological nitrogen fixation

BNF occurs via the nitrogenase enzyme complex:

$$N_2 + 8H^+ + 8e^- + 16Mg - ATP \xrightarrow{\text{nitrogenase}} 2NH_3 + H_2 + 16Mg - ADP + 16\,P_i. \qquad \text{(Eq. 14.1)}$$

NH_3 is toxic to cells in high concentrations, thus in free-living diazotrophs (archaea and bacteria capable of BNF), NH_3 is quickly assimilated into glutamate through the glutamine synthetase/glutamate synthase pathway (Bottomley and Myrold, 2015):

$$\text{Glutamate} + NH_3 + ATP \xrightarrow{\text{glutamine synthetase}} \text{Glutamine} + ADP + P_i \qquad \text{(Eq. 14.2)}$$

$$\text{Ketoglutarate} + \text{Glutamine} + \text{NADH} \xrightarrow{\text{glutamate synthetase}} 2\ \text{Glutamate} + \text{NAD}_{ox}. \quad \text{(Eq. 14.3)}$$

In symbiotic and associative diazotrophs the NH_3 is excreted and quickly assimilated by plant enzymes into amino acids. NH_3 can also escape into the soil solution, where it can be nitrified or otherwise assimilated by heterotrophs.

The nitrogenase complex is comprised of both dinitrogenase plus dinitrogenase reductase and requires metallic cofactors – most commonly molybdenum (Mo) but, in some nitrogenases, vanadium (V) and iron (Fe). Complete assembly of nitrogenase requires multiple *nif* genes: *nif*H encodes dinitrogenase reductase, *nif*D and *nif*K Mo nitrogenase, and *nif*K Fe nitrogenase. A variety of other *nif* genes (including B, F, I, J, L, LA, N, Q, S, U, V, W, X, Y, and Z) serve to regulate oxygen (O_2), sense inorganic N, assemble iron-sulfur clusters, and perform various other functions crucial for effective N_2 fixation (de Bruijn, 2015). At least nine *nif* genes are required for complete BNF.

Equations 14.1 to 14.3 show that BNF is energetically taxing. A total of 16 ATP molecules are necessary to reduce 1 molecule of N_2 to $2NH_3$, although some diazotrophs possess uptake hydrogenase, so the 4 ATP molecules needed to reduce $2H^+$ to H_2 can be recycled back to protons and electrons to power additional N_2 reduction. Energy is also required to maintain a large number of genes and their products necessary to synthesize and support a completely functional N_2-fixing enzyme system. Remarkably, nitrogenase can comprise ~10% of total cell protein in a functioning diazotroph (Bottomley and Myrold, 2015). With all said, the energetic cost of assimilating N_2 via nitrogenase vs. NH_4^+ free in the soil solution ranges from a factor of 1.8 to 5.4 (Hill, 1992).

Additional to the cost of maintaining and activating nitrogenase are the costs of maintaining an anaerobic microenvironment for BNF. Nitrogenase is exquisitely sensitive to O_2, which irreversibly denatures it; thus diazotrophs or their symbionts must invest in ways to exclude O_2. In the legume-rhizobia symbiosis this cost is largely borne by the plant, which creates a novel root nodule designed specifically to house diazotrophs. O_2 is actively excluded by transporting it away from the nodule via leghemoglobin – a red-colored protein with a high affinity for O_2, similar to hemoglobin in human blood. In actinorhizal symbioses (see below), although the plant creates nodules, O_2 exclusion appears mostly achieved by the vesicle wall of the bacterium itself.

In most free-living cyanobacteria (photosynthetic bacteria formerly known as blue-green algae) BNF occurs in vegetative cells called heterocysts, where thick walls and respiration protect nitrogenase from O_2, and light is used to directly power BNF. Cyanobacteria, with or without heterocysts, can also perform BNF at night when respiration consumes unwanted O_2. Free-living and associative diazotrophs in soil may be at the greatest O_2 disadvantage, which may limit BNF activity to episodic bursts following significant rainfall (Roley et al., 2019) or to high-respiration, low-O_2 soil microsites.

14.2.2 The diversity of biological nitrogen fixers

Genomic tools have massively expanded our knowledge of the number of organisms that can perform BNF. The ability to synthesize nitrogenase and fix N_2 remains exclusively with microbes – primarily bacteria, but also a few methanogenic archaea. Legume symbionts, collectively called rhizobia, are predominantly α-proteobacteria lineage proteobacteria. These include the well-studied genera *Rhizobium* and *Bradyrhizobium*, especially important in legumes cultivated for food such as soybean (*Glycine max* L.) and common bean (*Phaseolus vulgaris* L.). Since 1990, we have discovered scores of new genera that can partner with these and other legumes (Roy et al., 2020), almost all understudied, which has provided a wealth of opportunities to better understand this important association.

Symbiotic BNF also arose in a different bacterial phylum, the actinobacteria (Fig. 14.2). Approximately eight different nonleguminous plant families contain genera capable of hosting actinobacteria, such as *Frankia* and *Parasponia* (Table 14.3). So-called actinorhizal plants are globally distributed and tend to be woody shrubs (e.g., *Myrica* and *Ceanothus*) or trees (e.g., *Alnus* and *Casuarina*) that colonize early successional forests or shrublands following ecological disturbance but can also persist in aggrading forests (Binkley et al., 1992), where rates of N_2 fixation can rival those in leguminous field crops.

Free-living diazotrophs belong to a wide phylogenetic range of bacteria that includes the α-proteobacteria, β-proteobacteria (e.g., *Burkholderia*, *Nitrospira*), δ-proteobacteria, γ-proteobacteria (e.g., *Pseudomonas*, *Xanthomonas*), firmicutes, and cyanobacteria (Gaby and Buckley, 2015). Free-living diazotrophs also include symbiotic and associative N fixers that must persist in soil prior to associating with their plant hosts. Metagenomic analysis based on the presence of *nif*H genes in native switchgrass (*Panicum virgatum* L.) rhizospheres (Bahulikar et al., 2020) revealed the cooccurrence of diazotrophs from at least five phyla. Renewed interest in switchgrass and other native grasses for bioenergy production is fueling new interest in associative nitrogen fixation as a means for providing N to crops without the economic and environmental costs of N fertilizers (Robertson et al., 2017).

A phylogenetic map of *nif*H genes (Fig. 14.2) illustrates the broad diversity of BNF among archaea and bacterial phyla. The presence of *nif*H in archaea and bacteria worldwide is striking, with about as many representatives in marine as in terrestrial habitats. While most legume symbionts (rhizobia) are in the α-proteobacteria and nonlegume symbionts like *Frankia* in the Frankia, Paenibacillus, and

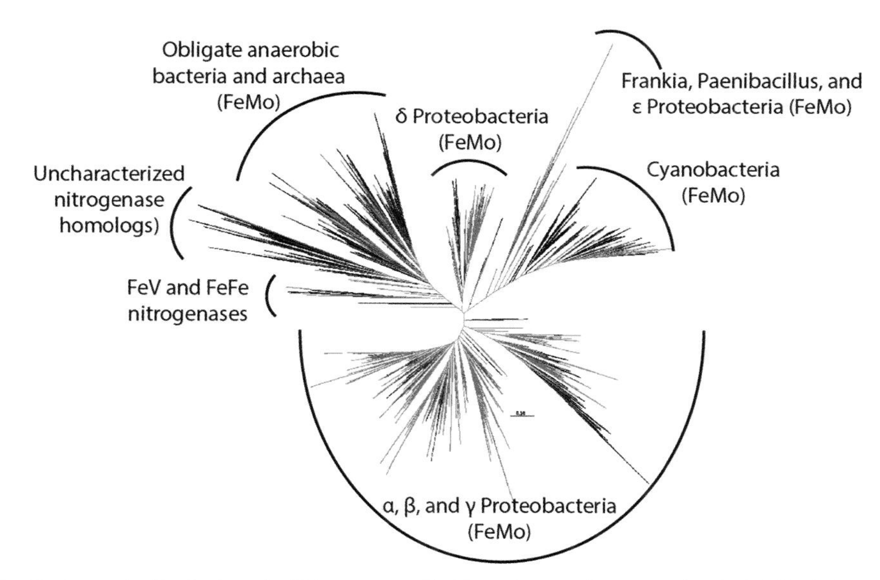

FIGURE 14.2 Phylogenetic distribution of *nif*H sequences. Red branches are those associated with N_2-fixing soil bacteria and archaea; black branches are from marine environments. Most legume symbionts (Rhizobia) are α-proteobacteria; nonlegume symbionts, such as actinobacteria, are in the *Frankia*, *Paenibacillus*, and ε-proteobacteria lineage. Free-living diazotrophs cut across all groups. *(Redrawn with permission from Gaby and Buckley, 2011.)*

TABLE 14.3 Families and genera of N_2-fixing plant-*Frankia* actinorhizal associations.

Family	N_2-fixing genera
Betulaceae	*Alnus*
Casuarinaceae	*Casuarina*
Coriariaceae	*Cariaria*
Datiscaceae	*Datisca*
Elaeagnaceae	*Elaeagnus, Hippophae*
Myricaceae	*Myrica and Comptonia*
Rhamnaceae	*Ceanothus*
Rosaceae	*Cercocarpus and Purshia*

From Bottomley and Myrold (2007)

ε-proteobacteria lineages, free-living diazotrophs cut across all groups: α-, β-, δ-, γ-, and ε-proteobacteria, as well as the firmicutes and cyanobacteria. While the presence of *nif*H in DNA does not necessarily reflect its use, it seems unlikely that coding for a protein this complex would be genetically conserved without consistent selective pressure. Notwithstanding, linking metagenomic knowledge to functional activity — in this case, quantitative N_2 fixation — remains a significant exciting challenge in soil microbial ecology.

14.2.3 Environmental control of biological nitrogen fixation

The conservative nature of the nitrogenase complex means that all diazotrophs are potentially constrained by the same set of ecological factors (Vitousek et al., 2013): low available energy, abundant inorganic N and O_2, and an insufficient supply of key resources, such as phosphorus (P), Fe, potassium, and Mo. Perhaps the most important environmental control on BNF is N availability. Given the high energetic cost of maintaining and using nitrogenase, it makes little evolutionary sense for a microbe to fix N_2 for assimilation into amino acids and proteins when more available forms of N are available. It is better to use energy for growth and reproduction than to fix N_2 when N is available in other useable forms. In fact, NH_4^+ is well known to inhibit nitrogenase synthesis and therefore BNF in pure culture, and in plant symbioses the plant takes the same tack. For example, BNF in soybeans declines to almost nil when soil inorganic N is available (Fig. 14.3), relieving the plant of much of the C cost of N assimilation. Conversely, BNF is especially important in infertile soils, whether naturally low in N or degraded. Exceptions to BNF N inhibition appear in some symbiotic plants and lichens (Binkley et al., 1992; Drake, 2011; Menge and Hedin, 2009), suggesting opportunities for a better understanding of the genetic and physiological basis for this functional trait.

Carbon or available energy can be an equally important constraint to BNF, especially in free-living and associative diazotrophs. Root exudates, known to be significant signaling compounds for legume nodulation, may also stimulate associative N_2 fixers (Coskun et al., 2017), whether intentionally or not. Evidence

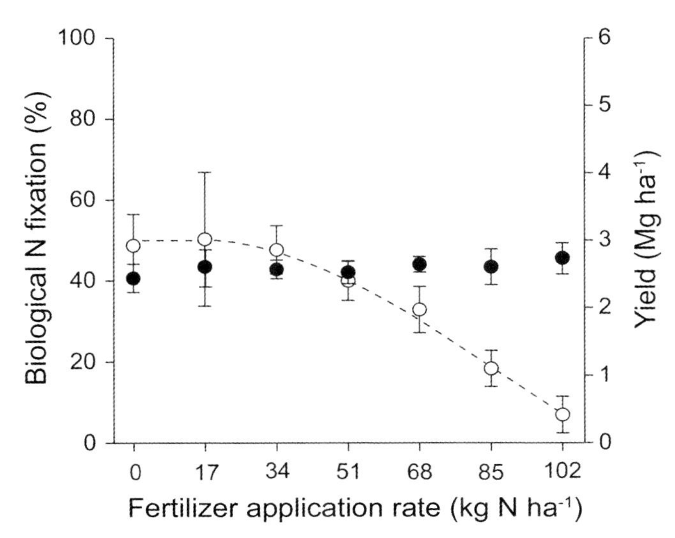

FIGURE 14.3 Biological N_2 fixation (open circles) and grain yield (filled circles) in soybeans grown in the field at different fertilizer rates. *(Redrawn with permission from Gelfand and Robertson, 2015.)*

for resource constraints other than N and energy primarily comes from resource addition experiments. BNF in alfalfa (*Medicago sativa* L.), for example, can be responsive to added Mo. Phosphorus has been observed to increase BNF in herbaceous legumes and forests (both in trees and epiphytic cyanolichens), although for nutrients that also limit plant productivity, it can be difficult to separate nutrient limitations on plant productivity and subsequent delivery of fixed C to diazotrophs from direct diazotrophic limitations on BNF per se. However, from a management standpoint, it may not matter.

Given that N more often than any other nutrient limits plant productivity in most terrestrial ecosystems, it can be interesting to ask why BNF, especially symbiotic BNF, is not more widespread. Shouldn't plants that partner with diazotrophs have a competitive advantage over plants that rely exclusively on atmospheric N deposition and SOM mineralization? Several hypotheses have been advanced to explain this apparent paradox (Vitousek et al., 2013): (1) in forests N_2-fixing plants tend to be shade intolerant due to the high energetic demands of BNF and thus are able to establish only following canopy-clearing disturbances when plant competitors are temporarily suppressed; (2) legumes and other N_2-fixing plants have higher leaf N and protein contents, making them more palatable than nonfixing plants to herbivores, especially in grazed systems; and (3) higher non-N nutrient demands of symbiotic N_2 fixers (Mo, Fe, perhaps P) place them at a competitive disadvantage. It is curious, however, that despite the higher energetic costs of BNF to symbiotic plants, at least in moderately fertile soil there seems to be no agronomic yield penalty for a greater reliance on BNF than on N fertilizer (Fig. 14.3).

14.3 Nitrogen mineralization and immobilization

A critical process in any nutrient cycle is the conversion of organic forms of nutrients in dead biomass (detritus) into simpler, soluble forms that can be taken up again by plants and the soil organisms (Chapter 12). This conversion is carried out by microbes and other soil organisms that release or mineralize nutrients as a by-product of their detritus consumption. Although microbes consume detritus primarily as sources of energy and C to support the growth of new microbial biomass, they also have a need for nutrients, especially N, to assemble proteins, nucleic acids such as DNA, and other cellular components. If plant detritus is rich in N, microbial needs are easily met, and net N release or net N mineralization occurs. If plant detritus is low in N, microbes must scavenge additional inorganic N from their surroundings, leading to its net immobilization into microbial biomass.

The key to understanding mineralization-immobilization is to "think like a microbe": that is, think about a microbe's attempt to make a living by obtaining energy and C from detritus. Sometimes the detritus has all the N needed, so as detritus is consumed for its C, any extra N is released (mineralized) to the soil solution. Sometimes the detritus does not have enough N to meet microbial growth needs, so as detritus is consumed, additional N must be immobilized from the soil solution. Likely these two scenarios are happening simultaneously within even relatively small volumes of soil. While one group of microbes might be consuming a protein-rich and therefore N-laden bit of organic matter (think legume leaves), another group, perhaps <100 μm away, might be consuming detritus rich in C but low in N (think plant stalk). The first group is mineralizing N, while the second is immobilizing N, perhaps even immobilizing the same N that is being mineralized by the first.

As a result of the simultaneous nature and small scale of these processes, it is worth making a distinction between gross and net mineralization and immobilization. Gross N mineralization is the total amount of soluble N released by soil biota, and gross N immobilization is the total amount of soluble N consumed. Net N mineralization is the balance between the two. When gross mineralization exceeds gross immobilization, inorganic N in the soil increases (i.e., there is net mineralization). When gross immobilization exceeds gross mineralization, inorganic N in the soil decreases (i.e., there is net immobilization). This effect is readily apparent in compost management. Compost that has a high C-to-N (C:N) ratio, such as decomposing wood or wheat straw (Table 14.4), will lead to net N immobilization when applied to soil, whereas compost with a low C:N ratio, such as decomposing clover or lawn-grass clippings, will lead to net N mineralization. The difference will strongly affect plant N availability.

There is also an energetic cost to decomposition. Microbes invest more energy in the synthesis of enzymes to obtain nutrients (e.g., amidases to acquire N and phosphatases to acquire P) when decomposing substrates of low quality. Microbial N uptake is also affected by organism growth efficiency or the proportion of metabolized C that becomes microbial biomass (Chapter 9). Fungi have higher C:N ratios in their tissues than bacteria and archaea and so can grow more efficiently on low N substrates.

Traditionally, NH_4^+ has been viewed as the immediate product of mineralization, and in the older literature mineralization is often referred to as ammonification. More recently, recognition of the fact that plants from a variety of habitats can take up simple, soluble organic forms of N leads us to broaden our definition of mineralization products to include any simple, soluble form of N that can be taken up by plants (Moreau et al., 2019). Mycorrhizae can play a role in this uptake by absorbing amino acids, amino sugars, peptides, proteins, and chitin that are then used by their hosts as an N source (Chapter 4).

Soil fauna also play an important role in mineralization and immobilization (Chapter 5). They are responsible for much of the preliminary decomposition of detritus; feed on and can regulate populations of archaea, bacteria, and fungi; and can create or modify habitats for a wide array of organisms. For example, isopods shred leaf litter, earthworms create castings and burrows, and termites macerate wood. Often their own consumption is aided by gut microbes — wood-feeding termites, for example, rely on protozoan, bacterial, and fungal symbionts to digest cellulose. All heterotrophic soil organisms consume organic materials for energy and C and, at the same time, immobilize and mineralize N.

TABLE 14.4 C:N ratios of various organic materials.

Organic material	C:N ratio
Soil microorganisms	8:1
Sewage sludge	9:1
Soil organic matter	10:1
Alfalfa residues	16:1
Farmyard manure	20:1
Corn stover	60:1
Grain straw	80:1
Oak litter	200:1
Pine litter	300:1
Crude oil	400:1
Conifer wood	625:1

From Tisdale et al. (1993) and Hyvönen et al. (1996).

The widely distributed nature of mineralization and immobilization means that the environmental regulation of these processes is relatively straightforward. Rates of activity increase with temperature and are optimal at intermediate soil water contents, similar to rates of respiration (see Fig. 14.4), yet it is important to recognize that significant activity often occurs at extremes of both temperature and moisture. Globally, in most soils the quantity and quality of detrital inputs are the main factors that control the rates and patterns of mineralization and immobilization (Li et al., 2019). When moisture and temperature are favorable, large inputs of organic matter lead to high rates of microbial activity and the potential for high rates of mineralization and immobilization. However, in soils that are waterlogged or very cold (think wetlands or Arctic tundra), moisture or temperature can limit microbial activity, and SOM and the organic N it contains will accumulate due to low rates of mineralization.

Water-filled pore space is a useful measure to examine moisture's influence on soil biological activity because it includes information about the impact of soil water on aeration in addition to information on water availability. The calculation of the percent of water-filled pore space is:

$$\frac{\text{soil water content} \times \text{bulk density} \times 100}{1 - (\text{bulk density}/2.65)} \quad \text{(Eq. 14.4)}$$

Soil water content is determined gravimetrically (g H_2O g^{-1} dry soil), bulk density (g cm^{-3}) is the dry mass of a given soil volume, and the value 2.65 is the density (g cm^{-3}) of rock, which by definition has no pores. In most soils microbial activity (respiration) tends to be highest at a water-filled pore space of ~60% as first documented by Linn and Doran (1984) (Fig. 14.4).

What controls the balance between N mineralization and immobilization? The answer is primarily organic matter quality — the availability of C in the material relative to its available N. Consider the effects of

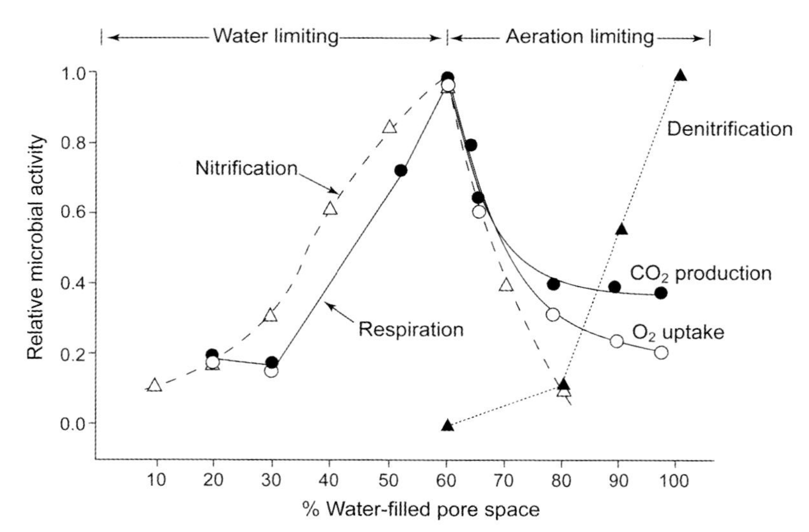

FIGURE 14.4 The relationship between water-filled pore space (a measure of soil moisture availability) and the relative amount of microbial activity in soil. *(Redrawn with permission from Linn and Doran, 1984.)*

adding various organic materials with different C:N ratios to soil (Table 14.4). When one adds manure to soil, with its relatively low C:N ratio of about 20:1, microbes have little trouble obtaining N, and as a result, mineralization dominates over immobilization, and plant-available N increases in soil. This is why manure is frequently used as a fertilizer. On the other hand, were one to add sawdust with its high C:N ratio (625:1) to soil, the microbes could not degrade this material without additional N because the sawdust has only 1 g of N for every 625 g of C, well below the amount of N needed to build proteins or other biomass constituents. Thus the microbes must acquire N from soil, resulting in a decrease in plant-available soil N. If there is no N to immobilize, microbial growth is slowed.

The balance between mineralization and immobilization is also affected by organism N use efficiencies or C:N ratios. As noted earlier, fungi have wider C:N ratios in their tissues than bacteria and will therefore have a lower need for N and, subsequently, will mineralize N more readily. As a general rule of thumb, materials with a C:N ratio >25:1 stimulate immobilization, whereas those with a C:N ratio <25:1 stimulate mineralization (Table 14.4). Highly decomposed substances, such as SOM, humus, and compost in which labile C and N have been depleted, are the exception to this rule. Even though these substances may have a low C:N ratio, the undecomposed C is in complex forms and inherently resistant to decomposition; thus mineralization also proceeds slowly.

There is a wide variety of methods for measuring mineralization and immobilization (Hart et al., 1994, Robertson et al., 1999). Measurement of net mineralization and immobilization rates is much easier and more common than the measurement of gross rates. Gross rates are measured using isotope dilution methods, whereby small amounts of ^{15}N-labeled NH_4^+ are added to soil, and the subsequent dilution of the ^{15}N with natural ^{14}N from mineralized organic matter is used as a basis for calculating the gross production and consumption of NH_4^+.

Measurement of net rates usually involves measuring changes in inorganic N levels in some type of whole soil incubation. The accumulation of N during incubation is considered net mineralization, whereas the loss of N is net immobilization. In most cases these incubations are conducted in well-aerated containers with no plant uptake or leaching losses. Changes in inorganic N levels are measured by periodic extractions of incubated soil. Incubation methods vary widely, from short (10 day) incubations of intact soil cores buried in the field to long (>52 week) incubations of sieved soils in the laboratory. Net N mineralization assays are a powerful means for understanding a soil's capacity for meeting plant N needs and are a common way to compare soil N availability through time and across ecosystems, landscapes, and even continents.

14.4 Nitrification

Nitrification is the microbial oxidation of NH_3 to less reduced forms, principally NO_2^- and NO_3^-. Nitrifying bacteria, first isolated in the late 1800s (Frankland and Frankland, 1890; Winogradsky, 1890), gain as much as 440 kJ of energy per mole of NH_3 oxidized when NO_3^- is the end product. The discovery of archaeal nitrifiers in 2005 (Könneke et al., 2005) and in 2015 a bacteria capable of oxidizing both NH_3 and NO_2^- in a single cell (Daims et al., 2015; van Kessel et al., 2015) has had a paradigm-shifting impact. We now know that nitrifiers are much more ubiquitous and diverse than earlier imagined, including bacteria, fungi, archaea, in addition to heterotrophs as well as autotrophs.

The importance of nitrifiers to ecosystem function is considerable. Even though some NO_3^- enters ecosystems as atmospheric N deposition or fertilizer, in most ecosystems NO_3^- is formed in situ via nitrification. This includes fertilized agricultural systems, insofar as the vast majority of chemical fertilizers are NH_3 based and organic fertilizers are first mineralized to soil NH_4^+. Because NO_3^- is an anion, in most soils it is substantially more mobile than NH_4^+, the ionized source of NH_3 in soil water:

$$NH_4^+\ (aq) \rightleftarrows NH_3\ (aq) + H^+\ (aq). \quad \text{(Eq. 14.5)}$$

As a positively charged ion, NH_4^+ can be held on cation-exchange sites associated with SOM, clay surfaces, and variable-charge minerals. Nitrate, on the other hand, is mostly free in the soil solution and can be easily transported out of the rooting zone by water when precipitation or irrigation exceeds evapotranspiration. An exception occurs in highly weathered soils, such as in much of the tropics, where variable charge minerals at low pH have anion exchange sites that can hold NO_3^-.

Nitrification is also a major source of soil acidity (Coleman and Thomas, 1967), which can have multiple effects on ecosystem health, including the mobilization of toxic metals and the hydrologic loss of base cations due to hydrogen ions' displacing other cations on exchange sites. In soils dominated by variable-charge minerals, which include most highly weathered tropical soils, soil acidity largely controls cation-exchange capacity (CEC) (Sollins et al., 1988), which can be driven to very low levels by nitrifier-generated acidity. Additionally, many plants (Moreau et al., 2019) and heterotrophic microbes (Jones and Richards, 1977) prefer one form of inorganic N over the other, implying a potential effect of nitrifiers on plant and microbial community composition. Finally, nitrifiers themselves can be direct and important sources of the atmospheric gases NO_x and N_2O through nitrifier denitrification when O_2 is low (Zhu et al., 2013) or via by-product formation.

14.4.1 The biochemistry of autotrophic nitrification

Autotrophic nitrifiers obtain their C from CO_2 or bicarbonate (HCO_3^-) rather than from organic matter and are obligate aerobes. Until recently, it was thought that autotrophic nitrification is necessarily a two-step process, carried out by separate groups of bacteria and archaea called NH_3 and NO_2^- oxidizers. We are now aware that so-called comammox bacteria (combined or complete ammonia oxidizers) in the genus *Nitrospira* can perform complete nitrification – both NH_3 and NO_2^- oxidation – within the same cell, although canonical nitrification, carried out sequentially by two separate taxa, appears to be far more common.

Ammonia oxidation is characterized as:

$$NH_3 + 1\frac{1}{2} O_2 \rightarrow NO2^- + H^+ + H_2O. \quad \text{(Eq. 14.6)}$$

The first step in this oxidation is mediated by the membrane-bound enzyme ammonia monooxygenase, which can also oxidize a wide variety of organic, nonpolar low-molecular-weight compounds, including phenol, methanol, methane, and halogenated aliphatic compounds such as trichloroethylene:

$$NH_3 + 2H^+ + O_2 + 2e^- \xrightarrow{\text{ammonia monooxygenase}} NH_2OH + H_2O. \quad \text{(Eq. 14.7)}$$

This reaction is irreversibly inhibited by small quantities of acetylene, which inhibits ammonia monooxygenase and provides a straightforward means for experimentally differentiating autotrophic from heterotrophic nitrification in soil.

Until recently, hydroxylamine was thought to be further oxidized directly to NO_2^- by hydroxylamine oxidoreductase. It has now been shown for ammonia-oxidizing bacteria (AOB) (Caranto and Lancaster, 2017) that hydroxylamine dehydrogenase (HAO) oxidizes hydroxylamine to NO, which is then oxidized to NO_2^- by an unidentified nitric oxide oxidoreductase (NOO) (Fig. 14.5):

$$NH_2OH + H_2O \xrightarrow{NH_2OH \text{ dehydrogenase}} NO + 3e^- + 3H^+ \quad \text{(Eq. 14.8)}$$

$$NO + H_2O \xrightarrow{\text{nitric oxide oxidoreductase}} NO_2^- + 1e^- + 2H^+. \quad \text{(Eq. 14.9)}$$

NO is likewise an essential metabolite in ammonia-oxidizing archaea (AOA), but whether it is similarly produced or is a cosubstrate with hydroxylamine for the production of NO_2^- is as yet unclear (Stein, 2019).

Two of the four electrons released in these reactions replace the two used in the first oxidation reaction, leaving a net of two electrons to generate energy for cell growth and metabolism via electron transport:

$$2H^+ + 1\,{}^1\!/\!{}_2O_2 + 2e^- \xrightarrow{\text{terminal oxidase}} H_2O. \quad \text{(Eq. 14.10)}$$

NO produced by AOB can escape into the atmosphere and influence the photochemical production of ozone (O_3) into the troposphere and the atmospheric abundance of hydroxyl (OH) radicals, primary oxidants for a number of important tropospheric trace gases, including methane. This is a good example of nitrification's indirect effect on global atmospheric chemistry.

The NO_2^- produced by AOB can also be used to produce N_2O, an important greenhouse gas that can then escape into the atmosphere by a process known as nitrifier denitrification. In O_2-stressed environments AOB can use NO_2^- as an electron acceptor rather than O_2. There is no current evidence that archaea or comammox nitrifiers can denitrify as they lack NO reductase genes (Prosser et al., 2020), but the NO_2^- produced by both AOA and AOB can, if not consumed by nitrite oxidizers, be abiotically converted to

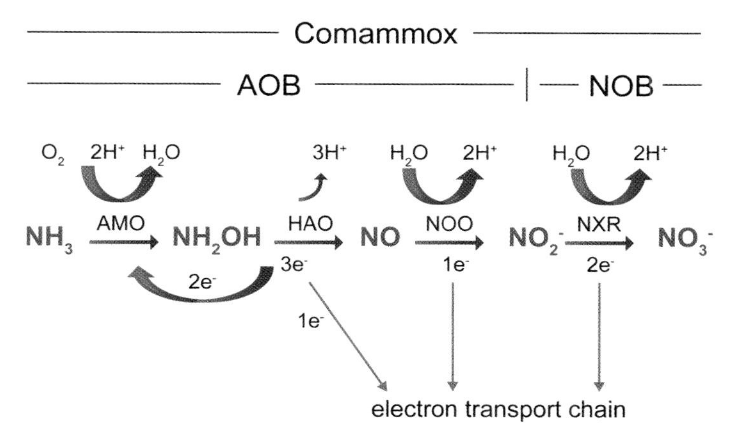

FIGURE 14.5 Autotrophic nitrification pathway for ammonia-oxidizing bacteria (AOB), nitrite-oxidizing bacteria (NOB), and comammox bacteria. AMO = ammonia monooxygenase, HAO = hydroxylamine dehydrogenase, NOO = nitric oxide oxidoreductase, NXR = nitrite oxidoreductase. Some NO can also escape into the atmosphere, as can N_2O, which is produced biotically from NH_2OH oxidation and the reduction of NO_2^- to NO then N_2O, termed nitrifier denitrification. N_2O can also be produced through abiotic reactions of NO, NH_2OH, and NO_2^-. The pathway for AOA is similar to that for AOB, but with different enzymes than HAO and NOO and an unclear role for NO. AOA also produce N_2O but only abiotically. *(Redrawn with permission from Stein, 2019.)*

reactive N gases by a variety of underappreciated chemical reactions resulting in NO_x and N_2O, and difficult to distinguish from biological sources (Heil et al., 2016).

In most soils the NO_2^- produced during canonical NH_3 oxidation is quickly oxidized to NO_3^- by NO_2^--oxidizing bacteria:

$$NO_2^- + H_2O \xrightarrow{\text{nitrite oxidoreductase}} NO_3^- + 2H^+ + 2e^-. \quad \text{(Eq. 14.11)}$$

These reactions are membrane associated, and because nitrite oxidoreductase is a reversible enzyme, the reaction can be reversed to result in NO_3^- reduction to NO_2^-.

Comammox bacteria differ from the canonical nitrifiers in that they can carry out complete nitrification within a single cell:

$$NH_3 + 2\,O_2 \rightarrow NO_3^- + 2H^+ + H_2O.$$

Up to 80% of the energy produced during nitrification is used to fix C. Growth efficiencies of all nitrifiers are correspondingly very low, which is especially the case for comammox bacteria (Koch et al., 2019).

14.4.2 The diversity of autotrophic nitrifiers

Our taxonomic understanding of nitrifiers has been fundamentally transformed over the past few years by new molecular techniques that have revealed considerable taxonomic diversity, whereas before, there was thought to be little. The development and use of 16S rRNA gene primers and subsequent metagenomic techniques targeting genes for ammonia monooxygenase (*amoA*) have demonstrated both a greater diversity among bacterial nitrifiers as well as the presence of nitrifiers in a completely different domain, the

Archaea. As first noted by Leininger et al. (2006), soil *amoA* gene abundance suggests that archaeal nitrifiers are far more abundant than bacterial nitrifiers in most soils. Comammox bacteria have now been documented in a wide variety of habitats (Koch et al., 2019), including agricultural soils (Orellana et al., 2018; Wang et al., 2020). Keep in mind, however, that greater abundance does not necessarily imply greater activity.

The ecological significance of these discoveries is slowly coming into focus. Inference enzyme kinetics for the few available isolates suggest that archaeal and comammox nitrifiers are favored in soil microenvironments with very low NH_4^+ concentrations (Koch et al., 2019; Prosser and Nicol, 2012), which by inference will have very little NH_3 available to nitrifiers. However, the discovery of AOA isolates with much higher NH_3 substrate affinities (e.g., Lehtovirta-Morley et al., 2016) complicates easy generalizations. Nevertheless, enzyme kinetic studies of whole soils with AOB inhibitors, which allow us to judge the relative importance of each group in different soils without the need to study hard-to-isolate individual populations (e.g., Liang et al., 2020; Taylor et al., 2012), corroborate the general trend of AOA's importance in soils with low NH_4^+ availability. AOA tend to be more active than AOB in soils with low NH_4^+ concentrations, such as low-pH forest soils (e.g., Liang et al., 2020), whereas AOB are more active than AOA in soils with an abundant NH_4^+ supply, such as fertilized agricultural soils. Evidence to date confirms the coexistence of all three groups (AOA, AOB, and comammox) in most soils, underscoring the importance of microsite heterogeneity for promoting microbial diversity.

Prior to 2000, the bacterial nitrifiers were viewed as the single family Nitrobacteraceae, defined by their characteristic ability to oxidize NH_3 or NO_2^-. Early work beginning with Winogradsky (1892) classified the NH_3-oxidizing genera of Nitrobacteraceae on the basis of cell shape and the arrangement of intracytoplasmic membranes. This yielded five genera: *Nitrosomonas*, *Nitrosospira*, *Nitrosococcus*, *Nitrosolobus*, and *Nitrosovibrio*. Recent work with isolates, based principally on 16S rRNA oligonucleotide and gene sequence analysis, places terrestrial NH_3-oxidizing bacteria in the beta subclass of the Proteobacteria (Fig. 14.6; Norton, 2011). *Nitrosolobus* and *Nitrosovibrio* are no longer considered distinct from *Nitrosospira*, and *Nitrosococcus* is being reclassified to *Nitrosomonas*. Today, we have nearly complete 16S rRNA gene sequences with >1000 nucleotides for the 14 described species of Betaproteobacteria NH_3 oxidizers, which have a gene sequence similarity of 89% (Fig. 14.7; Koops et al., 2006).

In arable soils the *Nitrosomonas communis* lineage is numerically dominant among culturable strains. Unfertilized soils usually also contain strains of the *Nitrosomonas oligotropha* lineage and strains of *Nitrosospira* and *Nitrosovibrio* (Koops and Pommerening-Röser, 2001). The latter two tend to be dominant in acid soils, which contain few if any *Nitrosomonas*. Culturable strains tell a very limited story, however.

Culture-free molecular techniques, such as 16S rRNA sequencing and the retrieval of *amoA* clones, have now been widely used to examine the diversity of NH_3 oxidizers in vivo. These techniques avoid the need for pure-culture cultivation and its bias toward those species that can be successfully separated from their native habitat. Although molecular techniques can themselves be biased because of their dependence on effective extraction of nucleic acid from soil and the bias associated with PCR amplification, primers, and cloning methods, they nevertheless suggest that most soils are dominated by archaeal species and *Nitrosospira*, not by *Nitrosomonas* (Prosser, 2011). Archaeal species are diverse and formally defined as class Nitrososphaeria in the phylum Thaumarchaeota, with four basal lineages: Ca. Nitrosocaldales, Nitrososphaerales, Ca. Nitrosotaleales, and Nitrosopumilales (Alves et al., 2018). Members of the Nitrososphaerales lineage appear to dominate soil environments. More than 80% of AOA sequenced from soils and sediments belong to this lineage, with a majority of taxa belonging to just two clades that lack

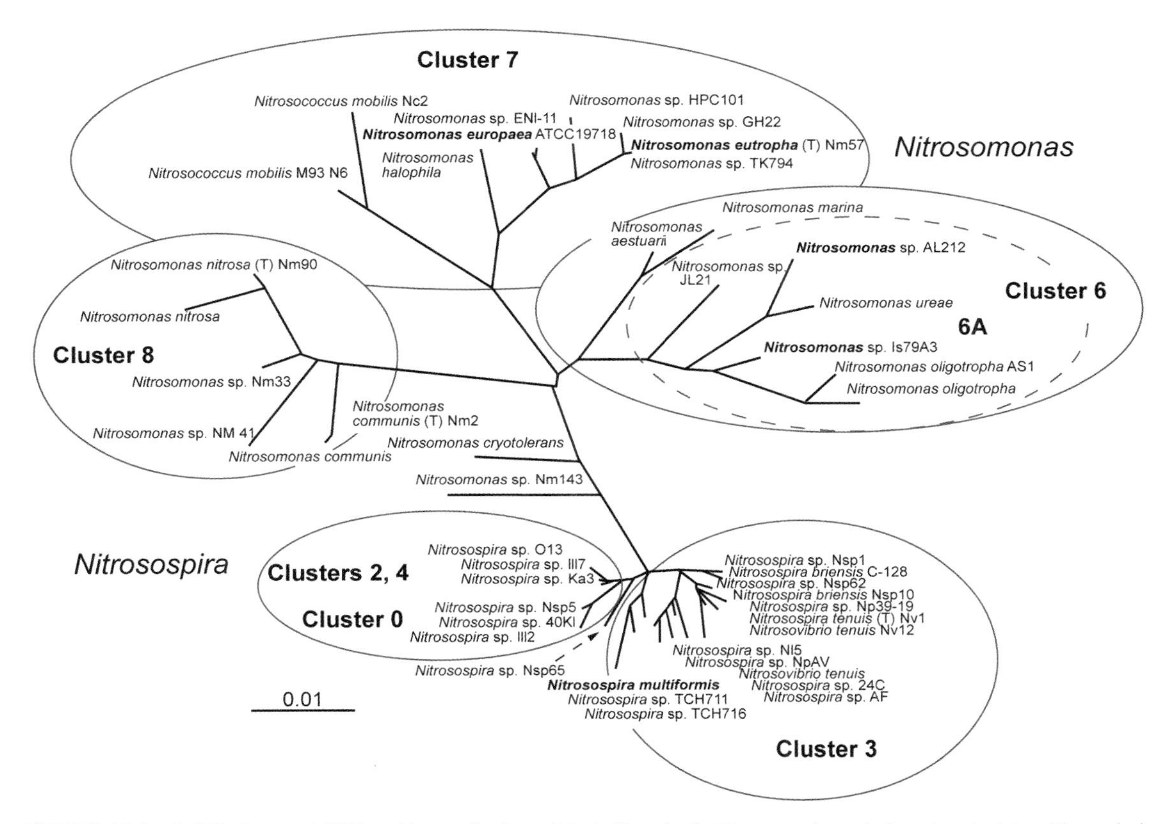

FIGURE 14.6 A 16S ribosomal RNA guide tree for bacterial nitrifiers in the Betaproteobacteria based on isolates. The scale is substitutions per site. *(Redrawn with permission from Norton, 2011.)*

cultivated species. Their recent discovery, ubiquity, numerical dominance in most soils, and unique physiology suggest additional surprises are in store.

Nitrite-oxidizing bacteria also appear in a broad array of phylogenetic groupings, but only the genus *Nitrobacter* and the candidate genus *Nitrotoga* have been cultured from soil (Daims et al., 2011). 16S rRNA analysis shows the presence of *Nitrospira* in most soils, which appear to be more diverse than *Nitrobacter* (Freitag et al., 2005). Members of *Nitrobacter* form an exclusive and highly related cluster in the Alphaproteobacteria. Though widely distributed in nature, pairwise evolutionary distance estimates are less than 1%, indicating little genetic diversity within the group, a finding supported by 16S rRNA sequence comparisons (Orso et al., 1994). The other NO_2^--oxidizing genera are in the delta (*Nitrospina* and *Nitrospira*), gamma (*Nitrosococcus*), and beta (*Candidatus* Nitrotoga) subclasses of the Proteobacteria.

All known comammox bacteria belong to *Nitrospira* lineage II, which is the most environmentally dispersed clade of this diverse genus. Based on 16S analysis, comammox appears comprised of two monophyletic sister clades A and B (Daims et al., 2015). All isolates thus far cultured are in clade A from

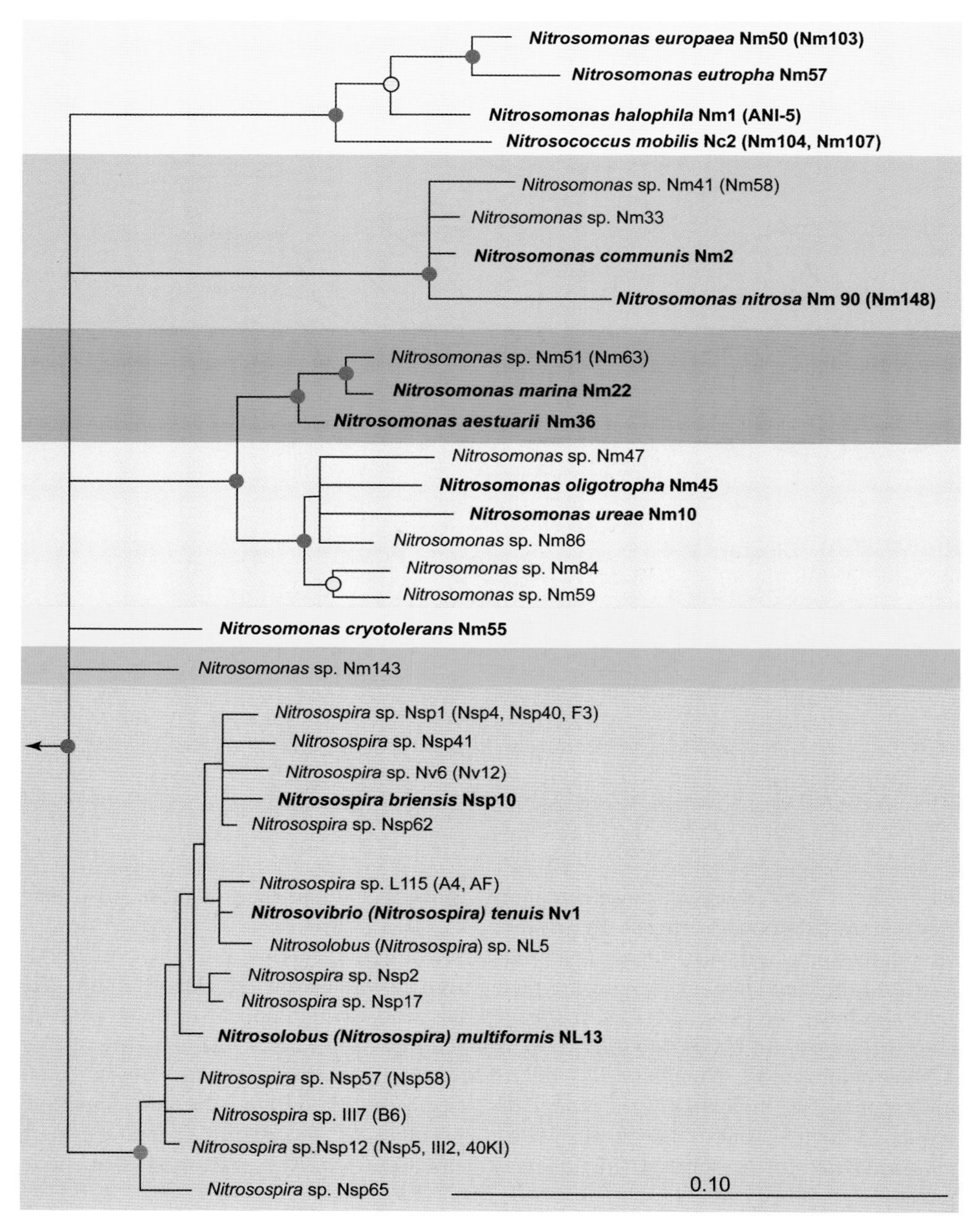

FIGURE 14.7 16S rRNA-based phylogenetic tree of the beta-proteobacterial ammonia oxidizers. The tree includes oxidizers of different genospecies (DNA-DNA similarity <60%) with available 16S rRNA gene sequences longer than 1000 nucleotides. Strains with DNA-DNA similarity >60% are in parentheses after the respective species name. Described species are depicted in bold. Scale bar represents 10% estimated sequence divergence. *(Redrawn with permission from Koops et al., 2006.)*

engineered systems (wastewater treatment plants and deep oil well biofilms), but comammox *Nitrospira* genes have been identified in a wide array of soil environments.

14.4.3 Heterotrophic nitrification

Wide varieties of heterotrophic bacteria and fungi have the capacity to oxidize NH_4^+. So-called heterotrophic nitrification is not linked to cellular growth as it is for autotrophic nitrification. There is evidence for two pathways for heterotrophic NH_3 oxidation. The first pathway is similar to that of autotrophic oxidation, in that the nitrifying bacteria have similar NH_3- and hydroxylamine-oxidizing enzymes. These enzymes can oxidize a number of different substrates, and it may be that NH_3 oxidation is only secondary to these enzymes' main purpose of oxidizing propene, benzene, cyclohexane, phenol, methanol, or any of a number of other nonpolar organic compounds.

The second heterotrophic pathway is organic and appears limited to fungi. It involves the oxidation of amines or amides to a substituted hydroxylamine followed by oxidation to a nitroso, and then a nitro compound with the following oxidation states:

$$\begin{array}{ccccccccc} RNH_2 & \rightarrow & RNHOH & \rightarrow & RNO & \rightarrow & RNO_3 & \rightarrow & NO_3^- \\ -3 & & -1 & & +1 & & +3 & & +5. \end{array} \qquad \text{(Eq. 14.12)}$$

These reactions are not coupled to ATP synthesis and thus produce no energy. Alternately, N compounds may react with hydroxyl radicals produced in the presence of hydrogen peroxide and superoxide, which may happen when fungi release oxidases and peroxidases during cell lysis and lignin degradation.

Heterotrophic nitrifying bacteria include *Arthrobacter globiformis*, *Aerobacter aerogenes*, *Thiosphaera pantotropha*, *Streptomyces griseus*, and various other species. The fungus *Aspergillus flavus* was first isolated as a nitrifier in 1954 and is the most widely studied of the nitrifying heterotrophs. Interest in heterotrophic nitrification increased substantially in the late 1980s when it became clear that accelerated inputs of atmospheric NH_4^+ to acid forest soils were being nitrified to NO_3^- with alarming effects on soil acidity, forest health, and downstream drinking water quality. Until recently, it was assumed that most of this nitrification was heterotrophic. We know now that most nitrification in acid soils is autotrophic (De Boer and Kowalchuk, 2001), perhaps chiefly performed by acidophilic (Lehtovirta-Morley et al., 2011) and archaeal (He et al., 2012) nitrifiers able to scavenge NH_3 under low-pH conditions. Heterotrophic nitrifiers thus appear to be important in some soils and microenvironments, perhaps where autotrophic nitrifiers are chemically inhibited (see following section), but they are thought now to rarely dominate the soil nitrifier community.

14.4.4 Environmental controls on nitrification

The single most important factor regulating nitrification in the majority of soils is NH_4^+ supply (Fig. 14.8). Where decomposition and thus N mineralization are low, or where NH_4^+ uptake and thus N immobilization by heterotrophs or plants are high, nitrification rates will be low. Conversely, any ecosystem disturbance that increases soil NH_4^+ availability will typically accelerate nitrification unless some other factor is limiting. Examples are tillage, fire, forest clear cutting, waste disposal, fertilization, and atmospheric N deposition — all of which have well-documented effects on NO_3^- production in soils, mostly due to their effects on soil NH_4^+ pools.

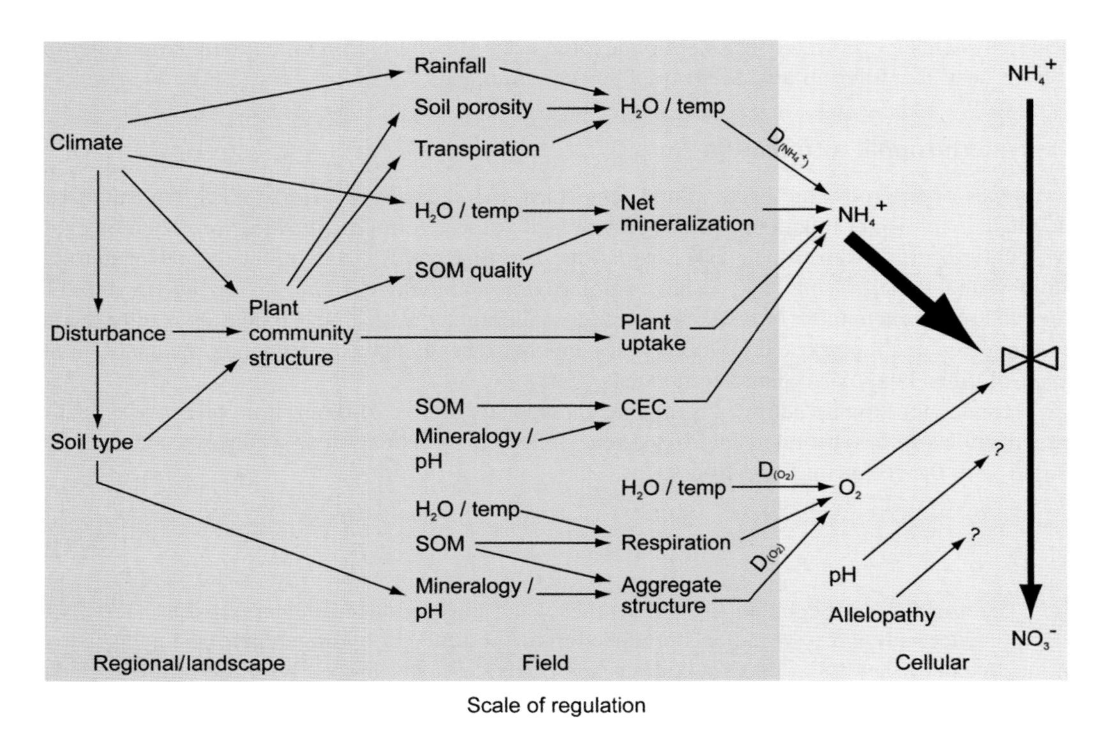

FIGURE 14.8 Environmental controls on nitrification at different scales. *(Redrawn with permission from Robertson, 1989.)*

Given that nitrification usually accelerates only when the NH_4^+ supply exceeds plant and heterotroph demand implies that nitrifiers are relatively poor competitors for NH_4^+ in soil solution. This is, in fact, the case: nitrification rates are typically low in midsuccessional communities and aggrading forests because of high plant demand for N. This also occurs following the addition of high C:N residues to agricultural soils because of high N demand by heterotrophic microbes (high immobilization; Fig. 14.1). In old-growth forests and mature grasslands, plant N demand has diminished, and consequently nitrification is usually higher than in midsuccessional communities where plant biomass is still accumulating, but not usually as high as in early successional and agricultural ecosystems, where N supply often greatly exceeds demand (Robertson and Vitousek, 1981).

Oxygen is another important regulator of nitrification in soil. All known nitrifiers are obligate aerobes, and nitrification proceeds very slowly, if at all, in submerged soils. In flooded environments, such as wetlands and lowland rice, nitrifiers are active only in the oxidized microzone around plant roots and at the water-sediment interface, which is usually only a few millimeters thick. Even though some canonical nitrifiers have the capacity to use NO_2^- rather than O_2 as an electron acceptor during respiration, O_2 is still required for NH_3 oxidation.

Nitrifiers are similar to other aerobic microbes with respect to their response to temperature, moisture, and other environmental variables (Fig. 14.4). Nitrification occurs slowly but readily under snow and in refrigerated soils. Soil transplant experiments (Mahendrappa et al., 1966) have demonstrated an apparent capacity for nitrifiers to adapt to different temperature and moisture regimes. For many decades, nitrifiers were thought to be inhibited in acid soils, probably because in many cases, especially in soils from cultivated fields, raising soil pH with calcium or magnesium carbonate stimulates nitrification, and culturable nitrifiers exhibit a pH optimum of 7.5 to 8 (Prosser, 2011). Rather, low pH often creates substrate limitation. The dissociation of NH_4^+ to NH_3 in soil solution (Eq. 14.5) is orders of magnitude lower in acid than in neutral or alkaline soils, creating an NH_3 limitation for nitrifiers. High rates of nitrification in very acidic forest soils (pH <4.5; Robertson, 1989) are likely due to either acidophilic nitrifiers or to archaea and comammox with very low substrate affinities.

14.4.5 Nitrifier inhibition

Nitrification is unaccountably slow in some soils and, in some circumstances, may be inhibited by natural or manufactured compounds. A wide variety of plant extracts can inhibit culturable nitrifiers in vitro (Rice, 1979), even though their importance in situ has long been challenged (Robertson, 1982). Likewise, commercial products, such as nitrapyrin and dicyandiamide, can be used to inhibit nitrification in soil with varying degrees of success (Beeckman et al., 2018). Neem oil, extracted from the Indian neem tree (*Azadirachta indica* A. Juss), has been used commercially to coat urea fertilizer pellets to slow its nitrification to NO_3^-.

Biological nitrification inhibition has been unequivocally documented in *Brachiaria* spp., a tropical perennial pasture grass where the root exudates appear to inhibit NH_3 oxidizers (Subbarao et al., 2009). It is also suspected in other species, such as sorghum (*Sorghum bicolor* [L.] Moench). However, documenting inhibition in situ is challenging, and to be truly effective, inhibitors must diffuse away from the rhizosphere, where nitrification may already be inhibited by low substrate supply, and must persist in the presence of heterotrophs into periods with little or no plant growth – most of the year in the case of annual crops. Both of these criteria are daunting.

That said, the potential value of managing nitrifiers in ecosystems can be easily seen from the position of nitrification in the overall N cycle (Fig. 14.1). Nitrogen is lost from ecosystems, primarily after its conversion to NO_3^- and prior to plant uptake. Thus keeping N in the NH_4^+ form prevents loss via NO_3^- leaching and denitrification, which in most ecosystems are the two principal pathways of unintentional N escape. Because many plants prefer to take up N as NO_3^-, it is not desirable to completely inhibit nitrification, even in intensively managed ecosystems such as fertilized row crops, but slowing nitrifiers or restricting their activity to periods of active plant growth is an attractive – if still elusive – management goal.

14.5 Denitrification

Denitrification is the reduction of soil NO_3^- to the gases NO, N_2O, and N_2. A wide variety of soil microorganisms can denitrify, whereby NO_3^- rather than O_2 is used as a terminal electron acceptor during respiration. Because NO_3^- is a less efficient electron acceptor than O_2, facultative denitrifiers will use NO_3^- only when O_2 is insufficient to meet respiratory demand, and obligate denitrifiers will be at a competitive disadvantage relative to O_2-utilizing heterotrophs. Significant denitrification in soil occurs only when O_2 availability is restricted. This primarily occurs following rainfall as soil pores fill with water and the diffusion of O_2 to microsites is drastically slowed. Denitrification is triggered at water-filled pore space concentrations of 60% and higher (Fig. 14.2) and O_2 concentrations below 2%. In wetlands and lowland rice paddies this may be the case most of the time, but even in well-aerated soils O_2 demand can exceed supply inside soil aggregates and around rapidly decomposing litter particles, microscale hotspots for denitrification (Kravchenko et al., 2017).

Denitrification was considered the only point in the N cycle where fixed N reenters the atmosphere as N_2 until the 1995 discovery of anaerobic ammonium oxidation (anammox), described later in this chapter. Notwithstanding anammox, denitrification remains the main process that serves to close the global N cycle. In the absence of denitrification N_2 fixers would eventually draw atmospheric N_2 to nil, and the biosphere would be awash in NO_3^- and other reactive forms of N. Denitrification is also the major source of atmospheric N_2O, an important greenhouse gas with ~300 times the global warming potential of CO_2. N_2O also consumes stratospheric ozone, important for shielding the Earth's surface from UV radiation.

Denitrification can also be employed to remove excess NO_3^- from soil prior to its movement as a pollutant to ground and surface waters. This can be accomplished by maintaining natural vegetation in riparian areas and wetlands adjacent to streams. As NO_3^--rich groundwater moves toward streams, it will encounter adjacent anaerobic zones where it can be denitrified to N_2O and N_2, keeping it from polluting downstream surface waters. Conceptually, this process is similar to urban wastewater treatment, which aims to move NO_3^-, produced by the oxidation of organic N in wastewater and its subsequent nitrification to NH_4^+, to anaerobic tanks containing denitrifiers. Anammox has recently been proposed for wastewater N removal (Hu et al., 2013).

Denitrification is less desirable in agricultural systems as it is better to conserve N for plant uptake. In regions with ample rainfall, N losses via denitrification can rival or exceed losses via NO_3^- leaching. Despite its importance, however, there are no technologies to directly inhibit denitrification. Denitrifiers are best managed indirectly by avoiding excess water (e.g., with drainage tiles in crop fields or levees in rice paddies) and excess NO_3^- (e.g., with N-efficient fertilizer technologies or nitrification inhibitors).

14.5.1 Denitrifier diversity and biochemistry

Denitrification is carried out by a broad array of microbes spanning all three domains of life. Best known are prokaryotic denitrifiers, because they have been isolated and described for over a century. Eukaryotes are a more recent addition to our knowledge of denitrifiers, as we've learned over the past 30 years that fungal denitrification is more common than originally thought (Mothapo et al., 2015). Although few archaea are known to denitrify, this is likely an artifact of the difficulty with which archaea are isolated for study. Emerging molecular evidence based on *nirK* gene abundance has revealed a number of archaea with the potential to denitrify.

Denitrifiers are in general quite cosmopolitan. Over 60 genera of bacterial denitrifiers have been identified (Coyne, 2018), including organotrophs, chemotrophs including nitrifiers, photolithotrophs, N_2 fixers, thermophiles, halophiles, and various pathogens. Some 14 different genera of aerobic denitrifiers — prokaryotes able to use O_2 and NO_3^- simultaneously as electron acceptors — have been identified from wastewater treatment systems (Ji et al., 2015). Within the eukaryotes, at least 70 species of fungi spanning 23 taxonomic orders possess denitrifying enzyme genes (Maeda et al., 2015). In soil most culturable denitrifiers are facultative anaerobes from only three to six genera, principally *Pseudomonas* and *Alcaligenes*, and to a lesser extent, *Bacillus*, *Agrobacterium*, and *Flavobacterium*. Typically, denitrifiers can constitute up to 20% of total microbial biomass (Tiedje, 1988).

Microbes denitrify to generate energy (ATP) by electron transport phosphorylation via the cytochrome system. The general pathway is:

$$2NO_3^- \underset{nar}{\rightarrow} 2NO_2^- \underset{nir}{\rightarrow} 2\overset{\uparrow}{NO} \underset{nor}{\rightarrow} \overset{\uparrow}{N_2O} \underset{nos}{\rightarrow} \overset{\uparrow}{N_2}\,. \qquad \text{(Eq. 14.13)}$$

Each step is enacted by individual enzymes: nitrate reductase (*nar*), nitrite reductase (*nir*), nitric oxide reductase (*nor*), and nitrous oxide reductase (*nos*). Each is inhibited by O_2. The organization of these enzymes in the cell membrane for gram-negative bacteria is described in Fig. 14.9. At any step in this process, intermediate products can be exchanged with the soil environment, making denitrifiers a significant source of NO_2^- in soil solution and very important sources of the atmospheric gases NO and N_2O.

Each denitrification enzyme is inducible, primarily in response to the partial pressure of O_2 and substrate (C) availability. Because enzyme induction is sequential and substrate dependent, there is usually a lag between the production of an intermediate substrate and its consumption by the next enzyme. In pure culture these lags can be on the order of hours (Fig. 14.10). In situ lags in soil can be substantially longer, and differences in lags among different microbial taxa may significantly affect the contribution of denitrifiers to fluxes of NO and N_2O into the atmosphere. For example, a dried soil recently wet may consume NO_3^- almost immediately, but until *nir* is induced, the NO produced by *nar* will accumulate and escape into the soil atmosphere (see Eq. 14.13). Until *nos* is induced, the N_2O produced by *nor* will escape. This can lead to the nonintuitive effect of drought's stimulating N_2O production.

The fact that induced enzymes also degrade at different rates, and more slowly than they are induced, also leads to a complex response to the environmental conditions that drive denitrification. Whether a soil has denitrified recently (whether denitrifying enzymes are present) may largely determine its response to newly favorable conditions for denitrification. Rainfall onto soil that is already moist, for example, will likely lead to a faster and perhaps stronger denitrification response than will rainfall onto the same soil when it is dry (Groffman and Tiedje, 1988) and will lead to a greater proportion of N product, that is N_2O vs. N_2, because of the presence of *nos* in soil that is already wet (Bergsma et al., 2002).

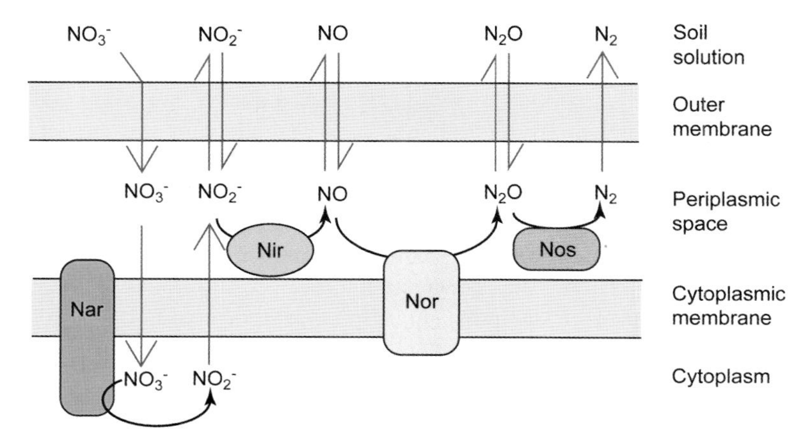

FIGURE 14.9 The organization of denitrification enzymes in gram-negative bacteria. *(Redrawn with permission from Ye et al., 1994.)*

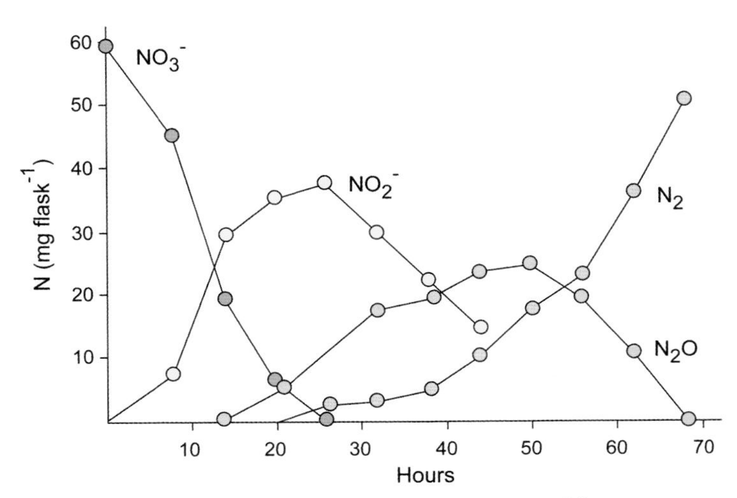

FIGURE 14.10 The sequence of products formed during denitrification in vitro as different enzymes are induced sequentially. *(Redrawn with permission from Cooper and Smith, 1963.)*

14.5.2 Environmental controls on denitrification

For over a century after its discovery as an important microbial process, denitrification was assumed to be important only in aquatic and wetland ecosystems. It was not until the advent of whole-ecosystem N budgets and the use of ^{15}N to trace the fate of fertilizer N that denitrification was found to be important in unsaturated soils. These studies suggested the importance of denitrification in agricultural soils, and with the development of the acetylene block technique in the 1970s, the importance of denitrification in even forest and grassland soils was confirmed. Acetylene selectively inhibits *nos* (see Eq. 14.13; Fig. 14.9), allowing the assessment of N_2 production by measuring N_2O accumulation in a soil incubated with acetylene. Unlike N_2, small changes in N_2O concentration are easily detected in air.

Today, denitrification is known to be an important N cycle process wherever O_2 is limiting and C and NO_3^- are not. In unsaturated soils this frequently occurs within soil aggregates, and in decomposing plant litter and rhizospheres. Soil aggregates vary widely in size but in general are composed of small mineral particles and pieces of organic matter <2 mm in diameter that are glued to one another with biologically derived polysaccharides or other binding agents. Like most particles in soil, aggregates are coated by a thin water film that impedes gas exchange. Oxygen diffuses through water ~10,000 times slower than through air. Quantitative models (Smith, 1980) suggested that the centers of these aggregates ought to be anaerobic owing to a higher respiratory demand in the aggregate's interior than could be satisfied by O_2 diffusion from soil air. This was confirmed experimentally in 1985 (Sexstone et al., 1985), and together with evidence for the importance of organic matter particles for providing anaerobic microhabitats (Kravchenko et al., 2017; Parkin, 1987), a logical explanation is provided for both active denitrification in soils that appear otherwise to be aerobic, as well as for the almost universal presence of denitrifiers and denitrification enzymes in soils worldwide.

The fact that denitrification — or at least N_2O production — occurs in well-aerated soils may also be due to fungal denitrifiers that can denitrify at higher O_2 concentrations than can archaeal or bacterial denitrifiers (Mothapo et al., 2015). In pure culture their NO_3^- consumption rates are much lower than those of prokaryotes, but their ability to denitrify in drier soils means they may be able to denitrify for longer periods. The fact that fungal denitrifiers also lack a functional *nos* enzyme means that N_2O is their dominant end product, providing added biogeochemical importance.

In addition to O_2, denitrification is also regulated by soil C and NO_3^-. Carbon is important because most denitrifiers are heterotrophs and require reduced C as the electron donor, although, as noted earlier, denitrifiers can also be chemolithotrophs like nitrifiers, and photolithotrophs like cyanobacteria. For all, NO_3^- serves as the electron acceptor and must be provided via nitrification, rainfall, or fertilizer. Oxygen, however, is a preferred electron acceptor because of its high energy yield and thus must be largely depleted before denitrification occurs. In most soils the majority of denitrifiers are facultative anaerobes that will simply avoid synthesizing denitrification enzymes until O_2 drops below some critical threshold that can differ substantially by taxa (Cavigelli and Robertson, 2001).

In the field O_2 is by far the dominant control on denitrification rates. Denitrification can be easily stimulated in an otherwise-aerobic soil by removing O_2 and can be inhibited in saturated soil by drying or otherwise aerating it. The relative importance of C and NO_3^-, the other major controls, will vary by ecosystem. In saturated soils, such as those in wetlands and lowland rice paddies, NO_3^- limits denitrification because the nitrifiers that provide NO_3^- are inhibited at low O_2 concentrations. Consequently, denitrification occurs only in the slightly oxygenated rhizosphere and at the sediment-water interface, places where there is sufficient O_2 for nitrifiers to oxidize NH_4^+ to NO_3^-, which can then diffuse to denitrifiers in the increasingly anaerobic zone away from the root surface or sediment-water interface. It is often difficult to find NO_3^- in persistently saturated soils, not only because of low nitrification but also because of the tight coupling between nitrifiers and denitrifiers. Nitrate may be more available in wetlands with fluctuating water tables or with significant inputs of NO_3^- from groundwater.

Under unsaturated conditions such as those in most upland soils, C availability may be a dominant control. For example, in these soils it is easy to stimulate denitrification simply by adding a readily oxidized C source. Carbon supports denitrification, both directly by providing donor electrons to denitrifiers, and indirectly by stimulating O_2 consumption by heterotrophs. It can be difficult to distinguish between these two effects experimentally, yet from a management perspective, there probably is no need to.

The multifactor control of denitrification often creates extraordinary spatial and temporal variation in its activity, which inhibits our ability to produce well-constrained estimates of just how much denitrification is occurring at field, landscape, or regional scales. One approach to addressing this variability is to focus on small areas of intensive activity where controlling factors converge to create episodic periods of high rates of activity that can account for a substantial percentage of overall denitrification at different scales. These spatial and temporal "control points" (Bernhardt et al., 2017) have been effectively targeted in denitrification studies at multiple scales and ecosystem types.

14.6 Other nitrogen transformations in soil

Several additional microbial processes transform N in soil, although none are thought to be as quantitatively important as mineralization, immobilization, nitrification, and denitrification. *Dissimilatory nitrate reduction to ammonium* (DNRA) refers to the anaerobic transformation of NO_3^- to NO_2^- and then to NH_4^+. Like denitrification, this process allows for respiration to go on in the absence of O_2 and is thought to be favored in environments where the ratio of C to NO_3^- is high because DNRA consumes more electrons than denitrification. A capacity for DNRA has been found in facultative and obligately fermentative bacteria and has long been thought to be restricted to high-C, highly anaerobic environments, such as anaerobic sewage sludge bioreactors, anoxic sediments, and the bovine rumen. However, DNRA has been identified in some tropical forest soils (Silver et al., 2001) and in a variety of freshwater sediments (Burgin and Hamilton, 2007). In these environments the flow of inorganic N through DNRA can be as large or larger than the flow through denitrification and nitrification and may help to retain N by shunting NO_3^- into NH_4^+ rather than into N_2O or N_2.

Nonrespiratory denitrification, like respiratory denitrification, also results in the production of N gas (mainly N_2O), but the reduction does not enhance growth and can occur in aerobic environments. A variety of NO_3^--assimilating bacteria, fungi, and yeast can carry out nonrespiratory denitrification, which may be responsible for some of the N_2O now attributed to nitrifiers in well-aerated soils (Robertson and Tiedje, 1987).

Anammox, in which NH_4^+ and NO_2^- are converted to N_2 (Mulder et al., 1995), is known to occur in sewage treatment plants and marine systems (Kuypers et al., 2005) where anammox can be the dominant source of N_2 flux. Anammox bacteria grow very slowly in enrichment culture and only under strict anaerobic conditions and are thus likely to be part of a significant soil process only in periodically or permanently submerged soils (Strous, 2011).

Bacteria capable of performing anammox occur within the single-order Brocadiales in the phylum Planctomycete. In these bacteria anammox catabolism occurs in a specialized organelle called the anammoxosome, wherein

$$NH_4^+ + NO_2^- \rightarrow N_2H_2 \rightarrow N_2, \quad \text{(Eq. 14.14)}$$

although much remains to be learned about the biochemistry and bioenergetics of the process, including intermediate compounds (Kartal et al., 2011).

Chemodenitrification occurs when NO_2^- in soil reacts to form N_2 or NO_x. This can occur through several aerobic pathways (Heil et al., 2016). In the Van Slyke reaction amino groups in the α position to carboxyls yield N_2:

$$RNH_2 + HNO_2 \rightarrow ROH + H_2O + N_2. \quad \text{(Eq. 14.15)}$$

In a similar reaction NO_2^- reacts with NH_4^+, urea, methylamine, purines, and pyrimidines to yield N_2:

$$HNO_2 + NH_4^+ \rightarrow N_2 + 2H_2O. \quad \text{(Eq. 14.16)}$$

Chemical decomposition of HNO_2 may also occur spontaneously:

$$3HNO_2 \rightarrow 3HNO_3 + H_2O + 2NO. \quad \text{(Eq. 14.17)}$$

Chemodenitrification is thought to be a minor pathway for N loss in most ecosystems. It is not easily evaluated in situ, however, and in the lab it requires a sterilization procedure that does not itself disrupt soil N chemistry.

14.7 Nitrogen movement in the landscape

Microbial transformations of reactive N (Table 14.5) have great importance for soil fertility, water quality, and atmospheric chemistry at ecosystem, landscape, and regional scales. It is at these scales that

TABLE 14.5 Forms of nitrogen of environmental concern.

N form	Sources	Dominant transport vectors	Environmental effects
Nitrate (NO_3^-)	Nitrification	Groundwater	Pollution of drinking water
	Fertilizer		Coastal eutrophication
	Disturbance that stimulates nitrification		
	Combustion (acid rain)		
Ammonia (NH_3, NH_4^+)	Fertilizer	Surface runoff	Pollution of drinking water
	Animal waste	Atmosphere	Eutrophication
Nitrous oxide (N_2O)	By-product of nitrification, denitrification, anammox	Atmosphere	Greenhouse gas
		Groundwater	Ozone destruction in stratosphere
Nitric oxide (NO)	By-product of nitrification, denitrification, anammox	Atmosphere	Ozone precursor in troposphere
Dissolved organic N	By-product of mineralization	Surface runoff	Eutrophication (?)
		Groundwater	

differences between what we have learned in the laboratory and what we observe in the environment (see Section 14.1) become most obvious.

One approach to thinking about microbial N cycle processes at large scales is to ask a series of questions that attempt to determine if a particular ecosystem is a source or sink of a particular N species (Table 14.6). Sites that are N rich either naturally or following disturbance have a high potential to function as sources of most of the reactive N forms identified in Table 14.1 because mineralization, nitrification, and denitrification occur at high rates.

Nitrogen sinks are defined as habitats that have a high potential to remove reactive N from the environment, preventing its movement into adjacent ecosystems. Ecosystems such as wetlands that are wet and rich in organic materials have high potential to function as sinks due to their ability to support denitrification. In many cases these sink areas retain N produced in source areas of the landscape. Riparian buffer zones next to streams can be managed to retain N moving as NO_3^- out of crop fields in groundwater (Lowrance et al., 1984). This NO_3^- can be stored in plant tissue or in SOM as organic N or can be denitrified to N_2O, or better, to N_2.

Humans have doubled the circulation of reactive N on Earth, creating N that then flows through the environment, leading to degraded air and water quality (Robertson and Vitousek, 2009). Solutions to landscape, regional, and global N enrichment problems often rely heavily on managing microbial N

TABLE 14.6 Criteria for determining if a site is a source or sink of nitrogen in the landscape.

Criteria	Determinants
Is the site N rich?	Fertilized
	Fine texture (clay)
	Legumes
	Wet tropics
Is the site highly disturbed?	Disturbance of plant uptake (e.g., harvest)
	Stimulation of mineralization (e.g., tillage)
	Disturbance of links between plant and microbial processes (e.g., tillage)
Does the site have a high potential for denitrification?	Wet soil
	Well-aggregated
	High available organic matter
Does the site have a high potential for NH_3 volatilization?	High pH (>8.0)

From Groffman (2000).

transformations. For example, coastal areas of the Gulf of Mexico suffer from eutrophication and hypoxia that have been linked to excess N from the Mississippi River Basin (Rabalais et al., 2002). Proposed solutions include better management of microbial N transformations in crop fields as well as the creation of wetland sinks to trap and remove N moving out of agricultural areas (Mitsch et al., 2001).

Source-sink dynamics of N ultimately depend on the juxtaposition of different ecosystems in the landscape and the hydrologic and atmospheric transport paths that link them — a complex topic that requires knowledge of hydrology and atmospheric chemistry in addition to soil ecology and microbiology. Because soil biota play a crucial role in forming and consuming reactive N in the environment, their management can be an important and even crucial means for regulating N fluxes at local, regional, and global scales. Better management at the ecosystem level is fundamental.

References and Further Reading

Alves, R.J.E., Minh, B.Q., Urich, T., von Haeseler, A., Schleper, C., 2018. Unifying the global phylogeny and environmental distribution of ammonia-oxidising archaea based on amoA genes. Nat. Commun. 9, 1–17.

Bahulikar, R.A., Chaluvadi, S.R., Torres-Jerez, I., Mosali, J., Bennetzen, J.L., Udvardi, M., 2020. Nitrogen fertilization reduces nitrogen fixation activity of diverse diazotrophs in switchgrass roots. Phytobiomes J. https://doi.org/10.1094/PBIOMES-09-19-0050-FI.

Beeckman, F., Motte, H., Beeckman, T., 2018. Nitrification in agricultural soils: impact, actors and mitigation. Curr. Opin. Biotechnol. 50, 166–173.

Bergsma, T.T., Robertson, G.P., Ostrom, N.E., 2002. Influence of soil moisture and land use history on denitrification end-products. J. Environ. Qual. 31, 711–717.

Bernhardt, E.S., Blaszczak, J.R., Ficken, C.D., Fork, M.L., Kaiser, K.E., Seybold, E.C., 2017. Control points in ecosystems: moving beyond the hot spot hot moment concept. Ecosystems 20, 665–682.

Binkley, D., Sollins, P., Bell, R., Sachs, D., Myrold, D., 1992. Biogeochemistry of adjacent conifer and alder-conifer stands. Ecology 73, 2022–2033.

Bottomley, P.J., Myrold, D.D., 2007. Biological N inputs. In: Paul, E.A. (Ed.), Soil Microbiology, Ecology, and Biochemistry, third ed. Academic Press, Boston, pp. 365–387.

Bottomley, P.J., Myrold, D.D., 2015. Biological N inputs. In: Paul, E.A. (Ed.), Soil Microbiology, Ecology and Biochemistry, fourth ed. Academic Press, Boston, pp. 447–470.

Burgin, A.J., Hamilton, S.K., 2007. Have we overemphasized the role of dentirification in aquatic ecosystems? A review of nitrate removal pathways. Front. Ecol. Environ. 5, 89–96.

Caranto, J.D., Lancaster, K.M., 2017. Nitric oxide is an obligate bacterial nitrification intermediate produced by hydroxylamine oxidoreductase. Proc. Natl. Acad. Sci. USA 114 (31), 8217–8222.

Cavigelli, M.A., Robertson, G.P., 2001. Role of denitrifier diversity in rates of nitrous oxide consumption in a terrestrial ecosystem. Soil Biol. Biochem. 33, 297–310.

Coleman, N.T., Thomas, G.W., 1967. The basic chemistry of soil acidity. In: Pearson, R.W., Adams, F. (Eds.), Soil Acidity and Liming. American Society of Agronomy, Madison, pp. 1–42.

Cooper, G.S., Smith, R., 1963. Sequence of products formed during denitrification in some diverse Western soils. Soil Sci. Soc. Am. J. 27, 659–662.

Coskun, D., Britto, D.T., Shi, W., Kronzucker, H.J., 2017. How plant root exudates shape the nitrogen cycle. Trends Plant Sci. 22, 661–673.

Coyne, M.S., 2018. Denitrification in soil. In: Lal, R., Stewart, B.A. (Eds.), Soil Nitrogen Uses and Environmental Impacts. CRC Press, Boca Raton, pp. 95–139.

Daims, H., Lücker, S., Le Paslier, D., Wagner, M., 2011. Diversity, environmental genomics, and ecophysiology of nitrite-oxidizing bacteria. In: Ward, B.B., Arp, D.J., Klotz, M.G. (Eds.), Nitrification. ASM Press, Washington, D.C., pp. 295–322

Daims, H., Lebedeva, E.V., Pjevac, P., Han, P., Herbold, C., Albertsen, M., et al., 2015. Complete nitrification by *Nitrospira* bacteria. Nature 528, 504–509.

De Boer, W., Kowalchuk, G.A., 2001. Nitrification in acid soils: micro-organisms and mechanisms. Soil Biol. Biochem. 33, 853–866.

de Bruijn, F., 2015. Biological nitrogen fixation. In: Lugtenberg, B. (Ed.), Principles of Plant-Microbe Interactions. Springer, Cham, pp. 215–224.

Drake, D., 2011. Invasive legumes fix N_2 at high rates in riparian areas of an N-saturated, agricultural catchment. J. Ecol. 99, 515–523.

FAO, 2019. World Fertilizer Trends and Outlooks to 2022. Food and Agriculture Organization of the United Nations, Rome. http://www.fao.org/3/ca6746en/CA6746EN.pdf?eloutlink=imf2fao.

Frankland, P.F., Frankland, G.C., 1890. V. The nitrifying process and its specific ferment – Part I. Philos. Trans. R. Soc. Lond. B Biol. Sci. 181, 107–128.

Freitag, T.E., Chang, L., Clegg, C.D., Prosser, J.I., 2005. Influence of inorganic nitrogen management regime on the diversity of nitrite-oxidizing bacteria in agricultural grassland soils. Appl. Environ. Microbiol. 71, 8323.

Gaby, J.C., Buckley, D.H., 2011. A global census of nitrogenase diversity. Environ. Microbiol. 13, 1790–1799.

Gaby, J.C., Buckley, D.H., 2015. Assessment of nitrogenase diversity in the environment. In: de Bruijn, F.J. (Ed.), Biological Nitrogen Fixation. Wiley, Hoboken, pp. 209–216.

Galloway, J.N., Townsend, A.R., Erisman, J.W., Bekunda, M., Cai, Z.C., Freney, J.R., et al., 2008. Transformation of the nitrogen cycle: recent trends, questions, and potential solutions. Science 320, 889–892.

Gelfand, I., Robertson, G.P., 2015. A reassessment of the contribution of soybean biological nitrogen fixation to reactive N in the environment. Biogeochemistry 123, 175–184.

Groffman, P.M., 2000. Nitrogen in the environment. In: Sumner, M.E. (Ed.), Handbook of Soil Science. CRC Press, Boca Raton, pp. C190–C200.

Groffman, P.M., Tiedje, J.M., 1988. Denitrification hysteresis during wetting and drying cycles in soil. Soil Sci. Soc. Am. J. 52, 1626–1629.

Hart, S.C., Stark, J.M., Davidson, E.A., Firestone, M.K., 1994. Nitrogen mineralization, immobilization, and nitrification. In: Bottomley, P.S., Angle, J.S., Weaver, R.W. (Eds.), Methods of Soil Analysis: Part 2 – Microbiological and Biochemical Properties. Soil Science Society of America, Madison, pp. 985–1018.

He, J.-Z., Hu, H.-W., Zhang, L.-M., 2012. Current insights into the autotrophic thaumarchaeal ammonia oxidation in acidic soils. Soil Biol. Biochem. 55, 146–154.

Heil, J., Vereecken, H., Brüggemann, N., 2016. A review of chemical reactions of nitrification intermediates and their role in nitrogen cycling and nitrogen trace gas formation in soil. Eur. J. Soil Sci. 67, 23–39.

Hill, S., 1992. Physiology of nitrogen fixation in free-living heterotrophs. In: Stacey, G., Burris, R.H., Evans, H.J. (Eds.), Biological Nitrogen Fixation. Chapman and Hill, New York, pp. 87–134.

Hu, Z., Lotti, T., de Kreuk, M., Kleerebezem, R., van Loosdrecht, M., Kruit, J., et al., 2013. Nitrogen removal by a nitritation-anammox bioreactor at low temperature. Appl. Environ. Microbiol. 79, 2807–2812.

Hungate, B.A., Dukes, J.S., Shaw, M.R., Luo, Y., Field, C.B., 2003. Nitrogen and climate change. Science 302, 1512–1513.

Hyvönen, R., Ågren, G.I., Andrén, O., 1996. Modeling long-term carbon and nitrogen dynamics in an arable soil receiving organic matter. Ecol. Appl. 6, 1345–1354.

Ji, B., Yang, K., Zhu, L., Jiang, Y., Wang, H., Zhou, J., et al., 2015. Aerobic denitrification: a review of important advances of the last 30 years. Biotechnol. Bioprocess Eng. 20, 643–651.

Jones, J.M., Richards, B.N., 1977. Effect of reforestation on turnover of N-15 labelled ammonium plus nitrate in relation to changes in soil microflora. Soil Biol. Biochem. 9, 383–392.

Kartal, B., Keltjens, J.T., Jetten, M.S.M., 2011. Metabolism and genomics of anammox bacteria. In: Ward, B.B., Arp, D.J., Klotz, M.G. (Eds.), Nitrification. ASM Press, Washington, D.C., pp. 181–200

Koch, H., van Kessel, M.A.H.J., Lücker, S., 2019. Complete nitrification: insights into the ecophysiology of comammox *Nitrospira*. Appl. Microbiol. Biotechnol. 103, 177–189.

Könneke, M., Bernhard, A.E., José, R., Walker, C.B., Waterbury, J.B., Stahl, D.A., 2005. Isolation of an autotrophic ammonia-oxidizing marine archaeon. Nature 437, 543–546.

Koops, H.-P., Pommerening-Röser, A., 2001. Distribution and ecophysiology of the nitrifying bacteria emphasizing cultured species. FEMS Microbiol. Ecol. 37, 1–9.

Koops, H.-P., Purkhold, U., Pommerening-Röser, A., Timmermann, G., Wagner, M., 2006. The lithoautotrophic ammonia-oxidizing bacteria. In: Dworkin, M., Falkow, S., Rosenberg, E., Schleifer, K.-H., Stackebrandt, E. (Eds.), The Prokaryotes: A Handbook on the Biology of Bacteria. Springer, Berlin, pp. 778–811.

Kravchenko, A.N., Toosi, E.R., Guber, A.K., Ostrom, N.E., Yu, J., Azeem, K., et al., 2017. Hotspots of soil N_2O emission enhanced through water absorption by plant residue. Nat. Geosci. 10, 496–500.

Kuypers, M.M.M., Lavik, G., Woebken, D., Schmid, M., Fuchs, B.M., Amann, R., et al., 2005. Massive nitrogen loss from the Benguela upwelling system through anaerobic ammonium oxidation. Proc. Natl. Acad. Sci. 102, 6478–6483.

Kuypers, M.M.M., Marchant, H.K., Kartal, B., 2018. The microbial nitrogen-cycling network. Nat. Rev. Microbiol. 16, 263–276.

Lehtovirta-Morley, L.E., Stoecker, K., Vilcinskas, A., Prosser, J.I., Nicol, G.W., 2011. Cultivation of an obligate acidophilic ammonia oxidizer from a nitrifying acid soil. Proc. Natl. Acad. Sci. U. S. A. 108, 15892–15987.

Lehtovirta-Morley, L.E., Ross, J., Hink, L., Weber, E.B., Gubry-Rangin, C., Thion, C., et al., 2016. Isolation of "*Candidatus* Nitrosocosmicus franklandus," a novel ureolytic soil archaeal ammonia oxidiser with tolerance to high ammonia concentration. FEMS Microbiol. Ecol. 92. https://doi.org/10.1093/femsec/fiw057.

Leininger, S., Urich, T., Schloter, M., Schwark, L., Qi, J., Nicol, G.W., et al., 2006. Archaea predominate among ammonia-oxidizing prokaryotes in soils. Nature 442, 806–809.

Li, Z., Tian, D., Wang, B., Wang, J., Wang, S., Chen, H.Y.H., et al., 2019. Microbes drive global soil nitrogen mineralization and availability. Glob. Change Biol. 25, 1078–1088.

Liang, D., Ouyang, Y., Tiemann, L.K., Robertson, G.P., 2020. Niche differentiation of bacterial versus archaeal soil nitrifiers induced by ammonium inhibition along a management gradient. Front. Microbiol. 11, 568588.

Linn, D.M., Doran, J.W., 1984. Effect of water-filled pore space on CO_2 and N_2O production in tilled and non-tilled soils. Soil Sci. Soc. Am. J. 48, 1267−1272.

Löhnis, F., 1913. Vorlesungen Uber Landwortschaftliche Bacterologia. Borntraeger, Berlin.

Lowrance, R.R., Todd, R.L., Fail, J., Hendrickson, O., Leonard, R., Asmussen, L., 1984. Riparian forests as nutrient filters in agricultural watersheds. Bioscience 34, 374−377.

Maeda, K., Spor, A., Edel-Hermann, V., Heraud, C., Breuil, M.-C., Bizouard, F., et al., 2015. N_2O production, a widespread trait in fungi. Sci. Rep. 5, 9697.

Mahendrappa, M.K., Smith, R.L., Christiansen, A.T., 1966. Nitrifying organisms affected by climatic region in western United States. Soil Sci. Soc. Am. J. 30, 60−62.

Menge, D.N., Hedin, L.O., 2009. Nitrogen fixation in different biogeochemical niches along a 120,000-year chronosequence in New Zealand. Ecology 90, 2190−2201.

Mitsch, W.J., Day Jr., J.W., Gilliam, J.W., Groffman, P.M., Hey, D.L., Randall, G.W., et al., 2001. Reducing nitrogen loading to the Gulf of Mexico from the Mississippi River Basin: strategies to counter a persistent ecological problem. Bioscience 51, 373−388.

Moreau, D., Bardgett, R.D., Finlay, R.D., Jones, D.L., Philippot, L., 2019. A plant perspective on nitrogen cycling in the rhizosphere. Funct. Ecol. 33, 540−552.

Mothapo, N., Chen, H., Cubeta, M.A., Grossman, J.M., Fuller, F., Shi, W., 2015. Phylogenetic, taxonomic and functional diversity of fungal denitrifiers and associated N_2O production efficacy. Soil Biol. Biochem. 83, 160−175.

Mulder, A., van de Graaf, A.A., Robertson, L.A., Kuenen, J.G., 1995. Anaerobic ammonium oxidation discovered in a denitrifying fluidized bed reactor. FEMS Microbiol. Ecol. 16, 177−184.

Norton, J.M., 2011. Diversity and environmental distribution of ammonia-oxidizing bacteria. In: Ward, B.B., Arp, D.J., Klotz, M.G. (Eds.), Nitrification. ASM Press, Washington, D.C., pp. 39−55

Orellana, L.H., Chee-Sanford, J.C., Sanford, R.A., Löffler, F.E., Konstantinidis, K.T., 2018. Year-round shotgun metagenomes reveal stable microbial communities in agricultural soils and novel ammonia oxidizers responding to fertilization. Appl. Environ. Microbiol. 84, 01617 e01646.

Orso, S.M., Navarro, M., Normand, P., 1994. Molecular phylogenetic analysis of *Nitrobacter* spp. Int. J. Syst. Bacteriol. 44, 83−86.

Parkin, T.B., 1987. Soil microsites as a source of denitrification variability. Soil Sci. Soc. Am. J. 51, 1194−1199.

Prosser, J.I., 2011. Soil nitrifiers and nitrification. In: Ward, B.B., Arp, D.J., Klotz, M.G. (Eds.), Nitrification. ASM Press, Washington, D.C., pp. 347−383

Prosser, J.I., Nicol, G.W., 2012. Archaeal and bacterial ammonia-oxidisers in soil: the quest for niche specialisation and differentiation. Trends Microbiol. 20, 523−531.

Prosser, J.I., Hink, L., Gubry-Rangin, C., Nicol, G.W., 2020. Nitrous oxide production by ammonia oxidizers: physiological diversity, niche differentiation and potential mitigation strategies. Glob. Change Biol. 26, 103−118.

Rabalais, N.N., Turner, R.E., Scavia, D., 2002. Beyond science into policy: Gulf of Mexico hypoxia and the Mississippi River. Bioscience 52, 129−142.

Rice, E.L., 1979. Allelopathy − an update. Bot. Rev. 45, 15−109.

Robertson, G.P., 1982. Factors regulating nitrification in primary and secondary succession. Ecology 63, 1561−1573.

Robertson, G.P., 1989. Nitrification and denitrification in humid tropical ecosystems: potential controls on nitrogen retention. In: Proctor, J. (Ed.), Mineral Nutrients in Tropical forest and savanna Ecosystems. Blackwell Scientific, Cambridge, pp. 55−69.

Robertson, G.P., Vitousek, P.M., 1981. Nitrification potentials in primary and secondary succession. Ecology 62, 376−386.

Robertson, G.P., Tiedje, J.M., 1987. Nitrous oxide sources in aerobic soils: nitrification, denitrification, and other biological processes. Soil Biol. Biochem. 19, 187−193.

Robertson, G.P., Vitousek, P.M., 2009. Nitrogen in agriculture: balancing the cost of an essential resource. Annu. Rev. Environ. Resour. 34, 97−125.

Robertson, G.P., Wedin, D.A., Groffman, P.M., Blair, J.M., Holland, E.A., Nadelhoffer, K.J., et al., 1999. Soil carbon and nitrogen availability: nitrogen mineralization, nitrification, and soil respiration potentials. In: Robertson, G.P., Bledsoe, C.S., Coleman, D.C., Sollins, P. (Eds.), Standard Soil Methods for Long-Term Ecological Research. Oxford University Press, New York, pp. 258−271.

Robertson, G.P., Hamilton, S.K., Barham, B.L., Dale, B.E., Izaurralde, R.C., Jackson, R.D., et al., 2017. Cellulosic biofuel contributions to a sustainable energy future: choices and outcomes. Science 356, eaal2324d.

Roley, S.S., Xue, C., Hamilton, S.K., Tiedje, J.M., Robertson, G.P., 2019. Isotopic evidence for episodic nitrogen fixation in switchgrass (*Panicum virgatum* L.). Soil Biol. Biochem. 129, 90–98.

Roy, S., Liu, W., Nandety, R.S., Crook, A., Mysore, K.S., Pislariu, C.I., et al., 2020. Celebrating 20 years of genetic discoveries in legume nodulation and symbiotic nitrogen fixation. Plant Cell 32, 15.

Sexstone, A.J., Revsbech, N.P., Parkin, T.P., Tiedje, J.M., 1985. Direct measurement of oxygen profiles and denitrification rates in soil aggregates. Soil Sci. Soc. Am. J. 49, 645–651.

Silver, W.L., Herman, D.J., Firestone, M.K., 2001. Dissimilatory nitrate reduction to ammonium in upland tropical forest soils. Ecology 82, 2410–2416.

Smith, K.A., 1980. A model of the extent of anaerobic zones in aggregated soils, and its potential application to estimates of denitrification. J. Soil Sci. 31, 263–277.

Sollins, P., Robertson, G.P., Uehara, G., 1988. Nutrient mobility in variable- and permanent-charge soils. Biogeochemistry 6, 181–199.

Stein, L.Y., 2019. Insights into the physiology of ammonia-oxidizing microorganisms. Curr. Opin. Chem. Biol. 49, 9–15.

Strous, M., 2011. Beyond denitrification: alternative routes to dinitrogen. In: Moir, J.W.B. (Ed.), Nitrogen Cycling in Bacteria: Molecular Analysis. Caister Academic Press, Norfolk, pp. 123–133.

Subbarao, G.V., Nakahara, K., Hurtado, M.P., Ono, H., Moreta, D.E., Salcedo, A.F., et al., 2009. Evidence for biological nitrification inhibition in *Brachiaria* pastures. Proc. Natl. Acad. Sci. U. S. A. 106, 17302–17307.

Taylor, A.E., Zeglin, L.H., Wanzek, T.A., Myrold, D.D., Bottomley, P.J., 2012. Dynamics of ammonia-oxidizing archaea and bacteria populations and contributions to soil nitrification potentials. ISME J. 6, 2024–2032.

Tiedje, J.M., 1988. Ecology of denitrification and dissimilatory nitrate reduction to ammonium. In: Zehnder, A.J.B. (Ed.), Biology of Anaerobic Microorganisms. John Wiley, New York, pp. 179–244.

Tisdale, S.L., Nelson, W.L., Beaton, J.D., Havlin, J.L. (Eds.), 1993. Soil Fertility and Fertilizers, fifth ed. MacMillan, New York.

van Kessel, M.A.H.J., Speth, D.R., Albertsen, M., Nielsen, P.H., Op den Camp, H.J.M., Kartal, B., et al., 2015. Complete nitrification by a single microorganism. Nature 528, 555–559.

Vitousek, P.M., Menge, D.N.L., Reed, S.C., Cleveland, C.C., 2013. Biological nitrogen fixation: rates, patterns and ecological controls in terrestrial ecosystems. Philos. Trans. R Soc. B 368, 20130119.

Wang, X., Wang, S., Jiang, Y., Zhou, J., Han, C., Zhu, G., 2020. Comammox bacterial abundance, activity, and contribution in agricultural rhizosphere soils. Sci. Total Environ. 727, 138563.

Winogradsky, S., 1890. Recherches sur les organisms de la nitrification. Annales de l'Institut Pasteur 4, 760–771, 213–231, 257–275.

Winogradsky, S., 1892. Contributions a la morphologie des organismes de la nitrification. Arch. Sci. Biol. 1, 86–137.

Ye, R.W., Averill, B.A., Tiedje, J.M., 1994. Denitrification: production and consumption of nitric oxide. Appl. Environ. Microbiol. 60, 1053–1058.

Zhu, X., Burger, M., Doane, T.A., Horwath, R.W., 2013. Ammonia oxidation pathways and nitrifier denitrification are significant sources of N_2O and NO under low oxygen availability. Proc. Natl. Acad. Sci. U. S. A. 110, 6328–6333.

Chapter 15

Biological transformations of mineral nutrients in soils and their role in soil biogeochemistry

Michael A. Kertesz* and Emmanuel Frossard†

*School of Life and Environmental Sciences, The University of Sydney, Australia; †Department of Environmental Systems Science, ETH Zürich, Switzerland

Chapter outline

15.1 Introduction

Soil biota play a major role in element cycling in terrestrial systems (Ehrlich and Newman, 2009; Gadd et al., 2012). The bacteria, archaea, and fungi inhabit a wide range of environmental niches in soils, populating oxygen gradients from fully aerobic conditions in surface soils to entirely anaerobic sites within soil aggregates, sediments, or in the deep subsurface. They tolerate a broad range of moisture contents, salinities, temperatures, and chemical compositions, providing 80% to 90% of the life support systems for many different types of soil (Pereira e Silva et al., 2013). Most microbes that inhabit the soil are able to access nutrients both in the soil solution and from mineral and organic substrates and can then interconvert them. They form multispecies biofilms on soil surfaces where organisms with different capabilities can interact, thus forming the basis of the food web for many different soil fauna. The redundancy of soil microbial functions means that nutrient transformations and soil fertility are largely resilient to local environmental

Soil Microbiology, Ecology, and Biochemistry. https://doi.org/10.1016/B978-0-12-822941-5.00015-6

stresses (Griffiths and Philippot, 2013), but there is increasing evidence that global change (especially increased temperature and drought) affects ecological C:N:P stoichiometry at multiple trophic levels (Maaroufi and De Long, 2020) and that ecosystem P availability is essential for C sequestration in soil (Sun et al., 2020). The importance of soil microorganisms in carbon (C) and nitrogen (N) transformations has been explored in other chapters of this volume. This chapter focuses on the role of microbes in transformations of other elements in soil systems, including phosphorus (P), sulfur (S), iron, and other trace elements. The transformations of these mineral nutrients have been subdivided into those carried out by bacteria, archaea, and fungi, primarily because the ratios of the elements are fundamentally different in these groups of organisms.

15.2 Nutrient needs of soil biota

Because microorganisms provide the lowest trophic level in soil food webs, their elemental composition clearly indicates their production dynamics, and the biogeochemical cycles in which they are involved. Microbial cells are predominantly made up of C and N. They also have structural requirements for P and S that are used to produce nucleic acids, phospholipids, and proteins. These elements are enriched in the microbial biomass above that of their content in the soil (Table 15.1).

The preferred forms of P and S for microbial growth are free inorganic phosphate and inorganic sulfate, respectively. Both these forms are uncommon in most soils because almost all of the P and S are either sequestered as organic compounds or are strongly sorbed on soil surfaces. The microbial preference for the simple anionic forms of these nutrients is clearly seen in the networks of "phosphate starvation-induced" and "sulfate starvation-induced" genes that control the uptake and assimilation of alternative P- and S-containing compounds (Kertesz, 1999; Santos-Beneit, 2015). A data compilation of the organic P and S contents of nearly 1000 different soils from around the world revealed that: (1) C:N ratios varied between 9.8 and 17.5 (598 soils), (2) C:P ratios were in the range of 44 to 287 (408 soils), and (3) C:S ratios were between 54 and 132 (527 soils) (Kirkby et al., 2011; Tipping et al., 2016). Soil bacteria have a C:P ratio of about 60, while fungi contain much less P, with C:P in the range of 300 to 1190 (Kirchman, 2012).

Many other elements are essential for microbial growth but are required in smaller amounts than the major biogenic elements. Their primary role is in biochemical processes, either in redox processes or as a component of specific enzymes. The most important of these elements is iron (Fe), which is abundant in the Earth's crust and is required in small amounts in microbial cells, primarily in electron transfer reactions. Additional roles for other essential elements include: (1) copper in some redox enzymes, (2) zinc in alkaline phosphatase and carbonic anhydrase, (3) molybdenum in the enzyme that catalyzes N fixation in bacteria, and (4) nickel in urease and in several enzymes that are important in anaerobic environments, including hydrogenase, methyl coenzyme M reductase, and carbon monoxide dehydrogenase. Other microbes require a range of less common metals, including tungsten and vanadium (Table 15.2).

Attempts have been made to define "typical" elemental compositions of soil microorganisms in vitro, and to link these measurements to the compositions observed in natural environments. This has been largely unsuccessful because the actual composition of cells in situ reflects both the biochemical requirements of the cell under the given conditions and the availability of particular nutrients in the soil environment (Khan and Joergensen, 2019). A corollary of this is that by measuring the ratios of C to N and P in soil microbial biomass, it should be possible to extract direct information about which nutrient limitations the microbes are experiencing in soils (Cleveland and Liptzin, 2007). However, this would

TABLE 15.1 Elemental composition of a representative soil bacterium, *Pseudomonas putida* (Passman and Jones, 1985); a representative soil archaeon, *Methanosarcina barkeri* (Scherer et al., 1983); a representative soil fungus, *Rhizophagus irregularis* (Olsson et al., 2008); and median elemental soil concentration (Strawn et al., 2015).

Element	*Pseudomonas putida* (% w/w)	*Methanosarcina barkeri* (% w/w)	*Rhizophagus irregularis* (young hyphae) (% w/w)	*Rhizophagus irregularis* (spores) (% w/w)	Soil median concentration (total % w/w)
C	51–53	37–44	nm	nm	2.0
N	41–42	9.5–12.8	nm	nm	0.2
P	1.3–2.2	0.5–2.8	0.24–0.96	0.13–0.8	0.08
S	0.5–0.54	0.56–1.2	0.26–0.92	0.008–0.02	0.07
Mg	0.26–0.53	0.09–0.53	nm	nm	0.5
Na	0.16–0.39	0.3–4	nm	nm	0.5
K	0.18–0.3	0.13–5	1.2–3.6	0.09–0.28	1.4
Fe	0.012–0.023	0.07–0.28	0.06–0.13	0.03	4.0
Ca	0.19–0.33	0.0085–0.055	0.57–2.2	0.23–0.6	1.5
Zn	0.006–0.013	0.005–0.063	0.07	0.008–0.04	0.009
Cu	0.002–0.003	0.001–0.016	nm	0.004	0.003
Mo	nm	0.001–0.007	nm	nm	0.00012
Mn	nm	0.0005–0.0025	0.02	0.003–0.04	0.1

nm, not measured.

also require an assessment of microbial activity (Ehlers et al., 2010). Most studies have focused on C:N:P ratios in fungi and bacteria, but coupling effects mean that the demand for P and other nutrients is also influenced by organisms at higher trophic levels. Imbalances in C:N:P stoichiometry between plant material, soil fauna, and soil microbes cause changes in the key ecosystem processes responsible for plant litter decomposition, soil organic matter (SOM) turnover, and associated nutrient mineralization. Thus organisms may be forced to change their metabolism and lifestyle to adapt (Bardgett and van der Putten, 2014; Maaroufi and De Long, 2020). Soil fauna play an important role in element transformations, primarily due to their influence on microbial community diversity and activity rather than their direct transformation. Through soil bioturbation, soil fauna create new microsites (pores, aggregate surfaces) where microbial activity is stimulated and nutrient turnover is accelerated (Kuzyakov and Blagodatskaya, 2015; Briones, 2018). In addition, bacterivorous and fungivorous soil fauna preferentially feed on specific microbial groups, and in doing so they exert selective pressure on microbial community structure and thereby indirectly control rates of element transformation by the microbial community (Briones, 2018).

TABLE 15.2 Major and trace elements used by bacteria, archaea, and fungi, with examples of their roles in cell metabolism.

Element	Inorganic form taken up	Role in cell metabolism
Major elements		
C	HCO_3^-	All organic compounds (bacteria/archaea/fungi)
N	N_2, NO_3^-, NH_4^+	Proteins, nucleic acids (bacteria/archaea/fungi)
P	PO_4^{3-}	Nucleic acids, phospholipids (bacteria/archaea/fungi)
S	SO_4^{2-}	Proteins, coenzymes (bacteria/archaea/fungi)
K	K^+	Cofactor for enzymes (bacteria/archaea/fungi)
Ca	Ca^{2+}	Intercellular signaling (bacteria/archaea/fungi)
Si	$Si(OH)_4$	Diatom frustules
Trace elements		
Fe	Fe^{3+} and Fe^{2+} complexes	Electron transfer systems (bacteria/archaea/fungi)
Mn	Mn^{2+}, MnO_2	Superoxide dismutase (bacteria/archaea/fungi), oxygenic photosynthesis in cyanobacteria
Mg	Mg^{2+}	Chlorophyll (bacteria/archaea)
Ni	Ni^{2+}	Urease, hydrogenase (bacteria/archaea/fungi)
Zn	Zn^{2+}	Carbonic anhydrase, alkaline phosphatase, RNA/DNA polymerase (bacteria/archaea/fungi)
Cu	Cu^{2+}	Electron transfer system, superoxide dismutase (bacteria/archaea/fungi)
Co	Co^{2+}	Vitamin B_{12} (bacteria/archaea)/antimicrobial compounds (fungi)
Se	SeO_4^{2-}	Formate dehydrogenase, betaine reductase, sarcosine reductase (bacteria/archaea)
Mo	MoO_4^{2-}	Nitrogenases, sulfite oxidases, nitrite reductase (bacteria/archaea/fungi)
Cd	Cd^{2+}	Carbonic anhydrase in diatoms
W	WO_4^{2-}	Hyperthermophilic enzymes (bacteria/archaea)
V	VO_4^{3-}	Nitrogenases (bacteria/archaea/fungi)

Several inorganic nutrients are used by bacteria to provide energy for metabolism and growth. Different redox states of S-containing compounds (sulfate, sulfide, sulfite, and elemental sulfur) can be used by a range of organisms either as electron donors or as terminal electron acceptors in redox pathways that supply reducing equivalents to the cell. Metals such as Fe and Mn are used as energy sources by chemoautotrophic bacteria and as terminal electron acceptors by a range of heterotrophic bacteria. The transformations of these metals between different redox states often have a significant effect on their solubility, and hence on their mobility within the soil environment and their bioavailability to microbes and plants. Ferric (Fe^{3+}) compounds, for example, are much less soluble and therefore less bioavailable than ferrous (Fe^{2+}) compounds. However, Fe^{3+} predominates in all aerobic environments, and the levels of Fe required for optimal microbial growth (10^{-7} to 10^{-5} molar; Robin et al., 2008) are rarely present. Ferrous iron is nonetheless the preferred form of Fe for soil microbes (Lemanceau et al., 2009) and is the active form controlling iron metabolism within the cell.

15.3 Effect of soil biota on inorganic nutrient transformations

15.3.1 Phosphorus

Phosphorus ultimately derives from phosphate-containing minerals, such as apatite, which are present in bedrock and are progressively released into the soil by chemical and biological weathering. The levels of P found in soils vary considerably. Young soils on bedrocks containing high levels of apatite are often quite rich in total P, while many highly weathered, tropical soils are low in total P and are particularly deficient in soluble P. The total P content of agricultural soils ranges from 150 to 2000 μg P g^{-1}. Chemical weathering releases inorganic phosphate (Pi) into the soil solution (Fig. 15.1), but the levels of soluble orthophosphate are low in most soils (often below 0.1 mg P L^{-1} and in highly weathered soils

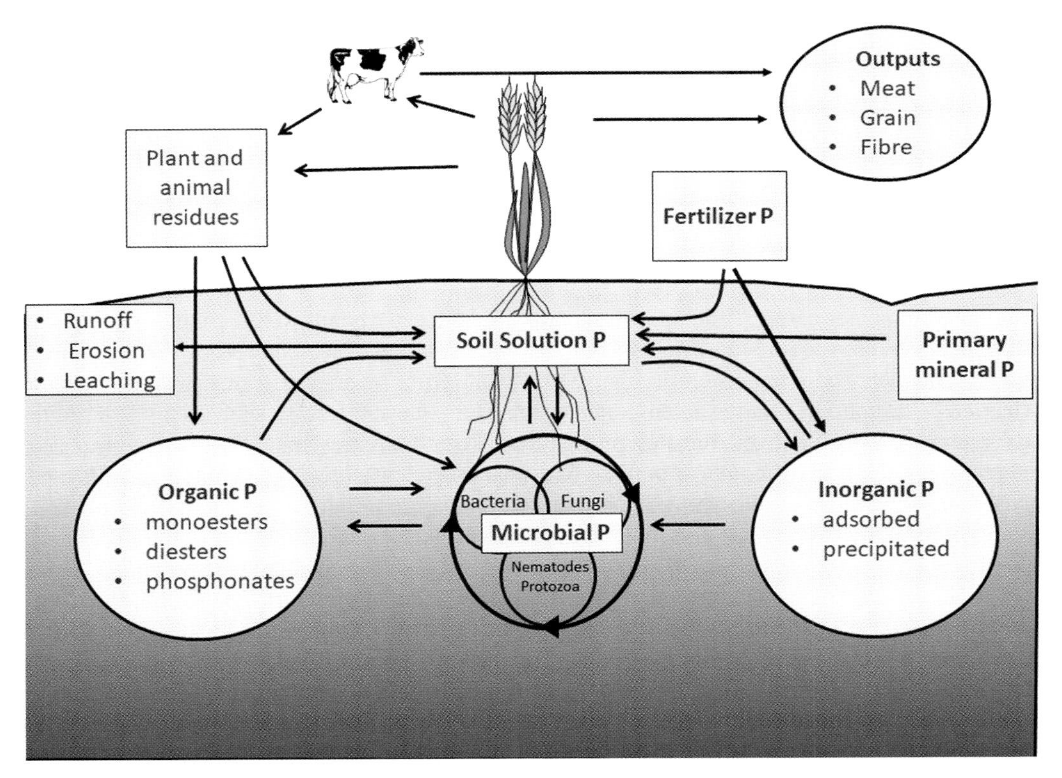

FIGURE 15.1 Major components of the phosphorus cycle in agricultural systems. Transformations between major P components are indicated.

below 0.01 mg P L^{-1}; (Randriamanantsoa et al., 2013). Orthophosphate sorbs to soil surfaces and forms precipitates with Ca salts in calcareous soils, reducing the phosphate available both from mineral weathering and from applications of phosphate-containing chemical fertilizer. In general, only 10% to 30% of applied fertilizer P is taken up by crop plants in the year following application (Doolette and Smernik, 2011), and the remainder is transferred to less soluble pools.

Soils also contain a large pool of organic P. These P forms are partly derived from biological compounds released into the soil as animal waste, plant litter, and microbes. The exact nature of organic P compounds is difficult to assess, but they have been classified (in decreasing order of abundance) as: (1) monoester phosphates, such as phytate (myo-inositol-hexakisphosphate), (2) diester phosphates, such as nucleic acids, and (3) phosphonates, containing a direct C-P bond. The phosphonates are partly derived from phosphonolipids, which replace phospholipids in some microbes, but they are also synthesized directly by many soil actinobacteria (Yu et al., 2013). In marine systems, methylphosphonate is synthesized by the Thaumarchaeota (Metcalf et al., 2012). This major group of versatile chemoorganotrophs in soils (Pester et al., 2011; Adam et al., 2017) may play a similar role in phosphonate synthesis. Phosphorus speciation has largely been conducted in recent years by two spectroscopic methods; nuclear magnetic resonance (NMR) and X-ray absorption near edge spectroscopy (XANES). Each of these approaches characterizes the chemical environment surrounding the P nucleus to provide information about the nature of the P atom. The NMR may be performed either on solid samples or soil extracts and is particularly useful for differentiating organic P forms (McLaren et al., 2020), while XANES is usually done with solid samples, providing useful information on inorganic P forms (Acksel et al., 2019).

Phosphorus availability in the soil is mediated by mineralization and immobilization from organic P fractions and also by sorption/desorption or precipitation/solubilization processes from inorganic P fractions (Frossard et al., 2000). The P in SOM can be mineralized either by microbes in search of C which thereby release excess P associated with the metabolized C or when microbes specifically scavenge P through the release of extracellular phosphatase enzymes (McGill and Cole, 1981). These microbial processes dominate organic P mineralization. Typical basal mineralization rates are 0.1–2.5 mg P kg^{-1} d^{-1}, but can be as high as 12.6 mg P kg^{-1} d^{-1} in grassland and forest soils (Bünemann, 2015).

Phosphatase enzymes catalyze the hydrolysis of phosphate ester bonds to release inorganic phosphate. These phosphatases are specific to forms of phosphate esters, many of which can only be found in certain compounds. Thus, phosphomonoesterases cleave phosphate from monoester forms such as phospholipids or nucleotides. Phosphodiesterases release phosphate from diester forms, such as nucleic acids, and phytases release phosphate from inositol phosphates. Their activities may be divided into acid and alkaline phosphatases with different pH optima. Phosphatases released by plant and fungal cells tend to be acid phosphatases, while bacteria synthesize primarily alkaline phosphatases and some acid phosphatases (Nannipieri et al., 2011). The most common bacterial phosphatases in soil belong to the *phoD* and *phoX* families (Ragot et al., 2015, 2017), with the corresponding genes detected in a wide range of soil taxa (15 bacterial phyla for *phoD*; and 16 bacterial, 1 archaeal, and 2 fungal phyla for *phoX*) (Bergkemper et al., 2016; Ragot et al., 2017). Most of these enzymes are targeted for extracellular localization, though it is not always clear whether the secreted enzymes are free in soil solution, bound to soil solids, or associated with the external surface of microbial/plant cells or cell debris. Their extracellular activities can potentially hydrolyze a broad range of phosphate-containing substrates outside the cell (Annaheim et al., 2013), and the phosphate released is taken up by specific high-affinity phosphate transporters in the cell membrane. This allows the cell to scavenge phosphate from a range of soil substrates without needing to synthesize specific transporters for multiple different compounds, many of which are of high molecular weight. By contrast, phosphonates are almost certainly degraded within the bacterial cell, as the enzymes responsible, the C-P lyases, are unstable and require several cofactors (Hove-Jensen et al., 2014). Because cells also require other nutrients for growth, the rate of incorporation of released phosphate into microbial biomass depends on the availability of C and N. Generally, if the C:P ratio is $> \sim 300:1$, net immobilization of P into microbial biomass occurs, while for C:P ratios $< \sim 200$, microbial growth yields a surplus of P, and net mobilization of orthophosphate into the soil solution is observed.

Although the process described above immobilizes P into soil microbial biomass, this P is also readily remineralized. Bacterial cell turnover in the soil can be rapid, especially in regions surrounding plant roots (the rhizosphere) where there is organic C for growth. Microbial turnover in soils is mediated through predation by the fauna (e.g., protists, nematodes) and by viral attack (Trap et al., 2016; Emerson, 2019), and the resulting cell lysis releases the bacterial cell contents (including P) into the soil solution. This mobilization is also affected by soil drying and rewetting cycles, since these cause changes in cell turgor which can both release cell contents during the drying process and cause cell lysis through osmotic shock on rewetting (Bünemann et al., 2013). This leads to the release of a significant proportion of the P from microbial biomass during rainfall events after a dry period (Blackwell et al., 2010). However, the effect will change with repeated cycles of wetting/drying as these environmental stresses also lead to selective changes in microbial community composition (Chen et al., 2021; Gschwend et al., 2020). Similar effects are also caused by cycles of freezing and thawing (Blackwell et al., 2010).

Bacteria play an important role in both the solubilization of phosphate from the precipitated inorganic soil P fraction and in weathering of minerals to release P (Uroz et al., 2009). This process is mediated by the release of organic acids (gluconic, oxalic, malonic, succinic, lactic, isovaleric, isobutyric, and acetic acids (Alori et al., 2017), which causes a localized decrease in soil pH. The phosphate-solubilizing microbes that mediate this process include a broad variety of bacterial taxa (phosphate-solubilizing bacteria [PSB]), but also many fungi (Alori et al., 2017). The best characterized PSB species is *Priestia (Bacillus) megaterium* which is commercially applied in biofertilizers. PSBs are readily isolated from soil by screening for the ability to solubilize Ca or Fe phosphates in an agar-plate test, but PSB diversity in soils has also been studied by directly using specific gene probes. The organic acid most frequently associated with phosphate mobilization by PSB is gluconic acid, which is produced in the bacterial periplasm by the membrane-bound quinoprotein glucose dehydrogenase (Alori et al., 2017). This enzyme is encoded by the *gcd* gene which has been used as a marker for PSB in soil environments (Bergkemper et al., 2016). Optimal expression of the *gcd* gene and of the genes encoding biosynthesis of the associated pyrroloquinoline quinone cofactor (PQQ) in the soil bacterium *Pseudomonas putida* KT2440 is observed under conditions of low phosphate (An and Moe, 2016).

Fungi take up P as orthophosphate ions using both low and high-affinity transporters (Nehls and Plassard, 2018; Plassard et al., 2019). Arbuscular mycorrhizal fungi (AMF) take up orthophosphate ions from the soil solution which have limited direct effect on soil P solubility (Smith and Smith, 2011). However, collaboration between AMF and PSB allows bacterial mobilization of organic P for the fungal partner in exchange for C supply to the PSB by the fungal hyphae. When soil P availability is low, the PSB compete with the AMF for P, but in the presence of additional organic P, the PSB enhance AMF hyphal growth (Zhang et al., 2016; Ezawa and Saito, 2018). Ectomycorrhizal fungi and fungal saprotrophs can synthesize and release phosphatase enzymes able to cleave Pi from organic P sources (Becquer et al., 2014). For instance, some fungi (e.g., *Aspergillus fumigatus*) release high quantities of phytase that cleave Pi from phytate which can be present in high concentration in soils. Fungi can also release low-molecular-weight organic acids, such as oxalate, which release P from inorganic forms by acidification and chelation of the P-binding cations (Plassard et al., 2011). When grown under conditions of low available P in symbiosis with *Pinus sylvestris*, the ectomycorrhizal *Paxillus involutus* was able to accelerate the rate of P release from apatite grains by utilizing plant C (Smits et al., 2012). Similarly, *Aspergillus niger*, which is known to produce large amounts of citric acid, can dissolve large amounts of apatite (Bojinova et al., 2008). Finally, siderophores produced by fungi for Fe acquisition can also significantly increase phosphate solubility (Reid et al., 1985), probably by displacing phosphate associated with Fe. Phosphate taken up in excess of growth requirements is stored in the fungal vacuole as polyphosphates, and compared to plants and bacteria, fungi therefore contain more polyphosphate and less diester P (Bünemann et al., 2011). Glomeromycota have been widely investigated as a means of improving the efficiency of soil P use by agricultural plants (Delavaux et al., 2017), including in the subsoil (Sosa-Hernández et al., 2019). However, their efficacy is still debated (Ryan and Graham, 2018), and their reliable application in the field on a large scale is dependent on a range of variables (Verbruggen et al., 2013).

15.3.2 Sulfur

Most global reserves of S are tied up in the lithosphere from which they are slowly released by weathering. The weathering process provides sulfate input to soils and surface water, with 27 mM sulfate eventually finding its way to the ocean. Agricultural soils contain 2 to 2000 μg S g^{-1}, most of which is bound in SOM in a range of organic S forms. Groundwater, however, contains considerable amounts of inorganic sulfate (Miao et al., 2012), as, unlike phosphate, it is relatively soluble at soil pH values. The availability of inorganic sulfate in surface soils therefore varies during the year with changes in the groundwater table.

The bulk of the S bound in SOM is found in the high-molecular-weight fraction (MW >100,000 kDa). Much of this is physically protected within soil microaggregates and is not readily mobilizable in the short term (Eriksen, 2009). The precise chemical structure of larger S-containing molecules in SOM is not known, but the chemical environment of the S atoms has been defined in a number of ways, with S speciation within SOM having been defined using X-ray spectroscopy (K-edge XANES) which provides specific information on the oxidation state of S functional groups within the sample. XANES is used to differentiate reduced S (e.g., cysteine), oxidized S (sulfonates, sulfate esters, and also inorganic sulfate), and S with intermediate redox state, such as sulfones and sulfoxides (Prietzel et al., 2011). XANES analysis of SOM can be done using either bulk soils directly (Prietzel et al., 2011), humic extracts of soils (Zhao et al., 2006), or the organic S fraction extracted from soil with acetylacetone (Boye et al., 2011). This information is often combined with more traditional methods of determining S-speciation according to its chemical reactivity with reducing agents. Using this latter approach, organic S compounds are divided into three broad groups: sulfate esters (containing the C-O-S moiety), amino acid-sulfur (cysteine/methionine/peptides), and all other carbon-bound S compounds, usually regarded to be dominated by sulfonate-sulfur ($-SO_3H$) (Frossard et al., 2012).

The soil organic-S fraction is primarily composed of oxidized S forms, including sulfate esters (30–75% of total S) and sulfonates (20–50% of total S), with inorganic sulfate making up less than 5% of total S in most aerobic soils. There is considerable variation between different types of soil, but on the whole, pastures and grasslands tend to be higher in sulfate esters, and forest soils contain more of the sulfonated fraction (Autry and Fitzgerald, 1990; Chen et al., 2001). Organic S found in soil is derived from plant litter and animal waste inputs. For example, sheep urine contains 30% of its S as sulfate esters, sheep dung contains about 80% of its S as sulfonates and amino acid-S (Williams and Haynes, 1993), and plant litter contains 60% to 90% of its S in C-bound form (Zhao et al., 1996), much of it derived from sulfolipid, a common lipid in the thylakoid membranes of the plant chloroplast (Benning, 1998). Carbon-bound S also enters soil in rainfall as methanesulfonate, which is derived from atmospheric dimethylsulfide (Kelly and Murrell, 1999).

S-containing compounds in aerobic soils undergo a range of transformations, including immobilization into microbial biomass and SOM, and mineralization of soil organic S by sulfatase and sulfonatase enzymes (Fig. 15.2). Immobilization of sulfate into organic matter has been studied primarily by experiments with ^{35}S-sulfate, which is initially incorporated into the sulfate ester pool and then slowly transformed into C-bonded S (Ghani et al., 1993a). This immobilization of sulfate is thought to be microbially mediated, because rates of immobilization can be increased by addition of assimilable C or N as glucose, organic acids, or model root exudates (Ghani et al., 1993a; Vong et al., 2003; Dedourge et al., 2004) and by pre-incubation to stimulate microbial growth (Ghani et al., 1993a). Incorporation of sulfate into the organic S pool is also dramatically increased by the addition of cellulose as a C source (Eriksen, 1997).

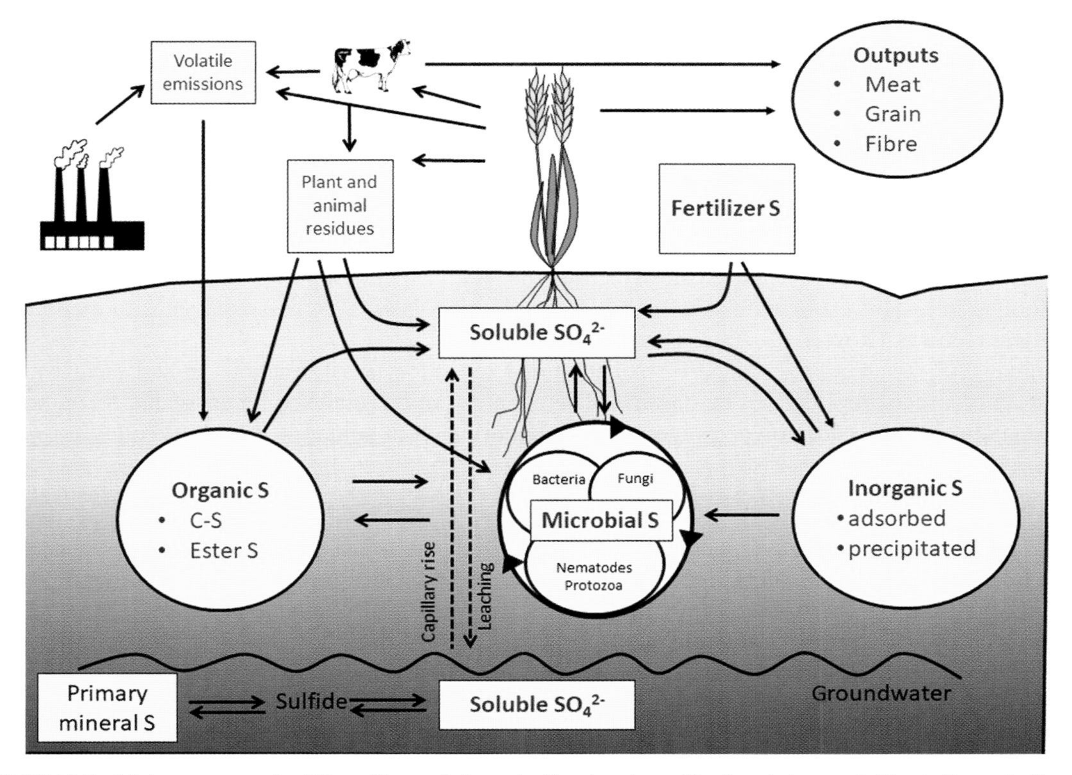

FIGURE 15.2 Major components of the sulfur cycle in agricultural systems. The S cycle is remarkably similar to the P cycle, with a small pool of available nutrient and large pools of the immobilized element in the soil. Unlike P, the S cycle includes enhanced sulfate exchange with groundwater and the redox transformations of S compounds.

Chemical-speciation studies have shown that almost all the S released in short-term incubation studies was derived from C-bonded S (Ghani et al., 1992, 1993b). This agrees with spectroscopic (XANES) data, which show a good correlation between S mineralization and the amount of peptide-S and sulfonate-S in the soil humic fraction (Zhao et al., 2006). The correlation with sulfate ester-S is not clear, but soil sulfatases are nonetheless thought to be important in soil S transformations. These enzymes catalyze the hydrolysis of sulfate esters, are easily measured in soils, and together with several other soil enzymes, have been widely used as a measure of soil health and soil microbial activity (Acosta-Martinez et al., 2019; Aponte et al., 2020). Moreover, they provide valuable markers for pesticides and heavy metals in soil (Kucharski et al., 2016; Aponte et al., 2020). Arylsulfatase activity is produced by a wide range of soil microbes (Stressler et al., 2016), and these enzymes are generally thought to be extracellular (Dunn et al., 2014). However, the arylsulfatase of one common soil genus, *Streptomyces,* is located in the cell membrane (Cregut et al., 2012), while those produced by several *Pseudomonas* species appear to be intracellular (Kertesz et al., 2007). Arylsulfatase activity is correlated with soil microbial biomass and the rate of S immobilization (Vong et al., 2003), as well as with pH and organic C levels (Goux et al., 2012; Piutti et al., 2015). The most common cultivable, sulfatase-producing bacteria in agricultural soils belong to the Actinobacteria and *Pseudomonas* clades (Cregut et al., 2009).

Sulfonates in soil can be desulfurized by a large proportion of bacteria (King and Quinn, 1997). The desulfurization of simple alkane and arylsulfonates is catalyzed by a family of flavin-dependent mono-oxygenase enzymes which cleave the sulfonate moiety to yield sulfite for assimilation into bacterial biomass. The key groups of bacteria responsible for this process in agricultural and grassland soils are members of the *Variovorax, Polaromonas,* and *Rhodococcus* genera, which have been studied using molecular methods to assess in situ diversity of the sulfonatase genes (Schmalenberger et al., 2008, 2009, 2010). The sulfonatase enzymes are thought to be entirely intracellular because they require flavin and nucleotide co-factors. However, it is not clear how high-molecular-weight organic substances carrying sulfonate moieties can enter the cell for mineralization to occur. Sulfonates of intermediate complexity, such as sulfoquinovose (6-deoxy-6-sulfoglucose, the sulfonated headgroup of plant sulfolipid), are transformed into short-chain sulfonates by bacterial sulfoglycolysis, a process analogous to the standard glycolytic pathway (Roy et al., 2003; Denger et al., 2014).

Unlike P, S is subject to a range of oxidation and reduction transformations in soil environments, almost all of which are mediated by bacteria. The main oxidation sequence for inorganic S compounds that is catalyzed by bacteria involves the sequential conversion of sulfide to sulfate as follows:

$$\underset{\text{Sulfide}}{S^{2-}} \rightarrow \underset{\text{sulphur}}{S^{0}} \rightarrow \underset{\text{thiosulfate}}{S_2O_3^{2-}} \rightarrow \underset{\text{tetrathionate}}{S_4O_3^{2-}} \rightarrow \underset{\text{sulfate}}{SO_4^{2-}}$$

Oxidation of these S-containing compounds is catalyzed by two main groups of microbes. Chemoautotrophic bacteria use the S-substrates as a source of energy while using inorganic compounds as their source of C (lithotrophs). These are primarily bacteria belonging to the *Thiobacillus* and *Acidithiobacillus* genera. Most of these are obligate aerobes and use oxygen as the terminal electron acceptor for growth, but *Thiobacillus denitrificans* can also grow anaerobically, using nitrate as an electron acceptor. Some of the thiobacilli are also able to oxidize inorganic S compounds while growing either heterotrophically or autotrophically (facultative heterotrophy) or with a mixture of organic and inorganic C sources (mixotrophy). Chemoautotrophic growth with inorganic S compounds is best known for organisms isolated from environments such as hot spring sediments, which have low levels of organic C (organic C <<< inorganic S). By contrast, levels of organic C in soils are typically much higher (organic C >>> inorganic S), and bacterial populations are dominated by heterotrophs. Many of these (e.g., *Arthrobacter, Bacillus, Micrococcus,* and *Pseudomonas* species) are also able to oxidize inorganic S compounds, although some of them only carry out partial oxidation (e.g., oxidize S to thiosulfate, or thiosulfate to sulfate, rather than the complete oxidation of S to sulfate). The role of these organisms is unclear, since they do not appear to gain any energy from the oxidation process, and it is possible that the transformation is co-metabolic. Populations of both thiobacilli and heterotrophic S-oxidizers are strongly stimulated by addition of elemental S to agricultural soils, but the increase in the thiobacilli population is only transient, suggesting that heterotrophic S-oxidation is more important in these environments (Yang et al., 2010).

Microbial oxidation of inorganic S compounds almost always requires oxygen. It is therefore not usually an important process in well-aerated soils (except in the interior of soil aggregates), but it becomes significant in sediments and in flooded soils, especially when large amounts of organic C are present. The reduction of sulfate to sulfide is catalyzed by the anaerobic sulfate-reducing bacteria (SRB), which are organotrophic organisms that use low-molecular-weight organic compounds (e.g., propionate, butyrate, or lactate) or H_2 as an electron donor and sulfate as a terminal electron acceptor. Marine sediments are particularly high in SRB due to the high levels of inorganic sulfate present in seawater. Dissimilatory sulfate reduction is known for five major groups of bacteria and two archaeal groups (Muyzer and Stams, 2008; Rabus et al., 2015), and many of these are also able to reduce sulfite and thiosulfate. Common SRB genera include *Desulfovibrio, Desulfobacter, Desulfococcus,* and *Desulfotomaculum*, while the best-known sulfate-reducing archaeal genus is *Archaeoglobus.*

The key genes involved in the dissimilatory sulfate reduction process are *dsrAB*, which encodes the dissimilatory sulfite reductase, and *aprA* which encodes adenosine phosphosulfate reductase. Populations of SRB are near-ubiquitous in anaerobic soils and sediments. They have been found in many different environmental habitats, including hydrothermal vents, mud volcanoes, acid mine drainage, soda lakes, oilfields, agricultural soils, and plant rhizospheres, and they play important biotechnological roles in treatment of wastewater, bioremediation of metals and hydrocarbons, and bioelectrochemistry and corrosion of iron and concrete (Rabus et al., 2015). Since the *aprA* gene is an essential part of the S oxidation mechanism, molecular analysis of *aprA* diversity yields detailed information on the activity of both S oxidizing and reducing pathways in soils, without the need to cultivate the bacteria in the laboratory (Meyer and Kuever, 2007). Analysis of the full genome sequences for many hundreds of SRB has revealed a great flexibility in energy metabolism and an ability to form syntrophic associations with other microbes, especially methanogens and H_2-producers. These genetic traits allow them to adapt to a wide range of anaerobic niches (Zhou et al., 2011; Rabus et al., 2015).

There is less information on S transformations by soil fungi, but many of the key bacterial enzymes involved in organic S transformations also have homologs in fungi (Linder, 2018). Wood-degrading fungi can utilize a large variety of different wood S sources for growth (Schmalenberger et al., 2011). Studies carried out on *Neurospora crassa*, *Aspergillus*, and *Penicillium* spp. show the existence of two types of transporters for sulfate uptake (Jennings, 1995; Marzluf, 1997) whose expression is regulated by sulfate availability (Jennings, 1995). After uptake, sulfate is reduced sequentially to 3′-phosphoadenosine 5′-phosphosulfate (PAPS), thiosulfate, and sulfide, from which it is incorporated into cysteine and homocysteine. When sulfate is taken up in excess, it can be stored as sulfate, choline sulfate, or glutathione in the vacuole. Fungi can also oxidize inorganic S, but it is not clear to what extent they gain energy from this process (Gadd and Raven, 2010; Gleason et al., 2019). Although the evidence for fungal soil S transformations is limited, S availability is important in determining the abundant and rare fungal subcommunities in soil (Jiao and Lu, 2020). Mycorrhizal fungi also modulate soil S transformations by selecting the sulfonate-degrading bacteria in the soil surrounding their hyphae (Gahan and Schmalenberger, 2015).

15.3.3 Iron

Iron, the fourth most abundant element in the Earth's crust, is poorly available to soil organisms because it is either bound up in primary minerals (either as Fe(II) or Fe(III)) or is in sparingly soluble form in Fe oxides and oxyhydroxides, which are often strongly sorbed to clays and organic compounds. In aerobic soils, the solubility of these oxides is governed by equilibria related to $Fe(OH)_3 \leftrightarrows Fe^{3+} + 3OH^-$ and is therefore strongly dependent on soil pH. In acidic soils (pH values ~3.5), the concentration of Fe^{3+} generated by this equilibrium is about 10^{-9} M, but in more alkaline soils (pH values ~8.5), this decreases to ~10^{-24} M, which is well below the levels required for biotic growth (Robin et al., 2008). Soil bacteria and fungi have therefore been forced to develop mechanisms to enhance the levels of bioavailable Fe in soil; accomplished with strategies based on acidification, chelation, and reduction (Fig. 15.3).

In the rhizosphere, microbial strategies for nutrient mobilization are often assisted by corresponding mechanisms in plants, with some evidence for synergism in Fe solubilization rather than competition for a limiting resource (Robin et al., 2008). Indeed, for aerobic soils, most studies of Fe and metal metabolism by bacteria have focused on the influence of soil bacteria in promoting uptake of Fe by plants for plant

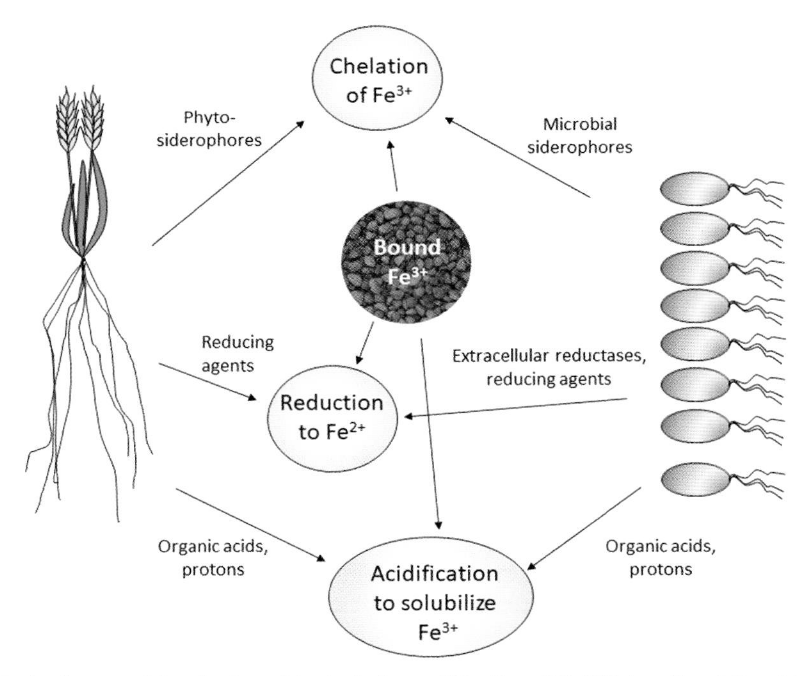

FIGURE 15.3 Mechanisms used by plants and bacteria for mobilization of Fe(III) in soil and the rhizosphere. Reducing agents include, for example, phenols and aliphatic acids. Both plants and bacteria preferentially take up Fe(II). Both plants and bacteria synthesize reductase enzymes to reduce Fe(III), but plant reductases are membrane bound and are not released into the soil solution.

health and nutrition (Robin et al., 2008) or in facilitating the sequestration of heavy metals by plants in phytoremediation applications (Glick, 2010).

The most common mechanism used by bacteria to enhance Fe solubility is the production of low-molecular-weight Fe-chelating compounds known as siderophores, which are released into the soil solution. These diverse molecules are able to sequester ferric iron from the soil environment even when the metal is tightly sorbed onto solid surfaces or metal oxides (Hider and Kong, 2010). They can also sequester iron by ligand exchange from other organic ligands, including organic acids, phenolates, and both soluble and insoluble complexes of iron and organic substances. The latter are especially important as they make up a large pool of plant-available iron in the soil (Colombo et al., 2014; Zanin et al., 2019). The nature of the chelating group varies between different siderophores which are classified according to the nature of the functional ligands as catecholates, hydroxamates, hydroxypyridonates, and hydroxyl- or amino-carboxylates (Albelda-Berenguer et al., 2019). Siderophore synthesis is strongly upregulated by low iron availability and is repressed under conditions where soluble iron is plentiful. Iron forms a complex with the siderophore via hydroxy and keto groups and this Fe-siderophore complex then binds to a siderophore-specific outer membrane receptor on the bacterium. In pseudomonads and other Gram-negative bacteria, this complex is then transported across the outer cell membrane and the metal is released from the siderophore by reduction to Fe^{2+}. The metal ion then enters the cell via a specific inner-membrane transporter, while the Fe-free siderophore molecule (or *apo-siderophore*) is released back into the extracellular environment using a specific exporter. In gram-positive bacteria, such as *Streptomyces*, the mechanism of Fe uptake is similar, except that the Fe-siderophore complex is reduced directly on binding to the cell. Plant roots also release Fe-binding molecules, called phytosiderophores, which are structurally distinct from bacterial siderophores (Robin et al., 2008) and have a lower affinity for Fe. Many bacteria also indulge in "siderophore piracy," taking up iron either from plant-derived phytosiderophores or from siderophores released by other microorganisms (Butaite et al., 2017; Kramer et al., 2020). Bacterial siderophores have considerably higher Fe-affinity than the siderophores released by many fungi, and thus, siderophore-synthesizing bacteria are often selected as biocontrol agents against fungal pathogens in agricultural applications (Laslo et al., 2012; Kramer et al., 2020).

Soil acidification increases the solubility of ferric salts and their bioavailability; therefore, plant roots have developed an active strategy of releasing protons to increase mobilization of Fe. Both microbes and roots also release organic acids in considerable quantities, contributing to decreases in soil pH. Carboxylic acids, such as citrate and malate, also help in the mobilization of Fe since they act as lower-affinity ligands for Fe and chelate the metal ion. A large proportion of organic acids in soil solution are complexed to metals including Fe.

Oxidation and reduction reactions also play a key part in Fe cycling in soils and sediments. Fe^{2+} is rapidly chemically oxidized to Fe^{3+} at neutral or alkaline pH. For many years, abiotic reactions were thought to dominate Fe redox reactions, but it has become clear that in many environments, microbial metabolism is extremely important in Fe redox transformations, almost all of which are also mediated by microbes (Fig. 15.4) (Melton et al., 2014). At acidic pH values, Fe^{2+} is oxidized to Fe^{3+} by chemoautotrophic bacteria, such as *Acidithiobacillus ferrooxidans*. At neutral pH values, Fe(II) functions as an electron donor for many lithotrophic Fe-oxidizing bacteria and archaea, including *Gallionella* and *Leptothrix* species (Emerson et al., 2010; Hedrich et al., 2011). Fe-oxidizing bacteria typically inhabit environments at redox boundaries, especially in wetlands, stream sediments, and waterlogged roots, where their activity is revealed by the presence of red Fe oxide precipitates (Emerson et al., 2010).

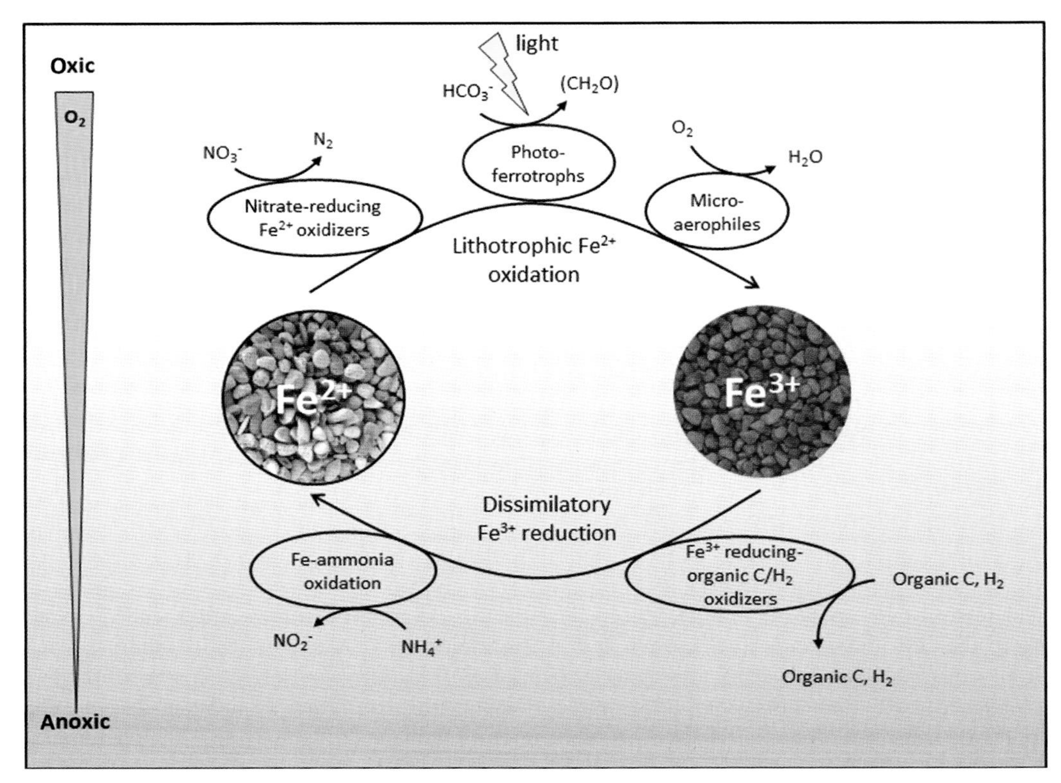

FIGURE 15.4 The microbially mediated iron redox cycle. Reduction of Fe(III) to Fe(II) occurs only under anoxic conditions. Lithotrophic Fe(II) oxidation is carried out by a range of organisms under different environmental conditions.

Insoluble Fe compounds may also be mobilized by dissimilatory Fe(III) reduction, in which Fe(III) is used as a terminal electron acceptor for microbial growth. Given the high levels of Fe(III) in the environment, this reaction is of great importance both for cell bioenergetics and for cycling of Fe between different redox states. The best studied organisms that carry out this process belong to the *Geobacter* and *Shewanella* genera, although Fe reducers are also known from many other bacterial and archaeal phyla (Weber et al., 2006). These microbes face a difficult problem in transferring reducing equivalents to an insoluble ferric mineral or salt. *Shewanella* species have solved this problem by using soluble quinones or flavins as electron shuttles between the cell and the substrate, which may be some distance from the cell itself, or by releasing redox cofactors within a surface-associated biofilm in a process called multistep electron hopping. By contrast, *Geobacter* species require direct contact between the cell and the substratum, allowing electrons to be passed from a bacterial biofilm directly to the metal. This contact may also be mediated by pili or microbial nanowires, which give metallic-like conductivity and facilitate electron transfer to the solid mineral phase (Lovley, 2012; Melton et al., 2014).

Iron is also needed in small amounts by fungi. It is essential as an acceptor and donor of electrons, e.g., in the cytochrome system, and is particularly important for wood-degrading fungi since it is required for the initial degradation of lignocellulosic polysaccharides and lignin (Comensoli et al., 2017). Like bacteria, fungi can acquire Fe by acidifying their environment or by releasing siderophores (Philpott, 2006). Fungal siderophores are almost exclusively of the hydroxamate type, where three families have been identified; the ferrichromes, fusarinines, and coprogens (Comensoli et al., 2017), with some organisms utilizing more than one of these systems (Voss et al., 2020). The fungus can take up the Fe^{3+}-siderophore as a complex or it can reduce Fe^{3+} to Fe^{2+}, which is then taken up by low or high-affinity Fe^{2+} transporters (Philpott, 2006). Some of these transporters need Cu to be functional, and some fungal Fe^{2+} transporters can also transport Mn^{2+} and Cu^{2+} into the cell (Philpott, 2006; Bolchi et al., 2011).

15.3.4 Other elements

Many other elements are subject to microbial transformations in soils and sediments. Microbes have high potassium (K) requirements, while elements like Ca, Zn, Cr, Se, or Mn are required in small amounts for specific biochemical functions (Table 15.2). In some cases, the uptake of higher concentrations of these elements may be toxic to the cell. Microbes have developed highly regulated mechanisms that control the bioavailability of metals in the soil environment to ensure that only appropriate amounts of the element are taken up, and guarantee that only the correct metal is incorporated into the relevant cofactor or enzyme (Sullivan and Gadd, 2019).

Bacteria and archaea use a range of mechanisms for this purpose. These include metal-binding proteins (metallochaperones) that selectively bind the metal ions on entry into the cell and systems that interact effectively with the target enzyme to insert the metal into its required site (Waldron and Robinson, 2009). Surprisingly, most metalloproteins do not have metal-binding preferences that directly match their function, and thus, metallation of an estimated 30% of metalloenzymes require specific metal delivery systems or preassembled metal cofactors (Foster et al., 2014; Capdevila et al., 2017). The desired metal selectivity is achieved by: (1) variation in the number and type of metal ligands on the binding protein, (2) binding the metal in a cellular compartment that selects for specific redox qualities (the cytoplasm is a more reducing environment than the periplasm), and (3) selective protein folding that depends on the type of metal ion bound (Waldron and Robinson, 2009). Comparative genomic studies of soil bacteria and archaea have revealed that these mechanisms are largely conserved, although there are distinct and dynamic evolutionary patterns in the utilization of different metals and metalloproteins (Zhang et al., 2019).

In many soils, trace elements (i.e., Cu, Ni, Zn, or Mn) are present at much higher levels than soil microbes can assimilate, either because the soils are derived from rocks containing these elements or because of industrial contamination with metal-containing effluents or wastes. Most of these elements exist in different oxidation states, and bacteria are involved in conversion between these redox forms. The bioavailability of metal salts often varies considerably between different oxidation states, although it is also affected by levels of organic matter, pH, and other factors. Microbial conversion of a contaminant metal compound to an insoluble form by oxidation or reduction can provide an effective method to immobilize it for bioremediation purposes (Antoniadis et al., 2017; Sullivan and Gadd, 2019). Conversely, solubilization of insoluble metals by redox transformation may also be used for "bio-mining" to mobilize valuable metals from low-grade ores (Shi et al., 2016). Examples of organisms involved in the oxidation or reduction of a range of soil elements are given in Table 15.3.

Manganese redox cycling is an extremely dynamic process in soil environments. Manganese oxidizing bacteria can be isolated from almost any soil or sediment sample, and the Mn oxides they generate are

TABLE 15.3 Examples of bacteria involved in the oxidation or reduction of manganese, chromium, arsenic, mercury, and selenium.

Element	Redox states	Reaction type	Examples of bacterial genera involved
Manganese (Mn)	Mn^{4+}, Mn^{2+}	Oxidation	*Arthrobacter, Bacillus, Pseudomonas, Leptothrix*
		Reduction	*Bacillus, Geobacter, Pseudomonas*
Chromium (Cr)	Cr^{6+}, Cr^{3+}	Reduction	*Aeromonas, Shewanella, Pseudomonas*
Arsenic (As)	As^{5+}, As^{3+}	Oxidation	*Alcaligenes, Pseudomonas, Thiobacillus*
		Reduction	*Alcaligenes, Pseudomonas, Micrococcus*
Mercury (Hg)	Hg^{2+}, Hg^{0}	Oxidation	*Bacillus, Pseudomonas*
		Reduction	*Pseudomonas, Streptomyces*
Selenium (Se)	Se^{6+}, Se^{4+}, Se^{0}, Se^{2-}	Oxidation	*Bacillus, Thiobacillus*
		Reduction	*Clostridium, Desulfovibrio, Micrococcus*

Data from (Sylvia et al., 1999; Tebo et al., 2005).

among the most reactive natural oxidizing agents, reacting rapidly with other reduced substrates, including Fe, S, and C compounds. Microbial Mn^{2+} oxidation occurs above pH 5, increases up to pH 8, and is catalyzed by a multi-copper oxidase system. It does not appear to provide energy for the bacteria (no Mn^{2+}-dependent chemoautotrophs are known), but does provide protection against damage by reactive oxygen intermediates such as hydrogen peroxide or superoxide (Waters, 2020). An excess of Mn affects intracellular pools of other ions and causes mis-metallation of important regulators and enzymes (Chandrangsu et al., 2017). In many soils that are subject to cycles of oxidizing and reducing conditions, due for example to flooding or a fluctuating water table, the products of Mn^{2+} oxidation can be seen in the form of manganese dioxide (MnO_2) nodules. Microbial dissimilatory reduction of Mn^{4+} is also a common reaction, and most microbes that can reduce Fe^{3+} are also able to reduce Mn^{4+} (Lloyd et al., 2003).

Chromium is required by bacteria as a trace element, but its common use in the metal and tanning industries has made it a priority pollutant in many countries since Cr^{6+} salts are readily mobile, toxic, and carcinogenic. A wide range of microbes can reduce the toxic Cr^{6+} form to the Cr^{3+} form. Cr^{3+} is 1000 times less toxic than Cr^{6+}, and its salts are insoluble at neutral pH. The reduction process normally requires anaerobic conditions and has been studied in both facultative anaerobes (e.g., *Shewanella* and *Aeromonas*) and in sulfate-reducing bacteria, yet has also been observed in aerobic species of *Pseudomonas*. The bacteria use the Cr as a terminal electron acceptor. The rate of reduction depends on various environmental conditions, including the availability of appropriate electron donors, pH, temperature, and the presence of other metals (Xia et al., 2019).

Arsenic is widely distributed in soils and groundwater and is commonly associated with pyrite and other minerals that contain sulfides. It is released in high concentrations into geothermal waters and groundwater, where it is a major health risk in several parts of the world, with over 40 million people considered to be in danger from drinking arsenic-containing water. For many years, arsenic-containing compounds were used as pesticides, and although their use has now been banned, they have left a legacy of contamination in agricultural soils. Arsenic exists as two major forms in soils: arsenate (As^{5+}) and arsenite (As^{3+}), with As^{3+} being more toxic and prevalent under reduced conditions. Arsenic is not associated with any essential intracellular microbial process, although its biochemistry resembles that of phosphate, and it can interfere with many aspects of phosphate metabolism. Bacterial oxidation and reduction of As occurs as a detoxification mechanism and provide electron donors and acceptors. Oxidation of As^{3+} as an energy source has been seen for bacteria in the Rhizobiaceae family, in which the *aox* genes encoding the arsenite oxidase enzyme are widespread. By contrast, reduction of As^{5+} as a terminal electron acceptor is more common and is carried out by a range of different bacteria. Functional genomic analysis of bacteria belonging to widely different phylogenetic groups suggests that arsenate-dissimilation occurs not only throughout the whole bacterial domain but also in the archaea (Andres and Bertin, 2016). The electron donors coupled with arsenate reduction vary between strains and environments. However, arsenic tolerance in bacteria is not linked to these redox processes, but is mediated primarily by active export of arsenic. The *ars* genes encoding these transporters are widespread in soil environments. Bacteria also carry out oxidative methylation of arsenic to produce methylarsenite, dimethylarsenate, dimethylarsenite, and trimethylarsine oxide. The ability to carry out these methylations has been described in a limited number of genera (*Clostridium, Desulfovibrio, Methanobacterium*), although the *arsM* gene associated with arsenic methylation has been found in over 120 different bacteria and archaea (Stolz et al., 2006; Slyemi and Bonnefoy, 2012).

Fungi can also strongly affect other elemental cycles in soil. Fungi have high requirements for potassium and take it up using both high- and low-affinity transport systems (Corratge et al., 2007; Haro and Benito, 2019). Potassium availability plays a role in mineral weathering by ectomycorrhizal fungi. *Pisolithus tinctorius* and *P. involutus* are both able to solubilize phlogopite (a mica) and access potassium trapped in the 2:1 layers (Paris et al., 1995). While *P. involutus* irreversibly transformed a fraction of the mica into hydroxy-aluminous vermiculite, the transformations caused by *P. tinctorius* remained reversible. When grown in symbiosis with *Pinus sylvestris, P. involutus* also mediates the weathering of biotite. The adherence of fungal hyphae to the mineral surface causes a mechanical distortion of the lattice structure, which delays the development of a potassium depletion zone and leads to the formation of vermiculite and Fe oxides (Bonneville et al., 2009).

The role of Ca in fungal growth has long been debated, but it is now recognized as an important secondary messenger within the fungal cell that can transmit a primary stimulus at the outer membrane to initiate intracellular events. The cell has both low-affinity and high-affinity Ca transporters that respond to different environmental conditions. Ca is stored at high concentrations in the vacuole (1 mM), but the concentration of Ca within the cytoplasm must be kept at a low level (100 nM), and its release from the vacuole in response to specific stimuli is affected by opening Ca^{2+} channels to permit passive flow into the cytosol (Lange and Peiter, 2020). Oxalate-Ca precipitates are often seen on fungal hyphae. In addition to the role of oxalate in mineral weathering and in softening cell walls, oxalate regulates the concentration of free Ca and the pH in the cytoplasm (Michael et al., 2014). Degradation of the Ca oxalate produced by fungi and plants by the so-called oxalate-carbonate pathway can result in the precipitation of calcite (Uren, 2018). The oxalate-carbonate pathway has been shown to produce significant amounts of calcite under trees, such as iroko (*Milicia excelsa*) in tropical soils (Cailleau et al., 2011).

Essential heavy metals (Zn^{2+}, Cu^{2+}, Mn^{2+}) are required for specific enzymatic reactions, but when they are present in excess, they can cause toxicity; thus, their uptake by fungi uses tightly regulated transporters. This is also true for algae, where both uptake and efflux are tightly controlled. Candidate metal transporters for Cu, Zn, Fe, and Mn have been identified in the *Chlamydomonas reinhardtii* genome by several techniques, including sequence similarity to previously characterized transporters, differential expression, and reverse genetics (Blaby-Haas and Merchant, 2017). Mycorrhizal fungi have been shown to alleviate stress to plants growing on soils containing large amounts of heavy metals. The genes involved in metal homeostasis in mycorrhizae such as *Tuber melanosporum* have also been characterized (Bolchi et al., 2011). The mechanisms involved in metal homeostasis by fungi include extracellular precipitation (e.g., in the presence of oxalate), sorption on the cell wall and within the cell, binding to low-molecular-weight compounds (glutathione, phytochelatins, metallothioneins, nicotianamine), and transport into the vacuole or secretion back into the soil solution (Colpaert et al., 2011; Ferrol et al., 2016). Plant-associated fungi protect their host plants by both accumulating metals and triggering differential expression of the genes involved in metal uptake in both partners (Ferrol et al., 2016; Domka et al., 2019; Shi et al., 2019). They can potentially play an important role in the phytoremediation of metal-contaminated sites (Ferrol et al., 2016).

15.4 Applied examples of interconnections between biotic community/activity and mineral nutrient transformations

15.4.1 Interactions of soil biota and plants in inorganic nutrient transformations during early soil development

The role of living organisms on mineral weathering has been studied extensively (Finlay et al., 2020), but there is less information on element transformations during the early stages of soil formation. The process of soil formation is particularly important in the current era of environmental change caused by human activities. Because of climate change, glaciers are receding at unprecedented rates all over the globe, revealing new terrestrial habitats and exposing new parent materials (Donhauser and Frey, 2018; Cauvy-Fraunie and Dangles, 2019). An understanding of new soil development is also needed to restore functional soil ecosystems in post-mining environments, as mining activity has severely disrupted many landscapes and soils worldwide (Martins et al., 2020). This section compares biological soil development over time in these two systems, focusing on how soil biota-plant interactions affect nutrient cycling over time.

Glaciers cover about 10% of the Earth's surface. Their rapid retreat in response to increasing global temperature has been well-documented for alpine, Antarctic, and high Arctic glaciers. Glacial retreat has major effects on biodiversity within the exposed substrates, which change over time as the newly revealed substrate develops into soil (Bradley et al., 2017; Cauvy-Fraunie and Dangles, 2019). The Damma Glacier in the Swiss Alps provides one of the best studied examples of how changed diversity affects cycling of nutrients along a soil chronosequence in a glacier forefield (Bernasconi et al., 2011).

The Damma glacier forefield is located at 2000 m altitude on a granitic parent material. It includes a chronosequence of soil ages (10−150 years), with scarce and patchy vegetation at the youngest sites (soil age 6−14 years), partial to full vegetation coverage at intermediate sites (57−79 years old), and full plant cover at older sites (108−150 years old). The vegetation is characterized by the presence of woody plants (e.g., *Rhododendron* and *Salix* spp.), grasses, and some N-fixing plants (*Alnus viridis* and *Lotus alpinus*) (Bernasconi et al., 2011; Brankatschk et al., 2011). Plant biomass production increases with soil age, total clay, soil organic C, N, and total P content. Soil microbial biomass also increases with soil age (estimated by total DNA, phospholipid fatty acids, and microbial C content), as does the abundance and richness of testate amoebae (Bernasconi et al., 2011). High bacterial diversity was observed across the entire chronosequence, and the bacterial community was dominated by Proteobacteria, Actinobacteria, Acidobacteria, Firmicutes, and Cyanobacteria. Euryarchaeota were predominantly found in younger soils, while Crenarchaeota primarily colonized older soils. Ascomycota dominated the fungal community in younger soils, whereas Basidiomycota were more common in older soils (Zumsteg et al., 2011). The proportion of arbuscular mycorrhizal fungi in the soil biota decreased with soil age, suggesting that AMF are more important in younger soils, yet decrease in importance in older soils as they become more acidic and enriched in organic matter (Welc et al., 2012).

Soil microbial biomass greatly influences the N cycle in forefield soils. Nitrogen mineralization (chitinase and protease) was the main driver of soil N turnover in the youngest soils (10 years), but biological N fixation by free-living organisms was more important in the intermediate soils (50–70 years). In older soils (120–2000 years), genes associated with nitrification (*amoA*) and denitrification (*nosZ*) were most abundant (Brankatschk et al., 2011). High nitrogenase diversity and high rates of biological N fixation by free-living microorganisms were observed in the rhizosphere soil of several pioneer plant species (*Leucanthemopsis alpina*, *Poa alpina*, *Agrostis* sp.) in both a younger (8 years) and an older (70 years) soil (Duc et al., 2009). Bacterial growth rates are limited by the availability of both C and N in the youngest soils (Goeransson et al., 2011); nevertheless, sufficient exogenous C and N are present for initial microbial development.

The microbial population plays a direct role in P, S, and trace metal transformations, even in the absence of plants. Fungal strains isolated from unvegetated granitic sediments in the Damma forefield (*Mucor hiemalis*, *Umbelopsis isabellina*, and *Mortierella alpina*) exuded large amounts of citrate, malate, and oxalate, releasing significant amounts of Ca, Cu, Fe, Mg, Mn, and P from granite powder (Brunner et al., 2011). Bacterial isolates of the *Arthrobacter, Leifsonia, Janthinobacterium,* and *Polaromonas* genera also caused significant granite dissolution and release of metal ions, but did not increase P mineralization (Frey et al., 2010). The effects of these bacteria on weathering were related to their ability to attach to the granite surfaces, secrete high amounts of oxalic acid, and lower the pH of the surrounding solution. The P released by fungal weathering was taken up by soil microorganisms as soon as it was released from the apatite, with a turnover time for microbial P of only a few weeks at the youngest site (Tamburini et al., 2012). Bacterial growth is never limited by P (Goeransson et al., 2011). By contrast, sulfate availability is low in the younger forefield soils. This is reflected in high diversity of sulfonate-desulfurizing bacteria based on the sequences of the oxidoreductase *asfA* gene. *asfA* genes associated with *Polaromonas* were found in the younger soils, while unidentified species were found in the older soils (Schmalenberger and Noll, 2010).

The biological chronosequence in the glacier forefield can be summarized as follows (Fig. 15.5). The parent material is colonized by microorganisms that use C from exogenous inputs, fix N, and release P and trace elements from the parent material in forms that can be taken up by plants. This allows colonization by different plant species and an increase in plant biomass, with AMF mediating efficient capture of nutrients by plants at the early stages. At a later stage, C inputs from plant rhizodeposition and litter are assimilated by soil microorganisms and microfauna, and the elements are recycled in the microbial-plant loop. This accelerates the weathering of minerals and leads to deeper, more developed, soils. Finally, as the ecosystem becomes richer in available nutrients, there is an increased tendency to lose a fraction of these to leaching and into the atmosphere.

The process of soil development during restoration of mining sites has much in common with glacier forefield soil development. When mining activities cease, the resulting areas of mining spoil constitute a mineral surface that is essentially devoid of plant or microbial activity. If these waste heaps are left to regenerate naturally, they are progressively colonized and revegetated over time, providing a chronosequence similar to the glacier forefields previously described. However, in more engineered cases

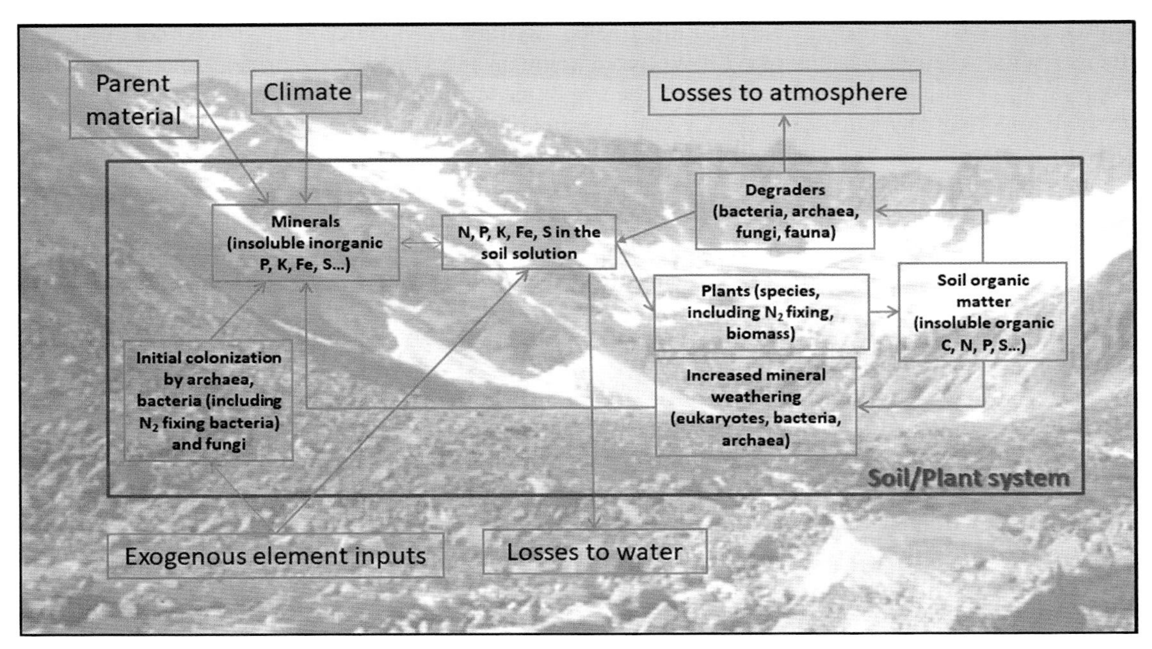

FIGURE 15.5 Element cycling during the early stages of soil development as affected by the interactions between soil bacteria, archaea, fungi, fauna, and plants.

(especially for open-cut mines), the topsoil removed at the start of the project is used to overlay the reconstructed landscape at the conclusion of the mining activity to accelerate rehabilitation. While this sounds as though it will greatly restore the soil, in practice, much of the topsoil microbial activity is often lost due to intense disturbance and desiccation during processing. Additional fertilization and inoculation treatments are typically also applied. Below, we highlight results from chronosequences at coal mine rehabilitation sites in the Czech Republic (Harantová et al., 2017; Ngugi et al., 2018). However, similar results have been seen elsewhere after extraction of iron, bauxite, and other metals (Daws et al., 2019; Gastauer et al., 2019; Martins et al., 2020; Wu et al., 2020).

The Sokolov region in the Czech Republic, located at 550 m altitude with annual precipitation of 650 mm, has been a center for open-cut, brown coal mining for over half a century. The mining waste is progressively deposited in spoil heaps, and predominantly consists of lacustrine clays, illite, kaolinite, and montmorillonite, with a high organic content of kerogens (2–18%). The oldest spoil heaps are 50 to 60 years in age. A linear chronosequence of microbial colonization and vegetation growth is available at a number of post-mining sites in the region. The initial vegetation cover (0–8 years) is extremely sparse, followed by development of low-stature grassland containing a range of grasses and forbs (8–16 years). After a transition period of grassland and trees, predominantly birch, aspen, and goat willow (16–21 years), a dense overstory of willow is developed with little understory (21–30 years). In the final period of the chronosequence (40–54 years), the understory cover increases to >60%, including a range of common grassland species and legumes, such as the N-fixing species *Trifolium* and *Melilotus* (Harantová et al., 2017). Physical treatment of the individual sites has a strong influence on the progression of revegetation, with leveled sites tending to remain as grassland. Sites left in undulating heaps show more tree growth, presumably because of increased moisture retained in the furrows between the heaps (Frouz et al., 2018).

Bacterial and fungal activity is low in the initial soils, with the bacterial community dominated by autotrophic taxa, such as *Thiobacillus* and *Desulfovibrio* which gain energy from S compounds remaining in the spoil. Microbial biomass increased more than 100-fold in the first 21 years, and was dominated by bacteria, with a bacterial:fungal ratio of 10-20 for the first 20 years, but decreasing to 5-10 thereafter (Harantová et al., 2017). Apart from the earliest samples, bacterial diversity remained constant throughout succession (Harantová et al., 2017), though functional diversity increased throughout the chronosequence, measured by changes in soil catabolic profiles and substrate-induced respiration patterns (Kaneda et al., 2019). Populations of Alpha- and Gammaproteobacteria, Acidobacteria, and Planctomycetes increased with time, while Actinobacteria, Cyanobacteria, Gemmatimonadetes, and Tenericutes decreased. The relative abundance of ascomycete fungi decreased over time, while basidiomycetes increased. Arbuscular mycorrhizal populations decreased over time (Harantová et al., 2017), though their diversity increased, primarily driven by changes in plant diversity rather than changes in soil chemistry (Krüger et al., 2017). Populations of soil mesofauna and macrofauna also increased with time, and although this could theoretically be due to increased soil organic C, it appears to be correlated with the amount of vegetation at individual sites (Moradi et al., 2017). The depth profile of these soil biota was different at sites where topsoil had been restored, in part because of compaction during leveling (Moradi et al., 2020), which may well affect development of the microbial community at these sites.

Post-coal mining soils in the Sokolov region contain substantial amounts of fossil organic matter (3–13%), i.e., C, S, and some N. This organic matter may slowly decompose, but it is not clear how quickly the constituent elements are incorporated into soil biogeochemical cycles (Čížková et al., 2018). The dominance of *Thiobacillus* and *Desulfovibrio* in the initial soils of the sequence suggests that at least the S portion of this fossil organic matter is bioavailable. Soil N levels increase slowly through the restoration process. Colonization of the initial spoil by free-living N-fixing bacteria facilitates the establishment of plants, and the N content is further increased by the presence of clover in the intermediate stages of succession. At later stages, available N is further increased by colonization with alder (*Alnus* sp.), one of the dominant pioneer tree species on these spoil heaps that actively fixes N through its symbiotic interactions with *Frankia* bacteria (Čížková et al., 2018). By contrast, the P content of these soils does not change significantly with time. Mine soils are low in P, and even though P is released by weathering, it is immediately bound by Fe oxides in the soils. However, the dynamics of this process have not been studied in detail for these soils. Reclamation by addition of topsoil did not change the P dynamics at these sites (Čížková et al., 2018), though in a similar post-coal mine study conducted in Australia, a progressive decrease in soil P was seen with reclamation (Ngugi et al., 2018). Because of this, it is often normal practice to fertilize restored topsoil with additional P, but care must be taken so that this does not affect the diversity of plant colonization of the site, especially in the long term (Daws et al., 2019).

15.4.2 Coupling of iron and sulfur transformations in acid mine drainage

Mining spoil from metalliferous ores often contains high concentrations of both ferric (Fe^{3+}) ions and more toxic heavy metals. The spoil usually also contains high levels of S, because metals such as Fe, Cu, Pb, and

Zn, commonly occur as sulfide minerals (e.g., pyrite, FeS_2). The spoil from coal mining can also contain up to 20% of sulfide S by weight, as well as metal contaminants. Leachate from the waste rocks and mineral tailings left behind after mining operations is known as *acid mine drainage* (AMD), since a combination of microbial Fe and S transformations leads to the *in-situ* production of sulfuric acid, with pH values as low as 2 to 3. The highly acidic pH of this leachate, together with the elevated levels of metals and metalloids and often high salinity in AMD, means that its release into the environment can have a disastrous impact on surrounding ecosystems (Rezaie and Anderson, 2020).

The overall transformation of pyrite minerals to generate sulfuric acid involves reaction of FeS_2 with oxygen (Eq. 15.1). However, this requires higher concentrations of oxygen than are commonly present in the subsurface environment. The reaction is therefore more efficient when the oxidant is not molecular oxygen, but Fe^{3+} (Eq. 15.2). The ferric ion (Fe^{3+}) is thereby converted to the ferrous form (Fe^{2+}), and under the acidic conditions of AMD, is then reoxidized to Fe^{3+} by Fe-oxidizing bacteria, such as the well-characterized species *Acidithiobacillus ferrooxidans* (Eq. 15.3).

$$FeS_2 + 3.5O_2 + H_2O \rightarrow Fe^{2+} + 2H^+ + 2SO_4^{2-} \quad \text{(Eq. 15.1)}$$

$$FeS_2 + 14Fe^{3+} + 8H_2O \rightarrow 15Fe^{2+} + 16H^+ + 2SO_4^{2-} \quad \text{(Eq. 15.2)}$$

$$14Fe^{2+} + 3.5O_2 + 14H^+ \rightarrow 14Fe^{3+} + 7H_2O \quad \text{(Eq. 15.3)}$$

Due to the nature of AMD, all the key organisms catalyzing this process are necessarily acidophiles, as they inhabit a highly acidic environment. These include autotrophs like *A. ferrooxidans* and *Leptospirillum ferrooxidans*, a range of acidophilic heterotrophs belonging to the *Acidophilum* genus, and heterotrophic archaea, such as *Ferroplasma acidiphilum*. Due to the minimal inputs of organic C and N to these systems, the population of heterotrophs is low. Most of the known acid-tolerant archaea are thermophilic, while the temperatures in AMD environments tend to be low, though there are a few mines where the energy released from rapid pyrite oxidation is sufficient to cause increased temperatures. The dominant organisms in these environments therefore tend to be psychrophiles (Hallberg, 2010).

In principle, the presence of high levels of sulfate in AMD might select for heterotrophic sulfate-reducing bacteria. Although evidence for sulfate reduction has occasionally been observed (e.g., the presence of blackened sulfide deposits in AMD sediment), no acidophilic SRB or archaea are known. Many members of the AMD microbial community can carry out S oxidation, using either sulfide, S, or reduced S compounds (e.g., trithionate or tetrathionate) as an electron donor. The common acidophile *A. ferrooxidans* can grow autotrophically with elemental S as the electron donor and Fe^{3+} as the electron acceptor. Thiobacilli, such as *Thiobacillus acidophilus*, use tetrathionate or trithionate for autotrophic growth. Overall, AMD ecosystems contain a limited range of bacterial and archaeal taxa, but there is significant variation within these taxa due to the availability of specialist niches that provide slightly different conditions of pH, temperature, and metal content (Fig. 15.6).

The high number of abandoned mine sites worldwide make AMD an economic and environmental issue of international significance. Remediation efforts include chemical strategies that focus on acid neutralization and metal oxidation/precipitation, and biological strategies involving sulfate reduction (Ayangbenro et al., 2018; Rezaie and Anderson, 2020). Metagenomic comparison of the bacterial species present in AMD water samples, AMD surface biofilms, AMD-impacted soils, and AMD treatment bioreactors have revealed significantly higher bacterial diversity in bioreactors than in natural environments. Significantly, while natural AMD samples were dominated by the Fe and S oxidizers that produce AMD,

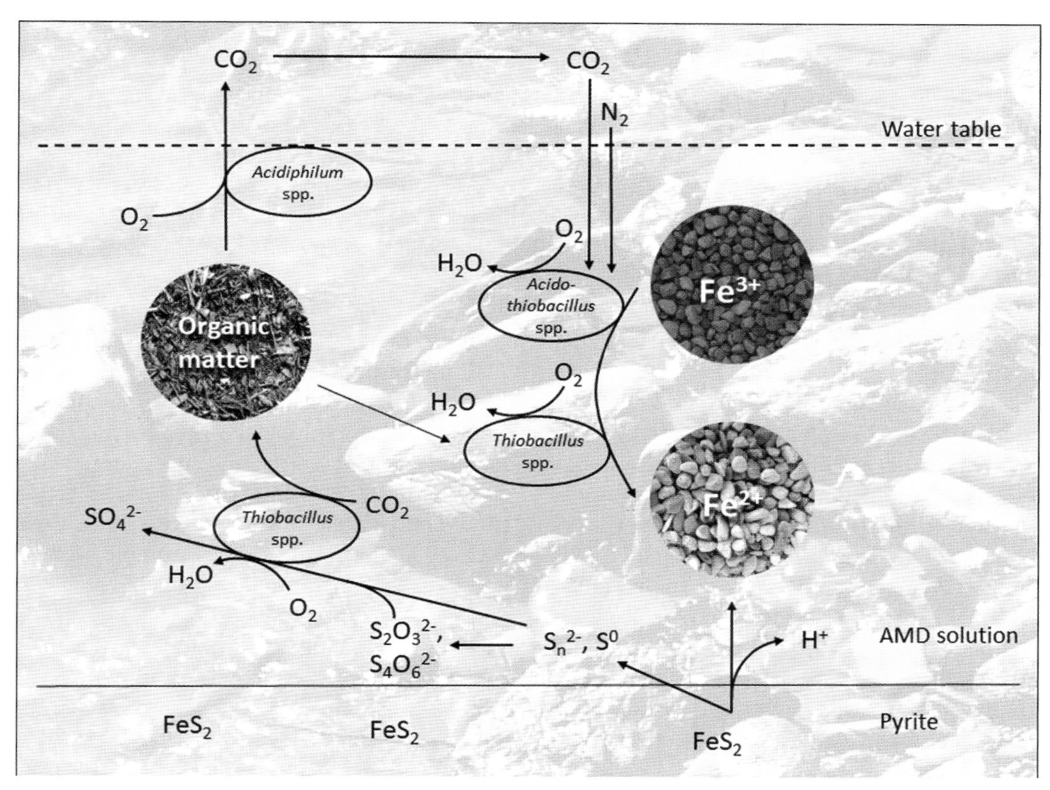

FIGURE 15.6 Simplified overview of Fe, S, and C transformations in acid mine drainage and the key groups of organisms that control these transformations at temperatures below 30°C. The lower bar represents pyrite crystals in equilibrium with acid mine drainage solution.

there was an enrichment of strictly anaerobic populations in bioreactors, including sulfate-reducing Deltaproteobacteria, such as *Desulfatirhabdium*, *Desulfovibrio*, *Desulfomicrobium*, *Desulfobulbus*, and *Desulfobacula*, which may be crucial in effective AMD bioremediation (Villegas-Plazas et al., 2019). These engineered systems provide great potential for successful bioremediation and restoration of AMD-impacted sites.

15.4.3 Element transformations in natural and artificial wetlands

Wetland ecosystems provide excellent examples of the interplay between plants, sediments, and different groups of microbes under a range of environmental conditions. Natural wetland systems are increasingly under threat due to climate change or anthropogenic inputs, including fertilizer runoff. Artificial wetland systems are often constructed to replace natural wetlands that have been lost, or for specific purposes such as sustainable wastewater management. Deeper sediments in wetlands tend to be anoxic, but water levels fluctuate throughout the year, leading to variable redox conditions. These fluctuations are influenced by changes in seasonal groundwater flow, by variations in inflow for constructed systems, and by irrigation events such as temporary flooding of rice fields in agricultural environments. The upper layers of the soils tend to be high in organic C from plant inputs and are dominated by interactions with plant roots. Plant roots also provide a partially aerobic redox environment by releasing oxygen into the rhizosphere. Understanding how interactions between redox changes and how the microbial community drives the transformation of other elements is critically important for wetland management. Specific microbial taxa have thus been proposed as biological indicators for wetland management (Urakawa and Bernhard, 2017).

In the anoxic environment of flooded wetland soils, ferric compounds are utilized by the resident microbial communities as electron acceptors, with reduced Fe compounds accumulating over time. However, when such soils are drained and become aerobic, Fe oxidizing bacteria become more active, and the ferric oxyhydroxides (FeOOH) they produce bind available phosphate and reduce its bioavailability. When the soil is flooded once more, the cycle is reversed. In flooded soil, oxygen is rapidly depleted as a microbial electron acceptor, and denitrification then dominates microbial respiration. Once nitrate is exhausted, the most energetically favorable alternative is dissimilatory Fe reduction, which converts the ferric-phosphate complexes previously mentioned into soluble Fe^{2+}, releasing inorganic phosphate into the groundwater. Because sulfate reduction and Fe^{3+} reduction are energetically not dissimilar in redox potential, Fe-reducing activity is quickly followed by an increase in the activity of heterotrophic sulfate-reducing bacteria, which convert inorganic sulfate in the groundwater into sulfide using organic matter as an electron donor. These sulfide products form highly insoluble iron sulfide precipitates (FeS_X) with reduced iron compounds, which are then stable until the system becomes aerobic once again (Fig. 15.7) (Burgin et al., 2011; Li et al., 2012).

There are several consequences of this complex combination of Fe, N, S, and P cycles. Once the water table falls in the wetland and the soil becomes aerobic, the FeS_X precipitates are oxidized by Fe-oxidizing bacteria, producing a pulse of sulfuric acid as a by-product of the process. The resultant acidification of the wetland soil can cause changes in both plant and microbial communities, which can be sufficient to

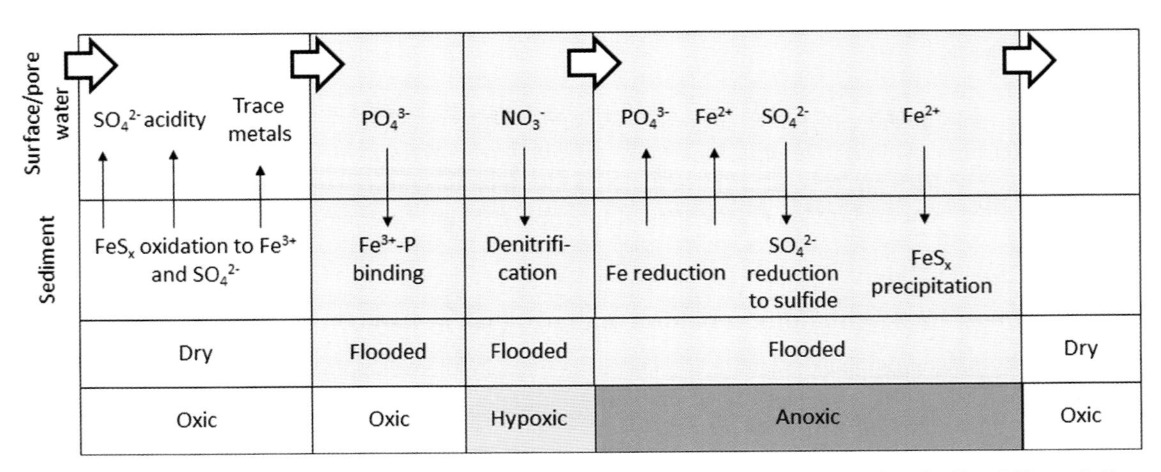

FIGURE 15.7 Hypothetical cycle of Fe and S transformations in wetlands and their implications for P and N metabolism. The figure represents a wetland that is subjected to periodic flooding and drying cycles. Once flooded, the sediment becomes progressively more anoxic, and the changes in redox state affect availability of nutrients in the porewater and in the overlying water. Note that the length of the Fe and sulfate reduction phases under anoxic conditions varies depending on nitrate inputs to the system.

mobilize toxic metals such as aluminum. Leachate from the wetland can then have damaging environmental effects on ecosystems downstream (Burgin et al., 2011).

In the past century, the application of synthetic fertilizers has had a dramatic effect on both natural and agricultural wetlands. Nitrate is one of the most common groundwater contaminants, with nitrate runoff contributing to surface water eutrophication. However, the impact of nitrate in wetlands goes beyond the N cycle. When nitrate concentrations are high, ferrous concentrations in the groundwater tend to remain low, since nitrate is energetically more favorable as an electron acceptor and therefore provides a redox buffer before ferric compounds are reduced. However, the presence of increased nitrate levels also stimulates what has been described as the "ferrous wheel" (Li et al., 2012), where chemolithoautotrophic nitrate reduction is coupled to the oxidation of Fe sulfide deposits, causing an increase in sulfate concentrations in the groundwater. This effect can be considerable. For example, between 1960 and 2000, sulfate concentrations in Dutch groundwater increased nearly threefold, with up to 70% thought to be derived from pyrite (FeS_x). SRB converted this sulfate to sulfide, which displaced even more phosphate from Fe-P complexes, causing a release of a large pulse of phosphate. Hence nitrate runoff and leaching from agricultural soils can cause phosphate-induced eutrophication within the wetland environment, and reduced phosphate-binding capacity of the soil in subsequent oxic/anoxic cycles (Smolders et al., 2010; van Dijk et al., 2019).

Nitrate pollution of wetland groundwater also affects the vegetation composition because of these coupled element cycles. Wetland plants are largely adapted to prefer ammonium (NH_4^+) as an N source, although this depends on the pH of the surface water. Changes in pH and in the availability of N sources can have a selective effect on the plant community. More importantly, the sulfide produced by SRB and archaea is toxic to plants and inhibits the growth of many plant species. Precipitation of iron-sulfide as a plaque around roots interferes with root physiology and P uptake, although this is partly alleviated by diffusion of oxygen from the roots. The released oxygen stimulates the activity of S-oxidizing bacteria in the immediate vicinity of the root, converting toxic sulfide to elemental S (Lamers et al., 2012; van Dijk et al., 2019).

References

Acksel, A., Baumann, K., Hu, Y.F., Leinweber, P., 2019. A critical review and evaluation of some P-research methods. Commun. Soil Sci. Plant Anal. 50, 2804–2824.

Acosta-Martinez, V., Perez-Guzman, L., Johnson, J.M.F., 2019. Simultaneous determination of beta-glucosidase, beta-glucosaminidase, acid phosphomonoesterase, and arylsulfatase activities in a soil sample for a biogeochemical cycling index. Appl. Soil Ecol. 142, 72–80.

Adam, P.S., Borrel, G., Brochier-Armanet, C., Gribaldo, S., 2017. The growing tree of Archaea: new perspectives on their diversity, evolution and ecology. ISME J. 11, 2407–2425.

Albelda-Berenguer, M., Monachon, M., Joseph, E., 2019. Siderophores: from natural roles to potential applications. Adv. Appl. Microbiol. 106, 193–225.

Alori, E.T., Glick, B.R., Babalola, O.O., 2017. Microbial phosphorus solubilization and its potential for use in sustainable agriculture. Front. Microbiol. 8. Article 971.

An, R., Moe, L.A., 2016. Regulation of pyrroloquinoline quinone-dependent glucose dehydrogenase activity in the model rhizosphere-dwelling bacterium *Pseudomonas putida* KT2440. Appl. Environ. Microbiol. 82, 4955–4964.

Andres, J., Bertin, P.N., 2016. The microbial genomics of arsenic. FEMS Microbiol. Rev. 40, 299–322.

Annaheim, K.E., Rufener, C.B., Frossard, E., Bunemann, E.K., 2013. Hydrolysis of organic phosphorus in soil water suspensions after addition of phosphatase enzymes. Biol. Fertil. Soils 49, 1203–1213.

Antoniadis, V., Levizou, E., Shaheen, S.M., Ok, Y.S., Sebastian, A., Baum, C., et al., 2017. Trace elements in the soil-plant interface: phytoavailability, translocation, and phytoremediation – a review. Earth Sci. Rev. 171, 621–645.

Aponte, H., Medina, J., Butler, B., Meier, S., Cornejo, P., Kuzyakov, Y., 2020. Soil quality indices for metal(loid) contamination: An enzymatic perspective. Land Degrad Dev 31, 2700–2719.

Autry, A.R., Fitzgerald, J.W., 1990. Sulfonate S – a major form of forest soil organic sulfur. Biol. Fertil. Soils 10, 50–56.

Ayangbenro, A.S., Olanrewaju, O.S., Babalola, O.O., 2018. Sulfate-reducing bacteria as an effective tool for sustainable acid mine bioremediation. Front Microbiol 9. Article 1986.

Bardgett, R.D., van der Putten, W.H., 2014. Belowground biodiversity and ecosystem functioning. Nature 515, 505–511.

Becquer, A., Trap, J., Irshad, U., Ali, M.A., Plassard, C., 2014. From soil to plant, the journey of P through trophic relationships and ectomycorrhizal association. Front. Plant Sci. 5. Article 548.

Benning, C., 1998. Biosynthesis and function of the sulfolipid sulfoquinovosyl diacylglycerol. Annu. Rev. Plant Physiol. Plant Mol. Biol. 49, 53–75.

Bergkemper, F., Kublik, S., Lang, F., Kruger, J., Vestergaard, G., Schloter, M., et al., 2016. Novel oligonucleotide primers reveal a high diversity of microbes which drive phosphorous turnover in soil. J. Microbiol. Methods 125, 91–97.

Bernasconi, S.M., Bauder, A., Bourdon, B., Brunner, I., Buenemann, E., Christl, I., et al., 2011. Chemical and biological gradients along the Damma glacier soil chronosequence, Switzerland. Vadose Zone J. 10, 867–883.

Blaby-Haas, C.E., Merchant, S.S., 2017. Regulating cellular trace metal economy in algae. Curr. Opin. Plant Biol. 39, 88–96.

Blackwell, M.S.A., Brookes, R.C., de la Fuente-Martinez, N., Gordon, H., Murray, P.J., Snars, K.E., et al., 2010. Phosphorus solubilization and potential transfer to surface waters from the soil microbial biomass following drying-wetting and freezing-thawing. Adv. Agron. 106, 1–35.

Boye, K., Almkvist, G., Nilsson, S.I., Eriksen, J., Persson, I., 2011. Quantification of chemical sulphur species in bulk soil and organic sulphur fractions by S K-edge XANES spectroscopy. Eur. J. Soil Sci. 62, 874–881.

Bojinova, D., Velkova, R., Ivanova, R., 2008. Solubilization of Morocco phosphorite by *Aspergillus niger*. Biores. Technol. 99, 7348–7353.

Bolchi, A., Ruotolo, R., Marchini, G., Vurro, E., di Toppi, L.S., Kohler, A., et al., 2011. Genome-wide inventory of metal homeostasis-related gene products including a functional phytochelatin synthase in the hypogeous mycorrhizal fungus *Tuber melanosporum*. Fungal Genet. Biol. 48, 573–584.

Bonneville, S., Smits, M.M., Brown, A., Harrington, J., Leake, J.R., Brydson, R., et al., 2009. Plant-driven fungal weathering: early stages of mineral alteration at the nanometer scale. Geology 37, 615–618.

Bradley, J.A., Anesio, A.M., Arndt, S., 2017. Microbial and biogeochemical dynamics in glacier forefields are sensitive to century-scale climate and anthropogenic change. Front. Earth Sci. 5. Article 26.

Brankatschk, R., Toewe, S., Kleineidam, K., Schloter, M., Zeyer, J., 2011. Abundances and potential activities of nitrogen cycling microbial communities along a chronosequence of a glacier forefield. ISME J. 5, 1025–1037.

Briones, M.J.I., 2018. The serendipitous value of soil fauna in ecosystem functioning: the unexplained explained. Front. Environ. Sci. 6. Article 149.

Brunner, I., Ploetze, M., Rieder, S., Zumsteg, A., Furrer, G., Frey, B., 2011. Pioneering fungi from the Damma glacier forefield in the Swiss Alps can promote granite weathering. Geobiology 9, 266–279.

Bünemann, E.K., 2015. Assessment of gross and net mineralization rates of soil organic phosphorus – a review. Soil Biol. Biochem. 89, 82–98.

Bünemann, E.K., Keller, B., Hoop, D., Jud, K., Boivin, P., Frossard, E., 2013. Increased availability of phosphorus after drying and rewetting of a grassland soil: processes and plant use. Plant Soil 370, 511–526.

Bünemann, E.K., Prusziz, B., Ehlers, K., 2011. Characterization of phosphorus forms in soil microorganisms. In: Bünemann, E.K., Oberson, A., Frossard, E. (Eds.), Phosphorus in Action: Biological Processes in Soil Phosphorus Cycling. Springer, Berlin, pp. 37–57.

Burgin, A.J., Yang, W.H., Hamilton, S.K., Silver, W.L., 2011. Beyond carbon and nitrogen: how the microbial energy economy couples elemental cycles in diverse ecosystems. Front. Ecol. Environ. 9, 44–52.

Butaite, E., Baumgartner, M., Wyder, S., Kummerli, R., 2017. Siderophore cheating and cheating resistance shape competition for iron in soil and freshwater *Pseudomonas* communities. Nat. Commun. 8. Article 414.

Cailleau, G., Braissant, O., Verrecchia, E.P., 2011. Turning sunlight into stone: the oxalate-carbonate pathway in a tropical tree ecosystem. Biogeosciences 8, 1755–1767.

Capdevila, D.A., Edmonds, K.A., Giedroc, D.P., 2017. Metallochaperones and metalloregulation in bacteria. Bioinorg. Chem. 61, 177–200.

Cauvy-Fraunie, S., Dangles, O., 2019. A global synthesis of biodiversity responses to glacier retreat. Nat. Ecol. Evol. 3, 1675–1685.

Chandrangsu, P., Rensing, C., Helmann, J.D., 2017. Metal homeostasis and resistance in bacteria. Nat. Rev. Microbiol. 15, 338–350.

Chen, C.R., Condron, L.M., Davis, M.R., Sherlock, R.R., 2001. Effects of land-use change from grassland to forest on soil sulfur and arylsulfatase activity in New Zealand. Aust. J. Soil Res. 39, 749–757.

Chen, H., Jarosch, K.A., Mészáros, É., Frossard, E., Zhao, X., Oberson, A., 2021. Repeated drying and rewetting differently affect abiotic and biotic soil phosphorus (P) dynamics in a sandy soil: a ^{33}P soil incubation study. Soil Biol. Biochem 153, 108079. https://doi.org/10.1016/j.soilbio.2020.108079.

Čížková, B., Woś, B., Pietrzykowski, M., Frouz, J., 2018. Development of soil chemical and microbial properties in reclaimed and unreclaimed grasslands in heaps after opencast lignite mining. Ecol. Eng. 123, 103–111.

Cleveland, C.C., Liptzin, D., 2007. C:N:P stoichiometry in soil: is there a "Redfield ratio" for the microbial biomass? Biogeochemistry 85, 235–252.

Colombo, C., Palumbo, G., He, J.Z., Pinton, R., Cesco, S., 2014. Review on iron availability in soil: interaction of Fe minerals, plants, and microbes. J. Soils Sediments 14, 538–548.

Colpaert, J.V., Wevers, J.H.L., Krznaric, E., Adriaensen, K., 2011. How metal-tolerant ecotypes of ectomycorrhizal fungi protect plants from heavy metal pollution. Ann. Forest Sci. 68, 17–24.

Comensoli, L., Bindschedler, S., Junier, P., Joseph, E., 2017. Iron and fungal physiology: a review of biotechnological opportunities. Adv. Appl. Microbiol. 98, 31–60.

Corratge, C., Zimmermann, S., Lambilliotte, R., Plassard, C., Marmeisse, R., Thibaud, J.-B., et al., 2007. Molecular and functional characterization of a Na^+-K^+ transporter from the Trk family in the ectomycorrhizal fungus *Hebeloma cylindrosporum*. J. Biol. Chem. 282, 26057–26066.

Cregut, M., Piutti, S., Slezack-Deschaumes, S., Benizri, E., 2012. Compartmentalization and regulation of arylsulfatase activities in *Streptomyces* sp., *Microbacterium* sp. and *Rhodococcus* sp. soil isolates in response to inorganic sulfate limitation. Microbiol. Res. 168, 12–21.

Cregut, M., Piutti, S., Vong, P.-C., Slezack-Deschaumes, S., Crovisier, I., Benizri, E., 2009. Density, structure, and diversity of the cultivable arylsulfatase-producing bacterial community in the rhizosphere of field-grown rape and barley. Soil Biol. Biochem. 41, 704–710.

Daws, M.I., Grigg, A.H., Tibbett, M., Standish, R.J., 2019. Enduring effects of large legumes and phosphorus fertiliser on jarrah forest restoration 15 years after bauxite mining. For. Ecol. Manag. 438, 204–214.

Dedourge, O., Vong, P.C., Lasserre-Joulin, F., Benizri, E., Guckert, A., 2004. Effects of glucose and rhizodeposits (with or without cysteine-S) on immobilized-S-35, microbial biomass-S-35 and arylsulphatase activity in a calcareous and an acid brown soil. Eur. J. Soil Sci. 55, 649–656.

Delavaux, C.S., Smith-Ramesh, L.M., Kuebbing, S.E., 2017. Beyond nutrients: a meta-analysis of the diverse effects of arbuscular mycorrhizal fungi on plants and soils. Ecology 98, 2111–2119.

Denger, K., Weiss, M., Felux, A.K., Schneider, A., Mayer, C., Spiteller, D., et al., 2014. Sulphoglycolysis in *Escherichia coli* K-12 closes a gap in the biogeochemical sulphur cycle. Nature 507, 114–117.

Domka, A.M., Rozpadek, P., Turnau, K., 2019. Are fungal endophytes merely mycorrhizal copycats? The role of fungal endophytes in the adaptation of plants to metal toxicity. Front. Microbiol. 10. Article 371.

Donhauser, J., Frey, B., 2018. Alpine soil microbial ecology in a changing world. FEMS Microbiol. Ecol. 94, fiy099.

Doolette, A.L., Smernik, R.J., 2011. Soil organic phosphorus speciation using spectroscopic techniques. In: Bünemann, E.K., Oberson, A., Frossard, E. (Eds.), Phosphorus in Action. Biological Processes in Soil Phosphorus Cycling. Springer, Berlin, pp. 3–36.

Duc, L., Noll, M., Meier, B.E., Buergmann, H., Zeyer, J., 2009. High diversity of diazotrophs in the forefield of a receding alpine glacier. Microb. Ecol. 57, 179–190.

Dunn, C., Jones, T.G., Girard, A., Freeman, C., 2014. Methodologies for extracellular enzyme assays from wetland soils. Wetlands 34, 9–17.

Ehlers, K., Bakken, L.R., Frostegard, A., Frossard, E., Buenemann, E.K., 2010. Phosphorus limitation in a Ferralsol: impact on microbial activity and cell internal P pools. Soil Biol. Biochem. 42, 558–566.

Ehrlich, H.L., Newman, D.K., 2009. Geomicrobiology. CRC Press/Taylor and Francis, Boca Raton.

Emerson, J.B., 2019. Soil viruses: a new hope. mSystems 4, e00120–00119.

Emerson, J.B., 2019. Soil viruses: a new hope. Msystems 4, e00120–19.

Eriksen, J., 1997. Sulphur cycling in Danish agricultural soils: turnover in organic S fractions. Soil Biol. Biochem. 29, 1371–1377.

Eriksen, J., 2009. Soil sulfur cycling in temperate agricultural systems. Adv. Agron. 102, 55–89.

Ezawa, T., Saito, K., 2018. How do arbuscular mycorrhizal fungi handle phosphate? New insight into fine-tuning of phosphate metabolism. New Phytol. 220, 1116–1121.

Ferrol, N., Tamayo, E., Vargas, P., 2016. The heavy metal paradox in arbuscular mycorrhizas: from mechanisms to biotechnological applications. J. Exp. Bot. 67, 6253–6265.

Finlay, R.D., Mahmood, S., Rosenstock, N., Bolou-Bi, E.B., Kohler, S.J., Fahad, Z., et al., 2020. Reviews and syntheses: biological weathering and its consequences at different spatial levels – from nanoscale to global scale. Biogeosciences 17, 1507–1533.

Foster, A.W., Osman, D., Robinson, N.J., 2014. Metal preferences and metallation. J. Biol. Chem. 289, 28095–28103.

Frey, B., Rieder, S.R., Brunner, I., Ploetze, M., Koetzsch, S., Lapanje, A., et al., 2010. Weathering-associated bacteria from the Damma glacier forefield: physiological capabilities and impact on granite dissolution. Appl. Environ. Microbiol. 76, 4788–4796.

Frossard, E., Bünemann, E.K., Oberson, A., Jansa, J., Kertesz, M.A., 2012. Phosphorus and sulfur in soil. In: Huang, P.M., Li, Y., Sumner, M.E. (Eds.), Handbook of Soil Sciences, second ed. CRC Press, Boca Raton. 26.22–26.15.

Frossard, E., Condron, L.M., Oberson, A., Sinaj, S., Fardeau, J.C., 2000. Processes governing phosphorus availability in temperate soils. J. Environ. Qual. 29, 15–23.

Frouz, J., Mudrák, O., Reitschmiedová, E., Walmsley, A., Vachová, P., Šimáčková, H., et al., 2018. Rough wave-like heaped overburden promotes establishment of woody vegetation while leveling promotes grasses during unassisted post mining site development. J. Environ. Manag. 205, 50–58.

Gadd, G.M., Raven, J.A., 2010. Geomicrobiology of eukaryotic microorganisms. Geomicrobiol. J. 27, 491–519.

Gadd, G.M., Rhee, Y.J., Stephenson, K., Wei, Z., 2012. Geomycology: metals, actinides and biominerals. Environ. Microbiol. Rep. 4, 270–296.

Gahan, J., Schmalenberger, A., 2015. Arbuscular mycorrhizal hyphae in grassland select for a diverse and abundant hyphospheric bacterial community involved in sulfonate desulfurization. Appl. Soil Ecol. 89, 113–121.

Gastauer, M., Vera, M.P.O., de Souza, K.P., Pires, E.S., Alves, R., Caldeira, C.F., et al., 2019. A metagenomic survey of soil microbial communities along a rehabilitation chronosequence after iron ore mining. Sci. Data 6. Article 19008.

Ghani, A., McLaren, R.G., Swift, R.S., 1992. Sulphur mineralisation and transformations in soils as influenced by additions of carbon, nitrogen and sulphur. Soil Biol. Biochem. 24, 331–341.

Ghani, A., McLaren, R.G., Swift, R.S., 1993a. The incorporation and transformations of sulfur-35 in soil: effects of soil conditioning and glucose or sulphate additions. Soil Biol. Biochem. 25, 327–335.

Ghani, A., McLaren, R.G., Swift, R.S., 1993b. Mobilization of recently-formed soil organic sulphur. Soil Biol. Biochem. 25, 1739–1744.

Gleason, F.H., Larkum, A.W.D., Raven, J.A., Manohar, C.S., Lilje, O., 2019. Ecological implications of recently discovered and poorly studied sources of energy for the growth of true fungi especially in extreme environments. Fungal Ecol. 39, 380–387.

Glick, B.R., 2010. Using soil bacteria to facilitate phytoremediation. Biotechnol. Adv. 28, 367–374.

Goeransson, H., Venterink, H.O., Baath, E., 2011. Soil bacterial growth and nutrient limitation along a chronosequence from a glacier forefield. Soil Biol. Biochem. 43, 1333–1340.

Goux, X., Amiaud, B., Piutti, S., Philippot, L., Benizri, E., 2012. Spatial distribution of the abundance and activity of the sulfate ester-hydrolyzing microbial community in a rape field. J. Soils Sediments 12, 1360–1370.

Griffiths, B.S., Philippot, L., 2013. Insights into the resistance and resilience of the soil microbial community. FEMS Microbiol. Rev. 37, 112–129.

Gschwend, F., Aregger, K., Gramlich, A., Walter, T., Widmer, F., 2020. Periodic waterlogging consistently shapes agricultural soil microbiomes by promoting specific taxa. Appl Soil Ecol 155. Article 103623.

Hallberg, K.B., 2010. New perspectives in acid mine drainage microbiology. Hydrometallurgy 104, 448–453.

Harantová, L., Mudrák, O., Kohout, P., Elhottová, D., Frouz, J., Baldrian, P., 2017. Development of microbial community during primary succession in areas degraded by mining activities. Land Degrad. Dev. 28, 2574–2584.

Haro, R., Benito, B., 2019. The role of soil fungi in K^+ plant nutrition. Int. J. Mol. Sci. 20. Article 3169.

Hedrich, S., Schlomann, M., Johnson, D.B., 2011. The iron-oxidizing proteobacteria. Microbiology 157, 1551–1564.

Hider, R.C., Kong, X.L., 2010. Chemistry and biology of siderophores. Nat. Prod. Rep. 27, 637–657.

Hove-Jensen, B., Zechel, D.L., Jochimsen, B., 2014. Utilization of glyphosate as phosphate source: biochemistry and genetics of bacterial carbon-phosphorus lyase. Microbiol. Mol. Biol. Rev. 78, 176–197.

Jennings, D.H., 1995. The Physiology of Fungal Nutrition. Cambridge University Press, Cambridge.

Jiao, S., Lu, Y.H., 2020. Abundant fungi adapt to broader environmental gradients than rare fungi in agricultural fields. Glob. Change Biol. 26, 4506–4520.

Kaneda, S., Krištůfek, V., Baldrian, P., Malý, S., Frouz, J., 2019. Changes in functional response of soil microbial community along chronosequence of spontaneous succession on post mining forest sites evaluated by Biolog and SIR methods. Forests 10. Article 1005.

Kelly, D.P., Murrell, J.C., 1999. Microbial metabolism of methanesulfonic acid. Arch. Microbiol. 172, 341–348.

Kertesz, M.A., 1999. Riding the sulfur cycle – metabolism of sulfonates and sulfate esters in Gram-negative bacteria. FEMS Microbiol. Rev. 24, 135–175.

Kertesz, M.A., Fellows, E., Schmalenberger, A., 2007. Rhizobacteria and plant sulfur supply. Adv. Appl. Microbiol. 62, 235–268.

Khan, K.S., Joergensen, R.G., 2019. Stoichiometry of the soil microbial biomass in response to amendments with varying C/N/P/S ratios. Biol. Fertil. Soils 55, 265–274.

King, J.E., Quinn, J.P., 1997. The utilization of organosulphonates by soil and freshwater bacteria. Lett. Appl. Microbiol. 24, 474–478.

Kirchman, D.L., 2012. Processes in Microbial Ecology. Oxford University Press, New York.

Kirkby, C.A., Kirkegaard, J.A., Richardson, A.E., Wade, L.J., Blanchard, C., Batten, G., 2011. Stable soil organic matter: a comparison of C:N:P:S ratios in Australian and other world soils. Geoderma 163, 197–208.

Kramer, J., Özkaya, Ö., Kümmerli, R., 2020. Bacterial siderophores in community and host interactions. Nat. Rev. Microbiol. 18, 152–163.

Krüger, C., Kohout, P., Janoušková, M., Püschel, D., Frouz, J., Rydlová, J., 2017. Plant communities rather than soil properties structure arbuscular mycorrhizal fungal communities along primary succession on a mine spoil. Front. Microbiol. 8. Article 719.

Kucharski, J., Tomkiel, M., Baćmaga, M., Borowik, A., Wyszkowska, J., 2016. Enzyme activity and microorganisms diversity in soil contaminated with the Boreal 58WG herbicide. J. Environ. Sci. Health. B. 51, 446–454.

Kuzyakov, Y., Blagodatskaya, E., 2015. Microbial hotspots and hot moments in soil: concept & review. Soil Biol. Biochem. 83, 184–199.

Lamers, L.P.M., van Diggelen, J.M.H., Op den Camp, H.J.M., Visser, E.J.W., Lucassen, E.C.H.E.T., Vile, M.A., et al., 2012. Microbial transformations of nitrogen, sulfur, and iron dictate vegetation composition in wetlands: a review. Front. Microbiol. 3. Article 156.

Lange, M., Peiter, E., 2020. Calcium transport proteins in fungi: the phylogenetic diversity of their relevance for growth, virulence, and stress resistance. Front. Microbiol. 10. Article 3100.

Laslo, E., Gyorgy, E., Mara, G., Tamas, E., Abraham, B., Lanyi, S., 2012. Screening of plant growth promoting rhizobacteria as potential microbial inoculants. Crop. Prot 40, 43–48.

Lemanceau, P., Bauer, P., Kraemer, S., Briat, J.F., 2009. Iron dynamics in the rhizosphere as a case study for analyzing interactions between soils, plants and microbes. Plant Soil 321, 513–535.

Li, Y.C., Yu, S., Strong, J., Wang, H.L., 2012. Are the biogeochemical cycles of carbon, nitrogen, sulfur, and phosphorus driven by the "Fe-III-Fe-II redox wheel" in dynamic redox environments? J. Soils Sediments 12, 683–693.

Linder, T., 2018. Assimilation of alternative sulfur sources in fungi. World J. Microbiol. Biotechnol. 34. Article 51.

Lloyd, J.R., Lovley, D.R., Macaskie, L.E., 2003. Biotechnological application of metal-reducing microorganisms. Adv. Appl. Microbiol. 53, 85–128.

Lovley, D.R., 2012. Electromicrobiology. Annu. Rev. Microbiol. 66, 391–409.

Maaroufi, N.I., De Long, J.R., 2020. Global change impacts on forest soils: linkage between soil biota and carbon-nitrogen-phosphorus stoichiometry. Front. For. Glob. Change. 3, UNSP16.

Martins, W.B.R., Lima, M.D.R., Barros, U.D., Amorim, L., Oliveira, F.D., Schwartz, G., 2020. Ecological methods and indicators for recovering and monitoring ecosystems after mining: a global literature review. Ecol. Eng. 145. Article 105707.

Marzluf, G.A., 1997. Molecular genetics of sulfur assimilation in filamentous fungi and yeast. Annu. Rev. Microbiol. 51, 73–96.

McGill, W.B., Cole, C.V., 1981. Comparative aspects of cycling of organic C, N, S and P through soil organic matter. Geoderma 26, 267–286.

McLaren, T.I., Smernik, R.J., McLaughlin, M.J., Doolette, A.L., Richardson, A.E., Frossard, E., 2020. The chemical nature of soil organic phosphorus: a critical review and global compilation of quantitative data. Adv. Agron. 160, 51–124.

Melton, E.D., Swanner, E.D., Behrens, S., Schmidt, C., Kappler, A., 2014. The interplay of microbially mediated and abiotic reactions in the biogeochemical Fe cycle. Nat. Rev. Microbiol. 12, 797–808.

Metcalf, W.W., Griffin, B.M., Cicchillo, R.M., Gao, J.T., Janga, S.C., Cooke, H.A., et al., 2012. Synthesis of methylphosphonic acid by marine microbes: a source for methane in the aerobic ocean. Science 337, 1104–1107.

Meyer, B., Kuever, J., 2007. Molecular analysis of the diversity of sulfate-reducing and sulfur-oxidizing prokaryotes in the environment, using *aprA* as functional marker gene. Appl. Environ. Microbiol. 73, 7664–7679.

Miao, Z., Brusseau, M.L., Carroll, K.C., Carreon-Diazconti, C., Johnson, B., 2012. Sulfate reduction in groundwater: characterization and applications for remediation. Environ. Geochem. Health 34, 539–550.

Michael, G., Bahri-Esfahani, J., Li, Q.W., Rhee, Y.J., Wei, Z., Fomina, M., et al., 2014. Oxalate production by fungi: significance in geomycology, biodeterioration and bioremediation. Fungal Biol. Rev. 28, 36–55.

Moradi, J., John, K., Vicentini, F., Veselá, H., Vicena, J., Ardestani, M.M., Frouz, J., 2020. Vertical distribution of soil fauna and microbial community under two contrasting post mining chronosequences: Sites reclaimed by alder plantation and unreclaimed regrowth. Glob Ecol Conserv 23, e01165.

Moradi, J., John, K., Vicentini, F., Veselá, H., Vicena, J., Ardestani, M.M., et al., 2020. Vertical distribution of soil fauna and microbial community under two contrasting post mining chronosequences: sites reclaimed by alder plantation and unreclaimed regrowth. Glob. Ecol. Conserv. 23. https://doi.org/10.1016/j.gecco.2020.e01165.

Muyzer, G., Stams, A.J.M., 2008. The ecology and biotechnology of sulphate-reducing bacteria. Nat. Rev. Microbiol. 6, 441–454.

Nannipieri, P., Giagnoni, L., Landi, L., Renella, G., 2011. Role of phosphatase enzymes in soil. In: Bünemann, E., Oberson, A., Frossard, E. (Eds.), Phosphorus in Action: Biological Processes in Soil Phosphorus Cycling. Springer, Berlin, pp. 215–243.

Nehls, U., Plassard, C., 2018. Nitrogen and phosphate metabolism in ectomycorrhizas. New Phytol. 220, 1047–1058.

Ngugi, M.R., Dennis, P.G., Neldner, V.J., Doley, D., Fechner, N., McElnea, A., 2018. Open-cut mining impacts on soil abiotic and bacterial community properties as shown by restoration chronosequence. Restor. Ecol. 26, 839–850.

Olsson, P.A., Hammer, E.C., Wallander, H., Pallon, J., 2008. Phosphorus availability influences elemental uptake in the mycorrhizal fungus *Glomus intraradices*, as revealed by particle-induced X-ray emission analysis. Appl. Environ. Microbiol. 74, 4144–4148.

Paris, F., Bonnaud, P., Ranger, J., Lapeyrie, F., 1995. In vitro weathering of phlogopite by ectomycorrhizal fungi: 1. Effect of K^+ and Mg^{2+} deficiency on phyllosilicate evolution. Plant Soil 177, 191–201.

Passman, F.J., Jones, G.E., 1985. Preparation and analysis of *Pseudomonas putida* cells for elemental composition. Geomicrobiol. J. 4, 191–206.

Pereira e Silva, M.C., Semenov, A.V., Schmitt, H., van Elsas, J.D., Salles, J.F., 2013. Microbe-mediated processes as indicators to establish the normal operating range of soil functioning. Soil Biol. Biochem. 57, 995–1002.

Pester, M., Schleper, C., Wagner, M., 2011. The Thaumarchaeota: an emerging view of their phylogeny and ecophysiology. Curr. Opin. Microbiol. 14, 300–306.

Philpott, C.C., 2006. Iron uptake in fungi: a system for every source. Biochim. Biophys. Acta 1763, 636–645.

Piutti, S., Slezack-Deschaumes, S., Niknahad-Gharmakher, H., Vong, P.C., Recous, S., Benizri, E., 2015. Relationships between the density and activity of microbial communities possessing arylsulfatase activity and soil sulfate dynamics during the decomposition of plant residues in soil. Eur. J. Soil Biol. 70, 88–96.

Plassard, C., Becquer, A., Garcia, K., 2019. Phosphorus transport in mycorrhiza: how far are we? Trends Plant Sci. 24, 794–801.

Plassard, C., Louche, J., Ali, M.A., Duchemin, M., Legname, E., Cloutier-Hurteau, B., 2011. Diversity in phosphorus mobilisation and uptake in ectomycorrhizal fungi. Ann. Forest Sci. 68, 33–43.

Prietzel, J., Botzaki, A., Tyufekchieva, N., Brettholle, M., Thieme, J., Klysubun, W., 2011. Sulfur speciation in soil by S K-edge XANES spectroscopy: comparison of spectral deconvolution and linear combination fitting. Environ. Sci. Technol. 45, 2878–2886.

Rabus, R., Venceslau, S.S., Wohlbrand, L., Voordouw, G., Wall, J.D., Pereira, I.A.C., 2015. A post-genomic view of the ecophysiology, catabolism and biotechnological relevance of sulphate-reducing prokaryotes. Adv. Microb. Physiol. 66, 55–321.

Ragot, S.A., Kertesz, M.A., Bünemann, E.K., 2015. *phoD* alkaline phosphatase gene diversity in soil. Appl. Environ. Microbiol. 81, 7281–7289.

Ragot, S.A., Kertesz, M.A., Mészáros, É., Frossard, E., Bünemann, E.K., 2017. Soil *phoD* and *phoX* alkaline phosphatase gene diversity responds to multiple environmental factors. FEMS Microbiol. Ecol. 93, fiw212.

Randriamanantsoa, L., Morel, C., Rabeharisoa, L., Douzet, J.M., Jansa, J., Frossard, E., 2013. Can the isotopic exchange kinetic method be used in soils with a very low water extractable phosphate content and a high sorbing capacity for phosphate ions? Geoderma 200, 120–129.

Reid, R.K., Reid, C.P.P., Szaniszlo, P.J., 1985. Effects of synthetic and microbially produced chelates on the diffusion of iron and phosphorus to a simulated root in soil. Biol. Fertil. Soils 1, 45–52.

Rezaie, B., Anderson, A., 2020. Sustainable resolutions for environmental threat of the acid mine drainage. Sci. Total Environ. 717. Article 137211.

Robin, A., Vansuyt, G., Hinsinger, P., Meyer, J.M., Briat, J.F., Lemanceau, P., 2008. Iron dynamics in the rhizosphere: consequences for plant health and nutrition. Adv. Agron. 99, 183–225.

Roy, A.B., Hewlins, M.J.E., Ellis, A.J., Harwood, J.L., White, G.F., 2003. Glycolytic breakdown of sulfoquinovose in bacteria: a missing link in the sulfur cycle. Appl. Environ. Microbiol. 69, 6434–6441.

Ryan, M.H., Graham, J.H., 2018. Little evidence that farmers should consider abundance or diversity of arbuscular mycorrhizal fungi when managing crops. New Phytol. 220, 1092–1107.

Santos-Beneit, F., 2015. The Pho regulon: a huge regulatory network in bacteria. Front. Microbiol. 6. Article 402.

Scherer, P., Lippert, H., Wolff, G., 1983. Composition of the major elements and trace elements of 10 methanogenic bacteria determined by inductively coupled plasma emission spectroscopy. Biol. Trace Elem. Res. 5, 149–163.

Schmalenberger, A., Noll, M., 2010. Shifts in desulfonating bacterial communities along a soil chronosequence in the forefield of a receding glacier. FEMS Microbiol. Ecol. 71, 208–217.

Schmalenberger, A., Hodge, S., Bryant, A., Hawkesford, M.J., Singh, B.K., Kertesz, M.A., 2008. The role of *Variovorax* and other *Comamonadaceae* in sulfur transformations by microbial wheat rhizosphere communities exposed to different sulfur fertilization regimes. Environ. Microbiol. 10, 1486–1500.

Schmalenberger, A., Hodge, S., Hawkesford, M.J., Kertesz, M.A., 2009. Sulfonate desulfurization in *Rhodococcus* from wheat rhizosphere communities FEMS Microbiol. Ecol. 67, 140–150.

Schmalenberger, A., Telford, A., Kertesz, M.A., 2010. Sulfate treatment affects desulfonating bacterial community structures in *Agrostis* rhizospheres as revealed by functional gene analysis based on *asfA*. Eur. J. Soil Biol. 46, 248–254.

Schmalenberger, A., Pritzkow, W., Ojeda, J.J., Noll, M., 2011. Characterization of main sulfur source of wood-degrading basidiomycetes by S K-edge X-ray absorption near edge spectroscopy (XANES). Int. Biodeterior. Biodegrad. 65, 1215–1223.

Shi, L., Dong, H.L., Reguera, G., Beyenal, H., Lu, A.H., Liu, J., et al., 2016. Extracellular electron transfer mechanisms between microorganisms and minerals. Nat. Rev. Microbiol. 14, 651–662.

Shi, W.G., Zhang, Y.H., Chen, S.L., Polle, A., Rennenberg, H., Luo, Z.B., 2019. Physiological and molecular mechanisms of heavy metal accumulation in nonmycorrhizal versus mycorrhizal plants. Plant Cell Environ. 42, 1087–1103.

Slyemi, D., Bonnefoy, V., 2012. How prokaryotes deal with arsenic. Environ. Microbiol. Rep. 4, 571–586.

Smith, S.E., Smith, F.A., 2011. Roles of arbuscular mycorrhizas in plant nutrition and growth: new paradigms from cellular to ecosystem scales. Annu. Rev. Plant Biol. 62, 227–250.

Smits, M.M., Bonneville, S., Benning, L.G., Banwart, S.A., Leake, J.R., 2012. Plant-driven weathering of apatite – the role of an ectomycorrhizal fungus. Geobiology 10, 445–456.

Smolders, A.J.P., Lucassen, E., Bobbink, R., Roelofs, J.G.M., Lamers, L.P.M., 2010. How nitrate leaching from agricultural lands provokes phosphate eutrophication in groundwater fed wetlands: the sulphur bridge. Biogeochemistry 98, 1–7.

Sosa-Hernández, M.A., Leifheit, E.F., Ingraffia, R., Rillig, M.C., 2019. Subsoil arbuscular mycorrhizal fungi for sustainability and climate-smart agriculture: a solution right under our feet? Front. Microbiol. 10. Article 744.

Stolz, J.E., Basu, P., Santini, J.M., Oremland, R.S., 2006. Arsenic and selenium in microbial metabolism. Annu. Rev. Microbiol. 60, 107–130.

Strawn, D.G., Bohn, H.L., O'Connor, G.A., 2015. Soil Chemistry. Wiley Blackwell, New York.

Stressler, T., Seitl, I., Kuhn, A., Fischer, L., 2016. Detection, production, and application of microbial arylsulfatases. Appl. Microbiol. Biotechnol. 100, 9053–9067.

Sullivan, T.S., Gadd, G.M., 2019. Metal bioavailability and the soil microbiome. Adv. Agron. 155, 79–120.

Sun, Y., Goll, D.S., Chang, J., Ciais, P., Guenet, B., Helfenstein, J., et al., 2020. Global evaluation of nutrient enabled version land surface model ORCHIDEE-CNP v1.2 (r5986). Geosci. Model Dev. Discuss. (GMDD) 2021. https://doi.org/10.5194/gmd-14-1987-2021.

Sylvia, D.M., Fuhrmann, J.F., Hartel, P.G., Zuberer, D.A., 1999. Principles and Applications of Soil Microbiology. Prentice-Hall, New Jersey.

Tamburini, F., Pfahler, V., Buenemann, E.K., Guelland, K., Bernasconi, S.M., Frossard, E., 2012. Oxygen isotopes unravel the role of microorganisms in phosphate cycling in soils. Environ. Sci. Technol. 46, 5956–5962.

Tebo, B.M., Johnson, H.A., McCarthy, J.K., Templeton, A.S., 2005. Geomicrobiology of manganese(II) oxidation. Trends Microbiol. 13, 421–428.

Tipping, E., Somerville, C.J., Luster, J., 2016. The C:N:P:S stoichiometry of soil organic matter. Biogeochemistry 130, 117–131.

Trap, J., Bonkowski, M., Plassard, C., Villenave, C., Blanchart, E., 2016. Ecological importance of soil bacterivores for ecosystem functions. Plant Soil 398, 1–24.

Urakawa, H., Bernhard, A.E., 2017. Wetland management using microbial indicators. Ecol. Eng. 108, 456–476.

Uren, N.C., 2018. Calcium oxalate in soils, its origins and fate – a review. Soil Res. 56, 443–450.

Uroz, S., Calvaruso, C., Turpault, M.P., Frey-Klett, P., 2009. Mineral weathering by bacteria: ecology, actors and mechanisms. Trends Microbiol. 17, 378–387.

van Dijk, G., Wolters, J., Fritz, C., de Mars, H., van Duinen, G.J., Ettwig, K.F., et al., 2019. Effects of groundwater nitrate and sulphate enrichment on groundwater-fed mires: a case study. Water Air Soil Pollut. 230. Article 122.

Verbruggen, E., van der Heijden, M.G.A., Rillig, M.C., Kiers, E.T., 2013. Mycorrhizal fungal establishment in agricultural soils: factors determining inoculation success. New Phytol. 197, 1104–1109.

Villegas-Plazas, M., Sanabria, J., Junca, H., 2019. A composite taxonomical and functional framework of microbiomes under acid mine drainage bioremediation systems. J. Environ. Manag. 251. Article 109581.

Vong, P.C., Dedourge, O., Lasserre-Joulin, F., Guckert, A., 2003. Immobilized-S, microbial biomass-S and soil arylsulfatase activity in the rhizosphere soil of rape and barley as affected by labile substrate C and N additions. Soil Biol. Biochem. 35, 1651–1661.

Voss, B., Kirschhofer, F., Brenner-Weiss, G., Fischer, R., 2020. *Alternaria alternata* uses two siderophore systems for iron acquisition. Sci. Rep. 10. Article 3587.

Waldron, K.J., Robinson, N.J., 2009. How do bacterial cells ensure that metalloproteins get the correct metal? Nat. Rev. Microbiol. 7, 25–35.

Waters, L.S., 2020. Bacterial manganese sensing and homeostasis. Curr. Opin. Chem. Biol. 55, 96–102.

Weber, K.A., Achenbach, L.A., Coates, J.D., 2006. Microorganisms pumping iron: anaerobic microbial iron oxidation and reduction. Nat. Rev. Microbiol. 4, 752–764.

Welc, M., Bünemann, E.K., Fließbach, A., Frossard, E., Jansa, J., 2012. Soil bacterial and fungal communities along a soil chronosequence assessed by fatty acid profiling. Soil Biol. Biochem. 49, 184–192.

Williams, P.H., Haynes, R.J., 1993. Forms of sulfur in sheep excreta and their fate after application on to pasture soil. J. Sci. Food Agric. 62, 323–329.

Wu, H., Chen, L., Zhu, F., Hartley, W., Zhang, Y.F., Xue, S.G., 2020. The dynamic development of bacterial community following long-term weathering of bauxite residue. J. Environ. Sci. 90, 321–330.

Xia, S.P., Song, Z.L., Jeyakumar, P., Shaheen, S.M., Rinklebe, J., Ok, Y.S., et al., 2019. A critical review on bioremediation technologies for Cr(VI)-contaminated soils and wastewater. Crit. Rev. Environ. Sci. Technol. 49, 1027–1078.

Yang, Z.H., Stoven, K., Haneklaus, S., Singh, B.R., Schnug, E., 2010. Elemental sulfur oxidation by *Thiobacillus* spp. and aerobic heterotrophic sulfur-oxidizing bacteria. Pedosphere 20, 71–79.

Yu, X.M., Doroghazi, J.R., Janga, S.C., Zhang, J.K., Circello, B., Griffin, B.M., et al., 2013. Diversity and abundance of phosphonate biosynthetic genes in nature. Proc. Natl. Acad. Sci. USA 110, 20759–20764.

Zanin, L., Tomasi, N., Cesco, S., Varanini, Z., Pinton, R., 2019. Humic substances contribute to plant iron nutrition acting as chelators and biostimulants. Front. Plant Sci. 10. Article 675.

Zhang, L., Xu, M.G., Liu, Y., Zhang, F.S., Hodge, A., Feng, G., 2016. Carbon and phosphorus exchange may enable cooperation between an arbuscular mycorrhizal fungus and a phosphate-solubilizing bacterium. New Phytol. 210, 1022–1032.

Zhang, Y., Ying, H.M., Xu, Y.Z., 2019. Comparative genomics and metagenomics of the metallomes. Metallomics 11, 1026–1043.

Zhao, F.J., Wu, J., McGrath, S.P., 1996. Soil organic sulphur and its turnover. In: Piccolo, A. (Ed.), Humic Substances in Terrestrial Ecosystems. Elsevier, Amsterdam, pp. 467–506.

Zhao, F.J., Lehmann, J., Solomon, D., Fox, M.A., McGrath, S.P., 2006. Sulphur speciation and turnover in soils: evidence from sulphur K-edge XANES spectroscopy and isotope dilution studies. Soil Biol. Biochem. 38, 1000–1007.

Zhou, J.Z., He, Q., Hemme, C.L., Mukhopadhyay, A., Hillesland, K., Zhou, A.F., et al., 2011. How sulphate-reducing microorganisms cope with stress: lessons from systems biology. Nat. Rev. Microbiol. 9, 452–466.

Zumsteg, A., Bernasconi, S.M., Zeyer, J., Frey, B., 2011. Microbial community and activity shifts after soil transplantation in a glacier forefield. Appl. Geochem. 26, S326–S329.

Chapter 16

Advancing quantitative models of soil microbiology, ecology, and biochemistry

Wally Xie*, Elizabeth Duan[†], Brian Chung[‡], and Steven D. Allison[†,‡]
*Center for Complex Biological Systems, University of California, Irvine, CA, USA; †Department of Ecology and Evolutionary Biology, University of California, Irvine, CA, USA; ‡Department of Earth System Science, University of California, Irvine, CA, USA

Chapter outline

16.1 Introduction

Soils host diverse biological communities, including plants, animals, and microbes. Together, these communities provide benefits essential for ecosystem functioning and human well-being. Decomposition of organic matter—primarily driven by microbes—regenerates nutrients that support plant growth in agricultural and unmanaged systems. In turn, plant growth and microbial transformations of organic matter enhance soil carbon (C) sequestration that mitigates greenhouse gas emissions from human activities.

At the same time, the biological services provided by soils are vulnerable to human-caused environmental change (Cavicchioli et al., 2019; Jansson and Hofmockel, 2020). For example, there is concern that global warming will stimulate metabolic activity in soils, weakening C sequestration and potentially turning soils into a net source of greenhouse gases (Davidson and Janssens, 2006). Given these concerns, soil microbes and biological processes are topics of intense research interest.

Improvements in sequencing technologies and other approaches for probing biological diversity and functioning have led to rapid advances in fundamental knowledge of soil ecology (Bahram et al., 2018). In parallel with these empirical advances, mathematical models of soil systems have recently blossomed (Allison, 2017; Wieder et al., 2015). Foundational models of soil biogeochemistry developed during the

Soil Microbiology, Ecology, and Biochemistry. https://doi.org/10.1016/B978-0-12-822941-5.00016-8

1980s and 1990s have been joined by a new generation of biologically inspired models starting in the early 2000s. Since then, these models have increased in scale and complexity.

Still, there is room for additional model improvement and intellectual development. Large-scale models fail to replicate fundamental patterns in soil biogeochemical pools and fluxes (Todd-Brown et al., 2013, 2014; Wu et al., 2018). Many of the most recent models with updated biological mechanisms have not been tested extensively. The field of soil ecological modeling has come a long way, but the pathway to addressing soil-relevant challenges with models remains uncertain.

In an effort to elevate the relevance and impact of soil modeling, this chapter aims to summarize the current state of the art while providing guidance for next steps to advance the field. We discuss some of the main reasons for engaging in soil modeling and then review selected modeling approaches from molecular to global scales. This review does not attempt to be exhaustive, and we focus our attention primarily on advances from the past 5 to 10 years, especially since the publication of Parton et al. (2015). The chapter concludes with recommendations for model-data integration and future intellectual development.

16.2 Justification for modeling

As with empirical approaches, soil scientists use models to address a range of different goals and questions. Models play an important role in advancing fundamental understanding of soil processes by representing concepts and mechanisms in a quantitative framework. For instance, the priming effect is a common biological mechanism in soil whereby addition of fresh organic matter stimulates, or "primes," the decomposition of existing soil C that may be older and more resistant to decay (Fontaine et al., 2004). Soil researchers have developed models that represent this mechanism, thereby quantifying the magnitude and impact of priming effects in soil systems (Guenet et al., 2016).

Models are also useful for generating hypotheses. Koven et al. (2015) used a depth-resolved version of the Community Land Model (CLM4.5BGC) to simulate permafrost thaw and its effects on ecosystem C balance. This version of the model is notable for incorporating fundamental understanding of how soil processes vary with depth, a crucial concept in frozen soils with seasonal changes in active layer thickness. Moreover, CLM4.5 represents nitrogen (N) dynamics which likely play into C-climate feedbacks. In response to climate warming, the Koven et al. (2015) modeling study suggested that the positive effects of N release on plant productivity and associated C storage would be outweighed by the negative effects of permafrost thaw and increased microbial metabolism with soil warming. This outcome is a testable hypothesis that can be addressed with laboratory, field, and global change experiments (Mack et al., 2004; Xue et al., 2016).

More broadly, models can help guide experimental work. A conceptual paradigm proposed by Blankinship et al. (2018) calls for better integration between theory, models, and measurements. This aim could be partially achieved by aligning modeled mechanisms and outcomes with experimental data. For example, models of soil biogeochemistry include a wide array of pools ranging from largely inert to mineral-associated organic matter and highly dynamic microbial biomass. Aligning these pools with the chemical composition of real soils provides a rationale for exploiting cutting-edge organic matter fractionation and characterization approaches, such as nuclear magnetic resonance (NMR), X-ray microspectroscopy, and pyrolysis gas chromatography-mass spectroscopy (Kalbitz et al., 2003; Lehmann et al., 2008; Quideau et al., 2005). Likewise, recent advances in modeling microbial diversity can drive new approaches for analyzing sequencing and other datasets that probe the functioning of microbial communities. Building a model can generate practical guidelines for distilling, organizing, and processing the information contained in complex omics datasets.

Scaling is another relevant application of soil models (Allison, 2017; Wieder et al., 2015). Nearly all the grand challenges facing soils at the global scale require knowledge of emergent properties arising from smaller spatial scales and shorter time scales. At the molecular level, cells exchange metabolites, enzymes catalyze reactions, and organic compounds interact with mineral surfaces. At cellular to ecosystem scales, these molecular processes combine into emergent biological properties such as growth and respiration. All the way up to the global scale, biological systems interact with soil physical properties to determine outcomes like C and nutrient balance. Modeling offers a quantitative, rational approach for representing key emergent properties at ever-increasing scales. Nested sets of models can, for example, provide insight into how Michaelis-Menten enzyme kinetics at the molecular level scale up to control organic matter decomposition rates at the community scale (Tang and Riley, 2013; Wang and Allison, 2019).

Models are also the primary tool available to scientists for making predictions, particularly in the context of global environmental change (Bradford et al., 2016; Todd-Brown et al., 2012). In many studies the goal of prediction complements other modeling aims, such as advancing fundamental understanding, generating hypotheses, and scaling up. Although predictions remain highly uncertain, soil models offer the potential to apply empirical and theoretical advances to simulate C and nutrient pools at the scale of the entire planet, decades or centuries into the future. Such models can provide answers to scientists and decision makers concerned about the future state of soils, including the capacity to store C in the face of climate and land use change (IPCC, 2019). The increasing prominence of model outputs in Intergovernmental Panel on Climate Change (IPCC) reports and policy making emphasizes prediction as a relevant, if not always singular, goal of model development.

16.3 Modeling approaches

Across scales, including the ecosystem scale, differential equation models are often applied to track soil biogeochemical pools and fluxes. Sierra and Müller (2015) described a general framework for this type of soil model based on first principles of mass balance, substrate dependence, heterogeneity of decomposition rates, chemical transformations, variation in environmental drivers, and interactions among soil pools. Nearly all existing models of soil biogeochemistry fit under this general framework, allowing for rigorous comparison of stability and mathematical properties across models.

Differential equation models like RothC and CENTURY emerged in the late 1970s and 1980s (Jenkinson and Rayner, 1977; Parton et al., 1988), embracing the principles of mass balance and substrate dependence as envisioned by Olson (1963) with organic matter decaying in proportion to its concentration. These models further included the principle of heterogeneity by representing different pools of organic matter with different decay rates. Transfers among the pools were allowed, following the principle of chemical transformations, and decay rates were functions of temperature and moisture levels, consistent with the principle of varying environmental drivers. Bosatta and Ågren (1985, 1999) generalized the principle of heterogeneous decomposition in their theory of continuous organic matter quality, which was intended to better reflect the complexity and diversity of soil organic compounds.

Models like RothC and CENTURY have some convenient mathematical properties, but they omit the fundamental principle of interacting soil pools in Sierra and Müller's (2015) framework. Commonly known as "linear" or "first-order," differential equation models without complex dependencies among pools can be readily represented in matrix form and solved analytically (Xia et al., 2013). They also tend to be mathematically stable, meaning that pool sizes and fluxes do not oscillate as the system returns to steady state following perturbation. Despite these advantages, linear models simplify or omit mechanisms of interaction among organic matter pools, such as enzymatic degradation driven by microbial decomposers. Rather, the biological roles of microbes in linear models are assumed to be "implicit" (Schimel, 2001).

An alternative approach to account for the principle of soil pool interactions is to make microbial mechanisms mathematically "explicit." The idea of microbial control over soil biogeochemical processes dates back to at least Waksman (1927). In the late 1970s O. L. Smith proposed a complex model of soil microbial biogeochemistry that included many of the features described by Sierra and Müller's (2015) general framework but that did not receive much attention (Smith, 1979a, b). More recently, there has been an explosion of microbially explicit model development and applications (Abramoff et al., 2018; Allison et al., 2010; Fontaine and Barot, 2005; Schimel and Weintraub, 2003). Although they attempt to represent biological mechanisms with higher fidelity, challenges remain with the stability, interpretability, and scaling of microbially explicit models (Wang et al., 2014). Efforts to analyze microbial processes with models at different scales could help address some of these challenges (Allison, 2012; Kaiser et al., 2014).

Dynamical differential equation models are valuable for representing fundamental processes, but predictive statistical models are a valuable alternative approach. Process-based models with many differential equations require careful parameterization; otherwise they may be mathematically unstable or generate inaccurate predictions. If accurate prediction is the goal, rather than representing mechanisms, statistical models can be very useful, assuming sufficient training data are available. Rapid development of machine learning techniques has made it possible to extrapolate soil properties across time and space based on training data and algorithms, such as neural networks and random forest. For example, this approach has been used to determine the global age of soil C based on radiocarbon profiles (Shi et al., 2020) and to map soil C stocks across Scotland (Aitkenhead and Coull, 2016).

New approaches have started to combine features of process-based and probabilistic modeling. Rather than representing explicit pools of C, Waring et al.'s (2020) PROMISE model tracks the flow of individual C molecules through a heterogeneous soil system. Molecules undergo transformations and movements based on soil parameters, proximity to microbes and enzymes, and stochastic processes. In this way molecules with different chemical properties vary in transit time such that the total soil C pool contains a distribution of residence times. This modeling framework requires relatively few assumptions and parameters while replicating emergent properties of soil C more accurately than pool-based models. It also incorporates mass balance and interactions among soil compounds, consistent with the six key principles identified by Sierra and Müller (2015).

16.4 Modeling across scales

16.4.1 Cellular/molecular

Molecular interactions, both within and outside cells, underlie all soil biotic and abiotic processes. Key interactions include metabolic pathways within microbial cells along with sorption/desorption, enzymatic catalysis, and molecular diffusion outside of cells. Molecular-scale interactions between organic molecules and soil minerals contribute to the physical protection of soil organic matter (Schmidt et al., 2011), whereas extracellular enzyme activity catalyzes decomposition of polymeric molecules (Burns et al., 2013). Many of these interactions are represented in models at larger scales.

Metabolic pathways can be represented with flux balance models that simulate how specific substrates are metabolized in microbial cells. In ^{13}C metabolic flux analysis (^{13}C-MFA), isotopic labeling experiments provide models with information to estimate intracellular metabolic fluxes. Together with ^{13}C fingerprinting to pinpoint central metabolic pathways and RNA-seq to complement the results of ^{13}C-MFA, Varman et al. (2016) uncovered the lignin degradation pathway of the bacterium *Sphingobium* sp. SYK-6. Environmental constraints and microbial community interactions must also be considered when modeling microbial metabolism. Jansson and Hofmockel (2018) defined the term *metaphenome* as the product of microbial functions that are expressed given abiotic and biotic environmental constraints. Flux balance models can be used to determine how microbial metaphenomes will respond to different environmental conditions and perturbations.

Information on molecular mechanisms can be used to quantify and better represent emergent properties in models. Carbon use efficiency (CUE) describes the proportion of C converted to microbial biomass and results from a combination of multiple metabolic processes. Hagerty et al. (2018) suggested modeling CUE explicitly to account for its dependence on microbial growth and C allocation processes, including costs of extracellular enzyme production and substrate assimilation. By representing these additional cellular processes, the accuracy of larger-scale models with static CUE parameters could be improved.

Enzymes are biochemical catalysts involved in many molecular transformations that occur in soil (Burns et al., 2013). Microbes secrete extracellular enzymes outside the cell to obtain resources from complex biopolymers, which are abundant in soils and litter. Given their role as biocatalysts targeting soil organic matter, extracellular enzyme activity represents a mechanism of interaction between soil pools, namely microbial biomass and organic polymers. The Michaelis-Menten equation describes this activity, which often represents the initial and rate-limiting step in microbial decomposition. The Michaelis-Menten equation predicts reaction velocity (dC/dt) as a function substrate concentration (C) based on two parameters: the maximum velocity (V_{max}) at unlimited substrate concentration and the half-saturation constant (K_M), which is the substrate concentration at $\frac{1}{2}$ V_{max}:

$$dC/dt = V_{max} \cdot C/(K_M + C), \qquad \text{(Eq. 16.1)}$$

V_{max} and K_M can be experimentally determined and used to parameterize models. German et al. (2012) used experimental data on Michaelis-Menten enzyme kinetics obtained from enzyme assays to build a decomposition model and determine the temperature sensitivity of extracellular enzymes. They found that both V_{max} and K_M are temperature sensitive and the level of sensitivity is enzyme specific.

Michaelis-Menten theory was extended in the Dual Arrhenius Michaelis-Menten (DAMM) model (Davidson et al., 2012). DAMM represents the interaction between Arrhenius and Michaelis-Menten kinetics at the scale of enzyme active sites to predict CO_2 production from soil. The model accounts for temperature, moisture, and oxygen limitation effects on the metabolism of soluble C substrates. Model predictions aligned well with laboratory measurements of extracellular enzyme activity at different temperatures and field measurements of soil respiration across seasons. DAMM was later extended to incorporate microsite variation in substrate concentrations and applied to predict not only soil respiration but also CH_4 and N_2O fluxes (Sihi et al., 2020).

The Reverse Michaelis-Menten (RMM) and Equilibrium Chemistry Approximation (ECA) equations have emerged as additional options to explicitly model enzyme kinetics (Moorhead and Weintraub, 2018;

Tang, 2015). The RMM equation describes the reaction velocity as a function of enzyme concentration (E), where K_E is the enzyme concentration at ½ V_{max}:

$$dC/dt = V_{max} \cdot E/(K_E + E). \quad \text{(Eq. 16.2)}$$

This equation is a better fit for situations in which substrate available for enzyme binding is limiting. Such situations may be common in soils, and therefore RMM was included in one of the first microbial-explicit models of soil C and N dynamics (Weintraub and Schimel, 2003).

The ECA considers both free substrate and enzyme limitations by accounting for mass balance constraints. Michaelis-Menten and RMM kinetics are special cases of the ECA (Tang, 2015):

$$dC/dt = k \cdot E \cdot C/(K_{ES} + C + E), \quad \text{(Eq. 16.3)}$$

where k is a rate constant, and $1/K_{ES}$ is the apparent substrate affinity of the enzyme. The ECA is more widely applicable than Michaelis-Menten and RMM kinetics because it handles a wider range of substrate-to-enzyme ratios. These ratios can shift in soil systems, and the ECA accounts for those changes by converging to either Michaelis-Menten or RMM kinetics (Wang and Allison, 2019). However, the ECA is more complex and requires additional data for parameterization, so the simpler Michaelis-Menten and RMM formulations may be better fits in some environmental contexts.

16.4.2 Population

As microbes consume substrates to obtain energy and nutrients, population size increases, resulting in changes in substrate demand and decomposition ability. Monod growth is an established model used to describe microbial growth given substrate availability (Parton et al., 2015). Analogous to Michaelis-Menten kinetics, the specific growth rate (μ') is a function of substrate concentration (S), where μ_{max} is the maximum potential growth rate and K_t is the Monod constant, or substrate concentration at ½ μ_{max}:

$$\mu' = \mu_{max} \cdot S/(K_t + S). \quad \text{(Eq. 16.4)}$$

Under the assumption that initial microbial biomass is much greater than initial substrate concentration, the Monod equation can be simplified to the Michaelis-Menten equation. The Monod equation does not account for density dependence, so other models, such as the logistic equation, may be more appropriate if resources limit microbial population growth.

16.4.3 Community

Moving up in scale, multiple models represent interacting populations of microbes, many of which include physical features of the environment. Georgiou et al. (2017) found that introducing density-dependent growth of microbial biomass in decomposition models of varying complexity reduced divergence between model predictions and experimental observations. Density-dependent growth accounts for community-level mechanisms, such as competition and spatial limitations, though the exact parameterization may vary across biomes and should be experimentally determined.

Multiple community-scale models have adopted trait-based approaches that focus on the physiological characteristics of microbes. Analogous to some vegetation models, the Guild Decomposition Model (GDM) represents three distinct microbial functional groups involved in litter decomposition: opportunists that process available organic matter, decomposers that break down holocellulose, and miners that degrade more chemically resistant lignin polymers (Moorhead and Sinsabaugh, 2006). The GDM is a differential equation model with explicit degradation of substrate pools by the microbial functional groups following Michaelis-Menten kinetics. The model also includes N, which is often a limiting nutrient for fresh litter decomposition. Overall, the GDM successfully simulated decomposition and successional patterns consistent with observations.

The MIcrobial-MIneral Carbon Stabilization (MIMICS) model also represents microbial functional groups along with mineral stabilization, making it suitable for application to soil systems (Wieder et al., 2014). The functional groups in MIMICS distinguish r- versus K-selected life histories, where r-strategists specialize in the degradation of low-molecular-weight compounds and K-strategists process structural litter and chemically protected compounds relatively more efficiently. Like the GDM, MIMICS assumes Michaelis-Menten kinetics and reproduces observed patterns, including litter decomposition rates and soil response to disturbance.

Building on the idea of functional traits, other community-scale models represent interacting populations and even individuals. The DEMENT model (Allison, 2012) assigns traits at random to tens or hundreds of individual microbial taxa that compete and interact on a spatial grid (Fig. 16.1). Rather than assigning taxa to functional groups a priori, taxa with favorable trait combinations for a given set of environmental conditions increase in abundance in the model simulations. The model is individual based, meaning that it tracks the locations of individual cells or colonies that grow, divide, and disperse according to model assumptions and parameters. DEMENT's unique structure allows for simulation of "virtual microbiome" composition and functioning, including the cycling of C, N, and phosphorus. Once assigned, the traits of individual taxa in DEMENT are fixed, but related models have allowed for trait evolution within taxa (Allison, 2005; Folse and Allison, 2012).

Other models also represent microbial traits at the community scale. For example, an individual-based model with trait-based functional groups interacting on a spatial grid predicted tight cycling of N during litter decomposition, allowing the microbial community to maintain CUE by overcoming stoichiometric imbalances (Kaiser et al., 2014). These findings, along with applications of DEMENT (Allison, 2014), show that community-scale models are useful for predicting emergent, and sometimes unexpected, properties of community functioning. At the same time, challenges remain in translating genomic and physiological datasets into the trait distributions required to parameterize these models.

Spatially explicit models like DEMENT are designed to represent enzyme kinetics and microbial interactions at appropriately small scales. Simulations with these models have provided insight into the

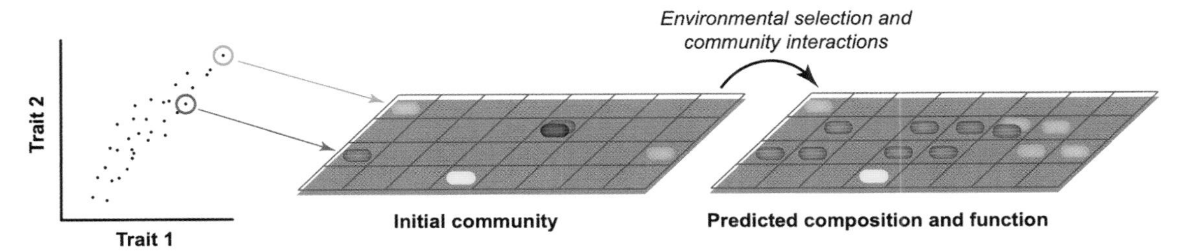

FIGURE 16.1 Schematic of the Decomposition Model of Enzymatic Traits (DEMENT). Traits are assigned to microbial taxa by drawing at random from empirically based distributions. Taxa are placed randomly on a spatial grid where they consume substrates, reproduce, disperse, and interact over time. The model predicts community composition and function as taxon abundances change due to environmental selection. *(Adapted from Allison, 2012.)*

emergent properties of heterogeneous enzyme-substrate interactions occurring at submicron scales, which could be useful for refining differential equation models operating at larger scales (Wang and Allison, 2019). Similarly, modeling the heterogeneous spatial structure of soil aggregates and associated microbial communities leads to more mechanistic prediction of trace gas fluxes (Ebrahimi and Or, 2016). Like individual-based models, aggregate-based models are valuable for determining the scaling rules needed to incorporate heterogeneous soil properties and microbial communities into larger-scale models (Wang et al., 2019).

16.4.4 Ecosystem

Ecosystem-scale models of soil microbial and biological processes often include community-level processes as well as inputs and outputs that interact with other ecosystem components such as plants and minerals. Classical models, such as RothC and CENTURY, have long been applied in an ecosystem context, and now, microbial-explicit models are also being used at ecosystem scales. Efforts to integrate these approaches are likewise gathering momentum. The Millennial model combines the best of both classical and microbial-explicit models, including microbial processes, mineral stabilization, aggregate dynamics, and soil pools that can actually be measured (Abramoff et al., 2018).

Compared to classical ecosystem models, the techniques for developing and analyzing microbial-explicit models are relatively similar. Like classical models, microbial-explicit models require technical expertise to formulate differential equations that represent soil pools, fluxes, and mechanisms of interest. For microbial-explicit models, those equations typically include nonlinear terms to represent the interaction between microbial or enzyme biomass and other soil pools (Sierra and Müller, 2015). Microbial-explicit models should be evaluated for stability and behavior across a range of relevant parameter values, much like classical linear models. For some models, the mathematics involved in these analyses may be more complicated, especially if there are no analytical solutions. However, complex microbial models can be solved numerically, much like their classical counterparts. Therefore researchers developing microbial-explicit models will likely find the process familiar if they have experience with classical models.

Microbial-explicit models represent key microbial traits, such as CUE, microbial turnover, and enzyme production, that lead to different behaviors and predictions compared to microbial-implicit models (Fig. 16.2). The Allison-Wallenstein-Bradford (AWB) model was proposed as a relatively simple microbial-explicit model of soil C cycling at the ecosystem scale. In contrast to the MIMICS model (as described in the Community section), the AWB model does not include functional groups. Instead, it represents average traits of the whole microbial community, such as CUE, enzyme kinetic parameters, and temperature sensitivities. Simulations with AWB showed that the soil C response to 5°C warming depends on the temperature sensitivity of CUE. Greater temperature sensitivity of CUE results in more stable soil C pools in response to warming due to reductions in the biomass of microbial decomposers.

The Microbial-Enzyme-mediated Decomposition (MEND) model, developed by Wang et al. (2013), is similar in structure to AWB but also accounts for mineral stabilization mechanisms. MEND splits soil organic C (SOC) into particulate organic C (POC) and mineral-associated organic C (MOC), both of which are converted into dissolved organic C (DOC) via enzyme activity. DOC can adsorb onto or desorb from MOC. The rate of breakdown into DOC is lower for MOC than POC, representing the physical protection of soil organic matter (Schmidt et al., 2011). Still, MOC and POC respond similarly to a step

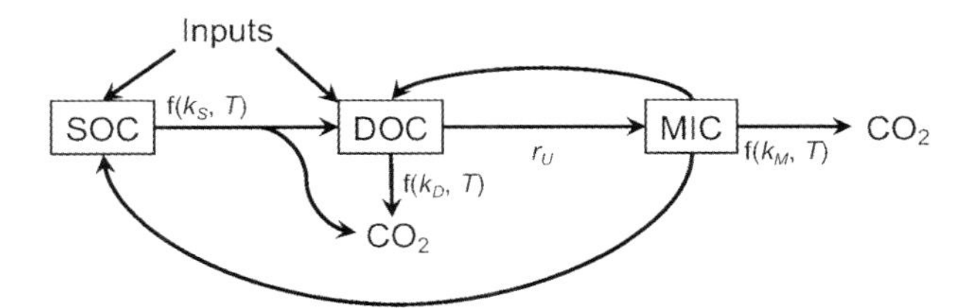

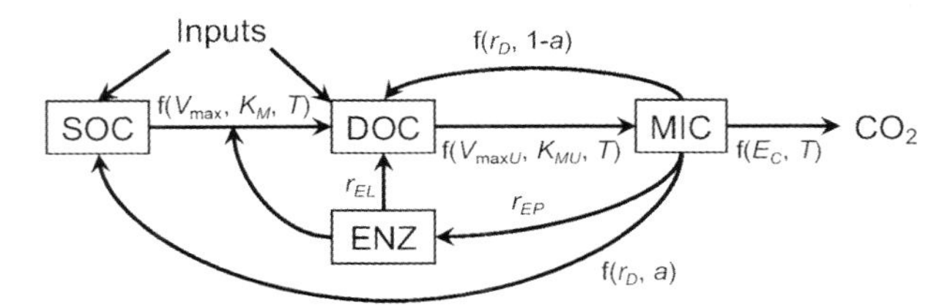

FIGURE 16.2 (a) Classical first-order linear model with microbial implicit transfers among pools. (b) Allison-Wallenstein-Bradford (AWB) model with microbial-explicit interactions among pools of soil organic carbon (SOC), dissolved organic carbon (DOC), microbial biomass (MIC), and extracellular enzymes (ENZ). In the classical model pool turnover depends on first-order decay constants (k_S for SOC, k_D for DOC, k_M for MIC) as well as DOC uptake by MIC (r_U). Turnover also depends on temperature (T). In the AWB model SOC turnover is represented as a Michaelis-Menten process dependent on T and ENZ with parameters $V_{\max}$ and K_M. An analogous process describes DOC uptake by MIC with parameters $V_{\max U}$ and K_{MU}. Carbon uptake is allocated to biomass versus respiration according to a carbon use efficiency parameter E_C. Enzymes are produced in proportion to MIC biomass at rate r_{EP} and are lost to the DOC pool at rate r_{EL}. MIC biomass dies at rate r_D and is partitioned into SOC versus DOC according to coefficient a. Partition coefficients are used in the conventional model but omitted from the figure for clarity. *(Adapted from Allison et al., 2010)*

increase in temperature, meaning that MEND and AWB end up predicting comparable SOC responses to warming.

Sulman et al. (2014) developed the Carbon, Organisms, Rhizosphere, and Protection in the Soil Environment (CORPSE) model, which explicitly represents microbes, but has a somewhat unique structure. Carbon in CORPSE can move between physically protected and unprotected pools, but unlike in MEND, only unprotected C pools can be decomposed. Another difference between CORPSE and MEND is that protected C pool sizes in CORPSE increase with clay content. These differences emphasize a need for additional empirical studies that quantify physical protection and the decomposition rates of protected SOC.

Soil models at the ecosystem scale differ substantially in their responses to plant C inputs. Microbial-explicit models like AWB and CORPSE represent the priming effect, or increased turnover of SOC in response to the addition of fresh plant C, documented in many empirical studies (Bernal et al., 2016; Perveen et al., 2019). For example, Sulman et al. (2014) fitted CORPSE to empirical data from free-air CO_2 enrichment experiments at Duke Forest and Oak Ridge National Laboratory (ORNL). They found that the priming effect almost completely offset increased litter input at Duke Forest. However, the model predicted that physical protection was stronger at ORNL, while the priming effect was much weaker, which corresponds with observations at ORNL showing increased protection of SOC in soil microaggregates.

Ecosystem model development remains a very active area of research. Although there are multiple microbial-explicit models available now, many of them still lack key mechanisms, such as spatial heterogeneity and cycling of N and other nutrients. When these mechanisms are incorporated, model outcomes may change substantially. For example, the Stoichiometrically Coupled, Acclimating Microbe-Plant-Soil (SCAMPS) model includes N dynamics and allows for variable C:N within the microbial community (Sistla et al., 2014). This stoichiometric flexibility allows the microbial community to acclimate to warming, resulting in greater losses of soil C through decomposition, especially in winter. The implication is that soil C dynamics likely depend on interactions with nutrients mediated by decomposers and plants.

16.4.5 Earth system

Most Earth system models (ESMs) do not explicitly represent microbial communities. Of the 11 ESMs in the 6th Coupled Model Intercomparison Project (CMIP6), only one ESM explicitly represents microbes (Arora et al., 2020). That model—GFDL-ESM4.1 from NOAA's Geophysical Fluid Dynamics Laboratory—represents soil C cycling using CORPSE.

Although they are not fully coupled, there have been efforts to run microbial-explicit models on a global grid, forced with output from ESMs. Wieder et al. (2013) created a microbial-explicit version of the Community Land Model (CLM) and compared its outputs with those from the Daily CENTURY (DAYCENT) model and CLM4cn, a version of CLM with N cycling. Compared to microbial-implicit CLM4cn and DAYCENT, microbial CLM predicted spatial patterns of steady-state soil C that better aligned with global observations. Furthermore, a 20% increase in litter inputs only increased global soil C temporarily due to priming effects in microbial CLM (Fig. 16.3). In contrast, soil C steadily increased in

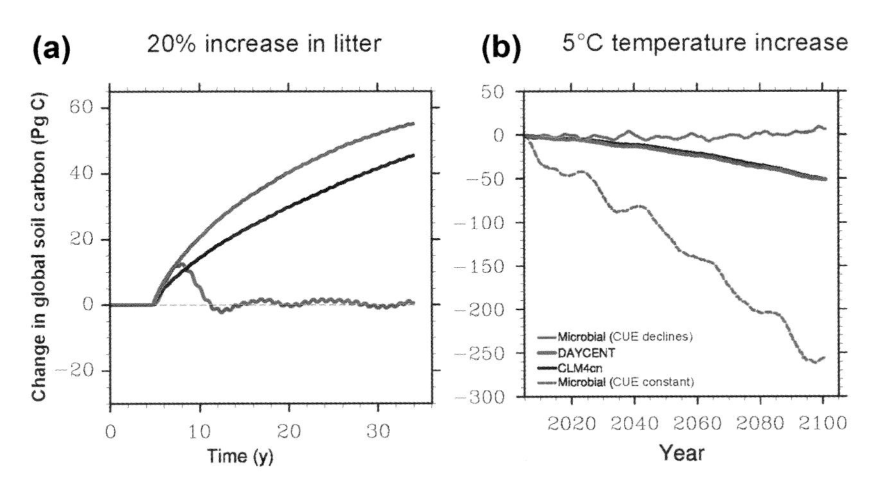

FIGURE 16.3 Soil carbon response of first-order and microbial-explicit models to (a) increased litter inputs and (b) warming at the global scale. Warming response in the microbial model depends on whether carbon use efficiency (CUE) declines or remains constant with increasing temperature. *(Adapted from Wieder et al., 2013.)*

the microbial-implicit models CLM4cn and DAYCENT. Global soil C responses to warming were also variable and mediated by the temperature sensitivity of CUE, as observed with the AWB model at the ecosystem level.

Hararuk et al. (2015) ran AWB and an ecosystem model by German et al. (2012)—which the study called the GER model—on a global grid. Both models simulated steady-state global soil C more accurately than the microbial-implicit CLM-CASA model. After calibrating the models using a global soil C database, AWB and GER predicted faster declines in soil C compared to CLM-CASA under the RCP 8.5 climate forcing scenario. The Hararuk et al. (2015) analysis also quantified the net outcome of decreasing CUE and the priming effect, allowing for key insights into how these opposing processes ultimately influence soil C predictions.

16.5 Model-data integration

16.5.1 Uncertainty quantification

As soil models continue to advance, they should be evaluated systematically for their effectiveness in achieving research goals (Fig. 16.4). The process of reviewing and stress-testing models against observations is termed "model validation" (Marzouk and Willcox, 2015). Uncertainty quantification is a core part of model validation that involves assessment of model variation, biases, limitations, and constraints that lead to deviations between the model and the true, underlying data-generating processes. Uncertainty may arise from unknown values and meanings of system parameters and inputs, potentially because parameters do not correspond to measurable quantities. Related to parameter uncertainty, parametric variability concerns the unknown effects of varying conditions on parameter and input values. Uncertainty also stems from model discrepancy, or the intentional and unintentional assumptions and simplifications separating a model from the actual processes it aims to represent.

Parameter uncertainty, parametric variability, and model discrepancy continue to be high for soil biogeochemical models (Shi et al., 2018). Some soil models have parameters that facilitate the functionality of the model but do not have clear biological interpretations. For instance, the AWB model assumes Arrhenius temperature dependence for SOC transformations, but the associated activation energy parameters are not easy to measure directly (Allison et al., 2010; Xie et al., 2020). Modeling temperature dependence also introduces parametric variability and model discrepancy. Empirical studies confirm that parameters, such as CUE and enzyme V_{max} and K_M, are temperature sensitive (Sinsabaugh et al., 2013, 2017), but the magnitude and functional form of temperature dependency is still an active area of investigation (Alster et al., 2020; Davidson et al., 2006).

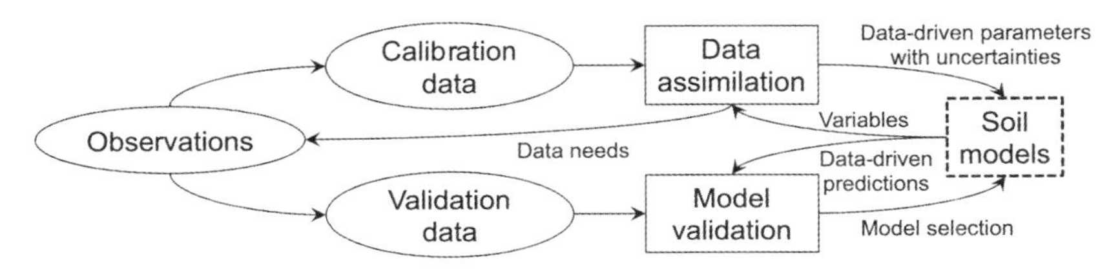

FIGURE 16.4 Framework for model-data integration. Observations are used for validating model outputs or calibrating model parameters via data assimilation. Bayesian approaches can be used for data assimilation and model validation to obtain posterior parameter distributions and calculate indices of model fit that aid in model selection. *(Adapted from Wieder et al., 2015.)*

Complex models have many parameters that may covary, making it difficult to constrain parameter uncertainty (Sierra et al., 2015). Reducing this uncertainty requires that model parameters are identifiable, such that change in parameter value causes an associated change in variables predicted by the model. Sierra et al. (2015) proposed a collinearity index to quantify the identifiability of a model—the higher the index, the lower the identifiability, and the more difficult it is to find the true parameter values. Increasing the number of datasets used to parameterize a model can increase identifiability of linear models and reduce overfitting, thereby improving predictive accuracy. For microbial-explicit models, additional datasets including microbial variables (e.g., soil enzyme activities, microbial biomass) might be needed to increase parameter identifiability and reduce uncertainty.

Bayesian probabilistic frameworks are increasingly applied to interpret uncertainty in soil models. Central to Bayesian uncertainty quantification and model validation are the processes of Bayesian parameter estimation and inference, also known as data assimilation and probabilistic/Bayesian inversion in the geosciences (Lahoz and Schneider, 2014). With these approaches, the likely distribution of model parameter values for a given data set is estimated and characterized. The numerical approximation of parameter distributions and model likelihood estimation is carried out through Markov chain Monte Carlo (MCMC) simulation methods (Christensen et al., 2006). Although the exact Monte Carlo simulation algorithm may vary, most data assimilation frameworks include the following steps:

1. Choose model types and specific models to evaluate. In the case of soil biogeochemistry, the assimilation of linear and nonlinear ordinary differential equation models can be compared (Xie et al., 2020).
2. Choose a dataset for comparison with model outputs.
3. Establish preinference probability density functions of model parameter values (known as the "prior distributions" or "priors").
4. Iteratively propose model parameter values to generate model outputs for computing model likelihood for the given data set.
5. Approximate the distributions and probability density functions of parameter values that correspond to better model fits to the data set (known as the "posterior distributions" or "posteriors").
6. Compare model likelihoods conditional on the data set with available and desired goodness-of-fit metrics. The specific Monte Carlo algorithm will dictate the options available for goodness-of-fit metrics.

"Exact" Bayesian Monte Carlo schemes comprehensively sample parameter values to compute posterior distributions. These methods include traditional Gaussian random walk Metropolis-Hastings MCMC and Gibbs samplers (McElreath, 2020), adaptive approaches derived from evolutionary optimization algorithms, such as differential adaptive evolution Metropolis (Vrugt, 2016), and the physics-inspired, momentum-driven family of Hamiltonian Monte Carlo algorithms (Neal, 2011).

Statisticians have also been investigating "nonexact" Bayesian inference schemes that seek to increase speed through approximation and simplification of parameter spaces. Nonexact approaches include the approximate Bayesian computation (Alahmadi et al., 2021; Csilléry et al., 2010) and variational Bayesian classes of methodologies (Blei et al., 2017; Ryder et al., 2018). Goodness-of-fit methods range from simpler frequentist computations, such as coefficient of determination and maximum likelihood estimation, to Bayesian metrics, including information criteria and cross-validation computations (Gelman et al., 2013). Fully Bayesian goodness-of-fit metrics can be more stable than their frequentist counterparts and provide more diagnostic information about overfitting and inference validity (Vehtari et al., 2016), though there may be higher computational resource demands.

There have been several powerful applications of Bayesian parameter estimation to soil biogeochemical models. Hararuk et al. (2014) integrated global soil C data into the C-only version of CLM coupled with the Carnegie-Ames-Stanford Approach submodel (CLM-CASA), while Ťupek et al. (2019) integrated respiration data from boreal forests in Finland into the Yasso07, Yasso15, and CENTURY models. Both studies compared model outputs before and after using a Bayesian data assimilation process to constrain model parameters. In all cases data integration resulted in model predictions that more closely matched observations.

However, each of these studies has caveats. The soil C database used by Hararuk et al. (2014) did not include time-series data, thereby necessitating a steady-state assumption about C pool sizes. If this assumption is not accurate, estimates of model uncertainty may be difficult to interpret. Ťupek et al. (2019) calibrated models with observed data but did not use an independent dataset to validate model predictions, which can lead to model overfitting. Maintaining separate training and validation datasets, a common practice in machine learning approaches, can help avoid this problem (Botu et al., 2017).

Approaches like Bayesian data assimilation are most effective when extensive, multivariate datasets are available for model calibration and validation across a range of ecosystems. For example, field measurements of dryland soils have improved biogeochemical models of ecosystem-specific C-cycling dynamics (Shen et al., 2016; Zhang et al., 2014). Going forward, rapid advancements in remote and in situ environmental sensing tools, such as light detection and ranging (LiDAR) (Kemppinen et al., 2018) and soil nutrient sensors (Burton et al., 2020), can increase the availability of ecosystem-specific measurements at lower cost, higher resolution, and greater sampling intensity than ever before.

16.5.2 Model intercomparison

Model intercomparison goes hand in hand with model selection and data assimilation to evaluate the behaviors and performance of different models relative to one another. For instance, Li et al. (2014) compared three microbial-explicit models with a classical first-order model and found that steady-state SOC was much more responsive to varying temperature sensitivity of CUE in the microbial-explicit models. In contrast, SOC stocks were largely independent of microbial CUE in the first-order model. This analysis points toward a need for additional empirical research on how microbial CUE varies with temperature and other factors.

The application of Bayesian approaches to model calibration and selection can readily be extended to model intercomparison. In their global analysis of soil C responses under RCP8.5, Hararuk et al. (2015) used a Bayesian approach to show that the microbial-explicit models AWB and GER better explained the spatial variation of steady-state soil C compared to the CLM-CASA model. However, at least with some parameter values, the microbial-explicit models simulated oscillations in soil C over time, which is an unrealistic behavior at ecosystem to global scales.

Xie et al. (2020) applied a Bayesian approach to compare AWB with a classical model (Fig. 16.2). Both models were fit to a meta-analysis dataset on soil respiration response to warming (Romero-Olivares et al., 2017) and compared using Bayesian goodness-of-fit metrics, such as the widely applicable information criterion (WAIC) and leave-one-out (LOO) cross-validation. These metrics account for the posterior distributions of parameter values after model fitting; LOO is a useful metric when limited data are available for model selection. Both models fit the meta-analysis data reasonably well, but the simpler structure of the classical model led to slightly better WAIC and LOO scores. These findings emphasize that model selection involves tradeoffs. Simple models with few parameters may be calibrated to match observational datasets with good validation scores, but these models may fall short in capturing the mechanistic details needed to make accurate predictions across a broader range of soil ecosystems.

16.6 Recommendations to advance soil models

Despite recent progress, substantial barriers still prevent the widespread application of models to grand challenges in soil biology. In particular, specialized language, expertise, and skill sets can make it challenging to integrate modeling with other scientific approaches. This specialization can be a barrier to information flow between modeling and empirical analyses. Such issues can exacerbate the challenge of collecting data in a form that supports model development, calibration, and validation. In addition, models can be difficult to access and apply if recent versions, adequate documentation, and user interfaces are not available. Scaling-up models, for example, to make Earth system predictions, can be limited by insufficient tools for model selection and intercomparison. Approaches for model validation are still under development and involve specialized knowledge of computational and statistical tools.

Overcoming these barriers would be beneficial. Predictive accuracy would increase for models applied to simulate future soil C stocks, nutrient cycling, and climate change. Given that models have multiple uses beyond prediction, broader community engagement in the science of modeling would also advance fundamental knowledge across the disciplines of soil science, biology, and biochemistry. To reap these benefits, we recommend the following steps:

1. Integrate modeling and empirical approaches. Rather than viewing modeling and empirical activities as separate, we recommend codeveloping models and empirical research. Operationally, this means reconfiguring science teams so that researchers with modeling expertise interact directly and frequently with empirical researchers. From the proposal writing stage through model development and manuscript publication, scientists creating models and collecting data should create spaces to develop a common language and align research goals. By cocreating models and experiments, researchers can ensure that models represent key processes, critical model parameters are measurable, and both model and experimental outcomes are relevant to one another. Such cooperation would be particularly helpful for incorporating complex datasets into trait-based community models.
2. Collect more data. Relatively few time-series datasets are available for some soil variables, such as C stocks, making it difficult to evaluate or avoid the steady-state assumptions often made in biogeochemical models. Sparse data can also limit the possibility of separating data into training versus validation subsets. Better integration between modeling and empirical research could help fill some of these data gaps.
3. Cross-train researchers in modeling. To enable the interactions necessary for integration, researchers should receive training in modeling perspectives and approaches. For example, training activities such as workshops, short courses, and online modules can help students acquire common vocabulary used in modeling. Conversely, students with a modeling background can benefit from training activities focused on theory and empirical work. If designed thoughtfully, seminars and courses can provide opportunities for students to get comfortable communicating and collaborating across the modeling-empirical divide.

4. Improve accessibility to model code and analysis tools. The principles of FAIR data should also apply to model code and outputs: findable, accessible, interoperable, and reusable (Wilkinson et al., 2016). Code repositories such as GitHub and platforms such as the Department of Energy's KBase can host code along with input/output files and user interfaces to make models accessible. For new models, writing and documenting code in widely used, open-source formats, such as R Markdown, Java, C++, and Jupyter Notebook for Python and other computer languages, can promote interoperability and reusability. A version control system is also important to ensure analyses from a prior model version can be replicated. Regardless of the model or platform, researchers should always strive to make model code and analyses publicly available with guidelines for reuse so that others may validate, build upon, and broaden applications of existing models.
5. Plug-and-play models and datasets. Taking the principle of interoperability to another level, we encourage the development of model testbeds that enable mixing and matching of different models and datasets (Wieder et al., 2018). Ideally, such testbeds should allow for modifications of model structure and input datasets. Testbeds can also facilitate standardization of input/output protocols and datasets to enhance interoperability, thereby avoiding tedious data reformatting procedures while also providing guidance on standards that could be adopted by the broader soil science community.
6. Develop improved model selection and intercomparison tools. Moving beyond testbeds, the research community would benefit from wider availability of model selection and intercomparison resources. For example, the *soilR* package enables users to run simulations with an array of differential equation models, including some that represent soil radiocarbon (Sierra et al., 2012). Global intercomparison initiatives, such as the CMIP, have also been tremendously valuable for comparing Earth system models by establishing a standardized set of simulation scenarios and output variables (Arora et al., 2020; Todd-Brown et al., 2013). As new tools for model inference become available, they should be incorporated into intercomparison projects to enable one-stop-shopping for model comparison and selection (Xie et al., 2020).

16.7 Conclusion

Within the last decade, models of soil systems have advanced substantially. There are now many new approaches for representing microbial and biochemical processes in soil models. As these new models came online, synthesis efforts placed them in the context of broad principles that guide quantitative soil science across scales and ecosystems. We anticipate that these advances will support further integration and unification of soil biological modeling in the next 5 to 10 years. Still, another modeling renaissance faces some significant challenges. Disciplinary silos, as well as difficulties in scaling models from genes to ecosystems, must be overcome to maximize the impact of recent model advances. Breaking down these barriers will require better integration of modeling approaches into all branches of soil science. Our recommendations to build computational infrastructure and train a new generation of researchers well-versed in modeling can serve as an initial roadmap for integration. Following our roadmap should help elevate models as powerful tools for tackling soil-related grand challenges facing society, from food security to climate change.

16.8 Acknowledgments

The work was supported by funding from the US National Science Foundation Ecosystem Studies Program (DEB-1900885) and the Department of Energy, Office of Science, BER, Genomic Sciences Program (DE-SC0020382).

References

Abramoff, R., Xu, X., Hartman, M., O'Brien, S., Feng, W., Davidson, E., et al., 2018. The millennial model: in search of measurable pools and transformations for modeling soil carbon in the new century. Biogeochemistry 137, 51−71.

Aitkenhead, M.J., Coull, M.C., 2016. Mapping soil carbon stocks across Scotland using a neural network model. Geoderma 262, 187−198.

Alahmadi, A.A., Flegg, J.A., Cochrane, D.G., Drovandi, C.C., Keith, J.M., 2021. A comparison of approximate versus exact techniques for Bayesian parameter inference in nonlinear ordinary differential equation models. R. Soc. Open Sci. 7, 191315.

Allison, S.D., 2005. Cheaters, diffusion, and nutrients constrain decomposition by microbial enzymes in spatially structured environments. Ecol. Lett. 8, 626−635.

Allison, S.D., 2012. A trait-based approach for modelling microbial litter decomposition. Ecol. Lett. 15, 1058−1070.

Allison, S.D., 2014. Modeling adaptation of carbon use efficiency in microbial communities. Front. Microbiol. 5, 571.

Allison, S.D., 2017. Building predictive models for diverse microbial communities in soil. In: Tate, K.R. (Ed.), Microbial Biomass: A Paradigm Shift in Terrestrial Biogeochemistry. World Scientific, pp. 141−166.

Allison, S.D., Wallenstein, M.D., Bradford, M.A., 2010. Soil-carbon response to warming dependent on microbial physiology. Nat. Geosci. 3, 336−340.

Alster, C.J., von Fischer, J.C., Allison, S.D., Treseder, K.K., 2020. Embracing a new paradigm for temperature sensitivity of soil microbes. Glob. Chang. Biol. 26, 3221−3229.

Arora, V.K., Katavouta, A., Williams, R.G., Jones, C.D., Brovkin, V., Friedlingstein, P., et al., 2020. Carbon−concentration and carbon−climate feedbacks in CMIP6 models and their comparison to CMIP5 models. Biogeosciences 17, 4173−4222.

Bahram, M., Hildebrand, F., Forslund, S.K., Anderson, J.L., Soudzilovskaia, N.A., Bodegom, P.M., et al., 2018. Structure and function of the global topsoil microbiome. Nature 560, 233−237.

Bernal, B., McKinley, D.C., Hungate, B.A., White, P.M., Mozdzer, T.J., Megonigal, J.P., 2016. Limits to soil carbon stability; Deep, ancient soil carbon decomposition stimulated by new labile organic inputs. Soil Biol. Biochem. 98, 85−94.

Blankinship, J.C., Berhe, A.A., Crow, S.E., Druhan, J.L., Heckman, K.A., Keiluweit, M., et al., 2018. Improving understanding of soil organic matter dynamics by triangulating theories, measurements, and models. Biogeochemistry 140, 1−13.

Blei, D.M., Kucukelbir, A., McAuliffe, J.D., 2017. Variational Inference: a review for statisticians. J. Am. Stat. Assoc. 112, 859−877.

Bosatta, E., Ågren, G., 1985. Theoretical analysis of decomposition of heterogeneous substrates. Soil Biol. Biochem. 17, 601−610.

Bosatta, E., Ågren, G.I., 1999. Soil organic matter quality interpreted thermodynamically. Soil Biol. Biochem. 31, 1889−1891.

Botu, V., Batra, R., Chapman, J., Ramprasad, R., 2017. Machine learning force fields: construction, validation, and outlook. J. Phys. Chem. C 121, 511−522.

Bradford, M.A., Wieder, W.R., Bonan, G.B., Fierer, N., Raymond, P.A., Crowther, T.W., 2016. Managing uncertainty in soil carbon feedbacks to climate change. Nat. Clim. Chang. 6, 751−758.

Burns, R.G., DeForest, J.L., Marxsen, J., Sinsabaugh, R.L., Stromberger, M.E., Wallenstein, M.D., et al., 2013. Soil enzymes in a changing environment: current knowledge and future directions. Soil Biol. Biochem. 58, 216−234.

Burton, L., Jayachandran, K., Bhansali, S., 2020. Review—the "real-time" revolution for in situ soil nutrient sensing. J. Electrochem. Soc. 167, 37569.

Cavicchioli, R., Ripple, W.J., Timmis, K.N., Azam, F., Bakken, L.R., Baylis, M., et al., 2019. Scientists' warning to humanity: microorganisms and climate change. Nat. Rev. Microbiol. 17, 569−586.

Christensen, O., Roberts, G., Sköld, M., 2006. Robust Markov chain Monte Carlo methods for spatial generalized linear mixed models. J. Comput. Graph Stat. 15, 1−17.

Csilléry, K., Blum, M.G.B., Gaggiotti, O.E., François, O., 2010. Approximate bayesian computation (ABC) in practice. Trends Ecol. Evol. 25, 410–418.

Davidson, E.A., Janssens, I.A., 2006. Temperature sensitivity of soil carbon decomposition and feedbacks to climate change. Nature 440, 165–173.

Davidson, E.A., Janssens, I.A., Luo, Y., 2006. On the variability of respiration in terrestrial ecosystems: moving beyond Q10. Glob. Chang. Biol. 12, 154–164.

Davidson, E.A., Samanta, S., Caramori, S.S., Savage, K.E., 2012. The Dual Arrhenius and Michaelis-Menten (DAMM) kinetics model for decomposition of soil organic matter at hourly to seasonal time scales. Glob. Chang. Biol. 18, 371–384.

Ebrahimi, A., Or, D., 2016. Microbial community dynamics in soil aggregates shape biogeochemical gas fluxes from soil profiles—upscaling an aggregate biophysical model. Glob. Chang. Biol. 22, 3141–3156.

Folse, H.J., Allison, S.D., 2012. Cooperation, competition, and coalitions in enzyme-producing microbes: social evolution and nutrient depolymerization rates. Front. Microbiol. 3, 338.

Fontaine, S., Barot, S., 2005. Size and functional diversity of microbe populations control plant persistence and long-term soil carbon accumulation. Ecol. Lett. 8, 1075–1087.

Fontaine, S., Bardoux, G., Abbadie, L., Mariotti, A., 2004. Carbon input to soil may decrease soil carbon content. Ecol. Lett. 7, 314–320.

Gelman, A., Carlin, J.B., Stern, H.S., Dunson, D.B., Vehtari, A., Rubin, D.B., 2013. Texts in Statistical Science, Bayesian Data Analysis, third ed. Chapman & Hall/CRC Press.

Georgiou, K., Abramoff, R.Z., Harte, J., Riley, W.J., Torn, M.S., 2017. Microbial community-level regulation explains soil carbon responses to long-term litter manipulations. Nat. Commun. 8, 1223.

German, D.P., Marcelo, K.R.B., Stone, M.M., Allison, S.D., 2012. The Michaelis-Menten kinetics of soil extracellular enzymes in response to temperature: a cross-latitudinal study. Glob. Chang. Biol. 18, 1468–1479.

Guenet, B., Moyano, F.E., Peylin, P., Ciais, P., Janssens, I.A., 2016. Towards a representation of priming on soil carbon decomposition in the global land biosphere model ORCHIDEE (version 1.9.5.2). Geosci. Model Dev. 9, 841–855.

Hagerty, S.B., Allison, S.D., Schimel, J.P., 2018. Evaluating soil microbial carbon use efficiency explicitly as a function of cellular processes: implications for measurements and models. Biogeochemistry 140, 269–283.

Hararuk, O., Xia, J., Luo, Y., 2014. Evaluation and improvement of a global land model against soil carbon data using a Bayesian Markov chain Monte Carlo method. J. Geophys. Res. Biogeosci. 119, 403–417.

Hararuk, O., Smith, M.J., Luo, Y., 2015. Microbial models with data-driven parameters predict stronger soil carbon responses to climate change. Glob. Chang. Biol. 21, 2439–2453.

IPCC, 2019. Climate Change and Land: An IPCC Special Report on Climate Change, Desertification, Land Degradation, Sustainable Land Management, Food Security, and Greenhouse Gas Fluxes in Terrestrial Ecosystems.

Jansson, J.K., Hofmockel, K.S., 2018. The soil microbiome—from metagenomics to metaphenomics. Curr. Opin. Microbiol. 43, 162–168.

Jansson, J.K., Hofmockel, K.S., 2020. Soil microbiomes and climate change. Nat. Rev. Microbiol. 18, 35–46.

Jenkinson, D.S., Rayner, J.H., 1977. The turnover of soil organic matter in some of the Rothamsted classical experiments. Soil Sci. 123, 298–305.

Kaiser, C., Franklin, O., Dieckmann, U., Richter, A., 2014. Microbial community dynamics alleviate stoichiometric constraints during litter decay. Ecol. Lett. 17, 680–690.

Kalbitz, K., Schwesig, D., Schmerwitz, J., Kaiser, K., Haumaier, L., Glaser, B., et al., 2003. Changes in properties of soil-derived dissolved organic matter induced by biodegradation. Soil Biol. Biochem. 35, 1129–1142.

Kemppinen, J., Niittynen, P., Riihimäki, H., Luoto, M., 2018. Modelling soil moisture in a high-latitude landscape using LiDAR and soil data. Earth Surf. Process. Landforms 43, 1019–1031.

Koven, C.D., Lawrence, D.M., Riley, W.J., 2015. Permafrost carbon−climate feedback is sensitive to deep soil carbon decomposability but not deep soil nitrogen dynamics. Proc. Natl. Acad. Sci. 112, 3752–3757.

Lahoz, W., Schneider, P., 2014. Data assimilation: making sense of Earth observation. Front. Environ. Sci. 2, 16.

Lehmann, J., Solomon, D., Kinyangi, J., Dathe, L., Wirick, S., Jacobsen, C., 2008. Spatial complexity of soil organic matter forms at nanometre scales. Nat. Geosci. 1, 238–242.

Li, J., Wang, G., Allison, S.D., Mayes, M.A., Luo, Y., 2014. Soil carbon sensitivity to temperature and carbon use efficiency compared across microbial-ecosystem models of varying complexity. Biogeochemistry 119, 67–84.

Mack, M.C., Schuur, E.A.G., Bret-Harte, M.S., Shaver, G.R., Chapin III, F.S., 2004. Ecosystem carbon storage in arctic tundra reduced by long-term nutrient fertilization. Nature 433, 440–443.

Marzouk, Y.M., Willcox, K.E., 2015. Uncertainty quantification. In: Higham, N.J. (Ed.), The Princeton Companion to Applied Mathematics, Vol. II. Princeton University Press, pp. 131–134.

McElreath, R., 2020. Statistical Rethinking: A Bayesian Course with Examples in R and STAN. CRC Press.

Moorhead, D.L., Sinsabaugh, R.L., 2006. A theoretical model of litter decay and microbial interaction. Ecol. Monogr. 76, 151–174.

Moorhead, D.L., Weintraub, M.N., 2018. The evolution and application of the reverse Michaelis-Menten equation. Soil Biol. Biochem. 125, 261–262.

Neal, R.M., 2011. MCMC using Hamiltonian dynamics. In: Brooks, S., Gelman, A., Jones, G., Meng, X.-L. (Eds.), Handbook of Markov Chain Monte Carlo. Chapman & Hall/CRC Press, pp. 113–162.

Olson, J.S., 1963. Energy storage and the balance of producers and decomposers in ecological systems. Ecology 44, 322–331.

Parton, W.J., Stewart, J.W.B., Cole, C.V., 1988. Dynamics of C, N, P, and S in grassland soils - a model. Biogeochemistry 5, 109–131.

Parton, W.J., Del Grosso, S.J., Plante, A.F., Adair, E.C., Lutz, S.M., 2015. Modeling the dynamics of soil organic matter and nutrient cycling. In: Soil Microbiology, Ecology and Biochemistry, fourth ed. Elsevier Inc., pp. 505–537

Perveen, N., Barot, S., Maire, V., Cotrufo, M.F., Shahzad, T., Blagodatskaya, E., et al., 2019. Universality of priming effect: an analysis using thirty five soils with contrasted properties sampled from five continents. Soil Biol. Biochem. 134, 162–171.

Quideau, S.A., Graham, R.C., Oh, S.-W., Hendrix, P.F., Wasylishen, R.E., 2005. Leaf litter decomposition in a chaparral ecosystem, Southern California. Soil Biol. Biochem. 37, 1988–1998.

Romero-Olivares, A.L., Allison, S.D., Treseder, K.K., 2017. Soil microbes and their response to experimental warming over time: a meta-analysis of field studies. Soil Biol. Biochem. 107, 32–40.

Ryder, T., Golightly, A., McGough, A.S., Prangle, D., 2018. 2018. Black-box variational inference for stochastic differential equations. 35th Int. Conf. Mach. Learn. ICML 10, 7021–7030.

Schimel, J., 2001. Biogeochemical models: implicit versus explicit microbiology. In: Schulze, E.D., Harrison, S.P., Heimann, M., Holland, E.A., Lloyd, J.J., Prentice, I.C., et al. (Eds.), Global Biogeochemical Cycles in the Climate System. Academic Press, pp. 177–183.

Schimel, J.P., Weintraub, M.N., 2003. The implications of exoenzyme activity on microbial carbon and nitrogen limitation in soil: a theoretical model. Soil Biol. Biochem. 35, 549–563.

Schmidt, M.W.I., Torn, M.S., Abiven, S., Dittmar, T., Guggenberger, G., Janssens, I.A., et al., 2011. Persistence of soil organic matter as an ecosystem property. Nature 478, 49–56.

Shen, W., Jenerette, G.D., Hui, D., Scott, R.L., 2016. Precipitation legacy effects on dryland ecosystem carbon fluxes: direction, magnitude and biogeochemical carryovers. Biogeosciences 13, 425–439.

Shi, Z., Crowell, S., Luo, Y., Moore, B., 2018. Model structures amplify uncertainty in predicted soil carbon responses to climate change. Nat. Commun. 9, 2171.

Shi, Z., Allison, S.D., He, Y., Levine, P.A., Hoyt, A.M., Beem-Miller, J., et al., 2020. The age distribution of global soil carbon inferred from radiocarbon measurements. Nat. Geosci. 13, 555–559.

Sierra, C.A., Müller, M., 2015. A general mathematical framework for representing soil organic matter dynamics. Ecol. Monogr. 85, 505–524.

Sierra, C.A., Müller, M., Trumbore, S.E., 2012. Models of soil organic matter decomposition: the SoilR package, version 1.0. Geosci. Model Dev. 5, 1045–1060.

Sierra, C.A., Malghani, S., Müller, M., 2015. Model structure and parameter identification of soil organic matter models. Soil Biol. Biochem. 90, 197–203.

Sihi, D., Davidson, E.A., Savage, K.E., Liang, D., 2020. Simultaneous numerical representation of soil microsite production and consumption of carbon dioxide, methane, and nitrous oxide using probability distribution functions. Glob. Chang. Biol. 26, 200–218.

Sinsabaugh, R.L., Manzoni, S., Moorhead, D.L., Richter, A., 2013. Carbon use efficiency of microbial communities: stoichiometry, methodology and modelling. Ecol. Lett. 16, 930–939.

Sinsabaugh, R.L., Moorhead, D.L., Xu, X., Litvak, M.E., 2017. Plant, microbial and ecosystem carbon use efficiencies interact to stabilize microbial growth as a fraction of gross primary production. New Phytol. 214, 1518–1526.

Sistla, S.A., Rastetter, E.B., Schimel, J.P., 2014. Responses of a tundra system to warming using SCAMPS: a stoichiometrically coupled, acclimating microbe-plant-soil model. Ecol. Monogr. 84, 151–170.

Smith, O.L., 1979a. An analytical model of the decomposition of soil organic matter. Soil Biol. Biochem. 11, 585–606.

Smith, O.L., 1979b. Application of a model of the decomposition of soil organic matter. Soil Biol. Biochem. 11, 607–618.

Sulman, B., Phillips, R.P., Oishi, A.C., Shevliakova, E., Pacala, S.W., 2014. Microbe-driven turnover offsets mineral-mediated storage of soil carbon under elevated CO_2. Nat. Clim. Chang. 4, 1099–1102.

Tang, J.Y., 2015. On the relationships between the Michaelis-Menten kinetics, reverse Michaelis-Menten kinetics, equilibrium chemistry approximation kinetics, and quadratic kinetics. Geosci. Model Dev. 8, 3823–3835.

Tang, J.Y., Riley, W.J., 2013. A total quasi-steady-state formulation of substrate uptake kinetics in complex networks and an example application to microbial litter decomposition. Biogeosciences 10, 8329–8351.

Todd-Brown, K.E.O., Hopkins, F.M., Kivlin, S.N., Talbot, J.M., Allison, S.D., 2012. A framework for representing microbial decomposition in coupled climate models. Biogeochemistry 109, 19–33.

Todd-Brown, K.E.O., Hoffman, F.M., Post, W.M., Randerson, J.T., Allison, S.D., 2013. Causes of variation in soil carbon simulations from CMIP5 Earth system models and comparisons with observations. Biogeosciences 10, 1717–1736.

Todd-Brown, K.E.O., Randerson, J.T., Hopkins, F., Arora, V., Hajima, T., Jones, C., et al., 2014. Changes in soil organic carbon storage predicted by Earth system models during the 21st century. Biogeosciences 11, 2341–2356.

Ťupek, B., Launiainen, S., Peltoniemi, M., Sievänen, R., Perttunen, J., Kulmala, L., et al., 2019. Evaluating CENTURY and Yasso soil carbon models for CO_2 emissions and organic carbon stocks of boreal forest soil with Bayesian multi-model inference. Eur. J. Soil Sci. 70, 847–858.

Varman, A.M., He, L., Follenfant, R., Wu, W., Wemmer, S., Wrobel, S.A., et al., 2016. Decoding how a soil bacterium extracts building blocks and metabolic energy from ligninolysis provides road map for lignin valorization. Proc. Natl. Acad. Sci. U. S. A. 113, E5802–E5811.

Vehtari, A., Mononen, T., Tolvanen, V., Sivula, T., Winther, O., 2016. Bayesian leave-one-out cross-validation approximations for Gaussian latent variable models. J. Mach. Learn. Res. 17, 1–38.

Vrugt, J.A., 2016. Markov chain Monte Carlo simulation using the DREAM software package: theory, concepts, and MATLAB implementation. Environ. Model. Softw. 75, 273–316.

Waksman, S.A., 1927. Principles of Soil Microbiology. The Williams and Wilkins Company.

Wang, B., Allison, S.D., 2019. Emergent properties of organic matter decomposition by soil enzymes. Soil Biol. Biochem. 136, 107522.

Wang, B., Brewer, P.E., Shugart, H.H., Lerdau, M.T., Allison, S.D., 2019. Building bottom-up aggregate-based models (ABMs) in soil systems with a view of aggregates as biogeochemical reactors. Glob. Chang. Biol. 25, e6–e8.

Wang, G., Post, W., Mayes, M., 2013. Development of microbial-enzyme-mediated decomposition model parameters through steady-state and dynamic analyses. Ecol. Appl. 23, 255–272.

Wang, Y.P., Chen, B.C., Wieder, W.R., Leite, M., Medlyn, B.E., Rasmussen, M., et al., 2014. Oscillatory behavior of two nonlinear microbial models of soil carbon decomposition. Biogeosciences 11, 1817–1831.

Waring, B.G., Sulman, B.N., Reed, S., Smith, A.P., Averill, C., Creamer, C.A., et al., 2020. From pools to flow: the PROMISE framework for new insights on soil carbon cycling in a changing world. Glob. Chang. Biol. 26, 6631–6643.

Weintraub, M.N., Schimel, J.P., 2003. Interactions between carbon and nitrogen mineralization and soil organic matter chemistry in arctic tundra soils. Ecosystems 6, 129–143.

Wieder, W.R., Bonan, G.B., Allison, S.D., 2013. Global soil carbon projections are improved by modelling microbial processes. Nat. Clim. Chang. 3, 909–912.

Wieder, W.R., Grandy, A.S., Kallenbach, C.M., Bonan, G.B., 2014. Integrating microbial physiology and physio-chemical principles in soils with the MIcrobial-MIneral Carbon Stabilization (MIMICS) model. Biogeosciences 11, 3899–3917.

Wieder, W.R., Allison, S.D., Davidson, E.A., Georgiou, K., Hararuk, O., He, Y., et al., 2015. Explicitly representing soil microbial processes in Earth system models. Global Biogeochem. Cycles 29, 1782–1800.

Wieder, W.R., Hartman, M.D., Sulman, B.N., Wang, Y.P., Koven, C.D., Bonan, G.B., 2018. Carbon cycle confidence and uncertainty: exploring variation among soil biogeochemical models. Glob. Chang. Biol. 24, 1563–1579.

Wilkinson, M.D., Dumontier, M., Aalbersberg, I.J., Appleton, G., Axton, M., Baak, A., et al., 2016. Comment: the FAIR Guiding Principles for scientific data management and stewardship. Sci. Data 3, 160018.

Wu, D., Piao, S., Liu, Y., Ciais, P., Yao, Y., 2018. Evaluation of CMIP5 earth system models for the spatial patterns of biomass and soil carbon turnover times and their linkage with climate. J. Clim. 31, 5947–5960.

Xia, J., Luo, Y., Wang, Y.-P., Hararuk, O., 2013. Traceable components of terrestrial carbon storage capacity in biogeochemical models. Glob. Chang. Biol. 19, 2104–2116.

Xie, H.W., Romero-Olivares, A.L., Guindani, M., Allison, S.D., 2020. A Bayesian approach to evaluation of soil biogeochemical models. Biogeosciences 17, 4043–4057.

Xue, K.,M., Yuan, M.,J., Shi, Z., Qin, Y., Deng, Y., Cheng, L., et al., 2016. Tundra soil carbon is vulnerable to rapid microbial decomposition under climate warming. Nat. Clim. Chang. 6, 595–600.

Zhang, X., Niu, G.-Y., Elshall, A.S., Ye, M., Barron-Gafford, G.A., Pavao-Zuckerman, M., 2014. Assessing five evolving microbial enzyme models against field measurements from a semiarid savannah—what are the mechanisms of soil respiration pulses? Geophys. Res. Lett. 41. https://doi.org/10.1002/2014GL061399.

Chapter 17

The application of knowledge in soil microbiology, ecology, and biochemistry (SMEB) to the solution of today's and future societal needs

John C. Moore*,† and Nathaniel Mueller*,†,‡

*Natural Resource Ecology Laboratory; †Department of Ecosystem Science and Sustainability; ‡Department of Soil and Crop Sciences, Colorado State University, Fort Collins, CO, USA

Chapter outline

17.1 Introduction

Developments in soil microbial ecology have prompted significant advances in agronomy and in soil science. New methods in identifying, isolating, and cultivating microbes and their by-products have led to

Soil Microbiology, Ecology, and Biochemistry. https://doi.org/10.1016/B978-0-12-822941-5.00017-X

innovative means of increasing agricultural production and nutrient use efficiency, increasing the production and stabilization of soil organic matter (SOM), and providing credible means to sequester carbon (C) (National Academies of Sciences and Medicine, 2019, Zegeye et al., 2019). These advances were not entirely the result of new technologies but were spurred on by three major shifts in thinking: (1) the transition from a focus on soil quality to soil health; (2) the emergence of sustainability science where human needs and challenges define the direction of science; and (3) a movement from the traditional agricultural extension model to one of codevelopment and engagement.

The reshaping of a practice or industry through advances in technology and changes in thinking like those cited above are not new. The technological advances of the early 20th century had a profound impact on agriculture. The development of the internal combustion engine, industrial-level advances in chemistry applications such as the Haber-Bosch process to generate inexpensive nitrogen-based (N) fertilizers (see Galloway et al., 2003, 2004; Smil, 2001), and the development of pesticides led to a mechanized and chemical subsidized agriculture able to increase production and lower costs. These advances, combined with the development and adoption of modern crop varieties, dramatically increased agricultural production, improved food security (von der Goltz et al., 2020), and avoided even greater cropland expansion (Burney et al., 2010). However, advances in production and efficiency that the innovations brought notwithstanding, they also came at a cost to several critical environmental outcomes, including air quality, water quality, and chemical use (Tilman et al., 2002). The conservation movements led to changes in the philosophy of agricultural production and resource, first in forestry, followed by the Dust Bowl era in soil and rangeland management, and finally by the push to lower pesticide and fertilizer use following the publication of *Silent Spring* (Carson, 1962).

Changes in agricultural practices were further spurred by the emergence of the agroecology movement during the late 1970s. Programs in the US and Europe (Andrén et al., 1990; Brussaard et al., 1988; Elliott et al., 1984; Hendrix et al., 1986; Robertson, 1991) found that low input and low-impact management practices could lead to declines or increases in SOM and greater retention of nutrients, often with similar yields when compared to their conventionally managed counterparts (Suefert et al., 2012). These studies shared an approach to soil biogeochemical processes, soil organisms (i.e., microbes and invertebrates), and SOM under different management practices. Elliott and Coleman (1988) proposed that we "let the soil work for us whereby biotic interactions in soils could be harnessed to eliminate the worst excesses of the high input approaches that fueled the green revolutions in mid- to late 20th century." In many ways this philosophy was a prophetic precursor to global initiatives based on the concept of planetary health — the idea that human health and civilization depend on natural ecosystems.

The application of knowledge in soil microbiology, ecology, and biochemistry (SMEB) to develop the solutions to societal needs is based on "let the soil work for us" philosophy, but with the three shifts in the thinking discussed above. The first major shift was the evolution of the concept of soil health from soil quality. Although soil quality and health have been used synonymously (see Doran, 2002; Laishram et al., 2012), an important distinction can be made that connects soils and SMEB science to sustainability and sustainable development. Soil quality focuses on the ability of soil to grow crops and support plants. Soil health views soil as a living resource that provides vital functions for plant growth and ecosystem services to society. Healthy soils are resilient, possessing the ability to maintain these functions and services in response to natural disturbances. The second major shift in thinking was driven by the emergence of sustainability science (Clark, 2007; Clark and Dickson, 2003) as a science that addresses challenges defined by human needs as well as the pursuit of foundational knowledge. For soils and soil microbiology, this meant balancing our mechanistic understanding of the role that microbes play in agricultural and natural ecosystems, with not only increasing food production, promoting conservation and reclamation, and mitigating soil degradation, but also the need to harness its ability to sequester C as one way to mitigate rising CO_2 levels in the atmosphere. The third major shift in thinking represents a fundamental change in the relationship between science and practitioners across multiple domains. We are witnessing a movement within policy and public engagement realms writ large that embraces the best of the agricultural extension model as a partnership between the scientists and stakeholders to co-develop and apply solutions (Clark et al., 2016). Together, these shifts in thinking support the idea that soils are essential to the collective health of the planet beyond their importance to food and fiber production (Rockström et al., 2009; Whitmee et al., 2015) and are key to achieving a more sustainable future.

17.2 SMEB science and the frameworks for sustainable development

Adopting and applying SMEB science to meet societal needs requires that we identify the challenges, the ultimate goals of the applications with clearly defined benchmarks, and the strategies taken to achieve them. The UN Sustainable Development Goals (SDGs) identify these challenges and provide a framework that encompasses interdependence of Earth system biophysical realms and processes with humankind and activities writ large (United Nations, 2015). Key to this framework is the reformulation of ecosystem processes into ecosystem services, as outlined in the Millennium Ecosystem Assessment (MEA, 2005), and the positioning of soils within them.

The concept of ecosystem services, as presented in the MEA (2005), reshaped our understanding of the relationship between ecosystem processes and their benefits to society. It identifies the supporting services (i.e., nutrient cycling, primary production, soil formation) as the foundational natural ecosystem functions that drive the provisioning (e.g., food, fiber, freshwater, fuel, biochemicals, and genetic resources), regulatory (e.g., climate, disease, water, pollination), and cultural services (e.g., spiritual and religious, recreation and tourism, aesthetic, place). Initially developed to capture the benefits that humans derive from nature (Ehrlich and Ehrlich, 1981; Westman, 1977), the concept of ecosystem services evolved and was refined over time as an outgrowth of the social-ecological perspective by focusing on understanding the linkages between natural and social systems and the value and benefits that the services (read functions) provide humans.

Doran (2002) distilled the interdependence on Earth system and human activities to humans and their relationship to soil through agriculture. He advocated that the importance of soil quality and health to global sustainability could serve to develop a framework to define sustainable land management practices. Soil quality and health are inextricably linked to sustainable agriculture for food, fiber, and energy; environmental quality of land, water, and the atmosphere; and the health of plants, animals, and humans. As such, soil quality and health are seen as measurable attributes and primary indicators that link land management practices with sustainability goals. Barrios (2007) presents an early treatise that connected soil biota to soil-based ecosystem services and their importance to sustainable crop production and soil biochemistry. Robinson et al. (2013, 2014) provided a critical assessment of the value of soils and soil processes as natural capital and their importance to ecosystem services. Timmis et al. (2017) provided a blueprint for such an alignment that focused on microbial technologies and their contributions to the SDGs that accommodate SMEB science. Taken together, these developments are a recognition that societal needs dictate that the grand challenge for soils and SMEB science encompasses finding solutions within the food, energy, and water nexus in a changing climate (Fig. 17.1; Liu et al., 2018). This will require the simultaneous application of strategies based on SMEB science to improve soils that optimize

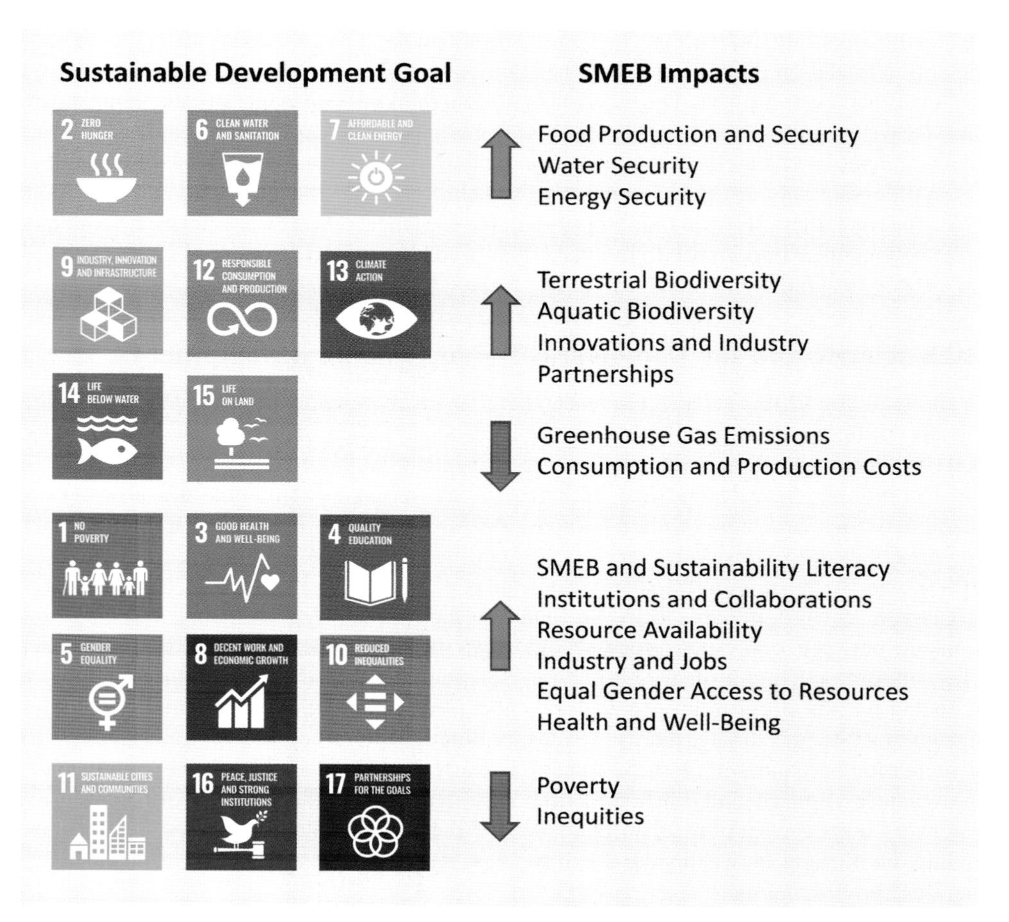

FIGURE 17.1 SMEB science applications as positioned within the Food-Water-Energy Nexus of the UN SDGs in a changing climate illustrating their direct and indirect effects on the ecosystem services that influence each of the SDGs. *(Adapted from Liu et al., 2018.)*

food production, sequester C to mitigate emissions, limit the release of reactive N to the atmosphere and water sources, and limit and mitigate pollutants. Finally, the SDGs embody the social-ecological ethos. As such, education will play an important role in the coproduction relationship among the generation of knowledge, decision-making, and policy implementation.

17.3 The grand challenge for global soils: food, energy, and water in a changing climate

Soils play an essential role in sustainability policy to meet human demands for food, energy, and water on a finite and warming planet. The combination of a growing human population, expected to reach nearly

10 billion people by 2050 (United Nations, 2015), and rising global affluence are rapidly increasing demand for resources (Steffen et al., 2015; Tilman et al., 2011). One recent synthesis of food demand estimates that demand for crop production will increase by 25–70% between 2014 and 2050 (Hunter et al., 2017), encompassing both increases in direct human consumption and consumption of feed to produce animal products. The importance of soil management for sustainable and productive landscapes goes beyond food production given the increasing pressure on global land resources to support bioenergy production, potentially linked with C capture and sequestration technology (Paustian et al., 2016). Soil health is also important for maximizing the supply of soil water available for food and bioenergy production, while judicious management of soil nutrients is critical for water quality.

The challenge of achieving sustainable production systems is magnified by the specter of climate change, which currently affects crop yields and water resources for global agricultural systems (Myers et al., 2017). Global land temperatures from 2006 to 2015 were 1.0°C warmer than the 20th-century average. Under a "moderate" emissions scenario (RCP 4.5), land temperatures would increase by 4.0°C by the end of the century (Myers et al., 2017). The projections also include increases in the frequency of extreme weather and weather-related events, e.g., storms, drought, and wildfires.

17.4 Soil biota and earth system processes

Soil biota are critical drivers of ecological and Earth system processes that operate across multiple scales. Soil communities are comprised of primary producers (e.g., some bacteria, cyanobacteria, algae, protozoa, and vascular and nonvascular plants) and heterotrophic species (e.g., other bacteria, fungi, protozoa, rotifers, tardigrades, nematodes, annelids, arthropods). The nature of the processes that soil microbes and fauna engage in have been discussed in detail throughout this book. For management and context purposes, we will focus on the roles soil organisms have in primary production and decomposition, nutrient cycling, and SOM formation. Next, we will compare soil, food web structure and function in native and managed lands, identifying common responses in terms of food web structure, nutrient fluxes among organisms and the environment, and dynamic stability. This will identify critical interactions and nodes within soil ecosystems that provide entry and control points for SMEB science to address pressing issues.

17.4.1 Primary production

Primary production is the biological transformation of inorganic C compounds to organic compounds. On a global scale, over 90% of terrestrial primary production is carried out by vascular plants, with the balance (<10%) being carried out by nonvascular plants and photosynthetic microbes (e.g., bacteria, cyanobacteria, algae, and protozoa; see Elbert et al., 2012). These estimates provide an incomplete picture of the importance of microbes as the primary production by vascular plants is mediated by direct and indirect interactions with microbes that provide plants with limiting nutrients (Coleman et al., 1983; Moore et al., 2003; Wall and Moore, 1999; Wardle et al., 1999, 2004).

17.4.2 The rhizosphere

The rhizosphere is a dynamic region of soil that includes plant roots and the surrounding soil that is influenced by plant roots, root products, and the trophic interactions of soil biota (Bardgett, 2005;

Coleman et al., 1983; Moore et al., 2003; Van der Putten et al., 2001; Wardle et al., 2004). It is characterized by rapid and prolific root growth, the sloughing of root cells, root death, and the exudation of simple C compounds. Trofymow and Coleman (1982) subdivided a growing root into a continuum of zones of activity from the root tip to the crown, where different microbial populations have access to a continuous flow of organic substrates derived from the root. The root tip is the site of growth with rapidly dividing cells and secretions or exudates that lubricate the tip as it passes through the soil. Exudates and sloughed root cells provide C for bacteria and fungi, which in turn immobilize N and P. Farther up the root is the region of nutrient exchange. This region is characterized by the birth and death of root hairs and lower rates of exudation (Bringhurst et al., 2001). The upper zones are characterized by symbiotic, mutualistic interactions (e.g., actinomycetes – *Frankia*; mycorrhizal fungi; Proteobacteria – *Bradyrhizobium*, *Rhizobium*) and structural support (Coleman et al., 1983).

A plant-microbial loop forms within each of the zones wherein there is an infusion of C into the rhizosphere by plants stimulating the activity of microbes (Bardgett et al., 1998; Foster, 1988; Grayston et al., 1996) and their invertebrate grazers (Lussenhop and Fogel, 1991; Parmelee et al., 1993), resulting in the mineralization of nutrients that become available for plant uptake and in the production of necromass that contributes to soil formation.

17.4.3 Mediators of C, N, and P cycles in soils

Soil microbes drive the cycling of C, N, and P. Plants shed leaves and structural materials (i.e., litter) onto the soil surface and release organic materials via rhizodeposition into soils from their roots that consist of plant metabolites (e.g., exudates) and plant debris (e.g., dead cells, mucilage, etc.). Microbes and invertebrates alter and metabolize the substrates for growth and, in the process of their growth and death, release excess N for plant uptake and necromass, contributing to SOM formation. Plant litter and rhizodeposition may change qualitatively and quantitatively with plant genotype, growth stages, and/or abiotic/biotic stresses, in turn affecting the microbial and invertebrate components and their activities. These biogeochemical processes form feedback loops that are foundational to understanding the C and N cycling in terrestrial systems, soil formation, and land-atmosphere exchanges, as well as in aquatic systems. The extent of these interactions can be conceptualized at the scale of an individual plant and the fate of plant production. Wall and Moore (1999) described these interactions as forms of mutualism (Table 17.1) that occur at the soil surface with plants providing C in the form of litter (leaves, stems, bark, and woody debris) within the rhizosphere and via leaching of substrates to the bulk soil (Fig. 17.2).

Mycorrhizal symbioses are interactions between different species of fungi and plants wherein the hypha of the fungus and the plant roots establish a site for nutrient exchange of C (from the plant to the fungus) and N and P (from the fungus to the plant). Arbuscular mycorrhizal (AM) fungi and plants are highly dependent on one another, while those involving ectomycorrhizal (EM) fungi forms are less so (Allen, 1991; Hoeksema et al., 2010). Several species of bacteria, archaea, actinomycetes, and cyanobacteria engage in the biological N fixation (BFN) in both agricultural (cropland and pastures) and native ecosystems (Herridge et al., 2008). These BFN-capable species form symbiotic (legume–rhizobia, nonlegume–*Frankia*, *Azolla* cyanobacteria) and nonsymbiotic associations with plants (cereal associative bacteria), while others are free-living (cyanobacteria, heterotrophic bacteria, and autotrophic bacteria). The importance of these direct and indirect mutualistic interactions extends beyond their impacts on plant growth and can influence the development of plant communities and nutrient dynamics of whole systems. Apart from the nutrient exchanges, many mycorrhizal fungi engage in the decomposition of organic

TABLE 17.1 A summary of the different forms of mutualisms in soil that include nutrient exchanges through direct and indirect means using the definition of Boucher et al. (1982): "An interaction between species that is beneficial to both."

Direct mutualism		Indirect mutualism
contact between species		no contact between species
Symbiotic		Nonsymbiotic
Intimate Physiological species link		No direct intimate Physiological species link
Examples		***Examples***
Plants – AM fungi Plants – EC fungi	Biota-microbes – grazing Biota-microbes – dispersal	Decomposition – facilitation Decomposition – priming
Legumes – *Rhizobia*	Cryptobiotic crusts	Decomposition – Comminution
Nonlegumes – *Frankia* *Azolla* – cyanobacteria	Cereal associative – bacteria	Decomposition – consortia Rhizosphere – microbial loop

Adapted from Moore (1988); Wall and Moore (1999).

materials directly; their necromass and hyphal networks are important to soil formation (Johnson et al., 2006; Ma et al., 2018). Mycorrhizal associations are critical drivers of post disturbance succession as well (Allen et al., 1987, 1992; Janos, 1980; Krüger et al., 2017). The actinorhizal association of the N-fixing actinomycete, *Frankia*, with *Myrica faya* in Hawaii was instrumental in establishing the invasiveness of the plant to the exclusion of native plant species and increased the rates of N mineralization and availability five-fold (Vitousek et al., 1987). At the global scale, species capable of BFN generate between 100 and 290 Tg N yr^{-1} (Elbert et al., 2012; Vitousek et al., 2013), which is significantly higher than the preindustrial estimates of 40 to 100 Tg N yr^{-1}.

17.4.4 Decomposition

Decomposition is the process of reducing complex dead organic material to simpler organic and inorganic forms. The decomposition of organic matter is a primary ecosystem service integral to soil formation, nutrient cycling, and plant growth. Decomposition involves highly interactive biotic and abiotic processes of physical fragmentation, chemical degradation, and leaching (Swift et al., 1979). The physical fragmentation of organic matter – comminution – involves invertebrates feeding directly on the material or on the microbes that have colonized the materials. Chemical degradation occurs via the actions of extracellular enzymes produced by bacteria, fungi, protozoa, and the gut enzymes of invertebrates during feeding. Hot, sunny areas such as deserts also experience chemical degradation of materials such as plant detritus. Leaching involves the physical movement of soluble forms of organic matter that are exposed during decomposition.

The rate of decomposition is a function of the effects that ecosystem type and abiotic (e.g., temperature, moisture, pH, nutrient availability) factors have on the activities of microbes and invertebrates,

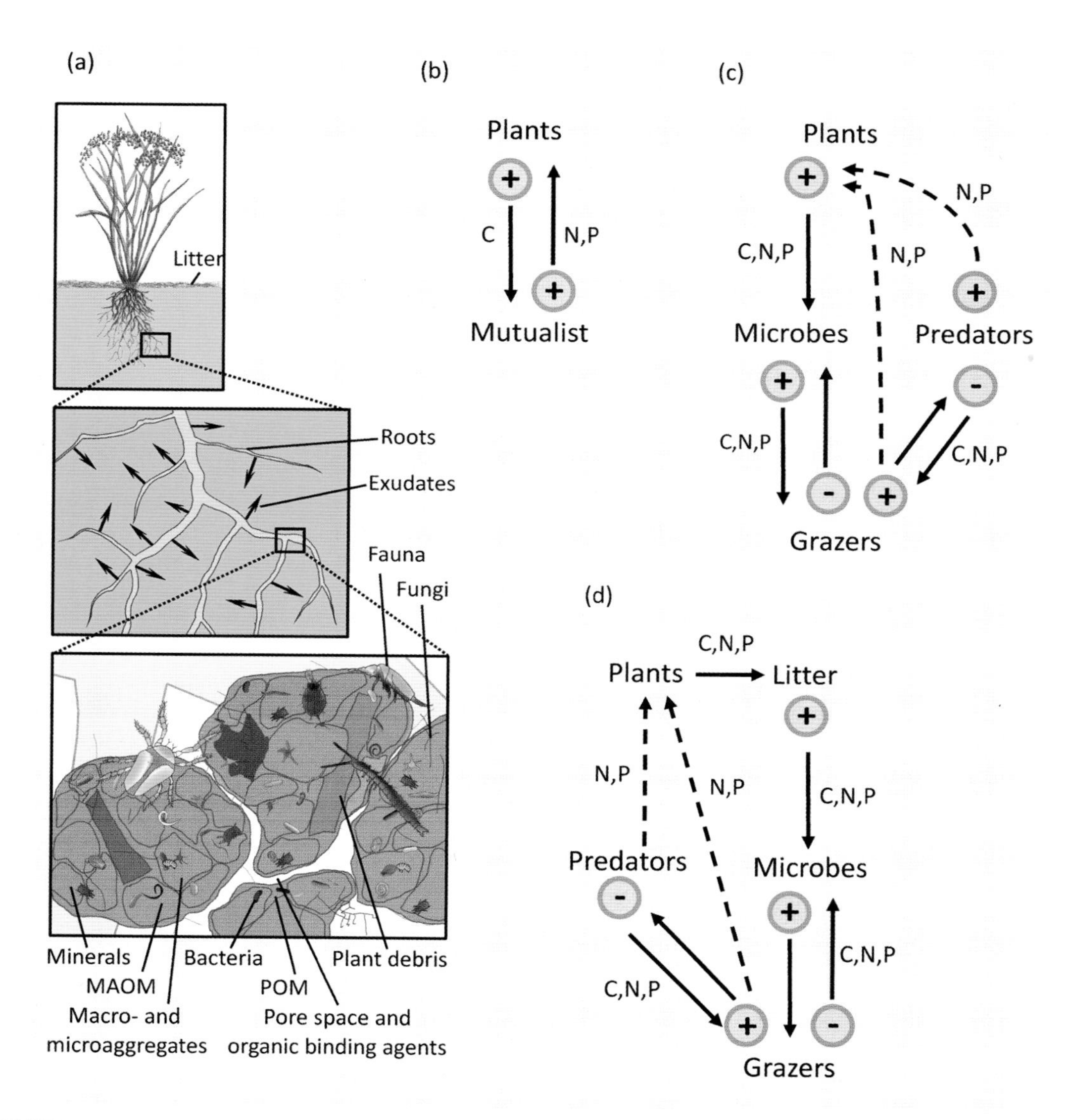

FIGURE 17.2 (a) Schematic of the multiscale nature of the interactions among plants and soil biota, highlighting the interactions with the rhizosphere. Wall and Moore (1999) characterized the interactions as forms of direct symbiotic mutualism, nonsymbiotic mutualisms, and indirect mutualisms that involve the exchange of C to soil microbes and biota, and N and P to plants. (b) Direct symbiotic mutualisms include mycorrhizae and N-fixing microbes. (c) Direct nonsymbiotic mutualisms include the release of C-rich exudates to soil, the utilization of the exudates by microbes, and the consumption of microbes by fauna that releases N for plant uptake. (d) Indirect mutualism that results from the utilization and consumption of plant litter by soil biota and the subsequent release of N for plant uptake. *Rhizosphere redrawn from Moore and R. Boone (unpublished).*

the quality (C:N, C:lignin) of the plant residues, and the biota that are present (Beare et al., 1992; Swift et al., 1979). Biotic community composition matters as microbes and invertebrates have adapted to utilize different substrates as food sources and appear to have coevolved with plants and the residues they produce. This is manifested in three distinct ways. First, several studies have documented patterns in the succession of microbes and invertebrates colonizing materials as decomposition proceeds. The type of litter determines the broad successional patterns when gauged in terms of broad taxonomic, functional groupings, and enzyme activity (Buresova et al., 2019). Early colonizers of leaf litter include fungi and their consumers, followed by bacteria and their consumers (Coleman et al., 1990; De Boer et al., 2005; Ingham et al., 1989; Purahong et al., 2016). Studies that excluded certain functional groups demonstrate lower rates of decomposition than those with the full complement of organisms present. Second, reciprocal transplants of plant litter types from one ecosystem to another reveal that litter tends to decompose faster in soil of its origin (home field) rather than soil outside of its origin (Gholz et al., 2000; Wall and Moore, 1999). This phenomenon of home-field advantage is ubiquitous across many ecosystem types but is most pronounced in forested systems where litter is complex and high in lignin and other recalcitrant materials (Ayres et al., 2009). Third, microbes affect decomposition and nutrient flow in a density-dependent manner by dint of the magnitude of active microbial biomass but also through adaptions in the nutrient use efficiency of existing populations and changes in the structure of the microbial community (Allison et al., 2010). Shifts in microbial community structure can change the quantity and quality of extracellular enzymes (Acosta-Martínez et al., 2003; Frey et al., 2004; Kourtev et al., 2002).

There is evidence that the dynamics of microbial communities are not entirely due to the additive responses of microbial species operating independently to environmental cues but rather represent consortia of species that respond to cues from one another in terms of their presence and resource use. Quorum, diffusion, and efficiency sensing and random awakenings are competing hypotheses proposed to explain the rapid growth of microbial populations and shifts in microbial community composition to other species, favorable changes in substrates, and environmental conditions (Epstein, 2009; Fuqua et al., 1994; Hense et al., 2007; Redfield, 2002). Each hypothesis is predicated on the conditions that microbes: (1) possess active and dormant states (Jones and Lennon, 2010; Lennon and Jones, 2011), (2) are sensitive to triggers that signal unfavorable and favorable conditions, and (3) produce autoinducer molecules whose fates either upregulate or downregulate gene expression and growth.

17.4.5 Soil formation

Different frameworks based on physical and chemical categorizations of organic materials and mineral factions have been developed to describe the dynamics of SOM. Tisdall and Oades (1982) presented a conceptual model for soil structure that was based on the association of SOM - categorized as mineral-associated organic material (MOAM) and particulate organic material (POM) derived from partially decomposed plant materials, roots, and their products (e.g., exudates and mucigels), and microbial necromass - with soil particles (i.e., sand silt and clay). At the finest scale, SOM interacts with the mineral soil and forms microaggregates (20–250 μm). Microaggregates interact with SOM and combine to form macroaggregates (250 μm to 2 mm). Elliott (1986) expanded this framework to include additional size classes of aggregates. Six et al. (2000, 2001, 2004) provided a dynamic framework for aggregate formation and degradation in natural and agricultural soils that included soil biota, establishing SOM as a measurable ecosystem property that is a structural component of soil and a resource of plant, microbial, and fauna origins utilized by soil biota (Coleman et al., 1983; Moore et al., 2003).

Contemporary science views SOM as a continuum of decomposing organic compounds of plant, microbial, and faunal origins (Cotrufo et al., 2013; Lehmann and Kleber, 2015; Ma et al., 2018) that can be binned into two distinct pools: POM and MAOM. Upward of half of SOM in soils originates from microbial necromass (Liang et al., 2019). A consensus has developed that recognizes that microbes, particularly fungi, produce stable SOM and are important drivers of SOM accumulation (Kallenbach et al., 2016). The basis for controls on long-term stability of SOM is still being debated (Cotrufo et al., 2013, 2015; Lavallee et al., 2019; Six and Paustian, 2014; von Lutzow et al., 2006).

17.4.6 Cryptobiotic crusts

Biological soil crusts (*aka* cryptobiotic, cryptogamic, microbiotic, or microphytic) are microbial consortia of different combinations of single-celled and filamentous bacteria, fungi, blue-green algae, green algae, lichens, and bryophytes that reside on the surface of soils, plants, and rocks. Exudates and necromass produced by these organisms serve as binding agents with soil particles to form a mat-like matrix that covers surfaces. Soil crusts structurally and functionally serve to stabilize soils and decrease wind and water erosion, increase surface temperatures by decreasing soil surface albedo, and influence soil water-holding relations while supporting a complex food web of soil microbes and invertebrates (Belnap et al., 2001). Crust communities can fix both atmospheric C via photosynthesis and N via biological nitrogen fixation (BNF). Biological crusts have a global distribution but are predominant in arid and semiarid regions, covering about 12% of the landmass with a C content of 3.0 to 8.2 Pg, representing about 1% of the 470 to 650 Pg C content estimates of terrestrial vegetation (Elbert et al., 2012). Global estimates of net C uptake by crusts from the atmosphere are 3.9 Pg C yr^{-1} (2.1−7.4 Pg C yr^{-1}), representing about 7% of the 56 C Pg yr^{-1} attributed to terrestrial vegetation. Estimate of BFN by crusts is 49 Tg N yr^{-1} (27−99 Tg N yr^{-1}), representing roughly 46% of the total global 107 Tg N yr^{-1} (100-90 Tg N yr^{-1}) (Elbert et al., 2012).

17.5 The organization of SMEB science concepts and information

An SMEB framework should be able to estimate foundational ecosystem processes and their influences on ecosystem services in ways that inform science and that can provide stakeholders with useable information to guide management and policy decisions (Barrios, 2007). The framework should be scalable to characterize soil communities and their contributions to ecosystem processes in ways that link presence and activity of soil biota to functional attributes (Cotrufo et al., 2013; Hall et al., 2018; Wieder et al., 2014). Global-scale biogeochemical models should include finer-scale soil biological processes linked to soils and land use to assess C capture and storage potential at local, regional, and global scales (Lavallee et al., 2019; Paustian et al., 2016). A gene to transcription to function approach would capture the dynamic nature of ecosystem function through the lens of the genetic make-up and activity of genes in soil microbial communities (Brussaard et al., 1997; Hall et al., 2018; Jochum and Eisenhauer, 2022). Ultimately this means possessing the ability to characterize which genes and traits are present and then measuring the dynamics of gene function in relation to ecosystem processes.

17.5.1 SMEB food web

An SMEB framework should integrate the structural, functional, and dynamic aspects of whole soil communities to address fundamental ecosystem processes and ecosystem services rather than those of specific taxa, processes, or biogeochemical pathways (Barrios, 2007; Brussaard et al., 1997; de Vries and Wallenstein, 2017; Hunt et al., 1987; Moore and de Ruiter, 2012; Moore 2021). The foundations of this approach are based on the seminal works by Odum (1969), May (1973), and Paine (1980) on the influences of the structure (diversity and architecture) of and energy flow within ecological communities on their stability and persistence.

The model of rhizosphere food web for the North American shortgrass steppe in Colorado presented by Hunt et al. (1987) provides a template for a SMEB food web (Fig. 17.3). The model was based on the three descriptions of food webs proposed by Paine (1980), offering the structural, functional, and

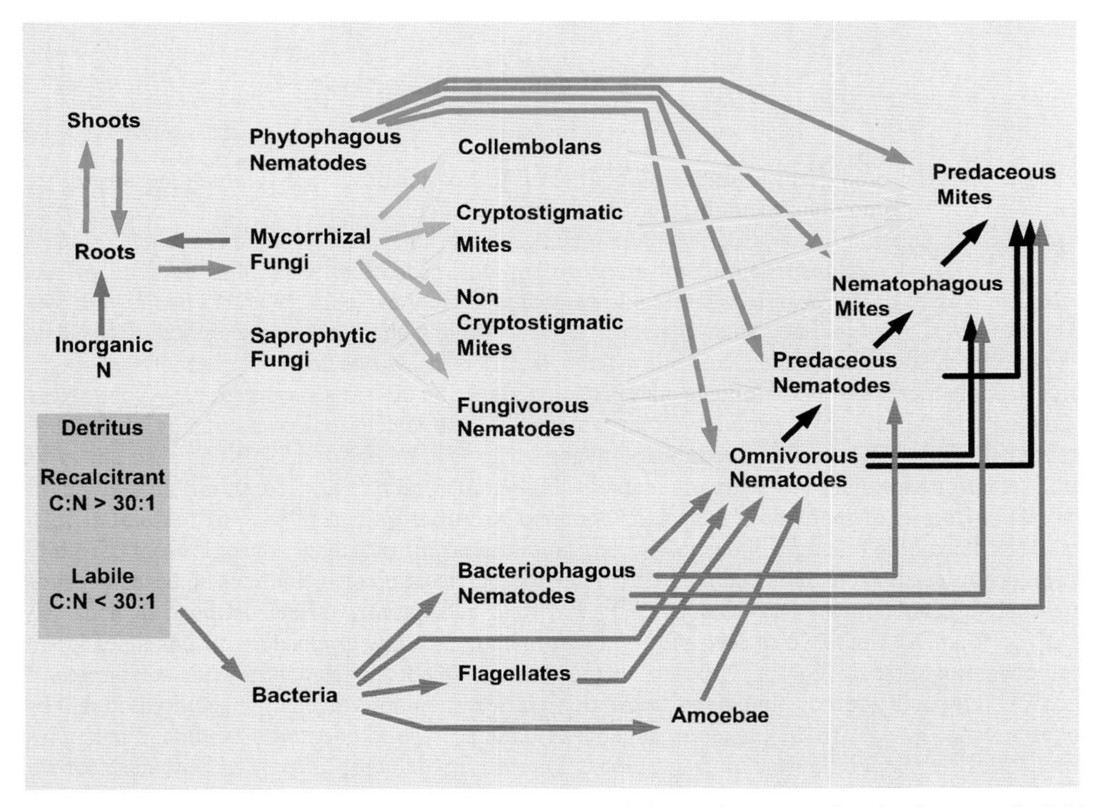

FIGURE 17.3 Template for an SMEB food web depicting the trophic interactions among functional groups of soil biota, plant (roots and shoots), detritus (labile and recalcitrant), and inorganic N within the rhizosphere of soils. The template is based on the rhizosphere food web description from the Shortgrass Steppe of Colorado. The arrows indicate the directional flow of C and N in a trophic interaction. All groups of soil biota contribute N to the inorganic N pool and respire C to the environment – C and N flows not represented for clarity. The colors represent flows/membership in the root (green), fungal (yellow), and bacterial (red) energy channels. Black flows indicate representation from all energy channels. Blue represents inorganic N. *(Adapted from Hunt et al., 1987.)*

dynamic aspects discussed above: (1) for structure, the connectedness web depicts the trophic interactions among organisms, (2) for function, the energy flow web represents the flow of nutrients among organisms and to the environment, and (3) for dynamics, the functional web depicts the influences of the dynamics of one group on another. This approach has been adopted by several research groups that have attempted to link the structure of soil food webs in relation to the decomposition of organic matter and the mineralization of nutrients (Andrén et al., 1990; Andrés et al., 2016; Brussaard et al., 1988, 1997; de Ruiter et al., 1993a,b, 1994; Hendrix et al., 1986; Hunt et al., 1987; Moore et al., 1988; Pressler et al., 2016; Schwarz et al., 2017). We provide a description of the approach in Fig. 17.4,

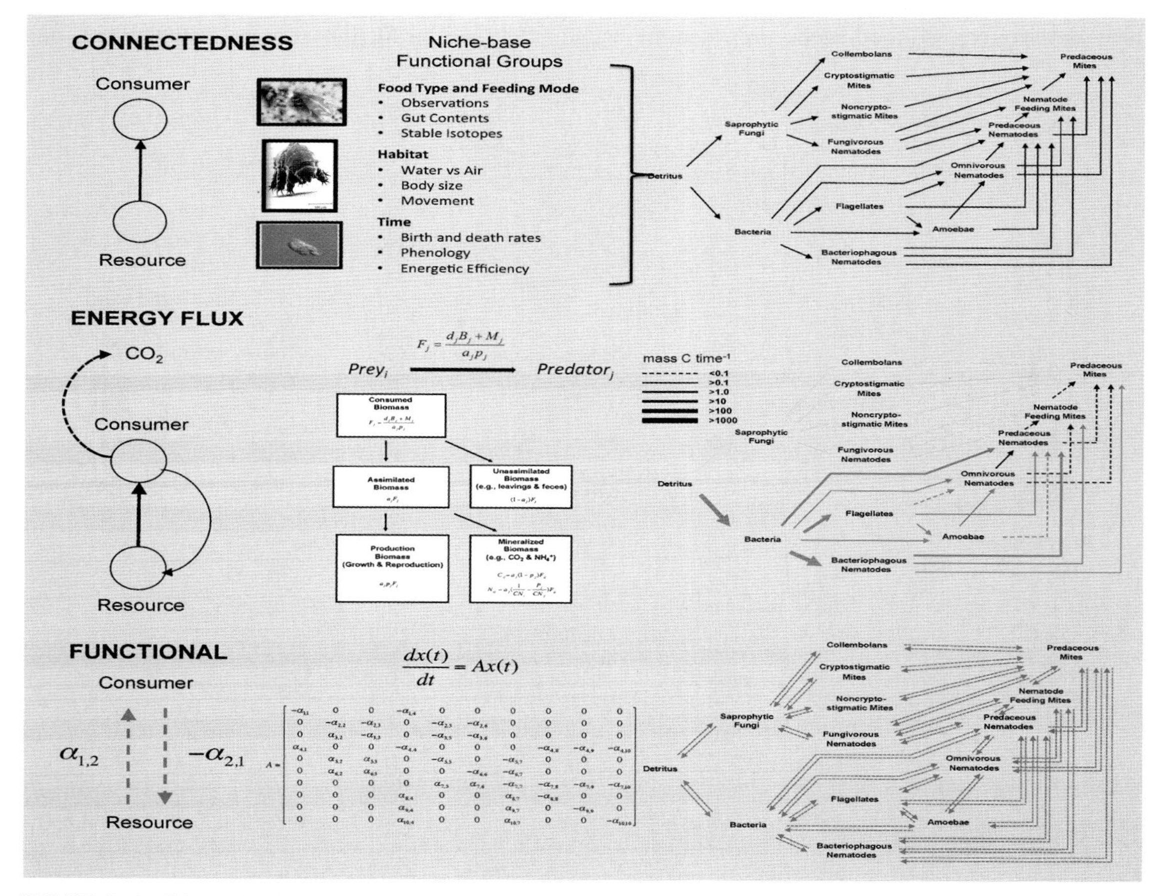

FIGURE 17.4 Schematic of the framework used to construct the Connectedness, Energy Flux, and Functional food webs showing the interrelationship between structure, function, and dynamics (Moore and de Ruiter, 2012). The webs include only the detritus portions for clarity here, but the synthesis will include plants and plant roots. The energy flux description highlights the fungal (yellow) and bacterial (red) energy channels. The elements of the Jacobian matrix A are estimated by equating the partial derivatives of the ODEs describing the dynamics of the functional groups ($\alpha_{i,j}$, $\alpha_{j,i}$) to the corresponding fluxes assuming steady-state *(de Ruiter et al. 1995; Moore et al., 1993).*

a presentation of patterns that the approach has revealed and areas of refinement that are being developed.

The connectedness web defines the model's basic structure. The description is based on functional groups of organisms that have similar resources, prey and predators, feeding modes, life history attributes and habitat preferences, different forms of detritus, and inorganic N. At the base of the web are plant roots, labile (C:N ratio $< 30:1$) and resistant (C:N ratio $> 30:1$) forms of detritus, and an inorganic N source. These basal resources are utilized by functional groups of microbes and invertebrates, terminating with predatory mites.

The energy flow web (see C and N flow) provides a functional description of the SMEB food web that provides estimates of nutrient dynamics derived from population sizes (biomass) and feeding rates. The estimates of flow are derived from estimates of population sizes, turnover rates, consumption rates, prey preferences, and energy conversion parameters (Table 17.2; see Hunt et al., 1987; de Ruiter et al., 1993b; O'Neill, 1969). Consumed matter is divided into a fraction that is immobilized by the consumer and one that is returned to the environment. From the immobilized fraction, material is either incorporated into new biomass (production) or is mineralized as inorganic material. The flow estimates start with the top predators under the assumption that the amount of consumed matter required to maintain the predator biomass must equal the sum of its steady-state biomass and loss due to death divided by its ecological efficiency:

$$F = (D_{\text{nat}} B + P)/e_{\text{ass}} e_{\text{prod}}, \qquad \text{(Eq. 17.1)}$$

where F is the feeding rate (biomass time^{-1}), D_{nat} is the specific death rate (time^{-1}) of the consumer, B (biomass) is the population size of the consumer, P is the death rate to predators (biomass time^{-1}), and e_{ass} and e_{prod} are the assimilation (%) and production (%) efficiencies, respectively. For predators that consume multiple functional groups, the total flux required to sustain the predator is parsed by its feeding preferences for each prey group. The estimation procedure moves from the top predators to the basal resources with fluxes to each prey, taking into account the estimated biomass lost to predation. A dynamic version accounts for changes in the biomasses of the organisms and resources over a specified time interval, i.e., add ΔB over the time interval to the numerator of Eq. 17.1. The death due to predation for the top predator is zero.

The functional web emphasizes the strengths of the interactions among the functional groups and the impacts of the interactions on the dynamic stability of the web. The process starts with the development series of differential equations that describe the dynamics of each functional group and basal resource from the connectedness description (May 1972). The Jacobian matrix is then constructed from the partial derivatives of the differential equations at or near equilibrium.

$$\alpha_{ij} = \left[\delta(\mathrm{d}X_i/\mathrm{dt})/\delta X_j\right], \qquad \text{(Eq. 17.2)}$$

where α_{ij} refers to the interaction strengths that designate the per capita (see biomass), effects of each functional group (X_i, X_j) and resource. Interaction strengths for the off-diagonal elements of the matrix can be derived from the equations used to model the population dynamics of the functional groups using their population densities, physiological parameters, and estimated feeding rates (Eq. 17.1) (de Ruiter et al., 1995; Moore et al., 1993; Moore and de Ruiter, 2012). The diagonal elements of the matrix cannot be derived from field data or estimates of energy fluxes but can be scaled to the specific death rates (de Ruiter et al., 1995) or levels tied to stability (Neutel et al., 2002).

TABLE 17.2 Physiological parameter values and population densities (g C m^{-2}) for the functional groups of the belowground food web of the Colorado shortgrass steppe as depicted in Fig. 17.3 (after Hunt et al., 1987 for soil biota and Lauenroth and Burke, 2008 for roots and detritus). The functional groups are loosely arranged by trophic level, with lower levels at the bottom and upper levels at the top. The food web processes a steep pyramid of biomass, with the majority dominated at the base of the web. The parameters and data are used as model input to estimate C and N fluxes among functional groups and resources, and C and N mineralized into the atmosphere and soil, respectively.

Functional group	C:N	Turnover rate (yr^{-1})	Assimilation efficiency %	Production efficiency %	Biomass (g C m^{-2} yr^{-1})
Predatory mites	8	1.84	60	35	1.60
Nematophagous mites	8	1.84	90	35	1.60
Predatory nematodes	10	1.60	50	37	4.30
Omnivorous nematodes	10	4.36	60	37	6.50
Fungivorous nematodes	10	1.92	38	37	4.10
Bacteriophagous nematodes	10	2.68	60	37	58.00
Collembola	8	1.84	50	35	4.64
Mycophagous prostigmata	8	1.84	50	35	13.60
Cryptostigmata	8	1.20	50	35	16.80
Amoebae	7	6.00	95	40	37.80
Flagellates	7	6.00	95	40	1.60
Phytophagous nematodes	10	1.08	25	37	29.00
AM-mycorrhizal fungi	10	1.20	100	30	70.00
Saprobic fungi	10	2.00	100	30	630.00
Bacteria	4	1.20	100	30	3040.00
Detritus	10	0.00	100	100	177.00
Roots	10	1.00	100	100	1149.00

17.5.2 Patterns within SMEB food webs

The SMEB food web possesses three sets of patterns that involve elements of their structure, function, and dynamics that are important to its stability. Understanding these patterns and how changes in them affect ecosystem processes and services should enhance their usefulness as tools for monitoring, management, and policymaking.

The first set of patterns deals with the relationship between number of basal resources and functional groups and complexity of trophic interactions within the food web. Theory asserts that as the diversity and complexity of systems increase, the likelihood of the system's stability decreases (Gardner and Ashby, 1970; May 1972), leading to an inverse relationship between diversity and complexity among stable systems. Theory also asserts that compartmentalized or "blocked" arrangements of interactions within a system enhance stability. Empirical evidence supports these assertions (see Fig. 17.3) as the SMEB food web possesses distinct pathways of C and N flows — aka energy channels — resulting in a compartmentalized structure and falls within the predicted relationship between diversity and complexity within the compartments when compared to other food web descriptions (Moore and Hunt, 1988).

A second set of patterns emerges from the arrangement of the estimates of biomass and nutrient flux within the web by trophic position (Table 17.2). The SMEB food web forms pyramids of biomass and nutrient fluxes due in large part to the dominance of microbes at the base of the web. Bacteria and fungi account for upward of 95% of living biomass, protozoa, and invertebrates, accounting for 5% at the upper trophic positions (de Ruiter et al., 1993a; Hunt et al., 1987). The pyramidal structures have shown to be more stable than alternatives, wherein higher levels of biomass occur at upper trophic levels (Moore and de Ruiter, 2000; Neutel et al., 2002). Alterations to the climate or in plant-limiting nutrients (fertilization, movement of reactive N) can lead to increased productivity, adding trophic positions or increases in biomass at higher trophic positions, both of which can induce instability (Moore et al., 1993; Oksanen et al., 1981; Rosenzweig, 1971).

The nutrient fluxes applied to the trophic interactions among functional groups highlight the distinct subassemblages, or energy channels referenced above, within the web. The SMEB web includes a root-energy channel based on living plant roots and detritus-energy channel based on nonliving materials, which include plant litter, root exudates, and corpses of microbes and consumers (Moore and Hunt, 1988). The detritus pathway can be subdivided into a "fast cycling" bacterial energy channel that is comprised of relatively low C:N labile detritus, bacteria and their consumers, and a "slow cycling" fungal pathway comprised of high C:N recalcitrant detritus, fungi, and their consumers (Fig. 17.5).

The linkages among energy channels tend to be weak at the lower trophic levels occupied by roots, bacteria, and fungi and strongest at the upper trophic levels occupied by predators. The dominance of the root, bacterial, and fungal energy channels within a system varies by ecosystem type and changes within a growing season and is affected by disturbance and nutrient turnover rates. For example, the fungal energy channel tends to be dominant in forest and shrubland systems that possess detritus with high C:N ratios, while the bacterial channel is more important in grassland and sedge systems that possess detritus with narrow ratios.

The final set of patterns deals with the strengths of the trophic interactions with trophic position (Fig. 17.6). Recall that there are two strengths for each trophic interaction — the effect of the prey at the lower trophic position on the dynamics of the predator at the trophic position — a positive "bottom-up" effect — and the effect of the predator at the upper trophic position on the dynamics of the prey at the lower trophic position — a negative "top-down" effect. de Ruiter et al. (1995) demonstrated that soil food webs exhibit strong top-down effects relative to bottom-up effects at the lower trophic position that switches to the strong bottom-up effects relative to top-down effects at upper trophic positions. This asymmetry is aligned with the pyramid of nutrient flows and biomass that were introduced above (Fig. 17.6).

17.5.3 Moving forward

Taken together, these patterns lead to the general conclusion that the energetic organization of soil communities (arguably ecological communities, see Rooney et al., 2006) defines their function and forms

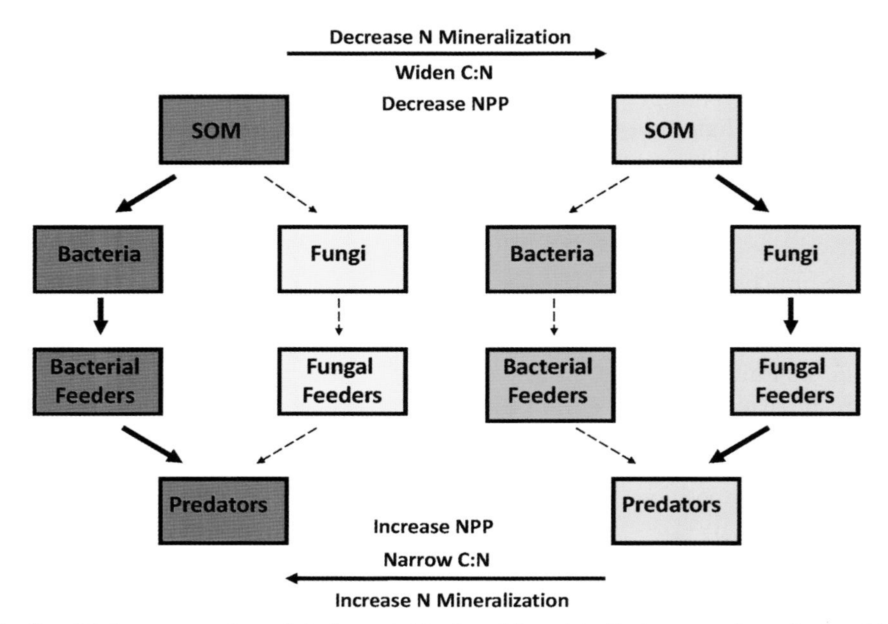

FIGURE 17.5 Simplified representations of the bacterial (red) and fungal (yellow) energy channels of a rhizosphere food web presented in Fig. 17.3. The bacterial energy channel represents the "fast cycle" while the fungal energy channel represents the "slow cycle" due to the higher turnover rates of bacteria and their consumers relative to those of fungi and their consumers. The energy channels coexist in soils, with the relative dominance of one channel to the other affected by the C:N ratio of the detritus and NPP of the plant communities. Changes in the rates of C and N mineralization have been associated with shifts in the channels. *(Adapted from Doles, 2000 and Moore et al., 2003.)*

the basis of their stability and resilience. There is an inextricable link between the distribution of biomass, the pattern of nutrient flow, the patterning of interaction strength affect, and ultimately stability. A "let the soil work for us" philosophy would start by recognizing not only the contributions of individual groups of soil biota to ecosystem processes and services but also include the relationships and patterns discussed above to shape management, mitigation, adaptation practices, and policy. For this to happen, improvements are needed to better align the models with ecosystem processes and services (Treseder et al., 2012).

A movement in the direction of characterizing communities in ways that link presence and activity to function has developed (Hall et al., 2018; Wieder et al., 2014). Global-scale biogeochemical models are now being reimagined in ways that include biological processes that operate in soil at finer scales, linked to soils and land-use to assess C capture and storage potential at local, regional, and global scales (Lavallee et al., 2019; Paustian et al., 2016). Traditional models have focused on the fate of plant-derived substrates and microbial and microbial-faunal interactions at various scales (Moore and de Ruiter, 2012; Parton et al., 1994; Schimel and Weintraub, 2003). Ecosystem-scale biogeochemical models (Parton et al., 1994; Sulman et al., 2014, Tang and Riley, 2015; Wang et al., 2015, Wieder et al., 2014) often include detailed pools of organic material of different types and quality yet lack detailed information about microbes, representing them implicitly within rate constants or explicitly in simplified, uniform pools. Soil food web models operate on smaller scales, possess copious details on microbivores and

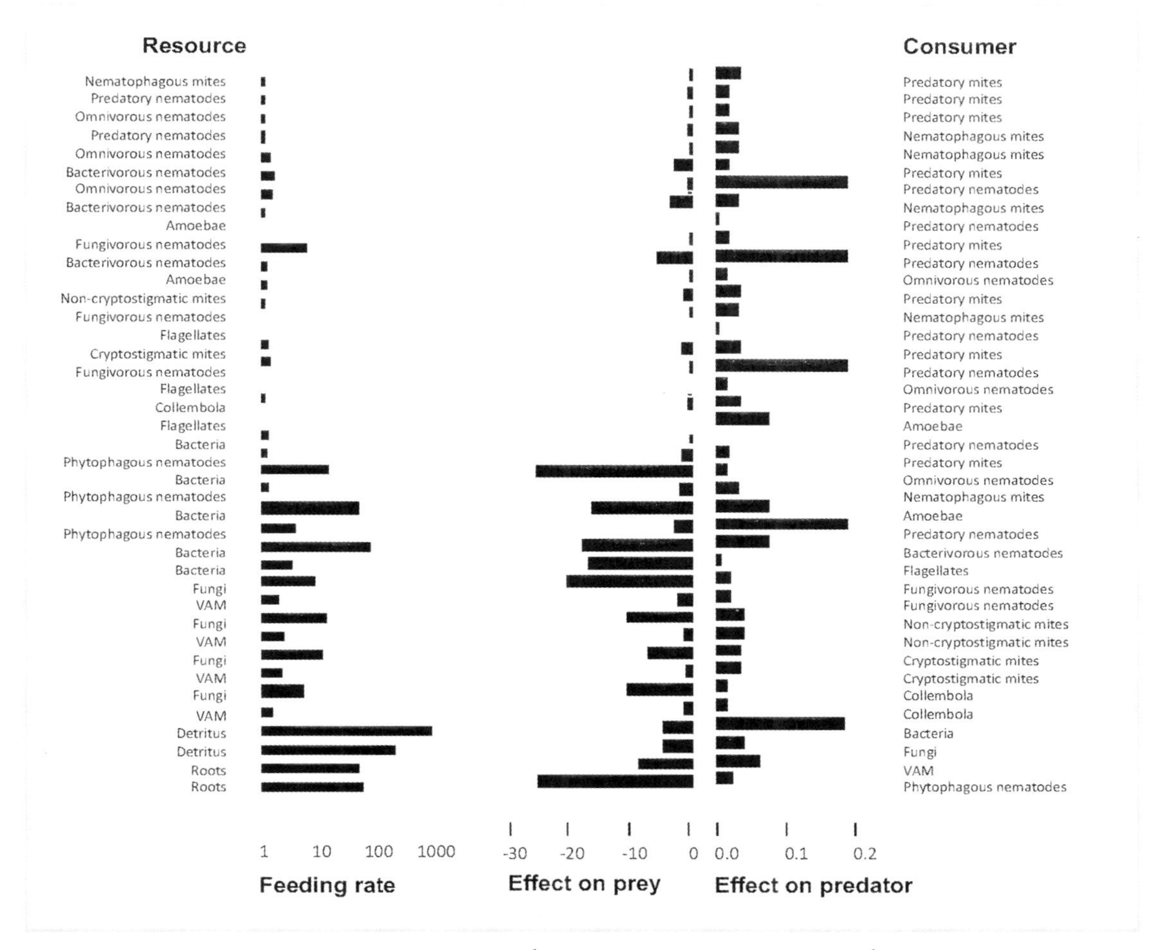

FIGURE 17.6 Estimates of feeding rate (kg C ha yr^{-1}) and the interaction strengths (yr^{-1}) by trophic position for the rhizosphere food web from the Shortgrass Steppe in Colorado (de Ruiter et al., 1995). The bars correspond to the feeding rates and strengths for pair-wise trophic interactions in a food web arranged from resources (*left*) to consumers (*right*), starting with interactions at the base of the food web (basal resources) to the top of the food web (top predators). The estimates of the feeding rates were obtained using Eq. 17.2. The elements of the interaction strengths include the impact of a consumer on a resource (predator on prey) and the impact of the resource on the consumer (prey on predator). The asymmetry in the patterning of interaction strengths (yr^{-1}) was observed for food webs from the Shortgrass Steppe in Colorado (presented here), and agricultural sites in the Netherlands, Sweden, and Georgia, USA *(de Ruiter et al., 1995).*

higher-order consumers, and often lack sufficient details about basal resources (i.e., roots and organic matter pools) and microbes (i.e., bacteria and fungi). These deficits in structural detail have prompted calls for greater resolution of basal resources, functional groups of microbes, and trophic interactions (Bradford, 2016; de Vries and Wallenstein, 2017; Grandy et al., 2016; Kallenbach et al., 2016; Liang et al., 2019; Soong and Neilson, 2016) given the major role of soil microbiota and fauna in SOM formation and nutrient cycling. Many have called for reflecting differences in microbial and faunal traits in biogeochemical models as the best path forward (de Long et al., 2018; Malik et al., 2020).

The SMEB framework provides middle-scale granularity for a gene to function–based approach with some refinements. At present, the descriptions separate microbes into crude functional groups – saprophytic fungi, mycorrhizal fungi, and bacteria – but not at the resolution envisioned by microbial ecologists (Hall et al., 2018). Modeling-based research has included microbes and the exoenzymes they produce to study nutrient dynamics and decomposition (Allison et al., 2010; Folse and Allison, 2012; Moore et al., 2014; Moorhead et al., 2012; Resat et al., 2012; Schimel and Weintraub, 2003); however, a consensus has yet to be reached on which microbes and enzymes to include. Development of a core microbiome defined by microbial functional groups and inclusion of exoenzymes into the mix could also be of benefit (de Vries and Wallenstein, 2017; Shade et al., 2012). Research on soil formation has identified the need to include measurable forms of soil organic fractions such as POM tied to plant residues, microbes, and invertebrates (Cotrufo et al., 2015; Lavallee et al., 2019), as well as the more resistant, mineral-associated materials that can be thousands of years old.

17.6 Anthropogenic impacts on the SMEB food web

Human activity has directly affected over 70% of land surfaces, with agricultural lands representing nearly 40% of the Earth's ice-free land surface (Foley et al., 2011; IPCC, 2018). With the exclusion of lands converted to hardscapes, conversion of native landscapes for human use structurally and chemically affects soils in ways that alter SOM and nutrient stores and dynamics (Don et al., 2011; Smith et al., 2015; Wei et al., 2014). Upper soil horizons are homogenized by the removal of surface vegetation and the incorporation of the remaining plant residue, surface litter, and roots into the soil in the preparation of planting. The introduction of crop species and accompanying management practices (e.g., liming, biochar, and fertilizer applications) to optimize plant growth can result in changes in pH, macro- and micronutrients, and hydrology. Land conversion leads to major above- and belowground C loss into the atmosphere on the order of 30–50% until a new steady-state is reached (Burke et al., 1989; Don et al., 2011; Paustian et al., 2016; Wei et al., 2014).

Following land conversion, natural seasonal fluctuations remain but with variations of impact based on vegetation type. For example, seasonal signatures of surface albedo and hydrology are different for native forests and grasslands, managed pastures, orchards, or croplands. Land conversions for agricultural use also change the disturbance regime of the system. Agricultural systems receive frequent disturbances that vary in intensity depending on the cropping system and management practices. Tillage, fertilizer, and pesticide applications; irrigation; and crop removal affect soil communities to varying degrees.

Conversion of native land to agriculture for annual grain and row crop production and the ways in which grazing affects soils and soil biota with the frameworks for SMEB science will be further discussed in this chapter. In doing so, some common responses that soil and soil communities undergo with land conversion will be highlighted, focusing on agricultural practices that underpin the management decisions to meet SDGs.

17.6.1 Agricultural management practices

Annual row crop and grain production relies on the frequent tilling of soils to prepare seed beds for planting and to control weeds. Tillage changes the soil structure and chemical composition, the community structure of the soil biota, and the ensuing biogeochemical pathways. For example, tillage practices have led to a soil C loss of 133 Pg across the top 2 m of soil (Sanderman et al., 2017). The production and use of inorganic fertilizers and the release of reactive forms of N exceed 121 Tg N into other ecosystems (Galloway et al., 2003, 2004), well above ambient levels in nonagricultural systems and preindustrial agriculture. These alterations to soils and biochemical cycles are linked to changes in the structure and function of the SMEB food web.

Cultivation directly kills soil biota or indirectly affects their abundances and activities through the changes it causes to the soil environment (see Wardle, 1995). The extent of the effects depends on the intensity and frequency of tillage and on the morphology, physiologies, and life histories of the soil biota. The sizes of soil biota span several orders of magnitude from submicrons to microns for bacteria and fungi (microbiota) to microns to millimeters for tardigrades, nematodes, mites (mesobiota), and millimeters to centimeters for larger mites, collembola, macroarthropods, and annelids (macrobiota). Surface active macrobiota, particularly annelids, microarthropods, and macroarthropods, are apt to be killed directly by tillage, while smaller mesobiota and microbiota typically survive (Wardle, 1995).

The indirect effects of tillage on soil biota are more pervasive than direct effects through the changes in soil structure, moisture, and the distribution of organic matter. Tillage incorporates surface residues, altering the surface albedo, habitable pore space within soils, moisture retention, and soil moisture and temperature. Bacteria, protozoa, and micro invertebrate taxa (rotifers, tardigrades, and nematodes) are essentially aquatic organisms residing in soil water and water films, while fungi, annelids, and arthropods reside on water films and within the high humidity of air-filled pore spaces. Soil microbes and micro-invertebrates possess cryptobiotic states wherein biota form spores, cysts, or inactive forms when conditions are unfavorable (e.g., extreme cold or heat, or drought conditions), while annelids and arthropods can migrate to more favorable environs.

Cultivation such as plowing physically alters soil structure by disrupting soil aggregates exposing occluded forms of POM and MOAM and incorporates surface residues into the soils, effectively homogenizing the upper 20 to 30 cm soil profiles via mixing organic materials. The incorporation of surface litter into soil through tillage effectively alters the habitat of litter microbiota and fauna while turning the soil leads to wind and water erosion and desiccation. Several studies have associated these changes with declines in SOM, both in terms of declines in POM and MAOM (Camberdella and Elliott, 1992; Grandy and Robertson, 2007; Tiessen and Stuart, 1983). Much of this is driven by the enhanced heterotrophic activity of soil biota that is initiated by changes in the availability of stored SOM.

Nitrogen additions to native and agricultural ecosystems alter the C:N of the system writ large and of plants and the litter they produce. These changes in turn alter the structure and dynamics of the SMEB food web, flux rates within the N cycle, decomposition of plant materials, and the formation and loss of SOM. The magnitudes and direction of these system responses are dependent on the N limitations of the system (Aber et al., 1998; Galloway et al., 2003). For example, a meta-analysis of N additions and plant litter decomposition by Knorr et al. (2005) found decomposition to be stimulated in sites with low ambient N-deposition rates and high-quality litter (low lignin:N) and inhibited decomposition in sites with high N addition rates or ambient N deposition rates and low-quality litter (high lignin:N). Similarly, meta-analyses on the impacts of nutrient limitations on mycorrhizal associations align well with the framework and draw a similar conclusion (Hoeksema et al., 2010; Johnson et al., 2006; Treseder, 2004). As mycorrhizal benefits were greater in N-limited plants, N additions reduced mycorrhizal benefits to plants and mycorrhizal biomass.

Long-term fertilization studies have provided some insight into the relationships among nutrient availability, soil community structure, and the SOM content of soils. Mack et al. (2004) found significant increases in annual net primary productivity (ANPP), shifts in Arctic tundra plant communities, and declines in SOM content of Arctic peat and soils after 20 years of annual additions of N. Aboveground ANPP increased by 1,500 g m^{-2} and the communities transitioned from graminoid sedge community dominated by *Eriophorum vaginatum* to a shrub tundra community dominated by *Betula nana*. Belowground, the system lost nearly 2000 g C m^{-2}, yet no significant loss of N over the 20 years, attributed in large part to an increase in the decomposition mediated by soil biota. In a comparison of grassland soils in the US Central Great Plains after 3 to 5 years of annual additions of N, P, and K, Riggs et al. (2015) highlighted the complexity of how added nutrients interact with different plant types, SOM fractions, and soil texture. Nitrogen addition increased the infection rates of AM fungi, which tended to decrease overall microbial respiration of unoccluded OM, and stimulated aggregate formation, suggesting that N enrichments could lead to increased sequestration of soil C. The increase in AM fungal infection of roots led to speculation that greater hyphal biomass, exudates, and glomalin production associated with the fungi (Rillig, 2004) served as physical and chemical mechanisms for macroaggregate formation.

These and other studies reveal some general trends about the differential effects of agricultural practices on soil biota, biochemistry, and soils when viewed through the SMEB framework (Andrén et al., 1990; Beare et al., 1993; Didden et al., 1994; Frey et al., 1999; Hendrix et al., 1986; Holland and Coleman, 1987; Moore and de Ruiter, 1991). The first observation summarizes the impacts of practices on different taxa or segments of the SMEB food webs in that the diversity of soil communities in native systems tends to be higher than the diversity in the agricultural systems (Brinkmann et al., 2019; Jangid et al., 2008; Moore, 1994; Rodrigues et al., 2013; Wardle, 1995). The decline in the diversity from microbes to macrobiota within the agricultural systems is most striking in practices that rely on intensive and frequent tillage, inorganic fertilizer applications, and pesticide usage than in less intensive, more conservation-oriented practices.

The second observation is that agricultural practices alter the structure, function, and dynamics of the SMEB food web in ways that affect SOM and nutrient formation and depletion (Beare et al., 1997; Drijber et al., 2000; Fraser et al., 1988; Frey et al., 1999; Wardle, 1995). Agroecosystems tended to lose SOM, while native systems tend to accrue SOM or maintain SOM at a near-term steady state (Tiefenbacher et al., 2021). Declines in SOM are coincident changes in the aggregate structure of the soils (Beare et al., 1994; Camberdella and Elliott, 1992) and declines in soil fungi (Frey et al., 1999). It is further observed that agricultural management practices that used intensive tillage and commercial fertilizers stimulated activity within the fast-cycling bacterial energy channel relative to that of the slower-cycling fungal and root energy channels. Less intensive management practices that relied on no-till or other reduced-tillage techniques and organic fertilizers and amendments enhanced the slow-cycling fungal energy channel relative to the fast-cycling bacterial energy channel (Beare, 1997; Hendrix et al., 1986; Moore and de Ruiter, 1991; Wardle, 1995).

The final observation addresses the impacts of agricultural practices on the spatial arrangement of the SMEB food web within soils and its activity through time. Agricultural cropping system practices tend to support monocultures of plants, remove plant litter from the soil surface, or integrate it into the soil; for tilled systems, they homogenize the upper soil profiles. The spatial separation between activity within the rhizosphere and surface litter disappears. The seasonal dynamics of activity within and among the energy channels through time and the nutrients they generate become asynchronous with plant growth, particularly within fallow fields. These conditions and practices disrupt the spatial and temporal compartmentalization of soil communities with increased intensity and frequency of tillage and fertilization, leading to an increase in mineralization of SOM and nutrient loss.

We can also align the SMEB food web (Fig. 17.3) and the conceptual representation of the simplified version of the SMEB food with coupled fungal and bacterial energy channels (Fig. 17.4) with the stages outlined by Galloway et al. (2003) in terms of changes in the C:N ratio of organic material and N availability in soils (Fig. 17.7). The summary is very process oriented yet focuses on methods that are mediated to soil biota (e.g., mineralization, nitrification, and denitrification) and is linked to C and C-oriented processes (i.e., decomposition and C mineralization). Variations of this summary address the impacts of the placement of plant residues, organic matter, and how the availability of limiting nutrients interact in ways that affect organisms and the processes they engage in. For soil food webs operating within stages 0 to 2 discussed above, those dominated by high C:N ratio organic materials with low levels of inorganic N tend to be dominated by the "slow-cycling" fungal energy channel with low rates of N mineralization, leaching, and efflux. Systems with low C:N ratio organic materials and high levels of inorganic N tend to be dominated by the "fast-cycling" bacterial energy channel with high rates of N mineralization, leaching, and efflux. Referring back to the discussion of how tillage affects soils, soil processes, and adding fertilizers to the mix, repeated tillage disrupts aggregates exposing SOM as

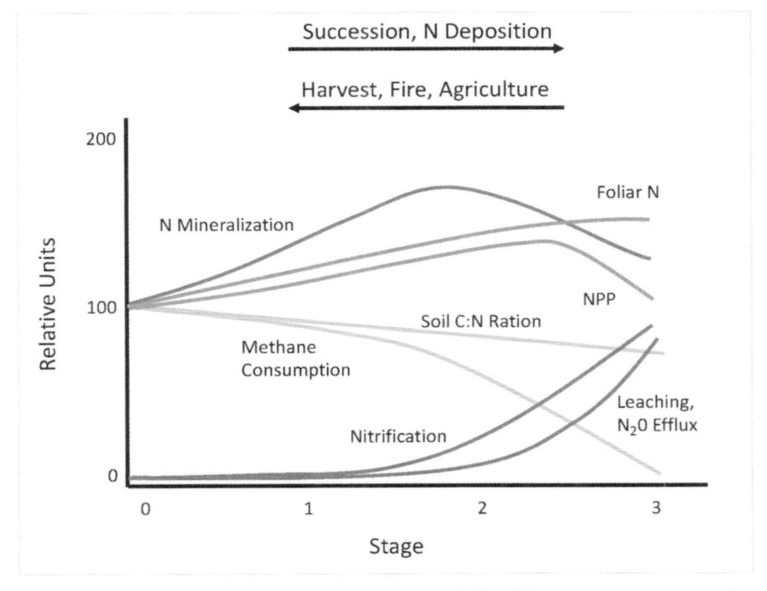

FIGURE 17.7 Summary of biogeochemical responses (C, plants, and N) of forest systems to reactive N additions that are mediated by soil biota. *Stage 0:* The system is N limited, the C:N of plant leaves and roots is high, and the interactions between plants and soil biota, particularly symbiotic interactions, are efficient with regards to N with little leaching or efflux. *Stage 1:* Added N results in increased plant productivity, higher N content in leaves and roots, and increased mineralization and flow rates. *Stage 2:* The system is N limited, N mineralization exceeds immobilization and net nitrification, and low but increased levels of leaching and efflux. *Stage 3:* The system is no longer N limited, symbioses decrease, many biotic interactions in soil are inhibited, and net nitrification increases, with high levels of N leaching and efflux from the system. *(Redrawn from Galloway et al., 2003; adapted from Aber et al., 1998.)*

unoccluded forms, fertilizers release N-limitation, and the fast-cycling bacterial channel responds resulting in net SOM loss.

17.6.2 Grazing

Grasslands and shrublands support diverse populations of indigenous flora and grazers which have co-evolved with one another. The impacts of introduced livestock grazing on soil communities and processes of grasslands and shrublands have resulted in losses of topsoil through wind and water erosion, loss in soil SOM, and changes in plant communities. The magnitude of these impacts depends largely on the grazing intensity, grazing history, and the geomorphology of the ecosystem in question (Andrés et al., 2017; Burke et al., 1999). A suite of studies from deserts, semiarid shrubs and grasslands, and the Arctic provide insights into the effects of grazing on soil communities.

Studies from desert and semiarid regions to the Arctic, regions that are characterized by low productivity, have shown that modifications in grazing can alter production and induce shifts or regime changes in the plant communities that impact soil biota. Soil biota in arid and semiarid ecosystems with high incidents of cryptobiotic crusts are extremely vulnerable to overgrazing by livestock (Belnap et al., 2001; Johansen, 1993). Gibbens et al. (2005) documented the dramatic shift in the plant community composition with an approximate 30% decline in grass cover and 30% increase in shrub cover within the Jornada Basin of the Chihuahuan Desert in the southwestern region of the US, following decades of cattle grazing. Trampling by native and introduced ungulates destroyed biological crusts, compacted the soils, leading to higher bulk density, decreased the rate of water infiltration, and increased wind and water erosion.

Gough et al. (2012) used exclosures to study the impacts of ungulate grazing on the plant and soil communities within the Arctic tundra of the north slope of Alaska. The landscape is dominated by moist acidic tussock (MAT) tundra communities comprised of sedges, mosses, and shrubs, and the lesser productive dry heath (DH) tundra communities dominated by lichens, diminutive evergreen shrubs, and mosses. The ANPP of the MAT communities was stimulated by grazing with little effect on the overall plant community structure but with an increased predator biomass in the soil food web. The exclusion of ungulates from the DH community induced a shift in the plant community from one dominated by lichens and mosses to one dominated by a palatable grass. With the shift in the DH plant community, the soil food web showed increases in phytophagous nematodes (root energy channel) and enchytraeids (detritus/fungal energy channel).

Site-specific factors include soil texture, plants, and legacy of grazing. For example, the shortgrass steppe of Colorado evolved with large ungulate herbivores, the most spectacular being bison. Following the extirpation of bison and the introduction of cattle, plant communities, and aboveground biomass showed effects locally, with minor impacts on soil biota, soil processes, and SOM (Lauenroth and Burke, 2008). Milchunas et al. (1998) compared native grassland plant communities with and without grazing exclosures. Their study found that grazing decreased above- and belowground plant biomass but did not induce marked shifts in plant species composition seen in more arid regions. Burke et al. (1999) report that topography, variation in microsites (resource islands), plant species, and grazing accounted for 70% to 90% of the variation in spatial distribution of the SOM in the topsoil. A large fraction of the variation in SOM was attributed to topography (recalcitrant SOM pools) and microsite (mineralizable SOM and POM), with plant species and grazing accounting for about 2%. Studies on grazing impacts on soil biota found changes in the densities of biota and restructuring within the energetic configuration of the soil food

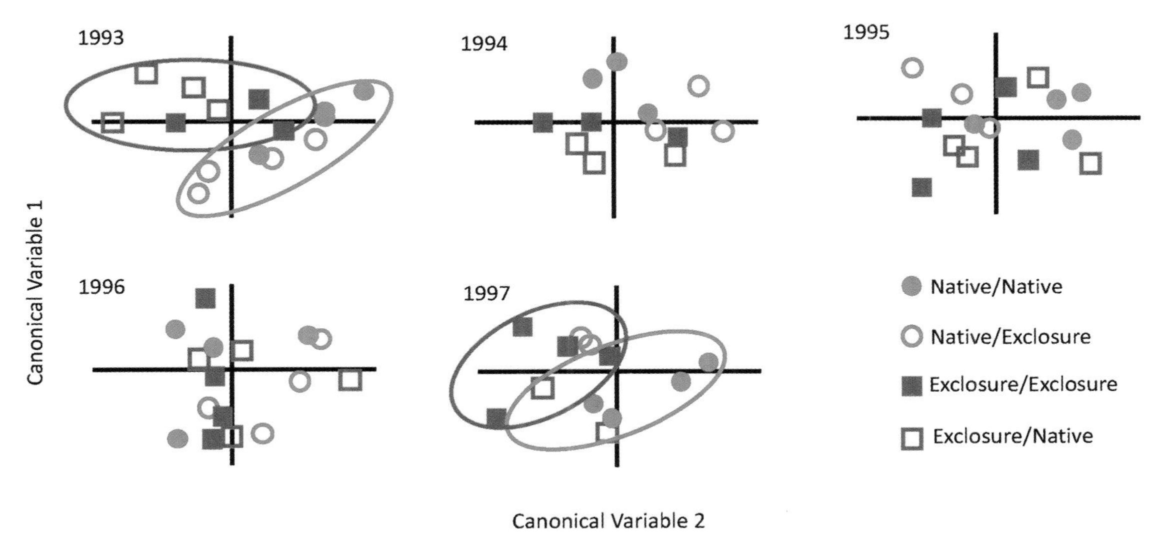

FIGURE 17.8 A case study on the impacts of grazing on the structure and energetic organization of the SMEB food web. The study manipulated the fencing used in a comparative study established by the USDA in 1938 during the Dust Bowl era to study the impact of cattle grazing on prairie plant communities and soils. The USDA study included a series of paired open native (N = native) shortgrass steppe and fenced regions to exclude the cattle (E = grazing exclosure). In 1993 portions of the areas fenced in 1938 were reopened to cattle and portions of the open native prairie were fenced to exclude cattle, creating four treatments: 1938 Native/1993 Native (N/N); 1938 Exclosure/1993 Exclosure (E/E); 1938 Native/1993 Exclosure (N/E); and 1938 Exclosure/1993 Native (E/N). *(Adapted from Moore et al., 2008.)*

web (Fig. 17.8). The soil communities of native systems grazed by both native and introduced ungulates were dominated by bacteria and their consumers (bacterial energy channel). The soil communities in systems where ungulates were excluded exhibited a marked shift toward fungi and their consumers (fungal energy channel).

17.6.3 Fire

Wildfires are integral to ecosystem development and function (Sugihara et al., 2006). Fire is also used as a prescribed management tactic in grasslands to increase forage quality, in forest and shrublands to clear understories or remove dominant vegetation for replanting, or for changes in land use. Changes in climate, when coupled with changes in land use and historical management practices to suppress fires, have altered fire regimes to include more frequent and intensive events.

Fire changes plant community structure and dynamics, surface vegetation, litter, albedo, soil water, soil organisms, and soil nutrient storage and dynamics (Hart et al., 2005). The impact of fire on soils, soil biota, and processes is varied but can be profound. In the extreme Mack et al. (2011) cataloged that the 2007 Anaktuvuk River fire on Alaska's North Slope vaporized 1039 km^2 of tundra vegetation and surface organic layer, resulting in a loss of about 2.1 Tg C ($\sim$ 2000 g C m^{-2}) into the atmosphere; equivalent to the annual net C sink for the entire Arctic tundra biome. In many cases the impacts of fire are locally intense but not catastrophic, with the systems exhibiting a high degree of both resistance and resilience as a response.

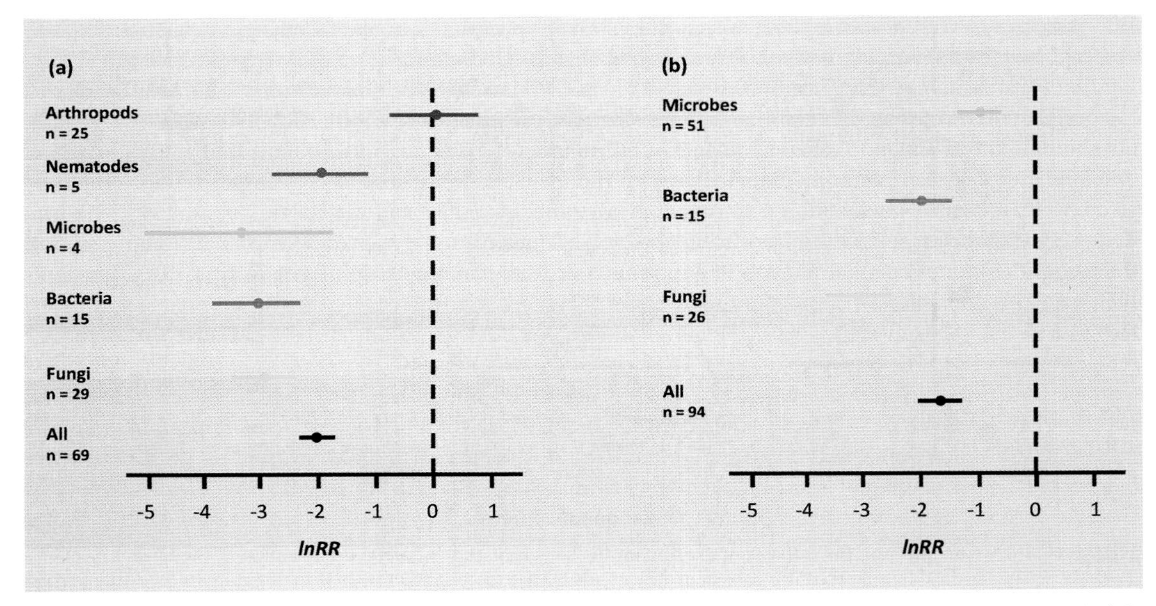

FIGURE 17.9 Fire impacts on the (a) biomass and (b) abundance soil biota expressed as effect size (*lnRR*) from a meta-analysis by Pressler et al. (2019). The effect size, *lnRR*, is the difference in the natural logs of the means for the burned and nonburn sites, and *n* is the number of studies included in the comparison. *(Redrawn from Pressler et al., 2019.)*

Several studies have addressed the impacts of fire on the linkages between above- and belowground communities (Hart et al., 2005; Neary et al., 1999; Pressler et al., 2019). In a comprehensive meta-analysis Pressler et al. (2019) summarized 1634 observations from 131 studies on the impact of fire on soil biota in terms of the biomass, abundance, and biodiversity of the dominant taxonomic groupings residing in the litter layer and topsoil (Fig. 17.9). Microbial populations were adversely affected by fire, with losses up to 96% of their biomass post fire. Fungi were more sensitive to fire than bacteria, which was directly attributed to differences in the thermal effect on these organisms and indirectly to losses of plant symbionts and surface litter. Surprisingly, mesofauna were less sensitive to fire than microbial communities. Given the nature of the reported losses of arthropods and nematodes, and their positioning within the SMEB food web (Fig. 17.9), it appears that the fungal energy channel is more sensitive to fires than the bacterial energy channel.

17.7 Applications of SMEB science

There are two strategies that SMEB science applications can adopt with major ramifications for meeting the UN SDGs: (1) optimize agricultural production in ways that minimize nutrient losses and (2) tune the management of natural and agricultural terrestrial ecosystems to increase nutrient levels (e.g., C sequestration, N retention) in soils. Achieving these goals will require a hybrid of mitigative and adaptive management approaches to readdress declines in soil health and requirements for increasing production (Pacala and Socolow, 2004). Hence, the application of SMEB science includes the management of the soils of native and agricultural ecosystems through the management of microbes and the monitoring of SMEB systems as sentinels of environmental change.

Changes in soil management would better equip the world to remain within the "safe operating space" of environmental conditions in which human civilization can develop and flourish. This safe operating space represents important thresholds or benchmarks that can be defined as a series of "planetary boundaries" that attempt to quantify dangerous degrees of disruption. Beyond these boundaries, we may find Earth's systems shifting into new states with harmful impacts on human societies (Rockström et al., 2009). Soils are particularly important for planetary boundaries that relate to climate change and ocean acidification, as well as the boundaries for disruption to the nitrogen and phosphorus cycles. Although defining global boundaries is difficult and a subject of ongoing scientific debate, the concept is nonetheless useful for articulating the need to reduce humanity's environmental impact along key axes.

17.7.1 Regenerative agriculture

Regenerative agriculture is a farming movement that embraces management practices that optimize production, crop quality, and climate resilience by restoring soils. As practiced, conservation tillage, no-till, and related integrative practices seek to maintain or increase production, reverse the legacy of soil C-loss in agricultural soils, and increase C sequestration (Fig. 17.10; Paustian et al., 2016). Actions that increase C inputs to the soil include improving crop rotations, increasing crop residues, cover-cropping, improving grazing practices, and adding external sources of C (e.g., compost, manure, or biochar;

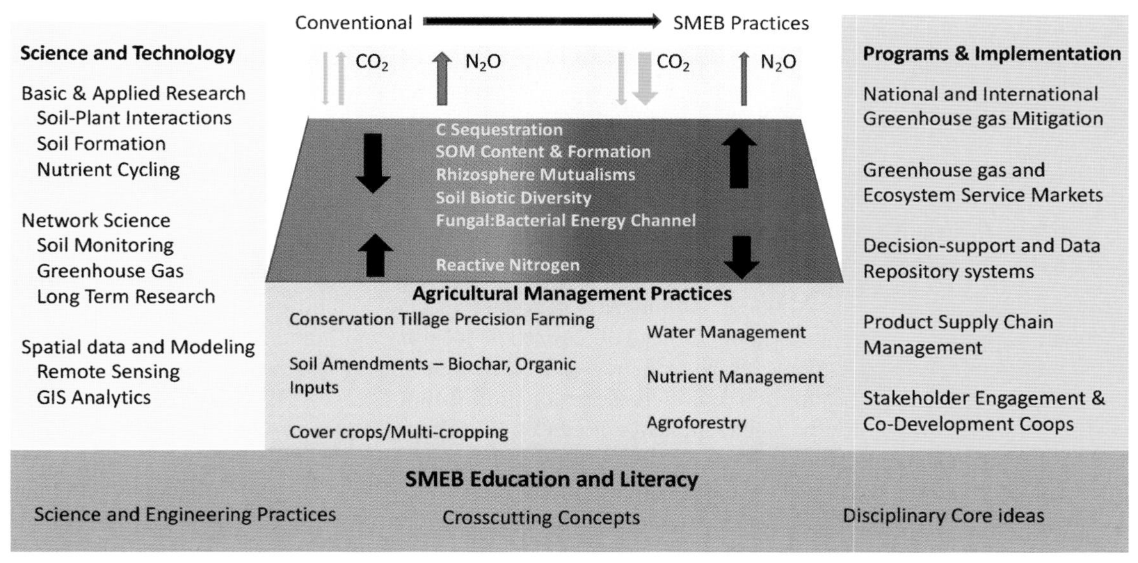

FIGURE 17.10 Regenerative agricultural practices provide a suite of ecologically based management practices to restore and sustain soils to meet the needs of increasing crop production and reducing greenhouse gas emissions (center panel). Paustian et al. (2016) cast the ecologically based practices as climate-smart practices. Management alone will likely not succeed, as science and technology support combined with stakeholder engagement, policy, and regulations as advocated by Clark et al. (2016) and improvements in SMEB literacy are needed (*right* and *left* panels). *(Adapted from Paustian et al., 2016.)*

Brussaard et al., 1997; Pressler et al., 2016). Conversely, other approaches seek to decrease C losses by utilizing conservation tillage approaches, defallowing, and rewetting organic soils (i.e., peat and muck). Converting lands to legumes and deep-rooted perennial grasses is a strategy that can both increase C inputs by enhancing interactions within the rhizosphere and reduce C losses by reducing cultivation events (Asbjornsen et al., 2013; Peixoto et al., 2022). In each case the strategies employed are ecologically based and rely on soil biota to mediate soil formation and nutrient dynamics within the cropping systems and reduce their impacts on adjacent systems.

17.7.2 Mitigating disruptions of the C, N, P cycles in soils

Can changes in management practices make meaningful increases in SOM in ways that can increase and sustain food production while significantly reducing greenhouse gas emissions and stabilizing their atmospheric concentrations? Recent studies suggest that at the global scale, soils have an immense capacity to mitigate a wide range of global change challenges by increasing soil organic C and reducing greenhouse gas emissions (Paustian et al., 2016; Tiefenbacher et al., 2021). Increasing SOM content can dramatically increase the sustainable productivity of agricultural soils (Chan, 2008; Smith et al., 2015). The adoption of management practices designed to increase the accumulation of SOM, coupled with those designed to decrease the loss of SOM, has the potential to significantly impact food production (Kell, 2012; Oldfield et al., 2022, Paustian et al., 2016; Schlesinger and Amundson, 2018; Tiefenbacher et al., 2021). Paustian et al. (2016) estimated that on a global scale, the management practices that reduce tillage, provide organic amendments, and shift toward perennial cultivars could increase SOM accumulation to 8 Pg CO_2(eq) yr^{-1}. The overall magnitude of C sequestration potential from these diverse strategies is a topic of considerable debate (Schlesinger, 2000; Schlesinger and Amundson, 2018). Claims of the potential for soils to sequester enough C using the aforementioned management practices to offset that which is being released through burning of fossil fuels have been deemed overly ambitious. Schlesinger and Amundson (2018) estimated that the management practices could offset about 5% of the amount released annually through burning fossil fuels.

Soil management is also central to reigning in anthropogenic impacts on N and P cycles (Lassaletta et al., 2014, 2016; Vitousek et al., 2013). Surplus N from agriculture drives degradation of water and air quality while also contributing to climate change and stratospheric ozone deposition (Sutton et al., 2011), while P losses from cropland are highly detrimental to water quality causing harmful algal blooms, eutrophication, and hypoxia. Although surplus N is generated in many countries around the world, more than half of all surplus N is attributable to just two countries — China and India. More efficient allocation of N resources — shifting inputs away from regions of surplus toward regions of N deficit — could simultaneously decrease N pollution and increase production (Mueller et al., 2012, 2017). Improved on-farm efficiency by applying the right type and amount of N at the right place and the time, the so-called "4Rs" of N management, could make a sizeable impact in reigning in N surplus (Robertson et al., 2008). Likewise, more efficient use of P resources would involve closer matching between soil supply and plant demand to minimize P buildup and losses and address geographic imbalances in P supplies (MacDonald et al., 2011). Phosphorus is also unique in that it is a scarce and geopolitically important element. Most P reserves are located in Western Sahara, a region occupied by Morocco with much smaller reserves in China, South Africa, the US, and elsewhere. Enhanced P recycling, more efficient use, innovative technologies, and thoughtful policies are required (Elser and Bennett, 2011). Enhancing P-efficiency within the rhizosphere (see Fig. 17.2) through inoculation with a broad spectrum of microorganisms

(Richardson et al., 2001) to boutique consortia of microbes (Baas et al., 2016) is being used to increase plant uptake of P (Richardson and Simpson, 2011). Other approaches have employed the genetic engineering of plant roots with microbial genes to produce phosphorolytic enzymes (Richardson et al., 2005).

17.7.3 Erosion control

Whether preventing erosion can be considered a climate adaptation is open for debate, but it is likely that the benefits of healthy soil may even be greater in a changed climate. Further, poor soil conservation contributes to poor growing conditions. Poor soils contribute to crop failure and increase aerosols in the environment. In the case of the 1930s Dust Bowl these factors contributed to amplified temperatures and drought conditions across much of the Central US and Great Plains region (Cook et al., 2009).

Although considerable efforts were made at reigning in soil loss, recent findings suggest cropland expansion (Lark et al., 2020) in the Great Plains is again contributing to enhanced dust, with negative consequences for human health, the environment, and long-term sustainability (Lambert et al., 2020). Mitigation incudes rebuilding topsoil using the regenerative practices previously discussed, maintaining the litter layer (O-horizon) through no tillage and minimum tillage and, where possible, maintaining cryptobiotic crusts, all of which will stabilize soil and reduce erosion (Belnap, 2013). For example, Kheirfam et al. (2017a,b) explored seeding erosion-prone, bare, and low-quality soils with slurries containing native N-fixing and photosynthetic cyanobacteria and bacteria isolated to promote crust-like activity to improve soil quality and control erosion with some success.

17.7.4 Managing soils to improve resilience (adaptation)

As climate change poses increasing threats to global agricultural systems, significant attention is being focused on the potential for soil management to help our food production systems adapt to changing conditions. Adaptation is "… the process of adjustment to actual or expected climate and its effects." In human-managed systems such as agriculture, "adaptation seeks to moderate or avoid harm or exploit beneficial opportunities" (IPCC, 2014). While changes in management practices may lead to increasing yields regardless of climate conditions, changes that are true adaptations will mitigate the harm (or exploit the benefit) caused by those changes (Lobell, 2014). Resilience is the ability of a system to cope with disturbance, persist, and recover. Agricultural adaptations help with the goal of resilience by allowing systems to adjust to changing climate (Davis et al., 2019).

Quantifying the impacts of climate change is a hotly debated topic of active research, yet despite uncertainties, it is recognized that climate change will pose an increasing headwind to global agriculture (Lobell and Gourdji, 2012), making it more difficult to continue historical trends in agricultural productivity (Ray et al., 2012, 2013). Rising temperatures can influence yields in several ways, including causing direct temperature stress and altering rates of crop development (Lobell and Gourdji, 2012). In addition, changes in temperature, precipitation, and carbon dioxide will interact to influence water stress on vegetation. As temperatures rise, atmospheric demand for water increases, which can dry out soils and lead to water stress for rainfed crops (Rigden et al., 2020; Swann et al., 2016). A rise in temperature is concurrent with the rise in carbon dioxide concentrations which increases plant water use efficiency (Leakey et al., 2009). These changes can mean changes in atmospheric aridity are partially decoupled from moisture stress to vegetation in future simulations of the Earth system (Swann et al., 2016). Even if the frequency of drought stays constant in the future, higher temperatures may lead to their severity.

Conversely, excess soil water can also damage crops by leading to waterlogging, creating harmful anaerobic conditions which may also promote pests and diseases (Rigden et al., 2020).

17.7.5 Land use and agricultural intensification

Meeting the SDGs will require addressing the twin needs of increasing food production to meet current and future demands in a secure and sustainable way while simultaneously curbing and ultimately mitigating soil C losses. Limiting land conversion from native to agricultural systems and increasing production on current agricultural lands – agricultural intensification – are two strategies in play. Implementing ecologically based approaches with the SMEB framework as underpinnings are key to these strategies (Elliott and Coleman, 1988; Matson et al., 1997).

Mitigating emissions from working landscapes must first focus on reducing deforestation and land conversion for agriculture, the largest source of emissions from the agriculture, forestry, and land use (AFOLU) sector (Hong et al., 2021). This is particularly important in the tropics, where land clearing produces approximately three times the carbon emissions per unit of crop production than temperate zones (West et al., 2020). Alternatively, limiting land use change and soil C loss will require investments in agricultural intensification – the increasing of crop yields on existing agricultural lands. A second focus would implement conservation and regenerative practices over more intensive conventional practices (Smith et al., 2008). Recall that intensive practices disturb soils and restructure the SMEB food web in ways that favor the bacterial energy channel resulting in increased mineralization of C and N, SOM loss, and leaching. Regenerative practices minimize soil disruption and incorporation of plant residues into soil, minimize the addition of commercial fertilizers through precision farming or eliminate it altogether with organic forms in croplands, and manage stocking rates in rangelands to reduce the physical disruption and maintain the plant communities. These practices favor the fungal and root energy channels within the SMEB food web that promote stabilizing or enhancing SOM formation, reducing erosion, and nutrient efflux.

The transition into the use of these management practices will only work if implemented alongside land use planning and regulations that decrease the incentive for agricultural land use expansion at the global scale (Mueller et al., 2012; Ray et al., 2012). In the absence of yield growth since the 1960s, contemporary crop area might have been about twice as large as the current global footprint (Burney et al., 2010), which would have had dramatic consequences for natural ecosystems and C losses. However, when increasing yields leads to greater profits and when lower prices cause increased food demand, then higher yields can still contribute to some degree of agricultural expansion. The SMEB-based strategies can temper but not entirely eliminate the need for further agricultural expansion. As a result, there is a need to couple increases in yields with land use zoning, economic incentives, and other strategies to ensure favorable outcomes at the local and global scales (Phalan et al., 2011). The SMEB framework needs to be part of the formulae that shape these regulations, incentives, and policies.

17.7.6 Managing water

Soil management practices that can promote adaptation to changing hydrologic conditions help promote optimal soil moisture conditions via infiltration enhancement and soil water holding capacity, maximizing the use of available moisture for crop growth, and preserving soil resources over the long run by preventing erosion (Kane et al., 2021; Stewart and Peterson, 2015). Such changes may be most important in semiarid and dry subhumid zones where dry spells are common due to high variability in rainfall across space and time (Rockström et al., 2010).

Agricultural practices that ensure continuous living cover on soils, i.e., cover crops, perennial grasses, and agroforestry, have been shown to substantially increase infiltration, water retention, and soil porosity (Basche and DeLonge, 2017, 2019). Utilizing no-till and residue retention can lead to increases in infiltration rates in wetter climates (Basche and DeLonge, 2019). In grazing lands infiltration rates can be increased by reducing stocking rates, adding complexity to grazing patterns, or providing an extended rest from grazing to minimize impacts on soil surfaces, biotic crusts, and plant cover (Belnap, 2013; DeLonge and Basche, 2018).

Maximizing the use of available moisture for crop growth requires limiting evaporation. If moisture is lost from the system, it is preferable that moisture is used productively for transpiration (Stewart and Peterson, 2015). A closed plant canopy helps limit soil evaporation, so efforts to address nutrient deficiencies, pest management, and other constraints on the least productive croplands across the globe could use soil water more productively (Brauman et al., 2013; Rockström and Barron, 2007; Sposito, 2013). No-till or limited-till systems, as well as the use of mulch, can also limit evaporation (Rockström and Barron, 2007). Optimal planting density and choice of crop varieties are also useful strategies for maximizing the use of moisture effectively, e.g., by rationing available moisture such that enough is available during critical periods of crop growth (Stewart and Peterson, 2015). Managing surface residues to conserve water also favors fungi over bacteria and the development of the fungal energy channel (Beare et al., 1993; Frey et al., 1999; Hendrix et al., 1986; Holland and Coleman, 1987).

17.8 Literacy, education, and engagement

The adoption of SMEB principles will depend on education and engagement networks and also the support systems currently in place. Ultimately, the level of stakeholder environmental science literacy (ESL) to implement these principles and the public's ability to accept them (Clark et al., 2016; Robertson and Swinton, 2005) will have the greatest effect in adopting future policy. The US National Research Council (2007) defined ESL as the capacity to understand and participate in evidence-based discussions of social-ecological systems and to make informed decisions about appropriate actions and policies. The definition applies to individuals, groups of individuals, and institutions.

Science and engineering are viewed as cultural ventures with specialized content knowledge and practices for participating in the construction, evaluation, and communication of scientific knowledge (Duschl, 2008; National Research Council, 2007, 2012). The environment, i.e., cultural, institutional, and disciplinary perspective, from which the knowledge of science and engineering is derived and applied is as important as the knowledge itself. As such, ESL represents a form of social capital that influences human behavior, and the nature and level of engagement in the decision-making process in social-ecological systems. Presented below are some tenets that emerge from this perspective.

1. The most adaptive social-ecological systems are distinguished by a high degree of general environmental literacy which may arise from a combination of different cognitive systems, including indigenous and scientific knowledge, and originate from formal and informal settings.
2. The resilience of social-ecological systems and the ability to manage them require a highly tuned understanding of resource and ecosystem dynamics, i.e., context-specific ESL is key.

3. Knowledge arises from many sources and contexts (i.e., a basic tenant of social constructivism in teaching and learning). Knowledge systems and associated management practices require strong social networks and an institutional framework (rules-in-use) to be effective. When social systems function in a way so that knowledge accumulates progressively through management and policy "experiments," resilience is enhanced.

The following sections present a general discussion about ESL and how it impacts SMEB science. Tasks or areas of action to facilitate the adoption of SMEB practices are then presented.

17.8.1 Literacy and SMEB science

Initiatives to elevate microbial literacy have been proposed (Timmis et al., 2019), but few studies have focused explicitly on literacy and SMEB topics. The Framework for K-12 Science Education (National Research Council, 2012) provides a guide that extends to higher education and professional career development. The Framework identifies three dimensions of science and engineering – core ideas, crosscutting concepts, and science and engineering practices (Table 17.3). The Framework further argues that curricula, instruction, and assessment include the three dimensions and that they are informed by empirically based learning progressions – descriptions of the successively, more sophisticated ways of thinking about a topic (Anderson et al., 2018; Duncan and Rivet, 2013; Table 17.4).

To date, learning progressions have not been developed for soils or microorganisms but have been created for concepts within life, physical, and Earth sciences. Learning progressions most germane to SMEB include those developed for the C cycle, water cycle, and biodiversity and for the ways in which

TABLE 17.3 The Next Generation Science Standards (NGSS, 2013) focus on three dimensions of learning: Practices, Crosscutting Concepts, and Disciplinary Core Ideas. The practices and crosscutting concepts readily align with the SMEB sciences. The disciplinary core ideas fall within broad areas of science. SMEB science is not covered explicitly within the core disciplinary ideas.

NGSS dimensions		
Practices	**Crosscutting concepts**	**Disciplinary core ideas**
Developing models	Patterns	***Physical Sciences***
Planning and carrying out investigations	Cause and effect	Matter and its interactions
Asking questions	Scale, proportion, and quantity	Energy
Analyzing and interpreting data	Systems and system models	***Life Sciences***
Mathematics and computational thinking	Energy and matter: flows, cycles, and conservation	From molecules to organisms
Constructing explanation	Structure and function	Ecosystems: interactions energy, dynamics
Engaging in argument from evidence	Stability and change	***Earth and Space Sciences***
Obtaining and communicating information		Earth systems
		Earth and human activity
		Engineering, Technology, and Applications of Science (ETAS)
		Linkages among ETAS

TABLE 17.4 Scientific model-based reasoning for SMEB science that aligns systems thinking with soil systems and SMEB science. The lower anchor represents how a novice or young child would reason. The upper anchor represents how experts in the field would define SMEB science to a literate person. An SMEB science learning progression would first establish the lower anchor, followed by the successive stages of reasoning to the upper anchor through assessments and interviews. The resulting learning progression would aid in the development of age-appropriate curricular materials, professional development training, and communication.

	Soil systems and SMEB science	
Systems thinking	**Lower anchor**	**Upper anchor**
Structure, function, and dynamics (SFD)	Macroscopic visible systems only; focuses on structure	Establishes boundaries; distinguishes SFD properties operating in soils
Scale	Macroscopic only	Molecular to larger scales – genes to function, C and N from microbes to atmosphere
Organization and representation	Physical and biological aspects of soils are studied in isolation	Nested hierarchical systems; traces C and N through connected systems
Principles (scientific, mathematical, and computational)	Force dynamics – describes the needs and wants of agents	Principle-based reasoning – identifies driving forces and constraints behind structural aspects of soils and soil processes
Control, regulation, constraints	Invokes other agents to act on soil processes or biota	Identifies driving forces and constraints that drive soil processes and soil biota

students learn with and from large-scale scientific data (Covitt et al., 2009; Duncan and Hmelo-Silver, 2009; Gunckel et al., 2012, 2022). Drawing on linguistics and psychology theory (Pinker, 2007; Talmy, 1988), discourse-based learning progressions are comprised of levels spanning from a lower anchor of informal, force-dynamic discourse to an upper anchor of scientific model-based discourse. Force-dynamic accounts describe events and phenomena as stories about actors with purpose, foregrounding short time frames and macroscopic scale. Phenomenological school science accounts focus on naming events and processes without attending to scientific principles. Model-based science invokes foundational principles (e.g., conservation of matter and energy, natural selection) to explain and predict phenomena at multiple scales in space and time. Studies have consistently found that most middle school and many high school students initially provide force-dynamic accounts, but with instruction, they have advanced to phenomenological reasoning. A small percentage of students transition further and achieve scientific model-based reasoning. K-12 teachers and college students generally provide phenomenological accounts, with a smaller percent providing scientific model-based reasoning (see Table 17.4).

17.8.2 Actions to facilitate the adoption of SMEB science

Aligning SMEB science within the frameworks used to meet the UN SDGs can be used to meet societal needs. The SDGs identify the challenges and provide a framework that encompasses interdependence of the Earth system, biophysical realms, and processes with humankind and activities. Timmis et al. (2017) provided a blueprint for such an alignment that focused on microbial technologies and their contributions to the SDGs. This effort should be expanded.

Next, K-16 education and postgraduate professional development should be reimagined to accommodate the transdisciplinary nature of SMEB topics and the UN SDGs. Clark et al. (2016) recognized that education was an important component of the level of engagement needed to address the SDGs at a high degree of ESL among all stakeholders. By extension, the complexities of the advances in SMEB and their application to meet current challenges reinforce the role of education in reaching solutions.

Finally, we need to align SMEB science with the current K-12 standards to promote SMEB literacy. Systemic changes in the education curricula at the K-12 and higher education levels are needed to elevate SMEB literacy and for practices to be adopted and effective. The Framework for K-12 Science Education (National Research Council, 2012) and a companion report, the Next Generation Science Standards (NGSS, 2013), identified discipline content, scientific practices, and crosscutting concepts as three important dimensions of science literacy. Introducing SMEB concepts into the K-12 curriculum will require that they be aligned along these dimensions within the existing NGSS-based science standards or their equivalent. These dimensions include: (1) an understanding of the content and underlying principles involved in SMEB; (2) the positioning of SMEB concepts and principles within social-ecological systems and, once positioned, within the system; and (3) the ability to apply reasoning that connects the content and principles in a system context across multiple temporal and spatial scales.

We advocate for the development of learning progressions that include SMEB science and their applications. The SMEB learning progressions will provide scientists, science educators, and K-12 teachers an understanding of and tools needed to advance instruction and to work with curriculum developers to ensure that SMEB concepts and practices are written into curricula. The transdisciplinary nature of SMEB science and the broad, discipline-based nature of NGSS science standards complicate the alignment process. SMEB concepts and principles are not explicitly covered in the NGSS standards. Soils are mentioned within the Earth systems science standard, with little to no discussion of soil formation or the importance of soils to plant growth. Microbes are presented in a manner that stresses their phylogeny and importance as mediators of biogeochemical cycling and plant growth as well as disease and human health. In addition, the principles of social-ecological systems are not highlighted.

17.9 Concluding remarks

This chapter focuses on the applications of SMEB science to meet the needs and challenges of society. A framework was presented for applying SMEB science to sustain ecosystem services operating within the food, water, and energy nexus articulated in the UN SDGs. The SMEB science framework operates on the premise that sustainable practices can be achieved by managing soils and soil biota in ways that emulate the structural, functional, and dynamic properties of natural ecosystems. In the process, we cataloged the different ways that information about soil biota is applied in the conservation management of natural ecosystems, the development of sustainable agricultural ecosystem practices, and to enhance the sequestration of C into soils.

The framework adapted recommendations by Robertson and Swinton (2005) for the development of sustainable agricultural production to apply to soils at the food, water, and energy nexus in a changing climate. First, ecosystem services provided by soils operating in different settings and applied at different scales are identified. Framing SMEB to address ecosystem services brings together basic and applied science, the arts, business and industry, education, and policy within the social-ecological construct (Collins et al., 2011) to facilitate the model of codevelopment to meet the UN SDGs (Clark et al., 2016). Second, the importance of a better understanding of the contributions that microbes and soil biota make to the services that soils provide and the trade-offs and synergies provided by SMEB applications is discussed. Ecosystem services are a human construct, and as such, there will be various degrees of negotiation and prioritization of resources depending on whether the service is supporting or regulatory and provisional or cultural. Understanding on an ecological basis provides a common forum from which to operate and a common currency to work within, connecting science to management and policy. Third, a middle-scale SMEB framework was sketched out that bridges large-scale geophysical processes with small-scale microbial processes operating within the nexus that provides information about these services to ensure SMEB science is prioritized and linked to both policy and market mechanisms. Fourth, a focus on SMEB science literacy as part of the broader discussion of ESL was strongly encouraged so that stakeholders can participate in the discussions and policy decision processes required to implement important changes.

Currently, food, timber, and bioenergy production are net sources of greenhouse gas emissions into the atmosphere, constituting roughly 24% of global emissions (Hong et al., 2021; IPCC, 2018). Global soils have an immense capacity to help mitigate a wide range of global change challenges. Adopting practices that increase the organic matter content of soils can stem emissions of trace gases into the atmosphere, dramatically increase the sustainable productivity of agricultural soils, and mitigate the export of reactive forms of N into the atmosphere and freshwater and marine environments (Chan, 2008; Galloway et al., 2003; Smith et al., 2015). Improved soil management tactics that include the application of SMEB science to increase production and mitigate losses of soil organic C and the export of reactive nutrients are part of the solution. Changes in economic incentives, regulatory policies, and dietary shifts must become part of the climate change solution if food production is to be substantially increased and emissions are to be simultaneously mitigated to meet international climate targets (Clark et al., 2016; Hasegawa et al., 2018; Hayek et al., 2020). In democratic societies changes of this magnitude will require input and debate from an informed public. The application of SMEB science to develop the solutions for societal needs is an extension of the "let the soil work for us" philosophy.

References

Aber, J.D., McDowell, W.H., Nadelhoffer, K.J., Magill, A., Berntson, G., Kamakea, M., et al., 1998. Nitrogen saturation in temperate forest ecosystems: hypotheses revisited. Bioscience 48, 921–934.

Acosta-Martínez, V., Zobeck, T.M., Gill, T.E., Kennedy, A.C., 2003. Enzyme activities and microbial community structure in semiarid agricultural soils. Biol. Fertil. Soils 38, 216–227.

Allen, E.B., Chambers, J.C., Connor, K.F., Allen, M.F., Brown, R.W., 1987. Natural reestablishment of mycorrhizae in disturbed alpine ecosystems. Arct. Alp. Res. 19, 11–20. https://doi.org/10.2307/1550995.

Allen, M.F., 1991. The Ecology of Mycorrhizae. Cambridge Studies in Ecology. Cambridge University Press, Cambridge, UK, p. 184.

Allen, M.F., Crisafulli, C., Friese, C.F., Jeakins, S.L., 1992. Re-formation of mycorrhizal symbioses on Mount St Helens 1980–1990: interactions of rodents and mycorrhizal fungi. Mycol. Res. 96, 447–453. https://doi.org/10.1016/S0953-7562(09).81089-7.

Allison, S.D., Wallenstein, M.D., Bradford, M.A., 2010. Soil-carbon response to warming dependent on microbial physiology. Nat. Geosci. 3, 336–340.

Anderson, C.W., de los Santos, E.X., Bodbyl, S., Covitt, B.A., Edwards, K.D., Hancock, J.B., et al., 2018. Designing educational systems to support enactment of the next generation science standards. J. Res. Sci. Teach. 55, 1026–1052.

Andrén, O., Linberg, T., Boström, U., Clarholm, M., Hanson, A.C., Johnansson, G., et al., 1990. Organic carbon and nitrogen flows. Ecol. Bull. 40, 85–125.

Andrés, P., Moore, J.C., Simpson, R.T., Selby, G., Cotrufo, M.F., Denef, K., et al., 2016. Soil food web stability in response to grazing in a semi-arid prairie: the importance of soil heterogeneity. Soil Biol. Biochem. 97, 131–143. https://doi.org/10.1016/j.soilbio.2016.02.014.

Andrés, P., Moore, J.C., Cotrufo, M.F., Denef, K., Haddix, M.L., Molowny-Horas, R., et al., 2017. Grazing and edaphic properties mediate soil biotic response to altered precipitation patterns in a semiarid prairie. Soil Biol. Biochem. 113, 263–274. https://doi.org/10.1016/j.soilbio.2017.06.022.

Asbjornsen, H., Hernandez-Santana, V., Liebman, M., Bayala, J., Chen, J., Helmers, M., et al., 2013. Targeting perennial vegetation in agricultural landscapes for enhancing ecosystem services. Renew. Agric. Food Syst. 29, 101–125. https://doi.org/10.1017/S1742170512000385.

Ayres, E., Steltzer, H., Simmons, B.L., Simpson, R.T., Steinweg, J.M., Wallenstein, M.D., et al., 2009. Home-field advantage accelerates leaf litter decomposition in forests. Soil Biol. Biochem. 41, 606–610.

Baas, P., Bell, C., Mancini, L.M., Lee, M.N., Conant, R.T., Wallenstein, M.D., 2016. Phosphorus mobilizing consortium Mammoth P™ enhances plant growth. PeerJ 4, e2121.

Bardgett, R., 2005. The Biology of Soil: A Community and Ecosystem Approach. Oxford University Press, Oxford. https://doi.org/10.1093/acprof:oso/9780198525035.001.0001.

Bardgett, R.D., Wardle, D.A., Yeates, G.W., 1998. Linking above-ground and below-ground food webs. How plant responses to foliar herbivory influences soil organisms. Soil Biol. Biochem. 30, 1867–1878.

Barrios, E., 2007. Soil biota, ecosystem services, and land productivity. Ecol. Econ. 64, 269–285.

Basche, A., DeLonge, M., 2017. The impact of continuous living cover on soil hydrologic properties: a meta-analysis. Soil Sci. Soc. Am. J. 81, 1179–1190. https://doi.org/10.2136/sssaj2017.03.0077.

Basche, A., DeLonge, M., 2019. Comparing infiltration rates in soils managed with conventional and alternative farming methods: a meta-analysis. PLoS One 14, e0215702. https://doi.org/10.1371/journal.pone.0215702.

Beare, M.H., 1997. Fungal and bacterial pathways of organic matter decomposition and nitrogen mineralization in arable soils. In: Brussaard, L., Ferrera-Cerrato, R. (Eds.), Soil Ecology in Sustainable Agricultural Systems. Lewis Publishers, Boca Raton, FL, pp. 37–70.

Beare, M.H., Parmelee, R.W., Hendrix, P.F., Cheng, W., Coleman, D.C., Crossley, D.A., 1992. Microbial and fauna interactions and effects on litter nitrogen and decomposition in agroecosystems. Ecol. Monogr. 62, 569–591.

Beare, M.H., Pohlad, B.R., Wright, D.H., Coleman, D.C., 1993. Residue placement and fungicide effects on fungal communities in conventional and no-tillage soils. Soil Sci. Soc. Am. J. 57, 392–399.

Beare, M.H., Hendrix, P.F., Coleman, D.C., 1994. Water-stable aggregates and organic matter fractions in conventional and no-tillage soils. Soil Sci. Soc. Am. J. 58, 777–786.

Beare, M.H., Hu, S., Coleman, D.C., Hendrix, P.F., 1997. Influences of mycelia fungi on aggregation and organic matter storage in conventional and no-tillage soils. Appl. Soil Ecol. 5, 211–219. https://doi.org/10.1016/S0929-1393(96)00142-4.

Belnap, J., 2013. Cryptobiotic Soils: Holding the Place in Place. U.S. Geological Survey. https://geochange.er.usgs.gov/sw/impacts/biology/crypto/.

Belnap, J., Kaltenecker, J.H., Rosentreter, R., Williams, J., Leonard, S., Eldridge, D., 2001. Biological Soil Crusts: Ecology and Management. U.S. Department of the Interior: Bureau of Land Management and U.S. Geological Survey. Technical Reference 1730-2.

Boucher, D.H., James, S., Keeler, K.H., 1982. The ecology of mutualism. Annu. Rev. Ecol. Syst. 13, 315–347.

Bradford, M.A., 2016. Re-visioning soil food webs. Soil Biol. Biochem. 102, 1–3.

Brauman, K.A., Siebert, S., Foley, J.A., 2013. Improvements in crop water productivity increase water sustainability and food security – a global analysis. Environ. Res. Lett. 8, 024030. https://doi.org/10.1088/1748-9326/8/2/024030.

Bringhurst, R.M., Cardon, Z.G., Gage, D.J., 2001. Galactosides in the rhizosphere: utilization by *Sinorhizobium meliloti* and development of a biosensor. Proc. Natl. Acad. Sci. 98, 4540–4545.

Brinkmann, N., Schneider, D., Sahner, J., Ballauff, J., Edy, N., Barus, H., et al., 2019. Intensive tropical land use massively shifts soil fungal communities. Sci. Rep. 9, 3403. https://doi.org/10.1038/s41598-019-39829-4.

Brussaard, L., van Veen, J.A., Kooistra, M.J., Lebbink, G., 1988. The Dutch Programme on soil ecology of arable farming systems. I. Objectives, approach, and some preliminary results. Ecol. Bull. 39, 35−40.

Brussaard, L., Behan-Pelletier, V.M., Bignell, D.E., Brown, V.K., Didden, W., Folgarait, P., et al., 1997. Biodiversity and ecosystem functioning in soil. Ambio 26 (8), 563−570.

Buresova, A., Kopecky, J., Hrdinkova, V., Kamenik, Z., Omelka, M., Sagova-Mareckova, M., 2019. Succession of microbial decomposers is determined by litter type, but site conditions drive decomposition rates. Appl. Environ. Microbiol. 85. https://doi.org/10.1128/AEM.01760-19. e01760-19.

Burke, I.C., Yonker, C.M., Parton, W.J., Cole, C.V., Flack, K., Schimel, D.S., 1989. Texture, climate, and cultivation effects on soil organic matter content in U.S. grassland soils. Soil Sci. Soc. Am. J. 56, 777−783.

Burke, I.C., Lauenroth, W.K., Riggle, R., Brannen, P., Madigan, B., Beard, S., 1999. Spatial variability of soil properties in the shortgrass steppe: the relative importance of topography, grazing, microsite, and plant species in controlling spatial patterns. Ecosystems 2, 422−438.

Burney, J.A., Davis, S.J., Lobell, D.B., 2010. Greenhouse gas mitigation by agricultural intensification. Proc. Natl. Acad. Sci. USA 107, 12052−12057.

Camberdella, C.A., Elliott, E.T., 1992. Particulate soil organic-matter changes across a grassland cultivation sequence. Soil Sci. Soc. Am. J. 56, 777−783. https://doi.org/10.2136/sssaj1992.03615995005600030017x.

Carson, R., 1962. Silent Spring. Houghton Mifflin, Boston, Cambridge, MA, p. 368.

Chan, Y., 2008. Increasing soil organic carbon of agricultural soils. Primefact 735, 1−5. http://www.dpi.nsw.gov.au/primefacts.

Clark, W.C., 2007. Sustainability science: a room of its own. Proc. Natl. Acad. Sci. USA 104, 1737−1738.

Clark, W.C., Dickson, N.M., 2003. Sustainability science: the emerging research program. Proc. Natl. Acad. Sci. USA 100, 8059−8061.

Clark, W.C., van Kerkhoff, L., Lebel, L., Gallopin, G.C., 2016. Crafting useable knowledge for sustainable development. Proc. Natl. Acad. Sci. USA 113, 4570−4578.

Coleman, D.C., Reid, C.P.P., Cole, C.V., 1983. Biological strategies of nutrient cycling in soil systems. In: MacFyaden, A., Ford, E.D. (Eds.), Advances in Ecological Research. Academic Press, London, pp. 1−55 vol. 13.

Coleman, D.C., Ingham, E.R., Hunt, H.W., Elliot, E.T., Reid, C.P.P., Moore, J.C., 1990. Seasonal and faunal effects on decomposition in semiarid prairie, meadow and lodgepole pine forest. Pedobiologia 34, 207−219.

Collins, S.L., Carpenter, S.R., Swinton, S.M., Orenstein, D.E., Childers, D.L., Gragson, T.L., et al., 2011. An integrated conceptual framework for long-term social−ecological research. Front. Ecol. Environ. 9, 351−357. https://doi.org/10.1890/100068.

Cook, B.I., Miller, R.L., Seager, R., 2009. Amplification of the North American "Dust Bowl" drought through human-induced land degradation. Proc. Natl. Acad. Sci. 106, 4997−5001. www.pnas.orgcgidoi10.1073pnas.0810200106.

Cotrufo, M.F., Wallenstein, M.D., Boot, C.M., Denef, K., Paul, E., 2013. The microbial efficiency-matrix stabilization (MEMS) framework integrates plant litter decomposition with soil organic matter stabilization: do labile plant inputs form soil organic matter? Glob. Chang Biol. 19, 988−995.

Cotrufo, M.F., Soong, J.L., Horton, A.J., Campbell, E.E., Haddix, M.L., Wall, D.H., et al., 2015. Formation of soil organic matter via biochemical and physical pathways of litter mass loss. Nat. Geosci. 8, 776−779.

Covitt, B.A., Gunckel, K.L., Anderson, C.W., 2009. Students' developing understanding of water in environmental systems. J. Environ. Educ. 40 (3), 37−51. https://doi.org/10.3200/JOEE.40.3.37-51.

Davis, K.F., Dalin, C., DeFries, R., Galloway, J.N., Leach, A.M., Mueller, N.D., 2019. Sustainable pathways for meeting future food demands. In: Ferranti, P., Berry, E.M., Anderson, J.R. (Eds.), Encyclopedia of Food Security. Elsevier, Oxford, UK, pp. 14−20 vol. 3.

De Boer, W., Folman, L.B., Summerbell, R.C., Boddy, L., 2005. Living in a fungal world: impact of fungi on soil bacterial niche development. FEMS Microbiol. Rev. 29, 795−811. https://doi.org/10.1016/j.femsre.2004.11.005.

De Long, J.R., Jackson, B.G., Wilkinson, A., Pritchard, W.J., Oakley, S., Mason, K.E., et al., 2018. Relationships between plant traits, soil properties and carbon fluxes differ between monocultures and mixed communities in temperate grassland. J. Ecol. 107, 1704−1719. https://doi.org/10.1111/1365-2745.13160.2018.

de Ruiter, P.C., Moore, J.C., Zwart, K.B., Bouwman, L.A., Hassink, J., Bloem, J., et al., 1993a. Simulation of nitrogen mineralization in belowground food webs of two winter wheat fields. J. Appl. Ecol. 30, 95–106.

de Ruiter, P.C., Van Veen, J.A., Moore, J.C., Brussaard, L., Hunt, H.W., 1993b. Calculation of nitrogen mineralization in soil food webs. Plant Soil 157, 263–273.

de Ruiter, P.C., Neutel, A.M., Moore, J.C., 1994. Modelling food web and nutrient cycling in agro-ecosystems. Trends Ecol. Evol. 9, 378–383.

de Ruiter, P.C., Neutel, A.M., Moore, J.C., 1995. Energetics, patterns of interaction strength, and stability in real ecosystems. Science 269, 1257–1260.

de Vries, F.T., Wallenstein, M.D., 2017. Below-ground connections underlying above-ground food production: a framework for optimizing ecological connections in the rhizosphere. J. Ecol. 105, 913–920. https://doi.org/10.1111/1365-2745.12783.

DeLonge, M., Basche, A.D., 2018. Managing grazing lands to improve soils and promote climate change adaptation and mitigation: a global synthesis. Renew. Agric. Food Syst. 33, 267–278. https://doi.org/10.1017/S1742170517000588.

Didden, W.A.M., Marinissen, J.C.Y., Vreeken-Buijs, M.J., Burgers, S.L.G.E., de Fuiter, R., Geurs, M., et al., 1994. Soil meso- and macrofauna in two agricultural systems: factors affecting population dynamics and evaluation of their role in carbon and nitrogen dynamics. Agric. Ecosyst. Environ. 51, 171–186.

Doles, J., 2000. A Survey of Soil Biota in the Arctic Tundra and Their Role in Mediating Terrestrial Nutrient Cycling. Thesis. University of Northern Colorado, Greeley, Colorado, USA.

Don, A., Schumacher, J., Freibauer, A., 2011. Impact of tropical land use change on soil organic carbon stocks – a meta-analysis. Glob. Chang Biol. 17, 1658–1670. https://doi.org/10.1111/j.1365-2486.2010.02336.x.

Doran, J.W., 2002. Soil health and global sustainability: translating science into proactive. Agric. Ecosyst. Environ. 88, 119–127.

Drijber, R.A., Doran, J.W., Parkhurst, A.M., Lyon, D.J., 2000. Changes in soil microbial community structure with tillage under long-term wheat-fallow management. Soil Biol. Biochem. 32, 1419–1430.

Duncan, R.G., Hmelo-Silver, C.E., 2009. Learning progressions: aligning curriculum, instruction, and assessment. J. Res. Sci. Teach. 46 (6), 606–609. https://doi.org/10.1002/tea.20316.

Duncan, R.G., Rivet, A.E., 2013. Science learning progressions. Science 339, 396–397.

Duschl, R., 2008. Science education in three-part harmony: balancing conceptual, epistemic, and social learning goals. Rev. Res. Educ. 32 (1), 268–291.

Ehrlich, P.R., Ehrlich, A.H., 1981. Extinction. The Causes and Consequences of the Disappearance of Species. Random House, New York, NY.

Elbert, W., Weber, B., Burrows, S., Steinkamp, J., Büdel, B., Andreae, M.O., et al., 2012. Contribution of cryptogamic covers to the global cycles of carbon and nitrogen. Nat. Geosci. 5 (7), 459–462. https://doi.org/10.1038/ngeo1486.

Elliott, E.T., 1986. Aggregate structure and carbon, nitrogen, and phosphorus in native and cultivated soils. Soil Sci. Soc. Am. J. 50, 627–633.

Elliott, E.T., Coleman, D.C., 1988. Let the soil work for us. Ecol. Bull. Copenhagen 39, 23–32.

Elliott, E.T., Horton, K., Moore, J.C., Coleman, D.C., Cole, C.V., 1984. Mineralization dynamics in fallow dryland wheat plots, Colorado. Plant Soil 76, 149–155.

Elser, J., Bennett, E., 2011. A broken biogeochemical cycle. Nature 478, 29–31.

Epstein, S., 2009. Microbial awakenings. Nature 457, 1083.

Foley, J.A., Ramankutty, N., Brauman, K.A., Cassidy, E.S., Gerber, J.S., Johnston, M., et al., 2011. Solutions for a cultivated planet. Nature 478, 337–342. https://doi.org/10.1038/nature10452.

Folse, H.J., Allison, S.D., 2012. Cooperation, competition, and coalitions in enzyme-producing microbes: social evolution and nutrient depolymerization rates. Front. Microbiol. 3, 1–10.

Foster, R.C., 1988. Microenvironments of soil microorganisms. Biol. Fertil. Soils 6, 189–203.

Fraser, D.G., Doran, J.W., Sahs, W.W., Lesoing, G.W., 1988. Soil microbial populations and activities under conventional and organic management. J. Environ. Qual. 17, 585–590. https://doi.org/10.2134/jeq1988.00472425001700040011x.

Frey, S.D., Elliott, E.T., Paustian, K., 1999. Bacterial and fungal abundance and biomass in conventional and no-tillage agroecosystems along two climatic gradients. Soil Biol. Biochem. 31, 573–585.

Frey, S.D., Knorr, M., Parrent, J.L., Simpson, R.T., 2004. Chronic nitrogen enrichment affects the structure and function of the soil microbial community in temperate hardwood and pine forests. For. Ecol. Manag. 196, 159–171.

Fuqua, W.C., Winans, S.C., Greenberg, E.P., 1994. Quorum sensing in bacteria: the LuxR-LuxI family of cell density-responsive transcriptional regulators. J. Bacteriol. 176, 269–275.

Galloway, J.N., Aber, J.D., Erisman, J.W., Seitzinger, S.P., Howarth, R.W., Cowling, E.B., et al., 2003. The nitrogen cascade. Bioscience 53, 341–356.

Galloway, J.N., Dentener, F.J., Capnoe, D.G., Boyer, E.W., Howarth, R.W., Seitszinger, S.P., et al., 2004. Nitrogen cycles: past, present, and future. Biogeochemistry 70, 153–226.

Gardner, M.R., Ashby, W.R., 1970. Connectance of large, dynamical (cybernetic) systems: critical values for stability. Nature 228, 784–785.

Gholz, H.L., Wedin, D.A., Smitherma, S.M., Harmon, M.E., Parton, W.J., 2000. Long-term dynamics of pine and hardwood litter in contrasting environments: toward a global model of decomposition. Glob. Chang Biol. 6, 751–765.

Gibbens, R.P., McNeely, R.P., Havstad, K.M., Beck, R.F., Nolen, B., 2005. Vegetation changes in the Jornada Basin from 1858 to 1998. J. Arid Environ. 61, 651–668.

Gough, L., Moore, J.C., Shaver, G.R., Simpson, R.T., Johnson, D.R., 2012. Above- and belowground responses to increased nutrients in Arctic tundra: implications for understanding of carbon cycling. Ecology 93, 1683–1694.

Grandy, A.S., Robertson, G.P., 2007. Land-use intensity effects on soil organic carbon accumulation rates and mechanisms. Ecosystems 11, 59–74. https://doi.org/10.1007/s10021-006-9010-y.

Grandy, A.S., Wieder, W.R., Wickings, K., Kyker-Snowman, E., 2016. Beyond microbes: are fauna the next frontier in soil biogeochemical models? Soil Biol. Biochem. 102, 40–44.

Grayston, S.J., Vaughan, D., Jones, D., 1996. Rhizosphere carbon flow in trees, in comparison with annual plants: the importance of root exudation and its impact on microbial activity and nutrient availability. Appl. Soil Ecol. 5, 29–55.

Gunckel, K.L., Covitt, B.A., Salinas, I., Anderson, C.W., 2012. A learning progression for water in socio-ecological systems. J. Res. Sci. Teach. 49, 843–868. https://doi.org/10.1002/tea.21024.

Gunckel, K.L., Covitt, B., Berkowitz, A.R., Caplan, B., Moore, J.C., 2022. Computational thinking for using models of water flow in environmental systems: intertwining three dimensions in a learning progression. J. Res. Sci. Teach. 59, 1169–1203. https://doi.org/10.1002/tea.21755.

Hall, E.K., Bernhardt, E.S., Bier, R.L., Bradford, M.A., Boot, C.M., Cotnre, J.B., et al., 2018. Understanding how microbiomes influence the systems they inhabit. Nat. Microbiol. 3, 977–982.

Hart, S.C., DeLuca, T.H., Newman, G.S., MacKenzie, M.D., Boyle, S.I., 2005. Post-fire vegetative dynamics as drivers of microbial community structure and function in forest soils. For. Ecol. Manag. 220, 166–184.

Hasegawa, T., Fujimori, S., Havlík, P., Valin, H., Bodirsky, B.L., Doelman, J.C., et al., 2018. Risk of increased food insecurity under stringent global climate change mitigation policy. Nat. Clim. Chang. 8, 699–703. https://doi.org/10.1038/s41558-018-0230-x.

Hayek, M.N., McDermid, S.P., Jamieson, D.W., 2020. An appeal to cost undermines food security risks of delayed mitigation. Nat. Clim. Chang. 10, 418–419. https://doi.org/10.1038/s41558-020-0766-4.

Hendrix, P.F., Parmelee, R.W., Crossley Jr., D.A., Coleman, D.C., Odum, E.P., Groffman, P.M., 1986. Detritus food webs in conventional and no-tillage agroecosystems. Bioscience 36, 374–380.

Hense, B.A., Kuttler, C., Müller, J., Rothballer, M., Hartmann, A., Kreft, J.-U., 2007. Does efficiency sensing unify diffusion and quorum sensing? Nat. Microbiol. 5, 230–239.

Herridge, D.F., Peoples, M.B., Boddey, R.M., 2008. Global inputs of biological nitrogen fixation in agricultural systems. Plant Soil 311, 1–18. https://doi.org/10.1007/s11104-008-9668-3.

Hoeksema, J.D., Chaudhary, V.B., Gehring, C.A., Johnson, N.C., Karst, J., Koide, R.T., et al., 2010. A meta-analysis of context-dependency in plant response to mycorrhizal fungi. Ecol. Lett. 13, 394–407.

Holland, E.A., Coleman, D.C., 1987. Litter placement effects on microbial and organic matter dynamics in an agroecosystem. Ecology 68, 425–433.

Hong, C., Burney, J.A., Pongratz, J., Nabel, J.E.M.S., Mueller, N.D., Jackson, R.B., et al., 2021. Global and regional drivers of land-use emissions in 1961–2017. Nature 589, 554–561. https://doi.org/10.1038/s41586-020-03138-y.

Hunt, H.W., Coleman, D.C., Ingham, E.R., Ingham, R.E., Elliott, E.T., Moore, J.C., et al., 1987. The detrital food web in a shortgrass prairie. Biol. Fertil. Soils 3, 57–68.

Hunter, M.C., Smith, R.G., Schipanski, M.E., Atwood, L.W., Mortensen, D.A., 2017. Agriculture in 2050: recalibrating targets for sustainable intensification. Bioscience 67, 386–391. https://doi.org/10.1093/biosci/bix010.

Ingham, E.R., Coleman, D.C., Moore, J.C., 1989. An analysis of food-web structure and function in a shortgrass prairie, a mountain meadow, and a lodgepole pine forest. Biol. Fertil. Soils 8, 29–37.

IPCC, 2014. Climate change 2014: impacts, adaptation, and vulnerability. In: Field, C.B., Barros, V.R., Dokken, D.J., Mach, K.J., Mastrandrea, M.D., Bilir, T.E., et al. (Eds.), Part A: Global and Sectoral Aspects. Contribution of Working Group II to the Fifth Assessment Report of the Intergovernmental Panel on Climate Change. Cambridge University Press, United Kingdom, and New York, p. 1132.

IPCC, 2018. Summary for policymakers. In: Masson-Delmotte, V., Zhai, P., Pörtner, H.-O., Roberts, D., Skea, J., Shukla, P.R., et al. (Eds.), Global Warming of 1.5 °C. An IPCC Special Report on the Impacts of Global Warming of 1.5°C above Pre-industrial Levels and Related Global Greenhouse Gas Emission Pathways, in the Context of Strengthening the Global Response to the Threat of Climate Change, Sustainable Development, and Efforts to Eradicate Poverty. World Meteorological Organization, Geneva, Switzerland, p. 32.

Jangid, K., Williams, M.A., Franzluebbers, A.J., Sanderlin, J.S., Reeve, J.H., Jenkins, M.B., et al., 2008. Relative impacts of land-use, management intensity and fertilization upon soil microbial community structure in agricultural systems. Soil Biol. Biogeochem. 40, 2843–2853.

Janos, D.P., 1980. Mycorrhizae influence tropical succession. Biotropica 12, 56–64.

Jochum, M., Eisenhauer, N., 2022. Out of the dark: using energy flux to connect above- and belowground communities and ecosystem functioning. Eur. J. Soil Sci. 73, e13154. https://doi.org/10.1111/ejss.13154.

Johansen, J.R., 1993. Cryptogamic crusts of semiarid and arid lands of North America. J. Phycol. 29, 140–147.

Johnson, N.C., Hoeksema, J.D., Abbott, L., Bever, J., Chaudhary, V.B., Gehring, C., et al., 2006. From Lilliput to Brobdingnag: extending models of mycorrhizal function across scales. Bioscience 56, 889–900.

Jones, S.E., Lennon, J.T., 2010. Dormancy contributes to the maintenance of microbial diversity. Proc. Natl. Acad. Sci. 107, 5881–5886.

Kallenbach, C.M., Frey, S.D., Grandy, A.S., 2016. Direct evidence for microbial-derived soil organic matter formation and ecophysiological controls. Nat. Commun. 7, 13630. https://doi.org/10.1038/ncomms13630.

Kane, D.A., Bradford, M.A., Fuller, E., Oldfield, E.E., Wood, S.A., 2021. Soil organic matter protects US maize yields and lowers crop insurance payouts under drought. Environ. Res. Lett. 16, 044018. https://doi.org/10.1088/1748-9326/abe492.

Kell, D.B., 2012. Large-scale sequestration of atmospheric carbon via plant roots in natural and agricultural ecosystems: why and how. Philos. Trans. R. Soc. B. 367, 1589–1597. https://doi.org/10.1098/rstb.2011.0244.

Kheirfam, H., Sadeghi, S.H., Darki, B.Z., Homaee, M., 2017a. Controlling rainfall-induced soil loss from small experimental plots through inoculation of bacteria and cyanobacteria. Catena 152, 40–46.

Kheirfam, H., Sadeghi, S.H., Homaee, M., Darki, B.Z., 2017b. Quality improvement of an erosion-prone soil through microbial enrichment. Soil Till. Res. 165, 230–238.

Knorr, M., Frey, S.D., Curtis, P.S., 2005. Nitrogen additions and litter decomposition: a meta-analysis. Ecology 86, 3252–3257.

Kourtev, P.S., Ehrenfeld, J.G., Häggblom, M., 2002. Exotic plant species alter the microbial community structure and function in the soil. Ecology 83, 3152–3166.

Krüger, C., Kohout, P., Janoušková, M., Püschel, D., Frouz, J., Rydlová, J., 2017. Plant communities rather than soil properties structure arbuscular mycorrhizal fungal communities along primary succession on a mine spoil. Front. Microbiol. 8, 719. https://doi.org/10.3389/fmicb.2017.00719.

Laishram, J., Saxena, K.G., Maikhuri, R.K., Rao, K.S., 2012. Soil quality and soil health: a review. Int. J. Ecol. Environ. Sci. 38, 19–37.

Lambert, A., Hallar, A.G., Garcia, M., Strong, C., Andrews, E., Hand, J.L., 2020. Dust impacts of rapid agricultural expansion on the Great Plains. Geophys. Res. Lett. 47. https://doi.org/10.1029/2020GL090347. e2020GL090347.

Lark, T.J., Spawn, S.A., Bougie, M., Gibbs, H.K., 2020. Cropland expansion in the United States produces marginal yields at high costs to wildlife. Nat. Commun. 11, 4295. https://doi.org/10.1038/s41467-020-18045-z.

Lassaletta, L., Billen, G., Grizzetti, B., Anglade, J., Garnier, J., 2014. 50-year trends in nitrogen use efficiency of world cropping systems: the relationship between yield and nitrogen input to cropland. Environ. Res. Lett. 9, 105011. https://doi.org/10.1088/1748-9326/9/10/105011.

Lassaletta, L., Billen, G., Garnier, J., Bouwman, L., Velazquez, E., Mueller, N.D., et al., 2016. Nitrogen use in the global food system: past trends and future trajectories of agronomic performance, pollution, trade, and dietary demand. Environ. Res. Lett. 11, 095007. https://doi.org/10.1088/1748-9326/11/9/095007.

Lauenroth, W.K., Burke, I.C., 2008. Ecology of the Shortgrass Steppe: A Long-Term Perspective. Oxford University Press, New York.

Lauenroth, W.K., Milchunas, D.G., Sala, O.E., Burke, I.C., Morgan, J.A., 2008. Net primary production in the shortgrass steppe. In: Lauenroth, W., Burke, I. (Eds.), The Shortgrass Steppe Ecosystem. Oxford University Press, New York, pp. 270−305.

Lavallee, J.M., Soong, J.L., Cotrufo, M.F., 2019. Conceptualizing soil organic matter into particulate and mineral-associated forms to address global change in the 21st century. Glob. Chang Biol. 26, 261−273. https://doi.org/10.1111/gcb.14859.

Leakey, A.D.B., Ainsworth, E.A., Bernacchi, C.J., Rogers, A., Long, S.P., Ort, D.R., 2009. Elevated CO_2 effects on plant carbon, nitrogen, and water relations: six important lessons from FACE. J. Exp. Bot. 60, 2859−2876. https://doi.org/10.1093/jxb/erp096.

Lehmann, J., Kleber, M., 2015. The contentious nature of soil organic matter. Nature 528, 60−68. https://doi.org/10.1038/nature16069.

Lennon, J.T., Jones, S.E., 2011. Microbial seed banks: the ecological and evolutionary implications of dormancy. Nat. Rev. Microbiol. 9, 119−130.

Liang, C., Wulf Amelung, W., Lehmann, J., Kästner, M., 2019. Quantitative assessment of microbial necromass contribution to soil organic matter. Glob. Chang Biol. 25, 3578−3590. https://doi.org/10.1111/gcb.14781.

Liu, J., Hull, V., Godfray, H.C.J., Tilman, D., Gleick, P., Hoff, H., et al., 2018. Nexus approaches to global sustainable development. Nat. Sustain. 1, 466−476. https://doi.org/10.1038/s41893-018-0135-8.

Lobell, D.B., 2014. Climate change adaptation in crop production: beware of illusions. Glob. Food Sec. 3, 72−76.

Lobell, D.B., Gourdji, S.M., 2012. The influence of climate change on global crop productivity. Plant Physiol. 160, 1686−1697. https://doi.org/10.1104/pp.112.208298.

Lussenhop, J., Fogel, R., 1991. Soil invertebrates are concentrated on roots. In: Keiser, D.L., Cregan, P.B. (Eds.), The Rhizosphere and Plant Growth. Kluwer Academic Publishers, Dordrecht, The Netherlands.

Ma, T., Zhu, S., Wang, Z., Wang, Z., Chen, D., Dai, G., et al., 2018. Divergent accumulation of microbial necromass and plant lignin components in grassland soils. Nat. Commun. 9, 3480. https://doi.org/10.1038/s41467-018-05891-1.

MacDonald, G.K., Bennett, E.M., Potter, P.A., Ramankutty, N., 2011. Agronomic phosphorus imbalances across the world's croplands. Proc. Natl. Acad. Sci. 108, 3086−3091. https://doi.org/10.1073/pnas.1010808108.

Mack, M.C., Schuur, E.A.G., Bret-Harte, M.S., Shaver, G.R., Chapin III., F.S., 2004. Ecosystem carbon storage in Arctic tundra reduced by long-term nutrient fertilization. Nature 431, 440−443.

Mack, M.C., Bret-Harte, M.S., Hollingsworth, T.N., Jandt, R.R., Schuur, E.A.G., Shavers, G.R., et al., 2011. Carbon loss from an unprecedented Arctic tundra wildfire. Nature 475, 489−492. https://doi.org/10.1038/naturel0283.

Malik, A.A., Martiny, J.B., Brodie, E.L., Martiny, A.C., Treseder, K.K., Allison, S.D., 2020. Defining trait-based microbial strategies with consequences for soil carbon cycling under climate change. ISME J. 14, 1−9.

Matson, P.A., Parton, W.J., Power, A.G., Swift, M.J., 1997. Agricultural intensification and ecosystem properties. Science 277, 504−509.

May, R.M., 1972. Will a large complex system be stable? Nature 238, 399−418.

May, R.M., 1973. Stability and Complexity of Model Ecosystems. Princeton University Press, Princeton, New Jersey.

Milchunas, D.G., Lauenroth, W.K., Burke, I.C., 1998. Livestock grazing: animal and plant biodiversity of shortgrass steppe and the relationship to ecosystem function. Oikos 83, 65−74.

Millennium Ecosystem Assessment (MEA), 2005. Ecosystems and human well-being. General Synthesis: A Report of the Millennium Ecosystem Assessment. Island Press, Washington, DC.

Moore, J.C., 1988. Influence of soil microarthropods on belowground symbiotic and non-symbiotic mutualisms. Interactions between soil inhabiting invertebrates and microorganisms in relation to plant growth. Agric. Ecosyst. Environ. 24, 147−159.

Moore, J.C., 1994. Impact of agricultural practices on soil food web structure: theory and application. Agric. Ecosyst. Environ. 51, 239–247.

Moore, J.C., 2021. The re-imagining of a framework for agricultural land-use: a pathway for integrating agricultural practices into ecosystem services, planetary boundaries and sustainable development goals. Ambio 50 (7), 1295–1298. https://doi.org/10.1007/s13280-020-01483-w.

Moore, J.C., de Ruiter, P.C., 1991. Temporal and spatial heterogeneity of trophic interactions within belowground food webs: an analytical approach to understand multi-dimensional systems. Agric. Ecosyst. Environ. 34, 371–397.

Moore, J.C., de Ruiter, P.C., 2000. Invertebrates in detrital food webs along gradients of productivity. In: Coleman, D.C., Hendrix, P.F. (Eds.), Invertebrates as Webmasters in Ecosystems. CABI Publishing, Oxford, UK, pp. 161–184.

Moore, J.C., de Ruiter, P.C., 2012. Energetic Food Webs: An Analysis of Real and Model Ecosystems. Oxford University Press, Oxford, UK, p. 333.

Moore, J.C., Hunt, H.W., 1988. Resource compartmentation and the stability of real ecosystems. Nature 333, 261–263.

Moore, J.C., Walter, D.E., Hunt, H.W., 1988. Arthropod regulation of micro- and mesobiota in belowground detrital food webs. Annu. Rev. Entomol. 33, 419–439.

Moore, J.C., de Ruiter, P.C., Hunt, H.W., 1993. The influence of productivity on the stability of real and model ecosystems. Science 261, 906–908.

Moore, J.C., McCann, K., Setälä, H., de Ruiter, P.C., 2003. Top-down is bottom-up: does predation in the rhizosphere regulate aboveground production. Ecology 84, 84–857.

Moore, J.C., Sipes, J., Whittemore-Olson, A.A., Hunt, H.W., Wall, D.W., de Ruiter, P.C., et al., 2008. Trophic structure and nutrient dynamics of the belowground food web within the rhizosphere of the shortgrass steppe. In: Lauenroth, W., Burke, I. (Eds.), The Shortgrass Steppe Ecosystem. Oxford University Press, New York, pp. 248–269.

Moore, J.C., Boone, R., Koyama, A., Holfelder, K., 2014. Enzymatic and detrital influences on the structure, function, and dynamics of spatially-explicit model ecosystems. Biogeochemistry 117, 205–227.

Moorhead, D.L., Lashermes, G., Sinsabaugh, R.L., 2012. A theoretical model of C- and N-acquiring extracellular enzyme activities, which balances microbial demands during decomposition. Soil Biol. Biochem. 53, 133–141.

Mueller, N.D., Gerber, J.S., Johnston, M., Ray, D.K., Ramankutty, N., Foley, J.A., 2012. Closing yield gaps through nutrient and water management. Nature 490, 254–257. https://doi.org/10.1038/nature11420.

Mueller, N.D., Lassaletta, L., Runck, B., Billen, G., Garnier, J., Gerber, J.S., 2017. Declining spatial efficiency of global cropland nitrogen allocation. Glob. Biogeochem. Cycles. 31, 245–257. https://doi.org/10.1002/2016GB005515.

Myers, S.S., Smith, M.R., Guth, S., Golden, C.D., Vaitla, B., Mueller, N.D., et al., 2017. Climate change and global food systems: potential impacts on food security and undernutrition. Annu. Rev. Public. Health. 38, 259–277. https://doi.org/10.1146/annurev-publhealth-031816-044356.

National Academies of Sciences and Medicine, 2019. Science Breakthroughs to Advance Food and Agricultural Research by 2030. The National Academies Press, Washington, D.C. https://doi.org/10.17226/25059.

National Research Council, 2007. Taking science to school: learning and teaching science in grades K-8. In: Duschl, R.A., Schweingruber, H.A., Shouse, A.W. (Eds.), Committee on Science Learning Kindergarten through Eighth Grade. National Academies Press, Washington, D.C.

National Research Council, 2012. A Framework for K-12 Science Education: Practices, Crosscutting Concepts, and Core Ideas. National Academies Press, Washington, D.C.

Neary, D.G., Klopatek, C.C., DeBano, L.F., Folliot, P.F., 1999. Fire effects on belowground sustainability: a review and synthesis. For. Ecol. Manage. 122, 51–71.

Neutel, A.M., Heesterbeek, J.A.P., de Ruiter, P.C., 2002. Stability in real food webs: weak links in long loops. Science 296, 1120–1123.

Next Generation Science Standards (NGSS), 2013. Next Generation Science Standards: For States, by States. The National Academies Press, Washington, DC.

Odum, E.P., 1969. The strategy of ecosystem development: an understanding of ecological succession provides a basis for resolving man's conflict with nature. Science 164, 262–270.

Oksanen, L., Fretwell, S.D., Arruda, J., Niemel, P., 1981. Exploitative ecosystems in gradients of primary productivity. Am. Nat. 118, 240–261.

Oldfield, E.E., Eagle, A.J., Rubin, R.L., Rudek, R., Sanderman, J., Gordon, D.R., 2022. Crediting agricultural soil carbon sequestration. Science 375, 6586. https://www.science.org/doi/10.1126/science.abl7991.

O'Neill, R.V., 1969. Indirect estimation of energy fluxes in animal food webs. J. Theo. Biol. 22, 284–290.

Pacala, S., Socolow, R., 2004. Stabilization wedges: solving the climate problem for the next 50 years with current technologies. Science 305, 968–972. https://doi.org/10.1126/science.1100103.

Paine, R.T., 1980. Food webs: linkage, interaction strength and community infrastructure. J. Anim. Ecol. 49, 667–685.

Parmelee, R.W., Ehrenfeld, J.G., Tate, R.L., 1993. Effects of pine roots on microorganisms, fauna, and nitrogen availability in two soil horizons of a coniferous spodosol. Biol. Fertil. Soils 15, 113–119.

Parton, W.J., Ojima, D.S., Cole, C.V., Schimel, D.S., 1994. A general model for soil organic matter dynamics: sensitivity to litter chemistry, texture, and management. Soil Sci. Soc. Am. J. 39, 147–167.

Paustian, K., Lehmann, J., Ogle, S., Reay, D., Robertson, G.P., Smith, P., 2016. Climate-smart soils. Nature 532, 49–57.

Peixoto, L., Olesen, J.E., Elsgaard, L., Enggrob, K.L., Banfield, C.C., Dippold, M.A., et al., 2022. Deep-rooted perennial crops differ in capacity to stabilize C inputs in deep soil layers. Sci. Rep. 12, 5952. https://doi.org/10.1038/s41598-022-09737-1.

Phalan, B., Onia, M., Balmford, A., Green, R.E., 2011. Reconciling food production and biodiversity conservation: land sharing and land sparing compared. Science 333, 1289–1291.

Pinker, S., 2007. The Stuff of Thought: Language as a Window into Human Nature. Viking Press, New York.

Pressler, Y., Foster, E.J., Moore, J.C., Cotrufo, M.F., 2016. Coupled limited irrigation and biochar amendment strategies neither promote nor degrade soil food webs in a maize agroecosystem. Glob. Change Biol. Bioenergy. 9, 1344–1355. https://doi.org/10.1111/gcbb.12429.

Pressler, Y., Moore, J.C., Cotrufo, M.F., 2019. Belowground community responses to fire: meta-analysis reveals contrasting responses of soil microorganisms and mesofauna. Oikos 128, 309–327. https://doi.org/10.1111/oik.05738.

Purahong, W., Wubet, T., Lentendu, G., Schloter, M., Pecyna, M.J., Kapturska, D., et al., 2016. Life in leaf litter: novel insights into community dynamics of bacteria and fungi during litter decomposition. Mol. Ecol. 25, 4059–4074. https://doi.org/10.1111/mec.13739.

Ray, D.K., Ramankutty, N., Mueller, N.D., West, P.C., Foley, J.A., 2012. Recent patterns of crop yield growth and stagnation. Nat. Commun. 3, 1293. https://doi.org/10.1038/ncomms2296.

Ray, D.K., Ramankutty, N., Mueller, N.D., West, P.C., Foley, J.A., 2013. Yield trends are insufficient to double global crop production by 2050. PLoS One 8, e66428. https://doi.org/10.1371/journal.pone.0066428.

Redfield, R.J., 2002. Is quorum sensing a side effect of diffusion sensing? Trends Microbiol. 10, 365–370.

Resat, H., Bailey, V., McCue, L.A., Konopka, A., 2012. Modeling microbial dynamics in heterogeneous environments: growth on soil carbon sources. Microb. Ecol. 63, 883–897.

Richardson, A.E., Simpson, R.J., 2011. Soil microorganisms mediating phosphorus availability update on microbial phosphorus. Plant. Physiol. 156, 989–996. https://doi.org/10.1104/pp.111.175448.

Richardson, A.E., Hadobas, P.A., Hayes, J.E., O'Hara, C.P., Simpson, R.J., 2001. Utilization of phosphorus by pasture plants supplied with myo-inositol hexaphosphate is enhanced by the presence of soil microorganisms. Plant Soil 229, 47–56.

Richardson, A.E., George, T.S., Hens, M., Simpson, R.J., 2005. Utilization of soil organic phosphorus by higher plants. In: Turner, B.L., Frossard, E., Baldwin, D.S. (Eds.), Organic Phosphorus in the Environment. CABI, Wallingford, UK, pp. 165–184.

Rigden, A.J., Mueller, N.D., Holbrook, N.M., Pillai, N., Huybers, P., 2020. Combined influence of soil moisture and atmospheric evaporative demand is important for accurately predicting US maize yields. Nat. Food. 1, 127–133.

Riggs, C.E., Hobbie, S.E., Bach, E.M., Hofmockel, K.S., Kazanski, C.E., 2015. Nitrogen addition changes grassland soil organic matter decomposition. Biogeochemistry 125, 203–219. https://doi.org/10.1007/s10533-015-0123-2.

Rillig, M.C., 2004. Arbuscular mycorrhizae, glomalin, and soil aggregation. Can. J. Soil Sci. 84, 355–363. https://doi.org/10.4141/S04-003.

Robertson, G.P., 1991. The KBS LTER project. Long-Term Ecological Research in the United States, sixth ed. LTER Network Office, Seattle, Washington, pp. 86–92.

Robertson, G.P., Swinton, S.M., 2005. Reconciling agricultural productivity and environmental integrity: a grand challenge for agriculture. Front. Ecol. Environ. 3, 38–46.

Robertson, G.P., Allen, V.G., Boody, G., Boose, E.R., Creamer, N.G., Drinkwater, L.E., et al., 2008. Long-term agricultural research: a research, education, and extension imperative. Bioscience 58, 640–645.

Robinson, D.A., Hockley, N., Cooper, D.M., Emmett, B.A., Keith, A.M., Lebron, I., et al., 2013. Natural capital and ecosystem services, developing an appropriate soils framework as a basis for valuation. Soil Biol. Biochem. 57, 1023–1033.

Robinson, D.A., Fraser, I., Dominati, E.J., Davíðsdóttir, B., Jónsson, J.O.G., Jones, L., et al., 2014. On the value of soil resources in the context of natural capital and ecosystem service delivery. Soil Sci. Soc. Am. J. 78, 685–700.

Rockström, J., Barron, J., 2007. Water productivity in rainfed systems: overview of challenges and analysis of opportunities in water scarcity prone savannahs. Irrig. Sci. 25, 299–311. https://doi.org/10.1007/s00271-007-0062-3.

Rockström, J., Steffen, W., Noone, K., Persson, Å., Chapin III, F.S., Lambin, E., et al., 2009. Planetary boundaries: exploring the safe operating space for humanity. Ecol. Soc. 14 (2), 32.

Rockström, J., Karlberg, L., Wani, S.P., Barron, J., Hatibu, N., Oweis, T., et al., 2010. Managing water in rainfed agriculture: the need for a paradigm shift. Agric. Water. Manage. 97 (4), 543–550.

Rodrigues, J.L.M., Pellizari, V.H., Mueller, R., Baek, K., Jesus, E.D.C., Paula, F.S., et al., 2013. Conversion of the Amazon rainforest to agriculture results in biotic homogenization of soil bacterial communities. Proc. Natl. Acad. Sci. 110, 988–993. https://doi.org/10.1073/pnas.1220608110.

Rooney, N., McCann, K., Gellner, G., Moore, J.C., 2006. Structural asymmetry and the stability of diverse food webs. Nature 442, 265–269.

Rosenzweig, M.L., 1971. The paradox of enrichment: destabilization of exploitation ecosystems in ecological time. Science 171, 385–387.

Sanderman, J., Hengl, T., Fiske, G.J., 2017. Soil carbon debt of 12,000 years of human land use. Proc. Natl. Acad. Sci. 114, 9575–9580.

Schimel, J.P., Weintraub, M.N., 2003. The implications of extracellular enzymes activity on microbial carbon and nitrogen limitation in soil: a theoretical model. Soil Biol. Biochem. 35, 549–563.

Schlesinger, W.H., 2000. Carbon sequestration in soils: some cautions amidst optimism. Agric. Ecosyst. Environ. 82, 121–127. https://doi.org/10.1016/S0167-8809(00)00221-8.

Schlesinger, W.H., Amundson, R., 2018. Managing for soil carbon sequestration: let's get realistic. Glob. Chang Biol. 25, 386–389. https://doi.org/10.1111/gcb.14478.

Schwarz, B., Barnes, A.D., Thakur, M.P., Brose, U., Ciobanu, M., Reich, P.B., et al., 2017. Warming alters energetic structure and function but not resilience of soil food webs. Nat. Clim. Chang. 7, 895–900. https://doi.org/10.1038/s41558-017-0002-z.

Shade, A., Peter, H., Allison, S.D., Baho, D.L., Berga, M., Bürgmann, H., et al., 2012. Fundamentals of microbial community resistance and resilience. Front. Microbiol. 3, 417.

Six, J., Paustian, K., 2014. Aggregate-associated soil organic matter as an ecosystem property and a measurement tool. Soil Biol. Biochem. 68, A4–A9. https://doi.org/10.1016/j.soilbio.2013.06.014.

Six, J., Elliott, E.T., Paustian, K., 2000. Soil macroaggregate turnover and microaggregate formation: a mechanism for C sequestration under no-tillage agriculture. Soil Biol. Biochem. 32, 2099–2103. https://doi.org/10.1016/S0038-0717(00)00179-6.

Six, J., Guggenberger, G., Paustian, K., Haumaier, L., Elliott, E.T., Zech, W., 2001. Sources and composition of soil organic matter fractions between and within soil aggregates. Eur. J. Soil Sci. 52, 607–618. https://doi.org/10.1046/j.1365-2389.2001.00406.x.

Six, J., Bossuyt, H., Degryze, S., Denef, K., 2004. A history of research on the link between (micro) aggregates, soil biota, and soil organic matter dynamics. Soil Tillage Res. 79, 7–31. https://doi.org/10.1016/j.still.2004.03.008.

Smil, V., 2001. Enriching the Earth: Fritz Haber, Carl Bosch, and the Transformation of Food Production. Cambridge University Press and MIT Press, Cambridge, MA and London.

Smith, P., Martino, D., Cai, Z., Gwary, D., Janzen, H., Kumar, P., et al., 2008. Greenhouse gas mitigation in agriculture. Philos. Trans. R. Soc. B. 363, 789–813. https://doi.org/10.1098/rstb.2007.2184.

Smith, P., Cotrufo, M.F., Rumpel, C., Paustian, K., Kuikman, P.J., Elliott, J.A., et al., 2015. Biogeochemical cycles and biodiversity as key drivers of ecosystem services provided by soils. Soil 1, 665–685. https://doi.org/10.5194/soil-1-665-2015.

Soong, J.L., Nielsen, U.N., 2016. The role of microarthropods in emerging models of soil organic matter. Soil Biol. Biochem. 102, 37–39. https://doi.org/10.1016/j.soilbio.2016.06.020.

Sposito, G., 2013. Green water and global food security. Vadose Zone J. 12, 6.

Steffen, W., Ruchardson, K., Rockström, J., Cornell, S.E., Fetzer, I., Bennett, E.M., et al., 2015. Planetary boundaries: guiding human development on a changing planet. Science 247, 736, 1259855-1-9.

Stewart, B.A., Peterson, G.A., 2015. Managing green water. Agron. J. 107, 1544–1553.

Suefert, V., Ramankutty, N., Foley, J.A., 2012. Comparing yields of organic and conventional agriculture. Nature 485, 230–236. https://doi.org/10.1038/nature11069.

Sugihara, N.G., van Wagtendonk, J.W., Fites-Kaufman, J., 2006. Fire as an ecological process. In: Sugihara, N.G., van Wagtendonk, J.W., Fites-Kaufman, J., Shaffer, K.E., Thode, A.E. (Eds.), Fire in California's Ecosystems. University of California Press, Berkeley, pp. 58–74.

Sulman, B.N., Phillips, R.P., Oishi, A.C., Shevliakova, E., Pacala, S.W., 2014. Microb-driven turnover offsets mineral-mediated storage of soil carbon under elevated CO_2. Nat. Clim. Chang. 4, 1099–1102.

Sutton, M.A., Oenema, O., Erisman, J.W., Leip, A., van Grinsven, H., Winiwarter, W., 2011. Too much of a good thing. Nature 462, 159–161.

Swann, A.L.S., Hoffman, F.M., Koven, C.D., Randerson, J.T., 2016. Plant responses to increasing CO_2 reduce estimates of climate impacts on drought severity. Proc. Natl. Acad. Sci. 113, 10019–10024. https://doi.org/10.1073/pnas.1604581113.

Swift, M.J., Heal, O.W., Anderson, J.M., 1979. Decomposition in Terrestrial Ecosystems. Blackwell Scientific Publications, Oxford.

Talmy, L., 1988. Force dynamics in language and cognition. Cognitive Sci. 12, 49–100.

Tang, J., Riley, W.J., 2015. Weaker soil carbon climate feedbacks resulting from microbial and abiotic interactions. Nat. Clim. Chang. 5, 56–60.

Tiefenbacher, A., Sandén, T., Haslmayr, H.-P., Miloczki, J., Wenzel, W., Spiegel, H., 2021. Optimizing carbon sequestration in croplands: a synthesis. Agronomy 11, 882. https://doi.org/10.3390/agronomy11050882.

Tiessen, H., Stewart, J.W.B., 1983. Particle-size fractions and their use in studies of soil organic matter: II. Cultivation effects on organic matter composition in size fractions. Soil Sci. Am. J. 47, 509–514.

Tilman, D., Cassman, K.G., Matson, P.A., Naylor, R., Polasky, S., 2002. Agricultural sustainability and intensive production practices. Nature 418, 671–677.

Tilman, D., Balzer, C., Hill, J., Befort, B.L., 2011. Global food demand and the sustainable intensification of agriculture. Proc. Natl. Acad. Sci. 108, 20260–20264. https://doi.org/10.1073/pnas.1116437108.

Timmis, K., de Voss, W.M., Ramos, J., Vlaeminck, S.E., Prieto, A., Danchin, A., et al., 2017. The contribution of microbial biotechnology to sustainable development goals. Microb. Biotechnol. 10, 984–987. https://doi.org/10.1111/1751-7915.

Timmis, K., Cavicchioli, R., Garcia, J., Nogales, B., Chavarría, M., Stein, L., et al., 2019. The urgent need for microbiology literacy in society. Environ. Microbiol. 21, 1513–1528. https://doi.org/10.1111/1462-2920.14611.

Tisdall, J.M., Oades, J.M., 1982. Organic matter and water-stable aggregates in soils. Eur. J. Soil Sci. 33, 143–163. https://doi.org/10.1111/j.1365-2389.1982.tb01755.x.

Treseder, K.K., 2004. A meta-analysis of mycorrhizal responses to nitrogen, phosphorus, and atmospheric CO_2 in field studies. New Phytol. 164, 347–355. https://doi.org/10.1111/j.1469-8137.2004.01159.x.

Treseder, K.K., Balser, T.C., Bradford, M.A., Brodie, E.L., Dubinsky, E.A., Eviner, V.T., et al., 2012. Integrating microbial ecology into ecosystem models: challenges and priorities. Biogeochemistry 109, 7–18.

Trofymow, J.A., Coleman, D.C., 1982. The role of bacterivorous and fungivorous nematodes in cellulose and chitin decomposition in the context of a root [rhizosphere] soil conceptual model. In: Freckman, D.W. (Ed.), Nematodes in Soil Ecosystems. University of Texas Press, Austin, TX, pp. 117–137.

United Nations, 2015. Transforming our world: the 2030 agenda for sustainable development. Sustainabledevelopment.un.org.

Van der Putten, W.H., Vet, L.E.M., Harvey, J.A., Wäckers, F.L., 2001. Linking above- and belowground multitrophic interactions of plants, herbivores, pathogens, and their antagonists. Trends Ecol. Evol. 16, 547–554.

Vitousek, P.M., Walker, L.R., Whiteaker, L.D., Mueller-Dombois, D., Matson, P.A., 1987. Biological invasion by *Myrica faya* alters ecosystem development in Hawaii. Science 238, 802–804.

Vitousek, P.M., Menge, D.N.L., Reed, S.C., Cleveland, C.C., 2013. Biological nitrogen fixation: rates, patterns, and ecological controls in terrestrial ecosystems. Philos. Trans. R. Soc. B. 368, 20130119. https://doi.org/10.1098/rstb.2013.0119.

von der Goltz, J., Dar, A., Fishman, R., Mueller, M.D., Barnwal, P., McCord, G.C., 2020. Health Impacts of the green revolution: evidence from 600,000 births across the developing world. J. Health Econ. 74, 102373. https://doi.org/10.1016/j.jhealeco.2020.

von Lutzow, M., Kogel-Knabner, I., Ekschmitt, K., Matzner, E., Guggenberger, G., Marschner, B., et al., 2006. Stabilization of organic matter in temperate soils: mechanisms and their relevance under different soil conditions — a review. Eur. J. Soil Sci. 57, 426—445.

Wall, D.W., Moore, J.C., 1999. Interactions underground: soil biodiversity, mutualism and ecosystem processes. Bioscience 49, 109—117.

Wang, X., Piao, S., Xu, X., Philippe Ciais, P., MacBean, N., Ranga, B., et al., 2015. Has the advancing onset of spring vegetation green-up slowed down or changed abruptly over the last three decades? Glob. Ecol. Biogeogr. https://doi.org/10.1111/geb.12289 http://wileyonlinelibrary.com/journal/geb.

Wardle, D.A., 1995. Impacts disturbance on detritus food webs in agro-ecosystems of contrasting tillage and weed management practices. Adv. Ecol. Res. 26, 105—185.

Wardle, D.A., Bonner, K.I., Barker, G.M., Yeates, G.W., Nicholson, K.S., Bardgett, R.D., et al., 1999. Plant removals in perennial grassland: vegetation dynamics, decomposers, soil biodiversity, and ecosystem properties. Ecol. Monogr. 69, 535—568.

Wardle, D.A., Bardgett, R.D., Klironomos, J.N., Setälä, H., van der Putten, W., Wall, D.H., 2004. Ecological linkages between aboveground and belowground biota. Science 304, 1629—1633.

Wei, X., Shao, M., Gale, W., Li, L., 2014. Global pattern of soil carbon losses due to the conversion of forests to agricultural land. Sci. Rep. 4, 4062. https://doi.org/10.1038/srep04062.

West, T.A.P., Börner, J., Sills, E.O., Kontoleon, A., 2020. Overstated carbon emission reductions from voluntary REDD+ projects in the Brazilian Amazon. Proc. Natl. Acad. Sci. 117, 24188—24194. https://doi.org/10.1073/pnas.2004334117.

Westman, W.E., 1977. How much are nature's services worth? Science 197, 960—964. https://doi.org/10.1126/science.197.4307.960.

Whitmee, S., Haines, A.H., Beyrer, C., Boltz, F., Capon, A.G., de Souza Dias, B.F., et al., 2015. Safeguarding human health in the Anthropocene epoch: report of the Rockefeller Foundation-Lancet Commission on planetary health. Lancet 386, 1973—2028. https://doi.org/10.1016/S0140-6736(15)60901-1.

Wieder, W.R., Grandy, A.S., Kallenbach, C.M., Bonan, G.B., 2014. Integrating microbial physiology and physio-chemical principles in soils with the MIcrobial-MIneral Carbon Stabilization (MIMICS) model. Biogeosciences 11, 3899—3917. https://doi.org/10.5194/bg-11-3899-2014.

Zegeye, E.K., Brislawn, C.J., Farris, Y., Fansler, S.J., Hofmockel, K.S., Jansson, J.K., et al., 2019. Selection, succession, and stabilization of soil microbial consortia. mSystems 4, 4. https://doi.org/10.1128/mSystems.00055-19.

Index

Note: Page numbers followed by *f* indicate figures and *t* indicate tables.

A

B

C

D

E

G

H

I

K

L

N

O

P

Q

R

S